Anesthesia and Analgesia in Laboratory Animals

AMERICAN COLLEGE OF LABORATORY ANIMAL MEDICINE SERIES

Steven H. Weisbroth, Ronald E. Flatt, and Alan L. Kraus, eds.:
The Biology of the Laboratory Rabbit, 1974

Joseph E. Wagner and Patrick J. Manning, eds.:
The Biology of the Guinea Pig, 1976

Edwin J. Andrews, Billy C. Ward, and Norman H. Altman, eds.:
Spontaneous Animal Models of Human Disease, Volume I, 1979; Volume II, 1979

Henry J. Baker, J. Russell Lindsey, and Steven H. Weisbroth, eds.:
The Laboratory Rat, Volume I: Biology and Diseases, 1979; Volume II: Research Applications, 1980

Henry L. Foster, J. David Small, and James G. Fox, eds.:
The Mouse in Biomedical Research, Volume I: History, Genetics, and Wild Mice, 1981; Volume II: Diseases, 1982; Volume III: Normative Biology, Immunology, and Husbandry, 1983; Volume IV: Experimental Biology and Oncology, 1982

James G. Fox, Bennett J. Cohen, and Franklin M. Loew, eds.:
Laboratory Animal Medicine, 1984

G. L. Van Hoosier, Jr., and Charles W. McPherson, eds.:
Laboratory Hamsters, 1987

Patrick J. Manning, Daniel H. Ringler, and Christian E. Newcomer, eds.:
The Biology of the Laboratory Rabbit, 2nd Edition, 1994

B. Taylor Bennett, Christian R. Abee, and Roy Henrickson, eds.:
Nonhuman Primates in Biomedical Research: Biology and Management, 1995

Dennis F. Kohn, Sally K. Wixson, William J. White, and G. John Benson, eds.:
Anesthesia and Analgesia in Laboratory Animals, 1997

ANESTHESIA AND ANALGESIA IN LABORATORY ANIMALS

EDITED BY

Dennis F. Kohn
Institute of Comparative Medicine
College of Physicians and Surgeons
Columbia University
New York, New York

Sally K. Wixson
Animal Research Facility
BASF Bioresearch Corporation
Worcester, Massachusetts

William J. White
Professional Services
Charles River Laboratories, Inc.
Wilmington, Massachusetts

G. John Benson
College of Veterinary Medicine
University of Illinois
Urbana, Illinois

ACADEMIC PRESS
San Diego London Boston New York Sydney Tokyo Toronto

This book is printed on acid-free paper.

Academic Press
A Harcourt Science and Technology Company
525 B Street, Suite 1900, San Diego, California 92101-4495, U.S.A.
http://www.apnet.com

Academic Press
24-28 Oval Road, London NW1 7DX, UK
http://www.hbuk.co.uk/ap/

Library of Congress Cataloging-in-Publication Data

Anesthesia and analgesia in laboratory animals / edited by Dennis F. Kohn,
Sally K. Wixson, William J. White, G. John Benson
p. cm.
Includes bibliographical references and index.
ISBN 0-12-417570-8 (alk. paper)
1. Laboratory animals—Surgery. 2. Veterinary anesthesia.
3. Analgesia. I. Kohn, Dennis F.
SF996.5.A54 1997
636.08'578—dc21 96-47584
CIP

PRINTED IN THE UNITED STATES OF AMERICA
01 02 EB 9 8 7 6 5 4

Contents

Tech University Health Sciences Center, Lubbock, Texas 79430

Care, College of Veterinary Medicine, University of Tennessee, Knoxville, Tennessee 37901

KATHLEEN L. SMILER (165), Bioresources Department, Wyeth-Ayerst Research, Chazy, New York 12921

ALISON C. SMITH (313), Department of Comparative Medicine, Medical University of South Carolina, Charleston, South Carolina 29425

M. MICHAEL SWINDLE (313), Department of Comparative Medicine, Medical University of South Carolina, Charleston, South Carolina 29425

GEORGE A. VOGLER (105), Department of Comparative Medicine, St. Louis University School of Medicine, St. Louis, Missouri 63104

WILLIAM J. WHITE (149), Professional Services, Charles River Laboratories, Inc., Wilmington, Massachusetts 01887

SALLY K. WIXSON (165, 274), Animal Research Facility, BASF Bioresearch Corporation, Worcester, Massachusetts 01605

Preface

This book presents a comprehensive review of anesthesia and analgesia techniques as they apply to animals used in a research setting. It differs in approach from other veterinary anesthesia texts in that it takes into account the research environment as well as a wide range of species, and will be useful to those engaged in biomedical research, including veterinarians, investigators, and technical staff.

In the biomedical research setting it is important not only to choose agents and methods that ensure the well-being of the research animal, but also to select agents that allow for scientifically valid and reproducible observations to be obtained with minimal or no effect on research outcomes. Wherever possible, background information and appropriate references have been provided to alert the reader to important effects and interactions of various drugs on the physiology and metabolism of animals used in a research setting. Where appropriate, authors have indicated key references in bold type.

The first set of chapters provides the reader with an overview of general topics related to the use of anesthetics and analgesics. Each of these chapters has a lengthy bibliography since the scope of the topics far exceeds that which can be dealt with in individual chapters. Those who have limited experience in administering anesthetics and analgesics to laboratory animals should consult these chapters for the background information required to understand the nature of these agents, including their mechanism of action.

The second set of chapters provides basic information on the nature of surgical and postsurgical facilities, equipment used to administer and support anesthesia, as well as essentials of preoperative and postoperative care. The information provided is applicable to a wide range of species.

The last set of chapters is oriented toward individual or groups of species. They provide the reader with a comprehensive discussion of agents, methods, and procedures from peer-reviewed reports as well as unpublished observations of the authors. Since the amount of data in the peer-reviewed literature on many agents is quite limited, especially with respect to individual laboratory animal species, unpublished observations are included to give the reader a reference point for applying less commonly used agents. All such observations are clearly identified as such. To further aid those who have limited experience in anesthetizing a particular species, the authors' preferences for agents and dosages are placed in the text in a tabulated format or are otherwise noted to provide guidance in selecting such agents. Extensive reference lists are included in the species chapters, which the reader is encouraged to consult for more detail on specific agents and their use in specific research settings.

The final chapter reflects the editors' view that the science and art of anesthesiology are not static and that there is a need for continued research applicable to the anesthetics and analgesics used in laboratory animals. Those individuals who provide anesthesia and analgesia to research animals should keep abreast of newly reported agents, combinations of agents, and methods of administration to ensure that laboratory animals in their care, as well as the research protocols in which they are being used, can benefit from such advances. A thorough understanding of the agents and techniques available for anesthesia and analgesia, as well as careful planning and thorough clinical evaluation of research animals used in experimental protocols, will allow all those who use animals in a research setting to uphold the trust given to them by the public to ensure that animals in their care are free of unnecessary pain during the course of experimental procedures.

This volume is one of a series of texts that the American College of Laboratory Animal Medicine has sponsored since 1974 as a means to nurture continuing education for its diplomates and their colleagues who have responsibilities for the care and use of laboratory animals in the biological sciences. We are indebted to the chapter authors for the dedication and effort each extended during the development of this book. We thank the reviewers whose suggestions and comments were extremely helpful to the chapter authors and editors. Royalties from the ACLAM sponsored texts are used to help support future continuing education programs.

Dennis F. Kohn
Sally K. Wixson
William J. White
G. John Benson

REFERENCES

Benson, G. J., Thurmon, J. C., and Davis, L. E. (1990). Laboratory animal analgesia. *In* ''The Experimental Animal in Biomedical Research,'' Vol. 1 (B. E. Rollin and M. L. Kesel, eds.), pp. 319–329. CRC Press, Boca Raton, FL.

Booth, N. H. (1982a). Clinical stages of general anesthesia. *In* ''Veterinary Pharmacology and Therapeutics,'' 5th Ed. (N. H. Booth and L. E. McDonald, eds.), pp. 165–174. Iowa State University Press, Ames, IA.

Booth, N. H. (1982b). Neuroleptanalgesics, narcotic analgesics, and analgesic antagonists. *In* ''Veterinary Pharmacology and Therapeutics,'' 5th Ed. (N. H. Booth and L. E. McDonald, eds.), pp. 267–296. Iowa State University Press, Ames, IA.

Lumb, W. V., and Jones, E. W. (1984). ''Veterinary Anesthesia.'' Lea and Febiger, Philadelphia, PA.

Short, E. E. (1987). Neuroleptanalgesia and alpha adrenergic receptor analgesia. *In* ''Principles and Practice of Veterinary Anesthesia'' (E. E. Short, ed.), pp. 47–57. Williams and Wilkens, Baltimore, MD.

Chapter 1

Pharmacology of Injectable Anesthetics

Richard E. Fish

ANESTHESIA AND ANALGESIA IN LABORATORY ANIMALS

ISBN 0-12-417570-8

I. INTRODUCTION

Injectable anesthetics are preferred by many for use in laboratory animals for a variety of reasons: minimization of equipment, administration less difficult than endotracheal intubation, and fewer safety concerns than associated with the use of inhalants. However, there is also increased interest in injectable anesthetic agents which may approach inhalants in speed of recovery or which have the potential for pharmacologic antagonism (Nunn *et al.,* 1989). The use of continuous infusion techniques is receiving increasing attention in human medicine, not only to account for individual variation, but also to improve anesthetic stability by approaching a steady-state blood concentration (White, 1989).

A few general cautions are in order for those evaluating the research literature on anesthetic agents. These include the importance of recognizing how (if) the parameters of anesthetic duration are defined, and the distinction between depth of anesthesia and antinociceptive potency (Whelan and Flecknell, 1992; Wixson *et al.,* 1987a); shortcomings of control data such as absence of data on nonmedicated animals, and use of conscious animals that are restrained, not at rest, or are evaluated shortly after recovery from instrumentation surgery/anesthesia; and failure to maintain normal body temperature and ventilation status. It should be remembered also that, although a pharmacologic effect may be attributed to a particular anesthetic agent, many reported effects are a more generalized feature of the anesthetic state. As is true for other drugs, individual variation plays an important role in biodisposition and pharmacokinetics, as well as therapeutics. Components of this variation include genotype (breed, stock, strain), sex, age, body composition, and nutritional and disease status. Finally, all conclusions/comparisons about drug potency, safety, and side effects must follow careful consideration of drug dosage.

Abundant, detailed information on the pharmacology of injectable anesthetics is available in numerous texts and articles in both the human and veterinary literature (see Additional Reading list at the end of this chapter). While it is presumed that the cellular mechanisms underlying anesthetic effects are well-conserved among species, it is clear that pharmacokinetics (or biodisposition) and clinical effects differ markedly among species. The focus of this chapter, therefore, is on the pharmacologic aspects of injectable agents in experimental animals, with particular emphasis on nondomestic species, and the effects (and side effects) of these agents on the physiology and metabolism of the animal.

II. BARBITURATES

A. Description

The barbiturates may be classified as hypnotic sedatives, reflecting their dose-dependent ability to produce either sedation or a deeper hypnotic state (Heavner, 1986; see Section III). Barbiturates are derived from barbituric acid, which is nondepressant, but appropriate side-chain substitutions result in central nervous system (CNS) depressant activity that varies in potency and duration with carbon chain length, branching, and saturation. Most barbiturate use in anesthesia is limited to those drugs classified as short (pentobarbital) or ultrashort-acting (methohexital, thiopental, and thiamylal) (Booth, 1988a). Inactin, the product name for ethyl(1-methylpropyl)malonylthiourea (also referred to as thiobutabarbital), has been popular for use in rats in renal studies for its prolonged and stable state of anesthesia (Buelke-Sam *et al.,* 1978; Cupples *et al.,* 1982; Turner and Howards, 1977). In rabbits, it is apparently ineffective alone, resulting in both short periods of anesthesia and deaths (Hobbs *et al.,* 1991).

B. Biodisposition

Differences among species in response to barbiturates are directly related to pharmacokinetics rather than to differences in drug receptor sensitivity (Davis *et al.,* 1973; Dos Santos and Bogan, 1974). For example, pentobarbital half-life is 38, 85, 100, and 200 min in the mouse, rabbit, rat, and dog, respectively (Thurmon, 1985), and thiopental elimination half-life and steady-state volume of distribution in the rabbit are markedly lower than in the dog or sheep (Ilkiw *et al.,* 1991). Barbiturates in general are readily absorbed from most sites, including the gastrointestinal tract.

Ultrashort-acting barbiturates are highly lipid-soluble and pass the blood–brain barrier rapidly, resulting in a rapid onset of action compared to the less lipid-soluble short-acting agents

(Booth, 1988a). The short duration of action of these agents is due primarily to redistribution from CNS to other tissues (although pentobarbital blood levels also decrease, in part, due to redistribution) (Thurmon, 1985). With large or repeated doses of thiopental, recovery may be prolonged due to equilibration between blood and tissues and resultant slow metabolism (Booth, 1988a). Duration of action of thiopental also is affected by speed of injection; bolus administration results in high plasma levels that result in a relatively greater fraction of unbound drug (Kurz and Fichtl, 1981). Differences in plasma protein binding of barbiturates may contribute to both species and individual differences in drug disposition (Sharma *et al.*, 1970; Thurmon, 1985).

Pentobarbital is metabolized, by 3-hydroxylation or oxidation, principally by the cytochrome P450-dependent microsomal enzyme system of hepatocytes (Booth, 1988a). Excretory routes other than urine are negligible in the sheep (Dos Santos and Bogan, 1974), in contrast to the rat in which 28% of a dose is excreted in bile within 6 hr (Klaassen, 1971). Thiopental also is metabolized primarily by liver (Booth, 1988a).

Tolerance to barbiturate effects, due to hepatic enzyme induction, can be demonstrated following previous exposure to the same or different barbiturate drug. For example, rats treated with phenobarbital and later with hexobarbital were anesthetized only 5% as long as untreated rats (Conney *et al.*, 1960), and pentobarbital plasma half-life following chronic pentobarbital pretreatment was only 12% that of control rats (Commissaris *et al.*, 1982). In contrast to tolerance phenomena, drugs which depress hepatic microsomal enzyme function may prolong anesthetic effect. The antibiotic chloramphenicol prolongs pentobarbital anesthesia in the rat, mouse, dog, cat, and monkey (Adams, 1970; Adams and Dixit, 1970; Azadegan *et al.*, 1980; Teske and Carter, 1971).

Barbiturate sleep time in rats and mice has been used in pharmacologic and toxicologic studies as a noninvasive measure of liver function; sleep time is inversely proportional to the rate of drug metabolism (Lovell, 1986b). A variety of factors affect sleep time, including age, sex, strain, feed and nutritional status, bedding material, and temperature (Cunliffe-Beamer *et al.*, 1981; Hall *et al.*, 1976; Jondorff *et al.*, 1958; Lovell, 1986a,b,c; Quinn *et al.*, 1958; Taber and Irwin, 1969; Vesell, 1968; Westenberg and Bolam, 1981; Westfall *et al.*, 1964). Sleep time can be prolonged by the administration of sulfonamides, salicylates, doxycycline, and phenylbutazone, each of which acts by displacing barbiturates from serum protein-binding sites (Booth, 1988a; Chaplin *et al.*, 1973).

Glucose administration to animals recovering from barbiturate anesthesia may result in reanesthetization. This ''glucose effect'' can be induced also by fructose, lactate, pyruvate, and glutamate (Booth, 1988a; Lumb and Jones, 1984), and a similar reanesthetization effect may occur with administration of adrenergic agents (Heavner and Bowen, 1968; Lamson *et al.*, 1952). The glucose effect is of limited practical concern when these agents are administered properly (Hatch, 1966).

C. Mechanism of Action

Multiple mechanisms may contribute to both the anticonvulsant (Rall and Schleifer, 1990) and anesthetic (CNS depressant) effects (Way and Trevor, 1986). One important mechanism of barbiturate action involves enhancement of the γ-aminobutyric acid (GABA)-mediated inhibition of synaptic transmission. The action is apparently not exerted directly at the GABA receptor or the chloride channel but allosterically, resulting in increased chloride channel conductance. The GABA receptor affected by barbiturates is distinct from that for benzodiazepines (Concas *et al.*, 1990). Thiobarbiturates act preferentially in the reticular activating system, depressing polysynaptic pathways and resulting in depression of higher brain centers (Muir *et al.*, 1991).

D. Pharmacologic Effects

1. Nervous System

An excitement phase during intravenous anesthetic induction is commonly observed with too slow injection, especially in large animals (Booth, 1988a). Intraperitoneal pentobarbital administration in rats is associated with mild excitement both on induction and on recovery (Wixson *et al.*, 1987a).

Barbiturates, especially thiobarbiturates, are poor analgesics (Booth, 1988a; Tomemori *et al.*, 1981). Pentobarbital provides inadequate or inconsistent analgesia in mice (Erhardt *et al.*, 1984), rats (Wixson *et al.*, 1987b), and rabbits (Borkowski *et al.*, 1990). It is not associated with a rise in plasma β-endorphin-like immunoreactivity, in contrast to the significant increases seen following fentanyl–fluanisone, urethane, or ether (Ramirez-Gonzalez *et al.*, 1991). Although hyperalgesic properties have been attributed to the barbiturates, spinal cord analgesic effects can be demonstrated (Jewett *et al.*, 1992).

Barbiturates provide ''brain protection'' (amelioration of neuronal damage following an ischemic event) by a variety of mechanisms, but especially by decreasing the cerebral metabolic rate (Hall and Murdoch, 1990).

2. Respiratory System

Respiratory depression following pentobarbital occurs in the rat (Folle and Levesque, 1976; Seyde *et al.*, 1985; Svendsen and Carter, 1985; Wixson *et al.*, 1987c), mouse (Erhardt *et al.*, 1984), rabbit (Borkowski *et al.*, 1990; Flecknell *et al.*, 1983), and hamster (Reid *et al.*, 1989). Minimal respiratory effects reported in other studies may have been due to maintenance of body temperature (Collado *et al.*, 1987; Heys *et al.*, 1989). Pentobarbital and thiopental in the dog each decrease the hypercapnic and hypoxic drive of respiration and attenuate carbon dioxide augmentation of the hypoxic response (Hirshman *et al.*, 1975). In the rabbit, thiopental results in hypercapnia with decreased minute volume and respiratory rate (Bellman and Pleuvry, 1981).

3. Cardiovascular System

Pentobarbital anesthesia in the dog typically increases heart rate (Booth, 1988a; Buckley *et al.,* 1979; Manders and Vatner, 1976). The heart rate in pentobarbital-anesthetized rats (Wixson *et al.,* 1987c) and rabbits (Borkowski *et al.,* 1990; Flecknell *et al.,* 1983) is not significantly altered, although tachycardia is seen in rabbits with subanesthetic doses (Murthy *et al.,* 1982). Barbiturate administration is not associated with increased sympathetic activity (Zimpfer *et al.,* 1982).

Reduced blood pressure, stroke volume, pulse pressure, and central venous pressure are common findings in pentobarbital-anesthetized animals (Parker and Adams, 1978). Pentobarbital results in prolonged hypotension in the rat (Svendsen and Carter, 1985; Wixson *et al.,* 1987c), although some investigators have reported an increased arterial pressure (Folle and Levesque, 1976). Cardiac output in the rat is reduced (Gumbleton *et al.,* 1990a; Kawaue and Iriuchijima, 1984; Lee *et al.,* 1985; Seyde *et al.,* 1985) and cardiovascular reflex responses are altered (Aisaka *et al.,* 1991; Fluckiger *et al.,* 1985; Wang *et al.,* 1991). Thiamylal anesthesia in the rabbit has little effect on the cardiovascular response to bilateral carotid artery occlusion or mild hemorrhage (Yamazaki and Sagawa, 1984). Ethylmalonylthiourea (Inactin) decreases arterial pressure, renal blood flow, and glomerular filtration rate in rats (Walker *et al.,* 1983).

Impaired myocardial contractility follows pentobarbital anesthesia in the dog (Vatner *et al.,* 1971b) and has been demonstrated also *in vitro* (Parker and Adams, 1978). Similar effects occur with thiopental exposure in the paced, isolated guinea pigheart and in the isolated rabbit ventricular myocardium preparation (Frankl and Poole-Wilson, 1981; Stowe *et al.,* 1992). Barbiturates alter the contractility of isolated mesenteric lymphatics and vascular smooth muscle, apparently by interfering with calcium-dependent processes (Altura *et al.,* 1980; Altura and Altura, 1975; McHale and Thornbury, 1989). Thiopental inhibits rat atrial and portal vein contractions *in vitro* (Bunting *et al.,* 1989).

The arrhythmogenic potential of thiobarbiturates appears to be well-accepted (Booth, 1988a; Hayashi *et al.,* 1989; Lumb and Jones, 1984), but there is little information on comparative aspects of this effect in different species. Barbiturates decrease cerebral blood flow and result in a consequent decrease in elevated intracranial pressure; cerebral vascular reactivity to carbon dioxide is preserved (Ilkiw, 1992).

4. Other Pharmacologic Effects

Other reported effects of barbiturates include the following: progressive decrease in core temperature in rats (Commissaris *et al.,* 1982; Wixson *et al.,* 1987d) and mice (Johnson *et al.,* 1976); decreased renal blood flow and urine output, secondary to lowered blood pressure (Booth, 1988a); increased sensitivity to barbiturate anesthesia in uremia, probably due to reduced protein binding (Booth, 1988a); reduced renal blood flow and glomerular filtration rate in rats (Gumbleton *et al.,* 1990a; Walker *et al.,* 1986), apparently responsible for decreased elimination rate of aminoglycoside antibiotics (Higashi *et al.,* 1982); rapid increase in plasma renin (Barrett *et al.,* 1988); depressed antipyrine clearance, an indicator of intrinsic hepatic clearance (Gumbleton and Benet, 1991); reduced *in vivo* rate of protein synthesis in liver and lung (Heys *et al.,* 1989); profound and prolonged decrease in output of lymphocytes in the sheep lymphocyte traffic model (Hall and Morris, 1962); decreased leukocyte counts and packed cell volumes in dogs (Usenik and Cronkite, 1965); decreased hematocrit following pentobarbital or thiamylal in miniature swine (Sawyer *et al.,* 1971) and mice (Friedman, 1959); altered gonadotropin secretion patterns in the rat (Chappel and Barraclough, 1976), baboon (Hagino, 1979), and rabbit (Mills *et al.,* 1981); increased level of plasma growth hormone (GH), and accentuated GH response to GH releasing hormone administration due, at least in part, to reduced somatostatin release from the hypothalamus (Hosoi *et al.,* 1988); elevated serum prolactin in pentobarbital-anesthetized rats following decapitation (Nazian, 1988); age-dependent changes in serum testosterone of rats decapitated after pentobarbital, thiopental, or thiamylal (Nazian, 1988); inhibition of normal masculinization in male hamsters given pentobarbital on postnatal days 2–4 (Clemens *et al.,* 1979); altered morphology of Type I hair cells from the vestibular epithelium of the inner ear of guinea pigs after thiopental (Scarfone *et al.,* 1991); decreased intraocular pressure (Ilkiw, 1992); resting myopia in humans and nonhuman primates, due largely to changes in central parasympathetic neuronal tone (Crawford *et al.,* 1990); alterations in tumor biology, including reduced blood flow and decreased radiosensitivity, and altered tumor energetics determined by ^{31}P NMR (Johnson *et al.,* 1976; Steen *et al.,* 1989); increased metastasis in rats (Agostino and Cliffton, 1964); increased lethality in mice when barbiturates are used concurrently with cyclophosphamide (Rose *et al.,* 1973); altered proximal tubule reabsorption in rats given ethylmalonylthiourea (Inactin) (Elmer *et al.,* 1972); and hyperglycemia in hamsters, unrelated to duration of anesthesia or degree of surgical manipulation (Turner and Howards, 1977), not seen in rats (Hinton, 1982).

E. Antagonists

Although the use of preanesthetics substantially reduces the dosage of barbiturate needed for anesthesia (Booth, 1988a; Muir *et al.,* 1991), there are no specific pharmacologic antagonists for the barbiturates. Effects of thiopental anesthesia are reduced in dogs and cats following administration of nonspecific agents such as 4-aminopyridine, amphetamine, or yohimbine (Hatch, 1973; Hatch *et al.,* 1984).

III. HYPNOTICS

The sedative hypnotics are drugs that are relatively nonselective CNS depressants, producing a dose-dependent response ranging from calming (sedation) to sleep (hypnosis) to unconsciousness and surgical anesthesia, and eventually to coma and death from depression of the respiratory and/or cardiovascular control centers (Rall, 1990). There may be important species differences in the utility of these drugs, depending on the degree of analgesia achieved as well as the seriousness of side effects.

A. Chloral Hydrate

1. Description

Chloral hydrate (trichloroacetaldehyde monohydrate) is a hypnotic agent that has been used in veterinary medicine primarily as a premedication for general anesthesia in horses or in combination with other drugs (magnesium sulfate and/or pentobarbital). It has been a useful hypnotic in this species due to a wide margin of safety and has been used in physiological experiments in laboratory animals for the same reason (Booth, 1988a). There have been few, if any, controlled studies of the anesthetic or analgesic actions of chloral hydrate (Silverman and Muir, 1993).

2. Biodisposition

Chloral hydrate is readily absorbed from the gastrointestinal (GI) tract, but onset of full effect is slow (Green, 1979). The rate of metabolism varies with species but is generally rapid in mammals (Daniel *et al.,* 1992). Most of the drug is reduced to trichloroethanol, an active metabolite which accounts for most of its hypnotic action (Booth, 1988a; Sourkes, 1992). Trichloroethanol is metabolized primarily by hepatic conjugation with glucuronic acid to form the inactive metabolite urochloralic acid, which is excreted in the urine; a small portion of the drug is excreted unchanged (Booth, 1988a). In some species, trichloroethanol also is oxidized to trichloroacetic acid but, in the rat and mouse, chloral hydrate oxidation/reduction to trichloroacetic acid/trichloroethanol is direct (Daniel *et al.,* 1992).

3. Reported Pharmacologic Effects

In hypnotic doses, the depressant effect of chloral hydrate is limited to the cerebrum, with minimal effects on medullary centers. Motor and sensory nerves are not affected except at high doses, and there is minimal analgesia (Booth, 1988a; Flecknell, 1987). Cerebral depression occurs slowly, and use of chloral hydrate for euthanasia of small animals may be preceded by gasping, muscle spasms, and vocalization [American Veterinary Medical Association (AVMA), 1993]. In the rat, the basal activity of nigrostriatal dopamine-containing neurons is reduced compared with unanesthetized paralyzed controls (Kelland *et al.,* 1990).

Hypnotic doses of chloral hydrate have minimal effects on cardiorespiratory function (Booth, 1988a), but anesthetic doses may be severely depressive (Field *et al.,* 1993; Lumb and Jones, 1984). In the dog, intravenous administration results in respiratory depression and lowered blood pressure, and sensitization of the heart to sudden vagal arrest or arrhythmias (Soma, 1983; Strobel and Wollman, 1969).

Chloral hydrate solution may be irritating to the stomach mucosa and causes severe inflammation and necrosis with perivascular injection (Booth, 1988a; Ogino *et al.,* 1990). Concentrated solutions may produce hemolysis and hematuria, and hepatic and renal damage may follow large repeated doses (Soma, 1983; Strobel and Wollman, 1969). Adynamic ileus, with morbidity and death, has been attributed to intraperitoneal administration of high concentrations of chloral hydrate in the rat (Fleischman *et al.,* 1977) and hamster (Dada *et al.,* 1992). Transient adynamic ileus also has been observed following intraperitoneal administration in the calf and pig (Silverman and Muir, 1993); these authors concluded that this route cannot be recommended for survival procedures. Gastric acid secretion is decreased in both vagally intact and denervated rats; this effect can be demonstrated also with ether and pentobarbital anesthesia (Graffner *et al.,* 1991).

B. α-Chloralose

1. Description

Chloralose is the weakly water-soluble reaction product of glucose with anhydrous chloral (trichloroacetaldehyde). Although the β isomer can cause convulsions, the α isomer is a useful hypnotic. The product is solubilized for administration by several routes, including *per os,* by heating to 60° C or mixing with 25% urethane. Chloralose produces hypnosis of long duration (8–10 hr), with minimal effect on reflexes. Analgesia generally has been considered poor (Flecknell, 1987; Strobel and Wollman, 1969), although it may be adequate for some procedures (Silverman and Muir, 1993). Effectiveness as an anesthetic appears to vary among species; it is probably not effective in the dog (Holzgrefe *et al.,* 1987). Chloralose use alone for survival procedures is not recommended due to rough induction, prolonged recovery, and seizure-like activity in some species (Booth, 1988a; Silverman and Muir, 1993; see also Chloralose Combinations, Section IX,D). There have been few, if any, controlled studies of the anesthetic or analgesic actions of chloralose (Holzgrefe *et al.,* 1987; Silverman and Muir, 1993).

2. Biodisposition

Chloralose is metabolized to chloral which, in turn, is metabolized to trichloroethanol. Its actions, therefore, should resemble those of chloral hydrate (Booth, 1988a), although there may be important species differences in pharmacokinetics.

3. Reported Pharmacologic Effects

α-Chloralose is used especially for chemical restraint with minimal cardiac and respiratory depression (Flecknell, 1987), although assisted respiration may be required at the higher doses necessary when the agent is used alone (Holzgrefe *et al.,* 1987). Spinal reflexes may actually be increased, and strychnine-like convulsions have been reported in the dog and cat (Booth, 1988a). Medicated animals may respond to tactile and auditory stimuli (Soma, 1983). There is no universally accepted mechanism to explain the actions of chloralose (MacDonald and Virtanen, 1992), although, like the barbiturates, a GABA-mimetic action on chloride currents can be demonstrated in frog isolated sensory neurons (Ishizuka *et al.,* 1989).

Transient changes following chloralose administration in the dog include decreased mean arterial blood pressure and peripheral resistance, and increased heart rate (Boucher *et al.,* 1991; Cox, 1972). However, there appear to be relatively few maintained changes in cardiopulmonary variables when this agent is used alone. Arterial pressure and heart rate of chloralose-anesthetized rats are similar (Wang *et al.,* 1991) or elevated (Folle and Levesque, 1976) compared with values in conscious controls.

α-Chloralose is commonly considered to preserve normal autonomic reflex activity, including baroreceptor and chemoreceptor reflexes (see review by Holzgrefe *et al.,* 1987). However, chloralose alters the baroreceptor reflex in lambs (Covert *et al.,* 1988), rats (Fluckiger *et al.,* 1985; Wang *et al.,* 1991), and rabbits (Ishikawa *et al.,* 1984); the somatosympathetic adrenal reflex (Gaumann and Yaksh, 1990) and micturition reflex (Rudy *et al.,* 1991) in cats; and the response to carotid chemoreceptor stimulation in the dog (Zimpfer *et al.,* 1981). Myocardial contractility is impaired (Parker and Adams, 1978).

In the rat, subanesthetic doses of chloralose do not increase serum renin activity, in contrast to several other anesthetic agents (Pettinger *et al.,* 1975). Intraperitoneal administration of chloralose to guinea pigs, rats, pigs, and calves leads to a severe inflammatory response which may be related to the concentration of the preparation (Silverman and Muir, 1993).

C. Urethane

1. Description

Urethane (ethyl carbamate) is the ethyl ester of carbamic acid, readily soluble in water, alcohol, and lipids. Following intravenous administration, urethane has a wide margin of safety and produces long-lasting narcosis (8–10 hr) with minimal cardiovascular or respiratory depression and maintenance of spinal reflexes. It differs from chloralose especially by having analgesic properties sufficient to permit surgery in small rodents (Field and Lang, 1988; Flecknell, 1987; Maggi and Meli, 1986a). The frequent use of urethane in neurophysiologic studies derives from its relatively minor effects on neurotransmission (see Albrecht and Davidowa, 1989).

2. Biodisposition

Because urethane is a known animal (and potential human) carcinogen, there has been particular interest in its metabolism. Metabolic activation of the ethyl moiety is required for carcinogenic action, but the metabolic pathways leading to activation or detoxication remain incompletely characterized (Kurata *et al.,* 1991).

Urethane has been administered by most routes, including topical application in frogs (Strobel and Wollman, 1969). It distributes evenly to most body tissues, except fat (Nomeir *et al.,* 1989), and is metabolized eventually to carbon dioxide, ammonia, and ethanol (Skipper *et al.,* 1951). In both rat and mouse given [^{14}C] carbonyl-labeled urethane, there is almost complete recovery of radiolabel in expired CO_2, with small amounts found in feces or urine (Bryan *et al.,* 1949). At low urethane doses, recovery of labeled CO_2 may approach 100% within 12 hr. *In vitro* metabolism to CO_2 can be demonstrated in a variety of tissues, including liver, plasma, brain, muscle, and kidney (Nomeir *et al.,* 1989). Recovery of exhaled ^{14}C following administration of [^{14}C]methylene-labeled urethane is slower, but still exceeds 90% by 24 hr (Skipper *et al.,* 1951).

3. Mechanism of Action

There is no universally accepted mechanism to explain the actions of urethane (MacDonald and Virtanen, 1992). It is tempting to speculate that the actions of urethane may, in part, be the result of metabolism to ethanol. Ethanol has anxiolytic and sedative/hypnotic effects similar to the barbiturates and benzodiazepines, and also appears to act at the GABA receptor complex (Suzdak *et al.,* 1986). However, compared with other anesthetics, urethane produces minimal enhancement of GABAergic neurotransmission (Maggi and Meli, 1986a).

4. Reported Pharmacologic Effects

Urethane anesthesia in rats occurs concomitantly with a significant rise in plasma β-endorphin-like immunoreactivity; the analgesic effect of urethane is partially reversed by pretreatment with naloxone (Ramirez-Gonzalez *et al.,* 1991). The basal activity of nigrostriatal dopamine-containing neurons in the rat is reduced compared with unanesthetized paralyzed controls (Kelland *et al.,* 1990). There is, in general, an activated sympathetic outflow from the CNS to peripheral organs (Maggi and Meli, 1986a). Urethane attenuates expression of kindled seizures in rats and may not be an appropriate anesthetic for the study of epileptiform phenomena (Cain *et al.,* 1989).

Respiratory effects of urethane anesthesia are minimal (Maggi and Meli, 1986c), although changes in blood gas values have been reported in the rabbit (Collado *et al.,* 1987) and rat (Buelke-Sam *et al.,* 1978; Folle and Levesque, 1976). Significant hypercapnia and hypoxia occur in the hamster with urethane (Reid *et al.,* 1989). In contrast, Field *et al.* (1993) found that rats anesthetized with urethane had a severely depressed arterial pH, with an increased PaO_2 and decreased $PaCO_2$ suggestive of hyperventilation.

Arterial pressure, cardiac output, and renal, hepatosplancnic, and brain blood flows in rats were lowest with intraperitoneal urethane anesthesia compared with four other anesthetic regimens, and lower than published values for the conscious rat (Gumbleton, *et al.,* 1989; Gumbleton *et al.,* 1990a). Similar depressant effects on cardiac dynamics of the rat have been reported (Maggi *et al.,* 1984; Wang *et al.,* 1991), although others have found cardiorespiratory effects to be minimal (De Wildt *et al.,* 1983; Folle and Levesque, 1976), especially when the intraperitoneal route of administration is avoided and doses are kept to the minimum required (Maggi and Meli, 1986b). Heart rate and systolic pressure in rats are stable during prolonged (3 hr) anesthesia, although pulse pressure is consistently elevated (due to decreased diastolic pressures) (Buelke-Sam *et al.,* 1978). The baroreceptor reflex in rats is altered (Fluckiger *et al.,* 1985; Wang *et al.,* 1991).

In discussing the pharmacologic effects of urethane, Maggi and Meli (1986c) stress the importance of dose and route of administration, and the need to distinguish between normal resting function during anesthesia and the degree of response to physiopharmacologic stimuli. Other reported pharmacologic effects of urethane include depressed xenobiotic renal clearance in the rat (Gumbleton *et al.,* 1990b); depressed antipyrine clearance, an indicator of intrinsic hepatic clearance, in the rat (Gumbleton and Benet, 1991); activation of the pituitary–adrenal axis, increased circulating catecholamines, and activation of the renin–angiotensin system (Carruba *et al.,* 1987; Pettinger *et al.,* 1975; Spriggs and Stockham, 1964); elevated blood glucose in the rabbit (Collado *et al.,* 1987) and rat, due at least in part to elevated catecholamine levels (Hinton, 1982; Maggi and Meli, 1986a); blunted plasma GH response to GH-releasing hormone, due in part to enhanced somatostatin release from the hypothalamus (Hosoi *et al.,* 1988); and lowered basal gastric acid secretion, due in part to increased synthesis and release of endogenous (gastric) somatostatin (Yang *et al.,* 1990).

Pathologic effects following intraperitoneal administration of urethane have been reported in the rat (Gumbleton *et al.,* 1988; Severs *et al.,* 1981; Van Der Meer *et al.,* 1975). The toxic effect on the mesenteric vasculature results in peritoneal effusion and secondary impairment of renal function; resultant hypovolemia may explain observed increases in serum renin (Severs *et al.,* 1981). Hypertonic urethane administration in the rabbit, by either intravenous or intraperitoneal routes, causes hemolysis, increased serum potassium, and prolonged clotting time (Bree and Cohen, 1965).

Urethane is an immunosuppressive agent and has a demonstrated antineoplastic effect. It is, however, more commonly recognized for its carcinogenic and mutagenic properties (Field and Lang, 1988; Inai *et al.,* 1991; Iversen, 1991; Leithauser *et al.,* 1990; Sotomayor and Collins, 1990).

D. Metomidate and Etomidate

1. Description

These nonbarbiturate hypnotic drugs are potentially useful agents for long-term anesthesia, due to minimal cumulative effect and good preservation of cardiovascular function. They generally require addition of an opioid for surgical anesthesia (Flecknell, 1987).

Metomidate is a hypnotic used in a variety of animal species. It has strong central muscle relaxant effects, but no analgesic properties in larger animals; muscle tremors and involuntary movements may occur. Etomidate is a hypnotic developed and used primarily in humans. No analgesic effect is seen in humans, but cardiorespiratory effects are minimal. Despite the lack of analgesic properties in larger mammals, Green *et al.* (1981) found that short-term, light surgical anesthesia could be produced in the mouse with either metomidate or etomidate given intraperitoneally; side effects included jerking and twitching movements, but cardiorespiratory depression was minimal. Anesthetic duration and depth were improved by a metomidate–fentanyl combination given subcutaneously (Green *et al.,* 1981).

2. Biodisposition

Etomidate is rapidly distributed following intravenous administration, with peak brain levels reached in less than 1 min. Etomidate is rapidly metabolized by ester hydrolysis in liver (primarily) and plasma to inactive products that are excreted primarily in the urine (Heykants *et al.,* 1975; Lewi *et al.,* 1976). The therapeutic index in rats and mice is wide for loss of the righting reflex (LD_{50}:ED_{50} = 29:1) and the biologic half-life in the rat is 40 min (Green *et al.,* 1981).

3. Reported Pharmacologic Effects

Etomidate is a carboxylated imidazole that has GABA-mimetic effects in the CNS (Evans and Hill, 1978). Its neurophysiologic actions are similar in many respects to the barbiturates and other injectable anesthetic agents such as alphaxalone (Way and Trevor, 1986). Etomidate decreases the cerebral metabolic rate and intracranial pressure, and has anticonvulsant properties (Batjer, 1993; Robertson, 1992; Wauquier, 1983).

In dogs, anesthetic doses of etomidate have little or no effect on heart rate, blood pressure, myocardial performance, or respiratory function (see Muir and Mason, 1989; Pascoe *et al.,* 1992). In artificially ventilated rats, there is minimal effect on heart rate and blood pressure, but aortic flow is markedly decreased (De Wildt *et al.,* 1983). Rat atrial and portal vein contractions *in vitro* are inhibited by the addition of etomidate (Bunting *et al.,* 1989). Direct negative inotropic effects of etomidate are dose dependent and may not reflect concentrations encountered clinically (Stowe *et al.,* 1992).

Imidazole compounds, including etomidate (and, presumably, metomidate), inhibit adrenal steroidogenesis (Preziosi and Vacca, 1988). Although this effect can be demonstrated following a single dose, the clinical significance is unknown (Maze *et al.,* 1991; Robertson, 1992).

Side effects of etomidate use in humans include pain on injection, nausea, vomiting, and myoclonic movements during induction (Booth, 1988a). In dogs, these effects are reduced following premedication with acepromazine or xylazine (Muir and Mason, 1989).

E. Propofol

1. Description

Propofol is an alkylphenol (2,6-diisopropylphenol), chemically distinct from the barbiturates, steroids, and imidazoles. Clinical formulations of the oil have been with the solubilizing agent Cremophor EL (polyoxyethylated castor oil) or as an emulsion with soybean oil, glycerol, and purified egg phosphatide (Sebel and Lowdon, 1989). Cremophor is known to result in histamine release in the dog. In addition, the Cremophor preparation (but not the emulsion) has been associated with pain on injection and anaphylactoid reactions in the rat and pig (Glen and Hunter, 1984). The anesthetic properties of propofol are similar in most respects to those of thiopental, although recovery from a single dose of propofol is more rapid and there is greater potential as a continuous infusion agent due to minimal cumulative effect (Glen, 1980; Sebel and Lowdon, 1989).

2. Biodisposition

Pharmacokinetic profiles following intravenous administration of the Cremophor preparation are similar in the rat, pig, rabbit, and cat. Plasma concentrations follow a biexponential curve, with a very short distribution half-life (1–6 min) and an elimination half-life between 16 and 55 min; in each case, lowest values are observed in the rat and rabbit. The total apparent volume of distribution is large, consistent with that for a lipophilic drug. The plasma concentration of propofol associated with awakening ranges from 1 to 4 μg/ml in the pig, rat, and cat, but is 7 μg/ml in the rabbit; in the latter, awakening occurs during the distribution phase of the drug (Adam *et al.,* 1980). Cockshott *et al.* (1992) came to similar conclusions using bolus injection of the emulsion formulation, although found that data were better fitted by a triexponential function when the sampling period was adequate. Total body clearance was rapid in the dog, rat, and pig (50–80 ml/min/kg) and was even higher in the rabbit (340 ml/min/kg).

Although redistribution of propofol is extensive, rapid elimination of the drug during the second phase is due to metabolic clearance primarily by the liver (Cockshott *et al.,* 1992). Metabolism is sufficiently rapid that speed of intravenous injection can affect both time to induction and dose needed (Glen, 1980). In mice, induction time and time to recover coordination are shorter than with thiopental; "utilization rate" of propofol, the amount required to maintain prolonged anesthesia, is greatest in mice compared with several other species. When light anesthesia is maintained, recovery is rapid following discontinuation of infusion; only slight cumulative effects of the drug were demonstrated. Intramuscular administration to rabbits, even at relatively high doses, results only in sedation (Glen, 1980).

3. Mechanism of Action

Propofol is a positive modulator of central GABAergic transmission and appears to act by enhancing the function of the GABA-activated chloride channel. Although acting at the level of the $GABA_A$ receptor complex, the specific site of action apparently is distinct from those for other anesthetics (e.g., barbiturates, steroids), benzodiazepines, and GABA itself (Concas *et al.,* 1990).

4. Reported Pharmacologic Effects

The anesthetic properties and hemodynamic effects of propofol are similar in the Cremophor and emulsion formulations, although the emulsion formulation has slightly greater potency in mice and male rats (Glen and Hunter, 1984). In the rat, propofol produces electroencephalogram (EEG) changes similar to those following barbiturate administration (Glen, 1980). Propofol is generally considered to be poorly analgesic, but spinal cord analgesic effects can be demonstrated (Jewett *et al.,* 1992). Although there have been warnings about the potential seizure-inducing action of propofol, the drug has potent anticonvulsive properties in mice (Lowson *et al.,* 1990).

Apnea occurs commonly in mice, rabbits, cats, and pigs, especially at higher doses; in rhesus monkeys it is seldom seen. In rabbits, there is a dose-dependent decrease in minute volume (Bellman and Pleuvry, 1981) and respiratory arrest may follow doses high enough to prevent pain response (Glen, 1980).

Hypotension associated with propofol administration has led to warnings about its use in patients with cardiovascular disease or compromise; changes other than hypotension are more variably reported (see Ilkiw *et al.,* 1992). Heart rate is elevated following propofol in rats (Rocchiccioli *et al.,* 1989) and rabbits (Blake *et al.,* 1988). Blood pressure is maintained during light propofol anesthesia in the rabbit (Blake *et al.,* 1988), but is decreased at higher doses; cardiac output may be elevated, reduced, or remain unchanged (Aeschbacher and Webb, 1993; Blake *et al.,* 1988; Glen, 1980; Van Leeuwen *et al.,* 1990). Antinociceptive doses of propofol in the rat result in lowered arterial pressure and heart rate (Tan *et al.,* 1993). Carmichael *et al.* (1993) confirmed a dose-dependent decrease in blood pressure in rats. There is no effect on cardiac output or on coronary or renal blood flows, and splanchnic hemodynamics and liver oxygenation are not adversely affected.

Cardiovascular effects of propofol in swine include a dose-dependent decrease in myocardial contractility (Coetzee *et al.,* 1989; Glen, 1980; Glen and Hunter, 1984). A negative inotropic response has been demonstrated *in vitro* along with marked attenuation of coronary flow autoregulation (Stowe *et al.,* 1992). Rat atrial and portal vein contractions are inhibited *in vitro* by the addition of propofol (Bunting *et al.,* 1989).

The baroreceptor–heart rate reflex is depressed following propofol administration to rats (Rocchiccioli *et al.,* 1989) and rabbits. The depression is much less potent than with alphaxalone/alphadolone (Blake *et al.,* 1988), thiopentone, or ketamine (Blake *et al.,* 1982; Van Leeuwen *et al.,* 1990). In the rabbit and cat, carotid artery infusion of propofol depresses afferent impulses from single chemoreceptor fibers in response to hypoxia and hypercapnia (Ponte and Sadler, 1989).

F. Tribromoethanol

Tribromoethanol (formerly available as Avertin) produces a generalized CNS depression, including both the respiratory and the cardiovascular centers. It is metabolized in the liver by conjugation with glucuronic acid and is excreted in the urine as tribromoethanol glucuronate (Green, 1979).

The intraperitoneal use of tribromoethanol in mice and other small rodents results in the rapid induction of short-term surgical anesthesia. The agent has received widespread acceptance in the United States, especially for use in the various manipulations required for the production of transgenic mice (Papaioannou and Fox, 1993). However, there have been numerous reports of postanesthetic complications. High death losses in mice after recovery were associated with fluid distension of the stomach and small intestine, suggesting intestinal ileus as the cause of death (Tarin and Sturdee, 1972). Similar results have been reported in gerbils, although the effects could be reduced, but not eliminated, by dilution to 1.25% and reducing dosage to 300 mg/kg or less (Norris and Turner, 1983). A single dose of tribromoethanol to mice before pregnancy resulted in impaired fertility (Kaufman, 1977), and administration following ovulation resulted in postimplantation parthenogenetic development (Kaufman, 1975). These complications have been attributed to decomposition products of the drug (hydrobromic acid and dibromoacetaldehyde) but, even with fresh solutions, high mortality has been reported following a second injection (Flecknell, 1987; Green, 1979). In contrast to these reports, Papaioannou and Fox (1993), in a thorough review of the literature, retrospective analysis, and pathological evaluation, concluded that tribromoethanol was a safe and effective anesthetic for mice when prepared and stored properly.

IV. CYCLOHEXAMINES: DISSOCIATIVE ANESTHETICS

A. Description

The cyclohexamines, phencyclidine and its congeners ketamine and tiletamine, have been categorized as dissociative anesthetics and sympathomimetic anesthetics (Soma, 1983). "Dissociative" refers to "apparent dissociation of the patient from its environment, believed to be caused by interruption of CNS impulses and differential depression and activation of various areas of the brain" (Muir, 1985). Ketamine may be the most widely used injectable anesthetic in veterinary practice, probably due to its wide margin of safety and compatibility with other drugs. There are published reports of ketamine use in more than 80 species (Muir, 1985).

Ketamine commercial preparations consist of equal parts of the two enantiomers, which differ in anesthetic potency and clearance rate (Muir and Hubbell, 1988; Reich and Silvay, 1989; Ryder *et al.,* 1978). The 10% aqueous solution has a pH of 3.5, which may result in pain on injection and muscle necrosis in small animals (Wright, 1982). Tiletamine was developed in an effort to find an agent with potency and duration intermediate between phencyclidine and ketamine; however, there are marked species differences in drug response (Chen *et al.,* 1969). Its current use in veterinary medicine is primarily in combination with zolazepam as the product Telazol (see Section IX).

B. Biodisposition

Ketamine has a high lipid solubility, associated with a rapid induction and return to consciousness following redistribution from the CNS to other body tissues (Thurmon, 1985). It has a high bioavailability following intravenous or intramuscular administration (Reich and Silvay, 1989), although intramuscular injection may result in more variable peak plasma concentrations (Löscher *et al.,* 1990). Ketamine is metabolized by the liver in most species. The most important pathway is *N*-demethylation by cytochrome P450 to norketamine (metabolite I), an active metabolite with reduced activity. Norketamine is then hydroxylated and conjugated to water-soluble products that are excreted by the kidney. The cyclohexanone ring also undergoes oxidative metabolism. So-called metabolite II is apparently not a naturally occurring *in vivo* metabolite, but a by-product of the chromatographic process (Reich and Silvay, 1989).

Peak brain levels of ketamine are reached within 1 min following intravenous injection in the rat, and brain:plasma ratios remain at 6.5:1 for more than 10 min (Cohen *et al.,* 1973). Over half of an oral, intramuscular, or intraperitoneal dose of radiolabeled ketamine in the rat is recovered in urine and feces within 24 hr. By 72 hr, 53–75% is recovered in urine and 23–25% in feces. Ketamine is transferred readily across the placenta in dogs, monkeys, and humans (Chang and Glazko, 1974). There are significant species differences in the relative amounts of free ketamine and ketamine metabolites in the urine (Chang and Glazko, 1974) and protein binding in serum (Wright, 1982). Löscher *et al.* (1990) present a detailed analysis of pharmacokinetic data in swine and compare their data with that in other domestic species.

One- and 2-week-old rats have a prolonged sleep time following intraperitoneal ketamine administration compared with adults, and plasma ketamine levels are higher at recovery. In rats over 2 weeks of age, females have a greater sleep time than males (Waterman and Livingston, 1978).

Ketamine pretreatment decreases the plasma half-life of intravenously administered ketamine by induction of hepatic microsomal enzyme activity (Marietta *et al.*, 1977). Tolerance has been demonstrated in rats (Livingston and Waterman, 1978) and is recognized in human clinical practice (Reich and Silvay, 1989).

C. Mechanism of Action

The anesthetic action of ketamine requires a functioning cerebral cortex (Wright, 1982). There is depression or disorganization of the associative areas of the brain, while subcortical areas may be activated (Haskins, 1992; Oguchi *et al.*, 1982).

There is no universally accepted mechanism to explain the actions of ketamine (MacDonald and Virtanen, 1992), and mechanisms for anesthesia and analgesia may be different. The cyclohexamines have no effect on GABA, although they do block ion channels. They inhibit excitatory polysynaptic pathways mediated by acetylcholine and L-glutamate in the spinal cord and *N*-methylaspartate (NMA) in the brain (Haskins, 1992; Reich and Silvay, 1989).

Ketamine is generally considered to be a potent analgesic, blocking the conduction of pain impulses to thalamic and cortical areas. Analgesia is reportedly more potent for procedures involving the musculoskeletal system than the abdomen (Wright, 1982). The mechanism of ketamine analgesia remains controversial, particularly the relationship with opioid receptors. The so-called phencyclidine (PCP) site may represent a subgroup of sigma opioid receptors that block spinal nociceptive reflexes (Reich and Silvay, 1989). Cross-tolerance between ketamine and opiates has been demonstrated, but naloxone does not consistently affect analgesia (Finck *et al.*, 1988; France *et al.*, 1989). Other sites of ketamine action have been suggested (Reich and Silvay, 1989).

D. Pharmacologic Effects

1. Nervous System

Although ketamine is typically considered to increase cerebral blood flow and intracranial pressure (ICP) (Wright, 1982), direct cerebral vasodilation has not been a consistent finding (Reich and Silvay, 1989). When ventilation is not controlled, ICP may rise following vasodilation secondary to hypercapnia (Pfenninger *et al.*, 1985; Schwedler *et al.*, 1982; Tranquilli *et al.*, 1983). Cerebral blood flow in the rabbit is increased with ketamine without an elevated P_{CO_2} (Dhasmana *et al.*, 1984).

Ketamine or tiletamine is capable of inducing seizures in dogs and cats (Chen *et al.*, 1969; Garmer, 1969; Wright, 1982). In rats and mice, however, these agents actually raise the threshold to chemically or electrically induced seizures (Chen *et al.*, 1969; Myslobodsky *et al.*, 1981).

Tiletamine produces excitation in mice and rats at low doses; at higher doses, anesthesia is produced in these species, but not in rabbits or guinea pigs (Chen *et al.*, 1969).

2. Respiratory System

Respiratory effects of ketamine are relatively minor, although a dose-dependent depression can be demonstrated. A characteristic apneustic pattern of breathing is seen commonly (Muir, 1985; Wright, 1982). In rhesus monkeys, normal respiratory variables are maintained, except for transient elevations in arterial P_{CO_2} and venous P_{O_2} (Reutlinger *et al.*, 1980). Low doses of ketamine in the rabbit result in a decreased respiratory rate and P_{O_2} (Dhasmana *et al.*, 1984). Ketamine sympathomimetic actions promote bronchial muscle relaxation, and protective airway reflexes are generally maintained (Thurmon and Benson, 1987).

3. Cardiovascular System

In contrast to the cardiovascular depression seen with most anesthetic agents, ketamine administration to dogs and cats typically increases hemodynamic variables (Muir, 1985; Wright, 1982). However, stimulatory cardiovascular effects (or lack of depression) may disappear at higher doses (Muir and Hubbell, 1988). The effect of ketamine on circulating levels of catecholamines has been controversial (Wright, 1982).

Positive inotropic and chronotropic effects of ketamine may be mediated by a peripheral sympathomimetic action of ketamine to inhibit norepinephrine uptake at adrenergic nerve endings (Miletich *et al.*, 1973). Cardiovascular stimulation with ketamine can be blocked by α- and β-blocking agents and verapamil and, clinically, by concurrent use of other anesthetic agents or sedatives (Muir, 1985; Reich and Silvay, 1989).

Increases in heart rate and mean arterial pressure have been reported following ketamine administration in rats and rabbits (Kumar and Kumar, 1984; Wang *et al.*, 1991). Other studies in rabbits have found an elevated heart rate (Dhasmana *et al.*, 1984) and a decreased arterial pressure and cardiac output (Van Leeuwen *et al.*, 1990). The effect of ketamine on cardiac output of the rat has been variable (Idvall *et al.*, 1980; Miller *et al.*, 1980). In nonhuman primates, ketamine administration has resulted in an unchanged heart rate, mean arterial pressure, and rectal temperature (Reutlinger *et al.*, 1980), or cardiodepression (Chimoskey *et al.*, 1975; Ochsner, 1977). It has been suggested that ketamine can open arteriovenous shunts in rats (Miller *et al.*, 1980) and pregnant guinea pigs (Mårtensson and Carter, 1984).

Baroreceptor reflex response is altered in ketamine-anesthetized rats (Wang *et al.*, 1991) and rabbits (Blake *et al.*, 1982), and there is a depressed vasoconstrictor response to hemorrhage compared with conscious animals (Van Leeuwen *et al.*, 1990).

Decreased myocardial contractility with anesthetic doses of ketamine *in vivo* and in isolated cardiac preparations (Dowdy and Kaya, 1968; Parker and Adams, 1978) may explain negative inotropic and chronotropic effects seen in pithed rat and rabbit preparations (Clanachan *et al.*, 1976). Conflicting reports on the *in vitro* inotropic effects of ketamine may be due to differences in species, tissue preparation, or temperature or calcium concentration of the medium (Stowe *et al.*, 1992). A direct negative inotropic effect is usually overshadowed *in vivo* by central sympathetic stimulation. The arrhythmogenic potential of ketamine remains controversial (Reich and Silvay, 1989).

4. Other Pharmacologic Effects

Other reported effects of ketamine include dose-dependent hypothermia in rats (Lin *et al.*, 1978); maintained thermal balance in rhesus monkeys (Hunter *et al.*, 1981); marked (normal) salivation response to hyperthermia, in contrast to five other anesthetic regimens (Furuyama *et al.*, 1989); increased serum renin activity (Pettinger *et al.*, 1975); altered serum follicle-stimulating hormone, testosterone, and androstenedione in rats following decapitation (Nazian, 1988); small decreases in hematocrit and plasma protein, and a larger decrease in leukocyte (principally lymphocyte) count compared with values in manually restrained animals (Loomis *et al.*, 1980); inhibited platelet aggregation in baboons (Atkinson *et al.*, 1985); resting myopia in humans and nonhuman primates, due largely to central parasympathetic neuronal tone (Crawford *et al.*, 1990); enhancement of neuromuscular blockade in human studies (Reich and Silvay, 1989) and in rabbits (Bogdan *et al.*, 1974) and monkeys (Tsai and Lee, 1989); enhanced ulcer formation in restrained rats, probably mediated by splanchnic vasoconstriction (Cheney *et al.*, 1974); teratologic effects in rats (Kochhar *et al.*, 1986); muscle necrosis from intramuscular injection of ketamine–xylazine in hamsters (Gaertner *et al.*, 1987) and rabbits (Beyers *et al.*, 1991); and autoamputation of digits following ketamine–acepromazine injections in guinea pigs (Latt and Ecobichon, 1984).

A single intramuscular injection of high doses of tiletamine in rabbits results in significant elevations in BUN (blood urea nitrogen) and creatinine. At lower doses, there are no changes in serum chemistry values, but mild nephrosis is evident histologically in most animals (Doerning *et al.*, 1992).

E. Antagonists

Reich and Silvay (1989) discuss a number of agents which, based on proposed mechanisms of action, should antagonize cyclohexamine effects; results in most cases have been contradictory. Agents include metaphit (which acetylates NMA receptors), naloxone (opioid antagonist), norepinephrine and serotonin receptor blockers, and anticholinesterase agents (e.g., 4-aminopyridine or physostigmine) (Reich and Silvay, 1989; Wright, 1982).

V. STEROIDS

A. Description

Anesthetic effects of steroids have been recognized since the 1940s (Sutton, 1972). Currently, Saffan or Althesin, a combination of the steroids alphaxalone and alphadolone, is available for veterinary use (although not in the United States). This combination is licensed in the United Kingdom for use in cats and nonhuman primates in which its use is characterized by a rapid induction of short-term anesthesia, with rapid recovery, a wide safety margin, and minimal accumulation with repeated doses (Jones, 1985). New steroid preparations, such as pregnanolone emulsion, may provide improved alternatives to available products (Høgskilde *et al.*, 1991).

Alphaxalone/alphadolone has been used successfully in most domestic species. Greatest difficulties have arisen in the dog in which Cremophor leads to histamine release and hypotension (Booth, 1988a).

Rapid onset and high potency are associated with presence of a 3-hydroxy group in the steroid molecules. Alphaxalone is insoluble in water and only partially soluble in Cremophor EL. Alphadolone, a steroid with one-third the anesthetic potency of alphaxalone, has been added to increase the solubility of alphaxalone (Jones, 1985).

B. Biodisposition

Induction is almost immediate following intravenous administration of alphaxalone/alphadolone in the rat. Duration is dose dependent, but generally short, and recovery is rapid. Alphaxalone plasma half-life is approximately 7 min and serum protein binding is about 40%. Excretion of the steroids is rapid, through both feces and urine. Following intravenous injection, 76.5% of a radiolabeled dose can be recovered from bile within 3 hr of injection; fecal excretion continues for several days, suggesting enterohepatic circulation of the steroid metabolites. The principal metabolite in rats apparently is a glucuronide of 2-hydroxy-alphaxalone. A similar plasma half-life is seen in the mouse and monkey (Child *et al.*, 1972a).

C. Mechanism of Action

Alphaxalone is a positive modulator of central GABAergic transmission and appears to act by enhancing the function of the GABA-activated chloride channel (Oliver *et al.*, 1991). It has been confirmed that the CNS target of the anesthetic steroids is the $GABA_A$ receptor and that there may be a specific steroid-binding site. Although the mechanism is similar to that of the barbiturates, there is not a common binding site (Im *et al.*, 1990; Turner *et al.*, 1989). Anxiolytic effects also can be demonstrated and appear to have a site of action independent of the benzodiazepine site on the $GABA_A$ receptor complex (Britton *et al.*, 1991).

D. Pharmacologic Effects

Alphaxalone–alphadolone has anticonvulsant effects in rat models of epilepsy, but only at hypnotic/anesthetic doses. This pattern resembles that of pentobarbital and supports the view that steroids and barbiturates have similar mechanisms of action (Peterson, 1989). In contrast to anticonvulsant effects, a high incidence of myoclonic jerks occurs in mice following alphaxalone/alphadolone administration (File and Simmonds, 1988).

Severe hypotension may follow intravenous alphaxalone/alphadolone administration to cats, but not sheep, rabbits, pigs, or humans (Booth, 1988a); in the pig, there is a transient hypotension and tachycardia (Glen, 1980). In the rat, cardiac output is decreased (Gumbleton *et al.,* 1990a) and blood pressure is decreased (Wang *et al.,* 1991) or unchanged (Faber, 1989). Renal and hepatosplanchnic blood flows were highest among five anesthetic regimes evaluated and most nearly those in conscious animals (Gumbleton *et al.,* 1990a). There is an enhanced pressor response to administration of a nitric oxide inhibitor (Wang *et al.,* 1991).

In pigs, there are minimal changes in blood gas values following alphaxalone/alphadolone administration (Glen, 1980). Surgical anesthesia in squirrel monkeys results in reduced body temperature and respiratory rate (Lögdberg, 1988); in cynomolgus monkeys, there is a rapid decrease in temperature without a change in respiratory rate (Box and Ellis, 1973). Gentamicin clearance and glomerular filtration rate in rats are similar to conscious controls (Gumbleton *et al.,* 1990b).

Hormonal and reproductive effects of alphaxalone/alphadolone are minimal, although weak antiestrogenic activity can be demonstrated (Child *et al.,* 1972b).

VI. α_2-AGONISTS

A. General

These drugs are usually classified as sedative–analgesics and skeletal muscle relaxants. They are not anesthetics (although there may be species differences in this regard) nor are they tranquilizers in the strictest sense (Booth, 1988b). Three α_2-agonists are commonly used clinically: xylazine, detomidine, and medetomidine.

B. Mechanism of Action

Central noradrenergic neurotransmission is potently decreased by α_2-agonists, but neither the hypnotic–anesthetic nor the analgesic effect is explained by this action alone. The mechanism involves a pertussis-sensitive guanine nucleotide-binding (G) protein and a 4-aminopyridine-sensitive ion channel (Daunt and Maze, 1992; Maze and Regan, 1991). In the CNS, α_2-agonists have the effect of inhibiting presynaptic calcium influx and neurotransmitter release (e.g., dopamine, norepinephrine) and result in sedation, decreased sympathetic tone, and analgesia (Haskins, 1992; MacDonald *et al.,* 1989; Virtanen, 1986). Inhibition of noradrenergic neurons in the locus coeruleus by α_2-agonists may explain sedative effects, as these neurons are known to be important in the regulation of vigilance (Stenberg, 1989; Virtanen, 1989). Evidence also shows α_2-mediated antinociceptive effects at the level of the spinal cord (Takano and Yaksh, 1992; Virtanen, 1986).

α_2-Agonism in the periphery causes vasoconstriction, decreased insulin release, diuresis, decreased gastrointestinal motility, and thrombasthenia (Haskins, 1992). Other α_2-agonists such as clonidine are used clinically for the treatment of hypertension, migraine prophylaxis, and relief of heroin withdrawal symptoms (Scheinin and MacDonald, 1989).

C. Xylazine

1. Description

Xylazine is a thiazole drug (Salonen, 1992) that is used alone or with other agents for sedation, immobilization, and anesthesia; it is an approved drug in the United States for use in cats, dogs, and horses. There is an abundance of information on the clinical use of this agent in these (and other) domestic species (Benson and Thurmon, 1987; Greene and Thurmon, 1988). Considerable variation exists among species in response to xylazine.

2. Biodisposition

Xylazine is rapidly absorbed following intramuscular administration and is rapidly and extensively metabolized (Salonen, 1992); absorption rates are similar for xylazine, medetomidine, and detomidine. Elimination is relatively rapid (Salonen, 1992).

Radiolabeled xylazine is rapidly distributed following intravenous injection in the rat, and 70% of the radioactivity may be recovered in urine, only 8% of which represents unchanged xylazine. The rapid metabolism of xylazine yields about 20 different metabolic products (Duhm *et al.,* 1969, discussed by Garcia-Villar *et al.,* 1981). Hepatic metabolism includes oxidation of the aromatic moiety and cleavage of the heterocyclic ring; no active metabolites have been reported (Salonen, 1992).

3. Pharmacologic Effects

As in most species, xylazine alone in mice does not produce sleep or loss of the righting reflex (Hsu, 1992). Analgesia following xylazine has been demonstrated in a variety of species, including the rat and mouse (Browning *et al.,* 1982; Schmitt *et al.,* 1974).

There is, in general, minimal effect of xylazine on the respiratory system (Greene and Thurmon, 1988). In rhesus monkeys, xylazine does not affect respiratory rate or blood gas values (Reutlinger *et al.,* 1980).

Xylazine administration results in hypotension and bradycardia in conscious, spontaneously hypertensive and urethane-anesthetized rats (Savola, 1986) and in rhesus monkeys (Reutlinger *et al.,* 1980). Arterial pressure in guinea pigs is decreased at a relatively low dosage (Flynn *et al.,* 1988). Cardiovascular effects in domestic species include a significantly reduced heart rate and a variety of arrhythmias, and effects on cardiac output and blood pressure vary with species and route of administration (Greene and Thurmon, 1988).

Other reported pharmacologic effects of xylazine include prolonged gastrointestinal transit time in several species, including mice, possibly due to localized inhibition of acetylcholine release (Greene and Thurmon, 1988; Hsu, 1982); polyuria in rats (Hsu *et al.,* 1986) and domestic animals (Greene and Thurmon, 1988); decreased rectal temperature in rhesus monkeys (Reutlinger *et al.,* 1980); decreased plasma insulin and increased plasma glucagon and blood glucose (Greene and Thurmon, 1988), probably a generalized α_2 effect on the pancreas (MacDonald and Virtanen, 1992); increased GH and decreased serum antidiuretic hormone (Greene and Thurmon, 1988); increased serum prolactin and altered serum testosterone levels in rats following decapitation (Nazian, 1988); acute reversible corneal desiccation and lens opacity in rats and mice (Calderone *et al.,* 1986); and increased uterine pressure (Wheaton *et al.,* 1989).

4. Antagonists

There are numerous reports of pharmacologic reversal of xylazine sedation using specific α_2-receptor antagonists such as yohimbine, tolazoline, and idazoxan (see Thurmon *et al.,* 1992). Reduced sedation also may be produced by nonspecific CNS stimulants such as 4-aminopyridine or doxapram. These agents may also speed recovery when xylazine is used in combination with other agents such as ketamine (Greene and Thurmon, 1988). Both *in vivo* and *in vitro* evidence shows significant species differences in the action of the α_2-antagonists (Hsu, 1992).

D. Medetomidine

1. Description

Medetomidine is a highly selective α_2-agonist that was developed in Finland for use as a sedative–analgesic in dogs and cats. It is an imidazole derivative (Salonen, 1992) that is more lipophilic and potent than xylazine, and is eliminated faster. Medetomidine is pharmacologically more α_2-selective, although the significance of this with respect to the anesthetic effect is unknown (Scheinin and MacDonald, 1989). The hypnotic/analgesic actions are due to the *d*-enantiomer (dexmedetomidine) (MacDonald and Virtanen, 1992). A comprehensive series of articles on the veterinary use of medetomidine has been published (Lammintausta *et al.,* 1989).

2. Biodisposition

Absorption of medetomidine following subcutaneous administration in the rat is rapid, with peak plasma concentrations reached within 10 min. Peak levels in the brain are five times higher than plasma and are reached in 15–20 min. Approximately 85% of the drug in plasma is protein bound. Excretion of ^{3}H-labeled medetomidine is mainly in the urine: 41% over 72 hr compared with 18% in the feces. Five percent or less of the urine radioactivity is unchanged medetomidine. Elimination half-life is 1.6 hr (Salonen, 1989).

Medetomidine and detomidine are metabolized similarly, by hepatic monooxygenases. Hydroxylated products may be subsequently oxidized or conjugated with glucuronic acid. The principal metabolite in rats is hydroxydetomidine glucuronide. Simple hepatic hydroxylation can explain the rapid removal of the drug; metabolism is regulated primarily by hepatic blood flow (Salonen, 1992).

3. Pharmacologic Effects

In mice, very low doses of medetomidine are anxiolytic without obvious signs of sedation (MacDonald *et al.,* 1989). There is a dose-dependent decrease in spontaneous activity and prolongation of barbiturate sleep time, more potent than with either detomidine or xylazine (Virtanen, 1986). In rats, there is a dose-dependent sedation and loss of righting reflexes, but deep sedation is not produced in mice or rabbits. Both medetomidine and detomidine can induce loss of the righting reflex in young chicks (MacDonald and Virtanen, 1992).

Antinociceptive actions of medetomidine have been demonstrated in a variety of assays and species (Vainio, 1992), including subcutaneous formalin injection in rats (Pertovaara *et al.,* 1990) and the acetic acid-induced writhing test in mice (Virtanen, 1986). Antinociceptive and sedative actions of medetomidine in rabbits, guinea pigs, and hamsters are inconsistent (Vainio, 1992).

A common feature of medetomidine administration is a dose-dependent bradycardia (MacDonald and Virtanen, 1992; Ruskoaho and Leppäluoto, 1989; Venugopalan *et al.,* 1994). The mechanism involves central, sympathetic effects at lower medetomidine doses and peripheral vagal effects present at higher doses (MacDonald and Virtanen, 1992). Second-degree atrioventricular block has been observed in dogs (Vainio, 1989).

There is typically a transient increase in blood pressure following medetomidine administration, attributed to peripheral α_2 effects, and a subsequent decrease which is probably centrally mediated. This pattern has been observed in dogs, chloralose-anesthetized cats, pentobarbital-anesthetized rats, and conscious spontaneously hypertensive rats (Savola, 1989; Vainio, 1990; Venugopalan *et al.,* 1994). Others have reported unchanged blood pressure in cynomolgus monkeys at sedative doses (Mann *et al.,* 1991) and in SHR rats (Ruskoaho and Leppäluoto, 1989).

Other reported pharmacologic effects of medetomidine (or dexmedetomidine) include hypothermia (MacDonald and Virtanen, 1992; MacDonald *et al.,* 1989; Vainio, 1989); vomiting, especially in cats, and occasional muscle jerks (Vainio, 1989); depressed adrenal steroidogenesis both *in vitro* and *in vivo* with dexmedetomidine (Maze *et al.,* 1991); suppressed gastric secretion in rats (Savola *et al.,* 1989); and diuresis and natriuresis in rats, due in part to increased plasma levels of atrial natriuretic peptide (Ruskoaho and Leppäluoto, 1989).

4. Antagonists

Atipamezole is a highly selective and potent α_2-antagonist that rapidly reverses sedation as well as other behavioral and physiologic effects of medetomidine (MacDonald *et al.,* 1989; Virtanen, 1989), including bradycardia in cynomolgus monkeys (Mann *et al.,* 1991) and rats (MacDonald and Virtanen, 1992).

E. Detomidine

1. Description

Detomidine is a selective α_2-agonist that was developed in Finland for use in horses and cattle. It is an imidazole derivative closely related to medetomidine (Daunt *et al.,* 1993; Salonen, 1992; see series of articles edited by Lindberg, 1986). The *in vitro* α-adrenoceptor interactions of detomidine are similar to medetomidine, although there are differences in potency, both *in vitro* and *in vivo,* which are small relative to xylazine. Detomidine is a sedative and potent analgesic in rats and mice, but does not lead to loss of the righting reflex, even at high doses (Virtanen, 1986).

2. Biodisposition

Absorption of detomidine following subcutaneous administration in the rat is rapid, with peak plasma concentrations reached within 10 min. Peak levels in the brain are three times higher than plasma. Excretion of ^{3}H-labeled detomidine is mainly in the urine: 62% over 72 hr compared with 22% in the feces. Only a small amount of the urine radioactivity is unchanged detomidine. Because the plasma elimination half-life is 12.7 hr (compared with a much shorter duration of action), redistribution of the drug must be relatively important in termination of clinical effect. In the dog, horse, and calf, approximately 90% of the drug in plasma is protein bound. Detomidine metabolism is similar to medetomidine. In rats, equal amounts of medetomidine glucuronide and medetomidine carboxylic acid are found in urine (Salonen, 1986).

3. Pharmacologic Effects

Detomidine prolongs barbiturate sleep time in mice in a dose-dependent manner. Its effect on spontaneous activity of mice is biphasic, with a decrease in activity at lower doses, but an actual increase in activity at higher doses; righting reflex is not lost at any dose (Virtanen, 1986). Antinociceptive effects in rodents are potent, similar to medetomidine but much greater than xylazine. These effects, as with sedation, are α_2 specific and are probably mediated through central inhibition of noradrenergic neurons; it is unlikely that dopaminergic, histaminergic, opioidergic, or serotoninergic pathways are involved (Virtanen, 1986).

Blood pressure and heart rate changes with detomidine are similar to those seen with medetomidine, except that detomidine does not reduce the blood pressure of conscious SHR rats, except at high dosage (Savola, 1986). Low doses of detomidine in rats result in hypothermia which can be blocked by yohimbine. At higher doses, an increased rectal temperature is seen (Virtanen, 1986).

VII. SEDATIVES AND TRANQUILIZERS

Tranquilizers are psychopharmacological agents, drugs that may have a number of physiologic effects in addition to mental calming, decreased response to environmental stimuli, and muscular relaxation. In general, however, they do not produce sleep, analgesia, or anesthesia, even at increased dosage, and effects can be reversed with adequate stimulation (Soma, 1983). They are also referred to as ataractics, neuroleptics, or psychotropic agents (Booth, 1988c). In human medicine, this broad class of agents is used widely in the treatment of various psychiatric disorders, where the distinction between neuroleptic/antipsychotic agents (e.g., phenothiazines and butyrophenones) and antianxiety/sedative agents (benzodiazepines) is perhaps more relevant than in animals (Baldessarini, 1990).

Abundant information is available in the veterinary (in the case of phenothiazines) and human (benzodiazepines) medical literature, but very little exists on the pharmacology of these agents in rodents and rabbits. Use of these agents either for calming effect or as preanesthetics has received relatively little attention in laboratory animal medicine; the greatest use has been as an adjunct with other anesthetic agents, especially ketamine (see Section IX,B).

A. Phenothiazine Derivatives

Phenothiazines have a wide spectrum of activity, from sedation and drowsiness at lower doses to ataxia and somnolence, and even cataleptic symptoms at higher doses. There are essentially no analgesic effects. Phenothiazines used in veterinary medicine are limited primarily to chlorpromazine (Thorazine), promazine (Sparine), and acepromazine; the last is in greatest use at present (Booth, 1988c).

Central actions of the phenothiazines are mediated by dopamine blockade, but there are also important peripheral effects involving α-adrenergic antagonism (and "epinephrine reversal") (Booth, 1988c; Soma, 1983).

Acepromazine in the dog may decrease arterial pressure and heart rate, and generally lowers rectal temperature and respiratory rate; effects on blood gas values are minimal. The phenothiazines potentiate the hypotension associated with epidural anesthesia (Booth, 1988c).

Other reported effects of phenothiazines vary with drug, dosage, and method of assessment. These include hyperglycemia, reduced hematocrit due to splenic sequestration, gastrointestinal antisecretory activity, lowered CNS seizure threshold, and teratogenic effects in mice and rats. These drugs also may be antipyretic, hypometabolic, and antiemetic and may be weakly anticholinergic, antihistaminic, and antispasmodic (Booth, 1988c; Niemegeers *et al.,* 1974). Intramuscular chlorpromazine in the rabbit is not recommended due to severe myositis and paralysis (Bree *et al.,* 1971). Chlorpromazine stimulates hepatic microsomal enzyme activity in the rat (Aurori and Vesell, 1974).

B. Butyrophenone Derivatives

Butyrophenones and phenothiazines, although chemically dissimilar, share many properties. They are used widely for the symptomatic treatment of psychoses in humans and also have been used to treat nausea, vomiting, and hiccoughing. Most theories of their antipsychotic effects are based on antidopaminergic actions in the basal ganglia and limbic portions of the forebrain (Baldessarini, 1990). Veterinary use has been limited primarily to droperidol in combination with fentanyl (Innovar-Vet), fluanisone in combination with fentanyl (Hypnorm), and azaperone (Stresnil).

1. Droperidol

Droperidol is relatively short-acting compared with other butyrophenones. It is an especially potent antiemetic and antitraumatic shock agent, and has greater potency than chlorpromazine in its tranquilizing effect, and inhibition of learning behavior and amphetamine antagonism in rats (Booth, 1988c). The neuroleptic agents have no direct analgesic effects, but controversy remains about the effects of these agents on opioid-mediated antinociception. Studies in sheep (Kyles *et al.,* 1993) and mice (Greene, 1972) found that droperidol enhanced the antinociceptive activity of fentanyl.

Cardiovascular effects are similar to chlorpromazine and are dose dependent. There is mild hypotension at low doses and hypotension, bradycardia, reduced cardiac output, and reduced myocardial contractility at higher doses. Hypotension is mediated at least in part by α-adrenergic blockade and decreased peripheral resistance. Droperidol also is effective against epinephrine-induced arrhythmias (Yelnosky *et al.,* 1964) and increases plasma prolactin (Booth, 1988c).

2. Azaperone

Azaperone in the rat has been characterized as a sedative neuroleptic. Conditioned and exploratory behaviors are inhibited and there is a potent protective effect in traumatic shock (Niemegeers *et al.,* 1974); an analgesic effect is apparent (Olson and Renchko, 1988). Similar sedative and behavioral effects are seen in mice, plus a potent potentiating effect on pentobarbital hypnosis (Niemegeers *et al.,* 1974). Azaperone reverses dominant–subordinate relationships in rats (Desmedt *et al.,* 1975) and mice (Niemegeers *et al.,* 1974), as well as in swine (Porter and Slusser, 1985). Its approved use in swine is to control aggressiveness and fighting (Booth, 1988c).

In the rat, 90% of a radiolabeled azaperone dose is excreted within 48 hr. About 25% of the dose is recovered in the urine and the remainder in feces; 13% is excreted unchanged. Principal metabolic pathways are oxidative N-dearylation and N-dealkylation (Heykants *et al.,* 1971).

Azaperone administration to male mice and to rats of both sexes results in a transient elevation in the respiratory rate. In female mice, there is a progressive decline in respiration (Olson and Renchko, 1988). An increased respiratory rate has been reported in swine, horses, and dogs. In swine, hypotension is observed and is apparently dose and route dependent (Booth, 1988c).

Azaperone displays a potent α-adrenolytic effect on rabbit spleen strips *in vitro* and a less potent antihistamine effect on guinea pig ileum. In the isolated guinea pig atrium, azaperone does not decrease contractility, but does produce a negative chronotropic effect (Niemegeers *et al.,* 1974).

3. Fluanisone

Fluanisone is used most commonly in combination with fentanyl as a neuroleptanalgesic (Hypnorm; see Section IX).

C. Benzodiazepine Derivatives

1. Description

A variety of benzodiazepines are used in human medicine for their sedative, hypnotic, anxiolytic, skeletal muscle relaxant, and anticonvulsant properties. Uses include preanesthesia and anesthesia induction, and they may be part of a balanced anesthesia regimen, although not capable alone of general anesthesia (Rall, 1990). Current veterinary use is limited largely to diazepam, midazolam, and zolazepam and typically involves combination with another agent (e.g., a dissociative agent) for anesthesia or anesthetic induction. Diazepam is insoluble in water; parenteral solutions, for intravenous injection, include propylene glycol and ethanol and may cause thrombophlebitis (Booth, 1988c; Hall and Clarke, 1991a). Midazolam is a relatively new benzodiazepine drug that is chemically similar to diazepam, but water soluble. In humans, it has a short duration and rapid elimination that make it suitable for administration as an infusion. Zolazepam is a benzodiazepine that is combined with the dissociative agent tiletamine as the veterinary product Telazol (Ilkiw, 1992).

2. Biodisposition

Rall (1990) provides a detailed discussion of benzodiazepine metabolism in humans. Diazepam shows a biexponential decline in plasma following intravenous administration and is highly protein bound. There are species differences in elimination half-life from approximately 1 to 7 hr in the rat, guinea pig, rabbit and dog. By comparison, the value in humans is 33 hr. In the rat, rabbit, and dog, the blood clearance rate exceeds the liver blood flow, suggesting extrahepatic elimination (Klotz *et al.,* 1976). The major metabolite of diazepam is *N*-desmethyldiazepam (nordiazepam), an active metabolite (Booth, 1988c).

3. Mechanism of Action

Benzodiazepines potentiate GABA effects in the CNS via receptors distinct from barbiturate, GABA, and chloride channel-specific receptors; they interact specifically at the $GABA_A$ receptor subset (Goodchild, 1993). The pharmacologic effect is modulation of the release of excitatory neurotransmitters such as norepinephrine, dopamine, and serotonin (Muir *et al.,* 1991). Specific receptors also are present in the periphery (Booth, 1988c), but the greatest binding capacity is found in the cerebral cortex (Rall, 1990).

4. Pharmacologic Effects

For a given drug in this class, considerable differences exist among species in the pharmacologic profile (Rall, 1990). Tranquilizing effects of the benzodiazepines are species variable, and excitation may follow administration in dogs and cats (Booth, 1988c; Rehm and Schatzmann, 1986).

Benzodiazepines in general have minimal cardiovascular effects and mild respiratory effects (Booth, 1988c). Diazepam in rabbits and mice leads to respiratory depression only at lethal doses (Bradshaw and Pleuvry, 1971). In swine, midazolam results in a decreased respiratory rate, but maintains blood gas values (Smith *et al.,* 1991).

Increasing, cumulative doses of midazolam in the pig result in a progressive decline in heart rate, but arterial pressure is increased and cardiac output is maintained (Smith *et al.,* 1991). Diazepam does not decrease blood pressure in guinea pigs (Flynn *et al.,* 1988). Decreased myocardial contractility occurs in swine (Smith *et al.,* 1991) and dogs (Jones *et al.,* 1979) at higher doses of midazolam. Direct negative inotropic effects have been demonstrated *in vitro,* but are dose dependent and may not reflect concentrations encountered clinically (Stowe *et al.,* 1992).

Other reported pharmacologic effects of diazepam include hypothermia in squirrel monkeys (Clark and Lipton, 1981) and high fetal losses in pregnant mice with large intravenous doses (Rank and Jensen, 1989).

Climazolam has been used in several domestic species (Hall and Clarke, 1991a) and is an effective sedative in rats, although there is no loss of righting reflex or analgesic effect. There are significant strain and age differences in response (West and Green, 1987).

5. Antagonists

Flumazenil is a specific benzodiazepine antagonist that attenuates the central effects of the benzodiazepines. Studies of the antagonistic effects of flumazenil in humans and in rats and mice have been reviewed (Brogden and Goa, 1991).

VIII. LOCAL/REGIONAL ANESTHESIA

Local anesthetics (or analgesics) are drugs that block nerve conduction when applied locally at sufficient concentration. A variety of chemical types may have this effect, ranging from alcohols to cocaine to complex toxins such as tetrodotoxin (Åkerman, 1988). Numerous techniques have been developed utilizing these agents to produce localized, regional, or topical analgesia without need for systemic anesthesia (Booth, 1988d). Epidural anesthesia (or analgesia) now includes use not only of the typical local anesthetic agents, but also agents that interact with spinal cord opiate, phencyclidine, or α_2-adrenergic receptors (e.g., morphine, ketamine, and xylazine, respectively) (Klide, 1992).

Most clinically useful local anesthetics have a common chemical pattern that includes an aromatic, lipophilic end connected by an intermediate chain to a hydrophilic amine group. Molecules exist in both charged (ionic) and uncharged forms, the proportion of which varies with local pH. Chemical differences in pK_a, lipid solubility, and protein binding affect drug potency and onset and duration of action (Hall and Clarke, 1991b). Local analgesics enter the nerve cell membrane and, in the ionic form, physically block the sodium channel and prevent nerve conduction (Durham *et al.,* 1992). In peripheral nerves, the smallest, unmyelinated fibers are affected first. Local anesthetic agents affect excitable tissue other than nerve fibers, and systemic and side effects may follow entry into the systemic circulation. These agents also may cause local tissue injury (Hall and Clarke, 1991b) and have a variety of effects when used topically on the eye (Durham *et al.,* 1992). Methemoglobinemia following topical use of benzocaine has been well-described (Davis *et al.,* 1993; Severinghaus *et al.,* 1991).

The biodisposition of the various local anesthetics and the pharmacologic and toxic effects in humans and domestic animals have been discussed (see Booth, 1988d; Hall and Clarke, 1991b; Ritchie and Greene, 1990; also Additional Reading list). Unfortunately, there is little specific information for the typical laboratory animal species, although the use of these agents as an anesthetic adjunct to reduce side effects of general anesthetics has been proposed (Raman *et al.,* 1989).

Flecknell *et al.* (1990) have suggested that use of a eutectic mixture of local anesthetic (EMLA cream) to reduce the discomfort of venipuncture may be a useful refinement, especially with inexperienced staff. Durham *et al.* (1992) provide an excellant general review of topical ocular anesthetics. These agents have been proposed for use in ocular irritancy testing, but their use would require essentially continuous administration to be efficacious, and long-term use produces irreversible damage to the cornea.

MS-222 (tricaine; metacaine; ethyl *m*-aminobenzoate; 3-aminobenzoic acid ethyl ester; used as the methane sulfonate salt) is a soluble local anesthetic agent, chemically related to procaine, that is used commonly for sedation, immobilization, and anesthesia of fish and amphibians. It usually is delivered via the ambient water or in the respiratory stream and produces both central effects, following absorption, and a local anesthetic effect. Onset and recovery are generally rapid, although recovery time increases with increasing concentrations of the agent or prolonged exposure (Lemm, 1993; Späth and Schweickert, 1977).

IX. ANESTHETIC COMBINATIONS

Most of the drugs discussed earlier lack one or more of the properties of a general anesthetic: hypnosis, analgesia, or muscle relaxation. This has resulted in frequent attempts to improve the overall quality of anesthesia by the combination of two or more drugs. The following sections are a listing of some of the pharmacologic effects of the various anesthetic combinations used in laboratory animal experimentation and medicine. [*Note:* These effects are often dose dependent and the reader should therefore refer to the original reference.] Little information is available on the biodisposition of anesthetic drugs when used in combination.

A. Neuroleptanalgesia

Neuroleptanalgesia refers to a combination of an opioid analgesic and a tranquilizer, typically a phenothiazine or butyrophenone neuroleptic. The combination has synergistic effects on sedation of the animal and may be sufficient for surgical procedures. An additional drug(s) may be used to produce neuroleptanesthesia or balanced anesthesia, including anticholinergic premedication (Short, 1987). Pharmacology of the analgesics is discussed elsewhere in this text.

1. Fentanyl–Droperidol (Innovar-Vet)

Reported pharmacologic effects include hyperacusia and poor muscle relaxation in the rat, especially at low doses, and residual sedative effect (Wixson *et al.,* 1987a); hypercapnia hypoxia, and acidosis in the rat (Wixson *et al.,* 1987c); reduced respiratory rate and some evidence of respiratory acidosis and hypoxemia in rabbits (Guerreiro and Page, 1987); unchanged (Wixson *et al.,* 1987c) or elevated (Gummerlock and Neuwelt, 1990) heart rate in rats; severe and prolonged hypotension in rats (Wixson *et al.,* 1987c); no major changes in heart rate or arterial blood pressure in rabbits (Guerreiro and Page, 1987); no decrease in mean arterial pressure in the guinea pig premedicated with diazepam (Flynn *et al.,* 1988); no effect on antipyrine clearance, an indicator of intrinsic hepatic clearance, in contrast to three other anesthetic regimes (Gumbleton and Benet, 1991); dose-dependent decrease in core body temperature of rats (Wixson *et al.,* 1987d); rapid increase in plasma renin (Barrett *et al.,* 1988); and myositis, neuritis, and self-mutilation associated with intramuscular injection in guinea pigs (Leash *et al.,* 1973; Newton *et al.,* 1975).

2. Fentanyl–Fluanisone (Hypnorm)

Reported pharmacologic effects include a significant rise in plasma β-endorphin-like immunoreactivity in rats (Ramirez-Gonzalez *et al.,* 1991); respiratory rate depression in rabbits, rapidly reversed by a variety of opioid antagonists (Flecknell *et al.,* 1989); decreased arterial pressure, respiratory rate, and P_{O_2} in rabbits when diazepam is added (Flecknell *et al.,* 1983; Peeters *et al.,* 1988); near normal blood gas values, reduced but steady blood pressure, and elevated heart rate in rats when diazepam and atropine are added (Svendsen and Carter, 1985); progressive metabolic acidosis and hypotension, with near normal P_{O_2} and P_{CO_2} in pregnant guinea pigs, when diazepam and atropine are added (Carter, 1983); depressed respiratory rate in a variety of laboratory animal species when midazolam is added (Flecknell and Mitchell, 1984); unchanged heart rate and arterial pressure in rabbits when midazolam is added (Flecknell *et al.,* 1989); cardiac output almost double that of pentobarbital-anesthetized animals, and markedly higher muscle blood flow and reduced peripheral resistance when midazolam is added (Skolleborg *et al.,* 1990); partial antagonism in the rat by climazolam (West and Green, 1987); and striking increase in blood glucose in fed, but not fasted, rats (Johansen *et al.,* 1993).

3. Other Neuroleptanalgesic Combinations

Reported pharmacologic effects include unchanged blood gasses, but decreased *in vivo* protein synthesis in lungs of rats with fentanyl–midazolam (Heys *et al.,* 1989); increased mean arterial pressure in guinea pigs with fentanyl–diazepam (Brown *et al.,* 1989); markedly reduced aortic flow in rats with fentanyl–etomidate (De Wildt *et al.,* 1983); severe respiratory depression in rabbits with etorphine–methotrimeprazine (small animal Immobilon), but improved respiratory rate and blood gas values with reduction of dose and addition of diazepam (Flecknell *et al.,* 1983); prolonged hypercapnia, hypoxia, and acidosis, with hypotension and tachycardia in rats with etorphine–acepromazine–atropine (Svendsen and Carter, 1985); and reduced heart rate and arterial pressure, and respiratory depression, in rabbits with midazolam–xylazine–alfentanil (Borkowski *et al.,* 1990).

B. Ketamine Combinations

Goals of ketamine combinations include improved analgesia, muscle relaxation and sedation, prolonged duration, and decreased side effects (Muir, 1985). Use of adjunctive agents alters the pharmacokinetics of ketamine, in general, leading to a potentiated and prolonged effect (Pascoe, 1992; Waterman, 1983, 1984).

1. Ketamine–Xylazine

Reported pharmacologic effects include excessive urination, defecation, and salivation with intraperitoneal administration in rats (Hsu *et al.,* 1986; Wixson *et al.,* 1987a); decreased body temperature in rats (Wixson *et al.,* 1987d) and rabbits (Hobbs *et al.,* 1991); respiratory depression in rats (Hsu *et al.,* 1986; Wixson *et al.,* 1987c), mice (Erhardt *et al.,* 1984), and rabbits (Borkowski *et al.,* 1990; Marini *et al.,* 1991; Peeters *et al.,* 1988; Wyatt *et al.,* 1989); no effect on respiratory rate or arterial P_{O_2} or pH, but elevated P_{CO_2} and depressed venous P_{O_2} in the rhesus monkey (Reutlinger *et al.,* 1980); severe, prolonged hypotension with intraperitoneal administration in rats (Wixson *et al.,* 1987c) and mice (Erhardt *et al.,* 1984); decreased (Erhardt *et al.,* 1984; Sanford and Colby, 1980) or unchanged (Wixson *et al.,* 1987c) heart rate in rats; decreased heart rate and blood pressure in the rabbit (Borkowski *et al.,* 1990; Hobbs *et al.,* 1991; Peeters *et al.,* 1988; Popilskis *et al.,* 1991) and rhesus monkey (Reutlinger *et al.,* 1980); small decreases in heart rate and blood pressure in the rabbit even when atropine is included (Lipman *et al.,* 1990; Marini *et al.,* 1991); significantly depressed antipyrine clearance (Gumbleton and Benet, 1991); corneal epithelial calcium deposition in rat pups, prevented by yohimbine but not by application of eye lubricant (Guillet *et al.,* 1988); profound and prolonged depression in lymphocyte traffic (decreased output of lymphocytes) in the sheep lymphocyte traffic model and elevated levels of efferent lymph prostaglandin E_2 (Moore *et al.,* 1988); reversal of bradypnea and regained consciousness and righting reflexes in rats following yohimbine (Hsu *et al.,* 1986); reversal of respiratory depression and reduced time to return of reflexes in rats following yohimbine, tolazoline, or 4-aminopyridine (Komulainen and Olson, 1991); and anesthetic reversal following yohimbine in the rabbit (Keller *et al.,* 1988; Lipman *et al.,* 1987), rat (Hsu *et al.,* 1986), budgerigar (Heaton and Brauth, 1991), and ferret (Sylvina *et al.,* 1990).

2. Ketamine–Diazepam

Reported pharmacological effects include poor muscle relaxation and hyperacusia in the rat (Wixson *et al.,* 1987a); decreased core body temperature in the rat (Wixson *et al.,* 1987d); hypercapnia, hypoxia, and acidosis in the rat (Svendsen and Carter, 1985; Wixson *et al.,* 1987c); transiently lowered blood pressure and unchanged heart rate in the rat (Wixson *et al.,* 1987c); steady decline in blood pressure and heart rate with higher or repeated doses in the rat (Svendsen and Carter, 1985); and decreased arterial pressure and no blood pressure response to 115 dB (decibel) white noise in the guinea pig (Flynn *et al.,* 1988).

3. Other Ketamine Combinations

Reported pharmacologic effects include highest (compared with four other anesthetic regimens) fractional distribution of cardiac output to brain in rats given ketamine–midazolam (Gumbleton *et al.,* 1989); prolonged increase in respiratory rate in rats and transient increase followed by a decrease in rate in mice given ketamine–azaperone (Olson and Renchko, 1988); effective surgical anesthesia with decreased respiratory rate, but unchanged heart rate and arterial pressure, in the rabbit with guaifenesin and ketamine (Olson *et al.,* 1987); hypercapnia, hypoxia, and acidosis, and steady decline in blood pressure and heart rate in rats with ketamine–pentobarbital (Svendsen and Carter, 1985); slight elevation of P_{CO_2} and deceased hemoglobin in female rabbits with ketamine–chlorpromazine (Barzago *et al.,* 1992); and heavy urination in rats given ketamine–medetomidine (Nevalainen *et al.,* 1989).

C. Tiletamine–Zolazepam (Telazol)

The analgesic effectiveness of tiletamine–zolazepam varies significantly among species. Although useful for surgical anesthesia in rats (Silverman *et al.,* 1983), ferrets (Payton and Pick, 1989), gerbils (Hrapkiewicz *et al.,* 1989), dogs and cats (Tracy *et al.,* 1988), nociception is not eliminated in mice, hamsters (Silverman *et al.,* 1983), guinea pigs, and rabbits (Ward *et al.,* 1974), even at relatively high dosages associated with prolonged recumbency. Reported pharmacologic effects include decreased respiratory rate in juvenile rhesus macaques (Booker *et al.,* 1982); variable respiratory effect in the ferret, from shallow, rapid breathing at low dosage to appreciable decrease in rate at higher dosage (Payton and Pick, 1989); mild negative chronotropic and inotropic effects, and unchanged cardiac output and rhythm, in the juvenile rhesus macaque (Booker *et al.,* 1982); slightly decreased rectal temperature in the ferret (Payton and Pick, 1989) and juvenile rhesus macaque (Booker *et al.,* 1982); nephrotoxicity in the rabbit after normal anesthetic recovery, including both clinical signs and histopathologic lesions (Brammer *et al.,* 1991); decreased respiratory rate in hamsters when xylazine is added (Forsythe *et al.,* 1992); and significant hypotension and respiratory depression in rabbits when xylazine is added (Popilskis *et al.,* 1991). The pharmacology of Telazol has been thoroughly reviewed by Lin *et al.* (1993).

D. Chloralose Combinations

Urethane increases the solubility of α-chloralose and complements some of its shortcomings. Analgesia is provided, excess muscle activity is reduced, and exaggerated spinal reflex activity and CNS stimulation are suppressed (Green, 1979). Shortcomings of chloralose also may be counteracted by combination with other preanesthetic medications. Beneficial effects of chloralose combined with thiobarbiturate may be due to altered thiobarbiturate pharmakokinetics (Holzgrefe *et al.*, 1987; Silverman and Muir, 1993). This interaction may explain the successful use of chloralose with thiopental for repeated anesthesia of beagle puppies (Grad *et al.,* 1988). Chloralose infusion following droperidol and flunitrazepam premedication results in the stability of hemodynamic variables for 2 to 3.5 hr in swine (Benharkate *et al.,* 1993). Parker and Adams (1978) emphasize that the preservation of function attributed to chloralose may be altered when other agents are used concurrently or as preanesthetics.

Reported pharmacologic effects of chloralose–urethane include augmented pressor response to administration of a nitric oxide inhibitor (Aisaka *et al.,* 1991); respiratory depression (Hughes *et al.,* 1982) or unchanged respiratory variables in rats (Seyde *et al.,* 1985); respiratory depression in guinea pigs (Brown *et al.,* 1989); decreased arterial pressure, heart rate, and cardiac output in rats (Faber, 1989; Seyde *et al.,* 1985); normal blood gas values for 6 hr in hamsters, using induction with chloralose–urethane–pentobarbital and maintenance with supplemental doses of chloralose and urethane (Reid *et al.,* 1989); and increased plasma renin activity in rats (Faber, 1989).

E. Pharmacologic Effects of Other Anesthetic Combinations

Reported effects include decreased respiratory rate and arterial blood pressure, and elevated heart rate in rabbits with guaifenesin and pentobarbital (Olson *et al.,* 1987); decreased respiratory rate in rats with fentanyl–medetomidine, reversed completely with nalbuphine plus atipamezole (Hu *et al.,* 1992); minimal cardiorespiratory effects in rabbits with medetomidine–propofol or medetomidine–midazolam–propofol (Ko *et al.,* 1992); and respiratory acidosis, hypoxemia, and reduced hemoglobin oxygen saturation in rabbits with medetomidine–ketamine–diazepam (Mero *et al.,* 1989).

ACKNOWLEDGMENTS

The author gratefully acknowledges the professional support and advice offered by Dr. H. Richard Adams and the technical and secretarial assistance of Bonnie Bass, Clover Chenoweth, Jim Coyle, Margaret Hogan, and Tesha Tellez. This work was supported in part by NIH Grant RR07004.

REFERENCES

Adam, H. K., Glen, J. B., and Hoyle, P. A. (1980). Pharmacokinetics in laboratory animals of ICI 35 868, a new I.V. anaesthetic agent. *Br. J. Anaesth.* **52,** 743–746.

Adams, H. R. (1970). Prolongation of barbiturate anesthesia by chloramphenicol in laboratory animals. *J. Am. Vet. Med. Assoc.* **157,** 1908–1913.

Adams, H. R., and Dixit, B. N. (1970). Prolongation of pentobarbital anesthesia by chloramphenicol in dogs and cats. *J. Am. Vet. Med. Assoc.* **156,** 902–905.

Aeschbacher, G., and Webb, A. I. (1993). Propofol in rabbits. 2. Long-term anesthesia. *Lab. Anim. Sci.* **43,** 328–335.

Agostino, D., and Cliffton, E. E. (1964). Anesthetic effect on pulmonary metastases in rats. *Arch. Surg. (Chicago)* **88,** 735–739.

Aisaka, K., Mitani, A., Kitajima, Y., Ohno, T., and Ishihara, T. (1991). Difference in pressor responses to N^G-monomethyl-L-arginine between conscious and anesthetized rats. *Jpn. J. Pharmacol.* **56,** 245–248.

Äkerman, B. (1988). On the chemistry and pharmacology of local anaesthetic agents. *In* ''Local Anaesthesia and Regional Blockade'' (J. B. Löfström and U. Sjöstrand, eds.), pp. 1–22. Elsevier, Amsterdam.

Albrecht, D., and Davidowa, H. (1989). Action of urethane on dorsal lateral geniculate neurons. *Brain Res. Bull.* **22,** 923–927.

Altura, B. M., Altura, B. T., Carella, A., Turlapaty, P. D. M. V., and Weinburg, J. (1980). Vascular smooth muscle and general anesthetics. *Fed. Proc., Fed. Am. Soc. Exp. Biol.* **39,** 1584–1591.

Altura, B. T., and Altura, B. M. (1975). Pentobarbital and contraction of vascular smooth muscle. *Am. J. Physiol.* **229,** 1635–1640.

American Veterinary Medical Association (AVMA) (1993). 1993 Report of the AVMA Panel on Euthanasia. *J. Am. Vet. Med. Assoc.* **202,** 230–249.

Atkinson, P. M., Taylor, D. I., and Chetty, N. (1985). Inhibition of platelet aggregation by ketamine hydrochloride. *Thromb. Res.* **40,** 227–234.

Aurori, K. C., and Vesell, E. S. (1974). Comparative stimulatory effects of four phenothiazines on hepatic microsomal enzymes. *Drug Metab. Dispos.* **2,** 566–572.

Azadegan, A., Johnson, D. W., and Stowe, C. M. (1980). Influence of time interval between administration of chloramphenicol and thiamylal on the sleeping time of mice. *Am. J. Vet. Res.* **41,** 976–977.

Baldessarini, R. J. (1990). Drugs and the treatment of psychiatric disorders. *In* ''The Pharmacological Basis of Therapeutics'' (A. G. Gilman, T. W. Rall, A. S. Nies, and P. Taylor, eds.), 8th ed., pp. 383–435. Pergamon, New York.

Barrett, J. D., Eggena, P., and Krall, J. F. (1988). Independent change of plasma and tissue renin in response to anesthetics. *Clin. Exp. Hypertens.* **A10,** 757–765.

Barzago, M. M., Bortolotti, A., Omarini, D., Aramayona, J. J., and Bonati, M. (1992). Monitoring of blood gas parameters and acid-base balance of pregnant and non-pregnant rabbits *(Oryctolagus cuniculus)* in routine experimental conditions. *Lab. Anim.* **26,** 73–79.

Batjer, H. H. (1993). Cerebral protective effects of etomidate: Experimental and clinical aspects. *Cerebrovasc. Brain Metab. Rev.* **5,** 17–32.

Bellman, M. H., and Pleuvry, B. J. (1981). Comparison of the respiratory effects of ICI 35 868 and thiopentone in the rabbit. *Br. J. Anaesth.* **53,** 425–429.

Benharkate, M., Zanini, V., Blanc, R., Boucheix, O., Coyez, F., Genevois, J. P., and Pairet, M. (1993). Hemodynamic parameters of anesthetized pigs: A comparative study of farm piglets and Gottingen and Yucatan miniature Swine. *Lab. Anim. Sci.* **43,** 68–72.

Benson, G. J., and Thurmon, J. C. (1987). Species difference as a consideration in alleviation of animal pain and distress. *J. Am. Vet. Med. Assoc.* **191,** 1227–1230.

Beyers, T. M., Richardson, J. A., and Prince, M. D. (1991). Axonal degeneration and self-mutilation as a complication of the intramuscular use of ketamine and xylazine in rabbits. *Lab. Anim. Sci.* **41,** 519–520.

Blake, D. W., Blombery, P. A., and Korner, P. I. (1982). Effect of ketamine, althesin, and thiopentone on the valsalva-constrictor and heart rate reflexes of the rabbit. *J. Auton. Nerv. Syst.* **5,** 291–301.

Blake, D. W., Jover, B., and McGrath, B. P. (1988). Haemodynamic and heart rate reflex responses to propofol in the rabbit. *Br. J. Anaesth.* **61,** 194–199.

Bogdan, L. G., Glisson, S. N., and El-Etr, A. A. (1974). The effect of ketamine upon depolarizing and non-depolarizing neuromuscular blockade in rabbit. *Naunyn-Schmiederberg's Arch. Pharmacol.* **285,** 223–231.

Booker, J. L., Erickson, H. H., and Fitzpatrick, E. L. (1982). Cardiodynamics in the rhesus macaque during dissociative anesthesia. *Am. J. Vet. Res.* **43,** 671–675.

Booth, N. H. (1988a). Intravenous and other parenteral anesthetics. *In* ''Veterinary Pharmacology and Therapeutics'' (N. H. Booth and L. E. McDonald, eds.), 6th ed., pp. 212–274. Iowa State Univ. Press, Ames.

Booth, N. H. (1988b). Nonnarcotic analgesics. *In* ''Veterinary Pharmacology and Therapeutics'' (N. H. Booth and L. E. McDonald, eds.), 6th ed., pp. 329–362. Iowa State Univ. Press, Ames.

Booth, N. H. (1988c). Psychotropic agents. *In* ''Veterinary Pharmacology and Therapeutics'' (N. H. Booth and L. E. McDonald, eds.), 6th ed., pp. 363–395. Iowa State Univ. Press, Ames.

Booth, N. H. (1988d). Local anesthetics. *In* ''Veterinary Pharmacology and Therapeutics'' (N. H. Booth and L. E. McDonald, eds.), 6th ed., pp. 407–423. Iowa State Univ. Press, Ames.

Borkowski, G. L., Danneman, P. J., Russell, G. B., and Lang, C. M. (1990). An evaluation of three intravenous anesthetic regimens in New Zealand rabbits. *Lab. Anim. Sci.* **40,** 270–276.

Boucher, M., Dubray, C., Li, J. H., Paire, M., and Duchêne-Marullaz, P. (1991). Influence of pentobarbital and chloralose anesthesia on quinidine-induced effects on atrial refractoriness and heart rate in the dog. *J. Cardiovasc. Pharmacol.* **17,** 199–206.

Box, P. G., and Ellis, K. R. (1973). Use of CT1341 anaesthetic ('Saffan') in monkeys. *Lab. Anim.* **7,** 161–170.

Bradshaw, E. G., and Pleuvry, B. J. (1971). Respiratory and hypnotic effects of nitrazepam, diazepam and pentobarbitone and their solvents in the rabbit and the mouse. *Br. J. Anaesth.* **43,** 637–643.

Brammer, D. W., Doerning, B. J., Chrisp, C. E., and Rush, H. G. (1991). Anesthetic and nephrotoxic effects of Telazol ® in New Zealand white rabbits. *Lab. Anim. Sci.* **41,** 432–435.

Bree, M. M., and Cohen, B. J. (1965). Effects of urethane anesthesia on blood and blood vessels in rabbits. *Lab. Anim. Care* **15,** 254–259.

Bree, M. M., Cohen, B. J., and Abrams, G. D. (1971). Injection lesions following intramuscular administration of chlorpromazine in rabbits. *J. Am. Vet. Med. Assoc.* **159,** 1598–1602.

Britton, K. T., Page, M., Baldwin, H., and Koob, G. F. (1991). Anxiolytic activity of steroid anesthetic alphaxalone. *J. Pharmacol. Exp. Ther.* **258,** 124–128.

Brogden, R. N., and Goa, K. L. (1991). Flumazenil: A reappraisal of its pharmacological properties and therapeutic efficacy as a benzodiazepine antagonist. *Drugs* **42,** 1061–1089.

Brown, N. J., Thorne, P. R., and Nuttall, A. L. (1989). Blood pressure and other physiological responses in awake and anesthetized guinea pigs. *Lab. Anim. Sci.* **39,** 142–148.

Browning, S., Lawrence, D., Livingston, A., and Morris, B. (1982). Interactions of drugs active at opiate receptors and drugs active at α_2-receptors on various test systems. *Br. J. Pharmacol.* **77,** 487–491.

Bryan, C. E., Skipper, H. E., and White, L. (1949). Carbamates in the chemotherapy of leucemia. IV. The distribution of radioactivity in tissues of mice following injection of carbonyl-labeled urethane. *J. Biol. Chem.* **177,** 941–950.

Buckley, N. M., Gootman, P. M., Brazeau, P., Matanic, B. P., Frasier, I. D., and Gentles, E. L. (1979). Cardiovascular function in anesthetized miniature swine. *Lab. Anim. Sci.* **29,** 200–208.

Buelke-Sam, J., Holson, J. F., Bazare, J. J., and Young, J. F. (1978). Comparative stability of physiological parameters during sustained anesthesia in rats. *Lab. Anim. Sci.* **28,** 157–162.

Bunting, P., Ramsay, T. M., and Pleuvry, B. J. (1989). An in-vitro study of the interactions between intravenous induction agents and the calcium antagonists verapamil and nifedipine. *J. Pharm. Pharmacol.* **41,** 840–843.

Cain, D. P., Raithby, A., and Corcorah, M. E. (1989). Urethane anesthesia blocks the development and expression of kindled seizures. *Life Sci.* **44,** 1201–1206.

Calderone, L., Grimes, P., and Shalev, M. (1986). Acute reversible cataract induced by xylazine and by ketamine-xylazine anesthesia in rats and mice. *Exp. Eye Res.* **42,** 331–337.

Carmichael, F. J., Crawford, M. W., Khayyam, N., and Saldivia, V. (1993). Effect of propofol infusion on splanchnic hemodynamics and liver oxygen consumption in the rat. *Anesthesiology* **79,** 1051–1060.

Carruba, M. O., Bondiolotti, G., Picotti, G. B., Catteruccia, N., and Da Prada, M. (1987). Effects of diethyl ether, halothane, ketamine and urethane on sympathetic activity in the rat. *Eur. J. Pharmacol.* **134,** 15–24.

Carter, A. M. (1983). Acid-base status of pregnant guinea pigs during neuroleptanalgesia with diazepam and fentanyl-fluanisone. *Lab. Anim.* **17,** 114–117.

Chang, T., and Glazko, A. J. (1974). Biotransformation and disposition of ketamine. *Int. Anesthesiol. Clin.* **12,** 157–177.

Chaplin, M. D., Roszkowski, A. P., and Richards, R. K. (1973). Displacement of thiopental from plasma proteins by nonsteroidal anti-inflammatory agents. *Proc. Soc. Exp. Biol. Med.* **143,** 667–671.

Chappel, S. C., and Barraclough, C. A. (1976). The effects of sodium pentobarbital or ether anesthesia on spontaneous and electrochemically-induced gonadotropin release. *Proc. Soc. Exp. Biol. Med.* **153,** 1–6.

Chen, G., Ensor, C. R., and Bohner, B. (1969). The pharmacology of 2-(ethylamino)-2-(2-thienyl)-cyclohexanone•HCl (CI-634). *J. Pharmacol. Exp. Ther.* **168,** 171–179.

Cheney, D. H., Slogoff, S., and Allen, G. W. (1974). Ketamine-induced stress ulcers in the rat. *Anesthesiology* **40,** 531–535.

Child, K. J., Gibson, W., Harnby, G., and Hart, J. W. (1972a). Metabolism and excretion of Althesin (CT 1341) in the rat. *Postgrad. Med. J., Suppl.,* June, pp. 37–42.

Child, K. J., English, A. F., Gilbert, H. G., and Woollett, E. A. (1972b). An endocrinological evaluation of Althesin (CT 1341) with special reference to reproduction. *Postgrad. Med. J., Suppl.,* June, pp. 51–55.

Chimoskey, J. E., Huntsman, L. L., Gams, E., and Flanagan, W. J. (1975). Effect of ketamine on ventricular dynamics of unanesthetized baboons. *Cardiovasc. Res. Cent. Bull.* **14,** 53–57.

Clanachan, A. S., McGrath, J. C., and MacKenzie, J. E. (1976). Cardiovascular effects of ketamine in the pithed rat, rabbit and cat. *Br. J. Anaesth.* **48,** 935–939.

Clark, S. M., and Lipton, J. M. (1981). Effects of diazepam on body temperature of the aged squirrel monkey. *Brain Res. Bull.* **7,** 5–9.

Clemens, L. G., Popham, T. V., and Ruppert, P. H. (1979). Neonatal treatment of hamsters with barbiturate alters adult sexual behavior. *Dev. Psychobiol.* **12,** 49–59.

Cockshott, I. D., Douglas E. J., Plummer, G. F., and Simons, P. J. (1992). The pharmacokinetics of propofol in laboratory animals. *Xenobiotica* **22,** 369–375.

Coetzee, A., Fourie, P., Coetzee, J., Badenhorst, E., Rebel, A., Bolliger, C., Uebel, R., Wium, C., and Lombard, C. (1989). Effect of various propofol plasma concentrations on regional myocardial contractility and left ventricular afterload. *Anesth. Analg.* **69,** 473–483.

Cohen, M. L., Chan, S.-L., Way, W. L., and Trevor, A. J. (1973). Distribution in the brain and metabolism of ketamine in the rat after intravenous administration. *Anesthesiology* **39,** 370–376.

Collado, P. S., Pozo-Andrada, M. J., González, J., Jiménez, R., and Esteller, A. (1987). Effect of pentobarbital or urethane on bile secretion and chemical composition of blood in the rabbit. *Lab. Anim.* **21,** 11–17.

Commissaris, R. L., Semsyn, D. R., and Rech, R. H. (1982). Dispositional without functional tolerance to the hypothermic effects of pentobarbital in the rat. *J. Pharmacol. Exp. Ther.* **220,** 536–539.

Concas, A., Santoro, G., Mascia, M. P., Serra, M., Sanna, E., and Biggio, G. (1990). The general anesthetic propofol enhances the function of γaminobutyric acid-coupled chloride channel in the rat cerebral cortex. *J. Neurochem.* **55,** 2135–2138.

Conney, A. H., Davison, C., Gastel, R., and Burns, J. J. (1960). Adaptive increases in drug-metabolizing enzymes induced by phenobarbital and other drugs. *J. Pharmacol. Exp. Ther.* **130,** 1–8.

Covert, R. F., Drummond, W. H., and Gimotty, P. A. (1988). Chloralose alters circulatory response to α-receptor stimulation and blockade. *Am. J. Physiol.* **255,** H419–H425.

Cox, R. H. (1972). Influence of chloralose anesthesia on cardiovascular function in trained dogs. *Am. J. Physiol.* **223:** 660–667.

Crawford, K., Gabelt, B. T., Kaufman, P. L., and Bito, L. Z. (1990). Effects of various anesthetic and autonomic drugs on refraction in monkeys. *Curr. Eye Res.* **9,** 525–532.

Cunliffe-Beamer, T. L., Freeman, L. C., and Myers, D. D. (1981). Barbiturate sleeptime in mice exposed to autoclaved or unautoclaved wood beddings. *Lab. Anim. Sci.* **31,** 672–675.

Cupples, W. A., Veress, A. T., and Sonnenberg, H. (1982). Lack of effect of barbiturate and ketamine anesthesia on renal blood flow in chronically instrumented rats prepared for micropuncture. *Can. J. Physiol. Pharmacol.* **60,** 204–205.

Dada, M. O., Campbell, G. T., Horacek, M. J., and Blake, C. A. (1992). Intraperitoneal injection of chloral hydrate causes intra-abdominal adhesions and unilateral testicular atrophy in golden Syrian hamsters. *Life Sci.* **51,** 29–35.

Daniel, F. B., DeAngelo, A. B., Stober, J. A., Olson, G. R., and Page, N. P. (1992). Hepatocarcinogenicity of chloral hydrate, 2-chloroacetaldehyde, and dichloracetic acid in the male B6C3F1 mouse[1]. *Fundam. Appl. Toxicol.* **19,** 159–168.

Daunt, D. A., and Maze, M. (1992). α_2-adrenergic agonist receptors, sites, and mechanisms of action. *In* ''Animal Pain'' (C. E. Short and A. Van Poznak, eds.), pp. 165–180. Churchill-Livingstone, New York.

Daunt, D. A., Dunlop, C. I., Chapman, P. L., Shafer, S. L., Ruskoaho, H., Vakkuri, O., Hodgson, D. S., Tyler, L. M., and Maze, M. (1993). Cardiopulmonary and behavioral responses to computer-driven infusion of detomidine in standing horses. *Am. J Vet. Res.* **54,** 2075–2082.

Davis, J. A., Greenfield, R. E., and Brewer, T. G. (1993). Benzocaine-induced methemoglobinemia attributed to topical application of the anesthetic in several laboratory animal species. *Am. J. Vet. Res.* **54,** 1322–1326.

Davis, L. E., Davis, C. H., and Baggot, J. D. (1973). Comparative pharmacokinetics in domesticated animals. *In* ''Research Animals In Medicine'' (L. T. Harmison, ed.), DHEW Publ. No. (NIH) 72-333, pp. 715–733. U.S. Department Of Health, Education, and Welfare, Washington, D.C.

Desmedt, L. K. C., Van Bruggen, J. A. A., and Niemegeers, C. J. E. (1975). The effects of azaperone, a sedative neuroleptic of the butyrophenone series, on dominant-subordinate behavior in Wistar rats competing for food. *Psychopharmacologia (Berlin)* **41,** 285–289.

De Wildt, D. J., Hillen, F. C., Rauws, A. G., and Sangster, B. (1983). Etomidate-anaesthesia, with and without fentanyl, compared with urethane-anaesthesia in the rat. *Br. J. Pharmacol.* **79,** 461–469.

Dhasmana, K. M., Saxena, P. R., Prakash, O., and Van Der Zee, H. T. (1984). A study on the influence of ketamine on systemic and regional haemodynamics in conscious rabbits. *Arch. Int. Pharmacodyn. Ther.* **269,** 323–334.

Doerning, B. J., Brammer, D. W., Chrisp, C. E., and Rush, H. G. (1992). Nephrotoxicity of tiletamine in New Zealand white rabbits. *Lab. Anim. Sci.* **42,** 267–269.

Dos Santos, M., and Bogan, J. A. (1974). The metabolism of pentobarbitone in sheep. *Res. Vet. Sci.* **17,** 226–230.

Dowdy, E. G., and Kaya, K. (1968). Studies of the mechanism of cardiovascular responses to CI-581. *Anesthesiology* **29,** 931–943.

Duhm, V. B., Maul, W., Medenwald, H., Patzschke, K., and Wegner, L. A. (1969). Untersuchungen mit radioaktiv markietem BAY Va 1470 an ratten. *Berl. Münch. Tierärztl. Wochenschr.* **6,** 104–108.

Durham, R. A., Sawyer, D. C., Keller, W. F., and Wheeler, C. A. (1992). Topical ocular anesthetics in ocular irritancy testing: A review. *Lab. Anim. Sci.* **42,** 535–541.

Elmer, M., Eskildsen, P. C., Kristensen, L. O., and Leyssac, P. P. (1972). A comparison of renal function in rats anesthetized with Inactin and sodium amytal. *Acta Physiol. Scand.* **86,** 41–58.

Erhardt, W., Hebestedt, A., Aschenbrenner, G., Pichotka, B., and Blümel, G. (1984). A comparative study with various anesthetics in mice (pentobarbitone, ketamine-xylazine, carfentanyl-etomidate). *Res. Exp. Med.* **184,** 159–169.

Evans, R. H., and Hill, R. G. (1978). GABA-mimetic action of etomidate. *Experientia* **34,** 1325–1327.

Faber, J. E. (1989). Effects of althesin and urethan-chloralose on neurohumoral cardiovascular regulation. *Am. J. Physiol.* **256,** R757–R765.

Field, K. J., and Lang, C. M. (1988). Hazards of urethane (ethyl carbamate): A review of the literature. *Lab. Anim.* **22,** 255–262.

Field, K. J., White, W. J., and Lang, C. M. (1993). Anaesthetic effects of chloral hydrate, pentobarbitone and urethane in adult male rats. *Lab. Anim.* **27,** 258–269.

File, S. E., and Simmonds, M. A. (1988). Myoclonic seizures in the mouse induced by alphaxalone and related steroid anaesthetics. *J. Pharm. Pharmacol.* **40,** 57–59.

Finck, A. D., Samaniego, E., and Ngai, S. H. (1988). Morphine tolerance decreases the analgesic effects of ketamine in mice. *Anesthesiology* **68,** 397–340.

Flecknell, P. A. (1987). ''Laboratory Animal Anaesthesia.'' Academic Press, San Diego, California.

Flecknell, P. A., and Mitchell, M. (1984). Midazolam and fentanyl-fluanisone: Assessment of anaesthetic effects in laboratory rodents and rabbits. *Lab. Anim.* **18,** 143–146.

Flecknell, P. A., John, M., Mitchell, M., Shurey, C., and Simpkin, S. (1983). Neuroleptanalgesia in the rabbit. *Lab. Anim.* **17,** 104–109.

Flecknell, P. A., Liles, J. H., and Wootton, R. (1989). Reversal of fentanyl/fluanisone neuroleptanalgesia in the rabbit using mixed agonist/antagonist opioids. *Lab. Anim.* **23,** 147–155.

Flecknell, P. A., Liles, J. H., and Williamson, H. A. (1990). The use of lignocaine-prilocaine local anaesthetic cream for pain-free venepuncture in laboratory animals. *Lab. Anim.* **24,** 142–146.

Fleischman, R. W., McCracken, D., and Forbes, W. (1977). Adynamic ileus in the rat induced by chloral hydrate. *Lab. Anim. Sci.* **27,** 238–243.

Fluckiger, J.-P., Sonnay, M., Boillat, N., and Atkinson, J. (1985). Attenuation of the baroreceptor reflex by general anesthetic agents in the normotensive rat. *Eur. J. Pharmacol.* **109,** 105–109.

Flynn, A. J., Dengerink, H. A., and Wright, J. W. (1988). Blood pressure in resting, anesthetized and noise-exposed guinea pigs. *Hear. Res.* **34,** 201–206.

Folle, L. E., and Levesque, R. I. (1976). Circulatory, respiratory and acid-base balance changes produced by anesthetics in the rat. *Acta Biol. Med. Ger.* **35,** 605–612.

Forsythe, D. B., Payton, A. J., Dixon, D., Myers, P. H., Clark, J. A., and Snipe, J. R. (1992). Evaluation of Telazol-xylazine as an anesthetic combination for use in Syrian hamsters. *Lab. Anim. Sci.* **42,** 497–502.

France, C. P., Snyder, A. M., and Woods, J. H. (1989). Analgesic effects of phencyclidine-like drugs in rhesus monkeys. *J. Pharmacol. Exp. Ther.* **250,** 197–201.

Frankl, W. S., and Poole-Wilson, P. A. (1981). Effects of thiopental on tension development, action potential, and exchange of calcium and potassium in rabbit ventricular myocardium. *J. Cardiovasc. Pharmacol.* **3,** 554–565.

Friedman, J. J. (1959). Effect of Nembutal on circulating and tissue blood volumes and hematocrits of intact and splenectomized mice. *Am. J. Physiol.* **197,** 399–402.

Furuyama, F., Ishida, Y., Furuyama, M., Hashitani, T., Isobe, Y., Sato, H., Ohara, K., and Nishino, H. (1989). Thermal salivation in rats anesthetized with barbiturates, chloralose, urethane and ketamine. *Comp. Biochem. Physiol. C* **94C,** 133–138.

Gaertner, D. J., Boschert, K. R., and Schoeb, T. R. (1987). Muscle necrosis in Syrian hamsters resulting from intramuscular injections of ketamine and xylazine. *Lab. Anim. Sci.* **37,** 80–83.

Garcia-Villar, R., Toutain, P. L., Alvinerie, M., and Ruckebush, Y. (1981). The pharmacokinetics of xylazine hydrochloride: An interspecific study. *J. Vet. Pharmacol. Ther.* **4,** 87–92.

Garmer, N. L. (1969). Effects of 2-ethylamino-2-(2-thienyl) cyclohexanone HCl (Cl-634) in cats. *Res. Vet. Sci.* **10,** 382–388.

Gaumann, D. M., and Yaksh, T. L. (1990). Alpha-chloralose anesthesia inhibits the somato-sympathetic reflex response in cats more effectively than halothane. *J. Vet. Med.* **37,** 669–675.

Glen, J. B. (1980). Animal studies of the anaesthetic activity of ICI 35 868. *Br. J. Anaesth.* 52, 731–742.

Glen, J. B., and Hunter, S. C. (1984). Pharmacology of an emulsion formulation of ICI 35 868. *Br. J. Anaesth.* **56,** 617–625.

Goodchild, C. S. (1993). GABA receptors and benzodiazepines. *Br. J. Anaesth.* **71,** 127–133.

Grad, R., Witten, M. L., Quan, S. F., McKelvie, D. H., and Lemen, R. J. (1988). Intravenous chloralose is a safe anesthetic for longitudinal use in beagle puppies. *Lab. Anim. Sci.* **38,** 422–425.

Graffner, H., Ekelund, M., and Hakanson, R. (1991). Anaesthetic agents suppress basal and stimulated gastric acid secretion. Are intramural neurons involved? *Scand. J. Gastroenterol.* **26,** 1200–1204.

Green, C. J. (1979). ''Animal Anaesthesia,'' *Lab. Anim. Handb.* **8.** Laboratory Animals Ltd., London.

Green, C. J., Knight, J., Precious, S., and Simpkin, S. (1981). Metomidate, etomidate and fentanyl as injectable anaesthetic agents in mice. *Lab. Anim.* **15,** 171–175.

Greene, M. J. (1972). Some aspects of the pharmacology of droperidol. *Br. J. Anaesth.* **44,** 1272–1276.

Greene, S. A., and Thurmon, J. C. (1988). Xylazine—a review of its pharmacology and use in veterinary medicine. *J. Vet. Pharmacol. Ther.* 11, 295–313.

Guerreiro, D., and Page, C. P. (1987). The effect of neuroleptanalgesia on some cardiorespiratory variables in the rabbit. *Lab. Anim.* **21,** 205–209.

Guillet, R., Wyatt, J., Baggs, R. B., and Kellogg, C. K. (1988). Anesthetic-induced corneal lesions in developmentally sensitive rats. *Invest. Ophthalmol. Visual Sci.* **29,** 949–954.

Gumbleton, M., Taylor, G., Sewell, R. D. E., and Nicholls, P. J. (1989). Anaesthetic influences on brain haemodynamics in the rat and their significance to biochemical, neuropharmacological and drug disposition studies. *Biochem. Pharmacol.* **38,** 2745–2748.

Gumbleton, M., and Benet, L. Z. (1991). Drug metabolism and laboratory anesthetic protocols in the rat: Examination of antipyrine pharmacokinetics. *Pharm. Res.* **8,** 544–546.

Gumbleton, M., Nicholls, P. J., and Taylor, G. (1988). The renin-angiotensin system and intraperitoneal toxicity: Possible basis to urethane anaesthesia-induced reductions in renal clearance in the rat. *Arch. Int. Pharmacodyn. Ther.* **294,** 7–32.

Gumbleton, M., Nicholls, P. J., and Taylor, G. (1990a). Differential influence of laboratory anaesthetic regimens upon renal and hepatosplanchnic haemodynamics in the rat. *J. Pharm. Pharmacol.* **42,** 693–697.

Gumbleton, M., Nicholls, P. J., and Taylor, G. (1990b). Differential effects of anesthetic regimens on gentamicin pharmacokinetics in the rat: A comparison with chronically catheterized conscious animals. *Pharm. Res.* **7,** 41–45.

Gumerlock, M. K., and Neuwelt, E. A. (1990). The effects of anesthesia on osmotic blood-brain barrier disruption. *Neurosurgery* **26,** 268–277.

Hagino, N. (1979). Effect of Nembutal on LH release in baboons. *Horm. Metab. Res.* **11,** 296–300.

Hall, C. E., Ayachi, S., and Hall, O. (1976). Differential sensitivity of spontaneously hypertensive (SHR) and control rats to various anaesthetic agents. *Clin. Exp. Pharmacol. Physiol.* **3,** 83–86.

Hall, J. G., and Morris, B. (1962). The output of cells in lymph from the popliteal node of sheep. *J. Exp. Physiol.* **47,** 360–369.

Hall, L. W., and Clarke, K. W. (1991a). Principles of sedation, analgesia, and premedication. *In* ''Veterinary Anesthesia,'' 9th ed., pp. 51–79. Baillière Tindall, London.

Hall, L. W., and Clarke, K. W. (1991b). General principles of local analgesia. *In* ''Veterinary Anesthesia,'' 9th ed., pp. 173–188. Baillière Tindall, London.

Hall, R., and Murdoch, J. (1990). Brain protection: Physiological and pharmacological considerations. Part II: The pharmacology of brain protection. *Can. J. Anaesth.* **37,** 762–777.

Haskins, S. C. (1992). Injectable anesthetics. *Vet. Clinic. North Am.: Small Anim. Pract.* **22,** 245–260.

Hatch, R. C. (1966). The effect of glucose, sodium lactate, and epinephrine on thiopental anesthesia in dogs. *J. Am. Vet. Med. Assoc.* **148,** 135–140.

Hatch, R. C. (1973). Experiments on antagonism of barbiturate anesthesia with adrenergic, serotonergic, and cholinergic stimulants given alone and in combination. *Am. J. Vet. Res.* **34,** 1321–1331.

Hatch, R. C., Kitzman, J. V., Zahner, J. M., and Clark, J. D. (1984). Comparison of five preanesthetic medicaments in thiopental-anesthetized cats: Antagonism by selected compounds. *Am. J. Vet. Res.* **45,** 2322–2327.

Hayashi, Y., Sumikawa, K., Yamatodani, A., Tashiro, C., Wada, H., and Yoshiya, I. (1989). Myocardial sensitization by thiopental to arrhythmogenic action of epinephrine in dogs. *Anesthesiology* **71,** 929–935.

Heaton, J. T., and Brauth, S. E. (1991). Effects of yohimbine as a reversing agent for ketamine-xylazine anesthesia in budgerigars. *Lab. Anim. Sci.* **42,** 54–56.

Heavner, J. E. (1986). Anesthesia, analgesia, and restraint. *In* ''Methods of Animal Experimentation'' (W. I. Gay and J. E. Heavner, eds.), Vol. 7, Part A, pp. 1–35. Academic Press, Orlando, FL.

Heavner, J. E., and Bowen, J. M. (1968). Influence of adrenergic agents on recovery of dogs from anesthesia. *Am. J. Vet. Res.* **29,** 2133–2139.

Heykants, J., Pardoel, L., and Janssen, P. A. J. (1971). On the distribution and metabolism of azaperone (R 1929) in the rat and pig. Part I: Excretion and metabolism of azaperone in the Wistar rat. *Drug Res.* **21,** 982–984.

Heykants, J. J. P., Meuldermans, W. E. G., Michiels, L. J. M., and Lewi, P. J. (1975). Distribution, metabolism and excretion of etomidate, a short-acting hypnotic drug, in the rat. Comparative study of (R)−(+) and (S)−(−)-etomidate. *Arch. Int. Pharmacodyn. Ther.* **216,** 113–129.

Heys, S. D., Norton, A. C., Dundas, C. R., Eremin, O., Ferguson, K., and Garlick, P. J. (1989). Anaesthetic agents and their effect on tissue protein synthesis in the rat. *Clin. Sci.* **77,** 651–655.

Higashi, Y., Notoji, N., Yamajo, R., and Yata, N. (1982). Effect of anesthesia on drug disposition in the rat. *J. Pharmaco-Dyn.* **5,** 112–119.

Hinton, B. T. (1982). Hyperglycemia in urethane-anesthetized rats: Involvement of the adrenal gland. *Lab. Anim. Sci.* **32,** 251–252.

Hirshman, C. A., McCullough, R. E., Cohen, P. J., and Weil, J. V. (1975). Hypoxic ventilatory drive in dogs during thiopental, ketamine, or pentobarbital anesthesia. *Anesthesiology* **43,** 628–634.

Hobbs, B. A., Rolhall, T. G., Sprenkel, T. L., and Anthony, K. L. (1991). Comparison of several combinations for anesthesia in rabbits. *Am. J. Vet. Res.* 52, 669–674.

Høgskilde, S., Wagner, J., Strøm, J., Sjøntoft, E., Olesen, H. P., and Sørensen, M. B. (1991). Cardiovascular effects of pregnanolone emulsion: An experimental study in artificially ventilated dogs. *Acta Anaesthesiol. Scand.* **35,** 669–675.

Holzgrefe, H. H., Everitt, J. M., and Wright, E. M. (1987). Alpha-chloralose as a canine anesthetic. *Lab. Anim. Sci.* **37,** 587–595.

Hosoi, E., Saito, H., Yamasaki, R., Kimura, S., and Saito, S. (1988). Influence of anesthetic agents on the release of growth hormone induced by growth hormone-releasing hormone in rats. *Tokushima J. Exp. Med.* **35,** 85–90.

Hrapkiewicz, K. L., Stein, S., and Smiler, K. L. (1989). A new anesthetic agent for use in the gerbil. *Lab. Anim. Sci.* **39,** 338–341.

Hsu, W. H. (1982). Xylazine-induced delay of small intestinal transit in mice. *Eur. J. Pharmacol.* **83,** 55–60.

Hsu, W. H. (1992). Antagonism of pharmacologic effects of xylazine. *In* ''Animal Pain'' (C. E. Short and A. Van Poznak, eds.), pp. 225–236. Churchill-Livingstone, New York.

Hsu, W. H., Bellin, S. I., Dellmann, H.-D., and Hanson, C. E. (1986). Xylazine-ketamine-induced anesthesia in rats and its antagonism by yohimbine. *J. Am. Vet. Med. Assoc.* **189,** 1040–1043.

Hu, C., Flecknell, P. A., and Liles, J. H. (1992). Fentanyl and medetomidine anaesthesia in the rat and its reversal using atipamazole and either nalbuphine or butorphanol. *Lab. Anim.* **26,** 15–22.

Hughes, E. W., Martin-Body, R. L., Sarelius, I. H., and Sinclair, J. D. (1982). Effects of urethane-chloralose anaesthesia on respiration in the rat. *Clin. Exp. Pharmacol. Physiol.* **9,** 119–127.

Hunter, W. S., Holmes, K. R., and Elizondo, R. S. (1981). Thermal balance in ketamine-anesthetized rhesus monkey *Macaca mulatta. Am. J. Physiol.* **241,** R301–R306.

Idvall, J., Aronsen, K. F., and Stenberg, P. (1980). Tissue perfusion and distribution of cardiac output during ketamine anesthesia in normovolemic rats. *Acta Anaesth. Scand.* **24,** 257–263.

Ilkiw, J. E. (1992). Advantages and guidelines for using ultrashort barbiturates for induction of anesthesia. *Vet. Clin. North Am.: Small Anim. Pract.* **22,** 261–265.

Ilkiw, J. E., Benthuysen, J. A., Ebling, W. F., and McNeal, D. (1991). A comparative study of the pharmacokinetics of thiopental in the rabbit, sheep and dog. *J. Vet. Pharmacol. Ther.* **14,** 134–140.

Ilkiw, J. E., Pascoe, P. J., Haskins, S. C., and Patz, J. D. (1992). Cardiovascular and respiratory effects of propofol administration in hypovolemic dogs. *Am. J. Vet. Res.* **53,** 2323–2327.

Im, W. B., Blakeman, D. P., Davis, J. P., and Ayer, D. E. (1990). Studies on the mechanism of interactions between anesthetic steroids and γ-aminobutyric $acid_A$ receptors. *Mol. Pharmacol.* **37,** 429–434.

Inai, K., Arihiro, K., Takeshima, Y., Yonehara, S., Tachiyama, Y., Khatun, N., and Nishisaka, T. (1991). Quantitative risk assessment of carcinogenicity of urethane (ethyl carbamate) on the basis of long-term oral administration to B6C3F1 mice. *Jpn. J. Cancer Res.* **82,** 380–385.

Ishikawa, N., Kallman, C. H., and Sagawa, K. (1984). Rabbit carotid sinus reflex under pentobarbital, urethane, and chloralose anesthesia. *Am. J. Physiol.* **246,** H696–701.

Ishizuka, S., Sikdar, S. K., Yasui, S., Oyama, Y., and Akaike, N. (1989). α-chloralose opens the chloride channel of frog isolated sensory neurons. *Brain Res.* **498,** 181–184.

Iversen, O. H. (1991). Urethan (ethyl carbamate) is an effective promoter of 7,12-dimethylbenz[a]anthracene-induced carcinogenesis in mouse skin two-stage experiments. *Carcinogenesis (London)* **12,** 901–903.

Jewett, B. A., Gibbs, L. M., Tarasiuk, A., and Kendig, J. J. (1992). Propofol and barbiturate depression of spinal nociceptive neurotransmission. *Anesthesiology* **77,** 1148–1154.

Johansen, O., Vaaler, S., Jorde, R., and Reikeras, O. (1994). Increased plasma glucose levels after Hypnorm anaesthesia, but not after Pentobarbital anaesthesia in rats. *Lab. Anim.* **28,** 244–248.

Johnson, R., Fowler, J. F., and Zanelli, G. D. (1976). Changes in mouse blood pressure, tumor blood flow, and core and tumor temperatures following Nembutal or urethane anesthesia. *Radiology* **118,** 697–703.

Jondorff, W. R., Maickel, R. P., and Brodie, B. B. (1958). Inability of newborn mice and guinea pigs to metabolize drugs. *Biochem. Pharmacol.* **1,** 352–354.

Jones, D. J., Stehling, L. C., and Zauder, H. L. (1979). Cardiovascular responses to diazepam and midazolam maleate in the dog. *Anesthesiology* **51,** 430–434.

Jones, R. S. (1985). Steroid anaesthetics. *Proc. 2nd Int. Congr. Vet. Anesthesiol.,* pp. 15–17.

Kaufman, M. H. (1975). Parthenogenetic activation of mouse oocytes following avertin anaesthesia. *J. Embryol. Exp. Morphol.* **33,** 941–946.

Kaufman, M. H. (1977). Effect of anesthesia on the outcome of pregnancy in female mice. *J. Reprod. Fertil.* **49,** 167–168.

Kawaue, Y., and Iriuchijima, J. (1984). Changes in cardiac output and peripheral flows on pentobarbital anesthesia in the rat. *Jpn. J. Physiol.* **34,** 283–294.

Kelland, M. D., Chiodo, L. A., and Freeman, A. S. (1990). Anesthetic influences on the basal activity and pharmacological responsiveness of nigrostriatal dopamine neurons. *Synapse* **6,** 207–209.

Keller, G. L., Bauman, D. H., and Abbott, L. (1988). Yohimbine antagonism of ketamine and xylazine anesthesia in rabbits. *Lab. Animal* **17(3),** 28–30.

Klaassen, C. D. (1971). Biliary excretion of barbiturates. *Br. J. Pharmacol.* **43,** 161–166.

Klide, A. M. (1992). Anatomy of the spinal cord and how the spinal cord is affected by local anesthetics and other drugs. *Vet. Clin. North Am.: Small Anim. Pract.* **22,** 413–416.

Klotz, U., Antonin, K.-H., and Bieck, P. R. (1976). Pharmacokinetics and plasma binding of diazepam in man, dog, rabbit, guinea pig and rat. *J. Pharmacol. Exp. Ther.* **199,** 67–73.

Ko, J. C. H., Thurmon, J. C., Tranquilli, W. J., Benson, G. J., and Olson, W. A. (1992). A comparison of medetomidine-propofol and medetomidine-midazolam-propofol anesthesia in rabbits. *Lab. Anim. Sci.* **42,** 503–507.

Kochhar, M. M., Aykac, I., Davidson, P. P., and Fraley, E. D. (1986). Teratologic effects of d, 1-2-(o-chlorophenyl)-2-(methylamino) cyclohexanone hydrochloride (ketamine hydrochloride) in rats. *Res. Commun. Chem. Pathol. Pharmacol.* **54,** 413–416.

Komulainen, A., and Olson, M. E. (1991). Antagonism of ketamine-xylazine anesthesia in rats by administration of yohimbine, tolazoline, or 4-aminopyridine. *Am. J. Vet. Res.* **52,** 585–588.

Kumar, R., and Kumar, A. (1984). Ketamine with and without premedication in rats and rabbits. *Indian Vet. J.* **61,** 372–376.

Kurata, N., Hurst, H. E., Benz, F. W., Kemper, R. A., and Waddell, W. J. (1991). Studies on inhibition and induction of metabolism of ethyl carbamate by acetone and related compounds. *Drug Metab. Dispos.* **19,** 388–393.

Kurz, H., and Fichtl, B. (1981). Interrelation between plasma protein binding, rate of injection and the anaesthetic effect of thiopental. *Biopharm. Drug Dispos.* **2,** 191–196.

Kyles, A. E., Waterman, A. E., and Livingston, A. (1993). Antinociceptive effects of combining low doses of neuroleptic drugs and fentanyl in sheep. *Am. J. Vet. Res.* **54,** 1483–1488.

Lammintausta, R., Outi, V., Virtanen, R., and Vähä-Vahe, T., eds. (1989). Medetomidine, a novel α_2-agonist for veterinary sedative and analgesic use. *Acta Vet. Scand., Suppl.* **85,** 1–204.

Lamson, P. D., Greig, M. E., and Williams, L. (1952). Potentiation by epinephrine of the anesthetic effect in chloral and barbiturate anesthesia. *J. Pharmacol. Exp. Ther.* **106,** 219–224.

Latt, R. H., and Ecobichon, D. J. (1984). Self-mutilation in guinea pigs following the intramuscular injection of ketamine–acepromazine. *Lab. Anim. Sci.* **34,** 516.

Leash, A. M., Beyer, R. D., and Wilber, R. G. (1973). Self-mutilation following Innovar-vet injection in the guinea pig. *Lab. Anim. Sci.* **23,** 720–721.

Lee, S. S., Girod, C., Valla, D., Geoffroy, P., and Lebrec, D. (1985). Effects of pentobarbital sodium anesthesia on splanchnic hemodynamics of normal and portal-hypertensive rats. *Am. J. Physiol.* **249,** G528–G532.

Leithauser, M. T., Liem, A., Steward, B. C., Miller, E. C., and Miller, J. A. (1990). 1, N6-ethenoadenosine formation, mutagenicity and murine tumor induction as indicators of the generation of an electrophilic epoxide metabolite of the closely related carcinogens ethyl carbamate (urethane) and vinyl carbamate. *Carcinogenesis (London)* **11,** 463–473.

Lemm, C. A. (1993). "Evaluation of Five Anesthetics on Striped Bass," *Resour. Publ. 196.* U.S. Department of the Interior, Fish and Wildlife Service, Washington, D.C.

Lewi, P. J., Heykants, J. J. P., and Janssen, P. A. J. (1976). Intravenous pharmacokinetic profile in rats of etomidate, a short-acting hypnotic drug. *Arch. Int. Pharmacodyn. Ther.* **220,** 72–85.

Lin, H. C., Thurmon, J. C., Benson, G. J., and Tranquilli, W. J. (1993). Telazol—a review of its pharmacology and use in veterinary medicine. *J. Vet. Pharmacol. Ther.* **16,** 383–418.

Lin, M. T., Chen, C. F., and Pang, I. H. (1978). Effect of ketamine on thermoregulation in rats. *Can. J. Physiol. Pharmacol.* **56,** 963–967.

Lindberg, L.-A. (1986). Sedative and analgesic effects of detomidine in horses and cattle. *Acta Vet. Scand., Suppl.* **82,** 1–208.

Lipman, N. S., Phillips, P. A., and Newcomer, C. E. (1987). Reversal of ketamine/xylazine anesthesia in the rabbit with yohimbine. *Lab. Anim. Sci.* **37,** 474–477.

Lipman, N. S., Marini, R. P., and Erdman, S. E. (1990). A comparison of ketamine/xylazine and ketamine/xylazine/acepromazine anesthesia in the rabbit. *Lab. Anim. Sci.* **40,** 395–398.

Livingston, A., and Waterman, A. E. (1978). The development of tolerance to ketamine in rats and the significance of hepatic metabolism. *Br. J. Pharmacol.* **64,** 63–69.

Lögdberg, B. (1988). Alphaxolone-alphadolone for anesthesia of squirrel monkeys of different ages. *J. Med. Primatol.* **17,** 163–167.

Loomis, M. R., Henrickson, R. V., and Anderson, J. H. (1980). Effects of ketamine hydrochloride on the hemogram of rhesus monkeys (*Macaca mulatta*). *Lab. Anim. Sci.* **30,** 851–853.

Löscher, W., Ganter, M., and Fassbender, C. P. (1990). Correlation between drug and metabolite concentrations in plasma and anesthetic action of ketamine in swine. *Am. J. Vet. Res.* **51,** 391–398.

Lovell, D. P. (1986a). Variation in pentobarbitone sleeping time in mice. 1. Strain and sex differences. *Lab. Anim.* **20,** 85–90.

Lovell, D. P. (1986b). Variation in pentobarbitone sleeping time in mice. 2. Variables affecting test results. *Lab. Anim.* **20,** 91–96.

Lovell, D. P. (1986c). Variation in barbiturate sleeping time in mice. 3. Strain × environment interactions. *Lab. Anim.* **20,** 307–312.

Lowson, S., Gent, J. P., and Goodchild, C. S. (1990). Anticonvulsant properties of propofol and thiopentone: Comparison using two tests in laboratory mice. *Br. J. Anaesth.* **64,** 59–63.

Lumb, W. V., and Jones, E. W. (1984). ''Veterinary Anesthesia,'' 2nd ed. Lea & Febiger, Philadelphia.

MacDonald, E., and Virtanen, R. (1992). Review of the pharmacology of medetomidine and detomidine: Two chemically similar α_2-adrenoreceptor agonists used in veterinary sedatives. *In* "Animal Pain" (C. E. Short and A. Van Poznak, eds.), pp. 181–191. Churchill-Livingstone, New York.

MacDonald, E., Haapalinna, A., Virtanen, R., and Lammintausta, R. (1989). Effects of acute administration of medetomidine on the behaviour, temperature and turnover rates of brain biogenic amines in rodents and reversal of these effects by atipamezole. *Acta Vet. Scand., Suppl.* **85,** 77–81.

Maggi, C. A., and Meli, A. (1986a). Suitability of urethane anesthesia for physiopharmacological investigations in various systems. Part 1: General considerations. *Experientia* 42, 109–114.

Maggi, C. A., and Meli, A. (1986b). Suitability of urethane anesthesia for physiopharmacological investigations in various systems. Part 2: Cardiovascular system. *Experientia* **42,** 292–297.

Maggi, C. A., and Meli, A. (1986c). Suitability of urethane anesthesia for physiopharmacological investigations. Part 3: Other systems and conclusions. *Experientia* **42,** 531–537.

Maggi, C. A., Manzinim, S., Parlani, M., and Meli, A. (1984). An analysis of the effects of urethane on cardiovascular responsiveness to catecholamines in terms of its interference with Ca^{++}. *Experientia* **40,** 52–59.

Manders, W. T., and Vatner, S. F. (1976). Effects of sodium pentobarbital anesthesia on left ventricular function and distribution of cardiac output in dogs, with particular reference to the mechanism for tachycardia. *Circ. Res.* **39,** 512–517.

Mann, W. A., Weizel, G., and Kinter, L. B. (1991). Hemodynamic effects of medetomidine in conscious unrestrained cynomolgus monkeys using implanted telemetric transmitters. *FASEB J.* **5,** A868.

Marietta, M. P., Vore, M. E., Way, W. L., and Trevor, A. J. (1977). Characterization of ketamine induction of hepatic microsomal drug metabolism. *Biochem. Pharmacol.* **26,** 2451–2453.

Marini, R. P., Avison, D. L., Corning, B. F., and Lipman, N. S. (1991). Ketamine/xylazine/butorphanol: A new anesthetic combination for rabbits. *Lab. Anim. Sci.* **42,** 57–62.

Mårtensson, L., and Carter, A. M. (1984). Distribution of cardiac output in the late pregnant guinea pig during anaesthesia with ketamine. *Z. Versuchstierkd.* **26,** 175–179.

Maze, M., and Regan, J. W. (1991). Role of signal transduction in anesthetic action: α_2 adrenergic agonists. *Ann. N. Y. Acad. Sci.* **625,** 409–423.

Maze, M., Virtanen, R., Daunt, D., Banks, S. J. M., Stover, E. P., and Feldman, D. (1991). Effects of dexmedetomidine, a novel imidazole sedative-anesthetic agent, on adrenal steroidogenesis: In vivo and in vitro studies. *Anesth. Analg.* **73,** 204–208.

McHale, N. G., and Thornbury, K. D. (1989). The effect of anesthetics on lymphatic contractility. *Microvasc. Res.* **37,** 70–76.

Mero, M., Vainiopää, S., Vasenius, J., Vihtonen, K., and Rokkanen, P. (1989). Medetomidine-ketamine-diazepam anesthesia in the rabbit. *Acta Vet. Scand., Suppl.* **85,** 135–137.

Miletich, D. J., Ivankovic, A. D., Albrecht, R. F., Zahed, B., and Ilahi, A. A. (1973). The effect of ketamine on catecholamine metabolism in the isolated perfused rat heart. *Anesthesiology* **39,** 271–277.

Miller, E. D., Kistner, J. R., and Epstein, R. M. (1980). Whole-body distribution of radioactively labelled microspheres in the rat during anesthesia with halothane, enflurane, or ketamine. *Anesthesiology* **52,** 296–302.

Mills, T., Copland, A., and Osteen, K. (1981). Factors affecting the postovulatory surge of FSH in the rabbit. *Biol. Reprod.* **25,** 530–535.

Moore, T. C., Spruck, C. H., and Leduc, L. E. (1988). Depression of lymphocyte traffic in sheep by anaesthesia and associated changes in efferent-lymph PGE_2 and antibody levels. *Immunology* **63,** 139–143.

Muir, W. W. (1985). Cyclohexanone drug mixtures: The pharmacology of ketamine and ketamine drug combinations. *Proc. 2nd Int. Congr. Vet. Anesthesiol.,* pp. 5–14.

Muir, W. W., and Hubbell, J. A. E. (1988). Cardiopulmonary and anesthetic effects of ketamine and its enantiomers in dogs. *Am. J. Vet. Res.* **49,** 530–534.

Muir, W. W., and Mason, D. E. (1989). Side effects of etomidate in dogs. *J. Am. Vet. Med. Assoc.* **194,** 1430–1434.

Muir, W. W., Bednarski, L., and Bednarski, R. (1991). Thiamylal- and halothane-sparing effect of diazepam in dogs. *J. Vet. Pharmacol. Ther.* **14,** 46–50.

Murthy, V. S., Zagar, M. E., Vollmer, R. R., and Schmidt, D. H. (1982). Pentobarbital-induced changes in vagal tone and reflex vagal activity in rabbits. *Eur. J. Pharmacol.* **84,** 41–50.

Myslobodsky, M. S., Golovchinsky, V., and Mintz, M. (1981). Ketamine: Convulsant or anti-convulsant? *Pharmacol., Biochem. Behav.* **14,** 27–33.

Nazian, S. J. (1988). Serum concentrations of reproductive hormones after administration of various anesthetics to immature and young adult male rats. *Proc. Soc. Exp. Biol. Med.* **187,** 482–487.

Nevalainen, T., Pyhälä, L., Voipio, H., and Virtanen, R. (1989). Evaluation of anaesthetic potency of medetomidine-ketamine combination in rats, guinea-pigs and rabbits. *Act Vet. Scand., Suppl.* **85,** 139–143.

Newton, W. M., Cusick, P. K., and Raffe, M. R. (1975). Innovar-vet-induced pathologic changes in the guinea pig. *Lab. Anim. Sci.* **25,** 597–601.

Niemegeers, C. J. E., Van Nueten, J. M., and Janssen, P. A. J. (1974). Azaperone, a sedative neuroleptic of the butyrophenone series with pronounced anti-aggressive and anti-shock activity in animals. *Drug Res.* **24,** 1798–1806.

Nomeir, A. A., Ioannou, Y. M., Sanders, J. M., and Matthews, H. B. (1989). Comparative metabolism and disposition of ethyl carbamate (urethane) in male Fisher 344 rats and male B6C3F1 mice. *Toxicol. Appl. Pharmacol.* **97,** 203–215.

Norris, M. L., and Turner, W. D. (1983). An evaluation of tribromoethanol (TBE) as an anaesthetic agent in the Mongolian gerbil (*Meriones unguiculatus*). *Lab. Anim.* **17,** 324–329.

Nunn, J. F., Utting, J. E., and Brown, B. R., Jr. (1989). Introduction. *In* ''General Anaesthesia'' (J. F. Nunn, J. E. Utting, and B. R. Brown, Jr., eds.), 5th ed., pp. 1–6. Butterworth, London.

Ochsner, A. J. (1977). Cardiovascular and respiratory responses to ketamine hydrochloride in the rhesus monkey (*Macaca mulatta*). *Lab. Anim. Sci.* **27,** 69–71.

Ogino, K., Hobara, T., Kobayashi, H., and Iwamoto, S. (1990). Gastric mucosal injury induced by chloral hydrate. *Toxicol. Lett.* **52,** 129–133.

Oguchi, K., Arakawa, K., Nelson, S. R., and Samson, F. (1982). The influence of droperidol, diazepam, and physostigmine on ketamine-induced behavior and brain regional glucose utilization in rat. *Anesthesiology* **57,** 353–358.

Oliver, A. E., Deamer, D. W., and Akeson, M. (1991). Evidence that sensitivity to steroid anesthetics appears late in evolution. *Brain Res.* **557,** 298–302.

Olson, M. E., and Renchko, P. (1988). Azaperone and azaperone-ketamine as a neuroleptic sedative and anesthetic in rats and mice. *Lab. Anim. Sci.* **38,** 299–304.

Olson, M. E., McCabe, K., and Walker, R. L. (1987). Guaifenesin alone or in combination with ketamine or sodium pentobarbital as an anesthetic in rabbits. *Can. J. Vet. Res.* **51,** 383–386.

Papaioannou, V. E., and Fox, J. G. (1993). Efficacy of tribromoethanol anesthesia in mice. *Lab. Anim. Sci.* **43,** 189–192.

Parker, J. L., and Adams, H. R. (1978). The influence of chemical restraining agents on cardiovascular function: A review. *Lab. Anim. Sci.* 28, 575–593.

Pascoe, P. J. (1992). The case for maintenance of general anesthesia with an injectable agent. *Vet. Clin. North Am.: Small Anim. Pract.* **22,** 275–276.

Pascoe, P. J., Ilkiw, J. E., Haskins, S. C., and Patz, J. D. (1992). Cardiopulmonary effects of etomidate in hypovolemic dogs. *Am. J. Vet. Res.* **53,** 2178–2182.

Payton, A. J., and Pick, J. R. (1989). Evaluation of a combination of tiletamine and zolazepam as an anesthetic for ferrets. *Lab. Anim. Sci.* **39,** 243–246.

Peeters, M. E., Gil, D., Teske, E., Eyzenbach, V., Brom, W. E., and Lumeij, J. T. (1988). Four methods for general anaesthesia in the rabbit: A comparative study. *Lab. Anim.* **22,** 355–360.

Pertovaara, A., Kauppila, T., and Tukeva, T. (1990). The effect of medetomidine, an α_2-adrenoceptor agonist, in various pain tests. *Eur. J. Pharmacol.* **179,** 323–328.

Peterson, S. L. (1989). Anticonvulsant profile of an anesthetic steroid. *Neuropharmacology* **28,** 877–879.

Pettinger, W. A., Tanaka, K., Keeton, K., Campbell, W. B., and Brooks, S. N. (1975). Renin release, an artifact of anesthesia and its implications in rats. *Proc. Soc. Exp. Biol. Med.* **148,** 625–630.

Pfenninger, E., Dick, W., and Ahnefeld, F. W. (1985). The influence of ketamine on both normal and raised intracranial pressure of artificially ventilated animals. *Eur. J. Anaesthesiol.* **2,** 297–307.

Ponte, J., and Sadler, C. L. (1989). Effect of thiopentone, etomidate and propofol on carotid body chemoreceptor activity in the rabbit and the cat. *Br. J. Anaesth.* **62,** 41–45.

Popilskis, S. J., Oz, M. C., Gorman, P., Florestal, A., and Kohn, D. F. (1991). Comparison of xylazine with tiletamine-zolazepam (Telazol) and xylazine-ketamine anesthesia in rabbits. *Lab. Anim. Sci.* **41,** 51–53.

Porter, D. B., and Slusser, C. A. (1985). Azaperone: A review of a new neuroleptic agent for swine. *Vet. Med.* **80** (March), 88–92.

Preziosi, P., and Vacca, M. (1988). Adrenocortical suppression and other endocrine effects of etomidate. *Life Sci.* **42,** 477–489.

Quinn, G. P., Axelrod, J., and Brodie, B. B. (1958). Species, strain, and sex differences in metabolism of hexobarbitone, amidopyrine, antipyrine and aniline. *Biochem. Pharmacol.* **1,** 152–159.

Rall, T. W. (1990). Hypnotics and sedatives; ethanol. *In* "The Pharmacological Basis of Therapeutics" (A. G. Gilman, T. W. Rall, A. S. Nies, and P. Taylor, eds.), 8th ed., pp. 345–382. Pergamon, New York.

Rall, T. W., and Schleifer, L. S. (1990). Drugs effective in the therapy of the epilepsies. *In* "The Pharmacological Basis of Therapeutics" (A. G. Gilman, T. W. Rall, A. S. Nies, and P. Taylor, eds.), 8th ed., pp. 436–462. Pergamon, New York.

Raman, J., Stewart, R., Montano, M., Reginald, S. A., and Lord, M. D. (1989). Assisted local anaesthesia in rabbits. *Microsurgery* **10,** 75–76.

Ramirez-Gonzalez, M. D., Barna, I., Wiegant, V. M., and de Jong, W. (1991). Effect of anaesthetics on the release of beta-endorphin-immunoreactivity in rat plasma. *Life Sci.* **48,** 1371–1377.

Rank, J., and Jensen, A.-G. (1989). The value of anaesthetic steroids alphaxolone-alphadolone in pregnant mice. *Scand. J. Lab. Anim. Sci.* **16,** 115–117.

Rehm, W. F., and Schatzmann, U. (1986). Pharmacological properties of benzodiazepines in animals. *In* "Comparative Veterinary Pharmacology, Toxicology and Therapy" (A.S.J.P.A.M. Van Miert, M. G. Bogaert, and M. Debackere, eds.), pp. 13–23. MTP Press, Boston.

Reich, D. L., and Silvay, G. (1989). Ketamine: An update on the first twenty-five years of clinical experience. *Can. J. Anaesth.* 36, 186–197.

Reid, W. D., Davies, C., Pare, P. D., and Pardy, R. L. (1989). An effective combination of anaesthetics for 6-h experimentation in the golden Syrian hamster. *Lab. Anim.* **23,** 156–162.

Reutlinger, R. A., Karl, A. A., Vinal, S. I., and Nieser, M. (1980). Effects of ketamine HCl-xylazine HCl combination on cardiovascular and pulmonary values of the rhesus macaque (*Macaca mulatta*). *Am. J. Vet. Res.* **41,** 1453–1457.

Ritchie, J. M., and Greene, N. M. (1990). Local anesthetics. *In* "The Pharmacological Basis of Therapeutics" (A. G. Gilman, T. W. Rall, A. S. Nies, and P. Taylor, eds.), 8th ed., pp. 311–331. Pergamon, New York.

Robertson, S. (1992). Advantages of etomidate use as an anesthetic agent. *Vet. Clin. North Am.: Small Anim. Pract.* **22,** 277–280.

Rocchiccioli, C., Saad, M. A. A., and Elghozi, J.-L. (1989). Attenuation of the baroreceptor reflex by propofol anesthesia in the rat. *J. Cardiovasc. Pharmacol.* **14,** 631–635.

Rose, W. C., Munson, A. E., and Bradley, S. G. (1973). Acute lethality for mice following administration of cyclophosphamide with barbiturates (37241). *Proc. Soc. Exp. Biol. Med.* **143,** 1–5.

Rudy, D. C., Downie, J. W., and McAndrew, J. D. (1991). Alpha-chloralose alters autonomic reflex function of the lower urinary tract. *Am. J. Physiol.* **261,** R1560–R1567.

Ruskoaho, H., and Leppäluoto, J. (1989). The effect of medetomidine, an α_2-adrenoceptor agonist, on plasma atrial natriuretic peptide levels, haemodynamics and renal excretory function in spontaneously hypertensive and Wistar-Kyoto rats. *Br. J. Pharmacol.* **97,** 125–132.

Ryder, S., Way, W. L., and Trevor, A. J. (1978). Comparative pharmacology of the optical isomers of ketamine in mice. *Eur. J. Pharmacol.* **49,** 15–23.

Salonen, J. S. (1989). Pharmacokinetics of medetomidine. *Acta Vet. Scand., Suppl.* **85,** 49–54.

Salonen, J. S. (1992). Chemistry and pharmacokinetics of the α_2-adrenoreceptor agonists. *In* "Animal Pain" (C. E. Short and A. Van Poznak, eds.), pp. 191–200. Churchill-Livingstone, New York.

Sanford, T. D., and Colby, E. D. (1980). Effect of xylazine and ketamine on blood pressure, heart rate and respiratory rate in rabbits. *Lab. Anim. Sci.* **30,** 519–522.

Savola, J.-M. (1989). Cardiovascular actions of medetomidine and their reversal by atipamezole. *Acta Vet. Scand., Suppl.* **85,** 39–47.

Savola, J.-M., Ruskoaho, H., Puurunen, J., Salonen, J. S., and Kärki, N. T. (1986). Evidence for medetoimidine as a selective and potent agonist at α_2-adrenoreceptors. *J. Auton. Pharmacol.* **5,** 275–284.

Savola, M., Savola, J.-M., and Puurunen, J. (1989). α_2-Adrenoceptor-mediated inhibition of gastric acid secretion by medetomidine is efficiently antagonized by atipamezole in rats. *Arch. Int. Pharmacodyn. Ther.* **301,** 267–276.

Sawyer, D. C., Lumb, W. V., and Stone, H. L. (1971). Cardiovascular effects of halothane, methoxyflurane, pentobarbital, and thiamylal. *J. Appl. Physiol.* **30,** 36–43.

Scarfone, E., Ulfendahl, M., Figueroa, L., and Flock, A. (1991). Anaesthetics may change the shape of isolated Type I hair cells. *Hear. Res.* **54,** 247–250.

Scheinin, M., and MacDonald, E. (1989). An introduction to the pharmacology of α_2-adrenoceptors in the central nervous system. *Acta Vet. Scand., Suppl.* **85,** 11–19.

Schmitt, H., Le Douarec, J.-C., and Petillot, N. (1974). Antagonism of the antinociceptive action of xylazine, an -sympathomimetic agent, by adrenoceptor and cholinoceptor blocking agents. *Neuropharmacology* **13,** 295–303.

Schwedler, M., Miletich, D. J., and Albrecht, R. F. (1982). Cerebral blood flow and metabolism following ketamine administration. *Can. Anaesth. Soc. J.* **29,** 222–226.

Sebel, P. S., and Lowdon, J. D. (1989). Propofol: A new intravenous anesthetic. *Anesthesiology* **71,** 260–277.

Severinghaus, J. W., Xu, F. D., and Spellman, M. J., Jr. (1991). Benzocaine and methemoglobin. *Anesthesiology* **74,** 385–386.

Severs, W. B., Keil, L. C., Klase, P. A., and Deen, K. C. (1981). Urethane anesthesia in rats. *Pharmacology* **22,** 209–226.

Seyde, W. C., McGowan, L., Lund, N., Duling, B., and Longnecker, D. E. (1985). Effects of anesthetics on regional hemodynamics in normovolemic and hemorrhaged rats. *Am. J. Physiol.* **249,** H164–H173.

Sharma, R. P., Stowe, C. M., and Good, A. L. (1970). Studies on the distribution and metabolism of thiopental in cattle, sheep, goats, and swine. *J. Pharmacol. Exp. Ther.* **172,** 128–137.

Short, C. E., ed. (1987). Neuroleptanalgesia and alpha-adrenergic receptor analgesia. *In* ''Principles & Practice of Veterinary Anesthesia'' (C. E. Short, ed.), pp. 47–57. Williams & Wilkins, Baltimore, Maryland.

Silverman, J., and Muir, W. W., III (1993). A review of laboratory animal anesthesia with choloral hydrate and chloralose. *Lab. Anim. Sci.* 43, 210–216.

Silverman, J., Huhndorf, M., Balk, M., and Slater, G. (1983). Evaluation of a combination of tiletamine and zolazepam as an anesthetic for laboratory rodents. *Lab. Anim. Sci.* **33,** 457–460.

Skipper, H. E., Bennett, L. L., Jr., Bryan, C. E., White, L., Jr., Newton, M. A., and Simpson, L. (1951). Carbamates in the chemotherapy of leukemia. VIII. Over-all tracer studies on carbonyl-labeled urethan, methylene-labeled urethan, and methylene-labeled ethyl alcohol. *Cancer Res.* **11,** 46–51.

Skolleborg, K. C., Grönbech, J. E., Grong, E., Åbyholm, F. E., and Lekven, J. (1990). Distribution of cardiac output during pentobarbital versus midazolam/fentanyl/fluanisone anaesthesia in the rat. *Lab. Anim.* **24,** 221–227.

Smith, A. C., Zellner, J. L., Spinale, F. G., and Swindle, M. M. (1991). Sedative and cardiovascular effects of midazolam in swine. *Lab. Anim. Sci.* **41,** 157–161.

Soma, L. R. (1983). Anesthetic and analgesic considerations in the experimental animal. *Ann. N. Y. Acad. Sci.* **406,** 32–47.

Sotomayor, R. E., and Collins, T. F. (1990). Mutagenicity, metabolism, and DNA interactions of urethane. *Toxicol. Ind. Health* **6,** 71–108.

Sourkes, T. L. (1992). Early clinical neurochemistry of CNS-active drugs. *Molec. Chem. Neuropathol.* **17,** 21–30.

Späth, M., and Schweickert, W. (1977). The effect of Metacaine (MS-222) on the activity of the efferent and afferent nerves in the teleost lateral-line system. *Naunyn-Schmiedeberg's Arch. Pharmacol.* **297,** 9–19.

Spriggs, T. L. B., and Stockham, M. A. (1964). Urethane anaesthesia and pituitary-adrenal function in the rat. *J. Pharm. Pharmacol.* **16,** 603–610.

Steen, R. G., Wilson, D. A., Bowser, C., Wehrle, J. P., and Glickson, J. D. (1989). ^{31}P NMR Spectroscopic and near infared spectrophotometric studies of effects of anesthetics on in vivo RIF-1 tumors. Relationship to tumor radiosensitivity. *NMR Biomed.* **2,** 87–92.

Stenberg, D. (1989). Physiological role of α_2-adrenoceptors in the regulation of vigilance and pain: Effect of medetomidine. *Acta Vet. Scand., Suppl.* **85,** 21–28.

Stowe, D. F., Bosnjak, Z. J., and Kampine, J. P. (1992). Comparison of etomidate, ketamine, midazolam, propofol, and thiopental on function and metabolism of isolated hearts. *Anesth. Analg.* **74,** 547–558.

Strobel, G. E., and Wollman, H. (1969). Pharmacology of anesthetic agents. *Fed. Proc., Fed. Am. Soc. Exp. Biol.* **28,** 1386–1403.

Sutton, J. A. (1972). A brief history of steroid anaesthesia before Althesin (CT1341). *Postgrad. Med. J., Suppl.* **2,** 9–13.

Suzdak, P. D., Schwartz, R. D., Skolnick, P., and Paul, S. M. (1986). Ethanol stimulates γ-aminobutyric acid receptor-mediated chloride transport in rat brain synaptoneurosomes. *Proc. Natl. Acad. Sci. U.S.A.* **83,** 4071–4075.

Svendsen, P., and Carter, A. M. (1985). Influence of injectable anaesthetic combinations on blood gas tensions and acid-base status in laboratory rats. *Acta Pharmacol. Toxicol.* **57,** 1–7.

Sylvina, T. J., Berman, N. G., and Fox, J. G. (1990). Effects of yohimbine on bradycardia and duration of recumbency in ketamine/xylazine anesthetized ferrets. *Lab. Anim. Sci.* **40,** 178–182.

Taber, R., and Irwin, S. (1969). Anesthesia in the mouse. *Fed. Proc., Fed. Am. Soc. Exp. Biol.* **28,** 1528–1532.

Takano, Y., and Yaksh, T. L. (1992). Characterization of the pharmacology of intrathecally administered alpha-2 agonists and antagonists in rats. *J. Pharmacol. Exp. Ther.* **261,** 764–772.

Tan, P. C. P., Shyr, M. H., Yang, C. H., Kuo, T. B. J., Pan, W. H. T., and Chan, S. H. H. (1993). Power spectral analysis of the electrencephalographic and hemodynamic correlates of propofol anesthesia in the rat: Intravenous infusion. *Neuroscience Letters* **160,** 205–208.

Tarin, D., and Sturdee, A. (1972). Sugrical anaesthesia of mice: Evaluation of tribromo-ethanol, ether, halothane and methoxyflurane and development of a reliable technique. *Lab. Anim.* **6,** 79–84.

Teske, R. H., and Carter, G. C. (1971). Effect of chloramphenicol on pentobarbital-induced anesthesia in dogs. *J. Am. Vet. Med. Asoc.* **159,** 777–780.

Thurmon, J. C. (1985). Comparative pharmacokinetics of selected injectable anesthetic agents. *Proc. 2nd Int. Congr. Vet. Anesthesiol.,* pp. 21–26.

Thurmon, J. C., and Benson, G. J. (1987). Pharmacologic considerations in selection of anesthetics for animals. *J. Am. Vet. Med. Assoc.* **191,** 1245–1251.

Thurmon, J. C., Tranquilli, W. J., and Benson, G. J. (1992). α_2-Antagonists: Use in domestic and wild animal species. *In* ''Animal Pain'' (C. E. Short and A. Van Poznak, eds.), pp. 237–247. Churchill-Livingstone, New York.

Tomemori, N., Shingu, K., Komatsu, T., Urabe, N., and Mori, K. (1981). Anti-analegsic action of thiamylal sodium in cats. *Acta Anaesthiol. Scand., Suppl.* **25,** 523–525.

Tracy, C. H., Short, C. E., and Clark, B. C. (1988). Comparing the effects of intravenous and intramuscular administration of Telazol. *Vet. Med.,* January, pp. 104–111.

Tranquilli, W. J., Thurmon, J. C., and Benson, G. J. (1983). Organ blood flow and distribution of cardiac output in hypocapnic ketamine-anesthetized swine. *Am. J. Vet. Res.* **44,** 1578–1582.

Tsai, S. K., and Lee, C. (1989). Ketamine potentiates nondepolarizing neuromuscular relaxants in a primate. *Anesth. Analg.* **68,** 5–8.

Tsun, C., and Glazko, A. J. (1974). Biotransformation and disposition of ketamine. *J. Int. Anesthesiol. Clin.* **12,** 157–176.

Turner, D. M., Ransom, R. W., Yang, J. S.-J., and Olsen, R. W. (1989). Steroid anesthetics and naturally occurring analogs modulate the γ-aminobutyric acid receptor complex at a site distinct from barbiturates. *J. Pharmacol. Exp. Ther.* **248,** 960–966.

Turner, T. T., and Howards, S. S. (1977). Hyperglycemia in the hamster anesthetized with Inactin® [5-Ethyl-5-(1-Methyl Propyl)-2-Thiobarbiturate]. *Lab. Anim. Sci.* **27,** 380–382.

Usenik, E. A., and Cronkite, E. P. (1965). Effects of barbiturate anesthetics on leukocytes in normal and splenectomized dogs. *Anesth. Analg.* **44,** 167–170.

Vainio, O. (1989). Introduction to the clinical pharmacology of medetomidine. *Acta Vet. Scand., Suppl.* **85,** 85–88.

Vainio, O. (1990). Reversal of medetomidine-induced cardiovascular and respiratory changes with atipamezole in dogs. *Vet. Rec.* **127,** 447–450.

Vainio, O. (1992). Pain control with medetomidine in dogs, cats, and laboratory animals. *In* ''Animal Pain'' (C. E. Short and A. Van Poznak, eds.), pp. 213–219. Churchill-Livingstone, New York.

Van Der Meer, C., Versluys-Broers, J. A. M., Tuynman, H. A. R. E., and Buur, V. A. (1975). The effect of ethylurethane on hematocrit, blood pressure and plasma-glucose. *Arch. Int. Pharmacodyn., Ther.* **217,** 257–275.

Van Leeuwen, A. F., Evans, R. G., and Ludbrook, J. (1990). Effects of halothane, ketamine, propofol and alfentanil anaesthesia on circulatory control in rabbits. *Clin. Exp. Pharmacol. Physiol.* **17,** 781–798.

Vatner, S. F., Higgins, C. B., Patrick, T., Franklin, D., and Braunwald, E. (1971b). Effects of cardiac depression and of anesthesia on the myocardial action of a cardiac glycoside. *J. Clin. Invest.* **50,** 2585–2595.

Venugopalan, C. S., Holmes, E. P., Fucci, V., Keefe, T. J., and Crawford, M. P. (1994). Cardiopulmonary effects of medetomidine in heartworm-infected and noninfected dogs. *Am. J. Vet. Res.* **55,** pp. 1148–1152.

Vesell, E. S. (1968). Factors affecting the responsiveness of mice to hexobarbital. *Pharmacology* **1,** 81–97.

Virtanen, R. (1986). Pharmacology of detomidine and other α_2-adrenoceptor agonists in the brain. *Acta Vet. Scand., Suppl.* **82,** 35–46.

Virtanen, R. (1989). Pharmacological profiles of medetomidine and its antagonist, atipamezole. *Acta Vet. Scand., Suppl.* **85,** 29–37.

Walker, L. A., Buscemi-Bergin, M., and Gellai, M. (1983). Renal hemodynamics in conscious rats: Effects of anesthesia, surgery, and recovery. *Am. J. Physiol.* **245,** F67–F74.

Walker, L. A., Gellai, M., and Valtin, H. (1986). Renal response to pentobarbital anesthesia in rats: Effect of interrupting the renin-angiotensin system. *J. Pharmacol. Exp. Ther.* **236,** 721–728.

Wang, Y., Zhou, T., Chua, T. C., and Pang, C. C. Y. (1991). Effects of inhalation and intravenous anaesthetic agents on pressor response to N^G-nitro-L-arginine. *Eur. J. Pharmacol.* **198,** 183–188.

Ward, G. S., Johnson, D. O., and Roberts, C. R. (1974). The use of CI 744 as an anesthetic for laboratory animals. *Lab. Anim. Sci.* **24,** 737–742.

Waterman, A. E. (1983). Influence of premedication with xylazine on the distribution and metabolism of intramuscularly administered ketamine in cats. *Res. Vet. Sci.* **35,** 285–290.

Waterman, A. E. (1984). The pharmacokinetics of ketamine administered intravenously in calves and the modifying effect of premedication with xylazine hydrochloride. *J. Vet. Pharmacol. Ther.* **7,** 125–130.

Waterman, A. E., and Livingston, A. (1978). Effects of age and sex on ketamine anaesthesia in the rat. *Br. J. Anaesth.* **50,** 885–889.

Wauquier, A. (1983). Profile of etomidate; A hypnotic, anticonvulsant and brain protective compound. *Anaesthesia* **38,** 26–33.

Way, W. L., and Trevor, A. J. (1986). Pharmacology of intravenous nonnarcotic anesthetics. *In* ''Anesthesia'' (R. D. Miller, ed.), 2nd ed., pp. 799–833. Churchill-Livingstone, New York.

West, C. D., and Green, C. J. (1987). The sedative effects of climazolam and climazolam with fentanyl-fluanisone in rats (*Rattus norvegicus*). *Lab. Anim.* **21,** 143–148.

Westenberg, I. S., and Bolam, J. M. (1981). Lethal doses of pentobarbital sodium in albino versus pigmented rats: A within-strain comparison. *Lab. Anim. Sci.* **31,** 360–361.

Westfall, B. A., Boulos, B. M., Shields, J. L., and Garb, S. (1964). Sex differences in pentobarbital sensitivity in mice. *Proc. Soc. Exptl. Biol. Med.* **115,** 509–510.

Wheaton, L. G., Benson, G. J., Tranquilli, W. J., and Thurmon, J. C. (1989). The oxytocic effect of xylazine on the canine uterus. *Theriogenology* **31,** 911–915.

Whelan, G., and Flecknell, P. A. (1992). The assessment of depth of anaesthesia in animals and man. *Lab. Anim.* **26,** 153–162.

White, P. F. (1989). Clinical uses of intravenous anesthetic and analgesic infusions. *Anesth. Analg.* **68,** 161–171.

Wixson, S. K., White, W. J., Hughes, H. C., Jr., Lang, C. M., and Marshall, W. K. (1987a). A comparison of pentobarbital, fentanyl-droperidol, ketamine-xylazine and ketamine-diazepam anesthesia in adult male rats. *Lab. Anim. Sci.* 37, 726–730.

Wixson, S. K., White, W. J., Hughes, H. C., Jr., Marshall, W. K., and Lang, C. M. (1987b). The effects of pentobarbital, fentanyl-droperidol, ketamine-xylazine and ketamine-diazepam on noxious stimulus perception in adult male rats. *Lab. Anim. Sci.* **37,** 731–735.

Wixson, S. K., White, W. J., Hughes, H. C., Jr., Lang, C. M., and Marshall, W. K. (1987c). The effects of pentobarbital, fentanyl-droperidol, ketamine-xylazine and ketamine-diazepam on arterial blood pH, blood gases, mean arterial blood pressure and heart rate in adult male rats. *Lab. Anim. Sci.* **37,** 736–742.

Wixson, S. K., White, W. J., Hughes, H. C., Jr., Lang, C. M., and Marshall, W. K. (1987d). The effects of pentobarbital, fentanyl-droperidol, ketamine-xylazine and ketamine-diazepam on core and surface body temperature regulation in adult male rats. *Lab. Anim. Sci.* **37,** 743–749.

Wright, M. (1982). Pharmacologic effects of ketamine and its use in veterinary medicine. *J. Am. Vet. Med. Assoc.* 180, 1462–1471.

Wyatt, J. D., Scott, R. A. W., and Richardson, M. E. (1989). The effects of prolonged ketamine-xylazine intravenous infusion on arterial blood pH, blood gases, mean arterial blood pressure, heart and respiratory rates, rectal temperature and reflexes in the rabbit. *Lab. Anim. Sci.* **39,** 411–416.

Yamazaki, T., and Sagawa, K. (1984). Effect of thiamylal on the response to carotid occlusion and mild hemorrhage in rabbits. *Am. J. Physiol.* **246,** H806–H810.

Yang, H., Wong, H., Wu, V., Walsh, J. H., and Taché, Y. (1990). Somatostatin monoclonal antibody immunoneutralization increases gastrin and gastric acid secretion in urethane-anesthetized rats. *Gastroenterology* **99,** 659–665.

Yelnosky, J., Katz, R., and Dietrich, E. V. (1964). A study of some of the pharmacologic actions of droperidol. *Toxicol. Appl. Pharmacol.* **6,** 37–47.

Zimpfer, M., Sit, S. P., and Vatner, S. F. (1981). Effects of anesthesia on the canine carotid chemoreceptor reflex. *Circ. Res.* **48,** 400–406.

Zimpfer, M., Manders, W. T., Barger, A. C., and Vatner, S. F. (1982). Pentobarbital alters compensatory neural and humoral mechanisms in response to hemorrhage. *Am. J. Physiol.* **243,** H713–721.

ADDITIONAL READING

Aitkenhead, A. R. (1988). Anaesthesia and the gastro-intestinal system. *Eur. J. Anaesthesiol.* **5,** 73–112.

Allert, J. A., and Adams, H. R. (1987). Pharmacologic considerations in selection of tranquilizers, sedatives, and muscle relaxant drugs used in inducing animal restraint. *J. Am. Vet. Med. Assoc.* **191,** 1241–1244.

Atlee, J. L., and Bosnjak, Z. J. (1990). Mechanisms for cardiac dysrhythmias during anesthesia. *Anesthesiology* **72,** 347–374.

Biggio, G., and Costa, E., Symp. eds. (1988). Chloride channels and their modulation by neurotransmitters and drugs. *Biochem. Psychopharmacol.* **45,** 1–375.

Child, K. J., Currie, J. P., Davis, B., Dodds, M. G., Pearce, D. R., and Twissell, D. J. (1971). The pharmacological properties in animals of CT1341—A new steroid anaesthetic agent. *Br. J. Anaesth.* **43,** 2–13.

Conney, A. H., and Burns, J. J. (1972). Metabolic interactions among environmental chemicals and drugs. *Science* **178,** 576–586.

Coulsom, N. M., Januszkiewicz, A. J., Dodd, K. T., and Ripple, G. R. (1989). The cardiorespiratory effects of diazepam-ketamine and xylazine-ketamine anesthetic combinations in sheep. *Lab. Anim. Sci.* **39,** 591–597.

Covino, B. G., Fozzard, H. A., Rehder, K., and Strichartz, G., eds. (1985). ''Effects of Anesthesia.'' Waverly Press, Baltimore, Maryland.

Davis, B., Dundee, J. W., Forrester, A. C., Johnson, S. E., Lewis, A. A. G., Simpson, B. R. J., and Sutton, J. A. (1972). Steroid anaesthesia. *Postgrad. Med. J.* **48,** 1–139.

Domino, E. F., ed. (1990). ''Status of Ketamine in Anesthesiology.'' NPP Books, Ann Arbor, Michigan.

Doze, V., Chen, B., Li, Z., and Maze, M. (1989). Pharmacologic characterization of the receptor mediating the hypnotic action of dexmedetomidine. *Acta Vet. Scand., Suppl.* **85,** 61–64.

Dundee, J. W., and Wyant, G. M. (1987). ''Intravenous Anaesthesia,'' 2nd ed. Churchill-Livingstone, Edinburgh.

Dundee, J. W., Clarke, R. S. J., and McCaughey, W., eds. (1991). ''Clinical Anaesthetic Pharmacology.'' Churchill-Livingstone, Edinburgh.

Farver, B., Haskins, S. C., and Patz, J. D. (1986). Cardiopulmonary effects of acepromazine and of the subsequent administration of ketamine in the dog. *Am. J. Vet. Res.* **47,** 631–635.

Franks, N. P., and Lieb, W. R. (1990). Mechanisms of general anesthesia. *Environ. Health Perspect.* **87,** 199–205.

Goodchild, C. S., and Serrao, J. M. (1989). Cardiovascular effects of propofol in the anaesthetized dog. *Br. J. Anaesth.* **63,** 87–92.

Haskins, S. C., and Klide, A. M., eds. (1992). Opinions in small animal anesthesia. *Vet. Clin. North Am.: Small Anim. Pract.* 22 (March), 245–502.

Hellyer, P., Muir, W. W., Hubbell, J. A. E., and Sally, J. (1988). Cardiorespiratory effects of the intravenous administration of tiletamine-zolazepam to cats. *Vet. Surg.* **17,** 105–110.

Hellyer, P., Muir, W. W., Hubbell, J. A. E., and Sally, J. (1989). Cardiorespiratory effects of the intravenous administration of tiletamine-zolazepam to dogs. *Vet. Surg.* **18,** 160–165.

Hsu, W. H., Hanson, C. E., Hembrough, F. B., and Schaffer, D. D. (1988). Effects of idazoxan, tolazoline, and yohimbine on xylazine-induced respiratory changes and central nervous system depression in ewes. *Am. J. Vet. Res.* **50,** 1570–1573.

Huidobro-Toro, J. P., Bleck, V., Allan, A. M., and Harris, R. A. (1987). Neurochemical actions of anesthetic drugs on the γ-aminobutyric acid receptor-chloride channel complex. *J. Pharmacol. Exp. Ther.* **242,** 963–969.

Ingwersen, W., Allen, D. G., Dyson, D. H., Pascoe, P. J., and O'Grady, M. R. (1988). Cardiopulmonary effects of a ketamine hydrochloride acepromazine combination in healthy cats. *Can. J. Vet. Res.* **52,** 1–4.

Jarvis, N., and England, G. C. W. (1991). Reversal of xylazine sedation in dogs. *Vet. Rec.* **128,** 323–325.

Kay, B., ed. (1991). ''Total Intravenous Anaesthesia,'' Monog. Anaesthesiol., Vol. 21. Elsevier, Amsterdam.

Klide, A. M., Calderwood, H. W., and Soma, L. R. (1975). Cardiopulmonary effects of xylazine in dogs. *Am. J. Vet. Res.* **36,** 931–935.

Kolata, R. J., and Rawlings, C. A. (1982). Cardiopulmonary effects of intravenous xylazine, ketamine, and atropine in the dog. *Am. J. Vet. Res.* **43,** 2196–2198.

Kumar, A., and Thurmon, J. C. (1979). Cardiopulmonary, hemocytologic and biochemical effects of xylazine in goats. *Lab. Anim. Sci.* **29,** 486–491.

Lagutchik, M. S., Januszkiewicz, A. J., Dodd, K. T., and Martin, D. G. (1991). Cardiopulmonary effects of a tiletamine-zolazepam combination in sheep. *Am. J. Vet. Res.* **52,** 1441–1447.

Langley, M. S., and Heel, R. C. (1988). Propofol: A review of its pharmacodynamic and pharmacokinetic properties and use as an intranenous anaesthetic. *Drugs* **35,** 334–372.

Lemke, K. A., and Tranquilli, W. J. (1994). Anesthetics, arrhythmias, and myocardial sensitization to epinephrine. *J. Am. Vet. Med. Assoc.* **205,** 1679–1684.

Lin, H. C., Thurmon, J. C., and Benson, G. J. (1989). The hemodynamic response of calves to tiletamine-zolazepam anesthesia. *Vet. Surg.* **18,** 328–334.

Löfström, J. B., and Sjöstrand, U., eds. (1988). ''Local Anaesthesia and Regional Blockade.'' Elsevier, Amsterdam.

Longnecker, D. E., and Harris, P. D. (1980). Microcirculatory actions of general anesthetics. *Fed. Proc., Fed. Am. Soc. Exp. Biol.* **39,** 1580–1583.

Miller, R. D., ed. (1986). ''Anesthesia,'' 2nd ed. Churchill-Livingstone, Edinburgh.

Nunn, J. F., Utting, J. E., and Brown, B. R., eds. (1989). ''General Anaesthesia,'' 5th ed. Butterworth, London.

Oyama, T., and Wakayama, S. (1988). The endocrine responses to general anesthesia. *Int. Anesthesiol. Clin.* **26,** 176–181.

Pocock, G., and Richards, C. D. (1991). Cellular mechanisms in general anaesthesia. *Br. J. Anaesth.* **66,** 116–128.

Priano, L. L., Bernards, C., and Marrone, B. (1989). Effect of anesthetic induction agents on cardiovascular neuroregulation in dogs. *Anesth. Analg.* **68,** 344–349.

Reeves, J. G., Fragen, R. J., Vinik, H. R., and Greenblatt, D. J. (1985). Midazolam: Pharmacology and uses. *Anesthesiology* **62,** 310–324.

Reynolds, J. E. F., ed. (1989). ''The Extra Pharmacopoeia/Martindale.'' Pharmaceutical Press, London.

Rubin, E., Miller, K. W., and Roth, S. H. (1991). ''Molecular and Cellular Mechanisms of Alcohol and Anaesthetics,'' Ann. N.Y. Acad. Sci., Vol. 625. New York Academy of Sciences, New York.

Soma, L. R., and Klide, A. (1987). Steady-state level of anesthesia. *J. Am. Vet. Med. Assoc.* **191,** 1260–1265.

Stoelting, R. K. (1991). ''Pharmacology and Physiology in Anesthetic Practice,'' 2nd ed. Lippincott, Philadelphia.

Strichartz, G. R., ed. (1987). ''Local Anesthetics,'' Handb. Exp. Pharmacol., Vol. 81. Springer-Verlag, Berlin.

Swerdlow, B. N., and Holley, F. O. (1987). Intravenous anaesthetic agents: Pharmacokinetic-pharmacodynamic relationships. *Clin. Pharmacokinet.* **12,** 79–110.

Trim, C. M., and Gilroy, B. A. (1985). Cardiopulmonary effects of a xylazine and ketamine combination in pigs. *Res. Vet. Sci.* **38,** 30–34.

Turner, D. M., and Ilkiw, E. J. (1990). Cardiovascular and respiratory effects of three rapidly acting barbiturates in dogs. *Am. J. Vet. Res.* **51,** 598–604.

Wood, M. (1991). Pharmacokinetic drug interactions in anaesthetic practice. *Clin. Pharmacokinet.* **21,** 285–307.

Chapter 2

Pharmacology of Inhalation Anesthetics

David B. Brunson

I. INTRODUCTION

The anesthetic properties of ether and chloroform were discovered in the 1840s and remained the only significant inhaled anesthetics until 1925 (Reynolds, 1981). Diethyl ether became the first routinely used anesthetic for research animal anesthesia. Researchers recognized that ether produced a stable anesthetic state that could be predictably reversed. The wide margin of safety and relative ease of administration of ether has made researchers reluctant to switch to newer anesthetics. Because of hazards associated with the flammability of ether and the toxicity associated with chloroform, clinical anesthesiologists began searching for a potent nonflammable and nontoxic inhaled anesthetic.

Because the relationship of chemical formula to anesthetic action is not fully understood, the discovery of new inhalant

ANESTHESIA AND ANALGESIA IN LABORATORY ANIMALS

ISBN 0-12-417570-8

anesthetics has been largely by trial and error. The addition of fluorides and bromides to gas anesthetic molecules resulted in the development of a few new inhaled anesthetics (Jones, 1990). In particular, substitution of other halogens by fluorine resulted in increased molecular stability and lower toxicity (Jones, 1990).

In the mid-1950s, halothane was discovered to be an anesthetic characterized by nonflammability, low solubility, molecular stability, and high potency (Jones, 1990). The major drawbacks to halothane were that it depressed ventilation and circulatory functions, enhanced epinephrine-induced arrhythmias, and was associated rarely with hepatic necrosis (Eger, 1985).

Methoxyflurane (MOF), a methyl ethyl ether, was evaluated and accepted as an inhaled anesthetic in 1960 (Eger, 1985). Although it was shown to be less arrhythmogenic to the myocardium than halothane, methoxyflurane was more susceptible to biodegradation and produced renal and hepatic injury. Compared to most other modern inhalants, the greater solubility of methoxyflurane resulted in slower equilibration and longer anesthetic recoveries.

Since 1960, four derivatives of MOF have been synthesized that are less soluble and retain the lack of myocardial sensitization of MOF: enflurane, isoflurane, desflurane, and sevoflurane. Enflurane was introduced to veterinary medicine in the mid 1970s but offered only minor advantages over halothane. In particular, the greater molecular stability and thus lower metabolism decreased the likelihood of hepatic necrosis. However, the occurrence of seizures in dogs during deep anesthesia accompanied by hypocapnia resulted in the general disfavor of enflurane in veterinary medicine.

In 1979, isoflurane was approved for clinical use for humans, in 1985 for horses, and in 1989 for dogs (Brunson, 1990). Isoflurane has become a popular anesthetic for animals due to the rapid onset of anesthesia, greater cardiovascular stability, and lower organ toxicity when compared to halothane and methoxyflurane (Brunson, 1990).

Continued efforts to identify a volatile anesthetic suitable for out-patient surgery resulted in the development of desflurane and sevoflurane. Their advantages are principally in lower solubility and thus faster induction and recovery. Greater skill in monitoring anesthetic depth and delivery are needed with the use of these anesthetics when compared to currently used inhalants. Additionally, the need for specialized agent specific vaporizers will most likely limit the use of desflurane for animal anesthesia.

Despite the development of less toxic and safer volatile anesthetics, the ideal anesthetic has not yet been identified. New inhalants will undoubtedly be discovered in the future which will have all of the properties desired in an ideal inhaled anesthetic. The following is a list of the properties of an ideal inhalant anesthetic (Jones, 1990).

1. *Stable molecular structure:* The compound should not be broken down by light, alkali, or soda lime. It should not require preservatives and it should have a long shelf life.
2. *Nonflammable and nonexplosive:* It should be nonflammable when mixed with air, oxygen, or nitrous oxide.
3. *High potency:* It should be easily delivered so that high concentrations of oxygen can be administered during anesthesia.
4. *Low solubility in blood:* Blood/gas solubility should be low in order that induction and emergence from anesthesia are rapid. This enables rapid and precise control of anesthetic depth.
5. *Nonirritating:* It should not cause irritation to the respiratory system, bronchospasm, or breath-holding. Inductions should be smooth and rapid.
6. *Not metabolized:* High molecular stability is desired to eliminate toxic metabolites and effect of disease on drug elimination.
7. *Cardiovascular and respiratory systems unaffected:* The anesthetic should not cause depression of the myocardial contractility or induce arrhythmias. It should not cause sensitization to catecholamines. Ventilatory rate and tidal volume should be unchanged by the anesthetic.
8. *Central nervous system (CNS) effects reversible and nonstimulatory:* Cerebral blood flow should either remain unchanged or decrease during the inhaled anesthesia.
9. *Compatible with other drugs:* The ideal inhalant anesthetic would be compatible with all other medications, especially other cardiovascular supportive drugs.
10. *Easily delivered:* Vaporization and potency must enable delivery of effective concentrations without limiting the inspired oxygen concentrations. Delivery through open systems or precision vaporizers should be possible.
11. *Affordable:* Low cost per hour of anesthesia.

The most important advantage of using inhaled anesthetics for research is that concentrations can be measured on a continual basis to ensure that all animals are at a similar anesthetic depth. This is important because anesthetic effects vary with the depth of anesthesia. Because gas anesthetics are delivered continuously via the lungs, during normal ventilation/perfusion exchange states, alveolar anesthetic concentrations will closely approximate the arterial blood concentrations. Tissues such as the brain that have high blood flows equilibrate rapidly with the arterial blood. By measuring the end tidal gas anesthetic concentration, the depth of anesthesia can be compared throughout the anesthetic period. Multiple animals can thus be studied under the same anesthetic conditions.

In contrast, injectable anesthetics are difficult to measure and stable blood concentrations are not easily confirmed. Absorption, distribution, and elimination of drugs injected intravenously, intramuscularly, subcutaneously, and intraperitoneally result in comparisons of anesthetic depth based on subjective observations that do not correlate with actual drug concentrations in blood (Morris *et al.,* 1979). Injectable anesthetic techniques may be preferred when waste gas elimination is not practical, gas anesthetic delivery is not possible, and when precise control of anesthetic depth is not needed.

II. CHEMICAL AND PHYSICAL PROPERTIES

The physical and chemical properties of the major inhaled anesthetics are shown in Table I. Although no longer recommended as an anesthetic for research, ether is included for comparison with the modern potent inhaled anesthetics. Desflurane

TABLE I

CHEMICAL AND PHYSICAL PROPERTIES OF INHALANT ANESTHETICS

Property	Ether (CH_3OCH_3)	Methoxyflurane ($CHCl_2CF_2OCH_3$)	Halothane ($CHClBrCF_3$)	Isoflurane ($CF_3CHClOCF_2H$)	Enflurane ($CHFCF_2OCHF_2$)	Desflurane ($CH_2HOCFHCF_3$)	Sevoflurane ($CFH_2OCH(CF_3)_2$)	Nitrous oxide (N_2O)
Molecular weight	74	165.0	197.4	184.5	184.5	168.0	200.1	44
Specific gravity	0.71	1.65	1.86	1.50	1.52	1.45	1.52	1.23
Boiling point	36.5°C	104.7°C	0.2°C	48.5°C	56.5°C	23.5°C	58.5°C	−89°C
Vapor pressure at 20°C	450	22.8	244.1	239.5	171.8	664	159.9	39,500
Maximum percent concentration	49	3%	32.53	31.86	22.93	88.53	21.33	100
MAC (Dog)	3.04%	0.23%	0.87%	1.28%	2.2%	7.2%	2.36%	188%
Odor	Ethereal	Fruity	Organic solvent	Pungent etheral	Etheral	Pungent etheral		Sweet
Stability		>50%	20%	0.2%	20%	<0.2%	20%	No
Explosive	>1.8% in air >2.1% in O_2	No	No	No	No	No	No	(supports combustion)
Preservatives	Yes 3% Ethanol	Yes 0.01 % Hydrylated hydrotoluene	Yes Thymol	None	None	None	None	None

and sevoflurane are new potent inhalant anesthetics that have unique chemical and physical properties. They are included for reference.

A. Vapor Pressure

Vapor pressure is the maximum pressure that can be produced by the anesthetic at a given temperature and pressure (Table I). It is an expression of the volatility of the drug. This value can be converted into the maximum percentage or concentration that the anesthetic can develop. The presence of liquid anesthetic in a container (bottle, vaporizer, anesthetic chamber) indicates that the partial pressure (concentration) of the anesthetic is at the maximum level (i.e., vapor pressure). When using highly potent inhaled anesthetics, this is a lethal concentration.

The physical property of vapor pressure is used in the design of vaporizers such that the correct volume of dilutional gas and anesthetic vapor is mixed to produce a clinically useable concentration (Dorsch and Dorsch, 1994). Usually oxygen and/or nitrous oxide is mixed with anesthetic vapors to deliver a concentration that is high enough to anesthetize the animal but low enough to prevent rapid overdosage. Anesthetics such as halothane and isoflurane, which have boiling points and vapor pressures that are nearly identical, can be used in identically manufactured vaporizers (Steffey *et al.,* 1983). However, the practice of using an anesthetic in a vaporizer not designed for that particular anesthetic is not recommended. This practice can easily result in mixtures of two anesthetics and misidentifying the contents of the vaporizer. Vaporizers should always be clearly labeled and only contain the designated anesthetic (Dorsch and Dorsch, 1994).

B. Molecular Weight

All of the inhaled anesthetics have similar molecular weights with the exception of nitrous oxide and ether (Table I). Because of this fact the milliliters of vapor produced by 1 ml of liquid anesthetic is also similar (Table II). The volume of gaseous anesthetic formed from 1 ml of liquid anesthetic can be calculated from the gas laws (Linde, 1971).

$$PV = nRT$$
$$V = nRT/P,$$

where the volume of gas formed from 1 ml of liquid is equal to the number of moles of the liquid (density divided by the molecular weight) times a constant [$R = 0.082$ (R = ml × atm/K × moles)] times the temperature in K divided by the number of atmospheres of pressure (atm). Direct injection of a liquid anesthetic into closed delivery systems can be successfully performed using halothane and isoflurane by calculating the volume of anesthetic required to produce a desired concentration in the anesthetic system. Anesthetic chambers can be used without precision vaporizers if the volume of the system is

TABLE II

MILLILITERS OF GAS ANESTHETIC FORMED FROM VAPORIZATION OF 1ML OF LIQUID ANESTHETIC[a]

Anesthetic	Volume (ml)
Methoxyflurane	191.4
Halothane	211.1
Isoflurane	182.1

[a]Calculated at standard temperature (273K) and pressure (760 mm Hg).

known and the volume of liquid anesthetic needed to produce the desired concentration is calculated. Table III provides the number of milliliters of liquid anesthetic needed to produce anesthetic concentrations in the clinically useable range. By using Table III it is possible to produce anesthetic concentrations that will produce safe anesthesia with even highly volatile anesthetics like halothane and isoflurane.

For reasons of cost and simplicity, laboratory rodents are frequently anesthetized in bell jars or anesthetic chambers. For many years, ether was used in this manner without concern for the inspired concentration. The high vapor pressure of ether produces a maximum anesthetic partial pressures of 450 mmHg, which is equal to an inspired concentration of 59% (Price, 1975).

Two reasons permitted ether to be delivered without the need for controlling the inspired concentration. First, the effective dose to produce anesthesia in half of the animals (MAC, minimum alveolar concentration) is 1.9%, because of the low potency, a high inspired concentration is desirable. Second, ether has a high blood/gas solubility and thus the equilibration is relatively slow, providing a wider time period when overdosage can be prevented by removing the animal from the chamber.

Methoxyflurane is now frequently recommended for chamber inductions instead of ether. Methoxyflurane can also be used without precise control of the inspired concentration for some of the same reasons as ether. Like ether it has a high blood/gas solubility that results in a long equilibration time. However, MOF is a very potent inhalant anesthetic with a MAC value of 0.23%. Once equilibration has occurred, very little methoxyflurane is required for maintenance of the anesthetic level. Unlike ether, MOF has a low volatility which limits the maximum concentration that can be produced at room temperature to 3%. The slow equilibration and the low volatility balance the high potency to make methoxyflurane useable without precise control of the inspired concentration.

All of the remaining potent inhaled anesthetics require control of the inspired concentration to be used safely. They possess high volatility, low blood/gas solubility, and relatively high potency. Together these factors result in rapid equilibration and greater potential for overdosage.

C. Boiling Point

At room temperature all of the currently used inhalant anesthetics are liquids with the exception of nitrous oxide. Nitrous oxide is supplied in cylinders which contain both liquid and gas; at pressures greater than 800 psi nitrous oxide is a liquid. Pressure gauges on nitrous oxide cylinders will indicate a pressure of approximately 700 psi until all of the liquid nitrous oxide has vaporized. The pressure will then progressively decrease until the cylinder is empty (Dorsch and Dorsch, 1994).

Desflurane has a unique physiochemical property that causes it to boil at 23.5°C (room temperature) (Jones, 1990). Because small changes in temperature would produce uncontrollable changes in output, delivery of a constant known volume or concentration of desflurane becomes difficult. Desflurane re-

TABLE III

Amount of Liquid Anesthetic Needed to Produce Various Anesthetic Concentrations

	Internal volume of anesthetic chamber (ml)				
Concentration of anesthetic (%)	1000	2000	3000	4000	5000
Halothane					
1	0.04[a]	0.09	0.13	0.18	0.22
2	0.09	0.18	0.26	0.35	0.44
3	0.13	0.26	0.40	0.53	0.66
4	0.18	0.35	0.53	0.71	0.88
5	0.22	0.44	0.66	0.88	1.10
Isoflurane					
1	0.05	0.10	0.15	0.20	0.26
2	0.10	0.20	0.31	0.41	0.51
3	0.15	0.31	0.46	0.61	0.77
4	0.20	0.41	0.61	0.82	1.02
5	0.26	0.51	0.77	1.02	1.28
Methoxyflurane					
1	0.05	0.10	0.15	0.19	0.24
2	0.10	0.19	0.29	0.39	0.49
3	0.15	0.29	0.44	0.58	0.73

[a]Volume given in milliliters.
Calculations at 20°C and 760 mmHg.

quires a special vaporizer which can precisely control the vaporization temperature (Jones, 1990). Standard ''Tec-type'' or plenum vaporizers should not be used for desflurane.

D. Partition Coefficients

Solubility of inhaled anesthetics in blood and tissue provides an indication of the capacity of these compartments to hold anesthetic (Eger, 1985). Partition coefficients are ratios of the concentrations of the anesthetic in two compartments at equilibrium. The most important partition coefficient for inhaled anesthetics is the blood/gas partition coefficient (Table IV). Because blood is the first compartment to equilibrate with alveolar anesthetic gases, the blood/gas coefficient is an indication of the speed of induction. All other factors being equal, the greater the solubility in blood the longer the equilibration time. Although solubilities vary among tissues, tissue components, temperature, and species (Table IV), partition coefficients provide a consistent guideline for predicting how uptake and elimination will occur when comparing anesthetics (Wollman and Smith, 1975).

The primary objective of inhalant anesthesia is to produce a constant and sufficient partial pressure of the anesthetic in the brain to produce analgesia and unconsciousness. The vessel-rich group of tissues, which includes the brain, receives approximately 75% of the cardiac output (Eger, 1964). Because of the large blood flow to this small compartment, equilibration rapidly occurs between the anesthetic concentration in the blood and the brain. Because nitrous oxide, desflurane, and sevoflurane have exceptionally low blood/gas solubilities, they are associated with especially rapid uptake and elimination (Eger, 1992; Jones, 1990).

The rapid effects of insoluble anesthetics are further accentuated in very small animals. Smaller animals have correspondingly small lung, blood, and tissue volumes which equilibrate very rapidly. Large animals reach equilibrium slowly even when insoluble anesthetics are administered (Steffey, 1978).

TABLE IV

BLOOD/GAS AND OIL/BLOOD PARTITION COEFFICIENTS[a]

Anesthetic	Blood/gas coefficient	Olive oil/blood
Ether	2.8	58.0
Methoxyflurane	12.0	970.0
Halothane	2.54	224.0
Isoflurane	1.46	91.0
Enflurane	2.0	96.0
Desflurane	0.42	19.0
Sevoflurane	0.69	53.0
Nitrous oxide	0.47	1.4

[a]Data from Eger (1974) and Jones (1990).

Tissue/blood partition coefficients also influence the anesthetic uptake. However, tissue solubility is even more important during elimination of the inhaled anesthetics. Anesthetics with high lipid solubilities will accumulate in fat tissue during anesthesia. On recovery the anesthetic will leave the fat and slow the animal's return to consciousness (Table IV). Lipid solubility and potency seemingly have a reciprocal relationship which has led to the theory that the mode of action of gas anesthetics is related to a relative constant number of anesthetic molecules dissolved in a hydrophobic phase [the ''Unitary Theory'' (Kaufman, 1977)].

Solubility of anesthetic gases in the delivery apparatus can also affect uptake and elimination. Absorption of the anesthetic by rubber or plastic components of the anesthetic delivery apparatus slows the increase in concentration inhaled by the animal, thus slowing induction. Uptake of anesthetic molecules by the anesthetic device also affects recovery. Animals breathing from the anesthetic device will continue to receive anesthetic released from rubber and/or plastic machine components even after the vaporizer has been turned off. The newer inhalants have low solubilities and are thus not significantly affected (Eger, 1974).

E. Minimum Alveolar Concentration

MAC is the concentration at 1 atmosphere that produces immobility in 50% of animals exposed to a noxious stimulus (Eger, 1974). Thus MAC is synonymous with the effective dose or ED_{50} and is a measure of the potency of inhaled anesthetics. MAC values can be determined as the concentration where either 50% of a group of animals respond or where an individual animal responds 50% of the time to a standardized noxious stimulus.

MAC is important for three reasons. First, it is a measure of the potency of inhalation anesthetics in the context of clinical use. General anesthesia is defined as the loss of pain perception and the loss of reflex and spontaneous muscle activity. To be properly anesthetized, an animal must be unaware of noxious (painful) stimuli and be immobile. These factors are the end points used to determine MAC. Second, MAC can be applied to all inhalation anesthetics. MAC allows direct comparisons of multiple animals or sequential anesthetic episodes by ensuring that they are performed at the same depth of anesthesia. The ability to measure end tidal anesthetic concentrations on a breath to breath basis ensures that each animal is at approximately the same magnitude of anesthetic depth. Third, MAC is important because continuous monitoring of alveolar concentrations can be used to maintain the animal at a known and stable depth of anesthesia.

1. How MAC Is Determined

The standard bracketing procedures for MAC determination are as follows. When possible the animals should be anesthetized, via a mask or chamber, and intubated with only the inhalational anesthetic. Sedatives and injectable anesthetics can affect the required inhalational anesthetic concentration needed to cause immobility. The concentration of the anesthetic is measured and held constant for at least 15 min to ensure equilibration between the alveoli and the brain. Measurements of anesthetic concentrations at the end of expiration provide the most accurate estimation of the anesthetic concentration in the blood leaving the lungs. A noxious supramaximal stimulus is then applied to the animal and the animal is observed for gross purposeful movement or the absence of movement. During MAC determinations, stretching, increased ventilation, or spontaneous movement, which is an elicited reflex, is not considered evidence of movement. Classically, either a large hemostat is clamped to the tail or digits (usually for 1 min) of the animal or passage of an electrical current through subcutaneous tissue is used as the noxious stimulus (Quasha *et al.,* 1980). If no response is observed, the alveolar concentration is reduced by 20% of the previous setting, 15 min is allowed to elapse for equilibration, and the stimulus is repeated. The step-down procedure is continued until gross purposeful movement is observed. The alveolar concentration is then increased by 10% until the animal fails to respond to the noxious stimulus. MAC (ED_{50}) is the average of the concentrations where the animal responds and does not respond.

Using alveolar anesthetic concentration as a measurement of the anesthetic depth of an animal is based on the assumption that gases within the alveoli will be in equilibrium with the blood exiting the lung. Arterial blood is assumed to be equilibrated with the brain and other high blood flow tissues. These assumptions are true as long as ventilation and perfusion are closely matched and sufficient time has elapsed to allow equilibration between the alveolar gases and the pulmonary blood. Alveolar concentrations are assumed to represent the anesthetic concentration in the brain of the animal. The anesthetic concentration in the gases last to exit the lung (end tidal concentrations) are from the deepest areas of the pulmonary system (i.e., the alveoli). Normally, ventilation and perfusion are equally matched in the lung and alveolar gas is fully equilibrated with blood. Ventilation/perfusion mismatching and shunting of blood past alveoli will result in discrepancies between the alveolar anesthetic concentrations and the anesthetic concentrations in the blood and brain (Eger, 1974). This is not a frequent problem in healthy mammals that weigh less than 50 kg.

MAC has been shown to be remarkedly consistent between a wide variety of animal species (Table V). Dohm and Brunson (1993) determined an isoflurane ED_{50} for a reptilian species (desert iguana) that was markedly higher than mammalian or avian species.

2. Factors That Do Not Alter MAC

Biological variability within an animal species has been shown to be small, with all animals being affected within a narrow range. Variation among dogs from the average MAC equals 10 to 20% (Eger, 1974). Variation within individual dogs is even less.

MAC is not affected by the type or the intensity of the stimulus (Jones, 1990). Suppression of movement during skin incision, tail clamp, and electrical current all require the same alveolar concentrations (Eger, 1974). The duration of anesthesia does not alter MAC; the concentration required to suppress movement is constant from the beginning to the end of a procedure. The potency of inhalant anesthetics does not differ between males and females (Quasha *et al.,* 1980).

Changes in acid–base status have little effect on MAC values. Infusions of sodium bicarbonate and hydrochloric acid to produce a base excess of 7 and −20 mEq, respectfully, did not alter MAC (Eger, 1974). Furthermore, when CO_2 is greater than 10 mm Hg and below 90 mm Hg, MAC is not changed. Hypoventilation resulting in CO_2 greater than 90 mm Hg results in

TABLE V

Isoflurane Minimum Alveolar Concentrations (MAC) for different species

Species	MAC (ED_{50}) values	Reference
Humans	1.28 ± 0.04	Stevens *et al.* (1975)
Monkey	1.28 ± 0.18	Tinker *et al.* (1977)
Pig	1.75 ± 0.1	Tranquilli *et al.* (1983)
Sheep	1.58 ± 0.17	Palahniuk *et al.* (1974)
Goat	1.5 ± 0.3	Antognini and Eisele (1993)
Dog	1.28 ± 0.06	Steffey and Howland (1977)
Cat	1.63 ± 0.06	Steffey and Howland (1977)
Rabbit	2.05 ± 0.00	Patel and Mutch (1990)
Guinea pig	1.15 ± 0.5	Seifen *et al.* (1989)
Rat	1.38 ± 0.06	White *et al.* (1974)
Mouse	1.41 ± 0.03	Deady *et al.* (1980)
Cockatoo	1.44 ± 0.07	Curro *et al.* (1994)
Iguana *(Diposaurus dorsalis)*	3.14 ± 0.16	Dohm and Brunson (1993)

an increased anesthetic effect of halothane. It is believed that this effect is related to an increased cerebrospinal fluid (CSF) hydrogen ion concentration. Complete CO_2 narcosis occurs at a CSF pH of 6.8 to 6.9 (Eger, 1974).

MAC is not affected by arterial oxygen partial pressures above 40 mm Hg. When PaO_2 drops below 40 mm Hg, MAC decreases. Similarly, when the arterial oxygen content decreases below 5 ml/100 ml of blood due to anemia, MAC decreases to one-third of the original value. This is most likely related to the development of cerebral lactic acidosis during hypoxemia (Quasha *et al.,* 1980).

Blood pressure changes have no effect on the MAC values of inhalational anesthetics. Hypertension induced by phenylephrine does not alter MAC; however, severe hypotension secondary to hemorrhage reduces halothane MAC. This was speculated to be related to the concomitant hypoxia which occurred.

3. Factors That Alter MAC

MAC has been shown to be affected by several factors. Small changes in MAC have been measured due to circadian rhythms in rats. Anesthesia during high activity time periods resulted in higher MAC requirements than anesthesia during low activity periods. Although this was only a 5–10% difference, anesthesia should be performed during similar activity time periods to reduce variability (Quasha *et al.,* 1980).

Anesthetic requirements decrease directly with decreases in the body temperature of the animal. Halothane and methoxyflurane requirements decrease by 5% for each degree centigrade decrease, thus animals allowed to cool during anesthesia will require less inhalational anesthetics than normothermic animals. Cooling of anesthetized animals occurs for numerous reasons. Animals lose heat from moisture evaporation through the respiratory system, exposed tissues during surgery, and convective heat loss to the surgical table. In order to be able to compare data from similar anesthetic preparations, animals must be carefully maintained at a known temperature and the gas anesthetic concentration monitored. Decreased MAC is probably not due to an altered metabolic rate since an increased metabolic rate associated with hyperthyroidism only slightly increased MAC requirements, but is associated with direct CNS depression associated with hypothermia (Eger, 1974; Quasha *et al.,* 1980).

Age has been shown to affect the MAC requirements of animals (Quasha *et al.,* 1980). Very old animals have lower MAC values. This may be due to a lower neuronal density or to a lower cerebral metabolic rate. The highest MAC requirements occur in neonates. Animals of similar age should be used to minimize variations in sensitivity to inhalational anesthetics.

Synergistic effects are common between CNS depressant drugs and inhalant anesthetics. Opioids have been shown to reduce MAC (Curro *et al.,* 1994; Eger, 1974). Even drugs such as diazepam reduce MAC in humans, even though individually they have no analgesic properties (Mathews *et al.,* 1990).

Inhalational anesthetics have been shown to be additive and thus can be used in combinations to produce general anesthesia. Nitrous oxide is frequently used in anesthetic protocols patterned after human anesthetic techniques. Although highly volatile and highly potent anesthetics have been used together, they are usually administered as sole agents or with N_2O.

Central nervous system catecholamines and drugs that affect their release can alter the MAC of inhalant anesthetics. α-Methyldopa and reserpine decrease MAC. D-Amphetamine releases CNS catecholamines and increases MAC. Chronic administration can result in depletion of central catecholamines and a decreased MAC (Quasha *et al.,* 1980).

F. Stability

In studies determining the extent of metabolism of inhaled anesthetics, it has been shown that the hepatic biodegradation of isoflurane is less than 0.2% (Holaday and Smith, 1981) of the administered dose, where 20% of halothane and more than 50% of methoxyflurane are metabolized (Carpenter *et al.,* 1986). Molecular stability is closely correlated with the lack of tissue toxicity. Although inhalant anesthetics are not considered to be directly tissue toxic, the metabolites of the anesthetics may cause tissue injury. Much of the effort used to identify new inhalant anesthetics has been motivated by the need for less toxic anesthetics. Substitution of fluorine and bromine for chlorine has resulted in greater molecular stability and less toxic anesthetics.

The low rate of biodegradation of the new inhalant anesthetics decreases the potential for renal and hepatic injury in anesthetized animals as well as in personnel exposed to trace anesthetic levels (Section VII).

Molecular stability is also important as it relates to the need for preservatives (Table I). Ether, methoxyflurane, and halothane have preservatives added to prevent degradation of the anesthetics during storage. Isoflurane, enflurane, desflurane, and sevoflurane do not require the addition of preservatives to prevent spontaneous oxidative decomposition. Thymol is the preservative added to halothane; it vaporizes slowly and thus accumulates in the vaporizer and may cause the turnstiles to stick. Accuracy of the vaporizer may be affected which requires periodic servicing (Dorsch and Dorsch, 1994).

III. MODE OF ACTION

The primary target sites of general anesthetics are not known. It is now believed that the ultimate target sites are probably ion channels in nerve membranes, influenced by proteins and/or lipids (Franks and Lieb, 1991; Jones, 1990).

Studies of halothane, isoflurane, and enflurane suggest that the major depressant effect involves enhancement of the inhibitory neurotransmitter α-aminobutyric acid (GABA) (Moody *et al.,* 1988). GABA effects are believed to be mediated through release of intracellular calcium. In particular, volatile anesthetics are associated with the ligand-gated ion channel $GABA_a$. When activated, this channel causes an increase in the chloride permeability of neurons (Tanelian *et al.,* 1993).

Although desflurane may approach the characteristics of an ideal anesthetic, it still has undesirable cardiovascular side effects. Better anesthetics will be identified when we understand how inhaled anesthetics work at the molecular level. The development of anesthetics without unwanted side effects must await identification of those sites (Jones, 1990).

IV. UPTAKE, DISTRIBUTION, AND ELIMINATION

As previously discussed, the alveolar anesthetic concentration is usually a close approximation to brain anesthetic concentrations. The goal during induction and maintenance is to raise the alveolar (brain) anesthetic concentration to the level which causes anesthesia. Three factors determine when alveolar equilibration occurs: ventilation, uptake, and the inspired concentration.

If unopposed by uptake, the alveolar concentration would rapidly equal the inspired concentration. Increasing ventilation increases alveolar anesthetic concentration. Likewise, factors that decrease ventilation slow the equilibration time.

Uptake of anesthetic into blood and tissue opposes or limits the rise in alveolar concentration. The greater the uptake, the lower the alveolar concentration will be relative to the inspired concentration. Anesthetics with high blood or tissue solubilities (see Section II,D) thus take longer for the alveolar concentration to increase to that required for anesthesia.

Conditions that increase cardiac output increase the quantity of blood passing the alveolar membrane per unit time. The higher the cardiac output, the larger the quantity of anesthetic removed from the lungs and thus the alveolar concentration will be lower, delaying induction. Conversely, a lower cardiac output will result in faster equilibration because the alveolar anesthetic concentration increases rapidly while cerebral blood flow is preserved due to autoregulation. This potentially can result in rapid deep anesthesia. If ventilation is held constant, a low cardiac output will result in faster increases in alveolar anesthetic concentrations. If high inspired concentrations are used, care must be taken to avoid excessive anesthetic depression (Eger, 1964).

The concentration gradient between the alveolus and the blood entering the lungs affects the uptake of the anesthetic. The greater the difference, the greater the anesthetic uptake. As equilibration occurs, tissue uptake decreases and the concentration of anesthetic in the blood returning to the lung is nearly equal to the alveolar concentration.

In general, ventilation facilitates whereas tissue uptake and cardiac output oppose equilibration of the alveolar anesthetic concentration with the inspired concentration.

Metabolism of highly soluble inhaled anesthetics may have an effect on anesthetic uptake. The majority of anesthetic metabolism has been shown to occur after clinical anesthesia has been discontinued. Anesthetics that have high fat solubilities have the slowest elimination from the lungs and thus will have the greatest hepatic metabolism. The importance of metabolism is not related to anesthesia recovery but is related to the potential of tissue toxic metabolites.

V. INHALED ANESTHETICS

A. Nitrous Oxide

Nitrous oxide is occasionally included in animal research protocols. Nitrous oxide is a very weak anesthetic for nonprimate species and the benefits of its use are minimal. The principal advantage of nitrous oxide is that it has minimal cardiovascular and respiratory depressive effects. The low blood/gas solubility produces rapid equilibration and elimination. The rapid uptake of the high inspired concentration can be used to accelerate the uptake of other inhaled anesthetics (Eger, 1974). This is known as the second gas effect and is most useful during the induction of mammalian species that are larger than 2 kg. Animals smaller than 2 kg equilibrate rapidly and the second gas effect is of no advantage. It is important to remember that in order for the second gas effect to occur, nitrous oxide must be inhaled after the other inhaled anesthetic is present in the alveoli.

Nitrous oxide is the least potent of the inhaled anesthetics. The human MAC is close to 105% whereas the MAC for dogs is calculated to be 188 $\pm$ 35% (Eger, 1974). The low potency in dogs means that even inspired concentrations of 60–70% will reduce the requirements for other anesthetics by only one-third.

Because of its minimal depressive effects, nitrous oxide is frequently used as an adjunct to other more potent and depressive anesthetics. Care must be taken to ensure that adequate oxygen is delivered to the patient because the low potency requires high delivered concentrations. The oxygen in nitrous oxide is not available for cellular metabolic needs. In all cases, a minimum of 20% of the inspired gases must be oxygen. Oxygen flow rates should be based on the metabolic needs of the animal and the nitrous oxide added to this base flow.

The high flow rates using nitrous oxide result in increased heat loss from the animal. Warm, moist exhaled air is either diluted or washed away from the animal. The high total gas flow also increases the cost associated with anesthesia. Because nitrous oxide and oxygen become carrier gases for the volatile anesthetics such as halothane and isoflurane, the amount of anesthetic vaporized increases proportionally to the increase in flow. Additional costs of using nitrous oxide include the need for flow meters and regulators.

Nitrous oxide rapidly diffuses into gas-filled spaces such as intestinal segments or pneumothoraces (Steffey *et al.,* 1979). For this reason, nitrous oxide is not recommended for use in herbivores.

Human health risks are associated with the chronic exposure to trace concentrations of nitrous oxide and with its abuse. Effective scavenging systems must always be used when delivering nitrous oxide. Monitoring of anesthetic use and adequate security should be employed to minimize the inappropriate use of nitrous oxide by humans. Nitrous oxide has been associated with atmospheric ozone damage (Schmeltekopf *et al.,* 1975).

In summary, nitrous oxide is a weak anesthetic gas that should be used only where small reductions in other anesthetics are critical. Because the effectiveness of nitrous oxide in animals is much lower than for humans, it should not be used for the sole reason of simulating human anesthetic protocols.

B. Ether

Ether was the principal inhaled anesthetic in research prior to the development of nonflammable and more potent gas anesthetics. However, ether is no longer recommended as an anesthetic. The principal reason for the abandonment of ether is that it is explosive in concentrations within the range required to produce anesthesia and that are frequently present during recovery (Wollman and Smith, 1975). Even the bodies of animals that have been euthanized will continue to release ether and have been responsible for explosions when placed in refrigerators prior to disposal.

In addition to the hazard of explosion or fire, ether is highly irritating to the respiratory tract. Increased production of respiratory secretions and increased bronchoconstriction are common. The use of an anticholinergic, such as atropine, was commonly advocated to decrease the signs of respiratory irritation (Wollman and Smith, 1975).

Methoxyflurane is a nonflammable alternative to ether. It can be used in inexpensive, nonprecision vaporizers or bell-jar systems.

C. Methoxyflurane

The low volatility and high blood solubility of methoxyflurane results in an anesthetic that is safe to use in nonprecision systems. Methoxyflurane is the most potent of inhaled anesthetics with an ED_{50} of less than 0.25%. Although MOF vaporizers are relatively inexpensive, the cost of the drug has increased due to the lack of use of the drug in human anesthesiology and the limited use in clinical veterinary medicine. Researchers with limited training or experience in anesthesiology may find MOF easier to use safely because anesthetic depth changes slowly.

The major disadvantage of methoxyflurane is that it is extensively metabolized. Metabolites include inorganic fluoride, oxalates, and trifluoroacetic acid, which are nephrotoxic (Brunson *et al.,* 1979). Human exposure should be avoided. Methoxyflurane should only be used with a functioning scavenging system or inside a nonrecirculating hood.

Methoxyflurane is a respiratory depressant. During spontaneous ventilation, animals will be hypercarbic. Apnea will occur at deep anesthetic levels. Methoxyflurane is nonirritating and does not stimulate tracheobronchial secretions or cause bronchoconstriction (Wollman and Smith, 1975).

The effects of MOF on the circulatory system are characterized by mild-to-moderate hypotension with or without bradycardia. Arterial hypotension is due to reduced myocardial contractility and cardiac output. Methoxyflurane causes less sensitization of the myocardium to catecholamines than halothane.

In general, MOF is an excellent anesthetic but is decreasing in clinical usage due to the slow onset and recovery and the potential for toxicity due to metabolism.

D. Halothane

Halothane is an inhaled anesthetic that has good potency, low blood/gas solubility, and high volatility. It produces anesthesia rapidly. The high vapor pressure of halothane enables concentrations as high as 33% to be produced which, combined with the rapid speed of equilibration, can lead to lethal levels of anesthesia. Delivery of a controlled anesthetic concentration is necessary for safe anesthesia. Stable anesthetic levels are easily produced within 10 min of the start of anesthesia.

Halothane causes respiratory depression which is dose dependent. At deep anesthetic levels, ventilation becomes inadequate. This limits anesthetic delivery if the animal is spontaneously ventilating, but spontaneous respiration should not be used to regulate anesthetic depth.

Halothane causes direct depression of myocardial muscle and relaxation of vascular smooth muscle. Myocardial contractility, cardiac output, and total peripheral resistance decrease. Systemic arterial blood pressure is progressively decreased as anesthetic depth becomes deeper, thus arterial blood pressure is a good indicator of anesthetic depth.

Halothane increases cerebral blood flow due to direct vasodilation of vascular smooth muscle. This increased blood flow is not prevented by preanesthetic hyperventilation.

E. Isoflurane

One of the advantages of isoflurane is that it has been shown to maintain cardiovascular functions better than halothane. Like other inhaled anesthetics, isoflurane causes a dose-related depression of systemic arterial blood pressure (Merin, 1993). The decreased blood pressure is due to a combination of decreases in myocardial depression and systemic vascular resistance. Isoflurane has less of a depressant effect on myocardial contractility but causes greater vasodilatory effects when compared to halothane. Isoflurane, like other ether derivatives (methoxyflurane, desflurane, and sevoflurane), does not sensitize the heart to the arrhythmogenic effects of exogenously administered epinephrine (Eger, 1992).

In keeping with the lower myocardial depression and the sensitivity to catecholamines, isoflurane has a greater safety margin than halothane in rats. The ratio of the concentration producing cardiovascular collapse with the anesthetic dose (MAC) is called the cardiac anesthetic index. Wolfson *et al.* (1978) determined that the cardiac anesthetic index for isoflurane and halothane was 5.7 and 3.0, respectively.

Isoflurane has several other vasodilatory effects that are important. Isoflurane is frequently used in coronary perfusion studies because it has been shown to be a more potent coronary vasodilator than halothane or enflurane (Merin, 1993). Like other inhalant anesthetics, isoflurane causes an increase in cerebral blood flow. However, unlike halothane, enflurane, or methoxyflurane, increases in cerebral blood flow can be prevented if hyperventilation is instituted prior to the administration of isoflurane (Boarini, 1984). For this reason, isoflurane is preferred where cerebral blood flow must not increase.

Isoflurane is a ventilatory depressant. Increasing concentrations produce a progressive decrease in tidal volume and response to increases in arterial carbon dioxide concentrations (Merin, 1993).

The high molecular stability of isoflurane results in metabolism of less than 0.2% of the inspired dose. This reduces the potential for renal and hepatic injury (Fugita *et al.,* 1991). Human exposure to trace amounts of isoflurane is less hazardous than exposure to either methoxyflurane or halothane. Low-flow delivery techniques are recommended to reduce environmental pollution and the associated cost of using isoflurane. Calculated volumes for induction chambers (Section II,B) and precision vaporizers should be used to minimize wastage and to increase the control of delivered concentrations.

The low solubility and high potency of isoflurane can result in rapid deep anesthesia. In the research environment, isoflurane is an excellent anesthetic. The primary difficulty in using isoflurane is that greater knowledge and training in monitoring anesthetized animals are needed. Researchers must be vigilant in adjusting the delivery of isoflurane in order to avoid overdosage.

F. Desflurane and Sevoflurane

These two new inhaled anesthetics are pharmacokinetically different from isoflurane, halothane, or methoxyflurane because of their lower solubility in blood (Table IV). This results in very rapid equilibration enabling precise and rapid changes in anesthetic depth (Eger, 1992; Yashuda *et al.,* 1990). The principal advantage of these two new inhalant anesthetics is their rapid induction and recovery (Kazama and Ikeda, 1988). They may be useful in animal research where rapid full recovery from anesthesia is imperative.

Cardiovascular and respiratory effects are similar to isoflurane (Weiskopf *et al.,* 1989). Desflurane is somewhat more irritating to the respiratory tract than isoflurane, resulting in increased secretions, breath-holding, and difficult mask inductions. Rapid increases in inspired desflurane concentration are associated with increased sympathetic activity. Increasing anesthetic depth causes a progressive decrease of blood pressure and tidal volume (Merin, 1993). Cerebral blood flow can be reduced if hyperventilation is instituted prior to the onset of desflurane delivery (Young, 1992).

In the United States, sevoflurane was approved for human clinical anesthesia in 1995. Sevoflurane is only metabolized to a small amount, approximately 2%. Degradation in the presence of soda lime and baralyme has raised concerns relating to the potential toxicity of degradation products. Two degradation products have been identified as $CF_2{=}C(CF_3){-}O{-}CH_2F$ (an olefin called compound A) and $CH_3{-}O{-}CF_2CH(CF_3){-}O{-}CH_2F$ (compound B). Only compound A is produced in significant quantities. Baralyme, which is composed of more KOH than soda lime, produces significantly higher concentrations of compound A (Bito and Ikeda, 1994). The greatest concern is associated with low-flow delivery techniques where increased contact time and temperature of the CO_2 absorbent result in greater degradation. Studies have shown that clinically detectable nephrotoxicity or hepatotoxicity do not occur with normal anesthetic delivery (Bito and Ikeda, 1994; Frink *et al.,* 1994).

Unlike desflurane, sevoflurane can be used with conventional methods of vaporization. The mild odor and rapid onset will make sevoflurane an attractive inhalant anesthetic for mask and chamber inductions of animals. The availability of precision vaporizers and the cost per animal will be the principal determinants of sevoflurane use.

Desflurane, like isoflurane, has a high molecular stability and low toxicity (Koblin 1992). The low boiling point of desflurane requires a special heated vaporizer to control delivered concentrations. The high cost of specialized equipment and the limited need in animal research for faster anesthetic inductions or recoveries make the routine use of desflurane unlikely (Jones, 1990).

VI. MEASUREMENT AND QUANTIFICATION OF ANESTHETIC CONCENTRATION FOR RESEARCH

Measurement of the alveolar anesthetic concentration is important when research goals are to evaluate the effects of inhaled anesthetics or when background levels of anesthesia must be similar in multiple animals.

Multiple methods for measuring anesthetic gas concentrations have been developed and are available for research. Monitors that measure the concentration of gaseous anesthetics are noninvasive and cause minimal interference with research procedures. They are either incorporated into the anesthetic delivery apparatus or are designed to aspirate a gas sample from the apparatus. In order to assess alveolar anesthetic concentrations, the monitor must be placed either on the expiratory side of the anesthetic machine or between the junction of the delivery system and the airway of the animal. If the animal is intubated, samples are withdrawn from the region of the endotracheal tube connector.

Estimations of the depth of anesthesia based on either the vaporizer setting or the calculated concentration in an anesthetic chamber will result in wide variability among animals. Monitoring the expired anesthetic concentrations eliminates the problem of estimating the complex dynamic relationship between delivered and alveolar anesthetic concentrations (Morris *et al.,* 1979).

Additionally, direct measurement of the anesthetic concentration enables determination of when equilibration of the animal has occurred with the inspired anesthetic concentration. Other procedures used to ensure a stable anesthetic plane, such as initially using high anesthetic concentrations and flow rates, are simplified by actually measuring the anesthetic concentration (Morris *et al.,* 1979).

A. Nonspecific Methods of Measurement

Anesthetic gas concentrations have been measured with a variety of techniques that are based on some physical property of the gas whereby the components can be identified and the quantity of each determined. Numerous nonspecific techniques have been used, including molecular density, thermal conductivity, refractive index, and alterations in the transmission of sound. These techniques require rigid control of the composition of the gases analyzed and only two gases can be present. Additionally, nonspecific methods often require larger sample sizes and have a longer response time than other methods (Payne, 1971).

One of the first monitors was the Narko test[A]. It is an anesthetic analyzer that is activated by the solubility of halocarbons in a silicone elastomer (increase in elasticity based on concentration) (Lowe and Hagler, 1971). Correction must be made for water vapor and nitrous oxide. The response time is slow, requiring up to 5 min for stable measurements. The Narko-test analyzer can be positioned on either the inspiratory or the expiratory sides of an anesthetic machine (Moens, 1988).

The concentration of volatile anesthetic agents can be monitored by a piezoelectric crystal technique. The natural resonance frequency of a lipophilic-coated piezoelectric quartz crystal changes frequency proportionally with the anesthetic concentration (Humphrey *et al.,* 1991; Westenskow and Silva, 1991). Early piezoelectric analyzers had relatively slow response times and were sensitive to water vapor. Improvements in design have resulted in a piezoelectric analyzer that compares favorably with other techniques (Westenskow and Silva, 1991). The piezoelectric crystal has a short 2-min warmup time and a rapid response time of less than 0.05 sec. Interferences by other lipophilic gases, water vapor, and nitrous oxide must be considered when using this technique (Parbrook *et al.,* 1990; Westenskow & Silva, 1991).

B. Specific Methods of Measurement

Methods that specifically identify anesthetics are used when gas mixtures have varying composition and contain more than two gases.

Infrared spectroscopy is probably the most widely used technique for measuring volatile anesthetics (Nielsen *et al.,* 1993). All gases either absorb or emit electromagnetic radiation in the infrared, visible or ultraviolet light ranges (Payne, 1971). Because all polyatomic gases are excited by specific infrared wavelengths, they have characteristic absorption patterns in the range of 1–15 μ (Payne, 1971). The accuracy of infrared analyzers is affected by the presence of water vapor, presence of other gases with absorption spectra that overlap the anesthetic of interest, and the size of the sample. Interference by exhaled methane by animals has been shown to affect halothane measurements by infrared analyzers (Jantzen, 1990). Ethanol in expired air causes an infrared anesthetic analyzer to register a vapor concentration as though it were measuring halothane (Foley *et al.,* 1990). Guyton and Gravenstein (1990) evaluated the Datex Capnomac PB254[B] infrared analyzer. Ethanol had a slight but clinically insignificant effect on isoflurane and enflurane, but the effect was greater with halothane. Mixtures of volatile anesthetics resulted in measurements that were additive (Guyton and Gravenstein, 1990).

Mass spectrometry separates the actual components of a mixture and quantitates them (Gillbe *et al.,* 1981). The components and their quantities can then be calculated based on their charge/mass ratio and on the proportions of molecular weight (Jantzen, 1990). The rapid response time (<0.05 sec) enables the mass spectrometer to follow changes within the respiratory cycle. Additionally, mass spectrometry can analyze very small samples (Larach *et al.,* 1988). Larach *et al.* (1988) measured CO_2, halothane, and other respiratory gases simultaneously in rats that were as small as 250 g. Continuous sampling did not distort end tidal CO_2 or arterial oxygen measurements. These advantages are generally outweighed by the mass spectrometers large size, complexity, and high cost (Parbrook *et al.,* 1990; Payne, 1971).

Gas chromatography is a technique used to measure volatile anesthetics in both liquids or gases and allows identification and measurement of very low concentrations of gas mixtures. Response time is dependent on the passage of the sample through a silica gel column and only provides serial measurements rather than continuous measurements (Parbrook *et al.,* 1990; Payne, 1971). An additional disadvantage is that measurements of volatile anesthetics in tissue or fluid require extraction by organic solvents, which is time-consuming and costly. Janicki *et al.* (1990) describes an alternate method for analyzing fluid samples using a high-performance liquid chromatographic (HPLC) method.

VII. HEALTH RISKS ASSOCIATED WITH EXPOSURE TO WASTE ANESTHETIC GASES

By the late 1960s, studies were published documenting the metabolism of halothane and methoxyflurane and relating their metabolites to hepatic and renal damage. At the same time, clinical epidemiologic studies began to accumulate evidence indicating that chronic exposure to trace amounts of inhalant anesthetics was associated with adverse human health effects. Although a cause-and-effect relationship had not been established, the high correlation suggested the possibility that exposure to trace amounts of volatile anesthetics was harmful.

Chronic exposure to trace amounts of waste anesthetic gases has been associated with immune system and bone marrow suppression, pruritus, enzyme induction, liver and kidney disease, reproductive failure, cancer, and central nervous system disturbances (fatigue, headache, irritability, nausea, and cognitive and motor skill disruptions) (Gross and Smith, 1993).

Because elimination of waste anesthetic gases was easily instituted, safety standards were issued by the National Institute for Occupational Safety and Health (NIOSH) in 1977. These regulations state that the exposure to concentrations of anesthetic gases should not exceed 25 ppm for nitrous oxide and 2 ppm for the halogenated anesthetics based on a timed weighted average. Ultrastructural changes occur in rat offspring after exposure to 10 ppm of halothane during gestation, although abnormal behavior of newborn rats was seen at a much higher concentration. No safe level has been determined (Baeder and Albrecht, 1990).

Inhalant anesthetics should only be used where waste gases can be safely removed from the work environment. Low fresh gas flows will minimize the potential for exposure to trace gases. Scavenging systems that allow both the collection and the removal of the anesthetic vapors can be used. Anesthetic chambers or delivery systems such as masks that are not easily sealed should be used inside fume hoods. Only fume hoods that do not recirculate exhausted gases into other areas are acceptable. The principal concern is that scavenged gases be completely removed from the work environment. The system should not recirculate into other areas. Newer anesthetic machines are equipped with scavenging adapters whereas older machines must be retrofitted. Active and passive systems have been recommended and are acceptable if properly designed. Active scavenging systems result in significantly lower exposure than passive scavenging systems (Gardner, 1989). Absorption of halogenated anesthetics by activated charcoal systems are appropriate for transport machines but do not absorb nitrous oxide and have a limited useable life. These filters must be changed frequently depending on the concentration and the volume of gas scavenged.

Specific information on health hazards and methods of scavenging are available in review articles (Gross and Smith, 1993). If exposure to anesthetic vapors cannot be avoided, it is safer to use the new anesthetic gases that have minimal metabolism due to greater molecular stability. Isoflurane is much safer to breathe than either halothane or methoxyflurane.

REFERENCES

Antognini, J. F., and Eisele, P. (1993). Anesthetic potency and cardiopulmonary effects of enflurane, halothane, and isoflurane in goats. *Lab. Anim. Sci.* **13**, 607–610.

Baeder, C., and Albrecht, M. (1990). Embryotoxic/teratogenic potential of halothane. *Int. Arch. Occup. Environ. Health* **62**, 263–271.

Bito, H., and Ikeda, K. (1994). Long-duration, low-flow sevoflurane anesthesia using two carbon dioxide absorbents. Quantification of degradation products in the circuit. *Anesthesiology* **81**, 340–345.

Boarini, D. J. (1984). Comparison of systemic and cerebrovascular effects of isoflurane and halothane. *Neurosurgery* **15**, 400–409.

Brunson, D. B. (1990). "A Compendium on Aerrane for Canine and Equine Anesthesia." Anaquest, Inc., Madison, Wisconsin.

Brunson, D. B., Stowe, C. M., and McGrath, C. J. (1979). Serum and urine inorganic fluoride concentrations following methoxyflurane anesthesia in the dog. *Am. J. Vet. Res.* **40**, 197–203.

Carpenter, R. L., Eger, E. I., Johnson, B. H., Unadkat, J. D., and Sheiner, L. B. (1986). The extent of metabolism of inhaled anesthetics in humans. *Anesthesiology* **65**, 201–205.

Curro, T. G., Brunson, D. B., and Paul-Murphy, J. (1994). Determination of the ED_{50} of isoflurane and evaluation of the isoflurane sparing effect of butorphanol in cockatoos, *Cacatua spp. Vet. Surg.* **23**, 429–433.

Deady, J. E., Koblin, D. D., and Eger, E. I. (1980). Anesthetic potencies and the theories of narcosis. *Anesth Analg.* **59**, 535.

Dohm, L., and Brunson, D. B. (1993). Effective dose of isoflurane for the desert iguana *Diposaurus dorsalis* and the effect of hypothermia on effective dose. *Vet. Surg.* **22**, 547.

Dorsch, J. A., and Dorsch, S. E. (1994). "Understanding Anesthesia Equipment," 3rd ed., pp. 91–147. Williams & Wilkins, Baltimore, Maryland.

Eger, E. I. (1964). Respiratory and circulatory factors in uptake and distribution of volatile anaesthetic agents. *Br. J. Anaesth.* **36**, 155–171.

Eger, E. I. (1974). MAC. U. *In* "Anesthetic Uptake and Action" (E. I. Eger, ed.), pp. 1–25. Williams & Wilkins, Baltimore, Maryland.

Eger, E. I. (1985). "Isoflurane (Forane), A Compendium and Reference." Anaquest, Inc., Madison, Wisconsin.

Eger, E. I. (1992). Desflurane animal and human pharmacology: Aspects of kinetics, safety, and MAC. *Anesth. Analg.* **75**, 53–59.

Foley, M. A., Wood, P. R., Peel, G. M., and Lawler, P. G. (1990). The effect of exhaled alcohol on the performance of the datex capnomac. *Anaesthesia* **45**, 232–234.

Franks, N. P., and Lieb, W. R. (1991). Stereospecific effects of inhalational general anesthetic optical isomers on nerve ion channels. *Science* **254**, 427–430.

Frink, E. J., Isner, R. J., Malan, T. P., Morgan, S. E., Jr., Brown, E. A., and Brown, B. R., Jr. (1994). Sevoflurane degradation product concentrations with soda lime during prolonged anesthesia. *J. Clin. Anesth.* **6**, 239–242.

Fugita, Y., Kimura, K., Hamada, H., and Takaori, M. (1991). Comparative effects of halothane, isoflurane, and sevoflurane in the liver with hepatic artery ligation in the beagle. *Anesthesiology* **75**, 313–318.

Gardner, R. J. (1989). Inhalation anaesthetic—exposure and control: A statistical comparison of personal exposures in operating theatres with and without anaesthetic gas scavenging. *Ann. Occup. Hyg.* **33**, 159–173.

Gillbe, C. E., Heneghan, C. P. H., and Branthwaire, M. A. (1981). Respiratory mass spectrometry during general anaesthesia. *Br. J. Anaesth.* **53,** 103–109.

Gross, M. E., and Smith, J. (1993). Waste anesthetic gas: What veterinarians should know. *Vet. Med.* **88,** 331–341.

Guyton, D. C., and Gravenstein, N. (1990). Infra-red analysis of volatile anesthetics: Impact of monitor agent setting, volatile mixtures, and alcohol. *J. Clin. Monit.* **6,** 203–206.

Holaday, D. A., and Smith, F. R. (1981). Clinical characteristics and biotransformation of sevoflurane in healthy human volunteers. *Anesthesiology* **54,** 100–106.

Humphrey, S. J. E., Luff, N. P., and White, D. C. (1991). Evaluation of the lamtec anaesthetic agent monitor. *Anaesthesia* **46,** 478–481.

Janicki, P. K., Erskine, W. A. R., and James, M. P. M. (1990). High performance liquid chromatographic method for the direct determination of the volatile anaesthetics halothane, isoflurane, and enflurane in water and in physiological buffer solutions. *J. Chromatogr.* **518,** 250–253.

Jantzen, J. P. A. H. (1990). Does fresh gas flow tare affect monitoring requirements? *Acta Anaesthesiol. Belg.* **41,** 211–224.

Jones, R. M. (1990). Desflurane and sevoflurane: Inhalation anaesthetics for this decade? *Br. J. Anaesth.* **65,** 527–536.

Kaufman, R. D. (1977). Biophysical mechanisms of anesthetic action: Historical perspective and review of current concepts. *Anesthesiology* **46,** 49–62.

Kazama, T., and Ikeda, K. (1988). Comparison of MAC and the rate of rise of alveolar concentration of sevoflurane with halothane and isoflurane in the dog. *Anesthesiology* **68,** 435–437.

Koblin, D. D. (1992). Characteristics and implications of desflurane metabolism and toxicity. *Anesth. Analg.* **75,** S10–S16.

Larach, D. R., Shuler, H. G., Skeehan, T. M., and Deer, J. A. (1988). Mass spectrometry for monitoring respiratory and anesthetic gas waveforms in rats. *J. Appl. Physiol.* **65,** 955–963.

Linde, H. W. (1971). The physics and chemistry of general anesthetics. *In* "Textbook of Veterinary Anesthesia" (L. R. Soma, ed.), pp. 30–49. Williams & Wilkins, Baltimore, Maryland.

Lowe, H. J., and Hagler, K. (1971). Clinical and laboratory evaluation of an expired anesthetic gas monitor (Narko-test). *Anesthesiology* **34,** 378–382.

Matthews, N. S., Dollar, N. S., and Shawley, R. V. (1990). Halothane sparing effect of benzodiazepines in ponies. *Cornell Vet.* **80,** 259–265.

Merin, R. G. (1993). Are differences in cardiopulmonary effects among the inhalation agents clinically important? *Proc. ASA Refresher Course Lect., Washington D.C.,* Lect. 226, pp. 1–6.

Moens, Y. (1988). Introduction to the quantitative technique of closed circuit anesthesia in dogs. *Vet. Surg.* **17,** 98–104.

Moody, E. J., Suzdak, P. D., Paul, S. M., and Skolnick, P. (1988). Modulation of the benzodiazepine/gamma-aminobutyric acid receptor chloride channel complex by inhalation anesthetics. *J. Neurochem,* **51,** 1386–1393.

Morris, P., Tatnall, M. L., and West, P. G. (1979). Breath-by-breath halothane monitoring during anaesthesia: A study in children. *Br. J. Anaesth.* **51,** 979–982.

Nielsen, J., Kann, T., and Moller, J. T. (1993). Evaluation of three transportable multigas anesthetic monitors: The Bruel & Kjaer anesthetic gas monitor 1304, the datex capnomac ultima, and the Nellcor N-2500. *J. Clin. Monit.* **9,** 91–98.

Palahniuk, R. J., Shnider, S. M., and Eger, E. I. (1974). Pregnancy decreases the requirements for inhaled anesthetic agents. *Anesthesiology* **41,** 82–83.

Parbrook, G. D., Davis, P. D., and Parbrook, E. O. (1990). Gas chromatography and mass spectrometry. *In* "Basic Physics and Measurement in Anaesthesia," 3rd ed., pp. 257–264. Butterworth-Heinemann, Boston.

Patel, P. M., and Mutch, W. A. (1990). The cerebral pressure-flow relationship during 1.0 MAC isoflurane anesthesia in the rabbit: The effects of different vasopressors. *Anesthesiology* **72,** 118–124.

Payne, J. P. (1971). Measurement of concentration of inhalation agents and respiratory gases in gas and blood. *In* "General Anaesthesia" (Gray and J. F. Nunn, eds.), 3rd ed., pp. 677–689. Butterworth, London.

Price, H. L. (1975). General anesthetics (continued). *In* "The Pharmacological Basis of Therapeutics" (L. S. Goodman and A. Gillman, eds.), 5th ed., pp. 89–96. Macmillan, New York.

Quasha, A. L., Eger, E. I., and Tinker, J. H. (1980). Determination and application of MAC. *Anesthesiology* **53,** 315–334.

Reynolds, R. C. (1981). General anesthetic agents. *Otolaryngol. Clin. North Am.* **14,** 489–500.

Schmeltekopf, A. L., Goldan, P. D., Henderson, W. R., Harrop, W. J., Thompson, T. L., Fehenfeld, F. C., Schiff, H. I., Crutzen, P. J., Isaksen, I. S. A., and Ferguson, E. E. (1975). Measurement of stratospheric $CFCl_3$, SF_2, Cl_2, and N_2O. *Geophys. Res. Lett.* **2,** 393–396.

Seifen, A. B., Kennedy, R. H., Bray, J. P., and Seifen, E. (1989). Estimation of minimum alveolar concentration (MAC) for halothane, enflurane and isoflurane in spontaneously breathing guinea pigs. *Lab. Anim. Sci.* **39,** 579–581.

Steffey, E. P. (1978). Enflurane and isoflurane anesthesia: A summary of laboratory and clinical investigations in horses. *J. Am. Vet. Med. Assoc.* **172,** 367–373.

Steffey, E. P., and Howland, D. D. (1977). Isoflurane potency in the dog and cat. *Am. J. Vet. Res.* **38,** 1833–1836.

Steffey, E. P., Johnson, B. J., Eger, E. I., and Howland, D. (1979). Nitrous oxide: Effects on accumulation rate and uptake of bowel gases. *Anesth. Analg.* **58,** 405–408.

Steffey, E. P., Woliner, M. L., and Howland, D. (1983). Accuracy of isoflurane delivery by halothane specific vaporizers. *Am. J. Vet. Res.* **44,** 1072–1078.

Stevens, W. C., Dolan, W. M., and Gibbons, R. T. (1975). Minimum alveolar concentrations (MAC) of isoflurane with and without nitrous oxide in patients of various ages. *Anesthesiology* **42,** 197–200.

Tanelian, D. L., Kosek, P., Mody, I., and MacIver, M. B. (1993). The role of the GABAa receptor/chloride channel complex in anesthesia. *Anesthesiology* **78,** 757–776.

Tinker, J. K., Sharbrough, F. W., and Michenfelder, J. D. (1977). Anterior shift of the dominant EEG rhythm during anesthesia in the Java monkey: Correlation with anesthetic potency. *Anesthesiology* **46,** 242–259.

Tranquilli, W. J., Thurmon, J. C., Benson, G. J., and Steffey, E. P. (1983). Halothane potency in pigs *(Sus scrofa). Am. J. Vet. Res.* **44,** 1106–1107.

Weiskopf, R. B., Holmes, M. A., Rampil, I. J., Johnson, B. H., Yasuda, N., Targ, A. G., and Eger, E. I. (1989). Cardiovascular safety and actions of high concentrations of I-653 and isoflurane in swine. *Anesthesiology* **70,** 793–798.

Westenskow, D. R., and Silva, F. H. (1991). Laboratory evaluation of the vital signs (ICOR) piezoelectric anesthetic agent analyzer. *J. Clin. Monit.* **7,** 189–194.

White, P. F., Johnston, P. R., and Eger, E. I. (1974). Determination of anesthetic requirements in rats. *Anesthesiology* **40,** 52–57.

Wolfson, B., Hetrick, W. D., Lake, C. L., and Siker, E. S. (1978). Anesthetic indices: Further data. *Anesthesiology* **48,** 187–190.

Wollman, H., and Smith, T. C. (1975). Uptake, distribution, elimination, and administration of inhalational anesthetics. *In* "The Pharmacological Basis of Therapeutics" (L. S. Goodman and A. Gillman, eds.), 5th ed., pp. 71–83. Macmillan, New York.

Yashuda, N., Targ, A. G., Eger, E. I., Johnson, B. H., and Weiskopf, R. B. (1990). Pharmacokinetics of desflurane, sevoflurane, isoflurane and halothane in pigs. *Anesth. Analg.* **71,** 340–348.

Young, W. L. (1992). Effects of desflurane on the central nervous system. *Anesth. Analg.* **75,** S32–S37.

Chapter 3

Pharmacology of Analgesics

James E. Heavner

I. INTRODUCTION

Pain as defined by the International Association for the Study of Pain is "an unpleasant sensory and emotional experience associated with actual or potential tissue damage or described in terms of such damage. It is always subjective." Pain is a well-recognized stressor in animals. Maladaptive behavior is one sign of distress, and sometimes the only one (Quimby, 1991). Pain can be triggered by a number of different stimuli acting at a number of different sites within the peripheral (PNS) or central nervous system (CNS). As shown in Fig. 1, a relationship exists among pain, fear, anxiety, and sleep. Awareness of this relationship is important to the understanding of why some approaches to pain treatment are used. Pharmacological approaches to managing pain (preventing or stopping) may include the administration of analgesics and the administration of drugs to treat the consequences of the pain (e.g., anxiolytics). This chapter deals with three groups of drugs used in pain management: (1) nonsteroidal anti-inflammatory drugs (NSAIDs), (2) opioids, and (3) α_2-adrenergic agonists. General and local

ANESTHESIA AND ANALGESIA IN LABORATORY ANIMALS

ISBN 0-12-417570-8

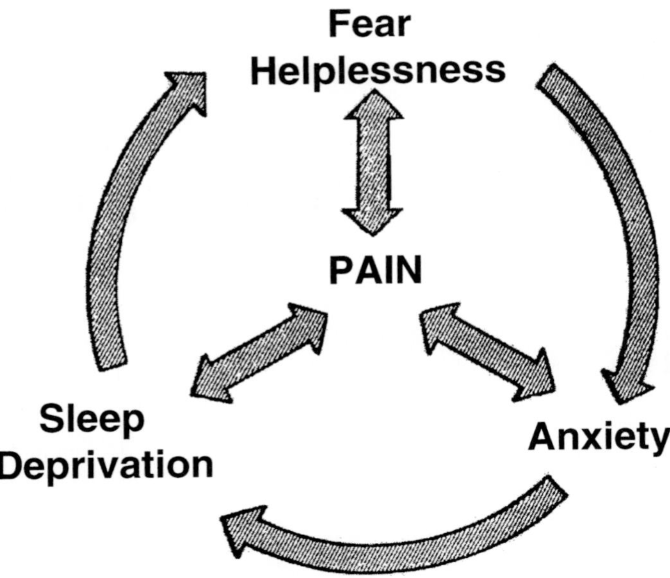

Fig. 1. The pain cycle. In humans, a sense of helplessness and fear add to the total suffering of the patient and exacerbate the pain, as does anxiety and sleep loss. It is reasonable to assume that the same thing happens in animals. Therefore, drugs that aid sleep, reduce anxiety, and produce a sense of well-being play a role in pain management.

anesthetics will not be discussed but they obviously are used in pain management, especially in the perioperative period. Analgesic adjuvants will be mentioned but not discussed in detail.

The chapter cites a significant amount of data regarding the pharmacology of analgesics in humans because this information is well-documented in a scientifically valid fashion. Data provide a solid base for illustrating important pharmacological properties that must be considered when using analgesics in laboratory animals. Pharmacodynamic effects of analgesics vary between animals and humans, and among animal species and even strains, but the greatest differences are in pharmacokinetic factors. Absorption, biotransformation, and elimination processes all vary rather remarkably, and examples of the species variability will be given. Detailed information about unique features of analgesics as they relate to each animal species is most appropriately discussed elsewhere in this volume in chapters describing the clinical use of analgesics in the different species.

II. NONSTEROIDAL ANTI-INFLAMMATORY DRUGS

A. Overview

The classes of aspirin-like analgesics are made up of *p*-Aminophenol derivatives, oxicam, salicylates, fenamates, propionic acids, pycazolon, acetic acids, and nicotinic acid derivatives. Nonsteroidal anti-inflammatory drugs are, for the most part, human drugs but they may become increasingly more important in the clinical practice of laboratory animal medicine. Details regarding the safety and efficacy of NSAIDs in laboratory animals, from a clinical perspective, are scarce. The drugs must certainly have been investigated in animals as part of the process to obtain approval to market them for human use, but the conditions of these studies are different from those encountered in the clinical practice of laboratory animal medicine. Some NSAIDs have been used in veterinary medicine for many years (e.g., phenylbutazone, flunixin, aspirin). A broad range of NSAIDs will be discussed, including those that are of potential use in laboratory animal medicine as well as those that are in relatively common use.

Aspirin is the prototype drug for this group. It has analgesic, antipyretic, and anti-inflammatory activity. The precise mechanism of action of drugs in this class is unknown, but inhibition of the enzyme, cyclooxygenase, is thought to be important for the intended pharmacologic effect.

1. History

Recognition of the antipyretic effects of willow bark led to the search for the active component. Leroux isolated salicin, which was found to be antipyretic, from willow bark in 1829. Hydrolysis of salicin yields glucose and salic alcohol. The latter can be converted to salic acid. In 1875, sodium salicylate was first used for the treatment of rheumatic fever and as an antipyretic. Acetylsalicylic acid was synthesized by Hoffman and introduced into medicine in 1899 under the name aspirin, after the demonstration of its anti-inflammatory effects.

The development of newer aspirin-type drugs began nearly 60 years later and included such drugs as phenylbutazone, indomethacin, ibuprofen, tolectin, naproxen, fenoprofen, and sulindac.

2. Mechanism of Action

In 1971, Vane and associates and Smith and Willis demonstrated (Insel, 1990) that low concentrations of aspirin and indomethacin inhibit the enzymatic production of prostaglandins (PG). Prostaglandins are among a number of chemicals that are released following tissue damage and affect nociceptors. Prostaglandins increase the sensitivity of nociceptors to other chemicals and/or mechanical and/or thermal stimuli.

Aspirin-like drugs inhibit the conversion of arachidonic acid to the unstable endoperoxide intermediate, prostaglandin G_2 (PGG_2), a reaction that is catalyzed by cyclooxygenase. Some agents are competitive inhibitors of cyclooxygenase, whereas others permanently modify the enzyme. Aspirin, for example, acetylates a serine residue at or near the active site of cyclooxygenase (Roth and Siok, 1978).

3. Therapeutic Activities and Side Effects of Aspirin-like Drugs

Although all aspirin-like drugs are antipyretic, analgesic, and anti-inflammatory, they differ in how they affect these clinical entities. For instance, acetaminophen is antipyretic and analgesic but is only weakly anti-inflammatory (Capetola *et al.*, 1983).

As a class, these drugs are considered to be effective against pain of low-to-moderate intensity. However, it is important to consider the type of pain as well as the intensity in assessing analgesic efficacy. These drugs are particularly effective in controlling pain related to nociceptors that have been sensitized to normally painless mechanical, thermal, or chemical stimuli. According to Insel (1990), aspirin-like drugs can be superior to the opioid analgesics is some forms of postoperative pain. Pain arising from the hollow viscera is usually not relieved by aspirin-like drugs.

All aspirin-like drugs are antipyretic, presumably via preventing PGE_2 production in the hypothalamus. PGE_2 apparently elevates the set point at which body temperature is maintained by the hypothalamus. Inhibition of PGE_2 production returns the set point to normal.

Common side effects of aspirin-like drugs are shown in Table I. All of the effects are generally thought to be mediated via the inhibition of cyclooxygenase. Although gastric ulceration may be caused by the local irritant effect of an orally administered drug, parenteral administration of aspirin-like substances can also cause gastric and intestinal ulceration. Prostaglandins inhibit acid secretion by the stomach and promote the secretion of cytoprotective mucus in the intestine. This protective function is blocked if prostaglandin synthesis is inhibited.

Synthesis in platelets of thromboxane A_2 (TXA_2), a platelet-aggregating agent, is inhibited by aspirin-like drugs, producing increased bleeding times.

Caution should be exercised in giving aspirin-like drugs to debilitated animals as they may affect renal function. In normal human subjects, aspirin-like drugs have little effect on renal function. However, renal effects may occur in patients with congestive heart failure, hepatic cirrhosis with ascites, chronic renal disease, or in those who are hypovolemic for any reason. In these conditions, renal perfusion is dependent on prostaglandin-induced vasodilation that opposes the norepinephrine- and angiotensin II-mediated vasoconstriction resulting from the activation of pressor reflexes. Chronic abuse of any aspirin-like drug or analgesic mixture may cause renal injury in susceptible individuals, although nephropathy is not commonly associated with the use of these drugs.

Aspirin-like drugs can be classified based on structural similarities as shown in Table II.

TABLE I

SIDE EFFECTS OF ASPIRIN-LIKE DRUGS

Gastric or intestinal ulceration
Disturbance of platelet function
Prolongation of gestation or spontaneous labor
Changes in renal function:
Decrease renal blood flow and glomerular filtration
Promote action of antidiuretic hormone
Increase absorption of chloride
Enhance K^+ reabsorption
Suppress renin secretion

B. Salicylates

Despite the introduction of many new drugs, aspirin still is widely used in humans for its antipyretic, analgesic, and anti-inflammatory action. Salicylates are derivatives of salic acid that are either esters of salicylic acid obtained by substitution in the carboxyl group or salicylate esters of organic acids in which the carboxyl group of salic acid is retained and substitution is made in the OH group. Aspirin is an ester of acetic acid.

In high doses, salicylates stimulate the CNS (including the production of seizures) followed by depression. Confusion, dizziness, tinnitus, high tone deafness, delirium, psychosis, stupor, and coma may occur.

Salicylates stimulate respiration directly and indirectly. Therapeutic doses increase CO_2 production and O_2 consumption by uncoupling oxidative phosphorylation. The increased CO_2 production stimulates respiration. Salicylates also directly stimulate the respiratory center in the medulla. Both rate and depth of respiration increase, producing marked hyperventilation. In human adults, salicylates produce a respiratory alkalosis with resultant renal compensation. This includes increased renal excretion of bicarbonate, accompanied by Na^+ and K^+ which restores blood pH via lowered plasma bicarbonate. After high doses of salicylates or after prolonged exposure, medullary depression occurs, producing circulatory collapse secondary to vasomotor depression and respiratory depression. Respiratory acidosis results not only from the respiratory depression but also from enhanced CO_2 production. No important cardiovascular changes are seen following ordinary therapeutic doses.

TABLE II

CLASSES OF ASPIRIN-LIKE ANALGESICS

Class	Generic names
Salicylates	Aspirin, diflunisal, choline magnesium trisalicylate, salsalate
Pycazolon	Phenylbutazone
p-Aminophenol derivatives	Acetaminophen
Acetic acids	Indomethacin, tolmetin, sulindac, suprofen, diclofenac, ketorolac
Fenamates	Mefenamic acid, meclofenamic acid
Propionic acids	Ibuprofen, naproxen, naproxen sodium, fenoprofen, ketoprofen, flurbiprofen
Oxicam	Piroxicam
Nicotinic acid derivatives	Flunixin

Salicylates may also induce nausea and vomiting, probably via an effect on the medullary chemoreceptor trigger zone. Epigastric distress, nausea, and vomiting may be produced by ingestion of salicylates. High-dose therapy may cause gastric ulceration, gastrointestinal hemorrhage, and erosive gastritis.

Salicylates may produce a dose-dependent hepatotoxicity most commonly in humans with connective tissue disorders. There usually are no symptoms and elevated serum transaminases are the principal laboratory findings. Hepatomegaly, anorexia, and nausea may be present in about 5% of humans with salicylate-induced hepatotoxicity.

The renal effects of salicylates were discussed earlier. Low doses of salicylates may decrease urate excretion and increase plasma urate concentrations. Moderate doses usually do not affect urate excretion, but higher doses may induce uricosuria. As discussed earlier, aspirin interferes with platelet aggregation, thereby prolonging bleeding time. Hyperglycemia, glycosuria, and depletion of muscle and liver glycogen may be produced by large doses of salicylates. Toxic doses cause a significant negative nitrogen balance. Salicylates reduce lipogenesis, inhibit epinephrine-stimulated lipolysis in fat cells, and displace long-chain fatty acids from binding sites on human plasma protein.

1. Pharmacokinetics and Metabolism

a. Absorption. When administered orally to humans, salicylates are absorbed rapidly, partly from the stomach but mostly from the upper small intestine. Absorption varies among animal species. Davis and Westfall (1972) demonstrated that orally administered sodium salicylate is rapidly absorbed by dogs and swine, less so in ponies, and slowly absorbed in goats.

b. Distribution and Fate. Salicylates are distributed throughout the body, primarily by pH-dependent passive processes. They readily cross the placenta and are actively transported into the cerebrospinal fluid (CSF) across the choroid plexus by a low capacity process. Up to 80–90% of salicylate is bound to plasma proteins, especially albumin.

Salicylate is biotransformed in many tissues, but especially in the hepatic endoplasmic reticulum and mitochondria. In humans, there are three primary metabolic products: salicyluric acid (the glycine conjugate), the ether or phenolic glucuronide, and the ester or acyl glucuronide. Salicylates are excreted in the urine as free salicylic acid, salicyluric acid, salicylic phenolic, acyl glucuronides, and gentisic acid. More than 30% of ingested drug may be eliminated as free salicylate in alkaline urine, but as low as only 2% may be excreted in acidic urine.

In humans, salicylate elimination is dose dependent because of the limited ability of the liver to form salicyluric acid and the phenolic glucuronide. The plasma half-life for salicylate is 2 to 3 hr at low dose and about 12 hr in higher doses; the half-life for aspirin is about 2 to 3 hr. The rates of elimination of sodium salicylate following IV injection of 44 mg/kg to goats, ponies, swine, dogs, and cats were measured by Davis and Westfall (1972). Clearance was fastest in ponies and goats and slowest in cats (~ 30 times slower). The clearance rate in dogs and pigs was about midway between cats and ponies and goats. Although differences in rates of biotransformation were a factor, so too was urine pH. The high clearance in ponies was due to rapid excretion in the alkaline urine of the species. The higher clearance in goats was associated with rapid biotransformation plus rapid clearance in alkaline urine.

According to Davis (1983), the half-life of IV sodium salicylate (44 mg/kg) is 37.6 hr in cats, 8.6 hr in dogs, 5.9 hr in swine, 1.03 hr in ponies, and 0.78 hr in goats. If the dosing interval is too short relative to the half-life of aspirin, toxicity will result. Dogs may be given aspirin twice daily, cats twice weekly. However, it may take 2 weeks to obtain effective blood levels in the cat (Hughes and Lang, 1983; Baggot, 1992).

2. Preparations and Routes of Administration

Aspirin and sodium salicylate are the two most commonly used preparations of salicylate for systemic effects. They usually are administered orally. Sodium salicylate is available for parenteral use. Because aspirin is poorly soluble and has many chemical incompatibilities, it should only be administered in dry form.

Diflunisal appears to be a competitive inhibitor of cyclooxygenase and is more potent than aspirin in anti-inflammatory tests in animals (Insel, 1990). It is used primarily as an analgesic in the treatment of osteoarthritis and musculoskeletal strains or sprains. The drug does not produce auditory side effects and has limited, if any, antipyretic activity. It is available as an oral preparation.

C. Pycazolon Derivatives

A number of drugs in this group have analgesic activity (e.g., antipyrine, oxyphenylbutazone, aminopyrine, dipyrone, apazone), but phenylbutazone is the most important one from a therapeutic point of view.

Phenylbutazone has anti-inflammatory effects similar to those of salicylates. Its analgesic activity for pain of nonrheumatoid origin is inferior to that of salicylates. Phenylbutazone causes agranulocytosis, has a mild uricosuric effect, and causes significant retention of Na^+ and Cl^- accompanied by a reduction in urine volume. It displaces other drugs bound to plasma protein.

1. Pharmacokinetics and Metabolism

Phenylbutazone is rapidly and completely absorbed from the gastrointestinal (GI) tract or the rectum. More than 98% of the drug is bound to plasma proteins. In humans, phenylbutazone undergoes extensive biotransformation. Hydroxylation of the phenyl rings or the butyl side chain and glucuronidation are the most significant primary reactions. Active metabolites with long plasma half-lives are formed. Half-lives of phenylbutazone in several species are shown in Table III. This drug, unlike salicylate, is slowly eliminated from ruminants. The reason for this is not known (Davis, 1983). As previously mentioned, dosing intervals must be adjusted relative to half-life of the drug: the longer the half-life, the greater the dosing interval. In turn, time to reach therapeutic blood concentration is lengthened.

2. Preparations and Route of Administration

Phenylbutazone is available as tablets or capsules for oral administration.

D. *p*-Aminophenol Derivatives

Acetaminophen is the drug in this class that is of therapeutic interest. Its analgesic and antipyretic effects do not differ significantly from those of aspirin. Its weak anti-inflammatory effect has been noted earlier in this chapter. Acetaminophen is an active metabolite of phenacetin. Therapeutic doses of acetaminophen have no effect on the cardiovascular and respiratory system and produce no acid–base changes. Acetaminophen has no effects on platelets, bleeding time, or excretion of uric acid and it does not produce gastric irritation, erosion, or bleeding.

As mentioned earlier, acetaminophen has analgesic and antipyretic activity but only weak anti-inflammatory action. This indicates that anti-inflammatory activity is not essential for the analgesic action of aspirin-like drugs. The inflammatory process involves a series of events that can be elicited by numerous stimuli, and many mediators of the inflammatory process have been identified. A central as well as peripheral mechanism for NSAID analgesic activity has been recognized. Malmberg and Yaksh (1993) demonstrated that NSAIDs modulate neuropathic pain at the spinal cord level.

TABLE III

HALF-LIVES OF PHENYLBUTAZONE IN SEVERAL SPECIES[a]

Species	$t_{1/2}$ (hr)
Human	72
Ox	55
	42
Goat	
Female	19
Male	14.5
Cat	18
Rat	6
Dog	6
Swine	4
Baboon	5
Horse	3.5
Rabbit	3

[a]From Davis, L.E. (1983). Species differences in drug disposition as factors in alleviation of pain. In ''Animal Pain: Perception and Alleviation'' (R. L. Kitchell and H. H. Erickson, eds.), pp. 161–178. *Am. Phys. Soc.*, Bethesda, Maryland.

1. Pharmacokinetics and Metabolism

Acetaminophen is rapidly and almost completely absorbed from the gastrointestinal tract. Binding of the drug to plasma protein is variable. It is excreted in the urine as a conjugate of glucuronic acid or cysteine; small amounts of hydroxylated and deacetylated metabolites have also been detected. Up to 80–90% of the therapeutic dose may be recovered in the urine within 24 hr of administration.

2. Toxic Effects

Therapeutic doses of acetaminophen usually are well tolerated. Skin rash and other allergic reactions occur occasionally. A dose-dependent potentially fatal hepatic necrosis is the most serious adverse effect of acute acetaminophen overdose. Acute hypoglycemia and renal tubular necrosis may also occur.

3. Preparations and Routes of Administration

Acetaminophen is available commercially in tablets, capsules, suppositories, chewable tablets, wafers, elixirs, solutions, and in combination with other drugs (e.g., opioids).

E. Acetic Acid Derivatives

Indomethacin and sulindac are representative drugs of this class. Indomethacin is widely used and is effective, but toxicity often limits its use. Its antipyretic, analgesic, and anti-inflammatory activities are similar to those of salicylates. This is also true for sulindac, which is less than half as potent as indomethacin.

Diclofenac is another drug in this family and has a pharmacologic profile similar to the others. It is available in enteric-coated tablet form for oral administration. Tolmetin appears to be approximately equal in efficacy to aspirin but it is usually better tolerated. It causes gastric erosion and prolongs bleeding time. In humans, it is rapidly absorbed following oral administration and is highly protein bound (99%) in plasma.

Ketorolac is similar pharmacologically to aspirin except it does not irreversibly interfere with platelet function. In chronic pain patients, the incidence of adverse gastrointestinal side effects is slightly higher with ketorolac than with aspirin (Physicians' Desk Reference, 1994, p.2373). Ketorolac is readily absorbed following oral or intramuscular administration and is highly bound to plasma protein (99%). In humans, the majority of ketorolac is excreted in the urine and may be accompanied by renal toxicity with repeated doses.

1. Pharmacokinetics and Metabolism

Indomethacin and sulindac are rapidly and almost completely absorbed from the gastrointestinal tract after oral administration. Indomethacin is 90% bound to plasma proteins. It is converted to inactive metabolites. Ten to 20% of the drug is excreted unchanged in the urine. Free and conjugated metabolites are eliminated in urine, bile, and feces. The metabolism and pharmacokinetics of sulindac are complex and vary enormously among species. It is a prodrug that must be metabolically changed to the active form (sulfide). About 90% of sulindac is absorbed in humans following oral administration.

2. Preparations and Route of Administration

Oral dosage forms of indomethacin include regular and sustained release capsules. It is also available as a suppository, as an oral suspension, and as an intravenous injection (the latter for inducing closure of patent ductus arteriosus). Sulindac is available in tablet form.

Tolmetin is supplied in the form of oral tablets and is approved in the United States for the treatment of osteoarthritis, rheumatoid arthritis, and the juvenile form of the disease in humans.

Ketorolac tromethamine is available in solution for parenteral injection and in tablets for oral administration.

F. Fenamates

The only drugs in this family available in the United States are mefenamic acid and meclofenamate. Meclofenamate is used in the treatment of rheumatoid arthritis and osteoarthritis, but is not recommended as initial therapy. Mefenamic acid is used for relief from symptoms of primary dysmenorrhea and for analgesia. The fenamates frequently cause side effects, especially diarrhea. They have no clear advantage over other aspirin-like drugs.

G. Propionic Acid Derivatives

Drugs in this family include ibuprofen, fenoprofen, ketoprofen, and naproxen. They usually are better tolerated by humans than are aspirin, indomethacin, and the pycazolon derivatives. The similarities between drugs in this class are more striking than are the differences. The compounds vary in their potency, but the difference is not of obvious clinical significance. All have anti-inflammatory, analgesic, and antipyretic activity. They can cause gastrointestinal erosions in experimental animals and produce gastrointestinal side effects in humans. They all alter platelet function and prolong bleeding time. Naproxen is particularly potent in inhibiting the migration and other functions of leukocytes.

Propionic acid derivatives generally are rapidly absorbed and highly bound to plasma protein. They generally undergo biotransformation in the liver, and the metabolites are excreted in the urine. The plasma half-life of naproxen is notably longer than that of other drugs in this family. These drugs are available in tablet form for oral administration.

H. Oxicams

The only clinical derivative that is available in the United States is piroxicam. Piroxicam is similar to aspirin and most of the other aspirin-like drugs in most respects. Its distinguishing feature is a relatively long half-life which permits once a day administration. Compounds are available for oral administration.

I. Nicotinic Acid Derivatives

Flunixin is an anti-inflammatory analgesic approved for use in horses. It is available in powder, pellet, and tablet form for oral administration and in liquid form for parenteral injection. It enjoys some popularity as a postoperative analgesic for a few laboratory animal species (e.g., dogs and calves). According to Ciofalo *et al.* (1977), flunixin meglumine is a potent analgesic agent after parenteral administration in mice, rats, and monkeys. It is significantly more potent than pentazocine, meperidine, and codeine in the rat yeast paw test after subcutaneous administration in saline. Vonderhaar and Salisbury (1993) presented reports of cases involving administration of flunixin meglumine (1.1 to 2.3 mg/kg) to dogs via various routes and for various durations associated with gastroduodenal ulceration and perforation. They concluded that the use of flunixin meglumine is appropriate in the management of many conditions in dogs, but clinicians must be aware of potential adverse effects.

III. OPIOIDS

A. Overview

The opioids are classified as morphine and related opioids, meperidine and congeners, methadone and congeners, opioids with mixed actions such as agonist–antagonists and partial agonists, and opioid antagonists. Opioid analgesics may be classified in a number of different ways, each providing useful distinguishing characteristics of the drugs. Table IV provides a listing of opioids grouped according to their relative analgesic potential. In this chapter opioids are classified by molecular structure. The structures of morphine, meperidine, and methadone, the prototypic drugs for the different groups, are shown in Fig. 2.

Abuse of opioids and addiction to them is of major concern. This concern in laboratory animal medicine primarily relates to

TABLE IV

OPIOIDS

''Weak'' opioids: Morphine-like agonists
Codeine
Oxycodone
Propoxyphene hydrochloride
Hydrocodone
Agonist–antagonists
Pentazocine
Nalbuphine
Butorphanol
''Strong'' opioids: Morphine and morphine-like agonists
Morphine
Hydromorphone
Oxymorphone
Meperidine
Heroin
Levorphanol
Methadone
Fentanyl transdermal system
Partial agonist
Buprenorphine

diversion of drugs intended for patient administration to personnel having access to the drugs. Substantial effort in the development of opioids is aimed at reducing the abuse and addiction potential of opioids, as well as reducing the side effects such as respiratory depression and constipation.

The abuse potential of opioids, the required permits to purchase and administer opioids, and the storage and record-keeping requirements imposed by federal and state governments often influence the decision of whether to administer these drugs to laboratory animals. Most opioids are classified as Schedule II Narcotics according to U.S. Drug Enforcement Agency documents (DEA Form 225a). Details of federal and state classifications and regulations of narcotics will not be discussed here.

The term ''narcotic'' as used in a legal sense includes opioids (as well as nonopioids, e.g., barbiturates) and the term generally has been used to refer to opioids. However, it has been pointed out that, for a number of reasons, the term narcotic is no longer useful in a pharmacological context. Opiate is a term once used to designate drugs derived from opium. Opioid is now used to designate those drugs plus all other drugs with morphine-like action.

1. History

The analgesic properties of poppy juice (opium) were recognized early in the history of mankind. *Papaver somniferum* is the plant from which poppy juice is extracted. Important to the modern developments in opioid pharmacology was the isolation of morphine from opium in 1806 by Sertürner and the relatively recent characterization and isolation of endogenous opioid substances and opioid receptors.

2. Mechanism of Action

Opioids produce their effects by binding to receptors. It is now generally accepted that there are at least three different

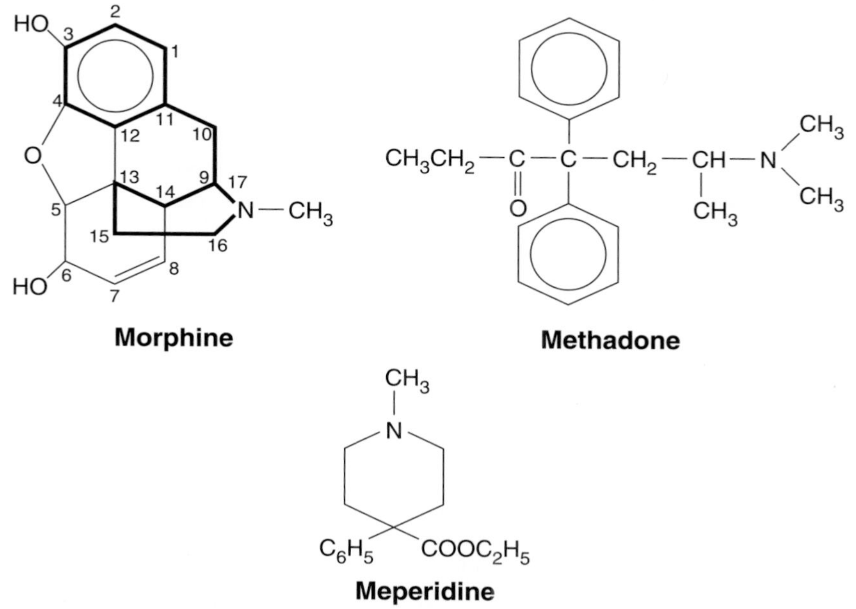

Morphine

Methadone

Meperidine

Fig. 2. Chemical structures of morphine, methadone, and meperidine.

receptors for opioids: mu, delta, and kappa (μ, δ, κ). As shown in Table V, the receptors differ to varying degrees in their anatomic location and in the effect elicited when stimulated by an agonist. Also shown in Table V are examples of agonists that bind to the different receptors. A given opioid may interact with all of the receptor types (Table VI). Analgesia is thought to involve the activation of μ receptors (largely at supraspinal sites) and κ receptors (mainly in the spinal cord); δ receptors may also be involved. A morphine receptor has been cloned (Randall, 1993).

Considerable interest exists in opioid receptors on or near nerve endings and the action of opioids on these receptors to produce analgesia. Evidence favors an analgesic effect of opioids mediated by these receptors, but the effect may be dependent on the presence of inflammation (Czlonkowski *et al.*, 1993). In humans, and with increasing frequency in animals, opioids are injected alone or with local anesthetics into the epidural or subarachnoid space to provide analgesia. The rationale for using this route obviously is to place the opioid as close as possible to the spinal cord opioid receptors, thereby reducing the effective analgesic dose and reducing systemic side effects.

Repeated use of opioids may include tolerance in some species. Development of tolerance will require adjustment of dosing to maintain efficacy.

B. Morphine and Related Opioids

Morphine is the standard by which new analgesics are measured. It is relatively difficult to synthesize and, therefore, it is obtained from opium or extracted from poppy straw. Many semisynthetic derivatives are made by relatively minor modifications of morphine. Important properties of opioids that can be altered by structural modification are affinity for various species of opioid receptors, agonistic versus antagonistic activity, lipid solubility, and resistance to metabolic breakdown.

1. Pharmacological Properties

Morphine and chemically related opioid agonists produce their predominate analgesic action via μ receptors. However, they have notable affinity for the δ and κ receptors. In addition to analgesia, the effects include drowsiness, alteration of mood, respiratory depression, decreased gastrointestinal motility, nausea, vomiting, and alterations of the endocrine and autonomic nervous systems. In humans, even large doses of morphine usually do not produce loss of consciousness, slurred speech, emotional lability, or significant motor incoordination.

This group of drugs relieves continuous dull pain more effectively than sharp intermittent pain (Jaffe and Martin, 1990). Often cited are reports that humans sometimes state that pain is still present after receiving an opioid but they are no longer uncomfortable. This is attributed, in part, to the fact that there is the sensory experience (nociception) as well as the reaction to the sensation that produces pain. Patients may report pain in the absence of nociception and vice versa. Pain evoked secondary to nerve damage (i.e., neuropathic pain) is poorly controlled by morphine and related drugs.

For reasons that are not clear, morphine does not appear to affect ruminant animals (Davis, 1983).

Opioids may increase locomotor activity, and large doses may produce seizures in animals. For example, buprenorphine, nalbuphine, and butorphanol all cause an increase in activity in normal rats (Liles and Flecknell, 1992). Relatively large doses of morphine will cause a rage reaction in cats. Morphine and related opioids depress respiration via a direct effect on the brain stem respiratory centers primarily by depressing responsiveness to carbon dioxide. These drugs also depress the cough reflex by a direct effect on the cough center in the medulla and may induce nausea and vomiting by direct stimulation of the chemoreceptor trigger zone in the area postrema of the medulla.

Morphine and morphine-like analgesics produce vasodilation, reduce peripheral resistance, inhibit baroreceptor reflexes, and blunt the reflex vasoconstriction caused by increased P_{CO_2}. These drugs may induce histamine release, but their cardiovascular effects are not dependent on histamine, although the effects are enhanced by histamine.

Morphine and other μ agonists usually decrease hydrochloric acid secretion in the stomach. Relatively low doses of morphine decrease gastric motility, thereby prolonging gastric emptying time. This drug decreases biliary, pancreatic, and intestinal secretions. The amplitude of nonpropulsive rhythmic, segmental

TABLE V

OPIOID RECEPTORS

Receptor	Location	Action	Drug
Mu (μ)	Cerebral cortex (lamina IV), thalamus, periaqueductal gray	Mu-1:analgesia, Mu-2:respiratory depression, physical dependence	Morphine, fentanyl, codeine, naloxone
Delta (δ)	Frontal cortex limbic system, olfactory tubercule	Analgesia, respiratory depression, dependence	Enkephalin
Kappa (κ)	Spinal cord	Spinal analgesia sedation, low physical dependence	Dynorphin, mixed agonist/antagonists, (e.g., butorphanol nalbuphine), phencyclidine

TABLE VI

ACTIONS OF PROTOTYPICAL AGONISTS, ANTAGONISTS, AND AGONIST–ANTAGONISTS AT OPIOID RECEPTORS[a]

Compound	Receptor type[b]		
	μ	δ	κ
Morphine	++	+	+
Fentanyl	+++	+	+
Pentazocine	−	NA	++
Butorphanol	−	NA	++
Nalbuphine	−	NA	++
Buprenorphine	P	NA	−
Naloxone	−	−	−
Nalorphine[c]	−	NA	+

[a]From Jaffe, J. H., and Martin, W. R. (1990). Opioid analgesics and antagonists. In "The Pharmacological Basis of Therapeutics" (A. G. Gilman, *et al.*, eds.), 8th ed., pp. 485–521. Pergammon, New York.

[b]+, Agonist; −, antagonist; P, partial agonist; and NA, data unavailable or inadequate.

[c]Produces dysphoric or psychotomimetic effects at relatively high doses that are poorly antagonized by naloxone.

contractions of the small and large intestines is usually enhanced, but propulsive contractions are markedly decreased in the small and large intestine. The tone of the anal sphincter is greatly augmented, and the reflex relaxation response to rectal distention is reduced. These actions, together with central actions of morphine, contribute to morphine-induced constipation. Morphine causes constriction of the sphincter of Oddi and pressure in the common bile duct may rise more than 10-fold. Fluid pressure in the gallbladder may also increase, producing signs that may vary from epigastric distress to biliary colic. Morphine inhibits the urinary voiding reflex, and both the tone of the external sphincter and the volume of the bladder are increased. Opioid-induced urinary retention and constipation do not appear to be clinically significant in rodents and rabbits (Flecknell, 1991).

Short-term (< 120 hr) morphine administration has been shown to reduce natural killer (NK) activity (Shavit *et al.,* 1984; Bayer *et al.,* 1990), impair immunoglobulin production (Bussiere *et al.,* 1992; Pruett *et al.,* 1992), suppress phagocytic activity (Levier *et al.,* 1993; Szabo *et al.,* 1993), and induce thymic hypoplasia (Fuchs and Pruett, 1993). In monkeys and humans, chronic morphine use is known to suppress NK activity (Novick *et al.,* 1989; Carr and France, 1993).

2. Absorption

Opioids generally are readily absorbed from the gastrointestinal tract. The more lipophilic opioids are readily absorbed through the nasal or buccal mucosa and those with the greatest lipid solubility can be absorbed transdermally. Opioids are readily absorbed after subcutaneous or intramuscular injection and can enter the spinal cord following epidural or intrathecal administration.

3. Distribution and Fate

About one-third of the morphine in the blood is bound to plasma proteins. Compared with other more lipid-soluble opioids such as codeine, heroin, and methadone, morphine crosses the blood–brain barrier at a slow rate; only small quantities pass the blood–brain barrier in adults. The major pathway for the metabolism of morphine is conjugation with glucuronide to form active as well as inactive products. Morphine-6-glucuronide is more potent than morphine and morphine-3-glucuronide has antianalgesic activity. Morphine is eliminated from the body primarily via glomerular filtration. There are considerable species differences in the elimination of morphine. The elimination half-life of morphine in dogs is about 75 min and about 3 hr in cats (Davis, 1983).

In this group of drugs, the ratio of the oral to IM effective dose of codeine, levorphanol, and oxycodone is substantially higher than is the ratio for morphine. The reason is that the former drugs undergo less first pass metabolism in the liver.

The analgesic effect of codeine is due to its conversion to morphine. Only a small fraction of codeine undergoes this conversion. Heroin is rapidly hydrolyzed to 6-monoacetylmorphine (6-MAM), which is further metabolized to morphine. 6-MAM and morphine produce the pharmacologic actions of heroin.

4. Side Effects

Morphine and morphine-related opioids produce respiratory depression, nausea, vomiting, dizziness, mental clouding, increased pressure in the biliary tract, urinary retention, and hypotension. Allergic reactions to opioids may occur but they are not common. Morphine may cause increased locomotor activity and produce mania in cats. Pruritis, especially in facial areas, is a common sequela to epidurally or intrathecally administered opioids.

5. Preparations and Route of Administration

Solutions of morphine sulfate are available for oral use and for injection. Tablets, controlled-release tablets, and rectal suppositories are also available. Preservative-free solutions are intended for intravenous, epidural, or intrathecal injection.

Codeine sulfate and codeine phosphate are available in tablet form for oral administration. Codeine phosphate is available for injection. Codeine is contained in numerous analgesic combinations (liquids, tablets, and capsules) and in various antitussive combinations (liquid and capsules). Codeine has no advantage over morphine when used parenterally. However, it is more advantageous when administered orally because of the high oral to IM effective dose ratio.

Oxymorphone is about 10 times as potent as morphine. According to Davis (1983), it has slight depressant effects on the CNS, which imparts advantages to this drug over morphine and methadone in certain circumstances. Hydromorphone hydrochloride is available in tablets, rectal suppositories, and solution for IV injection. Oxycodone hydrochloride is available in tablets, as a solution for injection, and in combination with other analgesics. It is about equipotent to morphine. Hydrocodone bitartrate is used in combination with other ingredients in antitussive and analgesic–antipyretic mixtures. Heroin is not available for therapeutic use in the United States. Levorphanol is available as levorphanol tartrate in tablets and as a solution for injection. Its clinical effectiveness in animals has not been established.

C. Meperidine and Congeners

Other drugs in this group are fentanyl, sufentanil, and alfentanil. Chemically, meperidine is classified as a piperidine. Its pharmacological activity is similar to, but not identical to, those of morphine. When given parenterally, meperidine is about one-tenth as potent as morphine. The oral to parenteral potency of meperidine is considerably higher than that of morphine. Meperidine has notable local anesthetic activity.

Meperidine differs from morphine in that toxic doses cause CNS excitation, characterized by tremors, muscle twitching, and seizures. The meperidine metabolite normeperidine is mainly responsible for these effects. The effects of meperidine on smooth muscle are less intense relative to its analgesic actions, as compared to morphine, and meperidine does not cause as much constipation.

1. Absorption and Fate

It is absorbed by all routes but absorption may be erratic following intramuscular injection. About 60% of the meperidine in blood is bound to plasma protein. It is metabolized in the liver with only a small amount excreted unchanged. The duration of analgesia provided by meperidine in the cat and dog is about 45 min (Davis, 1983). The half-life of meperidine in the cat is 0.7 hr because of rapid demethylaton to normeperidine.

2. Preparations and Routes of Administration

Meperidine hydrochloride is available for oral use in tablets and as a syrup and in solution for parenteral administration.

3. Side Effects

The side effects of meperidine are similar to those of morphine, except that a meperidine metabolite may produce CNS stimulation.

Fentanyl is approximately 80 times as potent as morphine. Higher doses produce marked muscular rigidity. It is usually used in combination with other drugs for anesthesia, but is also used for postoperative analgesia. Fentanyl citrate is available as a solution for injection and is supplied as a fixed-dose combination with droperidol for injection. Sufentanil citrate and alfentanil hydrochloride are also very potent and relatively selective μ receptor agonists. They are used, usually in combinations with other drugs, for general anesthesia. They also are injected intrathecally and epidurally for postoperative analgesia.

D. Methadone and Congeners

Methadone is primarily a μ agonist with pharmacologic properties similar to those of morphine. Notable properties of methadone include its oral efficacy, extended duration of action in suppressing withdrawal symptoms in physically dependent humans, and tendency to show persistent effects with repeated administration. About 90% of methadone is bound to plasma proteins. It undergoes extensive biotransformation in the liver. Methadone has a long plasma half-life (about 1–5 days in humans).

Propoxyphene binds to μ receptors with slightly less selectivity than morphine. As an analgesic, it is about one-half to two-thirds as potent as codeine given orally. The only recognized use of propoxyphene is for the treatment of pain that is not adequately relieved by aspirin.

1. Preparations and Routes of Administration

Methadone hydrochloride is available in tablets and in solutions for oral use and as a solution for parenteral administration.

Propoxyphene hydrochloride is available in capsules for oral administration. Propoxyphene napsylate is available in tablets or as a suspension for oral administration. Combinations of propoxyphene with aspirin or acetaminophen are marketed in tablets and capsules.

E. Opioids with Mixed Actions: Agonist–Antagonists and Partial Agonists

Some opioids are competitive antagonists at the μ receptor, but exert partial agonistic actions at other receptors, including the δ and κ receptors (e.g., nalorphine, cyclazine, and nalbuphine). Pentazocine is either a weak antagonist or a partial agonist of μ receptors and a relatively powerful κ receptor agonist. Buprenorphine is a partial μ agonist.

Pentazocine differs from typical μ agonists in that high doses cause an increase in blood pressure and heart rate. It does not antagonize respiratory depression produced by morphine, but it may produce withdrawal symptoms in patients who have been receiving μ-opioid receptor agonists. Pentazocine lactate is available as a solution for injection. Tablets containing pentazocine hydrochloride and naloxone hydrochloride are available for oral use. Naloxone is included to discourage the use of tablets as a source of injectable pentazocine. The naloxone absorbed after oral administration is rapidly removed by the liver.

Tablets containing pentazocine and aspirin or acetaminophen are also available. When given parenterally, three to six times as much pentazocine is required to equal the analgesic effect of a dose of morphine. Pentazocine is only slightly more potent than codeine when given orally. Ordinarily there is a ''ceiling'' on the respiratory depressant effects of pentazocine (i.e., beyond a certain doses, respiratory depression increases little if at all).

Nalbuphine is an agonist–antagonist with a spectrum of effects similar to those of pentazocine except nalbuphine is a more potent μ receptor agonist. Equal doses of nalbuphine and morphine given intramuscularly produce approximately equal analgesia. Nalbuphine has considerably less effect on the cardiovascular system than does pentazocine or butorphanol. Like pentazocine, there is a ceiling on the respiratory depressant effect of nalbuphine.

Butorphanol has a pharmacologic profile similar to that of pentazocine. On a milligram for milligram basis, butorphanol is about three to five times as potent as morphine when given parenterally. Butorphanol tartrate is available in solution for parenteral injection.

Buprenorphine is a partial μ agonist. It produces analgesia and other CNS side effects that are similar to those produced by morphine. In recommended therapeutic doses, buprenorphine is about 25 times as potent as morphine when given intramuscularly. However, the maximum analgesic effect of buprenorphine is less than that of morphine. Buprenorphine can produce withdrawal symptoms in patients who have been receiving μ receptor agonists. It antagonizes the respiratory depressant effects of some μ receptor agonists without completely preventing opioid pain relief. About 96% of the drug in the plasma is protein bound and it is mostly excreted unchanged in the feces.

Buprenorphine is available in solution for parenteral injection. It has a long duration of action in the rat (about 6 hr; Dum and Herz, 1981).

F. Opioid Antagonists

Naloxone usually causes no effect unless opioids with agonistic action have been given. Small intramuscular or intravenous doses of naloxone prevent or promptly reverse the effects of μ-opioid agonists. Naloxone hydrochloride is available in solution for parenteral injection. Other opioid antagonists are available, but naloxone is the most commonly used.

IV. α_2-ADRENERGIC AGONISTS

α_2-Adrenergic agonists can have diverse biological effects (MacDonald and Virtanen, 1992; Table VII). Of particular relevance to this chapter is that these agents produce sedation, loss of vigilance, and analgesia. Xylazine was introduced into veterinary medicine for sedation in the 1960s, but its mechanism of action was unknown. Later it was concluded that it was a rather specific α_2-adrenoreceptor agonist. Four α_2-adrenoreceptor agonists are commonly used to modify animal pain; clonidine, xylazine, detomidine, and medetomidine (Salonen, 1992). Detomidine and medetomidine are closely related imidazole derivatives, clonidine is an imidazoline, and xylazine is a thiazole.

TABLE VII

RESULTS OF α_2-ADRENORECEPTOR STIMULATION[a]

System or characteristic	Effect
CNS	Sedation, analgesia, hypotension, bradycardia
CVS	Peripheral vasoconstriction → initial hypertension, central bradycardia, and vasomotor depression → hypotension
Gut	Relaxation, decreased motility
Salivation	Decreased
Gastric secretion	Reduced
Uterus	Stimulation
Eyes	Mydriasis, decreased intraocular pressure
Hormones	Reduced release of insulin, renin, and antidiuretic hormone
Platelets	Aggregation

[a]From Hall, L. W. and Clarke, K. W. (1991). Principles of Sedation. In ''Veterinary Anaesthesia,'' 9th ed., p. 58. Baillière Tindall, London.

1. History

α_2-Adrenergic agonists were originally developed as antihypertensives and are currently being developed for intravenous administration for general anesthesia. Discovery of the role of a noradrenergic inhibitory system in nociception modulation at the spinal cord level provided a basis for the development of α_2-adrenoreceptor agonists as receptor-specific analgesics. Although receptor specific, α_2 receptor agonists developed to date are not effect specific. Clonidine was synthesized in the early 1960s, about the same time xylazine was introduced into veterinary medicine. The α_2-adrenergic agonist activity of xylazine was recognized in 1977 (Berthelsen and Pettinger, 1977; Hsu, 1981) and shortly thereafter, Farmos Group Ltd. (Turku, Finland) produced detomidine and medetomidine.

2. Mechanism of Action

Sedation and anesthesia from α_2-adrenergic agonists occur by actions in a small group of neurons in the brain stem, the locus coeruleus (Eisenach, 1993). The mechanism whereby these drugs produce analgesia is not completely understood, but the spinal cord is generally considered to be a primary action site. α_2-Adrenergic agonists appear to have a combined effect of presynaptic inhibition of afferent transmitter release from C-fibers and a postsynaptic inhibition of spinal cord dorsal horn transmission neurons.

3. Therapeutic Activities and Side Effects

As shown in Fig. 3, systemically administered α_2-adrenergic agonists produce a spectrum of behavioral effects, the most readily apparent being sedation and loss of vigilance. There is great variability in (1) the extent of sedation produced by different drugs and (2) responses of different animal species to the drugs. For example, detomidine and medetomidine can induce loss of the righting reflex in young chicks (Savola *et al.,* 1986), but only medetomidine does so in rats. Medetomidine does not produce such deep sedation in mice and rabbits. The dose response for this drug in rabbits is poor.

The analgesic action of α_2-adrenergic agonists is most clearly demonstrated following epidural or intrathecal injection. When administered systemically, they also have analgesic activity (e.g., inhibit acetic acid-induced writhing or tail-flick responses to heat), but there is difficulty in distinguishing true analgesia from inability to respond to painful stimuli. Clinically, the α_2-adrenergic agonists appear to have limited analgesic activity when used alone.

Common side effects of α_2-adrenergic agonists include inhibition of insulin secretion (and hence production of hyperglycemia). Low doses of these drugs decrease arterial blood pressure and heart rate, and may produce atrioventricular conduction block. Heart rate continues to slow as dose increases, but blood pressure may increase. α_2-Adrenergic agonists produce hypothermia and respiratory depression. They also produce emesis in a number of animal species (e.g., dog, cat).

Because of the wide range of effects of α_2-adrenergic agonists, there is potential for them to interact with many other drugs. The doses of barbiturates, inhalational anesthetics, and dissociative anesthetics should be reduced when given along with α_2-adrenergic agonists. It is common to administer α_2-adrenergic agonists with other drugs to take advantage of synergistic interactions (e.g., along with local anesthetic via epidural or subarachnoid route, with ketamine via IV or IM route).

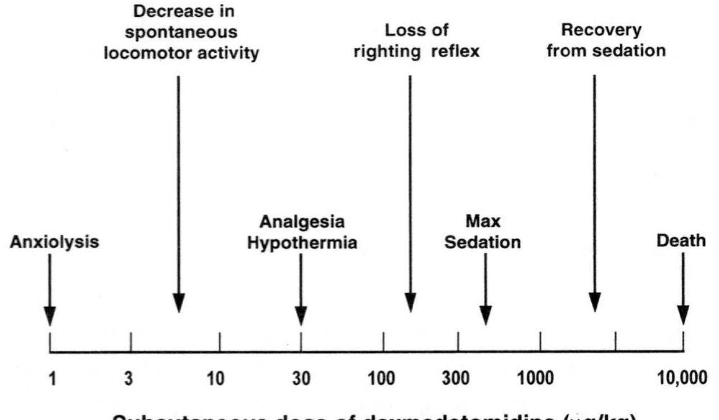

Fig. 3. Central actions of dexmedetomidine in rats after subcutaneous treatment. Dexmedetomidine is the dextro(+)isomer of medetomidine and is the active component of the racemic mixture. From MacDonald, E. and Virtanen, R. (1992). The effects of buprenorphine, nalbuphine and butorphanol alone or following halothane anesthesia on food and water consumption and locomotor movement in rats. *Lab. Anim.* 26, 180–189.

4. Pharmacokinetics and Metabolism

a. ABSORPTION. In general, clonidine is well-absorbed after oral administration and bioavailability is nearly 100%. However, the systemic availability of detomidine, medetomidine, and xylazine by the oral route is low. The few studies of the bioavailability of detomidine, medetomidine, and xylazine following IM administration revealed species variability (e.g., bioavailability good for detomidine in cattle and horses, good for xylazine in dog but not horse).

b. DISTRIBUTION AND FATE. α_2-Adrenergic agonists are lipophilic and hence are readily distributed into tissues. Approximately 85% of detomidine and medetomidine are bound to plasma protein and about 70% of xylazine and clonidine are bound.

Detomidine, medetomidine, and xylazine generally are eliminated relatively rapidly in all animal species. The half-life of elimination varies between 0.5 (xylazine, dog) and 1.5 hr (medetomidine and detomidine, horse and dog).

All four of the agonists are biotransformed primarily by the liver and then metabolites are excreted in the urine. With the exception of clonidine, renal clearance of unchanged drug is insignificant.

5. Preparations and Routes of Administration

Xylazine is available commercially as a 2 or 10% solution for intravenous or intramuscular injection. Detomidine is available in injectable form (1% solution) for use in the horse and injectable medetomidine is undergoing trials for approval for use in dogs and cats. No formulations are marketed for epidural or intrathecal administration by veterinarians. Agents marketed for IV or IM administration contain preservatives that may be unsuitable for epidural or intrathecal administration.

V. α_2-ADRENERGIC ANTAGONISTS

The α_2-adrenergic antagonists in clinical use today include yohimbine, tolazoline, idazoxan, and atipamezole. Yohimbine tolazoline and atipamezole are approved for veterinary use in the United States. The obvious value of antagonists is to reverse overdose or hasten recovery from the sedative effects of α_2-adrenergic agonists.

VI. ANALGESIC ADJUVANTS

A list of drug classes used as analgesic adjuvants is shown in Table VIII. These drugs usually are not noted as having analgesic activity but may be of benefit in the treatment of pain by,

TABLE VIII

ANALGESIC ADJUVANTS

Tricyclic antidepressants
Anticonvulsants
Antihistamines
Benzodiazepines
Sedative hypnotics
Steroids
Caffeine
Dextroamphetamines
Phenothiazines

for example, aiding sleep and countering depression and/or anxiety, muscle spasm, or inflammation. Drugs (e.g., antibiotics) used to treat conditions that produce pain such as bacterial infection may be classed as analgesic adjuvants. Evidence shows that sedative hypnotics may have antianalgesic activity at low doses (Archer *et al.,* 1994).

VII. FUTURE PROSPECTS

Efforts in four areas of research will strongly influence the management of pain in laboratory animals in the future. These areas are (1) understanding of pain mechanisms; (2) advances in optimization of alternative routes of drug administration, especially transmucosal, transcutaneous, epidural, and spinal; (3) improved methods for pain assessment; and (4) improved methods for assuring optimal dosing regimens, e.g., simple, rapid measurement of drug concentration in blood.

Of particular interest at this time are drugs that affect *N*-methyl-D-aspartate (NMDA) receptors at the spinal level. Activation of these receptors produces ''wind-up'' in dorsal horn neurons, a phenomenon linked to secondary algesia. NMDA receptor antagonists are viewed as being particularly useful as preemptive agents (i.e., administered in anticipation of injury instead of afterward).

ACKNOWLEDGMENT

The author thanks Dani Joyner for typing this chapter.

REFERENCES

Archer, B. P., Erven, A., Roth, S. H., and Samanani, N. (1994). Plasma, brain and spinal cord concentrations of thiopental associated with hyperanalgesia in the rat. *Anesthesiology* **80,** 168–176.

Baggot, J. D. (1992). Clinical pharmacokinetics in veterinary medicine. *Clin. Pharmocinet.* **22,** 254–273.

Bayer, B. M., Daussin, S., Hernandez, M., Irvin, L. (1990). Morphine inhibition of lymphocyte activity is mediated by an opioid dependent mechanism. *Neuropharmacology* **29,** 369–374.

Berthelsen, S., and Pettinger, W. A. (1977). A functional basis for classification of α_2-adrenergic receptors. *Life Sci.* **21,** 596.

Bussiere, J. L., Adler, M. W., Rogers, T. J., Eisenstein, T. K. (1992). Differential effects of morphine and naltrexone on the antibody response in various mouse strains. *Immunopharmacol. Immunotoxicol.* **14,** 657–673.

Capetola, R. J., Rosenthale, M. E., Dubinsky, B., and McGuire, J. L. (1983). Peripheral antialgesics: A review. *J. Clin. Pharmacol.* **23,** 545–556.

Carr, D. J. J., and France, C. P. (1993). Immune alterations in morphine-treated rhesus monkeys. *J. Pharmacol. Exp. Ther.* **267,** 9–15.

Ciofalo, V. B., Latranyi, M. B., Patel, J. B., and Taber, R. I. (1977). Flunixin meglumine: A non-narcotic analgesic. *J. Pharmacol. Ther.* **200,** 501–507.

Czlonkowski, A., Stein, C., and Herz, A. (1993). Peripheral mechanisms of opioid antinociception in inflammation: Involvement of cytokines. *Eur. J. Pharmacol.* **242,** 229–235.

Davis, L. E. (1983). Species differences in drug disposition as factors in alleviation of pain. *In* ''Animal Pain: Perception and Alleviation'' (R. L. Kitchell and H. H. Erickson, eds.), pp. 161–178. Am. Phys. Soc., Bethesda, Maryland.

Davis, L. E., and Westfall, B. A. (1972). Species differences in the biotransformation excretion of salicylate. *Am. J. Vet. Res.* **33,** 1253–1262.

Dum, J. E., and Herz, A. (1981). In vivo receptor binding of the opiate partial agonist buprenorphine, correlated with its agonistic and antagonistic actions. *Br. J. Pharmacol.* **74,** 627–633.

Eisenach, J. D. (1993). Overview: First international symposium on α_2-adrenergic mechanisms of spinal anesthesia. *Reg.-Anesth.* **18,** i–vi.

Flecknell, P. A. (1991). Post-operative analgesia in rabbits and rodents. *Lab. Anim.* **20,** 34–37.

Fuchs, B. A., and Pruett, S. B. (1993). Morphine induces apoptosis in murine thymocytes *in vivo* but not *in vitro*: Involvement of both opiate and glucocorticoid receptors. *J. Pharmacol. Exp. Ther.* **266,** 417–423.

Hall, L. W., and Clarke, K. W. (1991). Principles of Sedation. *In* ''Veterinary Anaesthesia,'' 9th ed., p. 58. Baillière Tindall, London.

Hsu, W. S. (1981). Xylazine-induced depression and its antagonism by α_2-adrenergic blocking agents. *J. Pharmacol. Exp. Ther.* **218,** 188.

Hughes, H. C., and Lang, C. M. (1983). Control of pain in dogs and cats. *In* ''Animal Pain: Perception and Alleviation'' (R. L. Kitchell and H. H. Erickson, eds.), pp. 207–216. Am. Physiol. Soc., Bethesda, Maryland.

Insel, P. A. (1990). Analgesic - antipyretics and antiinflammatory agents; drugs employed in the treatment of rheumatoid arthritis and gout. *In* ''The Pharmacological Basis of Therapeutics'' (A. G. Gilman, T. W. Rall, A. S. Nies, and P. Taylor, eds.), 8th ed., pp. 638–681. Pergamon, New York.

Jaffe, J. H., and Martin, W. R. (1990). Opioid analgesics and antagonists. *In* ''The Pharmacological Basis of Therapeutics'' (A. G. Gilman, T. W. Rall, A. S. Nies, and P. Taylor, eds.), 8th ed., pp. 485–521. Pergamon, New York.

Levier, D. G., Brown, R. D., McCay, J. A., Fuchs, B. A., Harris, L. S., and Munson, A. E. (1993). Hepatic and splenic phagocytosis in female B6C3F1 mice implanted with morphine sulfate pellets. *J. Pharmacol. Exp. Ther.* **267,** 357–363.

Liles, J. H., and Flecknell, P. A. (1992). The effects of buprenorphine, nalbuphine and butorphanol alone or following halothane anesthesia on food and water consumption and locomotor movement in rats. *Lab. Anim.* **26,** 180–189.

MacDonald, E., and Virtanen, R. (1992). Review of the pharmacology of medetomidine and detomidine: Two chemically similar α_2-adrenoceptor agonists used as veterinary sedatives. *In* ''Animal Pain'' (C. E. Short and A. Van Poznak, eds.), pp. 181–191. Churchill-Livingstone, New York.

Malmberg, A. B., and Yaksh, T. L. (1993). Pharmacology of the spinal action of ketorolac, morphine, ST-91, U50488H, and LPIA on the formalin test and an isobolographic analysis of the NSAID interaction. *Anesthesiology* **79,** 270–281.

Novick, D. M., Ochshorn, M., Ghali, V., Croxson, T. S., Mercer W. D., Chiorazzi, N., and Kreek, M. J. (1989). Natural killer cell activity and lymphocyte subsets in parenteral heroin abusers and long-term methadone maintenance patients. *J. Pharmacol. Exp. Ther.* **250,** 606–610.

Pruett, S. B., Han, Y. C., and Fuchs, B. A. (1992). Morphine suppresses primary humoral immune responses by a predominately indirect mechanism. *J. Pharmacol. Exp. Ther.* **262,** 923–928.

Quimby, F. W. (1991). Pain in animals and humans: An introduction. *ILAR News* **33,** 2–3.

Randall, T. (1993). Morphine receptor cloned-improved analgesics, addiction therapy expected. *JAMA, J. Am. Med. Assoc.* **270,** 1165–1166.

Roth, G. R., and Siok, C. J. (1978). Acetylation of the NH_2-terminal serine of prostaglandin synthetase by aspirin. *J. Biol. Chem.* **253,** 3782–3784.

Salonen, J. S. (1992). Chemistry and Pharmacokinetics of the α_2-Adrenoreceptor Agonist. *In* "Animal Pain" (C. E. Short and A. Van Poznak, eds.), pp. 191–200. Churchill-Livingstone, New York.

Savola, J.-M., Ruskoaho, H., Puurunen, J., Salonen, J. S., Kärki, N. T. (1986). Evidence for medetomidine as a selective and potent agonist of α_2-adrenoceptors. *J. Auton. Pharmacol.* **5,** 275–284.

Shavit, Y, Lewis, J. W., Terman, G. W., Gale, R. P., and Liebeskind, J. C. (1984). Opioid peptides mediate the suppressive effect of stress on natural killer cell cytolytic activity. *Science* **223,** 188–190.

Szabo, I., Rojivin, M., Bussiere, J. L., Eisenstein, T. K., Adler, M. W., and Rogers, T. S. (1993). Suppression of peritoneal macrophage phagocytosis of *Candida albicans* by opioids. *J. Pharmacol. Exp. Ther.* **267,** 703–706.

Vonderhaar, M. A., and Salisbury, S. K. (1993). Gastroduodenal ulceration associated with flunixin meglumine administration in three dogs. *J. Am. Vet. Med. Assoc.* **203,** 92–95.

Chapter 4

Paralytic Agents

Susan V. Hildebrand

I. INTRODUCTION

Paralytic agents, or muscle relaxants, are adjuncts to general anesthesia. Generally speaking, they provide superior muscle relaxation to that which can be achieved with general anesthesia alone. Improved muscle relaxation has a number of advantages. Endotracheal intubation and mechanical ventilation are facilitated. Surgical conditions can be improved, e.g., long bone fracture reduction or difficult exposure of abdominal structures. Anesthetic management, especially with respect to cardiovascular function, can be improved. Due to ongoing research, muscle relaxants developed in the early 1990s are easier to use, with fewer side effects, than the older relaxants.

That being said, there are important considerations that influence the decision to use paralytic agents. Because relaxants do not render the animal unconscious, particular care must be taken to ensure adequate anesthesia and analgesia (Gibbs *et al.,* 1989). Mechanical ventilation is necessary with most paralytic agents. A means of monitoring relaxation is necessary to ensure efficacy and to determine when and how much more relaxant to give. The drawbacks, then, of paralytic agents are not what they can offer, but rather the complexity of anesthetic delivery and monitoring that is required.

An important ethical concern is provision of adequate levels of anesthesia for the animal given muscle relaxants. Techniques for monitoring anesthetic depth are given in detail elsewhere in the text. Because the ability of the animals to move is eliminated by the use of paralytic agents, signs other than muscle movement must be used to monitor depth of anesthesia. Probably the best indicators are physiologic parameters such as heart rate, blood pressure, electrocardiogram, and arterial blood gas analysis. An animal that is inadequately anesthetized will demonstrate general stress responses such as tachycardia, arrhythmias, hypertension, and acidosis. Mydriasis and lacrimation are associated with light levels of anesthesia. Animals that sweat, e.g., equids, will sweat profusely if anesthesia is inadequate.

ANESTHESIA AND ANALGESIA IN LABORATORY ANIMALS

ISBN 0-12-417570-8

II. CLASSIFICATION OF PARALYTIC AGENTS

Muscle relaxants are classified by the anatomic location at which they act. There are two broad categories of relaxants.

A. Centrally Acting Relaxants

These drugs act on the central nervous system (CNS). Guaifenesin is the only centrally acting relaxant that is likely to be useful clinically.

1. Guaifenesin

Previously known as glyceryl guaiacolate, guaifenesin is a glyceryl ether of the myanesin group that affects the multisynaptic neuron pathways of the spinal cord and brain stem (Davis and Wolff, 1970; Booth, 1977). Tonic, or supporting, muscles are the most affected, whereas the muscles of respiration, including the diaphragm, are minimally affected. Because of this, the use of guaifenesin is different from other paralytic agents. Guaifenesin is most commonly used during induction of anesthesia, during parenteral anesthesia of short duration, or during inhalation anesthesia for brief periods when the inhalation agent must be discontinued (during extubation, for example).

There is no experimental evidence that guaifenesin provides surgical analgesia and unconsciousness (Davis and Wolff, 1970; Funk, 1970, 1973; Schatzman, 1974; Tavernor and Jones, 1970; Westhues and Fritsch, 1965; McLure and McLure, 1980).

a. CARDIOVASCULAR CHANGES. Depending on the species studied and the nature of the stimulus being applied, guaifenesin is associated with increased heart rate. Blood pressure may increase or decrease. No "pathologic deviations" in electrocardiogram were noted (Funk, 1970; Tavernor, 1970). Heart rate, right atrial pressure, pulmonary artery pressure, and cardiac output were not significantly different from baseline in horses given 134 mg/kg IV to produce lateral recumbency (Hubbell *et al.,* 1980). The horses were not subjected to surgical procedures or other drugs. Mean arterial blood pressure fell from 120 to 92 mm Hg after guaifenesin. Dogs given 200 mg/kg showed an elevated heart rate and blood pressure (Tavernor and Jones, 1970).

b. RESPIRATORY CHANGES. Laryngeal and pharyngeal relaxation increases upper airway resistance and facilitates intubation. $PaCO_2$ did not change, but PaO_2 fell after guaifenesin was given to horses (Hubbell *et al.,* 1980). In rabbits, 80 mg/kg produced no change in respiratory volume, whereas doses of 150–200 mg/kg reduced respiratory volume by 30% (Funk, 1970). Guaifenesin is an expectorant–antitussive component of cough medicines.

c. GASTROINTESTINAL CHANGES. Intestinal motility was unchanged in cats given 10–20 mg/kg guaifenesin. High (near lethal) doses of guaifenesin have been reported to cause diarrhea in guinea pigs. Rumen stasis and tympany have been reported in cattle, but the contribution of guaifenesin to this effect is unclear (Funk, 1970).

d. UROGENITAL CHANGES. High-dose guaifenesin causes crystalluria in horses and cattle (Funk, 1970). Crystals are metabolic products of guaifenesin. High dose or prolonged administration leads to glomerular damage. Hemoglobinuria was noted in dogs (Westhues and Fritsch, 1965) and cattle (Funk, 1970). Guaifenesin crosses the placenta (Hubbell *et al.,* 1980).

e. METABOLISM, EXCRETION, AND TOXICITY. Guaifenesin is metabolized by the liver and eliminated by the kidneys. One of its metabolites is catechol, which may be responsible for extensor rigidity, convulsions, and respiratory and cardiac arrest from an overdose. At three to four times the therapeutic dose, toxic manifestations include tetanus-like extensor spasm, cardiac arrest, and respiratory arrest (Funk, 1973; Westhues and Fritsch, 1965). Continuous infusion over several hours causes high mortality, renal glomerular lesions, and inflammatory renal interstitial lesions. Some degree of hemolysis occurs even at concentrations of less than 5% (Mostert and Metz, 1963). Other side effects include thrombophlebitis and tissue irritation.

f. PREPARATION. Guaifenesin is prepared as a 5 to 10% solution in water or 5% dextrose. A 10% solution in water best approximates plasma osmolarity (Grandy and McDonell, 1980). Solubility is improved by warming the solution. Solutions of 10% may precipitate at room temperature. Guaifenesin is compatible in mixture with barbiturates and ketamine.

g. SPECIES VARIATION. In horses, cattle, and small ruminants, the therapeutic (casting) dose is approximately 100 mg/kg IV. Duration from injection to standing is about 20 min. A gender difference in equids has been reported of 1.5 times longer duration in stallions than in mares (Davis and Wolff, 1970), with less potency in geldings than in mares (Hubbell *et al.,* 1980).

The usefulness of guaifenesin is limited to equine and ruminant species (Funk, 1973; Jackson and Lundvall, 1970, 1972). Guaifenesin has been described for small animal species, but the results have been unsatisfactory. In dogs, large volumes gave unsatisfactory relaxation with short duration and agitated recovery with prolonged ataxia (Tavernor and Jones, 1970; Westhues and Fritsch, 1965). In cats, 350 mg/kg gave approximately 60 min of relaxation, with prolonged recovery. Cats demonstrated salivation, emesis, respiratory depression and cyanosis, hypothermia, and cardiac arrhythmia. Rabbits required as much as 350 mg/kg for 30 min of relaxation, and birds showed less relaxation than mammals (Westhues and Fritsch, 1965).

2. Guaifenesin Combinations

Guaifenesin is often used in combination with other drugs. Guaifenesin modifies undesirable side effects and reduces the necessary doses of anesthetic drugs. A variety of sedative or tranquilizer drugs can be used to calm the animal prior to induction of anesthesia with guaifenesin and ketamine or with guaifenesin and a thiobarbiturate. These can all be given separately. For example, a pony could be sedated with xylazine, then 5–10 min later given guaifenesin until it was swaying slightly with relaxation (25–50 mg/kg), then given ketamine to render it unconscious, followed by another 25–50 mg/kg guaifenesin once the pony is recumbent. Guaifenesin can also be combined with other drugs and is compatible with ketamine, thiopental, thiamylal, pentobarbital, and α_2-agonists such as xylazine. For example, a combination known as "triple drip" consists of 5% guaifenesin containing 2 mg/ml of ketamine and xylazine (1 mg/ml for use in swine, 0.25 mg/ml for use in equids, and 0.1 mg/ml for use in ruminants). Anesthesia can be induced with triple drip and, if desired, can be given slowly IV (2 ml/kg/hr) to maintain anesthesia of 45 to 60 min duration.

B. Neuromuscular Blocking Agents

These drugs interact with receptors at the neuromuscular junction. During normal neuromuscular transmission, a stimulus to the motor nerve results in mobilization and release of acetylcholine from the motor nerve terminal (Fig. 1). The acetylcholine released crosses the synaptic cleft and attaches to receptors on the motor end plate of the muscle. Postjunctional receptors have been well defined (Standaert, 1990). Attachment of acetylcholine to these motor end plate receptors leads to muscle cell depolarization and muscle contraction. Acetylcholine–receptor attachment is brief, and the muscle relaxes again as acetylcholine detaches from the receptor and is broken down by acetylcholinesterase. Breakdown products are taken up by the motor nerve ending to be recycled in the production of more acetylcholine. This reuptake takes place via prejunctional receptors, which have not as yet been well defined (Standaert, 1990).

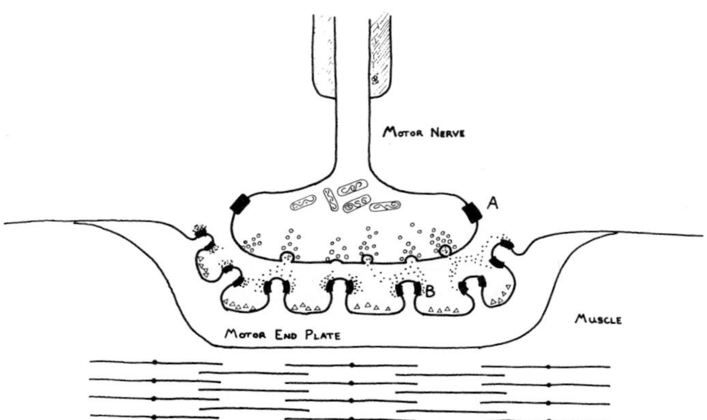

Fig. 1. A simplified diagram of events at the neuromuscular junction during normal neuromuscular transmission. The motor nerve loses its myelin sheath and branches before reaching the motor end plate, a specialized and distinct structure on skeletal muscle. A stimulus to the motor nerve initiates mobilization and release of acetylcholine, which attaches to receptors (B) in the motor end plate. Receptor attachment at B leads to depolarization and contraction of the muscle cell. Acetylcholine is hydrolyzed by acetylcholinesterase (Δ) in the clefts of the motor end plate. Breakdown products of acetylcholine are taken up by prejunctional receptors (A) on the motor nerve ending for recycling in the production of acetylcholine by the motor nerve ending. From Hildebrand, S. V. (1990). Neuromuscular blocking agents in equine anesthesia. *Vet. Clinics of North America* **6:** 587–606.

There are two categories of neuromuscular blocking agents: depolarizing and nondepolarizing.

1. Depolarizing Neuromuscular Blocking Agents

Depolarizing relaxants are known as leptocures because they are long, thin, flexible molecules. This structural feature contributes to a rapid onset of action. Depolarizing relaxants interact with postjunctional receptors and mimic the action of acetylcholine. They are called noncompetitive because their action is not antagonized by acetylcholine, i.e., depolarizing relaxants and acetylcholine do not compete for receptor sites. Increased amounts of succinylcholine at the neuromuscular junction will not influence magnitude or duration of a depolarizing block. Depolarizing relaxants cause initial depolarization of the muscle cells, but instead of separating quickly from the postjunctional receptor, receptor attachment of depolarizing relaxants persists for minutes to hours, depending on a number of factors including species, dose, and duration of administration. Even though the motor end plate remains depolarized, relaxation and transmission block persist due to differences in the receptors of the end plate and the muscle membrane.

a. SUCCINYLCHOLINE. Succinylcholine or suxemethonium represents the only depolarizing relaxant in use. Succinylcholine is a water-soluble powder available as succinylcholine chloride or iodide, the latter being two-thirds the potency of the former (Stowe *et al.,* 1958). Succinylcholine deteriorates rapidly in alkaline solutions. It loses potency in a warm environment and is therefore stored under refrigeration.

i. Action. Succinylcholine has a rapid onset and provides excellent relaxation, generally of brief duration, although duration depends on many factors. The neuromuscular blocking activity of succinylcholine is complex. Initially, postjunctional receptor occupation by succinylcholine results in depolarization of the muscle, causing marked fasciculation. This subsides within approximately 40 sec. Muscle relaxation follows, but muscle depolarization persists, as the neuromuscular junction remains refractory to acetylcholine and repolarization.

The action of succinylcholine can change with large doses, repeat doses, or prolonged administration. An initial depolarizing block may change to resemble a nondepolarizing block variously called phase II, dual, or desensitization block. This has been described as "complex and ever changing . . . difficult to predict" (Standaert, 1990). Desensitization block is incompletely understood and may occur due to an electrolyte imbalance that occurs when the continuous opening of ion channels allows an efflux of potassium and an influx of sodium and calcium (Standaert, 1990). The onset of desensitization block is seen as an apparent resistance to succinylcholine, then block characteristics gradually become nondepolarizing. The response of phase II block to relaxant antagonists is unpredictable.

Phase II block is an inherent risk associated with the use of succinylcholine, especially large doses or prolonged administration. Once phase II block occurs, it may or may not be reversible. Phase II block has the characteristic of the nondepolarizing block described in the following section. Phase II block can sometimes be reversed by the same antagonists used to reverse nondepolarizing block, but these are not always effective.

ii. Side Effects. A number of undesirable side effects are associated with succinylcholine.

1. *Musculoskeletal.* Initial depolarization causes muscle fasciculation and extensor rigidity, which leads to muscle pain and stiffness (scoline pain), potassium release, microscopic evidence of muscle damage, increased pressure within osteofascial compartments, and occasional myoglobinuria.
2. *Cardiovascular.* Depending on age, species, and depth of anesthesia, stimulation of nicotinic receptors in the sympathetic ganglia causes hypertension and tachycardia (Hildebrand and Howitt, 1983b). Stimulation of muscarinic receptors causes sinus bradycardia or cardiac arrest (Miller and Savarese, 1990). Increased serum potassium released from muscle cells can cause cardiac arrhythmias.
3. *Ocular.* Succinylcholine is associated with an intraocular pressure increase, although one equine study could not demonstrate an increase (Benson *et al.,* 1981). An intraocular pressure rise has been attributed to an increase in skeletal muscle tone, but in humans, a rise in intraocular pressure was noted even when the extraocular muscles were detached from the globe, suggesting that cycloplegia may contribute to a rise in intraocular pressure (Kelly *et al.,* 1991). In cats, contraction of the nictitating membrane can increase intraocular pressure.
4. *Potassium.* Concurrent with muscle cell damage, there is a release of potassium and increased serum potassium. This becomes life-threatening in certain conditions. A massive release of potassium with subsequent ventricular fibrillation can occur if succinylcholine is given to animals with renal failure, burns, or muscle trauma and neurological disorders such as paraplegia and hemiplegia (Miller and Savarese, 1990).

iii. Metabolism. Succinylcholine is broken down by nonspecific plasma cholinesterases to choline and succinylmonocholine.

iv. Factors Influencing Action of Succinylcholine. Because the breakdown of succinylcholine is dependent on plasma cholinesterases, decreased plasma cholinesterase activity prolongs its action. This can occur with hepatic disease, pregnancy, or exposure to anticholinesterase agents such as organophosphates. In humans, genetically abnormal plasma cholinesterases cause prolonged activity of succinylcholine.

Age influences the activity of succinylcholine. The immature neuromuscular junction of the neonate is resistant to succinylcholine (Goudsouzian and Standaert, 1986).

Inhalation anesthetics and local anesthetics can potentiate succinylcholine (Viby-Mogensen, 1990). Most experimental procedures that describe the use of succinylcholine in small animal species use noninhalational drugs such as pentobarbital, α-chloralose, ketamine, or urethane. These intravenous anesthetics have little effect on the neuromuscular blocking activity of succinylcholine.

Increased serum calcium due to hyperparathyroidism increases the succinylcholine dose requirement 1.5 to 2 times. Conversely, calcium antagonist drugs potentiate its action (Ocana *et al.,* 1992).

v. Contraindications to Succinylcholine Use. Succinylcholine should be avoided in the presence of penetrating wounds of the eye, glaucoma, aneurysm, and some fractures.

Succinylcholine, especially in combination with halothane, is one of the most potent initiators of malignant hyperthermia (MH) in genetically stress-susceptible animals. Malignant hyperthermia or MH-like reactions have been noted in pigs, dogs, cats, ponies, and horses. Because of its propensity to cause MH, succinylcholine in the presence of halothane has been used to determine MH susceptibility in swine.

vi. Use of Succinylcholine. There seems to be wide species variation in response to succinylcholine (Hansson, 1958). There is also considerable variation within species depending on the reference used. Cats, pigs, and ponies are said to be relatively ''resistant'' to the effects of succinylcholine, whereas dogs, sheep, and cattle are ''sensitive.'' This may in part be due to species variation in plasma cholinesterase activity. For example, sheep and cattle have relatively low cholinesterase activity, thus duration of action can be prolonged (Stowe *et al.,* 1958). The work of Muir and Marshall (1987), Muir *et al.* (1989a,b), and Hansson (1958) does not support resistance to succinylcholine in pigs as noted in veterinary texts. In ferrets, succinylcholine has been noted to have nondepolarizing activity (Tsai *et al.,* 1990). Succinylcholine doses and duration are noted in Table I.

2. Nondepolarizing Neuromuscular Blocking Agents

The nondepolarizing relaxants are large, bulky molecules called pachycures. These drugs compete with aceteylcholine for attachment to postjunctional receptors. However, these drugs do not cause muscle depolarization. Instead, they prevent the muscle from depolarizing. In addition, the nondepolarizing relaxants interact with prejunctional receptors to a varying degree, depending on the relaxant. The prejunctional receptor attachment of nondepolarizers is probably responsible for such phenomena as fade and posttetanic potentiation, which are described in Section III.

The nondepolarizing relaxants are grouped into two categories based on molecular structures: **steroid analogs,** which include pancuronium, vecuronium, pipecuronium, and rocuronium (ORG 9426); and **benzylisoquinoliniums,** which include tubocurarine, metocurine, gallamine, atracurium, doxacurium, and mivacurium.

Most current development of new muscle relaxants has focused on correcting the shortcomings of the earlier relaxants (Mirakhur, 1992). Undesirable features include slow onset of action, elimination dependent on hepatic and renal function, and autonomic side effects.

TABLE I

SUCCINYLCHOLINE DOSE/DURATION FOR ANIMAL SPECIES

Species	Dose (mg/kg)[a]	Duration (min)[b]	Ref.[c]
Dog	(ED_{80}) 0.1	(to 100%) 12	1
	(ED_{100}) 0.07	(to 100%) 8–9	2
Cat	(ED_{90-95}) 0.6–0.2	(complete) 5–11	3
Mouse	0.05–0.1	—	4
Rat	50	—	5
Rabbit	0.1	—	4
Pig	(ED_{90}) 0.75–1.3	2–3	6
	(infusion) 1 mg/kg/hr	—	7
Sheep/cattle	(ED_{100}) 0.04	>6–8	4
Pony	0.1	1–5	8
	(ED_{90-99} infusion) 2.2 mg/kg/hr	—	9
Horse	(ED_{100}) 0.6	6–7	2
Ferret	(ED_{95}) 0.15	—	10
Sparrow	5.0	—	4
Frog	2.5 (into ventral lymph sac)	—	4

[a]In parentheses, effective dose and percentage twitch reduction as noted in reference source.

[b]Reference descriptions of duration varied, but when indicated, were given as 100% or complete recovery.

[c]*Key to references:* (1) Hughes, 1970; (2) Hansson, 1958; (3) Ocana *et al.*, 1992; Muir and Marshall, 1987; Muir *et al.*, 1989a; van der Spek *et al.*, 1989; Hall and Clarke, 1983; (4) Lumb and Jones, 1973; (5) Yokoi *et al.*, 1992; (6) Ducey *et al.*, 1989; Muir and Marshall, 1987; Muir *et al.*, 1989a; (7) Matran *et al.*, 1991; (8) Hildebrand and Howitt, 1983a; (9) Benson *et al.*, 1981; (10) Tsai *et al.*, 1990.

The bulky nondepolarizing relaxants have a slow onset of action, with 95% effective (ED_{95}) doses typically requiring several minutes for onset. Onset is more rapid with large doses (three times ED_{95}), but doses larger than ED_{95} can have other undesirable side effects.

Some of the older relaxants have undesirable side effects at ED_{95} doses. The ideal neuromuscular blocking agent would interact specifically with the acetylcholine receptors of the neuromuscular junction and not affect other acetylcholine receptors. Most of the older nondepolarizers have some degree of action at other autonomic acetylcholine (nicotinic and muscarinic) receptors. The ''autonomic margin of safety'' describes the difference between doses needed for neuromuscular blockade and doses that cause undesirable effects (Bowman, 1982; Hackett *et al.*, 1989). For example, a therapeutic dose of gallamine has enough vagolytic activity to cause tachycardia. At therapeutic doses, curare causes hypotension due to sympathetic ganglion blockade. Curare also causes histamine release. These side effects, however, are negligible with the newer relaxants like vecuronium, which can be given in huge doses many times with therapeutic and no autonomic side effects noted.

Relaxants that depend to a great extent on organ function for metabolism and excretion have limitations. Prolonged duration of action is likely if hepatic or renal function is impaired (Agoston *et al.*, 1992). Similarly, large doses or administration over a prolonged period of time can result in cumulative effects with prolonged recovery of muscle strength. Doses of specific relaxants are noted in Table II.

a. STEROID ANALOGS. i. Pancuronium. Pancuronium has been used extensively for many years (Miller and Savarese, 1990) and is classified as a long-acting relaxant, with ED_{95} doses lasting 20 to 30 min. Pancuronium undergoes hepatic deacetylation to 3-OH, 17-OH, and 3,17-OH metabolites. The 3-OH metabolite has a similar duration of action to pancuronium and approximately one-half the neuromuscular blocking activity. Difficult reversal of pancuronium after prolonged administration may be due to the increased dose requirement of neostigmine for pancuronium metabolites (Booij *et al.*, 1979). Pancuronium depends in part on biliary excretion and substantially on renal excretion. Its action will be prolonged in hepatic or renal failure. Chronicity of renal failure may influence the degree of prolongation of the action of pancuronium (Booij *et al.*, 1982). Pancuronium is not protein bound to any extent and it causes increased heart rate due to vagal blockade at relaxant doses.

ii. Vecuronium. This is a more potent, shorter-acting monoquaternary homolog of pancuronium, having intermediate duration. The main metabolite, the 3-OH metabolite, of vecuronium has a neuromuscular blocking potency of 50 to 70% of vecuronium. Both unchanged vecuronium and its metabolite appear in the urine and bile. Renal or hepatic failure prolongs the action of vecuronium (Miller and Savarese, 1990). Vecuronium is highly protein bound and it has no autonomic side effects unless massive doses are given. Recovery of muscle strength from 25 to 75% of baseline takes approximately 5 min (Muir and Marshall, 1987; Bikhazi et al., 1988).

iii. Pipecuronium. A long-acting relaxant, approximately twice the duration of pancuronium, pipecuronium is 2 to 4 times as potent as pancuronium with a rapid onset of action. Pipecuronium undergoes hepatic metabolism and may be retained in the liver for many days. Approximately 50% is eliminated by the kidneys (Khuenl-Brady *et al.*, 1989). Pipecuronium is 30% protein bound. Doses of pipecuronium as great as 100 times the neuromuscular blocking dose do not influence cardiovascular function (Karpati and Biro, 1980).

iv. Rocuronium. Rocuronium (ORG 9426), a stable derivative of vecuronium, is approximately 20% of the potency of vecuronium. Its onset is rapid, similar to succinylcholine, with a duration of action of approximately 20 min (Miller and Savarese, 1990). An inverse relationship appears to exist between relaxant potency and speed of onset. This is apparently related to events at the motor end plate. It may be that a greater number of molecules available for receptor occupation hasten the onset of the less potent relaxants compared to more potent relaxants (Min *et al.*, 1992). Increasing the dose of rocuronium does not appreciably hasten onset, but it does prolong duration. Vagolytic side effects are noted with as low as two ED_{95}. Hepatic uptake

TABLE II

NONDEPOLARIZING RELAXANT ACTIVITY IN ANIMAL SPECIES

Agent and species	Dose (mg/kg)[a]	Onset (min) injection to maximum twitch reduction	Duration (min)[b] injection to 50% recovery	25 to 75%[c] recovery time (min)	Ref.[d]
Pancuronium					
Dog	(ED_{90}) 0.02	—	—	—	1
	3 × ED_{90}	—	108	—	1
Cat	(ED_{80-95}) 0.02	7	14	5	2
Rat	(ED_{80-95}) 0.1–0.19	2	4	1–4	3
Pig	(ED_{90}) 0.07	2	(injection to 90%) 15	(25–50%) 6	4
	3 × ED_{90}	1.3	60	15	4
Rabbit	0.1–0.5	—	—	—	5
	(infusion) 0.2 mg/kg/hr	—	—	—	6
Sheep	(fetal) 0.2–0.5	—	—	—	7
	(ED_{95} adult) 0.005	—	21	—	8
Calf	(ED_{90-99}) 0.04	—	26	—	9
Pony	(ED_{90-95}) 0.13	—	16	(injection to 100%) 24	10
Primate	(ED_{80-95}) 0.07	7	15	5	11
	(ED_{50} infusion): 0.3 μg/kg/min	—	—	—	12
Vecuronium					
Dog	(ED_{90}) 0.014	—	—	—	13
	3 × ED90	—	42	—	13
Cat	(ED_{80-95}) 0.025–0.04	5	9	2–5	14
Rat	(ED_{90-99}) 0.45	2	4	1	15
	(ED_{50} infusion) 0.1 mg/kg/min	—	—	—	16
Pig	(ED_{90-95}) 0.14–0.2	1.5	10	4	17
	3 × ED_{90}	1.3	23	6	18
Primate	(ED_{50}) halothane 0.01	8	16	7	20
	(ED_{50} infusion) 0.4 μg/kg/min	—	—	—	
Sheep	(ED_{80}) 0.005	—	(max to 50%) 14	—	21
Atracurium					
Dog	(ED_{95}) 0.125	5	(to 100%) 22	—	22
Cat	(ED_{99}) 0.25	3	(to 100%) 29	—	22
	(ED_{90-95} infusion) 3.7 μg/kg/min	—	—	—	23
Rat	(ED_{95}) 2.0	—	—	—	24
	(ED_{84-98} infusion) 0.01 mg/kg/min	—	—	—	25
	(ED_{50} infusion) 0.24 mg/kg/hr	—	—	—	26
Pig	(ED_{90-95}) 2.0	—	—	—	27
	(ED_{90-95} infusion) 0.12 mg/kg/min	—	—	—	27
Primate	(ED_{95-99}) 0.25–0.3	4	(injection to 25%) 20	—	28
	(ED_{50} infusion) 1.5 μg/kg/min	—	—	—	29

TABLE II

NONDEPOLARIZING RELAXANT ACTIVITY IN ANIMAL SPECIES (*continued*)

Agent and species	Dose (mg/kg)[a]	Onset (min) injection to maximum twitch reduction	Duration (min)[b] injection to 50% recovery	25 to 75%[c] recovery time (min)	Ref.[d]
Rabbit	(ED_{95}) 0.3	—	—	—	30
Pony	(ED_{98}) 0.11	9	(max to 10%) 24	(10 to 75%) 11	31
Chicken	(ED_{95}) 0.46	0.5	(injection to 75%) 31	10	32
Mivacurium					
Pig	(ED_{95}) 0.02–0.04	—	—	—	33
Primate	(ED_{95}) 0.04	1.7	9–12	3.4	34
	5 × ED_{95}	—	14–20	—	
Doxacurium					
Pig	(ED_{95}) 0.01	—	—	—	35
Rocuronium					
Cat	(ED_{90-95}) 0.16–0.3	1	(injection to 25%) 6.8	1.6	36
	6 × ED_{95}	0.8	26.4	6.7	36
Pig	(ED_{90}) 0.7	1.8	10	3.5	37
Pipecuronium					
Dog	(ED_{90}) 0.0056	—	26	—	38
Pig	(ED_{95}) 0.15	—	—	—	39
Cat/rabbit	(ED_{50}) 0.002–0.003	—	(of 50%) 24	—	38
	(ED_{85}) 0.004	rapid	28–32	—	38

[a]In parentheses, effective dose and percentage twitch reduction as noted in reference source.
[b]When relaxation time varied from injection to 50% recovery, the variation is noted in parentheses.
[c]When recovery index varied from 25 to 75%, the variation is noted in parentheses.
[d]*Key to references:* (1) Booij *et al.*, 1980a; (2) Bowman *et al.*, 1988; Durant *et al.*, 1980; Booij *et al.*, 1982; Bikhazi *et al.*, 1988; Durant *et al.*, 1980; (3) Bikhazi *et al.*, 1988; Durant *et al.*, 1980; (4) Muir and Marshall, 1987; (5) Baum *et al.*, 1989; Gouyon *et al.*, 1988; (6) Bhattacharya *et al.*, 1989; (7) Wilkening and Boyle, 1990; Wilkening *et al.*, 1989; Chestnut *et al.*, 1989; (8) Klein *et al.*, 1985; (9) Hildebrand and Howitt, 1983b; (10) Manley *et al.*, 1983; (11) Durant *et al.*, 1980; (12) Tsai and Lee, 1989; (13) Booij *et al.*, 1980a; (14) Durant *et al.*, 1980; van der Spek *et al.*, 1989; Baird *et al.*, 1982; (15) Bikhazi *et al.*, 1988; Durant *et al.*, 1980; (16) Walli *et al.*, 1989; (17) Motsch *et al.*, 1989; Muir and Marshall, 1987; Muir *et al.*, 1989b; Ducey *et al.*, 1989; (18) Muir *et al.*, 1989b; Durant *et al.*, 1980; (20) Tsai and Lee, 1989; (21) Klein *et al.*, 1985; (22) Hughes and Chapple, 1981; (23) Ilkiw *et al.*, 1990; (24) Pavlin *et al.*, 1988; (25) Law *et al.*, 1989; (26) Wali *et al.*, 1989; (27) Pittet *et al.*, 1990a; (28) Tsai *et al.*, 1987; (29) Tsai and Lee, 1989; (30) Hosking *et al.*, 1989; (31) Hildebrand *et al.*, 1986; (32) Nicholson and Ilkiw, 1992; (33) Sufit *et al.*, 1990; (34) Savarese *et al.*, 1984; (35) Muir *et al.*, 1989b; (36) Khuenl-Brady *et al.*, 1990, 1992; (37) Muir *et al.*, 1989b; (38) Karpati and Biro, 1980; (39) Pittet *et al.*, 1990a.

and biliary clearance are the dominant mechanisms for plasma clearance of rocuronium (Cason *et al.,*, 1990). Rocuronium is less dependent on renal clearance than vecuronium. Rocuronium metabolites have no neuromuscular blocking activity. Cumulative effects were not noted over the course of five ED_{95} doses in cats (Khuenl-Brady *et al.,* 1992). Recovery of muscle strength from 25 to 75% of baseline is rapid, approximately 5 min (Khuenl-Brady *et al.,* 1992; Muir *et al.,* 1989a,b). There may be species variations in the degree of fade or in the prejunctional effect of rocuronium. In rats, fade was found to be similar to other steroidal relaxants (Tian *et al.,* 1992), whereas rocuronium showed less fade than curare and pancuronium in cats (Shiraishi *et al.,* 1992).

b. BENZYLISOQUINILONIUMS. Griffith first reported the use of **curare,** or *d*-tubocurarine, in clinical anesthesia in 1942 (Griffith and Johnson, 1942). Curare is a long-acting muscle relaxant. It is highly protein bound and depends on hepatic metabolism and renal elimination. The benzylisoquinoliniums release histamine, and curare is the most potent histamine releasor in this group (Miller and Savarese, 1990). Curare also has a considerable amount of sympathetic ganlionic blockade.

As a result, curare is associated with vasodilation, hypotension, and tachycardia.

i. Metocurine. Methoxy substitution of curare produced **metocurine,** a long-acting relaxant with a much improved autonomic safety margin over curare. Metocurine is dependent on hepatic metabolism and renal elimination (Savarese *et al.,* 1977).

ii. Gallamine. A relatively impotent, long-acting relaxant, gallamine has pronounced vagolytic activity, with tachycardia as a consistent side effect. Gallamine is the only relaxant to depend entirely on renal elimination. Gallamine is also the only nondepolarizing relaxant that crosses the placenta.

iii. Atracurium. A recently released relaxant, atracurium has a unique method of degradation (Hughes and Chapple, 1981). The atracurium molecule is unstable and breaks down at a predictable rate by a process known as Hofmann elimination, a spontaneous chemical reaction. Atracurium must be stored at 4°C to prevent slow deterioration at room temperature. This breakdown is accelerated at body temperature and by changes in pH. Atracurium does not depend on cholinesterase activity, although some of its breakdown can be attributed to this mechanism. Atracurium does not utilize hepatic or renal function for metabolism and elimination. Metabolites of atracurium have no neuromuscular blocking activity, but one metabolite, laudanosine, has been the subject of much scrutiny. Laudanosine causes CNS excitation and seizures. Seizures have been noted at laudanosine plasma concentrations of 5 to 10μg/ml in rabbits, but could not be demonstrated at 100 μg/ml in cats (Lepage *et al.,* 1991). Laudanosine causes hypotension in dogs at 6 μg/ml. Laudanosine does not seem to be an issue for relaxation of several hours, but could potentially accumulate in animals receiving atracurium over several days. Because laudanosine is dependent on hepatic metabolism for clearance, hepatic insufficiency results in increased levels (Pittet *et al.,* 1990b). Potentially, laudanosine could increase the dose requirement of general anesthetics.

Atracurium is considered a mild histamine releaser, but has no autonomic side effects. At doses three times ED_{95}, histamine release can result in flushing, tachycardia, and hypotension. Histamine release is eliminated if that same dose is given slowly, over the period of 1 min (Scott *et al.,* 1985). Atracurium is considered an intermediate duration relaxant, with no cumulative effects from large or repeat doses.

iv. Mivacurium. Mivacurium is a new bischolinium ester that is broken down by plasma cholinesterase at a slightly slower rate than succinylcholine (Savarese *et al.,* 1988). The duration of action of mivacurium is slightly longer than succinylcholine and is approximately one-half that of vecuronium. Mivacurium is composed of three stereoisomers: trans–trans, cis–trans, and cis–cis. Ninety-five percent of mivacurium consists of the first two isomers (Belmont *et al.,* 1991; Maehr *et al.,* 1991). The cis–trans isomer has a longer elimination half-life than the other two isomers and could be responsible for prolonged activity with a repeat administration of mivacurium. The duration of action of mivacurium can also be prolonged if plasma cholinesterase activity is reduced, as is the case in hepatic disease, pregnancy, organophosphate exposure, or genetically abnormal cholinesterases. Mivacurium does cause a slight release of histamine, but does not have autonomic side effects.

v. Doxacurium. A long-acting relaxant that does not release histamine or cause autonomic side effects (Murray *et al.,* 1987), doxacurium is heavily dependent on renal elimination (Miller and Savarese, 1990).

c. Factors Influencing Neuromuscular Blockade by Nondepolarizers. i. Protein binding. The more highly protein bound a relaxant is, the less readily it will be excreted by glomerular filtration. Decreased plasma protein means that more of a highly protein-bound relaxant is available for activity at the neuromuscular junction.

ii. Hepatic and renal disease. Hepatic and renal disease will prolong the effects of relaxants that rely on those organs for metabolism and elimination.

iii. Hypothermia. This condition results in more profound, prolonged blockade due to increased sensitivity of the neuromuscular junction and to delayed urinary and biliary excretion (Miller *et al.,* 1978).

iv. Age. Neonatal animals will be sensitive to nondepolarizing relaxants because of immature neuromuscular junction and increased volume of distribution (Costarino and Polin, 1987). Relaxant activity may be prolonged in the elderly due to altered renal and hepatic function.

Fetal anesthesia and surgery are receiving more attention. Muscle relaxation may be required to perform fetal surgery. Because most of the neuromuscular blocking agents do not cross the placenta, relaxants must be administered to the fetus directly. A possible exception would be gallamine. Using muscle relaxants for fetal surgery does raise the issue of providing adequate anesthesia for the fetus. Muscle relaxants have also been administered to fetal animals to prevent trauma due to fetal movement during sampling from chronic catheters. Several investigators have determined that relaxants do alter fetal physiology (Wilkening and Boyle, 1990; Wilkening *et al.,* 1989; Rurak and Guber, 1983; Chestnut *et al.,* 1989). Changes include decreased fetal oxygen consumption, decreased uterine oxygen uptake, increased oxygen saturation and PO_2 of uterine venous, umbilical venous, and arterial blood flow, and increased uterine blood flow. Changes in fetal oxidative metabolism were also noted.

v. Drug Interactions. Inhalation anesthetics increase the potency of relaxants; enflurane, isoflurane, and desflurane more so than sevoflurane and halothane (Kobayashi *et al.,* 1990; Lee *et al.,* 1990). Ketamine potentiates nondepolarizing relaxants (Tsai and Lee, 1989), but barbiturates do not. It should be noted that the majority of animal studies describing the use of relaxants use intravenous anesthesia. Aminoglycoside antibiotics augment and prolong neuromuscular blockade (Viby-Mogensen, 1985; Forsyth *et al.,* 1990). Immune-suppressant drugs can also augment muscle relaxants, although this was not the case in a study done with cats receiving cyclosporin and atracurium relaxation (Ilkiw *et al.,* 1990). Calcium antagonists augment neuromuscular blockade (Law *et al.,* 1989; Ocana *et al.,* 1992; Hill *et al.,* 1988; Bikhazi *et al.,* 1988), whereas increased ionized calcium, as in hyperparathyroidism, for example, leads to resistance to block (Roland *et al.,* 1990; Gramstad *et al.,* 1990). Magnesium augments neuromuscular blockade by competing with calcium at the nerve ending to inhibit production and release of acetylcholine. Blocks with magnesium are difficult to reverse.

vi. Thermal Injuries. Injuries of this type cause resistance to atracurium block. In rats, the greatest effect was at 30 days after a 30% burn (Pavlin *et al.,* 1988). However, with smaller burns, muscle under the burned area was most resistant at 1 week postinjury (Pavlin *et al.,* 1991).

vii. "Priming." The giving a small dose of relaxant prior to the full dose, i.e., priming, can hasten the onset of block (Foldes, 1984).

viii. Muscle use. Muscle use or lack thereof influences the action of relaxants. Disuse atrophy causes resistance, whereas exercise produces sensitivity (Gronert *et al.,* 1984, 1989).

TABLE III

NEUROMUSCULAR BLOCKING AGENT ANTAGONISTS

Species	Doses of reversal agents			Ref.[b]
	Drug		Dose (mg/kg)[a]	
Rat	Neostigmine	ED_{50}	0.005–0.015	1
	Pyridostigmine	ED_{50}	0.018	
	4-Aminopyridine	ED_{50}	0.47	
	Methylguanidine	ED_{50}	2.04	
Cat	Neostigmine	Reversal 10–100%	0.04–0.07[c]	2
	Edrophonium	Reversal 10–100%	0.25–0.5[c]	
Dog	Neostigmine	Reversal 10–100%	0.02–0.2	3
Pony	Neostigmine	Reversal 10–100%	0.04	4
	Edrophonium	Reversal 10–100%	1.0	
Primate	4 AP	ED_{50}	0.73	5

[a]Effective dose and percentage twitch reduction as noted in reference source.
[b]*Key to References:* (1) Watanabe *et al.,* 1990; Booij *et al.,* 1980b; (2) Baird *et al.,* 1982; Booij *et al.,* 1979; (3) Jones, 1990; Hall and Clarke, 1983; Lumb and Jones, 1973; (4) Hildebrand and Howitt, 1984; (5) Biessels *et al.,* 1985.
[c]Equipotent doses: neostigmine, 0.06; edrophonium, 0.5.

C. Antagonists of Neuromuscular Blockade

The action of nondepolarizing neuromuscular blocking agents can be antagonized, or reversed (Cronnelly, 1985). Reversal is best accomplished once spontaneous recovery of muscle strength has begun and is unlikely to be effective if a large dose of relaxant has just been given. Some muscle relaxants have cumulative effects over time, whereas some have metabolites that have neuromuscular blocking activity. Reversal may be difficult with those agents. Other factors that complicate reversal include acid–base and electrolyte disturbances, hypothermia, hypotension, renal and hepatic dysfunction, and other drugs given. Doses of antagonists are noted in Table III.

Like the relaxants they antagonize, reversal agents have pre- and postjunctional effects (Cronnelly, 1985). The strength of muscle contraction is restored by providing acetylcholine to compete with the relaxants for receptor site occupation by inducing repetitive firing in the motor nerve terminal as a response to single action potential and directly depolarizing the motor nerve terminal.

1. Anticholinesterases

Anticholinesterases are the substituted esters of alkyl carbamic acids. These include edrophonium, neostigmine, and pyridostigmine. Anticholinesterase drugs increase the availability of acetylcholine by inhibiting acetylcholinesterase and by promoting the mobilization and release of acetylcholine from the motor nerve ending. Onset of action is most rapid for edrophonium, slightly slower for neostigmine, and longest for pyridostigmine. Equipotent doses of edrophonium and neostigmine have comparable duration, whereas pyridostigmine has the longest duration by virtue of a longer elimination half-life. All three agents depend heavily on renal excretion and have delayed clearance in renal failure (Miller and Savarese, 1990).

Side effects of the anticholinesterases are due to muscarinic effects of acetylcholine and include bradycardia and bradyarrhythmias, salivation, increased airway secretions, and increased gastrointestinal (GI) motility. Edrophonium has the least muscarinic side effects. Treatment of muscarinic side effects is with atropine or glycopyrrolate given prior to or at the same time as the reversal agent. The slower onset of glycopyrrolate is similar to neostigmine and pyridostigmine. The longer duration of glycopyrrolate matches that of pyridostigmine (Cronnelly and Morris, 1982).

Species variation in response to anticholinesterases has been noted. In equids, edrophonium without atropine caused increased blood pressure, little or no change in heart rate, and minimal to no intestinal signs and salivation (Hildebrand and Howitt, 1984).

2. 4-Aminopyridine and Guanidine

These drugs increase both evoked and spontaneous release of acetylcholine from the motor nerve terminal (Lundh and

Thesleff, 1977). Side effects include central nervous system stimulation with restlessness, confusion, and convulsions. This can occur at clinical doses. Increased heart rate and blood pressure can also be noted. These drugs are probably most useful in combination with acetylcholinesterases and may be useful to antagonize neuromuscular blockade caused by the aminoglycoside antibiotics (Nagashima *et al.,* 1988).

3. Calcium

Calcium enhances the release of acetylcholine from the nerve terminal and stabilizes the postjunctional membrane. Calcium is only partially effective as an antagonist of neuromuscular blockade.

III. MEASURING NEUROMUSCULAR FUNCTION

A. Muscle Response to Nerve Stimulus

The neuromuscular blocking activity of relaxants is evaluated through evoked responses of muscle to motor nerve stimulus (Ali and Savarese, 1976; Miller, 1978). Muscle response can be noted either mechanically by measuring evoked tension generated or electrically by evoked electromyogram (EMG). The two types of measurement have linear correlation, but may be disparate with respect to actual depression of the muscle response.

The preparation being studied can be either an isolated nerve and the muscle it innervates or an animal with the limb intact. For either method of evaluation, movement artifact is minimized by securing the limb. EMG measurements are made by recording motor unit action potentials via surface or needle electrodes on the muscle. For tension measurements in nonsurvival animals, the tendinous insertion of the muscle to be studied is dissected away from the bone and attached to a force displacement transducer or strain gauge. Noninvasive measurement of muscle strength can be achieved with the limb intact. The limb is secured so as to be free of motion and the distal limb is placed under resting tension, which varies with the species being studied. As for the nonsurvival model, tension is measured with a force displacement transducer or a strain gauge. Muscle response to a stimulus is referred to as ''twitch.'' For clinical studies, observation of the twitch instead of recording is common.

Stimulating electrodes are placed on either side of the motor nerve. Ideally, square wave electrical stimulus should be delivered to the nerve. Supramaximal voltage (20 to 25% in excess of that which ensures maximum contraction of the muscle) is determined. This ensures full activation of the nerve and muscle fibers. The electrical stimulus is usually in the order of 0.2 to 0.3 msec duration, with the frequency of stimulus varied to define different responses to muscle relaxants.

Figures 2, 3, and 4 illustrate the placement of stimulating electrodes and methods of evaluating muscle response. The equine is the demonstrator species in the examples given, but a similar approach has been used in calves, sheep, llamas, dogs, and cats. The approach would be the same for any species being studied. Typically, muscles of the face or the distal fore- or hind limb would be used. Stimulating electrodes are placed on either side of the motor nerve selected. A bit of creativity is needed for species in which noninvasive monitoring has not been described. One needs to verify the anatomical landmark for the motor nerve, stimulate it, and watch the appendage twitch. Muscle response to stimulation of the nerve would be evaluated by measuring the force of muscle contraction, by observing

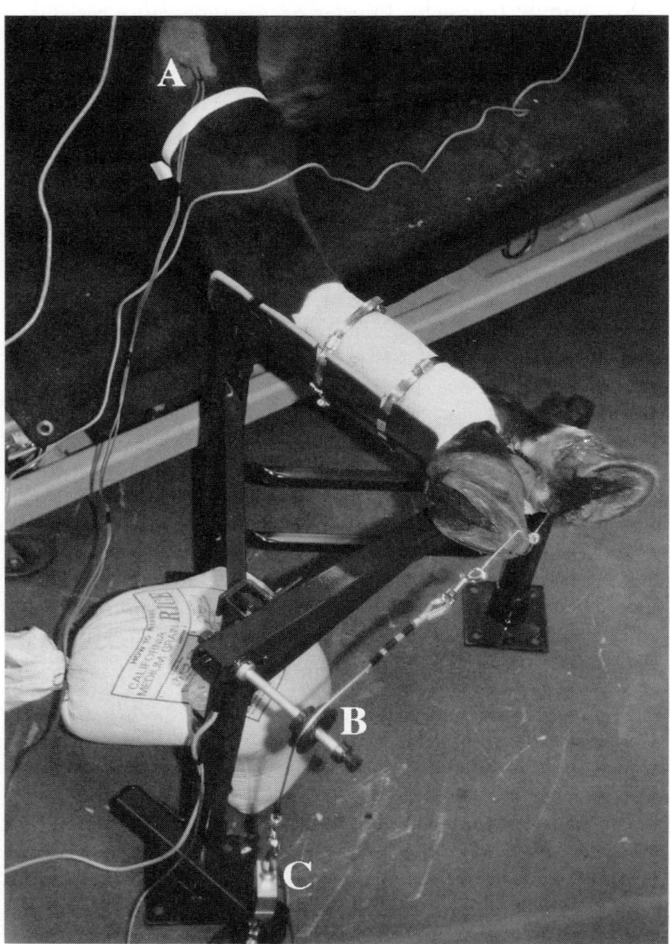

Fig. 2. Noninvasive quantitation of neuromuscular function. Stimulating electrodes are placed on either side of the superficial peroneal nerve (A) on the hind limb of a horse. The limb rests in a metal support frame and is held in place with a reusable cast. Tension is established on the hoof by attaching a wire from the insensitive hoof wall and extending it over a pulley (B) to attach to a strain gauge (C). Response of the extensor muscles of the hoof to peroneal nerve stimulus causes a ''twitch,'' the strength of which can be recorded. Variations of this type of measurement have been used for ponies, dogs, cats, llamas, and calves, and could likely be designed for any species. From Hildebrand, S. V. and Arpen, D. (1988). Neuromuscular and cardiovascular effects of atracurium administrated to healthy horses anesthetized with halothane. *Am. J. Vet. Res.* **49**: 1066–1071.

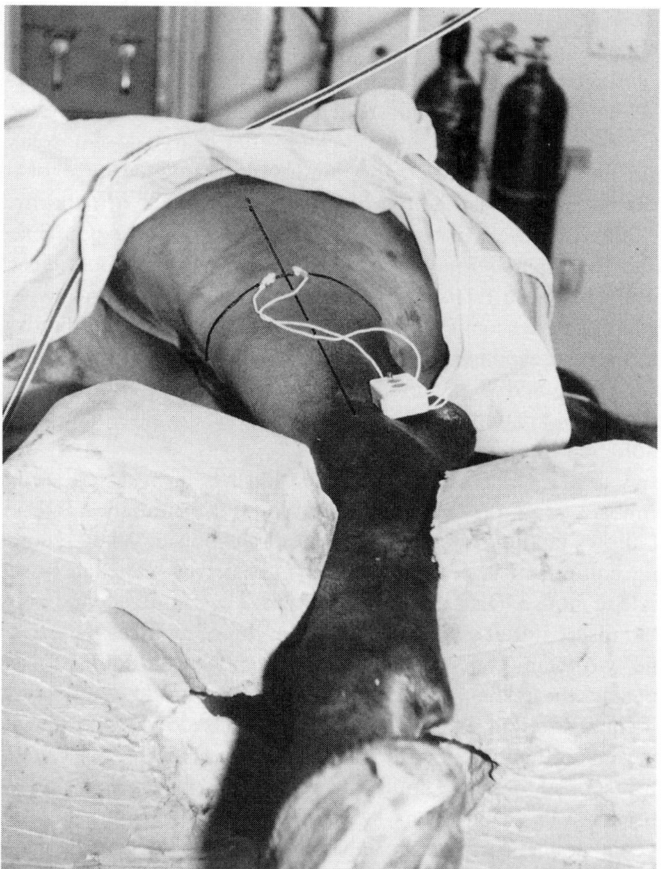

Fig. 3. Landmarks for placement of stimulating electrodes on either side of the superficial peroneal nerve of a horse's leg. The nerve is easily palpable as it crosses the lateral ridge of the tibia, just distal to the stife joint. The relaxant activity will be evaluated by observing the extensor response of the hoof. From Hildebrand, S. V. (1990). Muscle Relaxants. *In* ''Current Practice of Equine Surgery.'' (N. A. White and J. N. Moore, eds.). Lippincott-Raven, Philadelphia.

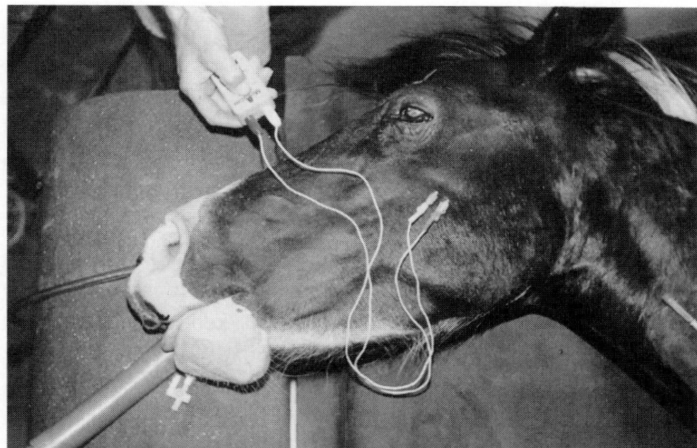

Fig. 4. An alternative site for stimulating electrode placement is shown on either side of the facial nerve. When the stimulus is given, muscle contraction causes the lips to curl upward. Because it is very difficult to quantitate the strength of muscle pull with this arrangement, the evaluation of relaxant activity is by observation. From Hildebrand, S. V., Howitt, G. A., and Arpin, D. (1986). Neuromuscular and cardiovascular effects of atracurium in ponies anesthetized with halothane. *Am. J. Vet. Res.* **47:** 1097.

movement of the appendage, or by palpating the appendage to detect movement.

A variety of equipment can deliver the appropriate stimulus to the motor nerve, from bulky stimulators that can deliver all modes of stimulus, including repetitive trains, to simple handheld stimulators that provide one or two functions. Figure 5 illustrates one very simple type of a handheld peripheral nerve stimulator.

B. Effect of Stimulus Frequency

The single twitch is the slowest frequency, ranging from 0.1 to 0.2 Hz, or a stimulus every 5 to 10 sec. The single twitch mode is the least demanding of neuromuscular function and thus the most resistant to the effects of relaxants. The single twitch indicates the postjunctional receptor attachment of relaxant molecules. Because of the tremendous margin of safety for neuromuscular transmission, a normal twitch height could mean that as many as 75% of the receptors at the motor end plate are occupied by relaxant molecules. The single twitch is unlikely to cause physiologic disruption of the neuromuscular junction.

A more rapid stimulus mode, the train of four (T4), was first described by Ali *et al.* (1970). The T4 consists of four separate stimuli given at a frequency of 2 Hz, or four stimuli in two sec. The train is repeated every 10 to 20 sec. The T4 is more demanding of neuromuscular function than the single twitch and provides more information. Because it is a more rapid stimulus, less relaxant is required for a comparable twitch reduction as for the single twitch mode (Ali, 1985). The T4 can also demonstrate fade, which is the inability of the muscle to sustain the strength of contraction. Fade is illustrated in Fig. 6. Fade is separate and distinct from reduction in twitch strength; the latter is due to postjunctional receptor occupation by the relaxant. Fade results from the prejunctional attachment of relaxant molecules (Bowman, 1980; Viby-Mogensen, 1990). The onset

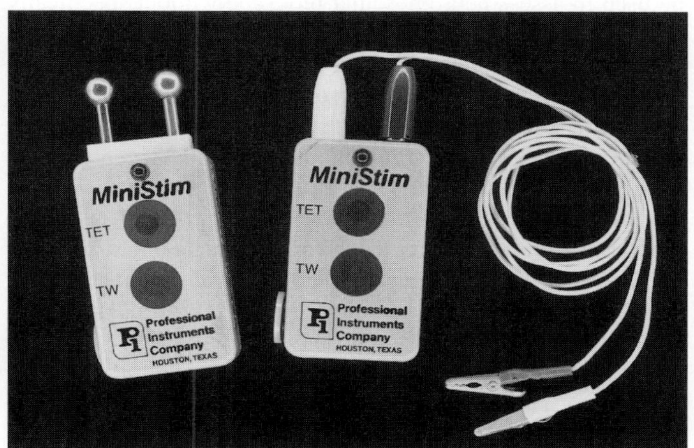

Fig. 5. A simple handheld peripheral nerve stimulator is illustrated. This stimulator can deliver train-of-four or 50 Hz tetanic stimulus. From Hildebrand, S. V. (1990). Neuromuscular blocking agents in equine anesthesia. *In* ''Veterinary Clinics of North America. Equine Practice.'' (T. W. Riebold, ed.). WB Saunders, Philadelphia.

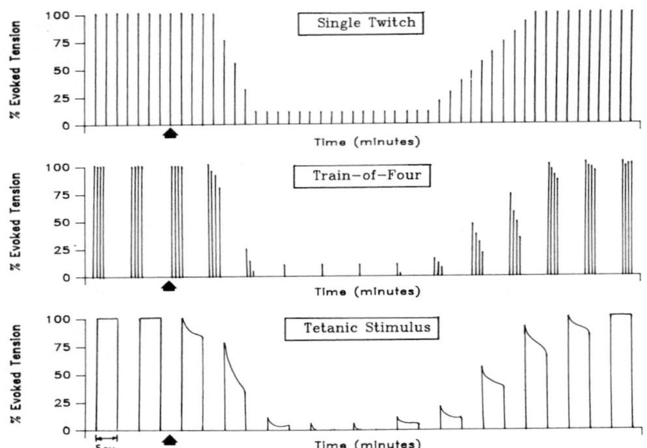

Fig. 6. Twitch response to different stimulus frequency. For each frequency, evoked strength of twitch is indicated on the left and twitch response is indicated over time. Relaxant is given at the arrow. Twitch response diminishes, then returns to prerelaxant strength. Single twitch can demonstrate only evoked tension response. Both train-of-four (T4) and tetanic frequencies demonstrate both evoked twitch tension and fade. Fade is seen as the progressive diminution of each twitch in the T4 and as rapid loss of twitch strength over the 5-sec tetanic stimulus period. This illustration is not to scale with respect to comparing onset and recovery of twitch and fade for the three frequencies. From Hildebrand, S. V., Muscle relaxants. In "Current Practice of Equine Surgery" (N. A. White and J. N. Moore, eds.), JB Lippincott, Philadelphia.

of fade is delayed compared to twitch reduction, and at recovery of muscle strenth, fade is more pronounced for any given level of relaxation compared to onset. Using the T4, fade is evaluated by comparing the magnitude of the last twitch in the train to the first. A T4 ratio of 70 to 75% correlates with the return of a single twitch to the prerelaxant baseline and the sustained mechanical response of a 50-Hz tetanic stimulus. Finally, T4 can be used to assess neuromuscular blockade without specifically quantitating the strength of the evoked response.

The most rapid stimulus mode used, tetanic stimulation, subjects the neuromuscular junction to the most demanding conditions. As a result, muscle is most sensitive to relaxants with this stimulus mode. Tetanic stimulation is generally given at a frequency of 50 Hz, held for 5 sec. Because this rate of stimulus is too rapid for the muscle to contract and relax between pulses, the muscle remains contracted as long as the tetanic stimulus is applied. The tetanic stimulus is painful and an inadequately anesthetized animal will respond to it. Tetanic stimulation will fade when nondepolarizing relaxants are used.

Tetanic stimulation can also provide useful information by virtue of the single twitch characteristics seen immediately after the tetanic stimulus (Viby-Mogensen, 1990). One such characteristic is called posttetanic potentiation (PTP). This refers to augmentation of the evoked single twitch tension by 100% after tetanic stimulus, as compared to before. As for fade, PTP is likely a prejunctional phenomenon, occurring as a result of increased mobilization and release of acetylcholine from the motor nerve terminal in response to the powerful tetanic stimulus. PTP can be used to monitor the posttetanic count. In the event of profound neuromuscular blockade after a large dose of relaxant, neither T4 nor tetanic stimulus will produce an evoked response. However, as the block begins to wear off, the first visible response will be PTP of single 1-Hz stimuli begun 3 sec after the tetanic stimulus. As intensity of the block continues to dissipate, more PTPs will be seen until T4 and tetanic twitch responses are seen. For any given relaxant, the time until return of the first response to T4 is related to the number of PTPs that can be counted. Posttetanic count can be useful when it is critical to avoid patient movement, as would be the case in ophthalmological surgery, for example.

A new method of nerve stimulation has been described by Engbaek *et al.* (1989). Called double-burst stimulation (DBS), it allows tactile evaluation of a muscle response. When recording equipment is not available, it is easier to feel than to see fade response. DBS consists of two 50-Hz stimuli separated by 750 msec. Initial studies indicate that three impulses in each of the two tetanic bursts give the best response. In nonparalyzed muscle, the response to DBS is two short muscle contractions of equal strength. In paralyzed muscle, the second response will be weaker than the first if fade is occurring. The absence of fade in response to DBS should mean that muscle strength has returned (Viby-Mogensen, 1990). Because DBS is a relatively recent concept in monitoring neuromuscular blockade, most older peripheral nerve stimulators cannot deliver this stimulus mode. DBS is available on at least one (Myotest, Biometer, Denmark) and may be available on others.

It should be noted that because fade and PTP are prejunctional phenomenon, they are features of nondepolarizing relaxants, but not of depolarizing blockade.

C. Clinical Features of Neuromuscular Blockade

Several features of relaxant activity are measured. First, onset of action. This may indicate the time taken from injection until the first reduction of twitch is noted or the time from injection to maximum reduction in twitch strength. This should be defined by those administering the relaxant.

The doses of relaxant that produce 50 and 95% reduction in twitch strength are often noted, although a range of relaxation may be reported by some investigators. A 95% reduction of evoked tension represents relaxation that is surgically useful. The ED_{95} is used as a reference for comparing potencies of relaxants. Multiples of ED_{95} are used as guidelines for hastening onset time and for evaluating nonneuromuscular side effects of relaxants. A 50% twitch reduction (ED_{50}) represents the steepest part of the dose–response curve (see Fig. 7). An infusion of relaxant to maintain 50% twitch reduction is a useful level of relaxation for evaluating the influence of other drugs on neuromuscular blockade.

Another feature of neuromuscular blockade noted is the duration of paralysis. Often this is noted as the time from injection

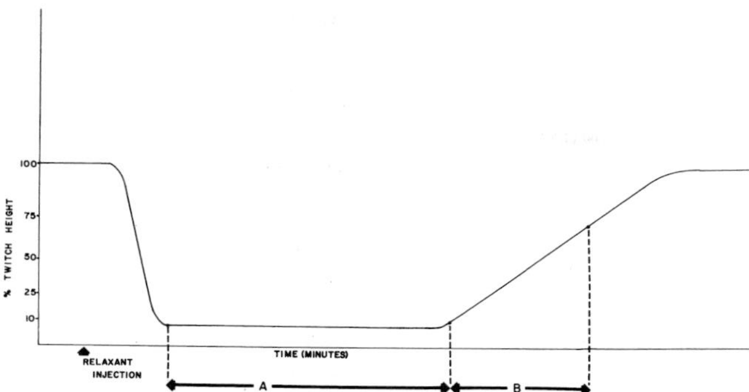

Fig. 7. Some important aspects of neuromuscular blockade used to evaluate and compare relaxants. The graph is generated by drawing a continuous line to represent evoked twitch tension. The percentage reduction of twitch is important, with 90% reduction (or greater) considered useful for surgical relaxation. Time intervals measured include onset of action from injection to maximum relaxation, time of relaxation (A), and recovery time for the muscle to regain its strength as relaxant activity diminishes (B). From Hildebrand, S. V., and Howitt, G. A. (1984). Dosage requirement of pancuronium in halothane anesthetized ponies: a comparison of cumulative and single dose administration. *Am. J. Vet. Res.* **45:** 2442.

to recovery of 10 or 20% of prerelaxant-evoked tension. This time period is important because it represents the time that muscle relaxation is likely to be profound enough to facilitate surgical procedures. However, some investigators will report duration of action to 100% recovery of muscle strength.

Finally, recovery of muscle strength as relaxant activity diminishes is an important aspect of neuromuscular blockade. Recovery is defined as the time taken for evoked tension to increase. The recovery slope is an important indicator of the cumulative action of a relaxant, or impact of interaction with other drugs. The steepest part of the slope of the recovery is generally used to define recovery of muscle strength. Thus recovery from 25 to 75% of baseline strength is often used, although investigators may vary the approach and use 10 to 50% or 10 to 75%, for example. In comparing the recovery aspects of relaxants, it is important to make sure the same time interval is noted.

One very important feature of recovery is the ability to ventilate. There is no one test to define adequacy of ventilation. Consider that an animal resting quietly and ventilating well is in a different situation than an animal that needs to overcome airway obstruction and be able to sustain effective coughing. Probably the best approach is to use as many tests as practically possible. The response to tetanic stimulus should be similar to prerelaxant strength, with no fade. Although T4 fade recovered to 75% is associated with the ability to breathe in humans, this has not been defined for animals. Observation of chest excursions, measurement of tidal volume, and evaluation of expiratory force are useful indicators. To determine expiratory force, the breathing bag can be removed from the anesthetic machine, the outlet occluded, and the expiratory force noted as the pressure generated on the circuit gauge. Arterial blood gas analysis is a valuable indicator of ventilation.

D. Response of Different Muscle Groups to Stimulus

Relaxant activity depends on the muscle studied (Muir *et al.*, 1989a). For example, because the tibialis muscle is more sensitive to the effects of relaxants than the soleus muscle, the tibialis muscle will require a lower dose of relaxant than will the soleus for the same degree of neuromuscular blockade. Onset of relaxation is slower and recovery is longer for the soleus muscle compared to the tibialis muscle. This is important for interpretation and comparison of data from relaxant studies. Practically, it is important in determining necessary dose of relaxant, and recovery of muscle strength. For example, the muscles of the face are more resistant to the effects of relaxants than are the limbs. By observation, facial twitch may be prerelaxant strength, yet hindlimb extensor muscles observed at the same time may not show any return of twitch. It follows that for the animal to be able to stand, additional time will be needed for recovery of the muscles of the limbs.

E. Species Differences

There are some species peculiarities of note with respect to response to relaxants and their antagonists. These are noted in Table IV.

TABLE IV

SPECIES VARIATION IN RESPONSE TO NEUROMUSCULAR BLOCKING DRUGS

Species	Comment	Ref.[a]
Dog	Severe histamine release with *d*-tubocurarine	
Cat	Good model (compared to humans) for onset of action and autonomic response to relaxants. Not representative of duration of action or hepatic uptake and distribution of relaxant in humans.	1
Pig	Good model (for humans) for time course data in pharmacokinetic studies of relaxants. Good model for hepatic uptake and distribution. Pipcuronium dose in pigs is twice that of humans; pancuronium dose is similar to humans.	2
Sheep	Adult sheep extremely sensitive to nondepolarizing neuromuscular blocking agents. Dose requirement as much as one-tenth that for other species.	3
Calves	Dose requirement of nondepolarizers similar to humans. Prolonged duration of action of succinylcholine.	4
Ponies	Dose requirement of pancuronium (and probably vecuronium) higher than most other species. Dose requirement of atracurium less than many other species.	5
Ferrets	Succinylcholine action similar to nondepolarizers.	6

[a]*Key to References:* (1) Motsch *et al.*, 1989; Muir *et al.*, 1989b; (2) Pittet *et al.*, 1990b; Motsch *et al.*, 1989; Muir *et al.*, 1989b; (3) Klein *et al.*, 1985; (4) Hildebrand and Howitt, 1983b; Hansson, 1958; (5) Hildebrand *et al.*, 1986; Manley *et al.*, 1983; Klein *et al.*, 1982; (6) Tsai *et al.*, 1990.

REFERENCES

Agoston, S., Vandenbrom, R. H. G., and Weirda, J. M. K. H. (1992). Clinical pharmacokinetics of neuromuscular blocking drugs. *Clin. Pharmacokinet.* **22,** 94–115.

Ali, H. H. (1985). Monitoring neuromuscular function. *In* ''Muscle Relaxants: Basic and Clinical Aspects'' (R. L. Katz, ed.), pp. 53–68. Grune & Stratton, Orlando, Florida.

Ali, H. H., and Savarese, J. J. (1976). Monitoring of neuromuscular function. *Anesthesiology* **45,** 216–249.

Ali, H. H., Utting, J. E., and Gray, T. C. (1970). Stimulus frequency in the detection of neuromuscular block in humans. *Br. J. Anaesth.* **42,** 967–978.

Baird, W. L. M., Bowman, W. C., and Ker, W. J. (1982). Some actions of edrophonium and neostigmine in the anaesthetized cat and in man. *Br. J. Anaesth.* **54,** 375–385.

Baum, M., Putz, G., Mutz, N., Putensen, C., Klima, G., and Benzer, H. (1989). Influence of high frequency ventilation at different end-expiratory lung volumes on the development of lung damage during lung lavage in rabbits. *Br. J. Anaesth.* **63,** 65S–70S.

Belmont, M. R., Wray, D. L., Maehr, R. B., Savarese, J. J., and Wastila, W. B. (1991). Comparative pharmacology of mivacurium isomers in cats. *Anesthesiology* **75,** A773.

Benson, G. J., Manning, J. P., Hartsfield, S. M., and Thurmon, J. C. (1981). Intraocular tension of the horse: Effects of succinylcholine and halothane anesthesia. *Am. J. Vet. Res.* **42,** 1831–1832.

Bhattacharya, S., Glucksberg, M. R., and Bhattacharya, J. (1989). Measurement of lung microvascular pressure in the intact anesthetized rabbit by the micropuncture technique. *Circ. Res.* **64,** 167–172.

Biessels, P. T. M., Houwertjes, M. C., Agoston, S., and Horn, A. S. (1985). A comparison of the pharmacological actions of 4-aminopyridine and two of its derivatives in the monkey. *Anesthesiology* **63,** A320.

Bikhazi, G. B., Leung, I., Flores, C., Mikati, H. M., and Foldes, F. F. (1988). Potentiation of neuromuscular blocking agents by calcium channel blockers in rats. *Anesth. Analg.* **67,** 1–8.

Booij, L. H. D. J., Miller, R. D., Jones, M. J., and Stanski, D. R. (1979). Antagonism of pancuronium and its metabolites by neostigmine in cats. *Anesth. Analg.* **58,** 483–486.

Booij, L. H. D. J., Edwards, R. P., Sohn, Y. J., and Miller, R. D. (1980a). Cardiovascular and neuromuscular effects of ORG NC45, pancuronium, metocurine, and d-tubocurarine in dogs. *Anesth. Analg.* **59,** 26–30.

Booij, L. H. D. J., van der Pol, F., Crul, J. F., and Miller, R. D. (1980b). Antagonism of ORG NC45 neuromuscular blockade by neostigmine, pyridostigmine, and 4-aminopyridine. *Anesth. Analg.* **59,** 31–34.

Booij, L. H. D. J., Vree, T. B., van der Pol, F., and Crul, J. F. (1982). Pharmacodynamics of vecuronium and pancuronium in cats with and without ligated renal pedicles.

Booth, N. H. (1977). Intravenous and other parenteral anesthetics. *In* ''Veterinary Pharmacology and Therapeutics'' (L. E. McDonald, ed.), pp. 283–285. Iowa State University Press, Ames.

Bowman, W. C. (1982). Non-relaxant properties of neuromuscular blocking drugs. *Br. J. Anaesth.* **54,** 147–160.

Bowman, W. C. (1980). Prejunctional and postjunctional cholinoceptors at the neuromuscular junction. *Anesth. Analg.* **59,** 935–943.

Bowman, W. C., Rodger, I. W., Houston, J., Marshall, R. J., and McIndewar, I. (1988). Structure:action relationships among some desacetoxy analogues of pancuronium and vecuronium in the anesthetized cat. *Anesthesiology* **69,** 57–62.

Cason, B., Baker, D. G., Hickey, R. F., Miller, R. D., and Agoston, S. (1990). Cardiovascular and neuromuscular effects of three steroidal neuromuscular blocking drugs in dogs (ORG9616, ORG 9426, ORG 9991). *Anesth. Analg.* **70,** 382–388.

Chestnut, D. H., Weiner, C. P., Thompson, C. S., and McLaughlin, G. L. (1989). Intravenous administration of d-tubocurarine and pancuronium in fetal limbs. *Am. J. Obstet. Gynecol.* **160,** 510–513.

Costarino, A. T., and Polin, R. A. (1987). Neuromuscular relaxants in the neonate. *Perinatal Pharmacol.* **14,** 965–989.

Cronnelly, R. (1985). Muscle relaxant antagonists. *In* ''Muscle Relaxants: Basic and Clinical Aspects'' (R. L. Katz, ed.), pp. 197–213. Grune & Stratton, Orlando, Florida.

Cronnelly, R., and Morris, R. B. (1982). Antagonism of neuromuscular blockade. *Br. J. Anaesth.* **54,** 183–194.

Davis, L. E., and Wolff, W. A. (1970). Pharmacokinetics and metabolism of glyceryl guaiacolate in ponies. *Am. J. Vet. Res.* **31,** 469–473.

Ducey, J. P., Deppe, S. A., and Foley, K. T. (1989). A comparison of the effects of suxemethonium, atracurium, and vecuronium on intracranial hemodynamics in swine. *Anaesth. Intensive Care* **17,** 448–455.

Durant, N. N., Houwertjes, M. C., and Crul, J. F. (1980). Comparison of the neuromuscular blocking properties of ORG NC45 and pancuronium in the rat, cat, and rhesus monkey. *Br. J. Anaesth.* **52,** 723–730.

Engbaek, J., Ostergaard, D., and Viby-Mogensen, J. (1989). Double burst stimulation (DBS): a new pattern of nerve stimulation to identify residual neuromuscular blockade. *Br. J. Anaesth.* **62,** 274–278.

Foldes, F. (1984). Rapid endotracheal intubation with non-depolarizing neuromuscular blocking drugs: The priming principle. *Br. J. Anaesth.* **56,** 663.

Forsyth, S. F., Ilkiw, J. E., and Hildebrand, S. V. (1990). Effect of gentamicin administration on the neuromuscular blockade induced by atracurium in cats. *Am. J. Vet. Res.* **51,** 1675–1678.

Funk, K. A. (1970). Glyceryl guaiacolate: A centrally acting muscle relaxant. *Equine Vet. J.* **2,** 173–178.

Funk, K. A. (1973). Glyceryl guaiacolate: Some effects and indications in horses. *Equine Vet. J.* **5,** 15–19.

Gibbs, N. M., Larach, D. R., and Schuler, H. G. (1989). The effects of neuromuscular blockade with vecuronium on hemodynamic responses to noxious stimuli in the rat. *Anesthesiology* **71,** 214–217.

Goudsouzian, N. G., and Standaert, F. G. (1986). The infant and the myoneural junction. *Anesth. Analg.* **65,** 1208–1217.

Gouyon, J. B., Torrado, A., and Guignard, J. P. (1988). Renal effects of d-tubocurarine and pancuronium in the newborn rabbit. *Biol. Neonate* **54,** 218–223.

Gramstad, L., and Hysing, E. S. (1990). Effect of ionized calcium on neuromuscular blocking actions of atracurium and vecuronium in the cat. *Br. J. Anaesth.* **64,** 199–206.

Grandy, J. L., and McDonell, W. N. (1980). Observations on concentrated solutions of guaifenesin for equine anesthesia. *J. Am. Vet. Med. Assoc.* **176,** 619–622.

Griffith, H. R., and Johnson, G. E. (1942). The use of curare in general anesthesia. *Anesthesiology* **3,** 418.

Gronert, G. A., Mateo, R. S., and Perkin, S. (1984). Canine gastrocnemius disuse atrophy:resistance to paralysis by dimethyltubocurarine. *J. Appl. Physiol.* **57,** 1502–1506.

Gronert, G. A., White, D. A., Schafer, S. L., and Matteo, R. S. (1989). Exercise produces sensitivity to metocurine. *Anesthesiology* **70,** 973–977.

Hackett, G. H., Jantzen, J. P., and Earnshaw, G. (1989). Cardiovascular effects of vecuronium, pancuronium, metocurine, and RGH-4201 in dogs. *Acta Anaesthesiol. Scand.* **33,** 298–303.

Hall, L. W., and Clarke, K. W. (1983). ''Veterinary Anaesthesia,'' pp. 115–140. Ballière Tindall, London.

Hansson, C. H. (1958). Studies on the effect of succinylcholine in domestic animals. *Nord. Veterinaer Med.* **10,** 201–216.

Hildebrand, S. V., and Howitt, G. A. (1983a). Succinylcholine infusion associated with hyperthermia in ponies anesthetized with halothane in oxygen. *Am. J. Vet. Res.* **44,** 2280–2284.

Hildebrand, S. V., and Howitt, G. A. (1983b). Neuromuscular and cardiovascular effects of pancuronium bromide in calves anesthetized with halothane. *Am. J. Vet. Res.* **45,** 1549–1552.

Hildebrand, S. V., and Howitt, G. A. (1984). Antagonism of pancuronium neuromuscular blockade in halothane anesthetized ponies using neostigmine and edrophonium. *Am. J. Vet. Res.* **45,** 2276–2280.

Hildebrand, S. V., Howitt, G. A., and Arpin, D. (1986). Neuromuscular and cardiovascular effects of atracurium in ponies anesthetized with halothane. *Am. J. Vet. Res.* **47,** 1096–1100.

Hill, D. C., Chelly, J. E., Dlewati, A., Abernathy, D. R., Dursout, M. F., and Merin, R. G. (1988). Cardiovascular effects of and interaction between calcium blocking drugs and anaesthetics in chronically instrumented dogs. IV. Verapamil and fentanyl-pancuronium. *Anesthesiology* **68,** 874–879.

Hosking, M. P., Lennon, R. L., and Gronert, G. A. (1989). Combined H1 and H2 receptor blockade attenuates the cardiovascular effects of high dose atracurium in rabbits. *Life Sci.* **44,** 347–353.

Hubbell, J. A. E., Muir, W. W., and Sams, R. A. (1980). Guaifenesin: Cardiopulmonary effects and plasma concentration in horses. *Am. J. Vet. Res.* **41,** 1751–1755.

Hughes, R. (1970). Hemodynamic effects of tubocurarine, gallamine, and suxamethonium in dogs. *Br. J. Anaesth.* **42,** 928–934.

Hughes, R., and Chapple, D. J. (1981). The pharmacology of atracurium: A new competitive neuromuscular blocking agent. *Br. J. Anaesth.* **53,** 31–44.

Ilkiw, J. E., Forsyth, S. F., Hill, T., and Gregory, C. R. (1990). Atracurium administration, as an infusion, to induce neuromuscular blockade in clinically normal and temporarily immune-suppressed cats. *J. Am. Vet. Med. Assoc.* **197,** 1153–1156.

Jackson, L. L., and Lundvall, R. L. (1970). Observations on the use of glyceryl guaiacolate in the horse. *J. Am. Vet. Med. Assoc.* **157,** 1093–1095.

Jackson, L. L., and Lundvall, R. L. (1972). Effect of glyceryl guaiacolate-thiamylal sodium solution on respiratory function and various hematological factors of the horse. *J. Am. Vet. Med. Assoc.* **161,** 164–168.

Jones, R. S. (1990). Reversal of atracurium with neostigmine in the dog. *Res. Vet. Sci.* **48,** 96–98.

Karpati, E., and Biro, K. (1980). Pharmacological study of a new competitive neuromuscular blocking steroid, pipecuronium bromide. *Arzneim. Forsch.* **30,** 346–354.

Kelly, R. E., Dinner, M., Turner, L. S., Abramson, D. H., Haik, B., and Daines, P. (1991). Intraocular pressure (IOP) effect of succinylcholine in the human eye with extraocular muscles (EOM) detached. *Anesthesiology* **75,** A804.

Khuenl-Brady, K. S., Sharma, M., Chung, K., Miller, R. D., Agoston, S., and Caldwell, J. E. (1989). Pharmacokinetics and disposition of pipecuronium bromide in dogs with and without ligated renal pedicles. *Anesthesiology* **71,** 919–922.

Khuenl-Brady, K. S., Castagnoli, K. P., Canfell, P. C., Caldwell, J. E., Agoston, S., and Miller, R. D. (1990). The neuromuscular blocking effects and pharmacokinetics of ORG 9426 and ORG 9616 in the cat. *Anesthesiology* **72,** 669–674.

Khuenl-Brady, K. S., Agoston, S., and Miller, R. D. (1992). Interaction of ORG 9426 and some of the clinically used intravenous anaesthetic agents in the cat. *Acta Anaesthesiol. Scand.* **36,** 260–263.

Klein, L., Beck, E., Hopkins, J., and Burton, B. (1982). Characteristics of neuromuscular blockade produced by gallamine, pancuronium, and vecuronium (ORG NC-45), and antagonism with neostigmine in anaesthetized horses. *Assoc. Vet. Anaesth.* **10,** 173–188.

Klein, L., Sylvina, T., and Beck, E. (1985). Neuromuscular blockade with d-tubocurarine pancuronium and vecuronium in halothane anesthetized sheep. *Proc. Int. Congr. Vet. Anesthesiol., 2nd,* p. 174.

Kobayashi, O., Ohta, Y., and Kosaka, F. (1990). Interaction of sevoflurane, enflurane, and halothane with non-depolarizing muscle relaxants and their prejunctional effects at the neuromuscular junction. *Acta Med. Okayama* **44,** 209–215.

Law, S. C., Ramzan, I. M., Brandom, B. W., and Cook, D. R. (1989). Intravenous ranitidine antagonizes intense atracurium induced neuromuscular blockade in rats. *Anesth. Analg.* **69,** 611–613.

Lee, C., Kwan, W. F., Chen, B., Tsai, S. K., Gyermek, L., Cheng, M., and Cantley, E. (1990). Desflurane (I-653) potentiates atracurium in humans. *Anesthesiology* **73,** A875.

Lepage, J. Y., Athouel, A., Vecherini, M. F., Malinovsky, J. M., and Cozian, A. (1991). Evaluation of proconvulsant effect of laudanosine in renal transplant recipient. *Anesthesiology* **75,** A781.

Lumb, W. V., and Jones, E. W., eds. (1973). "Veterinary Anesthesia," pp. 343–369. Lea & Febiger, Philadelphia.

Lundh, H., and Thesleff, S. (1977). The mode of action of 4-aminopyridine and guanidine on transmitter release from motor nerve terminals. *Eur. J. Pharmacol.* **42,** 411–412.

Maehr, R. B., Belmont, M. R., Wray, D. L., Savarese, J. J., and Wastila, W. B. (1991). Autonomic and neuromuscular effects of mivacurium and isomers in cats. *Anesthesiology* **75,** A772.

Manley, S. V., Steffey, E. P., Howitt, G. A., and Woliner, M. (1983). Cardiovascular and neuromuscular effects of pancuronium in ponies anesthetized with halothane. *Am. J. Vet. Res.* **44,** 1349–1353.

Matran, R., Alving, K., and Lundberg, J. M. (1991). Differential bronchial and pulmonary vascular responses to vagal stimulation in the pig. *Acta Physiol. Scand.* **143,** 387–393.

McClure, J. R., and McClure, J. J. (1980). Letter to the editor. *Vet. Surg.* **9,** 81–82.

Miller, R. D. (1978). Monitoring neuromuscular blockade. *In* "Monitoring in Anesthesia" (L. J. Saidman and N. T. Smith, eds.), pp. 127–144. Wiley, New York.

Miller, R. D., and Savarese, J. J. (1990). Pharmacology of muscle relaxants and their antagonists. *In* "Anesthesia" (R. D. Miller, ed.), Vol. 2, pp. 389–435. Churchill-Livingstone, New York.

Miller, R. D., Agoston, S., van der Pol, F., Booij, L. D. H. J., Crul, J. F., and Ham, J. (1978). Hypothermia and the pharmacokinetics and pharmacodynamics of pancuronium in the cat. *J. Pharmacol. Exp. Ther.* **207,** 532–538.

Min, J. C., Bekavac, M. I., Donati, F., and Bevan, D. R. (1992). Iontophoretic study of speed of action of various muscle relaxants. *Anesthesiology* **77,** 351–356.

Mirakhur, R. K. (1992). Newer neuromuscular blocking drugs. An overview of their clinical pharmacology and therapeutic use. *Drugs* **44,** 182–199.

Mostert, J. W., and Metz, J. (1963). Observations on hemolytic activity of guaiacol glyceryl ether. *Br. J. Anaesth.* **35,** 461–464.

Motsch, J., Hennis, P. J., Zimmermann, F. A., and Agoston, S. (1989). A model for determining the influence of hepatic uptake of nondepolarizing muscle relaxants in the pig. *Anesthesiology* **70,** 128–133.

Muir, A. W., and Marshall, R. J. (1987). Comparison of the neuromuscular blocking effects of vercuronium, pancuronium, ORG 6368, and suxamethonium in the anesthetized domestic pig. *Br. J. Anaesth.* **59,** 622–629.

Muir, A. W., Houston, J., Marshall, R. J., Bowman, W. C., and Marshall, I. G. (1989a). A comparison of the neuromuscular blocking and autonomic effects of two new short-acting muscle relaxants with those of succinylcholine in the anesthetized cat and pig. *Anesthesiology* **70,** 533–540.

Muir, A. W., Houston, J., Green, K. L., Marshall, R. J., Bowman, W. C., and Marshall, I. G. (1989b). Effects of a new neuromuscular blocking agent (ORG 9426) in anesthetized cats and pigs and in isolated nerve-muscle preparations. *Br. J. Anaesth.* **63,** 400–410.

Murray, D. J., Mehta, M. P., Forbes, F., Choi, W. W., Sokoll, M. D., Gergis, S. D., Krol, T., and Abou-Donia, M. (1987). Cardiovascular and neuromuscular effects of BWA938U: Comparison with pancuronium. *Anesthesiology* **67,** A367.

Nagashima, H., Salama, M. G., Goldiner, P. L., Chaudry, I. A., Duncalf, D., and Foldes, F. F. (1988). Reversal of the d-turbocurarine block by combinations of methylguanidine and cholinesterase inhibitors in vitro. *Anesthesiology* **69,** A511.

Nicholson, A., and Ilkiw, J. E. (1992). Neuromuscular and cardiovascular effects of atracurium in isoflurane-anesthetized chickens. *Am. J. Vet. Res.* **53,** 2337–2342.

Ocana, M., Del Pozo, E., Carlos, R., and Baeyens, J. M. (1992). Differential potential by calcium antagonists of neuromuscular blockade induced by pancuronium and succinylcholine in cats in vivo. *J. Neural Transm.* **88,** 223–224.

Pavlin, E. G., Haschke, R. H., Marathe, P., Slattery, J. T., Howard, M. L., and Butler, S. H. (1988). Resistance to atracurium in thermally injured rats. The roles of time, activity, and pharmacodynamics. *Anesthesiology* **69,** 696–701.

Pavlin, E. G., Howard, M. L., and Slattery, J. T. (1991). Resistance to atracurium in muscles under thermally injured skin in the rat. *Anesthesiology* **75,** A779.

Pittet, J. F., Tassonyi, E., Schopfer, C., Morel, D. R., Mentha, G., Fathi, M., LeCoultre, C., Steinig, D. A., and Benakis, A. (1990a). Plasma concentrations of laudanosine, but not of atracurium, are increased during an hepatic phase of orthotopic liver transplantation. *Anesthesiology* **72,** 145–152.

Pittet, J. F., Tassonyi, E., Schopfer, C., Morel, D. R., Leemann, P., Mentha, G., LeCoultre, C., Steinig, D. A., and Benakis, A. (1990b). Dose requirements and plasma concentrations of pipecuronium during bilateral renal exclusion and orthotopic liver transplantation in pigs. *Br. J. Anaesth.* **65,** 779–785.

Roland, E., Roupie, E., Wierda, J. M. K. H., Sarfati, E., Villiers, S., Dubost, C., and Eurin, B. (1990). Effects of hyperparathyroidism on the vecuronium induced neuromuscular blockade. *Anesthesiology* **73,** A901.

Rurak, D. W., and Guber, N. C. (1983). The effect of neuromuscular blockade on oyxgen consumption and blood gases in the fetal lamb. *Am. J. Obstet. Gynecol.* **145,** 258–262.

Savarese, J. J., Ali, H. A., and Antonio, R. P. (1977). The clinical pharmacology of metocurine: Dimethyltubocurarine revisited. *Anesthesiology* **47,** 277–284.

Savarese, J. J., Wastila, W. B., El-Sayad, H. A., Scott, R. P. F., Gargarian, M., Beemer, G., Basta, S. J., and Sunder, N. (1984). Comparative pharmacology of BW B1090U in the rhesus monkey. *Anesthesiology* **61,** A306.

Savarese, J. J., Ali, H. H., Basta, S. J., Embree, P. B., Scott, R. P. F., Sunder, N., Weakly, J. N., Wastila, W. B., and El-Sayad, H. A. (1988). The clinical pharmacology of mivacurium chloride (BW B1090U). *Anesthesiology* **68,** 723–732.

Schatzman, U. (1974). The induction of general anesthesia in the horse with glyceryl guaiacolate. *Equine Vet. J.* **6,** 164–169.

Scott, R. P. F., Savarese, J. J., Basta, S. J., Sunder, H. H., Ali, H. A., Gargarian, M., Gionfriddo, M., and Baston, A. G. (1985). Atracurium: Clinical strategies for preventing histamine release and attenuating the hemodynamic response. *Br. J. Anaesth.* **57,** 550–553.

Shiraishi, H., Suzuki, H., Suzuki, T., Katsumata, N., and Ogawa, S. (1992). Fading responses in the evoked EMG after rocuronium in cats. *Can. J. Anaesth.* **39,** 1099–1104.

Standaert. F. G. (1990). Neuromuscular physiology. *In* "Anesthesia" (R. D. Miller, ed.), Vol. 1, pp. 659–684. Churchill-Livingstone, New York.

Stowe, C. M., Bieter, R. N., and Roepke, M. H. (1958). The relationship between cholinesterase activity and the effects of succinylcholine chloride in the horse and cow. *Cornell Vet.* **48,** 241–259.

Sufit, R. L., Kreul, J. F., Bellay, Y. M., Helmer, P., Brunson, D. B., and Will, J. (1990). Doxacrium and mivacurium do not trigger malignant hyperthermia in susceptible swine. *Anesth. Analg.* **71,** 285–287.

Tavernor, W. D. (1970). The influence of guaiacol glyceryl ether on cardiovascular and respiratory function in the horse. *Res. Vet. Sci.* **11,** 91–93.

Tavernor, W. D., and Jones, E. W. (1970). Observations on cardiovascular and respiratory effects of guaiacol glyceryl ether in conscious and anesthetized dogs. *J. Small Anim. Pract.* **11,** 177–184.

Tian, L., Mehta, M. P., Prior, C., and Marshall, I. G. (1992). Relative pre- and post-junctional effects of a new vecuronium analogue, ORG 9426, at the rat neuromuscular junction. *Br. J. Anaesth.* **69,** 284–287.

Tsai, S. K., and Lee, C. (1989). Ketamine potentiates neuromuscular relaxants in a primate. *Anesth. Analg.* **68,** 1989.

Tsai, S. K., Lee, C., and Mok, M. S. (1987). Atracurium induced neuromuscular block is prolonged in hepatic vascular occlusion. *Anesthesiology* **67,** A604.

Tsai, S. K., Gyermek, L., Lee, C., and Nguyen, N. (1990). Neuromuscular pharmacology of succinylcholine in the ferret. *Anesthesiology* **73,** A873.

van der Spek, A. F., Reynolds, P. I., Ashton-Miller, J. A., Stohler, C. S., and Schork, M. A. (1989). Different effect of agonist and antagonist muscle relaxants on cat jaw muscles. *Anesth. Analg.* **69,** 76–80.

Viby-Mogensen, J. (1985). Interaction of other drugs with muscle relaxants. *In* "Muscle Relaxants: Basic and Clinical Aspects" (R. L. Katz, ed.), pp. 233–256. Grune & Stratton, Orlando, Florida.

Viby-Mogensen, J. (1990). Neuromuscular monitoring. *In* "Anesthesia" (R. D. Miller, ed.), Vol. 2, pp. 1209–1226. Churchill-Livingstone, New York.

Wali, F. A., Suer, A. H., Dark, C. H., and Tugwell, A. C. (1989). Assessment of neuromuscular blockade produced by atracurium in the rat diaphragm preparation. Measurements of tetanic fade, depression and recovery profile. *Pharmacol. Res.* **21,** 231–238.

Watanabe, K., Nagashima, H., Ohta, Y., and Foldes, F. F. (1990). Reversal of d-tubocurarine block by combination of methylguanidine and neostigmine in vitro. *Anesthesiology* **73,** A889.

Westhues, M., and Fritsch, R. (1965). "Animal Anesthesia," Vol. 2, pp. 181–187. Oliver & Boyd, London.

Wilkening, R. B., and Boyle, D. W. (1990). Effect of neuromuscular blockade on fetal oxidative metabolism. *Biol. Neonate* **58,** 57–60.

Wilkening, R. B., Boyle, D. W., and Meschia, G. (1989). Fetal neuromuscular blockade: Effect on oxygen demand and placental transport. *Am. J. Physiol.* **257,** H374–H378.

Yokoi, I., Akiyama, K., Kabuto, H., Fukuyama, K., Nishijima, Y., and Mori, A. (1992). Peripheral acting muscle relaxants alter the effects of baclofen on electrocorticograms in the rat. *J. Neural Transmission.* General Section. **87,** 37–47.

Chapter 5

Monitoring of Anesthesia

Diane E. Mason and Marilyn J. Brown

I. INTRODUCTION TO ANESTHETIC MONITORING

A. Importance of Monitoring

1. Anesthetics Alter Physiologic Homeostasis

The key body systems responsible for the short-term well-being of an animal during anesthesia and surgery consist of the central nervous system (CNS), the cardiovascular system, and the respiratory system. All three of these systems are markedly affected by anesthesia when the normal homeostatic mechanisms that influence their function are disturbed. Anesthetic monitoring allows one to recognize the extent of compromise to each body system during the course of a procedure and to make adjustments in the anesthetic protocol to prevent untoward long-term effects in the animal. For monitoring to be effective, three basic processes have to occur: early recognition, correct interpretation of changes, and appropriate intervention.

a. Early Recognition of Homeostatic Disturbances. The first aspect of monitoring anesthesia is to assure that the level of surgical anesthesia is consistent with the welfare of the animal. This requires monitoring anesthetic influence on the central nervous system, more commonly referred to as the depth of anesthesia. In addition to the danger of excessive depth of anesthesia, an animal in an adequate plane of anesthesia may become significantly compromised because of the direct depressant effects of particular anesthetic drugs on the cardiovascular or respiratory system. Many anesthetic-induced changes in the CNS, cardiovascular, or pulmonary system are gradually progressive over time rather than sudden events (with the exception of anesthetic induction); therefore, if one continually monitors the appropriate parameters it will allow early recognition of negative trends while physiologic disturbances are still reversible.

b. Correct Interpretation of Cause and Effect. When an animal is anesthetized, its physiologic processes are altered by the anesthesia itself, the surgery performed, and/or the research manipulations of the study. If an animal appears compromised during a procedure, one must be able to determine quickly the likely cause of the problem in order to choose

ANESTHESIA AND ANALGESIA IN LABORATORY ANIMALS

ISBN 0-12-417570-8

the appropriate steps to correct it. A thorough knowledge of the anticipated effects of different anesthetic drugs on each physiologic parameter will make this interpretation of cause and effect much easier.

c. Proper Intervention to Prevent Anesthetic Mishaps. Because regular anesthetic monitoring will indicate deterioration of the patient before it becomes irreversible, the anesthetist will be able to make adjustments in the anesthetic level, provide supportive therapy, or institute drug therapy that can positively affect outcome. The appropriate intervention is based on interpretation of the cause of the underlying problem.

B. Basic Principles of Monitoring

1. Choice of Parameters

The efficacy of monitoring depends on choosing the right variables to follow. Certain parameters, such as the withdrawal response to a toe pinch, are not very sensitive indicators of patient status because the presence of a withdrawal would indicate that an animal is too light under anesthesia, but the absence of withdrawal could be seen in an adequately anesthetized animal or in one that is dangerously overanesthetized. However, quantitative variables like blood pressure give numerical information that can be easily compared to previous readings to indicate trends in patient status.

2. Multiple Parameters

It is imperative that an anesthetist never relies on just a single parameter to monitor patient status. A single parameter, such as respiratory rate, may allow one to recognize trends during anesthesia; however, it can often be misleading when it comes to correct interpretation of cause and effect, therefore resulting in inappropriate treatment measures.

3. Power of Observation

The techniques used in monitoring range from the use of sophisticated electronic equipment to simple visual and tactile observation. Mechanical and electronic devices enhance patient monitoring and are very useful in the laboratory where the anesthetist often has other duties, including that of surgeon or surgical assistant. Further, some parameters can only be monitored using sophisticated equipment and these become especially important in high-risk patients or procedures. This equipment can also provide numerical data which may be necessary for the interpretation of research results. In light of all the advantages of such sophisticated equipment, it is crucial that anesthetists not neglect their senses as well as clinical observations. The senses are simple and reliable, rarely malfunction or require calibration, and do not need an emergency backup power generator. The importance of the use of the senses is a theme throughout this discussion of anesthetic monitoring. Electronic monitoring equipment should be used to *reinforce* the clinical observation of the anesthetist.

C. Unique Challenges of Laboratory Animals

1. Small Size

Monitoring anesthesia generally becomes more difficult the smaller the species being anesthetized. The powers of observation are impaired if an animal does not have a reasonable site to palpate pulse strength or if a surgical drape covers the animal entirely so that chest wall motion can no longer be used to monitor respiratory function. Creativity and the use of modified equipment are often necessary to be able to follow the status of very small patients under anesthesia.

2. Physiologic Stability

The necessity for physiologic stability is a prerequisite for valid experimental data. Another important aspect of anesthetic monitoring relates to the interpretation of research results. Anesthetics can have a profound effect on research results, and a well-documented anesthetic procedure is an important part of the data of an experiment. Data collected during anesthetic monitoring serve as the basis for design of subsequent anesthetic procedures.

3. Minimizing Pain and Stress

Minimizing pain and stress is an important aspect for *all* veterinary anesthesia; however, special emphasis is placed on this aspect in the research setting because of guidelines set forth by Institutional Animal Care and Use Committees. An approved anesthetic protocol is the first step toward this goal, but anesthetic monitoring plays a role as the indicator of an animal's response to the procedure while under the influence of anesthetic drugs. When monitored parameters indicate an excessive response to surgical manipulations, either additional anesthesia is indicated or the anesthetic protocol that was designed for that particular study should be modified.

II. PHYSIOLOGICAL SYSTEM OF IMPORTANCE

A. Cardiovascular System

1. Physiological Significance

Most anesthetic agents cause a dose-dependent depression of the cardiovascular system and therefore cardiovascular monitoring allows not only an assessment of adequate circulatory function, but also provides information on depth of anesthesia as well.

2. Important Cardiovascular Parameters: Observational Techniques

The information gathered through observational techniques used for monitoring the cardiovascular system includes heart rate and rhythm, pulse strength, and capillary refill time.

a. HEART RATE. Heart rate is important for the effect it has on overall cardiac output. Cardiac output is the product of heart rate and stroke volume. As a general rule, increases in heart rate will increase cardiac output whereas decreases in heart rate will decrease cardiac output. Heart rate is often influenced by depth of anesthesia, such that bradycardia frequently occurs as the anesthetic plane gets deeper and tachycardia occurs when the anesthetic plane is too light. The definition of bradycardia and tachycardia is species dependent, so at the outset of anesthesia one should be aware of the normal range of heart rate for the particular species that is being studied. The importance of not relying solely on one parameter while monitoring is well illustrated by heart rate. Tachycardia can be caused by painful surgical stimulation in a lightly anesthetized animal, and increasing the plane of anesthesia is often indicated in this case. However, hypotension, hypovolemia, hypoxia, hyperthermia, and hypercarbia are also potential causes of elevated heart rate and increasing the level of anesthesia would not be indicated for each of these causes. Determination of the actual cause of the tachycardia requires examining other monitored information. Bradycardia may be caused by specific anesthetic drugs (opioids, α_2-agonists), reflex activity (mesenteric traction, intubation, oculocardiac reflex, hypertension), hypothermia, hyperkalemia, or cardiac conduction disturbances. If anesthetic depth is determined to be appropriate, bradycardia may be treated by anticholinergics. Anticholinergics would be the treatment of choice for bradycardias caused by opioids, xylazine, vagal reflex activity, and certain cardiac conduction disturbances. Bradycardias secondary to hypothermia are often unresponsive to anticholinergics and require rewarming of the animal for improvement. Bradycardia as a result of hyperkalemia represents a serious disturbance in cardiac conduction and emergency steps to reduce serum potassium would need to be instituted.

b. PULSE STRENGTH. Pulse strength and regularity are usually determined by digital palpation of the pulse from an accessible site (e.g., femoral artery, lingual artery, auricular artery, tail artery). Palpation of the pulse indicates mechanical activity of the heart. Palpation can reveal the presence of an arrhythmia and, based on the subjective assessment of pulse "strength," can provide some information relating to adequacy of the cardiac output. It is important to realize when interpreting pulse strength that the intensity of the palpated pulse is a function of the magnitude of the pulse pressure (Hamlin, 1994). Pulse pressure is the numerical difference between systolic and diastolic arterial blood pressures. The larger the systolic/diastolic difference, the stronger the pulse feels. A large systolic/diastolic difference resulting in good pulse strength does not *always* indicate adequate tissue perfusion. Once again, relying only on a single parameter, in this case pulse strength, could be misleading and therefore other parameters need to be used to confirm adequate cardiovascular function. Although a good pulse is not always an absolute indication that all is going well during anesthesia, the absence of a palpable pulse should *always* be considered a sign of inadequate cardiovascular function.

c. CAPILLARY REFILL TIME. The time it takes for mucous membrane color to return to normal after releasing digital pressure applied to the membrane sufficient to blanch it of color is defined as capillary refill time (CRT). Normal CRT should be less than 2.5 sec. Capillary refill time is considered an indicator of peripheral perfusion, and a prolonged CRT is seen during hypotension or low cardiac output states. It is important to realize, however, that CRT is markedly influenced by arteriolar tone. A number of conditions causing peripheral arterial vasoconstriction will prolong CRT even though overall tissue perfusion is likely adequate. Pain, excitement, hypothermia, and certain drugs (e.g., xylazine) induce vasoconstriction and can increase CRT above the acceptable upper limit (Haskins, 1987). On detecting a prolonged CRT, other monitoring parameters must be examined in order to determine its significance.

3. Important Cardiovascular Parameters Requiring Instrumentation

a. ELECTROCARDIOGRAPHY. Electrocardiography (ECG) is a technique that provides important information during the course of anesthesia. An ECG is widely applicable to all species. Many ECG monitors provide a continuous readout of heart rate. When attached to a recording device, ECG monitors allow continuous or intermittent accumulation of heart rate and rhythm data at different stages of an experiment. In addition, an ECG is the only way to establish absolutely the diagnosis of arrhythmias that might occur as a result of anesthesia or surgical manipulations. The major shortcoming of electrocardiography as a monitoring tool is that it represents only cardiac electrical activity and does not provide an assessment of the cardiac output or tissue perfusion, which are the more important functional aspects of cardiovascular performance.

The ECG of most species consists of a set of waves: the p wave indicates atrial depolarization, the QRS complex occurs during ventricular depolarization, and the T wave represents ventricular repolarization. The shape, size, and timing of each of these waves are dependent on the species and the particular lead system that is being used. For the purpose of diagnostic electrocardiography in an individual with suspected cardiac disease, a multilead system (12 lead ECG) is frequently used. During monitoring of anesthesia, the ECG mostly serves as a heart rate and rhythm monitor and therefore a single lead system (e.g., lead II, with reference electrodes attached at right forelimb and left hindlimb) is often used. In small laboratory animal species, the electrical signal is small (low millivoltage), as are the amplitudes of the ECG waves. In addition, the heart rate is relatively rapid, sometimes making it difficult to discern anything other than the QRS complex on the ECG. Despite this situation, the ECG can still be useful in these species as an indicator of heart rate and an irregular P-R interval, or changes in the shape of the QRS complex can be a useful indicator of the development of arrhythmias. Recognition of the type of arrhythmia usually determines the appropriate treatment and consultation of a text on veterinary electrocardiography is recommended (Tilley, 1992).

b. Arterial Blood Pressure. The measurement of arterial blood pressure can be very helpful in determining the adequacy of cardiovascular function during anesthesia. The majority of anesthetic agents will cause a dose-dependent depression of blood pressure through their effects on cardiac output, vascular tone, or both and therefore a trend of progressively decreasing blood pressure may be an indication of excessive anesthetic depth. However, to accurately interpret the significance of changes, one must take into account the varied factors that influence blood pressure. Factors that contribute most notably to blood pressure (BP) are cardiac output (CO), peripheral vascular resistance (PVR), and blood volume. Analogous to Ohm's law, which applies to electrical current, pressure is equal to flow times resistance, such that $BP = CO \times PVR$. If cardiac output and/or peripheral vascular resistance increases, so will arterial blood pressure. If cardiac output decreases, by way of direct anesthetic depression of the myocardium or decreased filling pressures due to loss of blood volume, without compensatory increases in arteriolar tone, the blood pressure will fall.

The actual significance of arterial blood pressure is a component of tissue perfusion pressure, a determinant of tissue blood flow. The upstream (arterial) pressure must significantly exceed the downstream (venous) pressure to enhance the flow through tissue capillary beds. The perfusion pressure should be greater than 60 mm Hg for adequate tissue perfusion (Haskins, 1987). Therefore, in monitoring arterial blood pressure, a mean blood pressure less than 60 mm Hg is considered unacceptable for maintaining tissue blood flow. Mean blood pressure is not the mathematical mean of the systolic and diastolic blood pressures. Mean blood pressure can be estimated from the measured systolic and diastolic pressures by the following equation: mean BP = diastolic BP + 0.3 (systolic BP − diastolic BP). To obtain blood pressure ideally, a direct arterial catheter can be used, and through the use of a pressure transducer and electronic pressure monitor, accurate systolic, diastolic, and mean blood pressures are available. Because of the invasive nature of this technique, the difficulty in placing arterial catheters in many species, and the expense of the electronic monitor, direct arterial blood pressures are not always available. There are two reasonable noninvasive alternatives to indirectly measure arterial blood pressure: the oscillometric monitor and the Doppler ultrasonic blood pressure monitor. Although both of these devices are versatile and very useful in the clinical setting, there are some limitations to their use. Small patient size, extremes of heart rate (<40 or >200), significant hypotension, and hypothermia (through peripheral vasoconstriction) may all interfere with the accuracy of the pressures obtained with either of these devices. Oscillometric monitors generally report a measured systolic and mean blood pressure and extrapolate a diastolic blood pressure (Geddes, 1984a). Oscillometric devices report heart rate as well. The reported heart rate can be used as an index of the accuracy of the reported pressures. A rule of thumb for accuracy is that pressure values should be questioned when a pressure reading from the monitor is accompanied by a reported heart rate that varies more than 10% from the actual heart rate as determined by auscultation or palpation of pulses. The Doppler ultrasonic monitor gives systolic blood pressure only. As a rule, a systolic blood pressure >100 mm Hg is accompanied by adequate tissue perfusion. A major advantage of the Doppler monitor is that it also transmits a pulse sound signal through a speaker device. This audible signal can be used to count heart rate and indicates that there is functional cardiac activity adequate to generate a peripheral pulse. In addition, although subjective, changes in the intensity of the audible signal correlate with changes in cardiac output (Dyson *et al.*, 1985). If the audible Doppler signal becomes diminished, one should quickly examine other parameters of hemodynamic status. Doppler monitors are useful in virtually all species, although in small species (e.g., rats, mice, guinea pigs) they are primarily used as pulse monitors with the probe placed on the chest wall right over the cardiac apex.

c. Central Venous Pressure. Central venous pressure (CVP) reflects the pressure in the right atrium and is an index of cardiac filling pressure. The factors which influence CVP are the volume of blood in the central veins, the compliance of the right atrium during filling, central vein vascular tone, and intrathoracic pressure. Measurement of CVP requires placement of a catheter into the jugular vein with the tip of the catheter ideally placed into the right atrium or at the minimum past the thoracic inlet. Central venous pressure is useful when extensive blood loss, rapid fluid administration, or right-sided cardiac insufficiency is anticipated. Normal CVP is -1 to $+5$ cm H_2O, but during anesthesia the range of normal may increase to $+10$ cm H_2O. Low CVP indicates that the blood volume of the animal is too low for the capacity of its vascular system. This may result from a loss of blood volume (e.g., intraoperative hemorrhage) or an increase in vascular capacity (e.g., anesthetic-induced venodilation). In either case, expansion of the blood volume through blood or fluid administration is warranted. High CVP indicates either hypervolemia (e.g., fluid overload) or failure of the cardiac pump (e.g., anesthetic overdose). In either case the potential for the development of pulmonary edema exists and steps should be taken to improve myocardial function (decreased anesthetic, positive inotropes) and fluid administration should be limited until CVP returns to the normal range. Trends in CVP during a procedure are more helpful than relying on an isolated CVP value to determine volume or cardiovascular status. As always, CVP should be interpreted in the context of other indicators of hemodynamic function, such as blood pressure and CRT.

d. Pulmonary Artery Pressures and Cardiac Output. The Swan–Ganz catheter is a balloon-tipped, flow-directed catheter with a thermistor used for monitoring pulmonary artery pressures and passed into the pulmonary artery (PA) via the jugular vein–right atrium–right ventricle route. Confirmation of correct placement of a PA catheter is made by observation of characteristic changes in the pressure waveform during its passage (Geddes, 1984b). The Swan–Ganz catheter allows measurement of cardiac output, systolic, diastolic, and mean PA pressures as well as pulmonary artery occlusion pressure (PAOP, also called ''wedge'' pressure) which is obtained upon inflation of the balloon tip, impeding PA blood flow and reflecting left atrial pressure. Normal values in many species for PA pressures are: systolic, 20–40 mm Hg; mean, 10–20 mm Hg; diastolic, 5–10 mm Hg; and PAOP, 3–8 mm Hg. Pulmonary artery catheters are indicated for mixed venous blood gas sampling, measurement of cardiac output, monitoring for left-sided cardiac insufficiency, and pulmonary hemodynamics. Because much of the hemodynamic data obtained from PA catheters involves calculation of derived variables (e.g., vascular resistances, ventricular stroke work index), PA catheters are used more often in the experimental setting for the collection of study-specific data rather than for anesthetic monitoring per se; however, the information gained from the catheter is often useful for making future anesthetic determinations.

B. Respiratory System

1. Physiological Significance

The major function of the respiratory system under anesthesia is to act as a gas exchange organ. The lung is responsible for the introduction of oxygen into the arterial blood (oxygenation), the elimination of carbon dioxide from the body (ventilation), and the uptake and elimination of gas anesthetics when inhalation anesthesia is used. To evaluate respiratory function, both oxygenation (partial pressure of oxygen in arterial blood) and ventilation (partial pressure of carbon dioxide in arterial blood) must be assessed. It is important to monitor the respiratory system during anesthesia not only because of its life-sustaining importance, but also because many anesthetic agents depress respiratory control mechanisms and under anesthesia, respiratory arrest usually precedes cardiovascular collapse when excessive anesthesia occurs.

2. Important Respiratory Parameters: Observational Techniques.

Observational techniques allow one to assess respiratory frequency, rhythm, and volume as well as mucous membrane color. Each of these parameters alone provides little information of the adequacy of ventilation and oxygenation; however, when viewed together they may be very helpful in indicating states of respiratory insufficiency.

a. Respiratory Rate. This rate can be determined by visual observation of the animal, looking at chest wall motion, by auscultation of the thorax, or by observing the rebreathing bag on an anesthetic machine. There is species variation in the range of normal respiratory rates and one should be familiar with those values prior to anesthetizing an animal of a given species. Usually, the higher the basal metabolic rate (and CO_2 production) of an animal the higher their resting respiratory rate. Most anesthetics are respiratory depressants and as a general rule the respiratory rate decreases with increasing anesthetic depth.

b. Tidal Volume. The amount of gas entering the respiratory tract during one respiratory cycle is called the tidal volume. An average tidal volume for most species is about 10 ml/kg. Monitoring tidal volume is usually subjective based on the degree of chest wall motion with each breath or the amount of movement in the rebreathing bag of the anesthetic machine. It is important to realize that monitoring chest wall motion alone tells little about the actual volume of gas inhaled, as an animal with an airway obstruction or restrictive pulmonary disease often has marked chest wall motion with little air movement, with the chest wall motion reflecting a high work of breathing in this case. In certain species, such as ruminants, rabbits, and guinea pigs, with the potential to develop abdominal distension secondary to gas accumulation in their gastrointestinal tract, close attention should be paid to respiratory effort and tidal volume because abdominal distension can be a significant cause of respiratory insufficiency. It is therefore important to always evaluate other parameters, such as concurrent movement of volume in the rebreathing bag, to confirm adequate tidal volume. The volume of gas in each respiratory cycle can be measured by attaching an instrument called a respirometer to the endotracheal tube of the animal (Lumb and Jones, 1984).

c. Minute Ventilation. Respiratory rate and respiratory volume are determined as an indication of the minute ventilation of the animal. Minute ventilation is the product of respiratory rate and tidal volume. The adequacy of ventilation is judged by the effectiveness of the lung in removing carbon dioxide from the pulmonary capillary blood and is directly linked to minute ventilation. The partial pressure of carbon dioxide in the arterial blood is inversely proportional to minute ventilation, such that if metabolism (CO_2 production) remains constant, doubling the minute ventilation will decrease $PaCO_2$ by one-half. The partial pressure of CO_2 in arterial blood is the major stimulus for ventilation in the respiratory control center in the brain. It is there that anesthetics exert their respiratory depressant effect (increased threshold and decreased sensitivity to CO_2) (Nunn, 1987). A decreasing minute ventilation during anesthesia may reflect a decrease in CO_2 production (often seen with hypothermia) or it may reflect increased depression of the respiratory control centers from the anesthetic drugs. An elevation in $PaCO_2$ (above 60 mm Hg) indicates the need for increased minute ventilation, through an increase in respiratory rate, volume, or both, and may also indicate the need to decrease anesthetic depth.

d. MUCOUS MEMBRANE COLOR. Cyanosis is the discoloration of mucous membranes that results from reduced hemoglobin in the blood imparting a purplish cast to the tissue. It usually indicates severe hypoxemia; however, an animal may be severely hypoxemic and not exhibit cyanosis. Cyanotic mucous membranes are not seen until 5 g/dl of hemoglobin in the blood becomes unoxygenated (Allen, 1991). In anemic states, when the blood hemoglobin concentration is low, hypoxemia can be quite severe but an absolute value of 5 g/dl is not desaturated so cyanosis is not evident. Therefore the appearance of cyanotic mucous membranes indicates hypoxemia, but a lack of visible cyanosis does not mean that the animal is adequately oxygenated.

3. Important Respiratory Parameters: Noninvasive Techniques

a. NONINVASIVE MONITORS OF VENTILATION (CAPNOGRAPHY). The concentration of carbon dioxide in the inspired and expired gas can be continuously measured from the airway with capnography. The capnograph usually samples airway gas from a port located either at the endotracheal tube/breathing circuit interface or from a site at the distal end of the endotracheal tube. Inspired gas should contain virtually no carbon dioxide. As an animal exhales, carbon dioxide will appear in the exhaled gas as alveolar emptying occurs. The peak expired CO_2 is a reflection of the partial pressure of CO_2 in the alveolar gas which is equilibrated with arterial blood (Moens and Verstraeten, 1982). Therefore, increased values suggest hypoventilation whereas low peak expired CO_2 suggests hyperventilation. The peak expired (i.e., end-tidal) CO_2 in mm Hg approximates the $PaCO_2$ with reasonable accuracy unless ventilation–perfusion mismatching is significant within the lungs. In small-sized animals, the end-tidal CO_2 may tend to underestimate the $PaCO_2$ because the gas sampling flow rate of the machine is high enough that gas (containing virtually no CO_2) is entrained from the breathing circuit, diluting the end-tidal sample (Dorsch and Dorsch, 1984). In addition to ventilatory information, capnographs can be useful in the rapid detection of endotracheal tube malfunctions, such as a disconnect from the circuit or a kinked or obstructed tube. Capnographs have also proven to be sensitive early indicators of the development of malignant hyperthermia in susceptible species (primarily swine) in which a rapid rise in end-tidal CO_2 reflects the marked metabolic increase in CO_2 production that precedes the often fatal increase in body temperature (Bagshaw *et al,* 1978).

b. NONINVASIVE MONITORS OF OXYGENATION: PULSE OXIMETRY. A pulse oximeter is an instrument that by means of a light source and photodetector measures the light absorbance of tissues and indicates the level of oxygen saturation of hemoglobin in the blood. Oxygen saturation of hemoglobin (SaO_2) is related to the partial pressure of oxygen in arterial blood such that for each value of PaO_2 there is a corresponding percentage of hemoglobin that is saturated with oxygen. Adequate arterial oxygenation requires a minimum PaO_2 value greater than 60 mm Hg (and ideally $>$ 90 mm Hg). A PaO_2 of 60 mm Hg corresponds to a 90% saturation of the hemoglobin in the arterial blood. Using a pulse oximeter, changes in hemoglobin saturation can be monitored continuously. When the SaO_2 value falls below 90% the animal is becoming hypoxemic and steps should be taken to improve oxygenation, such as administering 100% oxygen, endotracheal intubation, and assisting ventilation as indicated (Clark *et al.,* 1992). There are also a number of conditions which will produce erroneously low oximeter values. Because the pulse oximeter is highly dependent on good peripheral perfusion, hypotension and hypothermia will impair accuracy. Motion artifacts (e.g., patient movement or shivering) and bright external light sources (e.g., fluorescent lights, surgery lights, heat lamps) will interfere with the probes ability to detect a signal (Severinghaus and Kelleher, 1992). In addition to reporting SaO_2, the pulse oximeter often gives a value for heart rate. If the heart rate value, as reported, varies significantly with the actual heart rate obtained by palpating a pulse or auscultation, then the accuracy of the reported SaO_2 should be questioned. Pulse oximeter probes work best when placed on nonpigmented, hairless tissues such as the tongue. Probes are now being designed specifically for veterinary application, including a probe inserted rectally like a thermometer, which may prove useful in small laboratory animal species.

4. Important Respiratory Parameters: Invasive Techniques

a. BLOOD GAS ANALYSIS. The only absolute way to effectively judge the adequacy of ventilation and oxygenation in any animal is through arterial blood gas analysis. The partial pressure of CO_2 in arterial blood should fall between 35 and 45 mm Hg. A $PaCO_2$ $<$35 mm Hg is defined as hyperventilation, which most often occurs iatrogenically in animals that are being ventilated. A $PaCO_2$ $>$45 mm Hg is by definition hypoventilation and, depending on the experimental conditions, may indicate the need for assisted ventilation. If an animal has a $PaCO_2$ above 60 mm Hg during anesthesia, steps should be taken to improve ventilation, which may include endotracheal intubation, manual or mechanical positive pressure ventilation, or decreasing the level of anesthesia as indicated.

The partial pressure of oxygen in the arterial blood $>$90 mm Hg assures adequate oxygenation (provided there is adequate tissue perfusion). Much higher levels of PaO_2 are often expected during anesthesia depending on the percentage of oxygen in the inspired gas. As a rule, the PaO_2 (in mm Hg) should be approximately five times the inspired oxygen concentration, so that when breathing 100% oxygen the arterial O_2 should approach 500 mm Hg. An animal may be relatively hypoxemic, in that its PaO_2 is less than would be expected for that particular inspired oxygen concentration. This can indicate concurrent hypoventilation (check the $PaCO_2$), ventilation–perfusion mismatching, shunt, or diffusion impairment. Absolute hypoxemia with significant impairment of tissue oxygenation occurs at PaO_2 values <60 mm Hg (Shapiro *et al.,* 1982). When PaO_2 falls to that level, measures should be taken to improve oxygenation, such as maximizing the inspired oxygen concentration and improving ventilation.

C. Central Nervous System

1. Physiological Significance

Anesthesia directly suppresses many of the homeostatic control systems within the body, most of which arise from or channel information through the central nervous system. Production of the state of anesthesia involves supressing the level of consciousness, pain perception, muscle tone, and reflexes; all of which result primarily from anesthetic drug activity on the CNS. The basis of our ability to control the effects of anesthesia depends on being aware of the impact the drugs are having within the CNS at any given moment. Awareness of the level of anesthetic-induced CNS depression requires monitoring signs in body systems that reflect the level of input from the CNS.

2. Important CNS Parameters: Depth of Anesthesia

Observational techniques are particularly important when assessing the depth of anesthesia. The depth of anesthesia required for an animal is determined by the type of procedure being performed and the response of the patient to the surgical stimulus. The depth of anesthesia required for a particular procedure also varies with the species of animal. Some species, such as sheep, are far more tolerant of manipulation than others, such as the rabbit. The type of procedure will determine the intensity of the stimulation. High-intensity painful procedures, such as joint capsule incision; periosteal stimulation; fracture manipulation; visceral or peritoneal traction; diaphragmatic stimulation; corneal manipulation; or the manipulation of inflamed tissue will require a deeper level of anesthesia than low-intensity pain procedures. Because different pain intensities occur within a procedure as different tissues are manipulated, the anesthetist must frequently reassess and adjust the depth of anesthesia as is appropriate. Experience with a procedure will allow the anesthetist to anticipate necessary changes in the depth of anesthesia. Observational techniques that help indicate depth of anesthesia include the level of muscle relaxation, reflex activities, and physiologic responses to surgical stimulation.

A good assessment of muscle relaxation can be made by monitoring jaw tone in certain species. The ease of monitoring jaw muscle tone varies with species due to the differences in jaw size and masseter muscle strength. Jaw tone is easily assessed in certain small animal species such as dogs and cats. It is much more difficult to evaluate in rodents, sheep, and swine. Although it is usually desirable to maintain some degree of muscle tone during anesthesia, if an animal attempts to close its mouth when gentle traction is placed on the mandible during a procedure, more anesthesia is needed.

Purposeful movement is an indication of light anesthetic depth and usually occurs in response to painful stimulus, but like most observations of anesthetic depth it must be assessed in relation to the anesthetic technique being used and the procedure being performed. Purposeful movement such as swallowing (often stimulated by the presence of an endotracheal tube) and head shaking (seen in small laboratory animal species such as the guinea pig) are typical indications of too light a plane of anesthesia. Purposeful movement in response to surgical manipulation must be differentiated from spontaneous movement which can be seen with certain anesthetic agents such as ketamine, opioids, enflurane, and methoxyflurane and which does not occur as a response to a surgical stimulus (although it may occur coincidentally with a stimulus). The pedal withdrawal reflex is commonly used to help determine the level of surgical anesthesia in small laboratory species. The foot is extended and the toe itself is pinched with the fingernails or gently with a hemostat. If the limb is withdrawn in response to the toe pinch, then the animal will need more anesthesia before a painful procedure can begin. The pinnae of rabbits or rodents are often tested in a similar manner, looking for head shaking in response to a painful stimulus.

Ocular reflexes can be used to indicate anesthetic depth. These include palpebral response, ocular position, and corneal reflex. The palpebral response is the blinking that occurs when the edge of the eyelid is lightly touched. There is species variation in this response under anesthesia. Most animals lose the palpebral response fairly early in surgical anesthesia; however, rabbits may maintain a palpebral response even at deeper planes of anesthesia. The intensity of the palpebral response is also influenced by the particular anesthetic agent used. The palpebral response is lost early with barbiturates and most inhalation agents; however, it is well-maintained with ketamine.

Ocular position is generally a reliable sign of changing anesthetic depth in many species. As anesthesia is induced and in light planes of anesthesia the eyeball remains central in the orbit and a palpebral response is present. Nystagmus and lacrimation are also indications of a light plane of surgical anesthesia. When a surgical plane of anesthesia is reached the globe rotates ventromedially. As anesthesia deepens (and muscle relaxation continues to increase) the globe will again rotate upward and return to a central position in the orbit. A centrally located globe during a deep plane of anesthesia can be distinguished from the centrally located globe of light anesthesia by the absence of the palpebral response during increased anesthetic depth. The corneal reflex is another ocular reflex that changes with the changing depth of anesthesia. To determine the presence of a corneal reflex, the surface of the cornea is lightly touched and the presence of a blinking response noted. A brisk corneal reflex is found in awake animals and its intensity begins to diminish as the plane of anesthesia deepens. Certain species, such as ruminants, will maintain a corneal reflex during a surgical plane of anesthesia, whereas the reflex is often absent at a surgical plane in other species, such as the dog and cat.

Various physiologic responses to stimuli are also used to assess the level of surgical anesthesia. Increases in heart rate, blood pressure, and respiratory rate can be seen in response to surgical stimulation when no purposeful movement has been observed. As discussed in previous sections, changes in heart rate, blood pressure, and respiratory rate are also affected by the specific drugs given and the physiologic state of the animal. To accurately interpret the cause of such autonomic responses requires an understanding of predicted cardiopulmonary responses to the pharmacologic agents being used in the animal, combined with an evaluation of anesthetic depth based on CNS signs. No single parameter that can be monitored is solely adequate to pinpoint accurately the plane of anesthesia in an experimental animal.

An adjunct to monitoring the depth of anesthesia is to monitor the end-expiratory (end-tidal) inhalation agent concentration during a procedure. As described with capnography, the end-expiratory concentration of an inhalation agent is assumed to be reflective of the arterial and brain concentration of the agent being monitored. Typically surgical anesthesia is achieved at 1.2 to 1.5 times the minimum alveolar concentration (MAC) of an inhalation agent (Steffey, 1994). As an example, the MAC of halothane in the dog or cat is 0.87%. Surgical anesthesia would therefore be achieved at an end-tidal halothane concentration of 1.04 to 1.31% halothane. An end-tidal inhalation agent monitor will indicate if the alveolar concentration of gas is in that range and, in combination with observational signs of depth of anesthesia, will indicate if the animal is suitably anesthetized for a surgical procedure. As with other parameters, using end-tidal agent concentration as the sole indicator of anesthetic depth can lead to erroneous conclusions. The administration of other anesthetic drugs, especially opioids or α_2-agonists that have significant analgesic properties, will decrease the MAC of inhalation agents and allow an animal to be adequately anesthetized at anesthetic concentrations much less than 1.5 times MAC. Hypothermia will also markedly decrease the MAC of inhalation agents, so relying on end-tidal concentration alone without observing additional signs in the animal may lead to inappropriately deep planes of anesthesia. It is important to recognize that the end-tidal gas concentration necessary to achieve surgical anesthesia is not the same thing as the concentration setting on the vaporizer. During the induction and the equilibration phase of inhalation anesthesia there can be a large difference between the vaporizer concentration and the end-tidal value (the vaporizer concentration being higher at the beginning of anesthesia). With time, the difference between these two concentrations narrows. Monitoring end-tidal agent concentrations can serve as a useful guide for when to decrease or increase the setting on the vaporizer. Monitoring end-tidal inhalant concentrations is often necessary in a research setting to quantitatively establish the dose of drug delivered.

3. Important CNS Parameters: Thermoregulation

Most anesthetic procedures cause a depression of the hypothalamic thermoregulatory mechanism, predisposing animals to hypothermia. This is an even greater problem in small laboratory species. These animals have a very large surface area relative to body mass which causes a correspondingly greater loss of body heat. This is further compounded by the removal of hair and wetting the remaining hair coat during aseptic preparation of the surgical site. When this small damp animal is then placed on a cold metal surface, the result can be fatal hypothermia. Opening a body cavity will further accelerate the loss of body heat. Body temperature is best monitored by a small thermistor placed into the esophagus to the level of the heart. This will provide a closer indication of core body temperature than a rectal thermistor because the temperature in the rectum falls more slowly than core body temperature (Lumb and Jones, 1984). Clinical thermometers do not register continuous decreases in body temperature and thus are not recommended. Hypothermia can be minimized by the warming of surgical preparation solutions, insulating the patient from both cool ambient temperatures and a cold restraining surface, using warmed fluids if supplemental fluids are provided, warming of inspired gases, and using supplemental heat provided by a circulating water warming blanket. Electric heating pads should be avoided as they can cause serious thermal burns and even hyperthermia. Anesthetic-induced hyperthermia can be seen in some breeds of dogs and pigs. Because both hypo- and hyperthermia can enhance the effects of central depressant drugs, as well as confound results in experimental data, monitoring body temperature of the patient is a particularly important parameter in laboratory animals.

III. SUMMARY AND CONCLUSIONS

Frequent evaluation of the physiologic status of an anesthetized animal leads to early recognition of homeostatic disturbances, facilitating timely and appropriate intervention to prevent anesthetic mishaps, and provides a stable physiologic state consistent with the welfare of the patient and sound experimental data. Accurate interpretation of monitoring data beings with a knowledge of what is normal for the animal species. Preanesthetic evaluation of the animal will provide the anesthetist with knowledge of what is normal for that individual patient.

It is usually not feasible to monitor all possible parameters in every patient so the anesthetist must choose those parameters that are the most likely to change in response to anesthesia and surgery and will have an impact on the well-being of the animal and on the research project. To make this determination, the anesthetist must have a basic knowledge of the physiologic effects of the anesthetics to be used, the procedures to be performed, and the experimental manipulations and data to be collected as part of the experimental protocol.

When a change in a monitored parameter occurs, the anesthetist must accurately determine the cause of that change. Monitoring several parameters, with serial data collected, allows trends to be observed that enable the anesthetist to tailor the anesthetic protocol to the response of the animal. Monitoring several parameters is important because systems interact with each other to create a "physiologic state" and because interventions will likely affect more than one system. Systems requiring at least some degree of monitoring include the cardiovascular system, the respiratory system, and the central nervous system.

Smaller laboratory animals present additional challenges to the anesthetist. Because of their small size and small blood volumes, frequent monitoring of certain parameters such as blood gases is not always feasible. Although newer equipment can determine blood gas using microliter quantities, limitations of serial sampling still exist. Other equipment, such as a blood pressure monitoring device, is often impractical. Modifications can sometimes be made in existing equipment to accommodate the smaller species used in laboratory animal research.

The most successful anesthetic protocols come with practice. To determine the best protocol for a given research or surgical procedure, it is helpful to have anesthetic monitoring records of past procedures. Such records assist the anesthetist in anticipating problems in future procedures. In addition, good monitoring records are always necessary for demonstrating consistency in experimental protocols and for eliminating variability that may invalidate the procedure performed.

REFERENCES

Allen, D. G. (1991). Cyanosis, pallor and erythema. *In* "Small Animal Medicine" (D. G. Allen, ed.), pp. 15–23. Lippincott, Philadelphia.

Bagshaw, R. J., Cox, R. H., Knight, D. M., and Detweiler, D. K. (1978). Malignant hyperthermia in a greyhound. *J. Am. Vet. Med. Assoc.* **172,** 61–62.

Clark, J. S., Votteri, B., Ariagno, R. L., Cheung, P., Eichhorn, J. H., Fallat, R. J., Lee, S. E., Newth, C. J. L., Rotman, H., and Sue, D. Y. (1992). Noninvasive assessment of blood gases. *Am. Rev. Respir. Dis.* **145,** 220–232.

Dorsch, J. A., and Dorsch, S. E. (1984). "Understanding Anesthesia Equipment," 2nd ed., pp. 136–181. Williams & Wilkins, Baltimore, Maryland.

Dyson, D. H., Allen, D. G., and McDonell, W. N. (1985). Evaluation of three methods for cardiac output determination in cats. *Am. J. Vet. Res.* **46,** 2546–2552.

Geddes, L. A. (1984a). "Cardiovascular Devices and Their Application," pp. 63–99. Wiley, New York.

Geddes, L. A. (1984b). "Cardiovascular Devices and Their Application," pp. 100–188. Wiley, New York.

Hamlin, R. L. (1994). Physical examination. *In* "Saunders Manual of Small Animal Practice" (S. J. Birchard, and R. G. Sherding, eds.), pp. 396–402. Saunders, Philadelphia.

Haskins, S. C. (1987). Monitoring the anesthetized patient. *In* "Principles and Practice of Veterinary Anesthesia" (C. E. Short, ed.), pp. 455–477. Williams & Wilkins, Baltimore, Maryland.

Lumb, W. V., and Jones, E. W. (1984). "Veterinary Anesthesia," pp. 493–520. Lea & Febiger, Philadelphia.

Moens, Y., and Verstraeten, W. (1982). Capnographic monitoring in small animal anesthesia. *J. Am. Anim. Hosp. Assoc.* **18,** 659–678.

Nunn, J. F. (1987). "Applied Respiratory Physiology," 3rd ed., pp. 350–378. Butterworth, London.

Severinghaus, J. W., and Kelleher, J. F. (1992). Recent developments in pulse oximetry. *Anesthesiology* **76,** 1018–1038.

Shapiro, B. A., Harrison, R. A., and Walton, J. R. (1982). "Clinical Application of Blood Gases," 3rd ed., pp. 131–142. Year Book Medical Publishers, Chicago.

Steffey, E. P. (1994). Inhalation anaesthesia. *In* "Anaesthesia of the Cat" (L. W. Hall and P. M. Taylor, eds.), pp. 157–193. Baillière Tindall, London.

Tilley, L. P. (1992). "Essentials of Canine and Feline Electrocardiography," 3rd ed. Lea & Febiger, Philadelphia.

Chapter 6

Monitoring of Analgesia

Peggy J. Danneman

I. INTRODUCTION

Pain hurts. Everyone can agree on that simple truth, but defining pain has proven considerably more challenging and measuring it more challenging yet. Although pain consists in part of purely physical sensation, the truly unpleasant aspects of it are all in the mind. As a result, pain is a uniquely individual experience and it is difficult, if not impossible, for us to fully understand the pain even of another human. When viewed in this light, the problems inherent in trying to measure pain in animals are evident. In fact, it is best to admit from the start that we cannot measure pain in animals with anything even approaching precision. Notwithstanding our limitations in this regard, veterinarians and investigators have both a moral and a legal obligation to prevent or minimize animal pain to the maximum extent consistent with scientific goals. To do so requires the ability to recognize or, better yet, predict the need for intervention as well as knowledge of the most effective forms of intervention. Because the effective treatment of pain requires recognition that there are many different types of pain which respond quite differently to specific therapeutic regimens, this chapter begins with a short review of the neurophysiological basis of the different types of pain. However, the emphasis of the chapter is on the recognition of pain in animals, the physiological effects of unrelieved pain, and approaches to predicting the severity of pain that may be associated with different experimental manipulations. Some general approaches to the prevention

ANESTHESIA AND ANALGESIA IN LABORATORY ANIMALS

ISBN 0-12-417570-8

or alleviation of pain are also reviewed (specific interventions are discussed in other chapters). Finally, for readers who must review protocols that involve the use of experimental animal models of pain, the types of models used, their advantages and disadvantages, and some of the problems and other considerations peculiar to these types of studies are reviewed at the end of the chapter.

II. NEUROANATOMY AND NEUROPHYSIOLOGY OF PAIN

Pain is a complex phenomenon consisting of both physical and psychological components. Although there is no single definition of pain that encompasses all of its manifestations, the most commonly used definition is that of the International Association for the Study of Pain: ''Pain is an unpleasant sensory and emotional experience associated with actual or potential tissue damage, or described in terms of such damage'' (IASP, 1986). The psychological (affective–motivational) component of pain will be discussed later (Section IV). This section reviews the physiological basis for the sensory component of pain, or *nociception.* Nociception refers to the physiological responses associated with the noxious stimulation of pain receptors (nociceptors) and processing of the information via specialized central nervous system (CNS) pathways.

Pain can arise from somatic tissues, the viscera, or injured nervous tissue. The neurological mechanisms underlying pain arising from these tissues differ, as do the characteristics of the pain and its response to specific therapeutic regimens. Pain also may be classified as acute or chronic and, again, these two types of pain often have different underlying causes and respond differently to treatment. All of these differences relate to the way in which pain is processed within the nervous system, and a basic understanding of the neurophysiology of pain is essential to both accurate diagnosis and effective treatment of different types of pain. The reader should note that the following discussion is intended as an overview and will not consider many of the complexities of the subject. Those wishing a more comprehensive review of the neurophysiological basis of pain can refer to Willis (1985), Light (1992), or Wall and Melzack (1994).

A. Somatic Pain

Acute superficial somatic nociception, or nociception arising from the skin, begins with the intense stimulation of specialized receptors called *nociceptors,* which are the peripheral terminations of finely myelinated [A-delta(δ) fibers] or unmyelinated (C-fibers) axons. Nociceptors respond to specific *noxious* stimuli (those which cause, or threaten to cause, tissue damage), but may also respond to innocuous stimuli under certain circumstances (Sections II, C and D). Most A-delta fibers respond to intense mechanical stimuli and many respond to intense heat. They subserve primarily well-localized sensations of sharp, stinging pain. Most C-fibers, or polymodal nociceptors, respond to intense mechanical stimuli, intense heat, and a variety of chemical irritants. They subserve sensations of dull or sharp burning pain. It should be noted that not all cutaneous C-fibers are nociceptors; some (the percentage varies from species to species) respond to benign stimuli such as warmth. Subclasses of both A-delta and C-fibers respond to intense cold. In general, the pain arising from stimulation of cutaneous nociceptors is well localized.

Acute deep somatic nociception is nociception arising from muscles, joints, tendons, bones, and associated fascia. Most of the pain receptors in these tissues are A-delta and C-fibers, although they are generally referred to as Group III and Group IV fibers, respectively. The Group III nociceptors respond primarily to intense mechanical stimulation (e.g., pressure, stretching), whereas the Group IV nociceptors may respond to intense mechanical stimulation, thermal stimulation, or chemical irritants (e.g., chemicals released during inflammation). As with cutaneous C-fibers, not all Group III or Group IV fibers are nociceptors. Pain arising from deep somatic tissues is less well localized than that arising from skin.

The majority of nociceptive afferents from the trunk and limbs enter the spinal cord via the lateral portion of the dorsal roots. After entering the dorsal root zone, the fibers divide into short ascending and descending branches that run longitudinally in the posterolateral fasciculus (Lissauer's tract) of the white matter. Fibers leave Lissauer's tract within one to two segments to make synaptic connections with neurons in the dorsal horn of the gray matter, which project to the ascending tracts of the anterolateral system (spinothalamic, spinoreticular, and spinomesencephalic tracts). Via these tracts, fibers project to various structures within the brain stem, including the reticular formation, thalamus, hypothalamus, and limbic system. Fibers from these areas relay nociceptive information to the primary and secondary somatosensory and limbic areas of the cerebral cortex. Nociceptive afferents from the face, cornea, sinuses, tooth pulp, and mucosa of the lips, cheeks, and tongue are carried into the spinal cord by way of the trigeminal nerve and trigeminal ganglion. Via the trigeminothalamic tract, nociceptive information is relayed to various structures within the brain stem and, ultimately, to the cerebral cortex.

Supraspinal processing is required for the conscious awareness of nociceptive stimulation, recognition that this stimulation is unpleasant, and the various affective–motivational responses (e.g., fear, rage, a desire to escape) or cognitive associations (e.g., anticipation, comparison with past experiences) that may accompany this recognition. In other words, it is through processing by higher centers that nociception becomes ''pain.''

B. Visceral Pain

Visceral pain differs from somatic pain in several respects: it cannot be evoked from all organs (e.g., liver and kidneys); it does not occur in association with certain types of injurious stimulation (e.g., burning, cutting); it is diffuse, poorly localized, and often referred to other locations; and it is accompanied by autonomic and motor reflexes (Cervero, 1988). Furthermore, whereas true nociceptors are present in some internal organs (Cervero, 1982; Kumazawa, 1986; Berkley *et al.*, 1988; Cervero and Sann, 1989), pain arising from other organs (e.g., bladder, colon) appears to be mediated by receptors that respond to both nonnoxious low-intensity stimulation and noxious high-intensity stimulation (Malliani and Lombardi, 1982; Bahns *et al.*, 1986; Jänig and Koltzenburg, 1991). In these latter organs, low-intensity stimulation produces nonpainful sensations such as fullness, whereas high-intensity stimulation produces pain. With prolonged high-intensity stimulation (e.g., resulting from hypoxia or inflammation), nociceptors become sensitized, previously unresponsive receptors (''silent nociceptors'') begin to respond, and the continued afferent barrage results in functional changes within the CNS (Cervero and Jänig, 1992; Cervero, 1994). These events lead to a state of chronic pain, which is discussed in more detail in Section II,C.

Several hypotheses have been proposed to explain the phenomenon of referred pain, which occurs when pain arising from the viscera is perceived to arise from somatic structures (e.g., pain in the arm following myocardial infarction). The most widely accepted of these are derived from the projection-convergence theory of Ruch (Willis, 1985; McMahon, 1994). According to this theory, many neurons within the spinothalamic tract receive convergent input from both visceral and somatic tissues arising from the same dermatotome. Activation of the spinal cord neurons by input from visceral afferents may be misinterpreted by the brain as coming from the convergent somatic afferents because these are the most common sources of nociceptive input.

C. Chronic Pain

Chronic pain has been variously defined as pain which persists beyond the normal time of healing (Bonica, 1953) and, more arbitrarily, pain which persists for more than 6 months (Gildenberg and DeVaul, 1985; Merskey and Bogduk, 1994). No matter how one defines chronic pain, it is important to understand that it is not simply a prolonged form of acute pain. Acute pain is associated with a diagnosable underlying physical pathology, generally responds to classical analgesic drugs, and can be expected to disappear once the underlying cause has been resolved. In contrast, chronic pain may not be associated with any recognizable active pathological process and it seldom responds more than temporarily to classical analgesics. These are critical differences because the clinician who does not distinguish chronic from acute pain or attempts to treat chronic pain as though it were acute pain is almost certainly doomed to failure.

Many different conditions and syndromes are associated with chronic pain, including malignancies, chronic inflammatory disorders (e.g., cholecystitis, colitis), and orthopedic disorders (e.g., spinal stenosis, osteophytes). Neuropathic pain, a particularly problematic type of chronic pain that occurs as a result of damage within the nervous system, is discussed separately in Section II,D. The common underlying feature that all forms of chronic pain share is abnormal activity within the peripheral and/or CNS. This abnormal activity is triggered by normal nociceptive input related to the initial acute insult but, due to functional changes within the nervous system, it becomes self-sustaining. As a result, the pain can continue even after the peripheral nociceptive input ceases. Types of abnormal function that may contribute to chronic pain are (1) sensitization of nociceptors (see later); (2) spread of excitation from damaged neurons to adjacent undamaged neurons (both nociceptors and low-threshold receptors that normally subserve mild sensations such as nonnoxious warmth or pressure); (3) expansion of the peripheral receptive fields of dorsal horn neurons; (4) increased excitability in populations of neurons within central nociceptive pathways; and (5) diminished activity of endogenous inhibitory mechanisms, leading to the reduced inhibition of excitability within nociceptive pathways (Bennett, 1994; Price *et al.*, 1994; Levine and Taiwo, 1994).

An important characteristic of nociceptors is their tendency to respond to less intense stimulation following tissue damage, inflammation, or repeated or prolonged stimulation. This phenomenon is called *sensitization* and it is at least partially responsible for the phenomenon of *hyperalgesia*. Hyperalgesia is an increased response to painful stimulation. It is associated with a lowered pain perception threshold, which results in perception of pain in response to previously innocuous stimuli (more properly referred to as *allodynia*), and in spontaneous pain which occurs in the absence of any external stimulation. Hyperalgesia occurs in both the injured/inflamed tissue (primary hyperalgesia) and in the surrounding ''normal'' tissue (secondary hyperalgesia). The neurophysiological basis for hyperalgesia is complex and incompletely understood, but it is at least partially due to changes in nociceptors mediated by chemicals released in injured and inflamed tissues. These include bradykinin, histamine, serotonin, prostaglandins, interleukins, and substance P. Studies suggest that functional changes within the CNS also contribute to the phenomenon of hyperalgesia (Dickenson and Sullivan, 1987; Coderre *et al.*, 1990; Malmberg and Yaksh, 1992; Vaccarino and Melzack, 1992).

D. Neuropathic Pain

Neuropathic pain is a clinically important type of pain that results from injury or dysfunction within the peripheral or CNS. The inciting trauma need not be extensive and may be minor or inapparent. A number of different neuropathic pain syndromes are recognized in humans, including *neuroma* (when injured peripheral nerve axons form sprouts at the site of damage), *neuralgia* (pain following the distribution of a nerve), *causalgia* (pain following traumatic peripheral nerve injury; it is associated with signs of sympathetic hyperactivity such as edema, sweating, and abnormalities of blood flow), postamputation pain (pain in the stump following amputation), *phantom pain* (pain in the missing limb following amputation), and *central pain* (pain caused by a primary lesion or dysfunction in the CNS). Features common to most of these syndromes include sensory deficit, *dysesthesia* (an unpleasant abnormal sensation), allodynia, *hyperpathia* (hyperalgesia associated with an increased threshold for sensory detection), and/or hyperalgesia. The pain associated with these syndromes is often characterized as burning or aching with sudden electric shock-like (''lancinating'') pains. It is not known whether animals experience similar sensations following nerve injury, but it is well known that animals engage in abnormal behaviors (e.g., self-mutilation) following nerve injury. In fact, various types of experimental nerve injury in animals are commonly used as models of neuropathic pain in humans. Therefore, we must assume that animals are subject to dysesthesia and/or pain in association with nerve damage.

A number of possible mechanisms of neuropathic pain have been hypothesized, including the following. (1) C-fibers that become sensitized following trauma may develop an increased sensitivity to norepinephrine (Wall and Gutnick, 1974a,b; Hu and Zhu, 1989), leading to exacerbation of pain in response to stress and other states involving increased sympathetic outflow. (2) Following transection or demyelination, the cell bodies and distal ends of damaged nerves may show spontaneous impulse generation in the absence of external stimulation and increased responsiveness to chemical, mechanical, and electrical stimulation (Wall and Gutnick, 1974a,b; Calvin *et al.*, 1982; Baker and Bostock, 1992; Kajander *et al.,* 1992). (3) Starting within minutes of peripheral nerve injury, there are permanent changes in the spontaneous activity and receptive fields of spinal cord neurons resulting in abnormal spontaneous discharges and amplification of input from the periphery (Palecek *et al.,* 1992, 1993; Laird and Bennett, 1991). Similar types of changes have been documented in higher centers, including thalamus and cortex (Guilbaud *et al.,* 1990; Guilbaud and Benoist, 1994).

III. MECHANISMS OF ANALGESIA

Analgesia, meaning the absence of pain in response to stimulation that would normally be painful (IASP, 1986), is produced by mechanisms that are as diverse as those of the different types of pain. Analgesic therapy is most effective when the analgesic is chosen based on understanding of both the nature and mechanism of the pain and the mechanism of action of the drug itself. The drug(s) should be chosen based on its ability to interfere with the specific physiological mechanisms that contribute to the pain. Effective analgesics can diminish nociceptor activation (e.g., nonsteroidal anti-inflammatory drugs), block peripheral nerve transmission (e.g., local anesthetics), interfere with nociceptive processing within the spinal cord (e.g., opioid agonists), and/or act on centers within the brain to reduce the emotional or cognitive component of pain, thereby making the sensation more tolerable (e.g., morphine, benzodiazepines).

Many of the most clinically useful drugs produce analgesia by enhancing the activity of natural descending pain inhibitory mechanisms within the CNS. The most notable examples are the opioid agonists (e.g., morphine), which interact with specific receptors within the CNS and activate pathways originating in the brain stem. There are several descending pathways originating within different areas of the brain that can modulate nociceptive transmission and pain-related behaviors. Important neurotransmitters in these pathways include the endogenous opioid peptides (e.g., β-endorphin, dynorphin, Met-enkephalin, and Leu-enkephalin), serotonin, and norepinephrine. These pathways project to the spinal cord dorsal horn, where nociceptive impulses may be inhibited or facilitated (Zhuo and Gebhart, 1992). An extensive body of experimental data suggests that these pathways are important in producing both (1) natural tonic modulation of nociceptive transmission in the dorsal horn (Hall *et al.,* 1982; Foong and Duggan, 1986) and (2) analgesia in response to pain, fear, and other stressors (Fields and Basbaum, 1994). Furthermore, there is good evidence that inhibition of dorsal horn neurons via descending inhibitory pathways is one of the mechanisms by which exogenous opioids produce analgesia (Gogas *et al.,* 1991; Hammond *et al.,* 1992). Other drugs which may produce analgesia by stimulation of endogenous pain-modulating mechanisms include tricyclic antidepressants such as amitriptyline and imipramine (Poulsen *et al.,* 1995) and α-adrenoreceptor agonists such as clonidine, medetomidine, and xylazine (Tasker, 1992; Daunt and Maze, 1992; Byas-Smith *et al.,* 1995).

IV. PAIN PERCEPTION

The perception of noxious stimulation requires initial detection of the stimulus by nociceptors, processing through peripheral and spinal cord pathways, and final processing through higher centers in the brain, including the cerebral cortex. However, mere conscious awareness of a stimulus is not equivalent to recognizing it as painful. At levels of stimulation just adequate to activate nociceptors, humans will report an awareness of sensation but will not describe the sensation as painful. This is referred to as the *nociceptive threshold.* Higher levels of stimulation are required to reach the *pain detection threshold,* which is the least experience of pain that a subject can recognize as painful (IASP, 1986). Under carefully controlled conditions, both nociceptive and pain detection thresholds are fairly consistent among different subjects. The highest experimentally defined level of pain perception is the *pain tolerance threshold.* This is the highest level of pain that a human subject is willing to tolerate (IASP, 1986). It is highly variable not only between subjects but even in the same subject at different times. The range of stimulus intensities between the pain detection and pain tolerance thresholds is the *pain sensitivity range* (Harris and Rollman, 1983). At the lower end of the pain sensitivity range, the cognitive and emotional responses that characterize the stimulus as *aversive* (provoking avoidance or escape) are weak [Chapman *et al.,* 1985; National Research Council (NRC), 1992; Dubner, 1994]. As the stimulus intensity increases, these responses become stronger. It is the cognitive and emotional components of pain that contribute to the wide variability in the pain tolerance threshold.

From the foregoing discussion, it is clear that pain perception is a broad concept, even in humans. So what does pain perception mean in an animal? Do animals feel pain in the same way as humans? The answer is that we do not know. We do know that the neurological mechanisms for processing nociceptive information are quite similar in humans and other mammals (Bonica, 1992; Gebhart, 1994). We also know that the stimulus intensities required to activate these mechanisms are remarkably similar for humans and other mammals (Kitchell, 1987). Furthermore, mammals exhibit the same behavioral and physiological responses to noxious stimulation as humans and, like humans, they will work harder to escape more intense stimulation (Kitchell, 1987; Dubner, 1994). However, because we cannot measure the all-important cognitive and emotional components of pain in animals, it is not possible to determine whether an animal perceives a noxious stimulus as unpleasant in the same way and to the same extent as a human would. Despite our inability to measure all aspects of the pain experience in animals, the assumption must nonetheless be made that nonhuman mammals have the capacity to perceive as unpleasant the same noxious stimuli that humans perceive as unpleasant.

The situation is less clear in phylogenetically more primitive animals (lower vertebrates and invertebrates). Even very primitive vertebrates have peripheral nociceptors and neural pathways that correspond to those involved in nociceptive processing in mammals. However, the cerebral cortex in these animals is either poorly developed or nonexistent, and the brain stem centers of the lower vertebrates lack the organization and complexity of these structures in mammals (Stevens, 1995). A functioning cerebral cortex is considered by many to be a basic requirement for noxious stimuli to be recognized as painful (AVMA Panel on Euthanasia, 1993; Kitchell, 1987). Thus, it is likely that nonmammalian animals are not able to perceive noxious stimuli as unpleasant in the same manner that mammals perceive them as unpleasant (Stevens, 1995). However, these animals do show physiological and behavioral responses to noxious stimulation, including avoidance or escape behaviors (Arena and Richardson, 1990; NRC, 1992). This indicates that, at the very least, noxious stimulation can be a potent stressor for more primitive animals and that the humane alternative is to minimize such stimulation or interfere with its processing with anesthetics or analgesics.

V. PAIN AND STRESS

Like pain, *stress* is a concept that continues to elude precise definition. Dorland's (1981) defines it as ''the sum of the biological reactions to any adverse stimulus, physical, mental, or emotional, internal or external, that tends to disturb the organism's homeostasis.'' The biological reactions that comprise the stress response include behaviors, changes in autonomic function, and changes in neuroendocrine function. These reactions allow the organism to adapt to the perceived adverse stimulus and return to a state of equilibrium. If the organism is unable to adapt, a state of *distress* ensues. Distress is characterized by maladaptive, often overtly harmful, behavioral and/or physiological changes. The adverse stimuli that provoke a stress response are called *stressors.* As implied in Dorland's definition of stress, almost anything can be a stressor, and a stressor for one animal may not be a stressor for another. Similarly, the response to a stressor may vary both qualitatively and quantitatively from one animal to another or in the same animal at different times.

Acute stressors trigger the release of corticotropin-releasing factor (CRF) from the hypothalamus. CRF causes the pituitary to secrete adrenocorticotropic hormone (ACTH) which, in turn, stimulates the adrenal cortex to release glucocorticoids (cortisol and corticosterone). These hormones cause (1) enhanced hepatic gluconeogenesis; (2) inhibition of glucose uptake by tissues; (3) enhanced lipid and protein catabolism; (4) altered function of macrophages, lymphocytes, and neutrophils; and (5) stimulation of tissue lipocortin production. Lipocortins inhibit the inflammatory response by, among other things, inhibiting the production of prostaglandins, thromboxanes, and leukotrienes (Breazile, 1987). The response to stress also involves increased sympathetic tone, which causes (1) reduced secretion of insulin by the pancreatic islet cells; (2) increased release of renin by the kidneys; (3) increased release of vasoactive intestinal peptide by the intestine; (4) increased release of substance P; and (5) increased secretion of catecholamines from the adrenal medulla and sympathetic nerve terminals. Changes in the plasma concentrations of these substances result in disturbances in cardiovascular, gastrointestinal, renal and immune function, and alterations in fluid, electrolyte, and acid–base balance. In addition to these and other measurable biochemical and physiological changes, clinical signs of stress may include delayed wound healing, muscle wasting, immune deficiencies, enhanced susceptibility to infection, stimulation or inhibition of feeding behavior, gastrointestinal bloating, and/or diarrhea (Breazile, 1987; NRC, 1992; Lewis *et al.,* 1994).

The general perception of stress is that it is undesirable and should be eliminated. In reality, the total elimination of stress is neither desirable nor even possible. However, many specific stressors can and should be eliminated. Among these is pain, especially pain that approaches or exceeds the tolerance threshold. Pain of this intensity is a known stressor that may have significant adverse effects on the general health and performance of the animal. Pain may also interact with other stressors (e.g., fear, cold, hunger, dirty environment) to produce an augmented stress response. The interaction between pain and other stressors is complex. In general, reducing pain can be expected to reduce the stress associated with other, unavoidable, stressors (e.g., the presence of human caretakers). Similarly, eliminating an important environmental or psychological stressor often will make pain more tolerable (see later). However, under certain circumstances, reduction of pain can also be achieved by just the opposite—exposure to another stressor. Two examples of this phenomenon are *stress-induced analgesia* (SIA) and *diffuse noxious inhibitory controls* (DNIC). SIA is characterized by decreased responsiveness to noxious stimulation following exposure to one of various stressors, including forced cold water swim, electric shock to the foot or tail, restraint, hypoglycemia, sudden loud noise, and defeat in aggressive encounters with conspecifics. The analgesia produced by some, but not all, of these stressors is reversible by naloxone, indicating that it is mediated by endogenous opioid systems. DNIC is characterized by decreased responsiveness to noxious stimulation following application of a distinctly noxious conditioning stimulus to another part of the body. The decrease in nociception is associated with a strong inhibition of convergent neurons within the dorsal horn, which is caused by the conditioning stimulus. It is hypothesized that the conditioning stimulus activates descending inhibitory controls originating in the brain stem (Le Bars *et al.,* 1992).

SIA and DNIC aside, veterinarians and investigators should make every effort to reduce significant stressors in the lives of laboratory animals. In particular, pain should be limited to levels not exceeding the animals' individual tolerance thresholds. This can be achieved by the administration of anesthetics or analgesics. However, reduction of the affective-motivational component of pain by reducing or eliminating environmental and psychological stressors is a desirable adjunct or alternative to drug treatment. With mild pain, decreasing nonpain stressors may be sufficient to bring pain within acceptable limits. With more severe pain, analgesics cannot be completely eliminated, but their effectiveness will be increased if nonpain stressors are minimized. In keeping with generally accepted husbandry practices, animals should be kept clean and dry, protected from extremes of temperature, and given sufficient quantities of clean water and palatable, nutritious food. Careful attention should be paid to the social environment of the animal. While social species may be seriously stressed by isolation, efforts must be made to protect vulnerable animals (e.g., those recovering from surgery) from aggression by cage or pen mates. A stable environment is particularly important. Changes in food, water, temperature, lighting, social groupings, or even position within the room all may be significant stressors and should be avoided when dealing with animals in pain. Investigators can do much to limit stress associated with experimental procedures. There are two basic approaches to minimizing experimentally induced stress: (1) plan studies with the likely responses of the animals in mind and, whenever possible, choose approaches and techniques that are less likely to cause significant pain or stress; and (2) allow time for the animals to acclimate to the research facility and the laboratory, and take the time required to accustom them to handling, restraint devices, and other specialized equipment prior to beginning the experiment.

VI. DETERMINING WHEN TO TREAT PAIN

Veterinarians and investigators have both a moral and a legal obligation to limit pain in experimental animals to that which is unavoidable in the conduct of valid scientific research [NRC, 1996; Public Health Service (PHS), 1986; Code of Federal Regulations (CFR)]. Beyond these obligations, however, control of pain in laboratory animals is both good medicine and good science. Aside from the general physiological changes associated with stress (Section V), pain has been shown to have specific effects on metabolism and on immune, pulmonary, cardiac, and gastrointestinal function. Postoperative pain in humans amplifies the stress-induced response to injury, resulting in an increased metabolic rate, accelerated loss of body protein, negative nitrogen balance, hyperglycemia, and disturbances in plasma electrolytes. The biochemical changes that may occur as a result of pain are discussed in more detail in Section VI,B, 2. Acute nonsurgical pain in experimental animals can accelerate the growth of tumors, increase mortality after experimental tumor inoculation, inhibit immune function, and inhibit cytotoxic activity of natural killer cells (Liebeskind, 1991). Chronic pain associated with experimentally induced mononeuropathy in rats was associated with increased delayed-type hypersensitivity and a decreased immunoglobulin G (IgG) response to keyhole limpet hemocyanin (KLH). A smaller decrease in the IgG response to KLH was observed in sham-operated rats, probably due to stress associated with surgery and acute pain (Herzberg *et al.,* 1994). Many, but not necessarily all, of these changes can be reduced or eliminated by the effective control of surgical pain (Brandt *et al.,* 1978; Yeager *et al.,* 1987; Lewis *et al.,* 1994). In humans, and undoubtedly in animals also, pain involving the upper abdomen or thorax causes a pattern of rapid, shallow breathing with reduced or absent deep breaths. This leads to reduced vital capacity, tidal volume, and functional residual capacity, which in turn predisposes the patient to atelectasis, hypoxemia, hypercarbia, retention of pulmonary secretions, and pulmonary infections (Cullen *et al.,* 1985; IASP, 1992; Lewis *et al.,* 1994). Pain-induced sympathetic stimulation causes vasoconstriction and increased heart rate, stroke volume, cardiac output, myocardial oxygen consumption, and blood pressure (IASP, 1992; Lewis *et al.,* 1994). Nociceptive impulses from viscera and somatic structures can lead to ileus, nausea and vomiting, and hypomotility of the urethra and bladder (IASP, 1992). Studies in humans have shown that superior pain control is associated with significantly reduced morbidity, a more rapid return to normal function, and a higher short-term survival rate following surgery (Brandt *et al.,* 1978; Cullen *et al.,* 1985; Yeager *et al.,* 1987; Scott and Kehlet, 1988; Lewis *et al.,* 1994).

There is no question that, except under unusual circumstances, every effort should be made to control excessive pain. However, there may be valid medical and scientific reasons for reserving the use of analgesic drugs for situations in which they are truly needed. The question is, how does one determine when intervention with analgesics is warranted? There are two basic approaches to making this determination. The simplest approach is to make the decision ahead of time that a particular procedure is likely to cause significant pain and to plan on treating all animals subjected to that procedure with an appropriate analgesic. With this approach, all animals that require analgesic treatment will receive it, but some, even many, that do not need it will also receive it. In addition, unless the animals are individually evaluated on a regular basis to assure adequacy of analgesia, the preselected treatment regimen may prove inadequate in some animals. The second approach is to evaluate each individual animal and, based on the presence or absence of specific signs indicative of pain, determine which animals should get analgesics and which should not. Depending on the knowledge, experience, and interpretations of the evaluator, there may be a relatively low or high incidence of incorrect assessments (i.e., animals falsely judged to be either in pain or pain free). In general, the best results will be obtained with a combination of these two approaches. Treatment of all animals subjected to a particular manipulation should be considered when the procedure is expected to cause moderate to severe pain and/or when trained, experienced personnel will not be available to evaluate the animals frequently (at least twice daily) for signs of discomfort. For animals that will be subjected to a procedure that is likely to cause moderate to severe pain, a preemptive approach to pain alleviation also should be considered (Section VI,A,2). Withholding treatment until it is deemed warranted by an experienced observer would be a reasonable approach in the case of procedures that are expected to cause milder pain.

A. Pain Prevention

1. Procedures that Cause Pain

It is generally possible to estimate the degree of discomfort that an animal is likely to experience during or after specific experimental procedures. Estimates of this nature are based on the incidence and severity of pain-related clinical signs that have been observed in other animals subjected to these procedures and on the degree of pain humans are known to experience following similar procedures. In general, it is valid to assume that procedures that cause pain in humans are likely to cause pain in animals. In fact, the U.S. Government Principles for the Utilization and Care of Vertebrate Animals Used in Testing, Research, and Training require investigators to make such assumptions in the absence of any evidence to the contrary (PHS, 1986). However, there are specific procedures which consistently appear to produce less postprocedural discomfort in animals than in humans. The most notable of these are surgical procedures involving midline abdominal incisions or thoracic intercostal incisions (Spinelli and Markowitz, 1987; Crane, 1987; Soma, 1987; NRC, 1992). It is hypothesized that animals experience less pain following abdominal incisions because ambulation in animals requires less use of abdominal musculature than it does in humans (Soma, 1987). Similarly, intercostal incisions may be less painful in animals because chest expansion during ventilation is normally not as great in animals as it is in humans (Soma, 1987). Thoracotomies involving sternal approaches usually are associated with severe postoperative pain in all species.

Table I lists the clinical signs and degree of pain that can be expected following surgical procedures on different tissues [Soma, 1987; Crane, 1987; Montgomery, 1990; Laboratory Animal Science Association (LASA) Working Party, 1990; NRC, 1992; FELASA, 1994]. Note that the clinical signs listed in Table I are primarily those that would be observed in dogs, so species-related differences in pain-related behavior should be taken into account when evaluating other laboratory animals. The degree of pain associated with these procedures is what would be expected if the procedures were performed by a gentle and skillful surgeon. Rough handling of tissues, allowing tissues to become dry, and/or generally poor surgical technique will result in increased postoperative pain. For example, although intercostal incisions appear to cause relatively little postoperative discomfort in animals, extensive retraction of the ribs will result in more severe pain. Increased inflammation, including that associated with infection, will also cause increased postoperative discomfort. Fortunately, these causes of increased postoperative pain can be prevented by assuring that the individuals who perform surgery on animals are well trained in both general surgical techniques and aseptic technique.

The Committee on Animal Research of the New York Academy of Sciences (NYAS) (1988) has published guidelines for determining the degree of discomfort or distress that would be expected to result from various surgical and nonsurgical experimental manipulations. These guidelines would be particularly useful for Institutional Animal Care and Use Committee (IACUC) members who must evaluate the severity of experimental procedures and determine what special precautions or stipulations should be considered prior to approving them.

2. Preemptive Analgesia

Preemptive analgesia refers to the reduction or prevention of pain by the administration of analgesics *prior to* the noxious insult (e.g., surgery). The preemptive approach is based on experimental evidence that the initial processing of nociceptive input results in changes within the CNS that augment the response to subsequent nociceptive input. It has long been known that repeated C-fiber stimulation causes a progressive increase in the responses of interneurons within the dorsal horn (Mendell,

TABLE I

Signs and Severity of Pain Associated with Experimental and Surgical Procedures

Tissue or site	Possible signs of pain	Severity of pain
Skin	Rubbing; licking; biting; scratching	Punctures/incisions: mild Burns/inflammation/scarification: moderate to severe
Muscle	Reluctance to move; lameness; rapid, shallow respirations	Mild to severe depending on site and degree of injury (e.g., may be severe following high amputation of hind leg)
Viscera	Reluctance to move; abnormal posture (e.g., hunched, stretched, "praying"); biting or kicking at abdomen; rolling or writhing; guarding	Mild to severe depending on organ injured and type of injury; greatest pain associated with distention or obstruction of hollow organ, ischemia, inflammation
Bones/joints	Reluctance to move; lameness; stiffness; guarding; licking; biting; self-mutilation	Bones (particularly humerus and femur): moderate to severe Joints: mild to severe (severity greatest with inflammation)
Spine	Cervical: reluctance to move, especially head; standing with head down; stiff gait Thoracic/lumbar: few signs	Cervical spine: moderate to severe Thoracic/lumbar: generally mild
Nervous system	Acute injury: none Postinjury neuropathy: licking, biting, self-mutilation	Acute injury: none Postinjury neuropathy: nonpainful abnormal sensations to moderate pain
Thorax	Reluctance to move; stiff gait; standing with front legs spread apart; rapid, shallow respirations	Lateral approach: mild to moderate Sternal approach: severe
Rectal area	Rubbing; licking; biting; abnormal excretory behavior	Moderate to severe
Abdomen	Arched back; hunched posture; abdomen tucked in; guarding; anorexia	Generally mild; moderate after extensive surgery
Eye	Rubbing; pawing or scratching at eye; blepharospasm; lids partially or completely closed	Intraocular or corneal injury: moderate to severe Enucleation: generally mild
Ear	Rubbing; pawing at ear; head shaking; head tilt	Moderate to severe
Mouth and teeth	Anorexia; abnormal mastication; excessive salivation; rubbing; pawing or scratching	Moderate to severe

1966). This phenomenon is referred to as *windup, central facilitation,* or *central sensitization.* More recent evidence shows that a similar phenomenon may occur within higher brain centers so that nociceptive input "primes" the brain to respond in an augmented fashion to subsequent nociceptive input (Vaccarino and Melzack, 1992). The net effect appears to be increased pain in response to noxious stimulation (hyperalgesia) and, perhaps, spontaneous pain in the absence of peripheral noxious stimulation. It has been hypothesized that administration of analgesics *before* the initial trauma (e.g., surgery) occurs will prevent the "priming" effect of nociceptive stimulation on the CNS and will reduce the amount of pain experienced in the posttrauma period. Furthermore, not only should pain be reduced by pretrauma administration of an analgesic, the reduction in pain should be greater than that obtained with the same regimen initiated after trauma.

Experimental studies of these phenomena have been performed using various animal models of protracted pain, including the formalin test. In this test, which is described in Section VII,B,1, pain occurs in two phases: (1) a short acute phase, which appears to be associated with direct activation of nociceptors; and (2) a more prolonged phase, which appears to be associated with the activation of nociceptors by inflammation. Numerous studies have shown analgesics given prior to the first phase are more effective in inhibiting both the neural and the behavioral responses to formalin than are the same analgesics given after the first phase has commenced (Dickenson and Sullivan, 1987; Coderre *et al.,* 1990; Malmberg and Yaksh, 1992). Such observations have led to the conclusion that the second phase is partially dependent on central sensitization, which enhances the nociceptive inputs associated with the inflammatory reaction and renders the pain less responsive to analgesics. Similarly, preinjury administration of a local anesthetic or morphine has been found to delay the onset and severity of pain-related behaviors in animal models of protracted neuropathic pain (Seltzer *et al.,* 1991; Puke and Wiesenfeld-Hallin, 1993).

Studies of preemptive analgesia in humans have yielded less consistent results. Most studies have shown that patients given analgesics prior to surgery had less postoperative pain or, at least, delayed onset of postoperative pain than those given no preoperative analgesics. These effects have been demonstrated with a variety of analgesics, including opioids, local anesthetics, nonsteroidal anti-inflammatory drugs (NSAIDs), and ketamine (Kiss and Kilian, 1992; Tverskoy *et al.,* 1994; Bartholdy *et al.,* 1994). However, although these studies demonstrated a beneficial effect of preoperative analgesia verus no analgesia, they did not investigate the effectiveness of preoperative analgesics versus the same analgesics given after surgery and thus did not demonstrate a true preemptive effect. In other studies, pain relief associated with the administration of NSAIDs and/or codeine in the preoperative *and* postoperative periods was compared with that obtained using the same drugs only postoperatively. Most, but not all, of these studies showed that patients had reduced or delayed postoperative pain when analgesics were started before surgery (Hill *et al.,* 1987; Dupuis *et al.,* 1988; Murphy and Medley, 1993). These studies were also inappropriately designed to demonstrate a true preemptive effect, however. Of the studies that did evaluate the effectiveness of preoperative analgesia versus the same intervention made after surgery, only a few have shown a preemptive effect, mostly with opioid analgesics. For example, both Katz *et al.* (1992) and Richmond *et al.* (1993) found that patient-controlled morphine use during the first 24 hr after abdominal or thoracic surgery was significantly less in patients treated preoperatively with opioid analgesics than in a similar group of patients given the same analgesics after the start of surgery. In contrast, most studies have shown no preemptive effect with either local anesthetics or NSAIDs (Hutchison *et al.,* 1990; Smith and Brook, 1990; Dahl *et al.,* 1992; Johansson *et al.,* 1994). Two studies did demonstrate a preemptive analgesic effect with local anesthetics, however. The time to first request for postoperative analgesia was significantly prolonged in abdominal surgery patients treated preoperatively versus postincisionally by wound infiltration with lignocaine (Ejlersen *et al.,* 1992) or epidural bupivacaine (Katz *et al.,* 1994). To date, only one study has demonstrated a preemptive effect with NSAIDs. Both postoperative pain and patient requests for postoperative analgesics were diminished when the NSAID diclofenac was given before as compared with after tonsillectomy (Nordbladh *et al.,* 1991).

Although clinical trials in humans have not provided strong support for the concept of preemptive analgesia, it is an approach that warrants further investigation. As more and better clinical trials are conducted in humans, analgesic regimens that can reliably preempt postoperative pain may be developed. Until then, veterinarians and investigators should keep in mind the repeatedly demonstrated ability of preoperatively administered analgesics to reduce or delay the onset of postoperative pain. These may not be true preemptive effects, but any measure that offers the potential for reducing pain in animals deserves serious consideration. The recommendation is to consider the use of preoperative analgesics for any experimental manipulation, including surgery, that is likely to cause moderate to severe discomfort. Based on available data, opioids would be the agents of choice for this purpose, as the evidence for a true preemptive analgesic effect is greatest for these drugs. However, it should be kept in mind that there is no evidence that preoperative analgesia can reliably eliminate the need for the postprocedural administration of analgesics. Animals should still be treated with, or at least monitored for the need for, analgesics starting about 4–8 hr after the end of the procedure.

B. Recognition and Assessment of Pain

It is generally recognized that pain consists of two components: sensation (nociceptive component) and the emotional or psychological response to that sensation (affective–motivational component). The physiological changes associated with the nociceptive component of pain can be measured, but the emotional component is not so easily quantitated. Even in humans, who can verbally describe their experience, pain defies precise measurement. In nonverbal beings such as animals, it is not possible to know their feelings and therefore not possible to measure pain. Instead, we must limit our efforts to studying nociception or *nociperception.* Nociperception refers to the conscious awareness of nociceptive stimulation and is often, although not always, accompanied by a behavioral response such as vocalization or withdrawal. It is the behavioral response that typically is measured as an indicator that the stimulus was perceived as unpleasant. This is not to say that animals experience no emotional response to the sensation. It is likely that they do, at least under some circumstances. But we must accept that it cannot be truly understood, let alone quantitated.

1. Behavioral Signs of Pain

The recognition of pain based on the behavior and appearance of an animal requires (1) careful observation; (2) knowledge of normal behavior and appearance; (3) knowledge of what signs or changes are associated with pain or distress; and (4) appropriate interpretation of the signs that are observed. Initial observations should be made from a distance, preferably in such a manner that the animal is unaware that it is being observed. Many animals—notably nonhuman primates—will make efforts to appear normal to observers even when they are sick or in pain, probably because obvious weakness or debilitation tends to have serious consequences in the wild. However, the same monkey that appears normal as you stand in front of its cage may show signs of pain if it can be viewed without it being aware of you (e.g., from outside the room). Observations should be made of the general appearance and condition, attitude, posture and movements, and interactions of the animal with cage/pen mates or other animals in the room. The animal can then be approached and, if appropriate, handled for a more detailed physical examination. Reactions of the animal to being approached and handled should be noted, as should responses to being palpated or manipulated. Food and water consumption and production of urine and feces are other important pieces of information. Accurate assessment of how much an animal is eating or drinking may require measuring food or taking an animal off of an automatic watering system and giving it water in a bottle or bowl. Placing a group-housed animal in a separate cage or pen may aid in the assessment of both food and water intake and production of urine and feces.

It is not possible to recognize abnormal appearance or behavior unless one is familiar with normal appearance and behavior. This requires specific knowledge of what is normal for that species and, in some cases, what is normal for that breed or strain. Ideally, one should know what is normal for the particular animal in question, although this is not always possible. It is useful to know how the animal normally behaves when approached or handled by humans. Is it quiet and docile? Does it vocalize and make vigorous attempts to escape? Does it become aggressive and try to bite? A normally quiet, docile animal that tries to hide or bite when approached should arouse suspicion, as should a normally wary or aggressive animal that does not resist handling. Similarly, it is important to know how a group-housed animal behaves with other animals. A mouse or pig that separates itself from cage mates probably is not normal. The normal feeding and drinking behavior of an animal is valuable information. Animals of some species (e.g., pigs) or breeds (e.g., hounds) are typically good eaters and will consume everything offered to them. However, it is not uncommon for animals of other species (e.g., monkeys) or breeds (e.g., huskies) to be picky eaters. Similarly, normal animals of some species will drink or waste a great deal of water (e.g., guinea pigs), whereas other animals will drink relatively little (e.g., gerbils, cats). If not aware of these differences, the observer will not be able to determine whether a partially full food hopper or water bottle is normal or a sign that something may be wrong.

Some of the signs of pain or distress may be observed in any species. These include dehydration, weight loss, behavioral depression, disturbances in sleep patterns, lameness, and self-mutilation. Some of the general signs of acute pain that may be observed following procedures on different tissues are listed in Table I. However, many of the potential signs of pain and distress are species specific and cannot be accurately interpreted without knowledge of the species in question. Tearing and accumulation of matter around the eyes may be seen in many species, but only in the rat is porphyrin staining around the eyes and nose a common indicator of abnormality. Sheep and goats with abdominal pain may kick at their bellies, but a dog or rabbit will not. An unkempt appearance is significant in an animal that grooms itself (e.g., a mouse or rat). Some animals, such as nonhuman primates, sheep, and gerbils, tend not to show obvious behavioral signs of pain, whereas others, such as cats, rabbits, and goats, are quicker to exhibit signs of pain or distress. For the former group, in particular, it is important to be alert for subtle signs of abnormality. The typical signs of pain and distress exhibited by some of the more common laboratory animals are listed in Table II [Association of Veterinary Teachers and Research Workers (AVTRW), 1986; Beynen *et al.,* 1987, 1988; Potthoff and Carithers, 1989; LASA Working Party, 1990; Montgomery, 1990; Molony and Wood, 1992; NRC, 1992; Canadian Council on Animal Care (CCAC), 1993; Choi *et al.,* 1994; FELASA, 1994]. In Table II, the signs of acute mild to moderate pain are differentiated from those of acute severe or chronic pain. Although these are useful guidelines, the reader should be aware that they are only guidelines and that an individual animal may deviate from them, e.g., by showing signs of ''mild'' pain in response to severe or chronic pain.

Interpretation of the signs that are observed is the most difficult aspect of assessing pain in an animal. Interpretation can be tricky for several reasons. (1) Although there are signs that are often associated with pain, there are no signs that are always exhibited by animals in pain or are never exhibited by animals that are not in pain. An example is vocalization. A cat that is in pain may growl and hiss when approached by a human, but a healthy cat may react in the same manner. A healthy guinea pig generally squeals and attempts to escape from humans, but it may also squeal in response to palpation of a painful area. However, a guinea pig in pain is more likely to be quieter and more docile than usual. Similar caveats apply to the evaluation of vocalizations in animals such as swine, rabbits, and nonhuman primates. (2) Along the same lines, no behaviors are absolutely indicative of pain versus nonpain-related distress. For example, vocalization associated with handling might be due to pain or fear, and anorexia and weight loss might be due to chronic pain or a nonpainful chronic illness. Based on behavior and/or appearance alone, the observer can do no more than conclude that the animal is or is not in distress. If the animal is judged to be in distress, its history and manipulations must be considered in deciding whether the distress is related to pain. (3) Pain related to a particular condition often is associated with particular changes in behavior or appearance, especially among animals of the same species. However, an animal may exhibit several of these changes or only one, and the signs that are shown will not be completely consistent even in animals of the same species with pain related to the same condition. The signs may not even be the same in the same animal examined at different times or by different people.

One of the approaches that has been used to render the assessment of pain more objective is the technique of scoring (Morton and Griffiths, 1985; AVTRW, 1986; Montgomery, 1990; Olfert, 1995). The first such system specifically designed for use with laboratory animals was proposed by Morton and Griffiths (1985). These authors suggested assessing five variables in each animal: body weight, appearance, clinical signs, unprovoked behavior, and responses to stimulation. Based on specified deviations from normal, each variable would be assigned a score of 0 (normal) to 3 (severely abnormal). The total score from all five variables could then be taken as an indication of the approximate degree of pain or distress the animal was experiencing and could be used to formulate an appropriate course of action (e.g., further observation to immediate intervention). Unfortunately, the practical application of such a system presents several problems, not the least of which is the time and effort that would be required to evaluate a rackful or roomful of animals. Furthermore, for a scoring system to be useful in assessing animals on a particular protocol, the parameters to be evaluated would have to be selected with care. Beynen *et al.* (1987, 1988) observed that the inclusion of parameters with poor predictive value for pain can result in lower overall pain scores and falsely suggest that animals in pain are experiencing no significant discomfort. These authors also observed a significant problem related to between-observer variation in the assignment of scores and suggested that the incidence of incorrect evaluations could be reduced by using more than one experienced, well-trained observer. Of course, these problems are not unique to the assessment of pain using a scoring system, but they suggest that the effort required to design and apply a scor-

TABLE II

Species-Typical Signs of Pain and Distress in Laboratory Animals

Species	Signs of pain or distress	
	Mild to moderate pain/distress	Severe or chronic pain/distress
Mouse	Eyelids partially closed; changes in respiration; rough hair coat; increased vibrissal movements; unusually apprehensive or aggressive; possible writhing, scratching, biting, self-mutilation; hunched posture; sudden running movements (escape); aggressive vocalization when handled or palpated; guarding	Weight loss; dehydration; incontinence; soiled hair coat; eyes sunken, lids closed; wasting of muscles on back; sunken or distended abdomen; decreased vibrissal movements; unresponsive; separates from group; hunched posture; ataxia; circling; hypothermia; decreased vocalization
Rat	Eyelids partially closed; porphyrin staining around eyes, nose; rough hair coat ± hair loss; increased aggression (toward humans and cage mates); reduced exploratory behavior; aggressive vocalization when handled; licking, biting, and/or scratching; guarding	Eyes closed; poor skin tone; muscle wasting along back; dehydration; weight loss; incontinence; soiled hair coat; depressed/unresponsive; sunken or distended abdomen; self-mutilation; recumbent position with head tucked into abdomen; decreased vocalization; hypothermia
Syrian hamster	Ocular discharge; increased aggression (toward humans and cage mates); hunched posture; reluctance to move	Loss of coat and body condition; increasing depression; extended daytime sleep periods; lateral recumbency; hypothermia; sores on lips, paws
Gerbil	Ocular discharge; eyelids partially closed and matted with dry material; may "faint" when handled; changes in activity and burrowing behavior; arched back; hunched posture	Loss of weight and condition; sores on face; hair loss on tail
Guinea pig	Eyes sunken and dull; changes in respiration; increased timidity; increased sleepiness; arched back; increased vocalization when handled	Weight loss; hair loss; scaly skin; dehydration; decreased timidity; unresponsive; excessive salivation (oral problems); increased barbering; loss of righting reflex; decreased vocalization; hypothermia
Rabbit	Ocular discharge; protruding nictitans; photophobia; constipation or diarrhea; depression; facing back of cage; excessive self-grooming; stretched posture; early failure to eat and drink; dull attitude or increased aggression when handled; possible vocalization when handled; tooth grinding	Tooth grinding; apparent sleepiness; dehydration; weight loss; fecal staining; wasting of lower back muscles; decreased production of night feces; unresponsive
Nonhuman primate	Generally very few signs, especially in the presence of humans; decreased activity; decreased food and water intake	Huddled or crouching posture, with hands folded over abdomen; clenching or grinding teeth; depression or increased restlessness; withdrawal from cage mates; increased (generally aggressive) attention from cage mates; anorexia; weight loss; decreased grooming
Dog	Decreased alertness; stiff posture; panting; biting, licking, or scratching; increased aggression; increased vocalization, especially when handled	Unwillingness to move; crouching posture; depression or increased aggression; crying when handled or moved; increased restlessness
Cat	Increased aggression when approached; decreased food intake; licking	Hunched, crouching or stretched posture; increased aggression, especially when palpated; anorexia; weight loss; vocalizing; wild escape behavior; unkempt appearance; pupillary dilation; stiff gait
Pig	Changes in gait or posture; increased efforts to avoid handling; increased squealing when approached or handled	Depression; unwillingness to move; attempts to hide; withdrawal from pen mates; anorexia
Sheep/goat	Sheep are more stoic than goats; lying with legs extended; stamping feet; swaying stance; mild ataxia; restlessness or depression; depressed food intake; increased aggression on handling; guarding; tooth grinding	Rolling; frequently looking or kicking at abdomen; falling over; walking backward; rapid, shallow respirations; weight loss; tooth grinding; grunting; vocalization on handling (goats especially); rigidity; unwillingness to move
Bird	Increased escape behavior and vocalization when approached or handled	Eyelids partially closed; anorexia; ruffled, drooping, unkempt appearance; immobility when approached

ing system might not pay off in terms of more accurate or reliable evaluations unless (1) care is taken to select only those variables that are highly likely to reflect pain in the particular species and situation in question and (2) the observers are well trained in its use. Otherwise, careful *subjective* evaluation by a trained and experienced observer is as likely, or more likely, to identify animals in pain or distress that require intervention. *Whether or not a scoring system is used, it is of the utmost importance that the evaluations be performed by well-trained, experienced personnel.*

2. Physiological and Biochemical Indicators of Pain

Injury and acute pain provoke a stress response that is characterized by stimulation of the sympathetic nervous system and activation of the pituitary–adrenal axis. This results in changes in the plasma concentrations of a wide variety of hormones and other substances. It also results in measurable changes in various physiological parameters. Physiological and biochemical changes that may occur in response to pain are listed in Table III (Brandt *et al.,* 1978; Breazile, 1987; Yeager *et al.,* 1987; IASP, 1992; Benson *et al.,* 1992; Lewis *et al.,* 1994). It is important to recognize that none of these changes is specific for pain, or even of nonpain-related stress. Alterations in any of these parameters must be evaluated in context before the conclusion is reached that the change(s) is due, at least in part, to pain.

3. Monitoring Pain during Surgery

Intraoperative monitoring is essential to assure that the animal is adequately anesthetized and feeling no pain. However, the need for adequate analgesia must be balanced against the need to avoid a dangerously deep plane of anesthesia. The anesthetist must be alert to signs indicating that the animal is perceiving pain while assuring that vital signs are being maintained at appropriate levels. Fortunately, many of the instruments that are routinely used to monitor the physiological status of the patient during surgery also can be used to monitor the adequacy of anesthesia.

The most obvious sign of inadequate depth of anesthesia is patient movement, either purposeful or in response to noxious stimulation. However, purposeful movements should never be seen intraoperatively, as this indicates a particularly inadequate plane of anesthesia. Movement in response to noxious stimulation is used routinely to titrate the depth of anesthesia before, and in the initial stages of, the surgical procedure. Similarly, this technique is used to determine when supplemental doses of injectable anesthetics are required during surgery. Initial titration of anesthetic depth is achieved by monitoring parameters such as the palpebral reflex, corneal reflex, and responses to pinching some part of the body, such as the foot, ear, tail, or abdominal skin. With the palpebral and corneal reflexes, inadequate anesthesia is indicated when the animal blinks in response to a light touch on the eyelid or cornea, respectively. The palpebral reflex is lost at a lighter plane of anesthesia than the corneal reflex. The pedal reflex is one of the most commonly used methods for monitoring depth of anesthesia. The hind leg of the animal is gently extended and the skin between the toes is pinched between the fingernails of the anesthetist or with an instrument such as a hemostat (it is important to be careful when using an instrument as tissue damage is neither necessary nor desirable). Responses that indicate inadequate anesthesia are vocalization, withdrawal of the leg, muscular twitching in the leg, or visible changes in respiration, such as breath-holding. Pinching other parts of the body will elicit similar responses if the animal is inadequately anesthetized. Once surgery begins, the anesthetist must watch for movement in response to more intense stimulation. An animal that has a negative pedal reflex may respond to incision of the skin, and an animal that does not respond to incision of the skin may respond to manipulation of deep somatic tissues or viscera. The usefulness of these different parameters for monitoring depth of anesthesia varies according to the species of animal and the specific anesthetic agent being used. All may be useful in larger animals such as dogs, cats, swine, ruminants, and nonhuman primates anesthetized with inhalant anesthetics or barbiturates. However, particular problems may be encountered with rodents and rabbits. For example, random spontaneous squirming movements may be seen in adequately anesthetized rodents, particularly guinea pigs. Similarly, in both guinea pigs and rabbits, the pedal reflex may not be lost until the animal is at a dangerously deep plane of anesthesia. For these species, the response to ear pinch (pinnae reflex) is a more useful index of anesthetic depth. The palpebral and corneal reflexes are not particularly useful in rodents and are notoriously unreliable indicators of anesthetic depth in rabbits. Even in the larger animals, the value of these reflexes is diminished when ketamine or a neuroleptanalgesic combination is used.

Physiological parameters that can be used to indicate depth of anesthesia and pain perception include respiratory rate, rhythm, and depth; heart rate; cardiac rhythm; and blood pressure. As indicators of pain, these parameters must be observed and interpreted with care by someone who has knowledge of both physiology and the specific agents being used for anesthesia. While pain is likely to cause an increase in respiratory rate, heart rate, or blood pressure, other variables may have the same

TABLE III

PHYSIOLOGICAL AND BIOCHEMICAL CHANGES OBSERVED IN ASSOCIATION WITH PAIN IN LABORATORY ANIMALS

	Biochemical changes	
Physiological changes	Increased plasma concentrations	Decreased plasma concentrations
Cardiovascular	Epinephrine, norepinephrine	Phosphorus
Vasoconstriction, increased blood pressure	Cortisol, corticosterone	Magnesium
Increased heart rate	Glucose	Testosterone
Increased stroke volume	Glucagon	Insulin
Increased cardiac output	Sodium	
Respiratory	β-Endorphin, enkephalins	
Rapid, shallow breathing	β-Lipotropin	
Decreased tidal volume	Substance P	
Decreased functional residual capacity	Amino acids	
Hypoxemia	Lipids	
Hypercapnia	Ketones	
	Vasoactive intestinal peptide	
Peripheral blood count	Renin, angiotensin II, aldosterone, vasopressin, antidiuretic hormone	
Lymphopenia		
Eosinopenia		
Neutrophilia		

effect or even mask the effect of pain. Other less commonly monitored parameters that may be useful indicators of pain perception include oxygen saturation, end-tidal CO_2, and the electroencephalogram (EEG). Quantitative electroencephalography, in particular, is attracting increasing attention as a method for monitoring pain perception during surgery in both humans and animals (Smith, 1987; Green *et al.,* 1992; Moore *et al.,* 1992). This technique, which involves computer processing of the raw EEG, requires specialized equipment as well as a highly skilled human interpreter. It clearly is not for every situation.

Monitoring the depth of anesthesia in an animal that has been paralyzed with a neuromuscular blocking agent is both critically important and challenging. It is vital that the animal be protected from feeling pain while paralyzed, but many of the techniques that may be used to monitor pain perception in an anesthetized animal are useless in a paralyzed animal. A paralyzed animal certainly cannot move in response to perception of painful stimulation, and most respiratory parameters cannot be assessed because the animal must be mechanically ventilated. Heart rate, blood pressure, and cardiac rhythm can be monitored in these animals, but the anesthetist must be aware that agents such as succinylcholine and gallamine can cause hypotension, arrhythmias, and changes in heart rate. Many investigators rely on end-tidal CO_2 to monitor pain in paralyzed animals. This is based on the premise that pain-related increases in catecholamines will cause an increase in CO_2 production (and a concomitant increase in oxygen consumption). To facilitate accurate interpretation of any of these parameters, it is best if the animal is at an adequate and stable plane of anesthesia prior to administration of the paralytic agent. Changes in physiological parameters associated with administration of the agent should be noted, then subsequent changes viewed as possible indicators of altered depth of anesthesia or pain perception. When consistent with the requirements of the experiment, allowing the anesthetized animal to recover periodically from the effects of the paralytic agent is desirable in that it provides opportunities for a more thorough evaluation of anesthetic depth.

VII. EXPERIMENTAL PAIN MODELS

Most of this chapter has dealt with pain that occurs as an undesirable side effect of experimental manipulation. However, IACUC members often must review protocols in which pain is the primary desired effect, and veterinarians must evaluate animals that have been used in such protocols. In either case, knowledge of the techniques used in, and expected outcomes of, pain research is required. This section reviews some of the more commonly used animal models of acute and chronic pain and highlights the major advantages and disadvantages of these models. This review will constitute neither a laundry list of all animal pain models nor a complete ''how-to'' description of the models that are discussed. References listed for every model can be consulted for those wishing a more in-depth description of that model.

A. Selected Models of Acute Pain

1. Simple Reflex Tests

With these tests, a reflex response to noxious stimulation is quantified as a measure of nociception. An advantage of most of them is that the pain involved is brief and generally mild. It is also escapable in that the stimulus is terminated when the reflex response is performed. A disadvantage to all of them is that these are spinal reflexes, not measures of pain perception, and they cannot be used to evaluate higher CNS functions involved in the pain experience. The tail flick response, for example, can be seen in animals with severed spinal cords (Irwin *et al.,* 1951). However, flexion reflexes in humans have been shown to be linearly correlated with subjective evaluations of pain (Chan and Dallaire, 1989).

a. TAIL FLICK. Since it was first described by D'Amour and Smith in 1941, the tail flick test has been widely used as an experimental model of nociception. It is most commonly used with rats and mice, but has also been used with larger animals such as monkeys and cats. It is a measure of acute cutaneous thermal pain and is generally considered to be a measure of nociceptive threshold. The test is based on withdrawal of the tail (tail flick) in response to noxious cutaneous thermal stimulation, usually from a focused beam of intense light or, less commonly, hot water into which the tail is dipped. An increase or reduction in the level of nociception is indicated by a shorter or longer latency to withdrawal of the tail, respectively. To protect against tissue damage, the test is terminated after a predetermined amount of time (e.g., 10–15 sec) if the animal does not withdraw its tail. The tail flick reflex has proven useful in predicting the analgesic efficacy of drugs, particularly opioid drugs, in humans. A particular disadvantage of this test is that the tail flick latency may vary with tail skin temperature (Berge *et al.,* 1988). A variation on this test involves the use of a CO_2 infrared laser as the noxious thermal stimulus (Danneman *et al.,* 1994). With this technique, the noxious stimulus is applied for less than 1 sec, resulting in synchronous activation of cutaneous nociceptors and generation of quantifiable, time-locked evoked potentials.

b. JAW-OPENING REFLEX. This test measures the reflex response to electrical stimulation of the tooth pulp. Depending on the species and the level of stimulation, this response may consist of licking or chewing, a head jerk, or dropping of the lower jaw. Electrodes are surgically implanted in the dentin of the incisor or canine tooth of the animal. A low-level current is passed through the electrodes to cause painful stimulation of the tooth. The current is increased incrementally until the appropriate response is seen. A decreased or increased level of nociception is indicated by an increase or decrease, respectively, in the amount of current needed to elicit the response. A more sensitive and sophisticated approach involves monitoring the electromyographic (EMG) activity in the digastricus muscle of the lower jaw. Nociception is measured by the current needed to elicit the EMG response or by the magnitude of the EMG response elicited by a current of fixed intensity and duration. This jaw-opening reflex has been studied in several species, including cat (Mitchell, 1964; Oliveras *et al.,* 1974), dog (Mitchell, 1964), and rabbit (Piercey and Schroeder, 1980). Tooth pulp stimulation (TPS) is generally regarded as an excellent noxious stimulus in that it is repeatable and quantifiable, and, at all but very low levels of stimulus intensity, it gives rise to purely painful sensations. Some investigators believe that this technique is not suitable for use in rodents (Hayashi, 1980; Engstrand *et al.,* 1983), but others have found them to be suitable subjects (Toda *et al.,* 1981; Sugimoto *et al.,* 1990; Danneman, 1994). Its major drawback in all species is the necessity of surgical instrumentation and postoperative maintenance of the animals.

In addition to the behavioral and EMG responses, TPS gives rise to a cortical evoked potential (CEP) in many species, including human (Chatrian *et al.,* 1975), monkey (Chudler *et al.,* 1986), cat (Roos *et al.,* 1982), dog (Chin and Domino, 1961), and rat (Hiranuma *et al.,* 1988; Danneman, 1994). This CEP can be measured precisely and its amplitude correlates well with stimulus intensity and subjective perception of pain. An increased or decreased level of nociception is indicated by an increase or decrease, respectively, in the amplitude of the evoked potential, particularly its early components. As a nonbehavioral, nonreflexive response, the CEP is not affected by nonnociceptive influences such as motor impairment or sedation (Danneman, 1994). However, it does require surgical placement of recording electrodes as well as the use of sophisticated recording equipment.

2. Organized Unlearned Behavioral Tests

These tests measure more complex responses to the types of stimuli that are used in the simple reflex tests. The required end point is a more complex behavior (e.g., vocalization or purposeful movement) that reflects supraspinal processing and, presumably, pain perception. As with the simple reflex tests, the pain involved is generally brief, mild, and escapable. There are three general disadvantages to these tests. (1) The end point is not as well defined as with the simple reflex tests. So, an animal that perceives the stimulus as noxious may react initially with the "wrong" response, thereby failing to terminate the test. (2) After repeated trials, the animal may learn the "correct" response and may perform it to terminate the test prior to actually perceiving pain (e.g., immediately upon being placed on the hot plate). (3) Performance on these tests can be affected by conditions such as sedation and abnormal motor function.

a. Hot Plate. The hot plate test was first described in 1944 by Woolfe and Macdonald and has since been modified (Cicero, 1974; O'Callaghan and Holtzman, 1975). This test is used with small laboratory animals, primarily rats and mice. The animal is placed on a heated (50–56° C) metal surface and the latency for the animal to jump or lick a hind paw is measured. An increase or reduction in the level of nociception is indicated by a shorter or longer latency, respectively, to jump or lick a hind paw. To protect against tissue injury, the test is terminated after a predetermined amount of time (e.g., 45 sec) if the animal does not either jump or lick a hind paw. A variant of the hot plate test involves noxious thermal stimulation of the face, which provokes a face-rubbing response (Rosenfeld *et al.,* 1983).

b. Flinch–Jump Test. With this test, electric shocks of fixed duration (e.g., 0.1–1 sec) are delivered at constant intervals (e.g., 30 sec) to the feet of an animal (usually a rodent) via the grid floor of a chamber in which the animal has been placed (Evans, 1961; Bonnet and Peterson, 1975; Crocker and Russell, 1984). Current levels generally vary between 0.025 and 0.55 mA. The shock intensity at which a behavioral reaction (reflexive flinch) is first observed is taken as the nociceptive threshold. With higher intensity stimuli, the animal responds by jumping off the floor. This is accompanied by other signs of discomfort such as vocalization or general agitation, indicating a level of nociception approximating the tolerance threshold. An increase or reduction in the level of nociception is indicated by a lower or higher current, respectively, required to elicit the flinch or jump response. It is important with this test that the same current intensity is delivered to all portions of the grid so that the animal will receive the same shock regardless of its position in the chamber. A variation of the flinch–jump test involves application of electric shock to the tail of a rodent (Levine *et al.,* 1984). The behavioral end point is vocalization. Higher shock intensities will cause vocalization that continues after termination of the stimulus, which is interpreted as an emotional response to the stimulus.

c. Colorectal Distention. The colorectal distention test measures the nociceptive threshold for acute visceral pain, which is produced by distention of the colon with a balloon catheter. It is based on the premise that pain is the only subjective sensory consequence of colorectal distention. The visceromotor response, which results from balloon distention of the colon, is quantifiable and reliably reproducible. This response consists of a characteristic abdominal contraction, which can be measured and recorded via a strain gauge around the abdomen of the animal. An increased or reduced level of nociception is indicated by a smaller or larger inflation pressure, respectively, required to elicit the characteristic nocifensive response. The stimulus itself is known to be noxious, based on human studies, but will not produce tissue damage if care is taken to avoid significantly exceeding the nociceptive threshold. The visceromotor response is supraspinally organized, is reduced by analgesic drugs such as morphine, and is eliminated by deep anesthesia. The colorectal distention test has been used in a variety of species, including rats (Ness and Gebhart, 1988;

Danzebrink and Gebhart, 1990), dogs (Sawyer *et al.,* 1991), cats (Sawyer and Rech, 1987), and horses (Pippi and Lumb, 1979).

d. RANDALL–SELITTO TEST. This test was first described in 1957 by Randall and Selitto. It is used primarily with rats and mice and involves placing the paw of an animal on a plinth. A rounded test pin is lowered so that it just touches the dorsal surface of the paw, after which pressure is applied and increased linearly at a constant rate. The test is terminated when the animal vocalizes or begins to struggle to remove its paw. The pressure at that point is recorded. An increase or reduction in the level of nociception is indicated by a lesser or greater pressure, respectively, needed to elicit vocalization or paw withdrawal. To prevent tissue damage, the test is terminated if the animal does not respond to a predetermined maximum pressure (e.g., 1000 *g*). The Randall–Selitto test is often used to measure hyperalgesia associated with experimentally induced inflammation (Winter and Flataker; 1965; Joris *et al.,* 1990). The pressure required to elicit vocalization or withdrawal of an inflamed paw is compared to that required to elicit vocalization or withdrawal of the same animal's uninflamed paw.

3. Tests Involving Complex Learned Behaviors

With these tests, animals are trained to modify their behavior to indicate recognition of, or unwillingness to tolerate, noxious stimulation (Chapman *et al.,* 1985; Vierck *et al.,* 1989; Dubner, 1994). These tests, especially the more complex paradigms, offer several advantages over other tests of nociception. The latency and vigor of the response are good indicators of the degree of aversiveness of the stimulus, especially when responses to stimuli of different intensities within the noxious range are recorded. With the more complex paradigms, confounding variables such as attention and motivation can be controlled. All of these tests permit the animal to escape the noxious stimulus and many of them give the animal a particularly high degree of control over stimulus intensity. A disadvantage of these tests is the time and effort required to train the animal. With some of the more complex paradigms, this may require months. Although the simplest of these tests are suitable for most species, the more complex paradigms are suitable only for the more intelligent species, such as nonhuman primates.

a. LEARNED ESCAPE/AVOIDANCE PARADIGMS. These are the simplest of the learned behavior paradigms. With the escape paradigms, the animal is taught to perform a task to terminate a noxious stimulus. For example, the animal must jump over a partition to another part of the cage to escape an electric shock delivered through a floor grid. A greater or lesser degree of nociception is indicated by a shorter or longer latency to perform the behavior required to terminate the stimulus. With the avoidance paradigms, the animal is taught to perform a task to prevent a noxious stimulus. For example, the animal might be placed in a chamber in which electric shocks of increasing intensity are delivered at regular intervals through a grid floor. The animal must press a bar to reduce the intensity of each successive shock. It is assumed that the animal will work to maintain the shock intensity well below the tolerance threshold, although just how far below that threshold cannot be determined. As with the organized unlearned behavioral tests, performance on the learned escape/avoidance tests may be affected by conditions such as sedation and abnormal motor function.

b. CONFLICT (APPROACH–AVOIDANCE) PARADIGMS. These are more sophisticated tests that require an animal to choose what level of nociception it will tolerate in order to receive a reward. The animal is taught to perform some task (e.g., bar press) to receive a reward (e.g., food or water), which is accompanied by a noxious stimulus (e.g., electric shock). At some level of stimulus intensity, the animal will refuse to work for the reward. Food or water restriction is used to reduce variability in performance associated with motivation. Parallel studies in humans and nonhuman primates have shown that the monkeys and humans will tolerate approximately the same level of noxious stimulation (Dubner, 1994).

c. DISCRIMINATION PARADIGMS. These are highly complex paradigms in which animals (nonhuman primates) are taught to distinguish between different stimuli (e.g., electric shocks), at least one of which is in the noxious range. Lineberry and Kulics (1978) described a model in which monkeys were taught to press a bar to indicate recognition of the more intense of two mildly noxious electric shocks and to refrain from bar pressing to indicate recognition of the less intense stimulus. In this model, the animals were punished for incorrect responses, whereas in other models they may be rewarded for correct responses. Once the animals are reliably distinguishing between the different stimuli, the effects of analgesics or other manipulations on their ability to distinguish can be measured (e.g., an analgesic might render the animals incapable of distinguishing a noxious stimulus from a nonnoxious stimulus).

B. Selected Models of Chronic Pain

Until fairly recently, most experimental pain research in animals has involved models of acute nociception. However, in humans, chronic pain is of far greater clinical importance than acute pain and, as discussed earlier, the mechanisms of, and effective therapies for, acute and chronic pain are different. As a result, animal models of protracted pain have proliferated and their use has increased tremendously. The primary drawback to all of these models is the necessity of causing the animal to experience inescapable pain which may last up to several weeks. Furthermore, unlike most models of acute pain, these models involve the production of tissue damage.

1. Models of Inflammatory Pain and Hyperalgesia

a. FORMALIN TEST. The formalin test was first described as a method for producing prolonged experimental pain in cats and rats by Dubuisson and Dennis in 1977. It employs a chemical stimulus (0.025–5% buffered formalin) that is injected into the subcutaneous tissues of the hind foot and causes pain of prolonged duration. Typically, the responses on the formalin test occur in two distinct phases which are believed to reflect two distinct processes. The first phase, which has an immediate onset and short duration (1–5 min), appears to reflect the nocifensive response to the direct activation of nociceptors. In humans, the sensation produced is one of sharp, burning pain. Rats and mice will typically react by holding the foot up and shaking, biting, and/or licking it. After a few minutes, most animals will cease to favor the foot and will ambulate normally. The second phase, which is of prolonged duration, starts approximately 10–20 min following injection of formalin. It is believed to reflect the nocifensive response to inflammation. In humans, the sensation is one of throbbing, aching pain. During this phase, the animals will once again start to elevate, shake, bite, and lick the injected foot. In general, this behavior will gradually diminish over about a 45-min period. Responses during both phases are reduced or eliminated by analgesic and anti-inflammatory drugs. The formalin test is valued because it involves both acute and protracted pain. The major problem with this test relates to determining the degree of discomfort experienced by the animal based on its behavioral responses. Different scoring systems have been devised to address this problem (Hunskaar *et al.,* 1985; Wheeler-Aceto and Cowan, 1991; Tjolsen *et al.,* 1992; Coderre *et al.,* 1993).

b. CARRAGEENAN TEST. Carrageenan is a mucopolysaccharide extracted from *Chondrus,* an Irish sea moss. Following SC injection into the plantar surface of the foot of a rat, it produces an inflammatory response characterized by localized hyperthermia and edema (Van Arman *et al.,* 1965; Planas *et al.,* 1995). The extent and duration of the inflammatory response depend on the amount of carrageenan injected. It may last as long as several days. For the first 7–10 min after the injection of carrageenan, rats respond by licking the injected foot. After that, there is little or no spontaneous behavioral response indicative of discomfort. However, throughout the period of inflammation and peaking at the point of peak inflammation, rats show an enhanced behavioral responsiveness to noxious thermal or mechanical stimulation (e.g., Randall–Selitto test). This hyperalgesic response is inhibited by opioid analgesics and nonsteroidal anti-inflammatory drugs. Inflammation in this test is measured by paw volume (edema) and skin surface temperature (hyperthermia). Variations on the carrageenan test involve the application or injection of other substances to produce inflammation and hyperalgesia. These include kaolin, acetic acid, yeast, Freund's complete adjuvant (CFA), mustard oil, and capsaicin (Van Arman *et al.,* 1965; Wheeler-Aceto *et al.,* 1990; LaMotte *et al.,* 1991; Iadarola *et al.,* 1988).

2. Freund's Complete Adjuvant-Induced Polyarthritis

Within 10–21 days after injection of CFA into the tail vein, rats develop a delayed hypersensitivity reaction that involves inflammation with swelling, edema and deformity, hyperalgesia, and destruction of bone and cartilage affecting multiple joints (De Castro Costa *et al.,* 1981; Butler *et al.,* 1985; Coderre and Wall, 1987). The animals show spontaneous behaviors suggesting pain, including licking and biting the affected limbs, decreased ambulatory and exploratory behavior, and weight loss. They vocalize when pressure is applied to the affected joints. They also show what has been described as a ''scratching'' behavior, where they vibrate the elevated hind foot near the ear. This is generally followed by biting the foot. Some investigators have reported a spontaneous decrease in some clinical signs 4–7 weeks after administration of CFA (De Castro Costa *et al.,* 1981), whereas others have seen no such decrease in signs (Butler *et al.,* 1985). A specific problem with this model is that the animals become systemically ill, with ulcerative skin lesions, impairment of hepatic function, and lymphadenopathy (Butler *et al.,* 1985; Coderre and Wall, 1987). Not only does this raise ethical issues beyond those associated with chronic pain, it makes it more difficult to determine which behavioral and physiological changes are due to pain and which are due to illness. A variation of this model involves the injection of sodium urate crystals into the ankle joint (Coderre and Wall, 1987). Within 24 hr, the ankle becomes swollen, inflamed, and hyperalgesic, and the animals become lame and lose weight. However, there is no destruction of bone or systemic illness. The arthritis peaks within 3 days, then begins to decline.

3. Nerve Injury Models

The nerve injury models are animal models of neuropathic pain. Among the many different models in this category are those which involve complete peripheral denervation (e.g., cutting, freezing, crushing, or ligating the sciatic nerve), partial peripheral denervation (e.g., loosely ligating or partially cutting the sciatic nerve), localized CNS injury (e.g., ischemic injury to the spinal cord, stereotaxic injection of toxic agents into specific brain centers), and nerve injury associated with systemic disease (e.g., streptozotocin-induced diabetic neuropathy). The different models are described by Zeltser and Seltzer (1994). Depending on the site and extent of injury, these models are associated with varying degrees and durations of abnormal behaviors, including autotomy (self-mutilation); guarding, licking, or biting a paw; excessive grooming; vocalization; and abnormal ambulation. Nonspecific signs of pain/distress may also be seen, such as weight loss, failure to groom, abnormal sleep patterns, and chromodacryorrhea. In addition, animals subjected to these treatments show stimulus-evoked responses such as hyperalgesia, hypoalgesia, allodynia, and sympathetic dependence (increased pain in association with sympathetic stimulation). These models are of considerable scientific importance because they permit the experimental investigation of some of the most troublesome and poorly understood forms of human pain. However, their use also raises significant ethical questions. The pain produced is prolonged, inescapable, and, to judge from the behavior of the animals, possibly severe. Ironically, the other major criticism of these models is that the behaviors of the animals, particularly autotomy, may reflect abnormal nonpainful sensations rather than nociception. This question was addressed in a review article by Coderre *et al.* (1986).

4. Abdominal Writhing

The writhing test was first described by Siegmund *et al.* in 1957. It has been used in rats, but is more often used in mice because the behavioral response is more readily observable in them. Within 3–10 min after intraperitoneal injection of an irritating substance (e.g., phenylbenzoquinone), the mouse begins to ''writhe.'' The writhing response consists of arching of the back, extension of the hind limbs, and contraction of the abdominal musculature. This response may last up to 60 min. Nociception is indicated by either (1) the presence (nociception) or absence (analgesia) of the writhing response or (2) the number of writhes occurring within a specified period of time (more writhes equal more pain). The writhing test is considered to be a reflexive test of protracted visceral nociception. The primary criticism of this test is that it may not truly reflect nociception. The test has not been performed in humans and because it has no clinical counterpart, the sensations involved are unknown. Furthermore, animals show no other signs (e.g., vocalization) to indicate that it is aversive. However, it may involve prolonged severe pain, which raises ethical concerns regarding its use.

C. Reviewing Experimental Pain Protocols

Among the most difficult experimental protocols to review are those involving the intentional infliction of pain and the deliberate withholding of analgesia. This section will offer some guidelines for the review of such protocols. Studies in which animals will be anesthetized during application of the painful stimulus present no problem other than assuring that the subjects will be adequately anesthetized. Studies of acute pain in awake animals make reviewers more uncomfortable. However, most of the experimental models of acute pain involve the use of noxious stimuli that are mild, brief, and escapable. Assuming that the tests are conducted properly and that the animals are acclimated to the experimental equipment, the animals used in such studies should experience less pain or distress than they would during recovery from a minor surgical procedure. The most serious concerns arise relative to the use of animals in chronic pain studies. The pain involved in these studies is by definition inescapable and of prolonged duration. The intensity of the pain varies according to the model and the perceptions of the individual animals, but may be severe. Although it might be tempting for an IACUC to simply disallow investigators to study chronic pain in animals, studies of this nature are essential. Knowledge regarding the pathogenesis and treatment of chronic pain is desperately needed. It has been estimated that chronic pain affects one in three Americans, for many of whom there is no effective treatment, and costs the U.S. economy about $90 billion each year (Gelb, 1994).

The Committee for Research and Ethical Issues of the International Association for the Study of Pain has published guidelines for the conduct of experimental pain research (Zimmerman, 1983). They can be summarized—with some interpretation and extrapolation by the author of this chapter—as follows. (1) The investigator must provide the IACUC with sufficient and continuing justification of the need for his/her studies, just as the IACUC must provide a thorough review of the proposed research. Both review and justification should be particularly rigorous in the case of chronic pain studies. (2) If possible, the investigator should try the pain stimulus on himself/herself and avoid using any stimulus on an animal that he/she would perceive as intolerable. Some stimuli cannot be tried by the investigator, but many of these have clinical counterparts in humans (e.g., colorectal distention, arthritis) and the degree of pain involved can be estimated. (3) To fully understand the effect of the stimulus on the behavior of the animal, both physiological and behavioral parameters should be measured and the data included in the published paper. (4) The animal should be exposed to the minimum amount of pain necessary to achieve the desired experimental results. (5) Animals in chronic pain studies should be treated with analgesics whenever this would not interfere with the purpose of the study. This might mean administering analgesics at some point(s) during the study, while withholding them at another time(s). (6) Neuromuscular blocking agents should be used only in animals that are anesthetized or have been subjected to a surgical procedure that eliminates sensory awareness (e.g., decerebration). (7) The duration of the experiment should be as short as possible, especially in the case of chronic pain studies, and the number of animals involved kept to the minimum required for statistical significance. Whenever possible, investigators should choose chronic pain models in which the duration of the pain is comparatively short (e.g., formalin test, sodium urate crystal-induced arthritis). Animals that are suffering debilitating pain or distress should be treated with appropriate analgesics or euthanized.

REFERENCES

Arena, P. C., and Richardson, K. C. (1990). The relief of pain in cold-blooded vertebrates. *ACCART News* **3,** 1–4.

AVMA Panel on Euthanasia (1993). Report of the AVMA Panel on Euthanasia. *J. Am. Vet. Med. Assoc.* **202,** 230–249.

Association of Veterinary Teachers and Research Workers (AVTRW) (1986). Guidelines for the recognition and assessment of pain in animals. *Vet. Rec.* **11,** 334–338.

Bahns, E., Ernsberger, U., Jänig, W., and Nelke, A. (1986). Functional characteristics of lumbar visceral afferent fibers from the urinary bladder and the urethra in the cat. *Pflügers Arch.* **407,** 510–518.

Baker, M., and Bostock, H. (1992). Ectopic activity in demyelinated spinal root axons of the rat. *J. Physiol. (London)* **451,** 539–552.

Bartholdy, J., Sperling, K., Ibsen, M., Eliasen, K., and Mogensen, T. (1994). Preoperative infiltration of the surgical area enhances postoperative analgesia of a combined low-dose epidural bupivacaine and morphine regimen after upper abdominal surgery. *Acta Anaesthesiol. Scand.* **38,** 262–265.

Bennett, G. J. (1994). Chronic pain due to peripheral nerve damage: An overview. *In* "Pharmacological Approaches to the Treatment of Chronic Pain: New Concepts and Critical Issues" (H. L. Fields and J. C. Liebeskind, eds.), Prog. Pain Res. Manage., Vol. 1, Chapter 5. IASP Press, Seattle, Washington.

Benson, G. J., Lin, H. C., Thurman, J. C., Olson, W. A., and Tranquilli, W. J. (1992). Assessment of analgesia by catecholamine analysis: Response to onychectomy in cats. *In* "Animal Pain" (C. E. Short and A. Van Poznak, eds.), pp. 436–439. Churchill-Livingstone, New York.

Berge, O-G., Garcia-Cabrera, I., and Hole, K. (1988). Response latencies in the tail-flick test depend on tail skin temperature. *Neurosci. Lett.* **86,** 284–288.

Berkley, K. J., Robbins, A., and Sato, Y. (1988). Afferent fibers supplying the uterus in the rat. *J. Neurophysiol.* **59,** 142–163.

Beynen, A. C., Baumans, V., Bertens, A. P. M. G., Havenaar, R., Hesp, A. P. M., and Van Zutphen, L. F. M. (1987). Assessment of discomfort in gallstone-bearing mice: A practical example of the problems encountered in an attempt to recognize discomfort in laboratory animals. *Lab. Anim.* **21,** 35–42.

Beynen, A. C., Baumans, V., Bertens, A. P. M. G., Haas, J. W. M., Van Hellemond, K. K., Van Herck, H., Peters, M. A. W., Stafleu, F. R., and Van Tintelen, G. (1988). Assessment of discomfort in rats with hepatomegaly. *Lab. Anim.* **22,** 320–325.

Bonica, J. J. (1953). "The Management of Pain." Lea & Febiger, Philadelphia.

Bonica, J. J. (1992). Pain research and therapy: History, current status, and future goals. *In* "Animal Pain" (C. E. Short and A. Van Poznak, eds.), pp. 1–29. Churchill-Livingstone, New York.

Bonnet, K. A., and Peterson, K. E. (1975). A modification of the jump-flinch technique for measuring pain sensitivity in rats. *Pharmacol., Biochem. Behav.* **3,** 47–55.

Brandt, M. R., Fernandes, A., Mordhorst, R., and Kehlet, H. (1978). Epidural analgesia improves postoperative nitrogen balance. *Br. Med. J.* **1,** 1106–1108.

Breazile, J. E. (1987). Physiologic basis and consequences of distress in animals. *J. Am. Vet. Med. Assoc.* **191,** 1212–1215.

Butler, S. H., Weil-Fugazza, J., Godefroy, F., and Besson, J.-M. (1985). Reduction of arthritis and pain behaviour following chronic administration of amitriptyline or imipramine in rats with adjuvant-induced arthritis. *Pain* **23,** 159–175.

Byas-Smith, M. G., Max, M. B., Muir, J., and Kingman, A. (1995). Transdermal clonidine compared to placebo in painful diabetic neuropathy using a two-stage 'enriched enrollment' design. *Pain* **60,** 267–274.

Calvin, W. H., Devor, M., and Howe, J. (1982). Can neuralgias arise from minor demyelination? Spontaneous firing, mechanosensitivity and afterdischarge from conducting axons. *Exp. Neurol.* **75,** 755–763.

Canadian Council on Animal Care (CCAC) (1993). Control of animal pain in research, teaching and testing. *In* "Guide to the Care and Use of Experimental Animals," 2nd ed., Vol. 1, pp. 115–127. Canadian Council on Animal Care, Ontario.

Cervero, F. (1982). Afferent activity evoked by natural stimulation of the biliary system in the ferret. *Pain* **13,** 137–151.

Cervero, F. (1988). Visceral pain. *In* "Proceedings of the Vth World Congress on Pain" (R. Dubner, G. F. Gebhart, and M. R. Bond, eds.), Pain Res. Clin. Manage., Vol. 3, pp. 216–226. Elsevier, Amsterdam.

Cervero, F. (1994). Sensory innervation of the viscera: Peripheral basis of visceral pain. *Physiol. Rev.* **74,** 95–138.

Cervero, F., and Jänig, W. (1992). Visceral nociceptors: A new world order? *Trends Neurosci.* **15,** 374–378.

Cervero, F., and Sann, H. (1989). Mechanically evoked responses of afferent fibers innervating the guinea-pig's ureter: An in vitro study. *J. Physiol. (London)* **412,** 245–266.

Chan, C. W. Y., and Dallaire, M. (1989). Subjective pain sensation is linearly correlated with the flexion reflex in man. *Brain Res.* **479,** 145–150.

Chapman, C. R., Casey, K. L., Dubner, R., Foley, K. M., Gracely, R. H., and Reading, A. E. (1985). Pain measurement: An overview. *Pain* **22,** 1–31.

Chatrian, G. E., Canfield, R. C., Knauss, T. A., and Lettich, E. (1975). Cerebral responses to electrical tooth pulp stimulation in man. *Neurology* **25,** 745–757.

Chin, J. H., and Domino, E. F. (1961). Effects of morphine on brain potentials evoked by stimulation of the tooth pulp of the dog. *J. Pharmacol. Exp. Ther.* **132,** 74–86.

Choi, Y., Yoon, Y. W., Na, H. S., Kim, S. H., and Chung, J. M. (1994). Behavioral signs of ongoing pain and cold allodynia in a rat model of neuropathic pain. *Pain* **59,** 369–376.

Chudler, E. H., Dong, W. K., and Kawakami, Y. (1986). Cortical nociceptive responses and behavioral correlates in the monkey. *Brain Res.* **397,** 47–60.

Cicero, T. J. (1974). The effects of alpha-adrenergic blocking agents on narcotic-induced analgesia. *Arch. Int. Pharmacodyn. Ther.* **208,** 5–13.

Code of Federal Regulations (CFR). Title 9 (Animals and Animal Products), Subchapter A (Animal Welfare), Parts 1–3. CFR, Washington, D. C.

Coderre, T. J., and Wall, P. D. (1987). Ankle joint urate arthritis (AJUA) in rats: An alternative animal model of arthritis to that produced by Freund's adjuvant. *Pain* **28,** 379–393.

Coderre, T. J., Grimes, R. W., and Melzack, R. (1986). Deafferentation and chronic pain in animals: An evaluation of evidence suggesting autotomy is related to pain. *Pain* **26,** 61–84.

Coderre, T. J., Vaccarino, A. L., and Melzack, R. (1990). Central nervous system plasticity in the tonic pain response to subcutaneous formalin injection. *Brain Res.* **535,** 155–158.

Coderre, T. J., Fundytus, M. E., McKenna, J. E., Dalal, S., and Melzack, R. (1993). The formalin test: A validation of the weighted-scores method of behavioural pain rating. *Pain* **54,** 43–50.

Crane, S. W. (1987). Perioperative analgesia: A surgeon's perspective. *J. Am. Vet. Med. Assoc.* **191,** 1254–1257.

Crocker, A. D., and Russell, R. W. (1984). The up-and-down method for the determination of nociceptive thresholds in rats. *Pharmacol., Physiol. Behav.* **21,** 133–136.

Cullen, M. L., Staren, E. D., El-Ganzouri, A., Logas, W. G., Ivankovich, A. D., and Economou, S. G. (1985). Continuous epidural infusion for analgesia after major abdominal operations: A randomized, prospective, double-blind study. *Surgery* **98,** 718–726.

Dahl, J. B., Hansen, B. L., Hjortso, N. C., Erichsen, C. J., Moinische, S., and Kehlet, H. (1992). Influence of timing on the effect of continuous extradural analgesia with bupivacaine and morphine after major abdominal surgery. *Br. J. Anaesth.* **69,** 4–8.

D'Amour, F. E., and Smith, D. L. (1941). A method for determining loss of pain sensation. *J. Pharmacol. Exp. Ther.* **72,** 74–79.

Danneman, P. J. (1994). Cortical potentials evoked by tooth pulp stimulation differentiate between the analgesic and sedative effects of morphine in awake rats. *J. Pharmacol. Exp. Ther.* **269,** 1100–1106.

Danneman, P. J., Kiritsy-Roy, J. A., Morrow, T. J., and Casey, K. L. (1994). Central delay of the laser-activated rat tail flick reflex. *Pain* **58,** 39–44.

Danzebrink, R. M., and Gebhart, G. F. (1990). Antinociceptive effects of intrathecal adrenoceptor agonists in a rat model of visceral nociception. *J. Pharmacol. Exp. Ther.* **253,** 698–705.

Daunt, D. A., and Maze, M. (1992). Alpha$_2$-adrenergic agonist receptors, sites, and mechanisms of action. *In* "Animal Pain" (C. E. Short and A. Van Poznak, eds.), pp. 165–180. Churchill-Livingstone, New York.

De Castro Costa, M., De Sutter, P., Gybels, J., and Van Hees, J. (1981). Adjuvant-induced arthritis in rats: A possible animal model of chronic pain. *Pain* **10,** 173–185.

Dickenson, A. H., and Sullivan, A. F. (1987). Subcutaneous formalin-induced activity of dorsal horn neurones in the rat: Differential response to an intrathecal opiate administered pre or post formalin. *Pain* **30,** 349–360.

Dorland's (1981). "Dorland's Illustrated Medical Dictionary," 26th ed. Saunders, Philadelphia.

Dubner, R. (1994). Methods of assessing pain in animals. *In* "Textbook of Pain" (P. D. Wall and R. Melzack, eds.), 3rd ed., pp. 293–302. Churchill-Livingstone, London.

Dubuisson, D., and Dennis, S. G. (1977). The formalin test: A quantitative study of the analgesic effects of morphine, meperidine, and brain stem stimulation in rats and cats. *Pain* **4,** 161–174.

Dupuis, R., Lemay, H., Bushnell, M. C., and Duncan, G. H. (1988). Preoperative flurbiprofen in oral surgery: A method of choice in controlling postoperative pain. *Pharmacotherapy* **8,** 193–200.

Ejlersen, E., Andersen, H. B., Eliasen, K., and Mogensen, T. (1992). A comparison between preincisional and postincisional lidocaine infiltration and postoperative pain. *Anesth. Analg.* **74,** 495–498.

Engstrand, P., Shyu, B.-C., and Andersson, S. A. (1983). Is a selective stimulation of the rat incisor tooth pulp possible? *Pain* **15,** 27–34.

Evans, W. O. (1961). A new technique for the investigation of some analgesic drugs on a reflexive behavior in the rat. *Psychopharmacology* **2,** 318–325.

FELASA Working Group on Pain and Distress (1994). Pain and distress in laboratory rodents and lagomorphs. *Lab. Anim.* 28, 97–112.

Fields, H. L., and Basbaum, A. I. (1994). Central nervous system mechanisms of pain modulation. *In* ''Textbook of Pain'' (P. D. Wall and R. Melzack, eds.), 3rd ed., pp. 243–257. Churchill-Livingstone, London.

Foong, F. W., and Duggan, A. W. (1986). Brainstem areas tonically inhibiting dorsal horn neurons: Studies with microinjection of the GABA analogue piperidine-4-sulphonic acid. *Pain* **27,** 361–372.

Gebhart, G. F. (1994). Pain and distress in research animals. *In* ''Research Animal Anesthesia, Analgesia and Surgery'' (A. C. Smith and M. M. Swindle, eds.), pp. 37–40. Scientists Center for Animal Welfare, Greenbelt, Maryland.

Gelb, R. L. (1994). Forward. *In* ''Pharmacological Approaches to the Treatment of Chronic Pain: New Concepts and Critical Issues'' (H. L. Fields and J. C. Liebeskind, eds.), Prog. Pain Res. Manage., Vol. 1, p. xi, IASP Press, Seattle, Washington.

Gildenberg, P. L., and DeVaul R. A. (1985). ''The Chronic Pain Patient: Evaluation and Management,'' Pain and Headache, Vol. 7. Karger, Basel.

Gogas, K. R., Presley, R. W., Levine, J. D., and Basbaum, A. I. (1991). The antinociceptive action of supraspinal opioids results from an increase in descending inhibitory control: Correlation of nociceptive behavior and c-fos expression. *Neuroscience* **42,** 617–628.

Green, S. A., Moore, M. P., Keegan, R. D., Gallagher, L. V., and Rosenthal, J. C. (1992). Quantitative electroencephalography for monitoring responses to noxious electrical stimulation in dogs anesthetized with halothane or with halothane and morphine. *In* ''Animal Pain'' (C. E. Short and A. Van Poznak, eds.), pp. 459–465. Churchill-Livingstone, New York.

Guilbaud, G., and Benoist, J.-M. (1994). Central transmission of somatosensory inputs in the thalamic ventrobasal complex and somatosensory cortex in rat models of clinical pain. *In* ''Touch, Temperature, and Pain in Health and Disease: Mechanisms and Assessments'' (J. Boivie, P. Hansson, and U. Lindblom, eds.), Prog. Pain Res. Manage., Vol. 3, pp. 339–354. IASP Press, Seattle, Washington.

Guilbaud, G., Benoist, J.-M., Jazat, F., and Gautron, M. (1990). Neuronal responsiveness in the ventrobasal thalamic complex of rats with an experimental mononeuropathy. *J. Neurophysiol.* **64,** 1537–1554.

Hall, J. G., Duggan, A. W., Morton, C. R., and Johnson, S. M. (1982). The location of brain stem neurones tonically inhibiting dorsal horn neurons of the cat. *Brain Res.* **244,** 215–222.

Hammond, D. L., Presley, R. W., Gogas, K. R., and Basbaum, A. I. (1992). Morphine or U-50,488H suppresses fos protein-like immunoreactivity in the spinal cord and nucleus tractus solitarii evoked by a noxious visceral stimulus in the rat. *J. Comp. Neurol.* **315,** 244–253.

Harris, G., and Rollman, G. B. (1983). The validity of experimental pain measures. *Pain* **17,** 369–376.

Hayashi, H. (1980). A problem in electrical stimulation of incisor tooth pulp in rats. *Exp. Neurol.* **67,** 438–441.

Herzberg, U., Murtaugh, M., and Beitz, A. J. (1994). Chronic pain and immunity: Mononeuropathy alters immune responses in rats. *Pain* **59,** 219–225.

Hill, C. M., Carroll, M. J., Giles, A. D., and Pickvance, N. (1987). Ibuprofen given pre- and post-operatively for the relief of pain. *Int. J. Oral Maxillofacial Surg.* **16,** 420–424.

Hiranuma, T., Kato, S., and Hachisu, M. (1988). Analgesic action of amfenac Na, a non-steroidal anti-inflammatory agent. *J. Pharmacobio-Dyn.* **11,** 612–619.

Hu, S., and Zhu, J. (1989). Sympathetic facilitation of sustained discharges of polymodal nociceptors. *Pain,* **38,** 85–90.

Hunskaar, S., Fasmer, O. B., and Hole, K. (1985). Formalin test in mice, a useful technique for evaluating mild analgesics. *J. Neurosci. Methods* **14,** 69–76.

Hutchison, G. L., Crofts, S. L., and Gray, I. G. (1990). Preoperative piroxicam for postoperative analgesia in dental surgery. *Br. J. Anaesth.* **65,** 500–503.

Iadarola, M. J., Douglass, J., Civelli, O., and Naranjo, J. R. (1988). Differential activation of spinal cord dynorphin and enkephalin neurons during hyperalgesia, *Brain Res.* **455,** 202–212.

International Association for the Study of Pain (IASP) (1986). Pain terms: A current list with definitions and notes on usage. *Pain, Suppl.* **3,** S217.

International Association for the Study of Pain (IASP) (1992). Adult postoperative pain. *In* ''Management of Acute Pain: A Practical Guide,'' A report of the Task Force on Acute Pain, pp. 22–33. IASP Press, Seattle, Washington.

Irwin, S., Houde, R. W., Bennett, D. R., Hendershot, L. C., and Seevers, M. H. (1951). The effects of morphine, methadone and meperidine on some reflex responses of spinal animals to nociceptive stimulation. *J. Pharmacol. Exp. Ther.* **101,** 132–143.

Jänig, W., and Koltzenburg, M. (1991). Receptive properties of sacral primary afferent neurons supplying the colon. *J. Neurophysiol.* **65,** 1067–1077.

Johansson, B., Glise, H., Hallerbäck, B., Dalman, P., and Kristoffersson, A. (1994). Preoperative infiltration with ropivacaine for postoperative pain relief after cholecystectomy. *Anesth. Analg.* **78,** 210–214.

Joris, J., Costello, A., Dubner, R., and Hargreaves, K. M. (1990). Opiates suppress carrageenan-induced edena and hyperthermia at doses that inhibit hyperalgesia. *Pain* **43,** 95–103.

Kajander, K. C., Wakisaka, S., and Bennett, G. J. (1992). Spontaneous discharge originates in the dorsal root ganglion at the onset of painful peripheral neuropathy in the rat. *Neurosci. Lett.* **138,** 225–228.

Katz, J., Kavanagh, B. P., Sandler, A. N., Nierenberg, H., Boylan, J. F., Friedlander, M., and Shaw, B. F. (1992). Preemptive analgesia. Clinical evidence of neuroplasticity contributing to postoperative pain. *Anesthesiology* **77,** 439–446.

Katz, J., Clairoux, M., Kavanagh, B. P., Roger, S., Nierenberg, H., Redahan, C., and Sandler, A. N. (1994). Pre-emptive lumbar epidural anesthesia reduces postoperative pain and patient-controlled morphine consumption after lower abdominal surgery. *Pain* **59,** 395–403.

Kiss, I. E., and Kilian, M. (1992). Does opiate premedication influence postoperative analgesia? A prospective study. *Pain* **48,** 157–158.

Kitchell, R. L. (1987). Problems in defining pain and peripheral mechanisms of pain. *J. Am. Vet. Med. Assoc.* **191,** 1195–1199.

Kumazawa, T. (1986). Sensory innervation of reproductive organs. *Prog. Brain Res.* **67,** 115–131.

Laboratory Animal Science Association (LASA) Working Party (1990). The assessment and control of the severity of scientific procedures on animals. *Lab. Anim.* 24, 97–130.

Laird, J. M. A., and Bennett, G. J. (1991). An electrophysiological study of dorsal horn neurons in the spinal cord of rats with an experimental peripheral neuropathy. *J. Neurophysiol.* **66,** 190–211.

LaMotte, R. H., Shain, C. N., Simone, D. A., and Tsai, E.-F. P. (1991). Neurogenic hyperalgesia: psychophysical studies of underlying mechanisms. *J. Neurophysiol.* **66,** 190–211.

Le Bars, D., Villanueva, L., Bouhassira, D., and Willer, J.-C. (1992). Diffuse noxious inhibitory controls (DNIC) in animals and in man. *Pathol. Physiol. Exp. Ther.* **4,** 55–65.

Levine, J. D., and Taiwo, Y. (1994). Inflammatory pain. *In* ''Textbook of Pain'' (P.D. Wall and R. Melzack, eds.), 3rd ed., pp. 45–56. Churchill-Livingstone, London.

Levine, J. D., Feldmesser, M., Tecott, L., Gordon, N. C., and Izdebski, K. (1984). Pain induced vocalization in the rat and its modification by pharmacological agents. *Brain Res.* **296,** 121–127.

Lewis, K. S., Whipple, J. K., Michael, K. A., and Quebbeman, E. J. (1994). Effect of analgesic treatment on the physiological consequences of acute pain. *Am. J. Hosp. Pharmacol.* **51,** 1539–1554.

Liebeskind, J. C. (1991). Pain can kill. *Pain* **44,** 3–4.

Light, A. R. (1992). ''The Initial Processing of Pain and Its Descending Control: Spinal and Trigeminal Systems,'' Pain and Headache, Vol. 12, Karger, Basel.

Lineberry, C. G., and Kulics, A. T. (1978). The effects of diazepam, morphine and lidocaine on nociception in rhesus monkeys: A signal detection analysis. *J. Pharmacol. Exp. Ther.* **205,** 302–310.

Malliani, A., and Lombardi, F. (1982). Consideration of the fundamental mechanisms eliciting cardiac pain. *Am. Heart J.* **103,** 575–578.

Malmberg, A. B., and Yaksh, T. L. (1992). Hyperalgesia mediated by spinal glutamate of substance P receptor blocked by spinal cyclooxygenase inhibition. *Science* **257,** 1276–1279.

McMahon, S. B. (1994). Mechanisms of cutaneous, deep and visceral pain. *In* ''Textbook of Pain'' (P. D. Wall and R. Melzack, eds.), 3rd ed., pp. 129–152. Churchill-Livingstone, London.

Mendell, L. M. (1966). Physiological properties of unmyelinated fiber projection to the spinal cord. *Exp. Neurol.* **16,** 316–332.

Merskey, H., and Bodguk, N., eds. (1994). ''Classification of Chronic Pain: Descriptions of Chronic Pain Syndromes and Definitions of Terms,'' 2nd ed. IASP Press, Seattle, Washington.

Mitchell, C. L. (1964). A comparison of drug effects upon the jaw jerk response to electrical stimulation of the tooth pulp in dogs and cats. *J. Pharmacol. Exp. Ther.* **146,** 1–6.

Molony, V., and Wood, G. N. (1992). Acute pain from castration and tail docking in lambs. *In* ''Animal Pain'' (C. E. Short and A. Van Poznak, eds.), pp. 385–395. Churchill-Livingstone, New York.

Montgomery, C. A., Jr. (1990). Oncologic and toxicologic research: Alleviation and control of pain and distress in laboratory animals. *Cancer Bull.* **42,** 230–237.

Moore, M. P., Greene, S. A., and Keegan, R. D. (1992). A method for assessing noxious stimuli in anesthetized dogs. *In* ''Animal Pain'' (C. E. Short and A. Van Poznak, eds.), pp. 439–446. Churchill-Livingstone, New York.

Morton, D. B., and Griffiths, P. H. M. (1985). Guidelines on the recognition of pain, distress and discomfort in experimental animals and an hypothesis for assessment. *Vet. Rec.* 116, 431–436.

Murphy, D. F., and Medley, C. (1993). Preoperative indomethacin for pain relief after thoracotomy: Comparison with postoperative indomethacin. *Br. J. Anaesth.* **70,** 298–300.

National Research Council (NRC) (1992). "Recognition and Alleviation of Pain and Distress in Laboratory Animals," A report of the Institute for Laboratory Animal Resources Committee on Pain and Distress in Laboratory Animals. National Academy Press, Washington, D.C.

National Research Council (NRC) (1996). ''Guide for the Care and Use of Laboratory Animals,'' A report of the Institute for Laboratory Animal Resources Committee on Care and Use of Laboratory Animals. U.S. Department of Health and Human Services, Washington, D.C.

Ness, T. J., and Gebhart, G. F. (1988). Colorectal distension as a noxious visceral stimulus: Physiologic and pharmacologic characterization of pseudoaffective reflexes in the rat. *Brain Res.* **450,** 153–169.

New York Academy of Sciences (NYAS) (1988). "Interdisciplinary Principles and Guidelines for the Use of Animals in Research, Testing, and Education," A report of the Ad Hoc Committee on Animal Research. New York Academy of Sciences, New York.

Nordbladh, I., Ohlander, B., and Björkman, R. (1991). Analgesia in tonsillectomy: A double-blind study on pre- and post-operative treatment with diclofenac. *Clin. Otolaryngol.* **16,** 554–558.

O'Callaghan, J. P., and Holtzman, S. G. (1975). Quantitation of the analgesic activity of narcotic antagonists by a modified hot-plate procedure. *J. Pharmacol. Exp. Ther.* **192,** 497–505.

Olfert, E. D. (1995). Defining an acceptable endpoint in invasive experiments. *AWIC Newsl.* **6,** 3–7.

Oliveras, J. L., Woda, A., Guilbaud, G., and Besson, J.-M. (1974). Inhibition of the jaw opening reflex by electrical stimulation of the periaqueductal grey matter in the awake, unrestrained cat. *Brain Res.* **72,** 328–331.

Palecek, J., Paleckovà, V., Dougherty, P. M., Carlton, S. M., and Willis, W. D. (1992). Responses of spinothalamic tract neurons to mechanical and thermal stimuli in an experimental model of peripheral neuropathy in primate. *J. Neurophysiol.* **67,** 1562–1573.

Palecek, J., Dougherty, P. M., Kim, S. H., Paleckovà, V., Lekan, H., Chung, J. M., Carlton, S. M., and Willis, W. D. (1993). Responses of spinothalamic tract neurons to mechanical and thermal stimuli in an experimental model of peripheral neuropathy in primates. *J. Neurophysiol.* **68,** 1951–1966.

Piercey, M. F., and Schroeder, L. A. (1980). A quantitative analgesic assay in the rabbit based on the response to tooth pulp stimulation. *Arch. Int. Pharmacodyn. Ther.* **248,** 294–304.

Pippi, N. L., and Lumb, W. V. (1979). Objective tests of analgesic drugs in ponies. *Am. J. Vet. Res.* **40,** 1082–1086.

Planas, M. E., Rodriguez, L., Sanchez, S., Pol, O., and Puig, M. M. (1995). Pharmacological evidence for the involvement of the endogenous opioid system in the response to local inflammation in the rat paw. *Pain* **60,** 67–71.

Potthoff, A., and Carithers, R. W. (1989). Pain and analgesia in dogs and cats. *Compend. Contin. Educ. Pract. Vet.* **11,** 887–896.

Poulsen, L., Arendt-Nielsen, L., Bronsen, K., Nielson, K. K., Gram, L. F., and Sindrup, S. H. (1995). The hypoalgesic effect of imipramine in different human experimental pain models. *Pain* **60,** 287–293.

Price, D. D., Mao, J., and Mayer, D. J. (1994). Central neural mechanisms of normal and abnormal pain states. *In* ''Pharmacological Approaches to the Treatment of Chronic Pain: New Concepts and Critical Issues'' (H. L. Fields and J. C. Liebeskind, eds.), Prog. Pain Res. Manage., Vol. 1, Chapter 5. IASP Press, Seattle, Washington.

Public Health Service (PHS) (1986). ''Public Health Service Policy on Humane Care and Use of Laboratory Animals.'' U.S. Department of Health and Human Services, Washington, D.C.

Puke, M. J. C., and Wiesenfeld-Hallin, Z. (1993). The differential effects of morphine and the alpha2-adrenoreceptor agonists clonidine and dexmedetomidine on the prevention and treatment of experimental neuropathic pain. *Anesth. Analg.* **77,** 104–109.

Randall, L. O., and Selitto, J. J. (1957). A method for measurement of analgesic activity on inflamed tissue. *Arch. Int. Pharmacdyn. Ther.* **111,** 409–419.

Richmond, C. E., Bromley, L. M., and Woolf, C. J. (1993). Preoperative morphine pre-empts postoperative pain. *Lancet* **342,** 73–75.

Roos, A., Rydenhag, B., and Andersson, S. A. (1982). Cortical responses evoked by tooth pulp stimulation in the cat. Surface and intracortical responses. *Pain* **14,** 247–265.

Rosenfeld, P. J., Pickrel, C., and Broton, J. G. (1983). Analgesia for orofacial nociception produced by morphine microinjection into spinal trigeminal complex. *Pain* **15,** 145–155.

Sawyer, D. C., and Rech, R. H. (1987). Analgesia and behavioral effects of butorphanol, nalbuphine, and pentazocine in the cat. *J. Am. Anim. Hosp. Assoc.* **23,** 438–446.

Sawyer, D. C., Rech, R. H., Durham, R. A., Adams, T., Richter, M. A., and Striler, E. L. (1991). Dose response to butorphanol administered subcutaneously to increase visceral nociceptive threshold in dogs. *Am. J. Vet. Res.* **52,** 1826–1830.

Scott, N. B., and Kehlet, H. (1988). Regional anaesthesia and surgical morbidity. *Br. J. Surg.* **75,** 299–304.

Seltzer, Z., Beilin, B. Z., Ginzburg, R., Paran, Y., and Shimko, T. (1991). The role of injury discharge in the induction of neuropathic pain behavior in rats. *Pain* **46,** 327–336.

Siegmund, E., Cadmus, R., and Lu, G. (1957). A method for evaluating both non-narcotic and narcotic analgesics. *Proc. Soc. Exp. Biol. Med.* **95,** 729–731.

Smith, A. C., and Brook, I. M. (1990). Inhibition of tissue prostaglandin synthesis during third molar surgery: Use of preoperative fenbufen. *Br. J. Oral Maxillofacial Surg.* **28,** 251–253.

Smith, N. T. (1987). New developments in monitoring animals for evidence of pain control. *J. Am. Vet. Med. Assoc.* **191,** 1269–1272.

Soma, L. R. (1987). Assessment of animal pain in experimental animals. *Lab. Anim. Sci.* **37,** 71–74.

Spinelli, J. S., and Markowitz, H. (1987). Clinical recognition and anticipation of situations likely to induce suffering in animals. *J. Am. Vet. Med. Assoc.* **191,** 1216–1218.

Stevens, C. W. (1995). An amphibian model for pain research. *Lab. Anim.* **24,** 32–36.

Sugimoto, T., Takemura, M., and Mukai, N. (1990). Histochemical demonstration of neural carbonic anhydrase activity within the mandibular molar and incisor tooth pulps of the rat. *Brain Res.* **529,** 245–254.

Tasker, R. A. R. (1992). Alpha-adrenergic receptors and their agonists. *In* ''Animal Pain'' (C. E. Short and A. Van Poznak, eds.), pp. 155–164. Churchill-Livingstone, New York.

Tjolsen, A., Berge, O-G., Hunskaar, S., Rosland, J. H., and Hole, K. (1992). The formalin test: An evaluation of the method. *Pain* **51,** 5–17.

Toda, K., Iriki, A., and Ichioka, M. (1981). Selective stimulation of intrapulpal nerve of rat lower incisor using a bipolar electrode method. *Physiol. Behav.* **26,** 307–311.

Tverskoy, M., Oz, Y., Isakson, A., Finger, J., Bradley, E. L., Jr., and Kissin, I. (1994). Preemptive effect of fentanyl and ketamine on postoperative pain and wound hyperalgesia. *Anesth. Analg.* **78,** 205–209.

Vaccarino, A. L., and Melzack, R. (1992). Temporal processes of formalin pain: Differential role of cingulum bundle, fornix pathway and medial bulboreticular formation. *Pain* **49,** 257–271.

Van Arman, C. G., Begany, A. J., Miller, L. M., and Pless, H. H. (1965). Some details of the inflammation caused by yeast and carrageenin. *J. Pharmacol. Exp. Ther.* **150,** 328–334.

Vierck, C. J., Jr., Cooper, B. Y., Ritz, L. A., and Greenspan, J. D. (1989). Inference of pain sensitivity from complex behaviors of laboratory animals. *In* ''Issues in Pain Management'' (C. R. Chapman and J. D. Loeser, eds.), Adv. Pain Res. Ther., Vol. 12, pp. 93–115. Raven Press, New York.

Wall, P. D., and Gutnick, M. (1974a). Properties of afferent nerve impulses originating from a neuroma. *Nature (London)* **248,** 740–743.

Wall, P. D., and Gutnick, M. (1974b). Ongoing activity in peripheral nerves: The physiology and pharmacology of impulses originating from a neuroma. *Exp. Neurol.* **43,** 580–593.

Wall, P. D., and Melzack, R., eds. (1994). "Textbook of Pain," 3rd ed. Churchill-Livingstone, London.

Wheeler-Aceto, H., and Cowan, A. (1991). Standardization of the rat paw formalin test for the evaluation of analgesics. *Psychopharmacology* **104,** 35–44.

Wheeler-Aceto, H., Porreca, F., and Cowan, A. (1990). The rat paw formalin test: Comparison of noxious agents. *Pain* **40,** 229–238.

Willis, W. D. (1985). ''The Pain System,'' Pain and Headache, Vol. 8. Karger, Basel.

Winter, C. A., and Flataker, L. (1965). Reaction thresholds to pressure in edematous hindpaws of rats and responses to analgesic drugs. *J. Pharmacol. Exp. Ther.* **150,** 165–171.

Woolfe, G., and MacDonald, A. D. (1944). The evaluation of the analgesic action of pethidine (Demerol). *J. Pharmacol. Exp. Ther.* **80,** 300–307.

Yeager, M. P., Glass, D. D., Neff, R. K., and Brinck-Johnsen, T. (1987). Epidural anesthesia and analgesia in high-risk surgical patients. *Anesthesiology* **66,** 729–736.

Zeltser, R., and Seltzer, Z. (1994). A practical guide for the use of animal models in the study of neuropathic pain. *In* ''Touch, Temperature, and Pain in Health and Disease: Mechanisms and Assessments'' (J. Boivie, P. Hansson, and U. Lindblom, eds.), Prog. Pain Res. Manage., Vol. 3, pp. 295–338. IASP Press, Seattle, Washington.

Zhou, M., and Gebhart, G. F. (1992). Characterization of descending facilitation and inhibition of spinal nociceptive transmission from the nuclei reticularis gigantocellularis and gigantocellularis pars alpha in the rat. *J. Neurophysiol.* **67,** 1599–1614.

Zimmerman, M. (1983). Ethical guidelines for investigations of experimental pain in conscious animals. *Pain* 16, 109–110.

Chapter 7

Anesthesia Equipment: Types and Uses

George A. Vogler

ANESTHESIA AND ANALGESIA IN LABORATORY ANIMALS

ISBN 0-12-417570-8

I. INTRODUCTION

Anesthesia equipment may be as simple as a syringe or as complex as the full panoply of anesthesia machines, ventilators, and invasive and noninvasive monitors and support equipment. In either case, careful selection of equipment based on an understanding of its use and limitations will ease the task of the anesthesia provider and improve the quality of the care provided. Although individual investigators may need to be familiar with only a limited range of species and techniques, managers of centralized surgical facilities must be prepared to offer a wide variety of services, involving a more extensive range of anesthesia, support, and monitoring equipment. The quality of anesthesia care has an immediate and long-term effect on the welfare of the animal and is important in both ethical and pragmatic terms. Modern equipment not only makes possible the use of new anesthetic agents and techniques, but also increases the safety and predictability of older agents which retain an important role in research. Budgetary constraints and prudence point to the importance of acquiring machines and instruments that are safe, flexible, and durable. This chapter attempts to provide an overview of the equipment commonly used to produce and support anesthesia in research animals. The interested reader is urged to consult the references cited for detailed discussions of individual topics.

II. COMPRESSED GASES

A variety of compressed gases may be encountered in anesthesia, but the most common are oxygen, nitrous oxide, and compressed air. However, carbon dioxide is often employed as an insufflation agent for laparoscopic procedures and nitrogen is the drive gas of choice for many pneumatically powered surgical instruments. Persons working in centralized or specialized surgical facilities may expect to encounter them as well.

Even in facilities having central gas supply systems, compressed gas cylinders are common in the operating room and are likely to remain so. Of the many sizes of cylinders available, the most common are E and H cylinders, and they are used as examples of small and large tanks in the following discussion.

A. Cylinders

1. E Tanks

E cylinders are small tanks usually connected directly to the anesthetic machine by means of a built-in yoke. The cylinder port forms a flush connection with the yoke and the Pin Index Safety System is used to prevent misconnection. The yoke is fitted with two steel pins which correspond to holes drilled into the cylinder valve assembly. The spacing of the pins and holes is specific to a particular gas, thus preventing a tank containing the wrong gas from being attached to the yoke. Damaged or missing pins will defeat the safety system, and care should be taken to assure that the pins are intact, inspected regularly, and maintained in good condition. No attempt should be made to bypass this safety system (Heavner *et al.,* 1989a).

A gasket is used to ensure a gas-tight seal when the cylinder is mounted to the yoke. E cylinders are usually supplied with a new gasket fitted over the port and covered by a plastic seal. If both the cylinder port and yoke are in good repair, only a single gasket is needed. Additional gaskets should not be used in an attempt to get a tight seal. Instead, the yoke should be repaired, if necessary, or the defective tank should be labeled and returned to the filler.

Special valve wrenches or handles are made to fit small tank valves and are often supplied with new anesthetic machines. There are a variety of designs and one common example is illustrated in Fig. 1. The two slots correspond to the tank valve, and the larger hexagonal slot is used to tighten the packing nut of the valve. Tank wrenches are convenient and are usually available from the compressed gas supplier at a minimal expense.

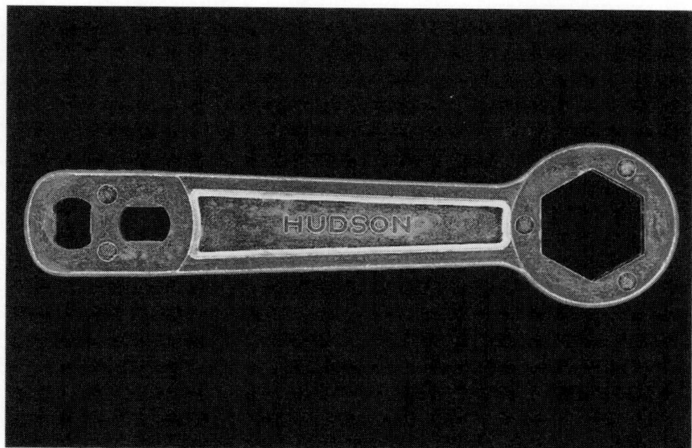

Fig. 1. Small tank wrench. Note the two small slots for the tank valve and the larger hexagonal slot for the valve packing nut.

A pressure relief device is located opposite and slightly below the level of the outlet port. Various types of relief devices are used, but all are designed to vent gas if the tank pressure reaches dangerous levels (Dorsch and Dorsch, 1994b).

2. H Tanks

These large tanks commonly are used to supply small central piping systems through tank manifolds or, in conjunction with a regulator, to supply gas directly to anesthetic machines. The valve assembly includes a handwheel, so a tank wrench is not needed for opening or closing the valve. A threaded cap is fitted over the valve assembly. Large tank outlet ports use specific styles and patterns of threaded connectors to prevent misconnection. As with small cylinder valve assemblies, a safety relief device is incorporated into the valve assembly (Dorsch and Dorsch, 1994b).

3. Identification

Cylinders are identified by a series of numbers and codes stamped into the metal near the neck. These numbers and symbols specify the manufacturing standard, maximum service pressure, the maker or owner of the cylinder, the serial number of the cylinder, and the test date and inspector's mark. If the cylinder has been in use for more than 10 years, the date of retesting also appears (Compressed Gas Association, Inc., 1989).

Gases suitable for medical use are clearly labeled, and the user should ensure that gases intended for industrial applications, which could contain toxic contaminants, have not been inadvertently supplied. The label should clearly identify the gas or gas mixture and provide information concerning the hazard classification (Dorsch and Dorsch, 1994b).

Cylinders containing medical gases are also color coded. Table 1 lists the color codes as well as the characteristics of commonly encountered medical gas cylinders and gases. Although the cylinder color is helpful in identifying cylinder contents, the label should always be read to verify tank contents.

4. Safety Precautions

Compressed gas cylinders are potentially hazardous and personnel responsible for handling them should be aware of the danger and be trained in safe handling practices. A few elements of safe practice are given below (Compressed Gas Association, Inc., 1989).

1. Cylinders should be stored in a manner which prevents exposure to extremes of temperature, weather, and chemicals or fumes that would cause corrosion of valves and valve caps.
2. Cylinders should never be subjected to rough or careless handling. Cylinder carriers or carts should always be used for transporting cylinders. Cylinders in use or in storage should be properly secured.
3. Cylinders should be inspected before use. Damaged cylinders or valve components should be tagged and the cylinder returned to the filler. The user should make no attempt to repair a cylinder.
4. Cylinder labels should be legible and the contents clearly identified. Cylinders having illegible labels should not be used, regardless of color coding.
5. Tank valves should be opened slowly to avoid damage or explosion caused by sudden pressure and temperature increases. Before connecting cylinders to yokes or regulators, the valve should be opened slightly to clear dust and debris from the port.
6. Cylinders, valves, connections, and regulators must be kept free of foreign materials. Never attempt to lubricate any gas fitting and avoid exposing any fitting to oils, greases, or any other flammable substance.
7. Regulator or yoke connections should never be forced. Regulators and manifolds are specific for a particular gas or gas mixture and are not interchangeable with other types.

B. Medical Gas Piping Systems

Central piping systems have long been the standard in hospitals and have become common in research settings. These systems are used to supply medical gases, medical vacuum, and waste gas evacuation to the operating rooms and medical support areas. For facilities that make regular use of pneumatically powered surgical instruments, common in orthopedic and neurosurgery, special control stations for centrally piped nitrogen are also available, eliminating the need for large cylinders in the operating room.

Central systems may supply wall or ceiling connections in a variety of configurations, ranging from relatively simple wall or ceiling hose stations to track-mounted hoses, booms, or elaborate retractable utility drops. The latter may integrate connection for anesthetic gases, waste gas evacuation, patient suction, and auxiliary electrical power outlets into a single centralized location. Even a simple system of wall-mounted outlets for oxygen, patient vacuum, and waste gas evacuation can greatly

TABLE I

MEDICAL GAS CYLINDERS[a]

Cylinder dimensions (inches) and color marking	Contents	Air	Carbon dioxide	Helium	Nitrogen	Nitrous oxide	Oxygen
B	psig	—	838	—	—	—	—
3.5 OD × 13	liters	—	370	—	—	—	—
	kg	—	0.68	—	—	—	—
D	psig	1900	838	1900	1900	745	1900
4.25 OD × 17	liters	380	940	350	370	940	400
	kg	—	1.73	—	—	1.73	—
E	psig	1900	838	1900	1900	745	1900
4.25 OD × 26	liters	620	1590	580	600	1590	650
	kg	—	2.92	—	—	2.92	—
M	psig	1900	838	1900	1900	745	1900
OD × 43	liters	2810	7570	2630	2760	7570	2970
	kg	—	13.9	—	—	13.9	—
G	psig	1900	838	1900	1900	745	1900
8.5 OD × 51	liters	4980	12300	4660	4890	13800	5270
	kg	—	22.7	—	—	25.4	—
H & K	psig	2217	838	2015	2015	745	2015
9.25 OD × 51	liters	6460	15800	5530	5800	15800	6270
	kg	—	29.1	—	—	29.1	—
Color marking							
United States		Yellow	Gray	Brown	Black	Blue	Green
Canada		Black and white	Gray	Brown	Black	Blue	White

[a]Adapted (1989) Compressed Gas Association, Inc.

enhance the convenience, safety, and function of operating and procedure rooms.

When planning a central supply system, it is important to work with architects and mechanical engineers experienced in operating room design. This topic is covered in Chapter 8 of this volume. Regardless of the system selected, the final connections will conform to the Diameter Index Safety System (DISS) or to one of several proprietary quick disconnect systems. In either case, the fittings are designed to minimize the hazards of misconnections. After installation, central systems should be thoroughly tested to detect possible leaks and cross-connections and to assure the purity of delivered gases at each outlet in the system.

III. ANESTHETIC MACHINES

Anesthetic machines used in research settings may be built especially for research use, may be designed for clinical veterinary use, or may be machines originally destined for human use. In contrast to the more complex veterinary clinical and human machines, research units made for rodent or other small animal use are often very simple and may lack even an absorber system. Any machine, properly designed and used, should provide for the safe delivery of breathing gas containing a controlled amount of anesthetic agent.

Anesthetic machines may be configured in a variety of ways. Modern human machines have large, heavy-duty frames and casters to support the multiple vaporizers, flowmeters, auxiliary tanks, ventilators, and monitors that comprise the machine. These machines usually provide a large, flat working surface and a monitor shelf, as well as one or more drawers for supplies and auxiliary equipment. In contrast, veterinary machines may offer little or no working surface, and the flowmeters, gauges, vaporizer, and absorber circuit may all be supported on a single central pole with casters. Some veterinary designs are even more compact, intended to be mounted on a table or wall. Such compact machines are often well-suited for use with rodents. They can be mounted on carts to provide mobile operating rooms for rodents, incorporating induction chambers and mask maintenance circuits using Bain circuits, as well as heating pads and other ancillary support equipment.

Although small mobile anesthetic machines are often ideal for use in induction and preparation areas, as well as for minor procedures, radiology, dentistry, etc., larger machines have other advantages. The additional working surface, storage, and monitor shelf can be very useful in organizing the anesthetic work space for long and complex cases. The flexibility of modern large machines makes them useful for rodents as well as for larger patients. They are well-suited for operating room use and, with their rugged stands and large casters, can be moved as needed. In an attempt to consolidate and simplify the task of the anesthesiologist, makers of human anesthetic machines have incorporated numerous safety features, monitors, alarms, and data acquisition systems into their products. This trend has resulted in large, complex, and very expensive anesthesia work stations. In contrast to this, veterinary machines tend to be quite simple, generally consisting of a stand, a set of flowmeters, a vaporizer, and a breathing circuit with a CO_2 absorber. Midway in complexity are the used or reconditioned human machines that are often found in research laboratories. These may range in age from obsolete equipment to newer machines still in service in human operating rooms.

Veterinary machines are sold directly by the manufacturer or through distributors of veterinary supplies. Companies that offer experimental physiology equipment often include anesthesia machines as part of their product line. Used human equipment may be obtained from a variety of sources, including firms that specialize in used medical equipment, vendors of new anesthesia machines who acquire used equipment in trade, and hospital anesthesia departments. Used equipment is often very suitable, but every effort should be made to assure that it is fully functional, properly maintained, and that it can be repaired. Factory-reconditioned human anesthesia machines are also available for research use from some companies. Regardless of the source, anesthetic machines share many common features and components, which are useful to understand in order to select and safely use them.

Human machines made in the United States since the early 1980s have complied with standards issued by the American National Standards Institute (Z79.8-1979) or, since 1989, with those of the American Society for Testing and Materials (ASTM F1161-88) (Eisenkraft, 1993a). The degree to which veterinary or research equipment conforms to these standards varies widely among manufacturers. The features and components discussed in this section are those of a relatively modern, generic anesthesia machine. Most veterinary and human machines will have many, but not all, of the features and characteristics described. Specialized veterinary machines used for very large domestic and exotic animals are beyond the scope of this chapter. Veterinary anesthesia texts and veterinary anesthesiologists accustomed to working with larger animals should be consulted by those in need of information concerning appropriate equipment. Earlier editions of veterinary anesthesia or human anesthesia equipment texts contain a wealth of information regarding older equipment.

A. Yoke Assemblies

Oxygen and other gases enter the anesthetic machine either from a small tank yoke (Fig. 2) or from a connection that accepts gas from a large tank or central supply system. The cylinder yoke assembly includes a flush-type port, a filter to remove debris from the gas, a retaining screw to tighten and support the cylinder, and a single-stage regulator. On all but the most obsolete equipment, the small cylinder yoke should conform with the Pin Index Safety System. If it does not, or if the index pins have been damaged, the machine should be modified or repaired to conform to this minimal safety standard.

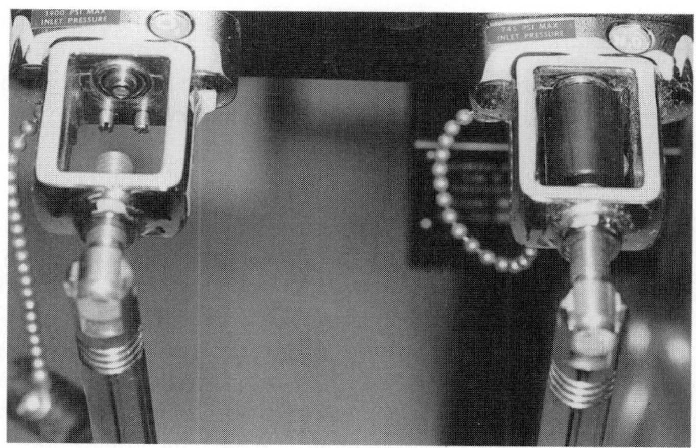

Fig. 2. Small tank yoke. The flush-type port, Pin Index Safety System pins, and retaining screw are shown on the left yoke. The right yoke is not in use and is protected by a yoke plug.

Small tanks are supported in the cylinder yoke by a retaining screw which mates with a conical depression on the valve assembly of the cylinder. A tight seal is attained by the use of a gasket, which is supplied with the cylinder. Only a single gasket should be needed: if a seal cannot be made with a single new gasket, either the tank or the yoke is faulty. The tank should be replaced or the yoke repaired by a qualified service technician.

Pressure in the cylinders is reduced to a safer and more constant level by regulators in the machine. Cylinder pressure is indicated by gauges, usually mounted on the front of the machine. Check valves limit the leakage of gas from the machine if a cylinder is not present. Yoke plugs are used to protect the port from damage and dirt when a cylinder is not mounted and to further limit the escape of gas from the machine. On most machines, the plug is chained to the yoke assembly. If yoke plugs are missing, as is sometimes the case with used equipment, they should be replaced (Dorsch and Dorsch, 1994c).

B. Yoke Blocks

Some very old anesthetic machines were equipped only for small gas cylinders and did not provide pipeline connections. In an attempt to extend the use of these machines and gain the advantages of central supply or large cylinder economy, adapters were made that duplicated the shape of small cylinder flush valves, including, in some cases, holes that corresponded to Pin Index Safety System pins. Although superficially appealing, yoke blocks or adapters have a number of disadvantages and their use is discouraged (Dorsch and Dorsch, 1984a).

C. Pipeline Connections

Pipeline inlets are male DISS connections that are usually located on the back of the machine. They may be connected directly to hoses using the corresponding DISS female connectors or fitted with quick-connect adapters. The pipeline inlets are protected by check valves which prevent back-flow into the central supply system. Some machines also use a filter, as with the small cylinder yoke assemblies. A pipeline connection is shown in Fig. 3. Pipeline pressures are usually regulated to about 45–55 psi (Dorsch and Dorsch, 1994c).

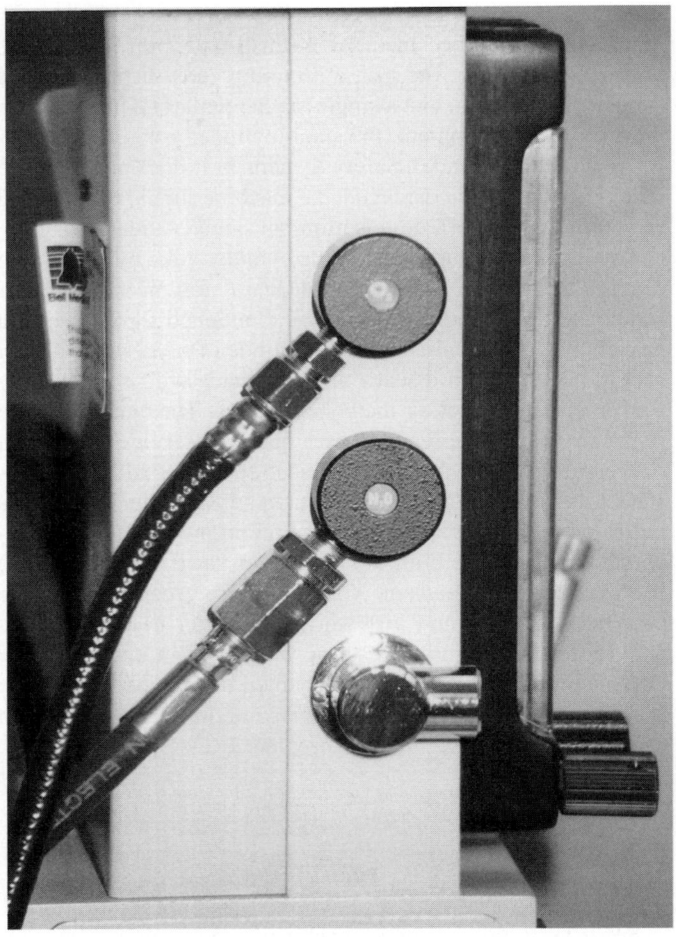

Fig. 3. DISS pipeline connections for oxygen (top hose) and nitrous oxide (lower hose). The chromed elbow fitting at the bottom of the photograph is the common gas outlet.

D. Pressure Gauges

In addition to small cylinder gauges, newer human machines have pipeline pressure gauges, located so that pipeline pressure is accurately displayed. This is not the case for most veterinary machines, which have pressure gauges only for small cylinders, or for certain older human machines, which may display the pressure within the intermediate pressure circuit of the machine instead of the pipeline itself. In the latter instance, if the small cylinder supply valves are open, the gauges will give no warning of central supply failure (Dorsch and Dorsch, 1994c).

It is useful to understand that gauge pressure is determined in part by the physical state of the compressed gases. Oxygen and nitrogen, for example, remain in the gas phase at the pressure used in medical cylinders. In their case, the gauge pressure will decrease as gas is used. However, in the case of carbon dioxide and nitrous oxide, which are partially liquefied in the cylinders, the gauge pressure is an unreliable indicator of gas content. Pressure will drop while gas is being withdrawn. However, if any liquid remains in the tank when the withdrawal is stopped, the gauge pressure will gradually return to a limit determined by ambient temperature. Once the liquid phase is exhausted, the gauge pressure will drop abruptly as the remaining gaseous tank contents are removed (Heavner *et al.,* 1989a). Small cylinders of nitrous oxide and carbon dioxide are commonly labeled with gross and net weights and can be weighed to determine the amount of gas remaining.

E. Flowmeters

Anesthetic machine flowmeters are precision instruments that control the delivery of gas to the vaporizers and to the breathing circuit. The most common design is a glass or plastic tube having a tapered bore in which an indicator bobbin or ball moves as gas flows through the tube from bottom to top. A needle valve controls the flow of gas entering at the bottom of the tube. Flow is indicated in liters per minute (lpm), usually read from a scale etched onto the tube itself. On some older machines, the scale may be mounted separately, adjacent to the flowmeter tube. The indicator may be a bobbin, which is read at the top, or a ball, which is read at the middle.

Flowmeters, calibrated for specific gases and flow rates, are not interchangeable. Dual flowmeters permit more precise control at the low flow rates used for low flow or closed circuit techniques. The two tubes are usually connected in series, with flow controlled by a single knob. The gas first enters the low flow tube, located to the left, and then the high flow tube on the right. Flows greater than 1 lpm, the usual limit of the low flow side, are read on the right-hand, or high flow, side (Dorsch and Dorsch, 1994c). Upper flow limits vary, with human machines typically permitting higher flows than small animal machines. Dual flowmeters for oxygen are increasingly common on veterinary machines. Newer human machines often have dual flowmeters for nitrous oxide and medical air as well.

The needle valve control knobs are usually color coded and labeled for the gas they control. In some cases the knobs have different shapes to further minimize confusion. Some human anesthetic machines do not permit oxygen flow to be turned off completely; rather, a continuous minimum flow of oxygen is established once the machine is connected to the pipeline supply or when the small cylinders are opened. This feature is intended to prevent accidental hypoxia and should not be mistaken for a faulty valve. Most machines, however, permit all gases to be discontinued at the flowmeter valve.

Flowmeters are deceptively simple in appearance and operation, but are subject to a number of problems. Unless specifically designed otherwise, they must be vertical in order to operate correctly; if the float ball or bobbin is in contact with the side of the tube, friction can cause erroneous readings. Dirt, debris, and static electricity can interfere with the free movement of the bobbin. Needle valves are subject to wear and may need to be serviced or replaced periodically (Davis *et al.*, 1995a). Flowmeter tubes are fragile and are usually protected by a plastic cover.

In addition to leaky or poorly functioning needle valves and dirty tubes, jammed indicator balls or bobbins are a common problem. This is usually the result of failing to turn the flowmeter off before disconnecting the pipeline supply or closing the small cylinder valve. When the system is abruptly repressurized, the ball or bobbin is forced to the top of the tube. The flowmeter must then be disassembled to free the bobbin, a process sufficiently complex, tedious, and expensive that the error is seldom repeated by the same person.

F. Oxygen Fail-Safe and Proportioning Systems

The ability to administer gases other than oxygen to the patient creates the risk of administering a hypoxic mixture. There are two ways in which this can happen: failure of oxygen pressure or inadvertent administration of a hypoxic mixture due to human error resulting in an inappropriate ratio of oxygen to other gas(es). The problem has been addressed in several different ways by means of safety devices added to some anesthetic machines. One approach, previously mentioned, sets a mandatory minimum oxygen flow. This does not prevent a hypoxic mixture from being given, however, as other gases, such as nitrous oxide, may be set at flows that still result in hypoxia. Oxygen:nitrous oxide proportioning devices, which maintain a constant ratio of oxygen to nitrous oxide, can prevent hypoxic mixtures of these gases, but will also limit the use of low flow techniques. The risk of oxygen pressure failure is addressed by yet another device, which prevents any gas flow if pressure is not sensed in the oxygen circuit of the machine. But oxygen pressure may be present without any oxygen flowing (Dorsch and Dorsch, 1994c). Each of these safety systems can help minimize the risk of hypoxia, but none replace operator vigilance.

G. Oxygen Flush Valve

The oxygen flush valve is usually located on the front of the machine, with some form of protection provided to prevent accidental activation. When the valve is activated, fresh oxygen is delivered to the common gas outlet at the rate of 35–75 liters per minute, bypassing the flowmeters and the anesthetic vaporizer. Except for some very old machines, the flush valves are spring-loaded and close as soon as pressure is released from the button.

Activating the flush valve will cause rapid dilution of the anesthetic gases in the patient circuit. Although the volume of the absorber circuit and breathing tubing usually acts to buffer pressure changes, a dangerous increase in the breathing circuit pressure may still occur. The adjustable pressure-limiting valve or pop-off valve should be fully open when the flush valve is used. Some breathing circuits, such as the Mapleson-D or Bain circuit, do not use the absorber system and *the flush valve should not be used because of the risk of barotrauma* (Hubble, 1993; Sawyer, 1982a).

H. Common Gas Outlet

The common gas outlet is the point at which the gas and anesthetic mixture produced by the flowmeters and the vaporizer exit the machine and, by means of a connecting hose, enter the breathing circuit. If the machine conforms to ASTM standard F1161-88, the outlet is a tapered 15-mm female fitting, usually located on the front or side of the anesthetic machine adjacent to the absorber pole. The female 15-mm fitting is the same size as an endotracheal tube connector. The common gas outlet and the corresponding male fitting that connects to the breathing circuit are carefully machined and should require little or no force to be gas tight. On modern human machines the common gas outlet is provided with a retaining mechanism in order to prevent accidental disconnection (Eisenkraft, 1993a). Some veterinary machines use a nonstandard outlet, which may be little more than a small hose connection leading to the breathing circuit.

I. Absorber Systems

Absorber systems remove carbon dioxide so breathing gases can be recirculated, conserving anesthetic agents, moisture, and heat. In most cases this is accomplished by means of a system of unidirectional breathing valves, which permit gas to flow in only one direction through the breathing circuit. Exhaled gas is directed through a canister assembly that contains a granular mixture of calcium (soda lime) or calcium and barium hydroxide (barium lime). The absorbent reacts with water and carbon dioxide, and an indicator is present that turns color when the absorbent is exhausted. The reaction is exothermic so the absorber canister will become warm with use. Soda lime or barium lime is available in large bulk cans, smaller sealed bags, or in plastic disposable prepacked inserts.

Absorber assemblies are usually located in the inspiratory limb of the breathing circuit, with the inspiratory valve mounted on or adjacent to the assembly. The absorbent is contained in a plastic housing, which accepts either loose granules or prepacked canisters. A dust and moisture trap or a moisture drainage valve is provided on some machines. Many anesthetic machines are equipped with dual canister absorbers, which are useful for larger animals and long procedures. A small canister used with a very large patient may not efficiently remove carbon dioxide. Recirculated gases enter the absorber assembly at the top of the canister and exit at the bottom. These gases, with fresh breathing gas added, are then returned to the patient via the inspiratory valve.

It is important to change the absorbent promptly when the indicator has turned color, denoting exhaustion of the carbon dioxide absorbing capacity. Some indicators will, upon standing and exposure to light, revert to colorless; this does not imply reconstitution of the absorbent. The canister assembly housing, gaskets, screen, and dust and moisture traps should be cleaned coincidentally with absorbent changes. Bulk absorbent should be added gradually and the housing tapped in order to promote even distribution and packing of the granules and to avoid ''channeling'' of gases. At the same time, the granules should not be overpacked or so agitated that the granules are broken down into smaller particles or dust (Dorsch and Dorsch, 1994g).

Large bulk cans of absorbent cost the least per unit weight, but are inconvenient and must be carefully sealed to retain moisture. Prepacked, disposable canisters are convenient, easy to change, and avoid the problems associated with over- or underpacking, but are more expensive.

J. Unidirectional Valves

Unidirectional valves direct the flow of gases around the patient breathing circuit and limit the rebreathing of expired gases. The most common valves are discs of mica, ceramic, plastic, or metal, which sit atop carefully machined metal ports. They are light, so they move with little resistance, and gas tight, to ensure directional control and prevent rebreathing. Dome valves are surmounted by clear plastic domes, which permit the observation of their function. The domes are made to permit easy removal and access for cleaning and maintenance. The inspiratory and expiratory unidirectional valve housings have standard 22-mm-diameter male connections for the breathing circuit.

Breathing valves are subject to accumulated moisture and should be cleaned and dried regularly, and carefully reassembled. Problems arising from sticking or incompetent inspiratory and expiratory valves have been described (Baxter *et al.,* 1991; Pyles *et al.,* 1984).

K. Adjustable Pressure-Limiting Valve

Pressure in the breathing circuit is controlled by an adjustable pressure-limiting valve (APL) or ''pop-off'' valve, located in the exhalation side of the circuit. On most machines, the APL valve has a 19-mm male fitting connecting to the waste gas scavenger system. The internal diameter of the fitting size differs from the standard 22-mm breathing circuit in order to prevent misconnections. Although the specific design of the valve varies with the manufacturer, all can be gradually adjusted from fully open to fully closed, providing progressive resistance to gas escape. When the valve is open, gas can escape the breathing circuit at very low pressures. As the valve is closed, the pressure at which gas escapes rises. The APL valve is fully or partially closed, and the breathing bag is compressed to permit manual ventilation of the patient. Otherwise, it is usually kept fully open to avoid excess circuit pressure.

Occasionally, attempts are made to connect a 22-mm breathing hose to the scavenger port, using tape or other means to achieve a seal. This practice defeats the safety advantage of the 19-mm port. Tubing of the correct size is inexpensive, readily available from anesthesia equipment suppliers, and should be used.

L. Reservoir Bag

The reservoir, or breathing bag, attaches to the bag port, usually close to, or part of, the APL valve assembly. It provides much of the gas needed for inspiration, limits pressure in the breathing circuit, and is used to assist, control, and monitor ventilation. With the reservoir bag, it is possible to use moderate fresh gas flow rates because the bag volume accommodates the high momentary flows that occur during normal inspiration (Dorsch and Dorsch, 1994e).

Rubber reservoir bags are designed to limit the pressure in the breathing system. Even when greatly overinflated, the pressure will reach only about 50 cm H_2O. Some plastic bags lack elasticity and permit rapid pressure rises (Davis *et al.,* 1995c; Dorsch and Dorsch, 1994e). Both disposable and reusable reservoir bags are available in sizes from 0.5 to 6.0 liter. It has been recommended that the bag volume should be three to five times the tidal volume of the patient (Bednarski, 1993). A bag that is too large will have smaller respiratory excursions, making monitoring more difficult and, by increasing circuit volume, will slow the response to changes in vaporizer settings at a given fresh gas flow rate. A bag that is too small risks inadequate reservoir volume for breathing and limits the safety provided by low pressure distention. Reusable bags are more robust and have heavier walls than the thinner disposable bags. A rubber nipple, or bag tail, is molded into the end of some bags opposite the neck. For certain applications, the bag tail is cut and connected to an exhaust tube, or valve.

M. Manometer

Pressure in the breathing circuit is usually expressed in centimeters of water, measured by an aneroid manometer located in the breathing system (Fig. 4). The manometer is an important safety component, used not only during assisted or controlled ventilation to avoid barotrauma, but also to detect and avoid dangerously high pressures due to operator error and equipment malfunction. Manometers indicate negative as well as positive pressure i.e., they register the degree to which circuit pressure is reduced below ambient, as when the patient attempts to inspire against an empty reservoir bag, or independently of a mechanical ventilator, or when there is a malfunction of an active waste gas scavenging system. Circuit manometers register relative, not absolute, pressure. The indicator needle should indicate zero at zero relative pressure and should move smoothly, without sticking. Airway manometers may be purchased separately through equipment suppliers as replacements.

N. Vaporizers

Vaporizers are used to convert liquid anesthetic agents into gases, which are then added to the breathing gas mixture. Vaporizers have been classified according to a variety of characteristics, including method of output regulation, method of vaporization, location in the breathing circuit, agent specificity, and temperature compensation. For the purposes of this discussion, vaporizers will be divided into precision and nonprecision types, and their further characteristics are described individually. If the vaporizers permit accurate selection of the final concentration of vapor, usually expressed as volume percent, they are referred to as precision vaporizers. If they do not, they are called nonprecision vaporizers. The underlying principles of vaporizer design and function, as well as extensive descriptions of individual vaporizer characteristics, are thoroughly covered in the references and other standard texts.

Fig. 4. Breathing system manometer. The gauge is located between the inspiratory valve and absorber assembly on the left and the expiratory valve and APL valve on the right.

1. Precision Vaporizers

Two types of precision vaporizers are described in the following section. The older, measured flow vaporizers are discussed briefly because they may be encountered in the research facility. On modern equipment, they have been supplanted by variable bypass vaporizers. In both cases, the vaporizers are located out of the breathing circuit, i.e., their output is blended into the total gas flow delivered by the anesthetic machine at the common gas outlet.

a. Measured Flow Vaporizers. Measured flow vaporizers are no longer made, but are still in service on older machines. Examples include Verni-Trol and Copper Kettle vaporizers. They use separate flowmeters to regulate the flow of carrier gas to the vaporizer, where it is bubbled through the anesthetic agent, becoming fully saturated. The degree of saturation depends largely on the vapor pressure of the agent and on the temperature. A thermometer indicates the temperature of the agent in the vaporizer. For agents such as halothane or isoflurane at room temperature, the gas leaving the vaporizer carries about 30 volume percent of agent. After exiting the vaporizer, the saturated gas is added to the basic fresh gas flow selected on the main flowmeters and is thus diluted to a safe and predictable concentration. These vaporizers are precise and can be used with many different agents, provided the physical properties of the agents and the principles of operation of the vaporizer are understood. In practice, the user must first select the total gas flow to be delivered to the patient and then decide what concentration of anesthetic agent is desired. The amount of fully saturated vapor needed to provide the desired final concentration is then calculated or read from a specialized slide rule. The correct use of these vaporizers requires detailed understanding. The inexperienced user should consult standard texts before using them (Hartsfield, 1987; Dorsch and Dorsch, 1984b).

Measured flow vaporizers can be demanding of the user when rapid changes in agent concentration and gas flows are needed and require more training to use correctly than the variable bypass vaporizers described later. A less apparent limitation is the restriction these vaporizers place on the use of low flow or closed circuit anesthesia techniques. The low flow techniques used in current practice require very low settings on the vaporizer flowmeter, frequently below the lower calibration limit. It is unsafe to extrapolate below the lower flowmeter limit unless agent analysis monitoring is used (Hartsfield, 1994).

b. Variable Bypass Vaporizers. Newer precision vaporizers, with few exceptions, are all of the variable bypass type. With these devices, the total output of the flowmeters is routed to the vaporizer, but only a portion is diverted through the vaporizing chamber, where wicks are used to maximize surface area for vapor saturation. The diverted gas, now saturated with vapor, rejoins the main flow before exiting the vaporizer. Agent concentration is controlled by varying the proportion of carrier gas diverted to the vaporizing chamber. These vaporizers are agent specific, calibrated by the manufacturer for a single anesthetic agent. A dial or hand wheel is used to select the concentration of agent delivered to the breathing circuit, with output indicated in volumes percent (Dorsch and Dorsch, 1994d). In order to prevent accidental use, some vaporizers have a positive ON/OFF mechanical lock, usually a button, which must be activated before the vaporizer dial can be turned.

The relationship between indicated output at the vaporizer dial and actual output at any given fresh gas flow rate is referred to as *linearity*. Output is affected by a number of physical factors, including temperature, pressure, fresh gas flow rate, and carrier gas composition. The vapor pressure of volatile anesthetic agents varies directly with temperature. Modern vaporizers have automatic temperature compensation mechanisms to maintain predictable output within the temperature range likely to be encountered in an operating room. Older vaporizers may require manual compensation. Pressure also has an effect on vaporizer output. Pressure variations transmitted from the breathing circuit during assisted or controlled ventilation will cause the output to vary. Again, modern vaporizers are designed to minimize this effect. The output of a vaporizer at any given setting can vary with the fresh gas flow rate, a phenomenon called flow dependence. This problem is also addressed in modern designs, which have a relatively linear output over a wide range of flows (Davis *et. al.,* 1995b). However, even modern vaporizers may not be linear at very low flow rates, and older vaporizers may deviate significantly from the indicated output at both high and low flow rates. Changes in carrier gas composition will affect output. Manufacturers calibrate precision vaporizers using either oxygen or air as the carrier gas. The addition of other gases, such as nitrous oxide or nitrogen, will affect output, but seldom to a significant extent (Eisenkraft, 1993b). Vaporizers used in human medicine have been extensively studied and their performance characteristics defined (Dorsch and Dorsch, 1984b, 1994d).

Vaporizers have filling ports, sight glasses to indicate fill level, and drain ports. Some filling ports are mechanically keyed to accept only the correct agent and require agent-specific adapters for the anesthetic agent bottles and the filling port. Many vaporizers have only a simple filling funnel. In addition to being more prone to spillage and personnel exposure to volatile agents, vaporizers without keyed filling ports are more liable to accidental misfilling. If this occurs, the vaporizer must be cleaned, and possibly serviced, before use. The manufacturer should be consulted for specific directions. If filling funnels are used, special bottle pouring adapters, shown in Fig. 5, greatly increase control while filling the vaporizer. Pouring adapters quickly pay for themselves in terms of reduced wastage and room contamination.

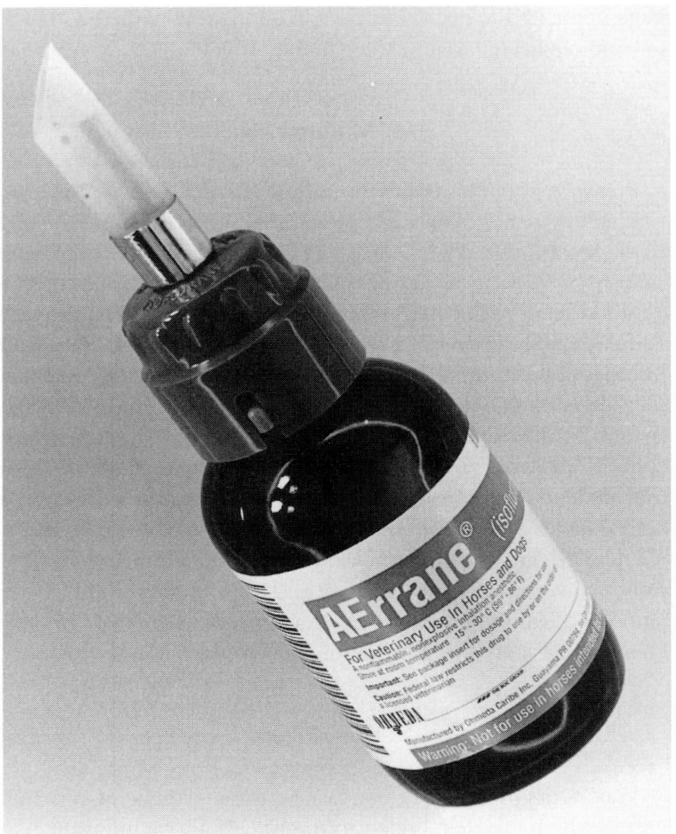

Fig. 5. A pouring adaptor replaces the original bottle cap. The adaptor minimizes spills during the filling of vaporizers with funnel fittings. The adaptor is color coded and will fit only the correct bottle of anesthetic agent.

To avoid the risk of dangerously high anesthetic agent levels, vaporizers should not be overfilled or tilted (Andrews, 1990). A maximum fill line on the sight glass indicates the safe capacity. Most vaporizers must be turned off and the fresh gas flow discontinued while being filled. Because this is an inconvenient process during use, the agent level should always be checked before starting anesthesia. It is important to secure the filling cap before turning the vaporizer on again or else pressure from the carrier gases may force the liquid anesthetic agent out of the filling spout, an expensive and potentially hazardous accident.

Drain ports permit the vaporizer to be emptied for servicing or storage. Because vaporizers are not completely vapor tight, a small amount of agent loss will occur upon standing. This loss may be greater with older designs. The vaporizer should be drained before storage, transport, or extended periods of disuse. When two or more vaporizers are mounted on the same machine, some form of lockout mechanism should be used to prevent simultaneous use. The practice of linking vaporizers in series, without the use of some positive lockout system, is risky. Safety lockouts are standard on modern machines made for use with humans. If a second vaporizer is added to a machine not equipped for it, it is safest to disconnect the vaporizer not in use from the fresh gas flow. If this is not done, extra caution is needed to avoid contaminating the vaporizers and administering an uncontrolled mixture of agents (Hartsfield, 1994). The position of a second vaporizer is important: it should be connected between the flowmeters and the oxygen flush valve. The temptation to connect the vaporizer downstream of the flush valve, between the common gas outlet and the breathing system, should be resisted because activation of the flush valve will direct high gas flows through the vaporizer, leading to increased agent levels (Andrews, 1990).

Although it is possible to use a volatile anesthetic in a precision vaporizer calibrated for another agent if the vapor pressures are very similar, no manufacturer recommends it. The advantages and disadvantages of this practice with respect to the use of isoflurane in halothane vaporizers have been discussed by Steffey *et al.* (1983). The practice becomes especially hazardous if the user intends to switch back and forth between two different agents. It is almost inevitable that errors in cleaning or filling will result in both agents being present in the vaporizer simultaneously, resulting in unknown and potentially lethal concentrations of mixed agents.

Modern agent-specific precision vaporizers are accurate, reliable, and simple to use, but do require care and regular maintenance. Vaporizer output should be checked periodically, at least yearly, either by the vendor or by qualified in-house service personnel. Vaporizers that fail to perform satisfactorily should be returned to the manufacturer or other qualified service center for repair and recalibration. They are not user serviceable.

The description given for precision vaporizers applies generally for vaporizers used with isoflurane, halothane, enflurane, and sevoflurane. Desflurane, however, has a much higher vapor pressure than the other agents and cannot be safely delivered from a conventional vaporizer. Rather, a heated, pressurized vaporizer designed specifically for the agent is required (Eisenkraft, 1993b).

2. Nonprecision Vaporizers

Nonprecision vaporizers remain popular because they are relatively inexpensive and simple. The Ohio No. 8 vaporizer is perhaps the most familiar example. Although originally used for ether, this "olive jar" vaporizer has long been used for methoxyflurane. It is essentially a glass jar with a paper wick and a lever-operated bypass valve that can divert a variable amount of gas through the jar and back into the circuit. When the lever is closed, gas bypasses the chamber. As the lever is opened, a greater proportion of gas is diverted through the chamber. Because it is located within the patient breathing circuit, the vaporizer output varies with respiratory rate and volume, use of positive pressure ventilation, vaporizer temperature, and fresh gas flow rates. The anesthetic agent is vaporized by gas circulating through the breathing circuit. Fresh gas provides oxygen, but also serves to dilute the agent in the circuit. The agent concentration rises with the use of low flow rates, increased ventilation, and positive pressure ventilation, and drops with increased fresh gas flows and decreased temperature (Soma and Klide, 1968). Although the variable output does not present an insurmountable problem with agents of low vapor pressure, such as methoxyflurane, this type of vaporizer can be dangerous if used with agents such as halothane or isoflurane, which have higher vapor pressures (Hartsfield, 1994).

During use, the paper wick becomes wet as moisture condenses from the breathing circuit, reducing the efficiency of vaporization. With poorly maintained vaporizers, it is not uncommon to see a layer of water on the surface of the methoxyflurane. The vaporizer should be emptied and the wick dried.

The Stephens universal vaporizer is a newer in-circuit variable bypass vaporizer marketed to the veterinary community, usually as a component of a compact small animal machine. It is designed for use with low flow anesthesia. As the name suggests, it is intended to be used for any volatile agent. Some degree of agent specificity is attained by varying the height of an internal metal sleeve above the agent level and inserting or removing a wick. Removing the wick reduces the efficiency of the vaporizer and permits the use of highly volatile agents such as halothane or isoflurane (Ludders, 1993). The Stephens vaporizer is subject to many of the same limitations cited for the Ohio No. 8.

The superiority of in-the-circuit nonprecision vaporizers versus out-of-the-circuit precision vaporizers remains an area of dispute. It has been argued that in-circuit vaporizers are superior for very low flow or closed circuit anesthetic techniques, despite the considerable variation in output due to fresh gas flow, temperature, and ventilation. Even proponents of in-circuit nonprecision vaporizers, however, allude to the need for thorough understanding of the functions and limitations of these units (Klide, 1992). Modern out-of-circuit precision vaporizers are more complex in construction and are more expensive, but are relatively precise and constant in their output and are easier to use. In contrast, the apparent simplicity, which makes nonprecision in-circuit vaporizers attractive to many users, demands a correspondingly higher level of sophistication and vigilance for optimal results—assets not always readily at hand in the experimental setting. Precision out-of-circuit vaporizers can be used for low flow and closed circuit anesthesia, are more economical with higher flow techniques, and are easily used with a variety of nonrebreathing systems commonly encountered with small laboratory animals. In terms of flexibility, ease of use, and safety, they are the preferred choice for laboratory animal anesthesia.

O. Scavenging Systems

Inhalation anesthesia produces waste gas, which must be safely removed. Scavenging systems accomplish this by means of collection devices, interfaces, and disposal systems. Both active or passive systems are used. Active systems use continuously applied suction from centrally piped vacuum or active duct systems to collect and dispose of gases. Passive systems depend on the pressure generated in the breathing circuit to propel waste gas into large-diameter tubing leading to the disposal.

The collection device used on rebreathing systems with carbon dioxide absorbers and on special adapters for nonrebreathing circuits is the APL valve. Although some very old APL valves do not include provisions for scavenging waste gas, all modern units should be of the scavenging type. Valves built to the standard used for human machines use a 19- or 30-mm OD (outer diameter) port for the transfer tubing to prevent the misconnection of 22-mm breathing circuit hoses (Dorsch and Dorsch, 1994e). Some breathing systems, such as Mapleson-F and Mapleson-E circuits, do not have an APL valve. The waste gas is exhausted via the reservoir bag tail (Mapleson-F) or directly from the expiratory limb (Mapleson-E). Special breathing bags are made for Mapleson-F systems, which permit convenient scavenging and single-handed bag compression by means of an enclosed scavenging line and valve.

Waste gas collected from the APL valve or other source is conducted to an interface for disposal. Either *closed* or *open* interfaces are used. Closed interface systems use a valve to provide a connection to the room atmosphere. If passive scavenging is used with a closed interface, a positive pressure relief valve will vent waste gas into the room if there is obstruction of the system downstream of the interface. If active scavenging is used, a closed interface will have both positive and negative pressure relief valves, and may also have a vacuum adjustment and a reservoir bag (Fig. 6). The positive pressure relief valve protects the breathing circuit from high pressures if the vacuum is obstructed or fails. The negative pressure relief valve prevents excess vacuum from being applied to the breathing circuit. If the interface is served by piped vacuum, a needle valve is incorporated to adjust the vacuum level. The reservoir bag is observed and the vacuum is adjusted so that it is neither overfull nor flattened by excess vacuum (Smith, 1993; Dorsch and Dorsch, 1994i).

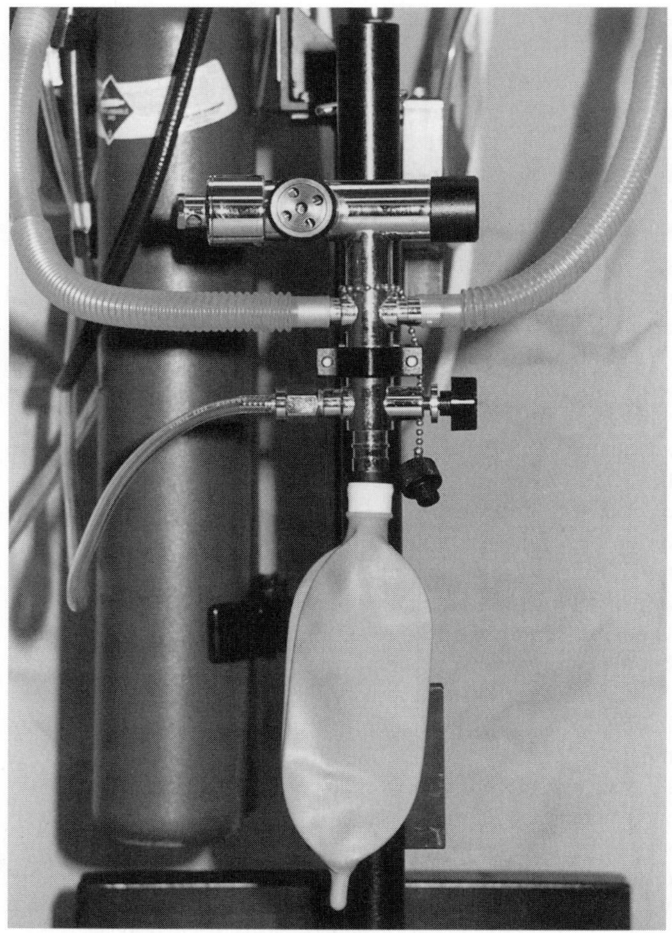

Fig. 6. A closed interface active scavenging system. The corrugated 19-mm hose to the right originates at the APL valve and to the left, from the ventilator relief valve. The positive and negative pressure relief valves are located in the crossbar assembly immediately above the corrugated hose connection. The small diameter clear tubing on the left leads to the central scavenging vacuum system. Opposite it is the vacuum adjustment needle valve. The reservoir bag is seen at the bottom.

Active scavenging open interface systems are simpler than closed interface systems. Because the system is open to the atmosphere through a series of vent holes, positive and negative relief valves are not necessary. If the interface is designed for piped vacuum, a needle valve is used to adjust the vacuum level. A reservoir bag or rigid reservoir is usually used to accommodate the volume of waste gas intermittently discharged from the APL valve (Fig. 7).

Suction for active scavenging systems is supplied by a piped vacuum system, or an active duct system. Piped vacuum systems use relatively narrow-gauge tubing to conduct gas from the interface to the vacuum connection. Active duct systems use specialized ducts and suction generators to move high volumes of gas at relatively low vacuum levels to the discharge site. This requires the use of relatively large-gauge (19 mm or larger) connecting tubing (Dorsch and Dorsch, 1994i). Both systems work well, but scavenger interfaces designed for one system may not work with another. The architects, engineers, and users should consult with each other closely to prevent a mismatch between the interface design and the conduit and disposal systems. They should also assure that the vacuum pump serving the scavenger system is itself safely vented and that the pump discharge does not contaminate equipment rooms or place maintenance personnel at risk.

Passive scavenging systems may simply lead the waste gas to nonrecirculating room air exhausts, chemical hoods, or

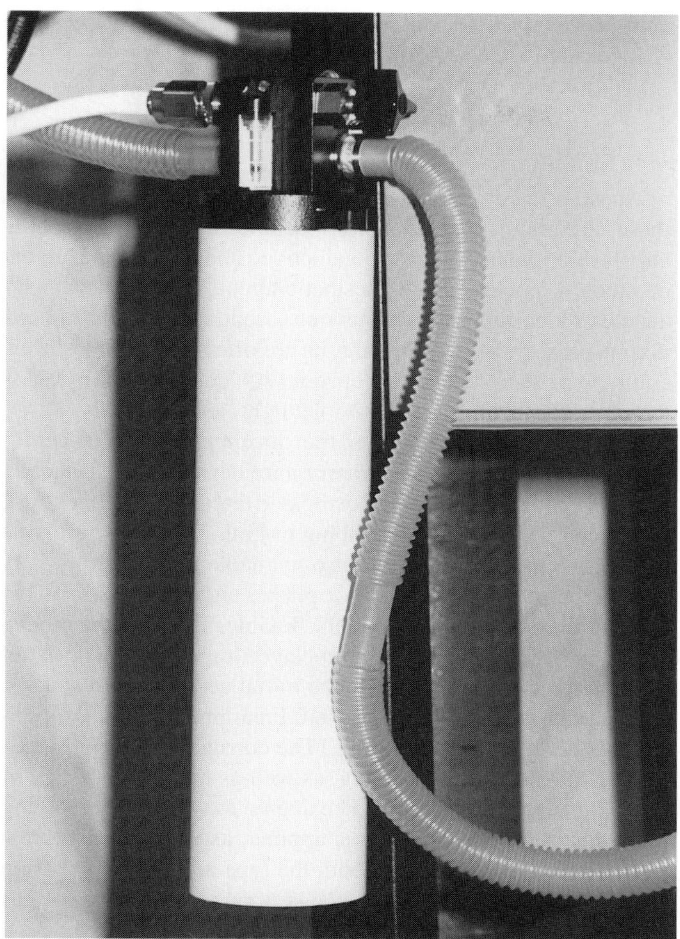

Fig. 7. An open interface scavenging system. This example is designed for a central vacuum with a vacuum adjustment needle valve and gauge seen at the top of the unit. The white hose to the top left leads to the central scavenger vacuum system. The corrugated hoses originate from the ventilator relief and APL valves (not shown). This design uses a rigid reservoir seen as the white cylinder below the tubing connections.

Fig. 8. Ventilator grill adaptor. The adaptor is made to fit a louvered room ventilation exhaust grill and to accept 19-mm waste gas tubing from the anesthetic machine. This interface must be used only with nonrecirculating air-handling systems.

through the wall, ceiling, or window to the outside. If a room ventilation grill is used, it is essential to check with the building engineer to assure that the room air is not recirculated (and that the correct grill is used). A 19-mm adaptor for the ventilation grill is shown in Fig. 8. Through the wall systems should be short and wide diameter to avoid back-pressure. Each passive duct system should serve a single station: they should not be conjoined. The exit must be located to prevent the pollution of adjacent window and air intakes, and the exit protected to avoid accidental blockage (Dorsch and Dorsch, 1994i). Narrow diameter plastic tubing cannot carry the high volumes needed without significant back-pressure and the risk of leaks or elevated circuit pressures.

In situations where no scavenging systems are available, activated charcoal filters can be used to absorb volatile anesthetic agents from the waste gas. Although charcoal canisters can remove halogenated anesthetic agents, they cannot remove nitrous oxide. They must be weighed regularly to determine the remaining absorption capacity. These disposable canisters are useful when anesthesia must be provided in unusual locations, but the need for regular replacement and careful monitoring to assure adequate performance renders them unsatisfactory for routine use.

Special considerations arise when dealing with the scavenging requirements for very small animals in which anesthesia is still frequently administered by a mask or an open "nose cone" with agents of low volatility, such as methoxyflurane. In these cases the simplest solution is to confine the procedure to a chemical hood or a similar high flow nonrecirculating hood. However, this often is not feasible. A more satisfactory solution is to use an actively scavenged mask in conjunction with a vaporizer and flowmeter controlled fresh gas delivery system. Designs of this type are described and illustrated by Glen *et al.* (1980) and Levy *et al.* (1980). Commercial rodent operating tables with integrated mask systems are offered by suppliers of experimental physiology equipment and veterinary equipment. These should be carefully evaluated before purchase to assure that they meet the needs of the user and provide adequate scavenging.

Use of appropriate scavenging devices and high standards of equipment maintenance are essential for the safe use of volatile anesthetics. Equally essential is good anesthetic practice, which minimizes the use of "dirty" techniques, such as mask and chamber induction.

IV. BREATHING CIRCUITS

A. Introduction

Breathing systems provide oxygen to the patient and remove waste products. If volatile anesthetic agents or gases such as nitrous oxide are used, they must be scavenged safely. The design and fabrication of breathing circuits and the study of their characteristics and performance have long been active areas of research, offering considerable scope for the inventive mind. The proliferation of circuits, valves, adapters, and related paraphernalia that are commercially available is so great that even experienced anesthesiologists are unlikely to be familiar with all of them. The principles of circuit design and function are well covered in standard texts (Dorsch and Dorsch, 1994a; Ehrenworth and Eisenkraft, 1993; Davis *et al.,* 1995c), supplemented by a large and growing literature. Those who are tempted to design *or modify* breathing circuits should consult the literature carefully: the subject is much more complex and subtle than is immediately apparent, and there are numerous pitfalls for the untutored and unwary. Fortunately, only a few types of breathing circuits are needed to accomplish a wide range of anesthetic tasks.

B. Dead Space

Inspired air that enters the respiratory passages but does not reach the alveoli or reaches poorly perfused alveoli does not participate in oxygen–carbon dioxide exchange. The total volume occupied by gas that is not losing oxygen or gaining carbon dioxide is called *physiologic dead space.* The volume accounted for by the upper respiratory structures—the nose, pharynx, trachea, and bronchi—is called *anatomic dead space.* The volume accounted for by under- or unperfused alveoli is called *alveolar dead space.* Anesthesia equipment can add to dead space. This *mechanical dead space* is undesirable and should be minimized. An endotracheal tube that is trimmed to the proper length does not increase dead space and, usually, decreases it. However, if the tube extends far beyond the mouth of the patient, dead space is increased. If an airway filter, elbow connection, or gas sampling connecter is added between the endotracheal tube and the breathing circuit, mechanical dead space is further increased. Properly designed breathing circuits, correctly used, do not significantly increase dead space (Gravenstein *et al.,* 1989).

C. Materials

Anesthesia circuits are made of rubber or plastic. Originally, these circuits, as well as other machine tubing and breathing bags, were made of heavy, conductive rubber. With the advent of modern, nonexplosive anesthetic agents and the disappearance of older, more hazardous ones, conductive tubing is seldom necessary. Still, these circuits are often supplied with new equipment. They are heavy, opaque, subject to attack by anesthetic agents, ozone, and ultraviolet light, and are costly.

Most users prefer the lighter, translucent plastic circuits ubiquitous in hospitals and respiratory care departments. Typically made of polyethylene, these corrugated circuits sometimes have reinforced walls, to resist crushing or kinking. Finally, silicone rubber breathing circuits are also available. Although initially more expensive than disposable plastic circuits and somewhat heavier, they are translucent, very flexible, and have the unique advantage that they can be autoclaved repeatedly. The latter characteristic may well justify the initial cost for some studies.

Regardless of the material used, breathing circuits are typically made of corrugated tubing. The corrugations confer flexibility and resistance to kinking, as well as a variable degree of compliance or distensibility. The degree to which the breathing hoses distend when pressure is applied, as with positive pressure ventilation, varies with both the type and the construction of the circuit. Circuit compliance is important because the gas that expands the circuit does not expand the lungs of the patient. For example, with mechanical ventilation of smaller patients, the proportion of the preset tidal volume lost to circuit compliance can be considerable, leading to hypoventilation, despite apparently adequate volume settings.

D. Classifications of Breathing Circuits

Terms such as *open, closed, semiopen,* and *semiclosed* are frequently used to describe and classify breathing circuits. Unfortunately, these terms are understood differently by various authors, a fact that has led to considerable confusion and misunderstanding. Only a few types of circuits are commonly used, and it seems more rational to simply describe their characteristics.

E. Rebreathing Systems

Although hardly the simplest design, the conventional circle system, using the unidirectional valves and carbon dioxide absorber described earlier, is the most common system (Fig. 9). The removal of carbon dioxide and the recirculation of the remaining gases reduce the use of both anesthetic agents and breathing gases, and with moderate flows, some heat and mois-

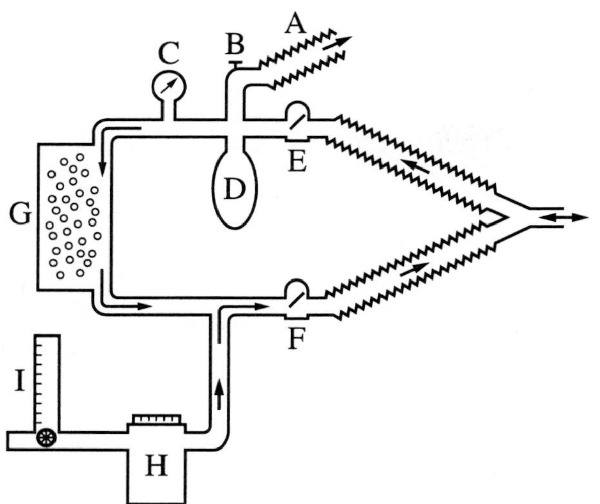

Fig. 9. Circle breathing system with the vaporizer outside the circle. (A) Scavenger tubing from the APL valve. (B) APL valve. (C) Circuit manometer. (D) Reservoir bag. (E) Unidirectional expiratory valve. (F) Unidirectional inspiratory valve. (G) Carbon dioxide absorber canister. (H) Vaporizer. (I) Flowmeter. Arrows indicate direction of gas flow. The patient Y connection is at the right.

ture are also retained. By reducing flow rates, the same circuit is used for very low flow or closed circuit anesthesia.

These systems have a fairly high internal volume, including the breathing bag, absorber canister(s), and tubing volume. As a result, if low flows are used with a precision vaporizer located out of the circuit, the agent concentration in the circuit will be very slow to change, despite changes in the vaporizer settings. If rapid changes in agent concentration are needed when using an out-of-the-circuit vaporizer, the flow rates must be increased accordingly. Aggressive use of the oxygen flush valve, with its high flow rates, will rapidly reduce the anesthetic concentration in the circuit. In contrast, if low resistance nonprecision vaporizers are used within the circle system, agent concentration at low flow rates will depend greatly on patient respiratory rate, volume, and the use of positive pressure ventilation.

Because of labor and liability considerations, most hospitals use disposable anesthetic circuits for each patient. These come preassembled and can be purchased with virtually any combination of breathing bags, elbow adapters, and bacterial filters. Typically, disposable circuits are available in 48- or 60-inch lengths. For research or veterinary clinical use, they may be washed and reused. As stated earlier, disposable breathing bags are available in a range of sizes and also can be reused, although they are less durable than the circuits. Some breathing circuits have reinforced cuffs at the machine ends, which makes the process of attaching and removing them much easier. Disposable circuits are available individually from veterinary suppliers and by the case from human anesthesia equipment vendors. Alternatively, it is very inexpensive to purchase 22-mm breathing hoses in bulk and to make circuits of any desired length with the addition of a Y connector.

Pediatric circle systems are available with the internal diameter of the connecting tubes reduced to 15 mm. Reduced volume circuits help conserve heat and moisture and can extend the advantages of circle systems to smaller patients, such as rabbits. If ventilation is closely monitored and supported, the adverse effects of increased resistance due to the absorber and unidirectional valves are minimal. These smaller circuits also are less compliant than larger ones, reducing volume losses with mechanical ventilation.

In contrast, large animal anesthesia machines have, for obvious reasons, larger valves and breathing ports, and use 50-mm ID connecting tubing, with correspondingly large Y connectors, endotracheal tube connectors, etc. These more specialized circuits are usually obtained through the makers of the large animal anesthetic machines.

Circle systems conveniently accept a variety of additional devices, including nebulizers, humidifiers, volumeters, and positive end expiratory pressure (PEEP) valves that are needed in special circumstances. A novel coaxial variation of the circle breathing circuit, called a Universal F Circuit, is now available.

F. Nonrebreathing Systems

''Nonrebreathing'' circuits are not, for the most part, nonrebreathing. Instead, they are circuits in which the degree of rebreathing is largely controlled by the fresh gas flow rate. Circuits that do not use unidirectional valves to direct the flow of gases to or from the patient have been described and classified as Mapleson types A through F (Dorsch and Dorsch, 1994f). They have been extensively studied, characterized, and modified, and four types, Mapleson-A, -D, -E, and -F, may be encountered in the research setting (Fig. 10).

The Mapleson systems are especially useful for rodents, and machines suitable for them can be constructed from a minimum of hardware. Typically, a gas cylinder with a pressure regulator, a suitable flowmeter, and a vaporizer are needed. The pressure regulator is required to control delivery pressure to the flow meter(s) at the usual 45–50 psig. The flowmeter then provides fresh gas to the vaporizer and then to the circuit. Some form of scavenging is essential—either a hood or a dedicated active or passive scavenging system. A flowmeter that is calibrated for lower flow rates should be used and the flow rates should take into account the minimum flow required for vaporizer accuracy. The alternative to fabricating a suitable machine for rodents is to purchase one of the increasing number of machines made for this purpose.

1. Mapleson-A Circuits

The Mapleson-A circuit is usually called a Magill system. In this configuration, fresh gas enters near the breathing bag,

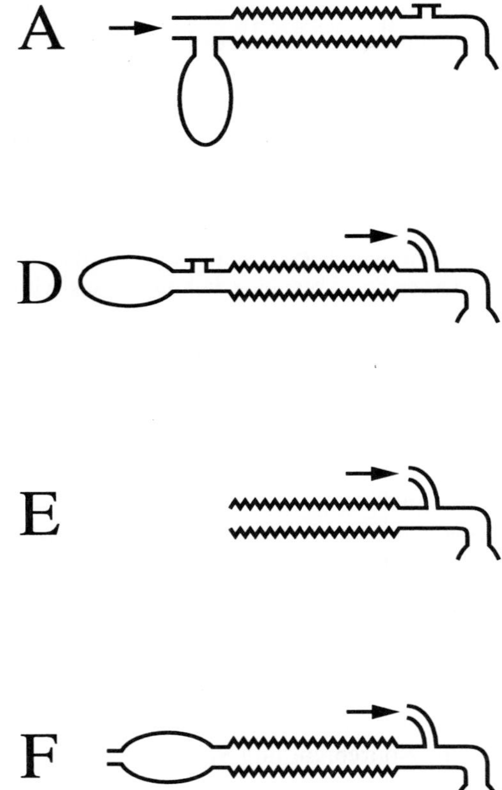

Fig. 10. Mapleson circuit types A, D, E, and F. The arrows indicate fresh gas flow from the anesthetic machine. Types A and D include relief valves, and types A, D, and E provide reservoir bags. The patient connection is to the right. Redrawn from Dorsch and Dorsch, (1984c).

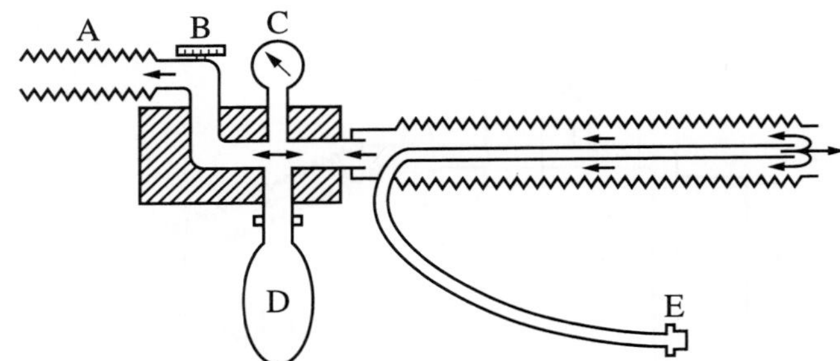

Fig. 11. Bain modification of a Mapleson-D circuit with adaptor. (A) Scavenger tubing from the APL valve. (B) APL valve. (C) Circuit manometer. (D) Reservoir bag. (E) Fresh gas connection to the common gas outlet. The patient connection is at the far right. Redrawn and modified from Dorsch and Dorsch (1984c).

passes toward the patient through a corrugated tube, and exits via a relief valve near the patient. The direction of fresh gas flow, which forces expired gas out of the circuit very close to the endotracheal tube or mask, makes this circuit very efficient and well suited for spontaneous ventilation. It is somewhat less satisfactory for assisted or controlled mechanical ventilation. The necessity of having a scavenging tube at the patient end of the circuit can make using the Magill circuit somewhat clumsy. Magill systems are less popular in the United States than elsewhere (Hartsfield, 1987; Bednarski, 1993).

2. Mapleson-D Circuits

The Bain modification is a common example of the Mapleson-D circuit (Fig. 11). In this version, the fresh gas supply tube is coaxially located within the larger exhalation limb. This circuit is popular because it is simple, offers little resistance to breathing, and works well for both spontaneous and positive pressure ventilation. The Bain circuit can be used by simply attaching a breathing bag to the exhaust limb and directing the overflow from the bag tail to an appropriate scavenging system (Hartsfield, 1987; Bednarski, 1993). However, it is much more convenient to use a Mapleson-D arm, or adaptor, as seen in Fig. 12. The adaptor is a drilled metal block that usually includes a circuit manometer, APL valve, and bag port. During positive pressure ventilation, the bag is removed and replaced with the ventilator hose. Some Mapleson adapters have selector levers permitting convenient and rapid switching from manual to mechanical ventilation. Various fresh gas flows have been recommended to prevent or minimize rebreathing, ranging from 100 to 130 ml/kg/min, with a minimum flow of 500 ml/min (Manley and McDonell, 1979; Hartsfield, 1987). It is generally accepted that positive pressure ventilation will permit the fresh gas flow to be reduced somewhat from that needed for spontaneous ventilation. Animals as small as rats have been successfully maintained on Bain circuits, using a mask or endotracheal

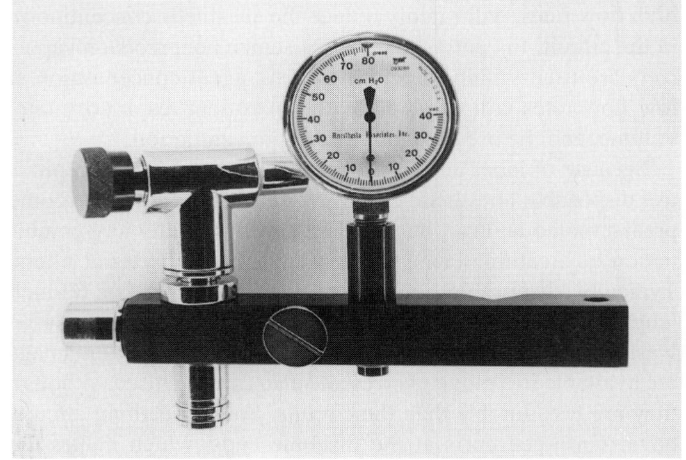

Fig. 12. Mapleson-D, or Bain adaptor. The circuit manometer, APL valve, circuit connection, and reservoir bag mount are shown counterclockwise from the top. The slotted metal covers a port for the addition of a ventilator/bag selector valve.

tube (Fig. 13). The circuit is not economical for larger patients due to the exorbitant fresh gas flows required.

As well as the advantages cited earlier, Mapleson-D circuits have some other characteristics that should be understood by the user. The fresh gas flow is delivered very near the endotracheal tube or mask connector by a narrow gauge, noncompliant tube. Changes in vaporizer settings are rapidly reflected in the circuit and the patient, a feature that may be disconcerting to those accustomed to circle systems. Whether this is an advantage or a disadvantage depends to a large extent on the disposition of the anesthetist. If a Mapleson-D control arm is used in conjunction with manual ventilation, the anesthetist must be scrupulous about returning the APL valve to the open position immediately after delivering a breath. Again, the rapid rise in pressure in the circuit can take an inattentive anesthetist by surprise, and can result in barotrauma to the patient. Modifications of the Mapleson-D arm to reduce the risk of barotrauma have been reported (McMurphy *et al.,* 1995). As previously mentioned, the oxygen flush valve is not needed and should not be used with circuits of this type due to the risk of barotrauma.

3. Mapleson-E Circuits

Mapleson-E circuits are extremely simple in concept and are well-suited to mask arrangements used for very small animals. A variety of ingenious, and sometimes complex, designs have been published (Glen *et al.,* 1980; Levy *et al.,* 1980; Ventrone *et al.,* 1982; Norris, 1981). It is important to remember that Mapleson-E circuits make it more difficult to employ assisted or controlled ventilation. For obvious reasons, proper waste gas scavenging also becomes problematic, a topic addressed in the designs of Glen *et al.* (1980), Levy *et al.* (1980), and Ventrone *et al.* (1982). Nonetheless, circuits of this type can be quite useful.

4. Mapleson-F Circuits

The version of the Ayre's T piece most commonly used today is the Jackson–Rees modification, which incorporates an expiratory limb and a breathing bag. In this configuration it is a Mapleson-F circuit (Dorsch and Dorsch, 1994f). When the bag is appropriately scavenged, the circuit is useful for small patients and is commonly used for the same reasons as is the Mapleson-D circuit. The use of a Jackson–Rees circuit in rats was described by Dardai and Heavner (1987).

Fig. 13. Bain circuit being used to provide inhalation anesthesia for a rat. Note the poor fit of this mask and the potential for anesthetic gas pollution.

Circuits can be designed to provide anesthesia to multiple small animals at once (Berryman and Sharp, 1995; Ventrone *et al.,* 1982). In designing and constructing such circuits, issues such as adequate fresh gas flows to prevent hypercapnia, resistance to breathing imposed by small diameter tubing or restrictive exhaust ports, and adequate scavenging must be carefully considered.

V. TESTING AND MAINTENANCE OF ANESTHETIC MACHINES

Before a patient is connected to an anesthetic machine, it is prudent to be sure that the machine is working properly. Modern equipment is reliable and designed with safety in mind. Still, equipment failure and human error are inevitable. In acknowledgment of that, an extensive checkout list for human anesthesia machines was published by the Food and Drug Administration in 1988 and has since been modified (Eichorn, 1993; Dorsch and Dorsch, 1994n). A discussion of checkout methods suited to veterinary and research use is given by Mason (1993). A standard checkout procedure, specific for the equipment used, should be developed and used. A thorough discussion of this very important issue is beyond the scope of this chapter. Only the most important and obvious points can be mentioned, and the reader should consult the references cited to develop an appropriate checkout procedure.

The machine is visually inspected to assure that the oxygen supply is present, adequate, properly connected, and functioning. The flowmeters and vaporizers are turned off, and a handheld suction bulb is connected to the common gas outlet. The bulb is compressed until empty and observed for 10 sec. If the bulb remains empty, the low pressure circuitry of the machine is not leaking. The test is repeated with the vaporizer turned on to detect internal vaporizer leaks. Reconnect the fresh gas hose.

Turn the flowmeters on and off while observing the float or bobbin. They should indicate zero at zero flow and move smoothly. Connect the breathing circuit and reservoir bag and check for correct position. Examine the absorber for proper filling and condition.

Close the APL valve and tightly occlude the breathing circuit at the patient end. An elbow connecter with the female fitting plugged by a rubber stopper is useful for this purpose. Use the oxygen flush valve to fill the bag to 30–40 cm H_2O pressure on the airway manometer. The pressure should not decrease more than 5 cm H_2O over 30 sec. A drop of more than 5 cm H_2O in that time indicates a leak of greater than 50 cm^3/min. *Release the pressure by opening the APL valve.* This checks the function of the valve and prevents dust from the absorber from being forced into the breathing circuit. See that the scavenging system is properly connected.

Bain circuits can be pressure checked, as described earlier, but also must be checked to ensure that the inner coaxial tube has not become disconnected: set the flowmeter to about 50 cm^3/min and use a small disposable syringe plunger to occlude the fresh gas delivery tube at the patient port. Oxygen flow should stop, and the flowmeter indicator should fall.

The anesthetic machine itself should be kept clean. Dust, hair, and spatters of saline and other fluids common to surgical procedures should be removed at the end of the case. Breathing circuits should be removed and washed between cases or, at least, at the end of the day. Unidirectional valves and absorber canisters can be cleaned by the user. Other parts of the machine, such as flowmeters and vaporizers, require a qualified service technician. Facilities with busy surgical caseloads may benefit from a maintenance contract so that machine function is checked and that parts subject to wear and deterioration, such as gaskets and O rings, are replaced on a regular basis.

VI. VENTILATORS

Because almost all anesthetic agents result in at least some respiratory depression, the anesthetist compensates by assisting or controlling ventilation. For short procedures, ventilatory support is usually provided manually by compressing the reservoir bag. For longer procedures, mechanical ventilation frees the anesthetist to attend to other monitoring and supportive duties.

Ventilators are classified in several different ways: the method used to initiate inspiration, the means used to transmit power, the pressure pattern produced, the nature of the safety limits, the power source, and the cycling control mechanism. Ventilators are available in almost every combination of power source, safety limit, and cycling control system imaginable. Fortunately for the anesthetist, anesthesia ventilators are generally much simpler than those designed for respiratory therapy and support. The general characteristics of the most common types of ventilators are described below (Schreiber, 1972).

Tidal volume is the amount of gas entering or leaving the patient with each breath. The sum of tidal volumes over 1 min is the **minute volume.** Many anesthesia ventilators use preset tidal volumes, i.e., the tidal volume is selected by the user. If the respiratory rate is also selected by the user, the minute volume is determined by these two independent settings. Some ventilators allow the user to set a desired minute volume and respiratory rate, and the ventilator itself determines the tidal volume. The **inspiratory flow rate** is the rate at which the gas enters the airway; in effect, the speed at which the tidal volume is delivered. Inspiratory flow rate has some effect on airway pressure but can also determine **inspiratory pause time,** the time between the end of inspiratory flow and the beginning of expiratory flow (Dorsch and Dorsch, 1994h). Ventilators in which the tidal volume is preset will tend to hold constant the volume delivered to the patient with each breath, but will allow the pressure to vary as the resistance to flow, compliance, etc., vary.

The **inspiratory phase time** is the time from the beginning of inspiratory flow to the beginning of expiratory flow. The **expiratory phase time** is the time from the beginning of expiratory flow to the beginning of inspiratory flow (Dorsch and Dorsch, 1994h). The inspiratory–expiratory phase time ratio, or **I:E ratio,** is fixed on some ventilators, usually at 1:2. On other machines it can be varied. It is possible, by varying the inspiratory flow rate and the I:E ratio, as well as tidal volume and respiratory rate, to gain considerable control over the ventilatory pattern of the patient.

Maximum inspiratory pressure is the highest pressure attained during the respiratory cycle (Schreiber, 1972). Some ventilators permit maximum inspiratory pressure to be preset. In this case, inspiration stops when the desired preset pressure is reached, regardless of the volume delivered. Pressure is constant, volume varies.

The relationship between the volume delivered and the pressure increase is called **compliance**—a measure of distensibility. Both patient and equipment compliance are important. A patient with high compliance, or high distensibility, has relatively less change in pressure for a given delivered volume of gas than does a patient with low compliance. Such a patient may be overdistended, and injured, at relatively low airway pressures. Conversely, a patient with low compliance may be underventilated at "normal" pressures. Young animals or animals with thin chest walls, such as rabbits, may be relatively compliant. In contrast, large animals or animals with certain diseases may have decreased compliance. Equipment compliance robs volume from the patient and can account for considerable losses, especially with small patients or with very long circuits. If this is not recognized and the delivered volume is not appropriately adjusted, the patient may be placed at risk of hypoventilation, despite apparently adequate tidal volume settings.

Anesthesia ventilators usually provide intermittent mandatory ventilation (IMV): they deliver each breath according to the preset frequency, although the patient can breathe independently of the ventilator. Some ventilators can also operate in the assist mode, in which the patient's attempt to breathe results in a low pressure in the circuit, which is sensed by the ventilator, triggering a breath cycle. Intermittent mandatory ventilation is often preferred in anesthetized patients because the attendant respiratory depression affects frequency as well as tidal volume. The assist mode is most useful during recovery and weaning from the ventilator. Numerous combinations of mandatory and assisted ventilation are possible and are used in respiratory therapy and critical care ventilators, but seldom in anesthesia. At the same time, some ventilators can be set to maintain positive pressure throughout some or all of the breathing cycle, resulting in a plethora of pressure patterns (and acronyms). In practice,

PEEP is occasionally useful. With this technique, the circuit pressure is kept above a certain preset value at all times during the breathing cycle instead of being allowed to fall to zero or near-zero during the end expiratory phase. PEEP may be obtained by a number of means, most of which involve some type of valve or spring mechanism in the expiratory limb of the circuit (Dorsch and Dorsch, 1994e).

Most new anesthesia ventilators are electronically controlled, requiring an electric power supply. Many older, and some newer, units use pneumatic or fluidic control systems and can function without electric power. Although the older systems are often relatively simple and reliable, they lack the alarms, convenience, and flexibility offered by newer designs.

A. Volume Ventilators

Ventilators that deliver a preset volume include bellows, or bag in a box designs that are common in clinical anesthesia, and the piston designs more frequently encountered in research laboratories.

1. Bellows in a Box

The most common design for an anesthesia ventilator is a bellows in a box, in which the bellows containing the breathing gas is intermittently compressed by high pressure gas admitted to the surrounding bellows housing, or box. This design, which separates the driving gas from the breathing mixture, is called a dual circuit. The tidal volume delivered by the bellows may be directly controlled by mechanically limiting the bellows volume, in which case the bellows is completely emptied with each breath. Alternatively, the tidal volume can be controlled indirectly by precisely limiting the volume of drive gas entering the housing. Because the bellows is compressed only to the extent that drive gas enters the housing, it need not be completely emptied with each breath. In both cases, exhalation is passive.

The pressure generated in the breathing circuit is primarily a function of the volume of gas delivered by the bellows. Excess gas in the patient circuit is exhausted by a spill valve, or an expiratory pressure relief valve. The spill valve is needed because, in most cases, the APL valve is isolated from the breathing circuit or is closed during ventilator operation (Dorsch and Dorsch, 1994h). In practice, the anesthetist usually adjusts the tidal volume to attain the desired peak inspiratory pressure, in effect using the volume ventilator as a pressure ventilator. Many ventilators include safety valves that limit the maximum delivered pressure. In this case, they can be described as volume preset, pressure-limited ventilators. With these ventilators, airway pressures below the upper safety limit are controlled by altering the delivered volume. A control allows selection of the number of breaths delivered per minute. Some ventilators permit the user to select the I:E ratio, whereas others have preset ratios. From the bellows, the gas is delivered to the patient breathing circuit. Pressure is usually monitored by the breathing system manometer.

There are two types of bellows. The older hanging or descending weighted bellows is forced upward to deliver the tidal volume, with the weight returning the bellows to its start position during exhalation (Fig. 14). Because of the weight used in the hanging or descending design, the bellows can continue to move even if the breathing circuit has become disconnected. A more recent variation is the ascending bellows, which is forced downward to deliver each breath. As the patient exhales, the exhaled volume displaces the bellows upward to its starting position. The principal advantage of the ascending design is that a leak or disconnection in the breathing circuit will be obvious because the bellows will not return to its starting position. The ascending bellows design results in slight positive end expiratory airway pressure, which is not harmful and may be useful (Dorsch and Dorsch, 1994h). With adequate alarms and vigilance,

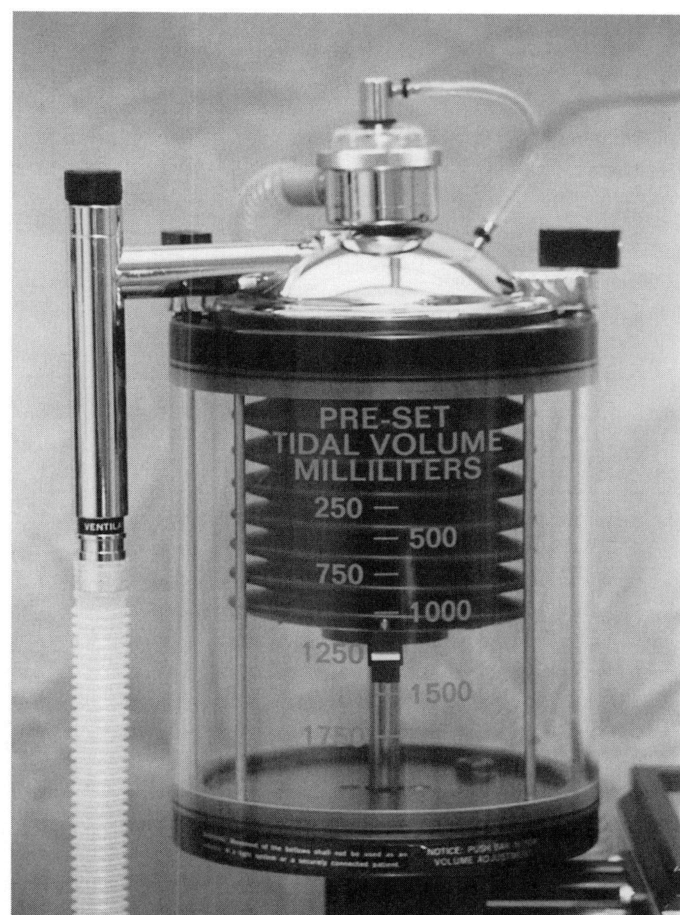

Fig. 14. Descending bellows. In this design the bellows is forced upward to deliver a breath. A mechanical stop is used to adjust bellows volume, and a weight is used to return the bellows to the start position during exhalation.

both designs are safe and effective. Reduced volume or pediatric bellows are available with many ventilators to accommodate smaller patients.

Volume ventilators can be used with Mapleson-D circuits by attaching the ventilator hose to the breathing bag port of a control arm or by using a special ventilator switching valve on a Mapleson-D control arm. If the breathing bag port is used, ventilator settings for Mapleson-D circuits can be complicated by the presence of the adjustable pressure limiting (APL) valve in the circuit. If the APL valve is partially opened, the pressures developed will be affected not only by the preset volume, but also by the resistance offered by the APL valve, as well as the rate of fresh gas flow into the circuit. The APL valve, used in this way, can limit the pressure developed in the circuit.

Volume ventilators of the type just described are usually integrated into new human anesthetic machines, but free-standing units are available as well. These units may be attached to an anesthetic machine as needed or can be permanently mounted. A modern veterinary ventilator of this type is seen in Fig. 15. Interchangeable bellows of various sizes make the ventilator adaptable to a variety of species.

Modern anesthesia ventilators are equipped with alarms to indicate low pressure or disconnection in the patient circuit. Other circuit and status alarms may be present, depending on the cost, age, and intended use of the ventilator, to warn of high pressure, negative pressure, and other conditions affecting performance and safety.

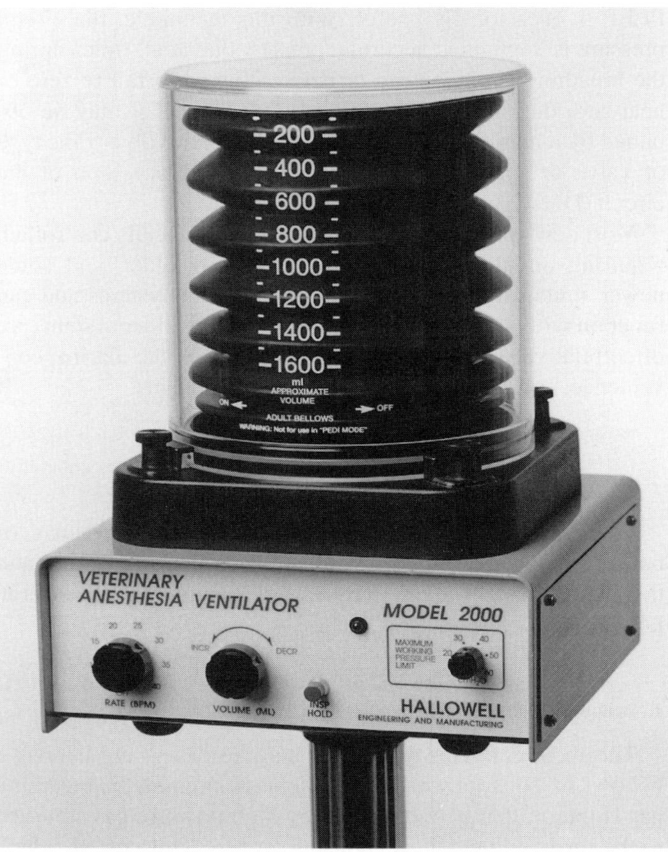

Fig. 15. Modern veterinary ventilator with an ascending bellows. Shown as a free-standing ventilator, the unit can also be mounted on an anesthetic machine. The bellows assembly can be changed to accommodate both large and small patients. [Courtesy of Hallowell Engineering and Manufacturing Corporation.]

2. Piston-Type Ventilators

These electrically driven mechanical ventilators are so ubiquitous in research institutions that ''Harvard pump'' has come to be used as a generic term for the class (Figs. 16 and 17). Piston ventilators deliver a preset volume, determined by syringe or cylinder assembly used and the stroke of the piston. Piston stroke is varied by mechanical linkage to alter the swept volume. Most ventilators of this type can accommodate several different sizes of syringes or cylinders, allowing considerable flexibility in the selection of tidal volume. The respiration rate is set by varying the motor speed or by changing drive belt and pulley combinations to produce preset rates with constant motor speeds. Some piston ventilators have adjustable I:E ratios.

A valve system is used that directs the fresh gas into the cylinder, to the animal, and then to an exhaust port via noncompliant tubing. Exhalation is usually passive. With some piston ventilators, the inlet valve can be connected to a reservoir containing an anesthetic gas mixture, and the exhaust valve output can be captured and directed into a scavenging system, thus permitting ventilators of this type to be used with inhalation anesthesia.

Piston ventilators may be used as a power source to drive a dual circuit bag-in-box ventilator. Setting up such an arrangement requires a thorough knowledge of the specific machines

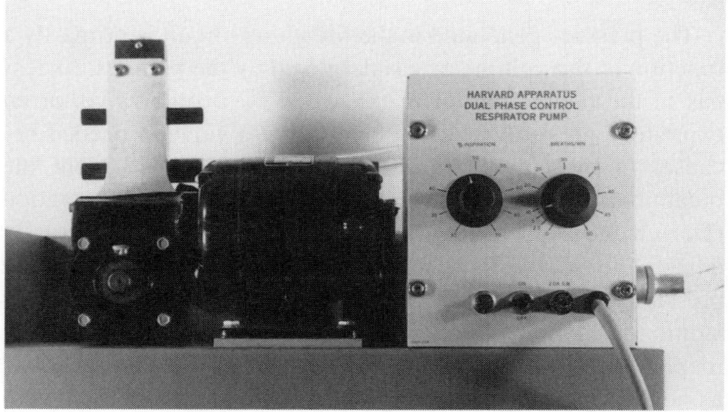

Fig. 16. Harvard pump. This view of the front of the instrument depicts the control panel, drive motor, and valve housing. Many styles and control configurations are or have been made for these respirators.

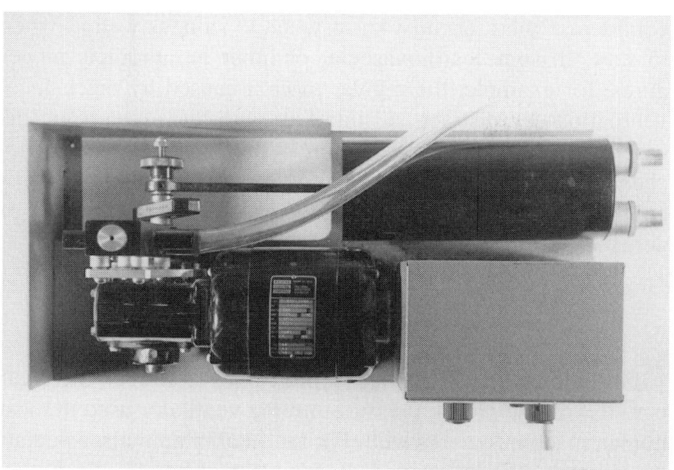

Fig. 17. Harvard pump. This view of the top of the respirator (in Fig. 16) shows the drive assembly, pump piston linkage, and cylinder assembly (top right).

used. Experienced professional help should be sought in designing and constructing such a system.

A number of small piston ventilators are marketed for use with rodents and other small animals. Because a variety of different sized syringes are available and the cycling speeds can easily reach the ranges typical for small animals, they are popular and often very satisfactory.

Historically, these ventilators were not designed for use with inhalation anesthesia and they lack the safety features and alarms usually associated with clinical anesthesia equipment. Nevertheless, they remain valuable and common in many experimental settings, especially when the precise control of volume is required. As mentioned earlier, they can be configured to deliver anesthetic gas mixtures. Safely connecting these ventilators to an anesthetic delivery system can be challenging, and the specific details depend on the model of the ventilator and the anesthesia machine or delivery system used. Because an improper connection may be dangerous to the animal, the users, or both, the ventilator manufacturer should always be consulted for precise directions. In practice, the principal source of injury associated with their use is probably barotrauma, arising from dependence on ''standard'' volume settings for a given species and weight of animal. If the animal is not ''standard'' by virtue of weight, conformation, or disease, over- or underventilation follows. This complication can be avoided, or at least diagnosed, by inserting an airway manometer into the breathing circuit, between the Y connector and the directional valves.

Piston ventilators tend to be simple and rugged in construction and appearance, traits that often lead the user to overlook the importance of regular inspection and maintenance. The cautions and recommendations of the manufacturer must be followed to assure safe and reliable performance.

B. Pressure Ventilators

Pressure ventilators are common in respiratory therapy and critical care applications; they are less often used in anesthesia. Pressure preset ventilators use valves to directly control the flow of pressurized gas to the breathing circuit. When the preset pressure is reached, inhalation stops, regardless of the delivered volume. Newer designs are usually electronically controlled, but pneumatic, fluidic, and combination control systems are also used. At the most basic, the respiratory rate and peak inspiratory pressure are selected by the operator. Beyond that, the number and sophistication of the controls and alarms vary greatly from one model to another and may include mandatory, control and control-assist cycle controls, variable I:E ratios, automatic sigh modes, and a variety of high pressure, low pressure, apnea, and other alarm settings (Banner and Lampotang, 1992).

Pressure limited ventilators have been designed for use with laboratory animals, including rodents (Fig. 18). Some of the ventilators have integral compressors and can be connected to an anesthetic machine. In addition to protection against barotrauma, the pressure-limiting mode of ventilation simplifies setup and adjustment and allows some flexibility with respect to patient size.

Because many of these ventilators depend on directly controlling high pressure gas, they are not well adapted to circle breathing systems. These ventilators require high tank or supply line inlet pressures. Ventilators of this type may subject the anesthetic vaporizer to intermittent high pressure, which may adversely affect vaporizer performance. Additionally, because the latter units cannot recirculate gas, they can be quite wasteful in terms of fresh gas and anesthetic agent if used for animals larger than rodents.

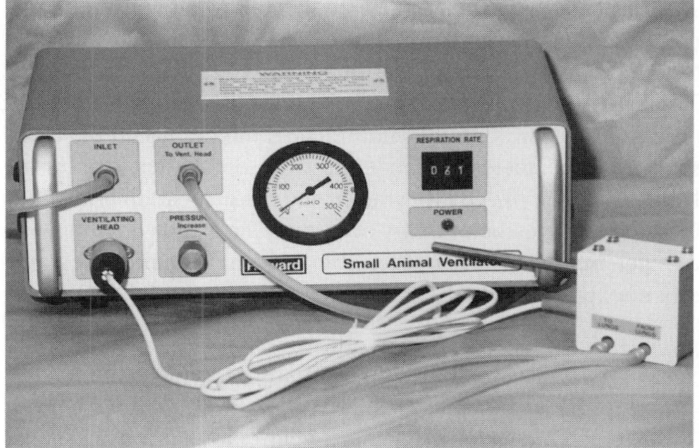

Fig. 18. Pressure-limited ventilator. In this older model the electrically operated inspiratory and expiratory valves are located in the small white box to the right of the main control unit. The hoses exiting from the front of the valve housing run to a patient Y connection.

C. Large Animal Ventilators

Large animal ventilators can accommodate adult horses and cattle. For nearly all animals encountered in biomedical research, ventilators designed for use with adult humans are large enough. However, if a large capacity volume ventilator is needed, the advice of a veterinary anesthesiologist experienced in the techniques and problems of ventilating large animals should be sought. The basic principles of these ventilators have already been described: they are very large bags in very large boxes.

D. Jet Ventilators

Jet ventilators are small diameter tubes, usually IV catheters, inserted into the trachea through which oxygen is given at relatively high pressures. The technique is used electively to provide adequate oxygenation during operations where an endotracheal tube cannot be used and a tracheostomy is not desired, e.g., head or neck surgeries, or as an emergency measure in situations where the patient cannot be intubated or maintained by mask, as might occur with laryngeal edema. Briefly, an IV catheter is inserted percutaneously into the trachea and is connected to a high pressure oxygen source. If an anesthetic machine is available, one method of using transtracheal ventilation is to connect a length of noncompliant tubing to the common gas outlet via an endotracheal connector, with the patient end of the tube connected to a Luer-lock adaptor. The Luer-lock connector is then fitted to the IV catheter, and the oxygen flush valve is used to deliver oxygen. A well-illustrated description of the equipment and techniques used in humans is given by Benumof *et al.* (1992).

A variation of this technique has been used by the author to supplement ventilation in rabbits in which an endotracheal tube was in place, but could not be connected to an anesthetic machine because of experimental considerations. The tube was connected to the common gas outlet, as described, except that a T tube was interposed in the supply line. A small gauge feeding tube threaded through the endotracheal tube replaced the transtracheal IV catheter. A moderate flow of oxygen was established using the oxygen flowmeter, and oxygen was intermittently directed to the catheter tip by occluding the T tube opening with a finger. Scavenging is not feasible, and the technique is generally not applicable with inhalation anesthesia.

E. Disadvantages of Mechanical Ventilation

The convenience of mechanical ventilation during anesthesia comes at a price. Greater technical expertise is required of the anesthetist to operate and maintain the equipment; the anesthetist may monitor the ventilator instead of the patient. Ventilators must be used carefully to minimize cardiovascular depression caused by increased intrathoracic pressure, as well as to avoid barotrauma. Convenience may lead to overreliance and overconfidence: the user must know when to stop, as well as when to start. In some cardiovascular or other intrathoracic procedures, for example, the regular motion caused by mechanical ventilation can distract and interfere with the surgeon: in this case attentive manual ventilation is preferred. However, mechanical ventilation is a requisite in many experimental protocols and is a valuable adjunct in the anesthetic management of demanding surgical cases.

F. Selection of Ventilators

For animals ranging in size from calves to ferrets, the dual-circuit, volume preset, pressure-limited ventilator used in most human machines works well. For the smaller animals, a pediatric bellows may be needed. Reconditioned human ventilators and new veterinary designs are available and will satisfy a wide range of needs. When selecting a ventilator, it is important to establish that the volume and breathing rate options will meet the anticipated needs. Some older human and veterinary models are not able to cycle quickly enough or, in some cases, slowly enough for veterinary patients. Ventilators having electronic controls are generally more flexible and somewhat easier to use than pneumatic or fluidic designs.

For rodents and other very small animals, specially designed volume (piston) or pressure ventilators may be the best choice. For anesthesia purposes, the modern pressure ventilators would seem to be most suitable in terms of controls and safety features.

As with all anesthesia equipment, used or reconditioned ventilators can be very satisfactory if carefully selected and well maintained. Ill-maintained, poorly functioning equipment is dangerous: it provides the illusion of function, but none of the advantages.

VII. INDUCTION CHAMBERS AND MASKS

A. Induction Chambers

Induction chambers are commonly used to produce brief anesthesia for very short procedures in rodents. For larger animals, chambers are frequently used to avoid restraint and the associated trauma to the animal and the handler. Depending on their design and the agent selected, anesthetic chambers may be used alone or in conjunction with an anesthetic machine.

Methoxyflurane is often used in closed chambers, particularly with rodents. The relatively slow induction time provides a margin of safety in the chamber, and a delayed recovery provides enough working time out of the chamber for many short procedures. In this case, no exit or entry ports are used and the agent is simply placed into the chamber, but out of contact with the animal. A typical method is to place the anesthetic onto a cotton pledget or gauze sponge below a grating or screen which serves as flooring for the animal. The animal must not come into skin contact with a volatile anesthetic.

When using volatile agents with homemade or commercial chambers, it is important to remember that most of them are powerful organic solvents that will attack many plastics, including many of the rodent cages used in laboratory animal facilities. Glass desiccator chambers are frequently used as anesthetic chambers for small rodents. They have the advantages that they can be cleaned easily, are not susceptible to damage by anesthetic agents, and some are supplied with a perforated floor, permitting easy use of the anesthetic agent. Unfortunately, they are heavy, expensive, and prone to breakage. If plastic chambers are used, the anesthetic agent should be placed in a metal or glass container that will permit vaporization, but prevent contact of the liquid agent with either the animal or the plastic.

Closed chambers present problems in terms of waste gas scavenging and should be used in conjunction with a properly vented chemical hood or other *nonrecirculating* hood. The HEPA filters commonly used in biohazard hoods do not remove volatile anesthetic agents and are not a suitable replacement for a nonrecirculating hood.

If the chamber is used with an anesthetic machine, there must be both an inlet for fresh gas and a scavenged outlet for waste gas. If a coaxial circuit is used, a single port into the chamber may serve both purposes, as seen in Fig. 19. The port has been machined to fit a standard coaxial Mapleson-D circuit.

The chamber volume may be quite large; a typical cat or rabbit chamber is commonly 18 liters. Relatively high gas flows are required if induction is not to be unduly prolonged. For the same reason, it is advisable to match the chamber size to the patient size. A rat can be anesthetized much more quickly and economically in a small chamber than in one used for larger animals. A variety of induction chambers are available from veterinary and experimental equipment vendors. A design for an induction chamber that is adjustable for a variety of species ranging from small dogs to mice has been published (Sawyer, 1982b).

Fig. 19. Rodent induction chamber. This chamber has a volume of about 4 liters and a single port, seen at the bottom left, for use with a Bain circuit. Although convenient and satisfactory for small volume boxes, such a design is inefficient and unsuitable for larger chambers.

Not all animals do well in induction chambers. Breath holding in rabbits has been described during mask and chamber induction with isoflurane and halothane and was associated with bradycardia, hypercapnia, and acidosis (Flecknell *et al.*, 1994).

Although induction chambers are valuable for the inhalation anesthesia of laboratory animals, they are also major sources of pollution. In contrast to closed chambers, those used with anesthetic machines have some scavenging via the system of the machine. Even so, contamination occurs as soon as the chamber is opened, primarily from the direct escape of anesthetic gases, but also from the anesthetic entrained in the fur of the animal and from exhalation.

B. Masks

Face masks are used to induce and maintain anesthesia and to provide supplementary oxygen to anesthetized animals or those recovering from anesthesia. A variety of masks are available from veterinary equipment suppliers for dogs and cats, which often work well for rabbits, pigs, and a number of other laboratory animals. Increasing interest in pet rodents and other small mammals, as well as birds, has led to the commercial development of masks and breathing circuits for these species as well. Masks intended for humans, especially for neonates, are sometimes usable with nonhuman primates. For many animals, however, custom or homemade masks are convenient and, if carefully made, provide the best fit.

Perhaps the most common example is the syringe-case mask commonly used for small rodents. In its simplest form, the mask may be little more than a syringe case with a cotton pledget to which methoxyflurane has been applied. Typically, the small rodent is anesthetized in an induction chamber, removed, and the mask applied to the head. No diaphragm is used: the air space around the head provides fresh air and is used to regulate anesthetic depth by pushing the head further into or withdrawing from the mask as needed. The system has the advantage of simplicity, but safe use requires practice and attention, as well as a suitable vented fume hood.

If a rodent mask is used in conjunction with an anesthetic machine, it is usually configured as part of a Mapleson-D or -E system. In this case, the mask should have a relatively good seal around the mouth and nose of the animal and should be designed to fit the appropriate circuit. The mask shown in Fig. 13 uses a machined metal adaptor which fits both a 35-cm^3 syringe case and the Bain circuit. The barrel of the syringe case has been cut short to minimize dead space, and the flared end is covered with the finger of a latex glove, fenestrated to fit the nose and mouth of the rodent. Other approaches include ingenious systems of padding in order to attain a comfortable and tight seal. Scavenging mask designs are also mentioned previously in the discussion of Mapleson-E circuits.

Similar ingenuity has been expended for larger animals, such as ruminants and domestic swine. In these cases, the favored raw material seems to be a plastic bottle or jug, which is cut and padded to fit the patient. Usually, the Y connector for a circle circuit is attached to the original threaded opening by copious amounts of tape.

As with breathing circuits, face masks should be washed and dried after use and inspected for deterioration and cracking of plastic, padding, tape, and rubber diaphragms.

VIII. FLUID DELIVERY SYSTEMS

Intravenous access is a part of anesthesia for all but the smallest patients or for the briefest procedures. Disposable intravenous catheters are available in a wide variety of sizes and lengths and are used for animals ranging in size from rats to ruminants. Numerous fluid administration and control devices are available, many for quite specialized needs.

The simplest device is a syringe attached to the catheter, sometimes by a T connector, or a reduced volume anesthesia extension set. Next in complexity are standard and pediatric fluid administration sets for bottles or bags of IV fluids, with their attendant volume and rate-limiting devices. Bag compression mechanisms increase the rate of fluid delivery and provide additional pressure sometimes needed for disposable blood pressure transducer continuous flush systems, adding another option. Electronic flow controllers, which use drop counters attached to the drip chamber of administration sets, provide more precise control of infusion rates. When coupled with peristaltic or disposable valve and piston pumps, flow controllers become infusion pumps. Finally, the need for precise administration of small volumes of potent drugs has led to the development of microprocessor controlled syringe pumps, the injectable analogs of precision vaporizers. Each of these is discussed in somewhat more detail below.

For reasons of convenience, control, or drug incompatibility with other IV fluids, the anesthetist may require a dedicated intravenous access site. Although a syringe can be taped to the catheter and limb, an anesthesia extension set is more convenient, especially if the patient is draped for a sterile procedure. Basic extension sets have a length of tubing with a male Luer connector on one end to fit the catheter and a female connector on the other end for a syringe, three-way stopcock, or syringe manifold. Extension sets with reduced lumens, to minimize dead space, are available as well, which permit better precision and accuracy for the administration of very potent agents.

Disposable intravenous administration sets are ubiquitous. The usual set consists of a hollow spike to penetrate the bag or bottle, a drip chamber, a clamp to regulate flow, and a line terminating in a Luer connection. One or more injection ports are also usually provided. Sets intended for use with bottles provide filtered air vents; those used with bags are not vented. Human adult administration sets deliver approximately 15–20 drops per milliliter. Pediatric sets usually deliver about 60 drops per milliliter, a more convenient rate for smaller patients. More precise control of rate is achieved with electronic devices, described below, or with mechanical valves, such as the dial-type controller seen in Fig. 20.

Bag compression devices are either pneumatic or spring-loaded mechanical types. The pneumatic infusors consist of a sleeve that encloses the bag and a bladder that is pumped up like a blood pressure cuff to apply pressure to the bag. A manometer is usually provided to indicate the pressure level.

Electronic flow controllers use an optical sensor to count the drops delivered through the drip chamber of the administration set per unit time. They depend on gravity to deliver the fluid and increase or decrease the rate to match a preset value. If a pump mechanism is added, the unit becomes an infusion pump. The pump mechanism may be a peristaltic roller, which may require specialized tubing, or a disposable piston and valve pump arrangement. Flow controllers and pumps can be set to deliver a given rate or volume. They usually have occlusion alarms and many have more elaborate alarms, indicators, and controls. Flow controllers are more sensitive to occlusion than pumps and work best with low viscosity, low specific gravity fluids (Holzman, 1992).

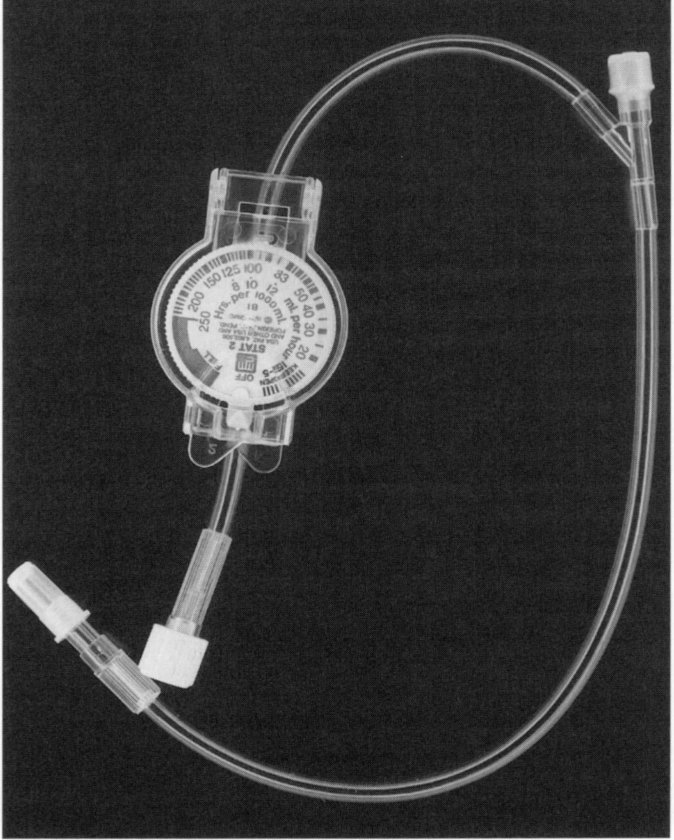

Fig. 20. Dial-type infusion rate controller used with an administration set to achieve finer control of the fluid infusion rate.

Syringe pumps have long been used in laboratories when the precise delivery of small volumes of fluid is required. More recently, microprocessor-controlled syringe pumps have been developed to assist in the delivery of newer, potent injectable anesthetic agents. Once certain data are entered, such as body weight and drug concentration, they can calculate the delivery rate and control the infusion of intravenous drugs. Some units are preprogrammed for agents and doses commonly used in the human operating room. In addition to continuous infusion, they permit bolus administration, monitor the amount of drug delivered, and greatly reduce the labor and detailed recording needed for the successful use of these drugs. Microprocessor-controlled syringe pumps are well suited to the delivery of agents such as propofol and the potent opioids, as well as other drugs used in anesthesia. It seems likely that infusion pumps will soon be a standard component of the anesthetic equipment found in experimental facilities. Indeed, their cost at this time is about the same as that of a new precision vaporizer.

IX. MONITORS

''As every anesthetist knows, the degree of analgesia or amnesia cannot be measured and often the degree of muscle relaxation is not measured. Instead, many variables that have little to do with the state of anesthesia but much to do with the side effects of anesthetics are recorded'' (Gravenstein and Paulus, 1982a).

Both the patient and the anesthesia machine are monitored. The patient is monitored to detect and correct the ill effects of the surgery and of the anesthesia itself. The equipment is monitored to detect failure and to prevent the potentially disastrous consquences of equipment malfunction or, more commonly, operator error.

The most important monitor is a trained and vigilant anesthetist. Training is essential in order to understand the effects of the anesthetic and surgery or other procedures on the patient, as well as to understand the appropriate use and pitfalls associated with the equipment. The anesthetic record is the primary tool used to promote vigilance on the part of the anesthetist, to detect subtle changes in the patient, and to provide the means to analyze and improve the safe administration of anesthesia. Other mechanical and electronic aids can enhance the senses of the anesthetist and increase the utility of the record, but cannot effectively replace either. Finally, we monitor what can be monitored, within the limitations imposed by training, equipment, and, often, convenience. The question of what to monitor, under what circumstances, is covered elsewhere in this and other texts.

Because most anesthetics depress ventilation and many adversely affect the cardiovascular system, most of the attention and equipment are directed to monitoring the adverse effects of anesthetics on these systems.

A. Cardiovascular Monitoring

1. Stethoscopes

With larger animals, the pulse can be palpated to assess rhythm, rate, and strength. This is difficult or impossible with smaller animals and is often inconvenient with larger animals fully draped for surgery. Perhaps the simplest instrument used to assess cardiovascular function is the stethoscope. Precordial stethoscopes, affixed to the thorax over the heart, are often used in human anesthesia, but the size and conformation of animals often limits their usefulness in veterinary anesthesia. Veterinary esophageal stethoscopes are available in a wide variety of sizes, and those marketed for human use can often be used for animals as well. The probe can be connected to a conventional binaural stethoscope or can be fitted to a molded monaural earpiece, permitting the anesthetist to monitor the stethoscope and still hear what is happening elsewhere in the room. Alternatively, the stethoscope probe can be connected to an amplifier, which will permit the heart and breath sounds to be heard by everyone in the operating room, freeing the anesthetist from being tethered to an earpiece. Disposable probes are available that have built-in temperature sensors with adapters to fit various types of electronic thermometers so that core temperature can also be measured.

The esophageal stethoscope detects mechanical heart sounds, which in some degree reflect the pumping of blood, unlike the electrical signals amplified by electrocardiography. They are useful in assessing rate and, to a lesser extent, rhythm, but cannot provide early warning of the electrical disturbances common in anesthetized patients and sound is lost or greatly diminished when the thorax is opened. The very high heart rates characteristic of many laboratory animals may defeat the quantitative assessment of rate and rhythm. Even so, the device is simple, easily used, inexpensive, and should probably be a part of the monitoring equipment for any animal in which it can be used.

2. Electrocardiography

Electrocardiography is used to assess cardiac rate and rhythm, both of which are affected by anesthesia, surgery, and disease. Electrocardiographs detect, amplify, and display the minute electrical impulses generated by the heart. In so doing, signal processing circuits and filters are used to determine rate and minimize artifact. Signals are detected at the skin surface by electrodes and are transmitted to the amplifier by wires (leads) configured as standard lead arrays. Of the leads most commonly used in anesthesia, leads I, II, and III are bipolar; they measure the voltage difference between two of three electrodes. Leads aVR, aVL, and aVF are augmented unipolar leads, which measure the voltage difference between a single exploring lead and the reference zero voltage obtained by connecting all of the four limb leads together. A cathode ray tube is commonly used to display the traces of one or more leads, usually with the heart rate also indicated. The rate, rhythm, and amplitude of the signals thus displayed permit assessment of the cardiac electrical activity and provide warning of dangerous trends or events. In most cases, because the change in rhythm and rate, rather than amplitude, most concerns the anesthetist, the precise placement of leads is not of primary importance (LeBlanc and Sawyer, 1993). Even in those species for which standard sites for electrode placement have been defined, it may be impossible to use them because of surgical or experimental reasons.

Free-standing ECG monitors are available, but they are more commonly found as components of multifunction monitors and defibrillators. In addition to selecting which electrode lead to monitor, many instruments also allow selection of ''monitoring'' or ''diagnostic'' modes. The monitoring mode is electronically filtered to remove electrical noise and artifact, but may potentially also remove some of the useful signal. The diagnostic mode is much ''noisier'' but presents more information to the trained eye. A useful feature of some monitors is the ability to freeze part of the display to permit closer examination of the ECG or to allow comparison of the trace at the beginning of surgery to that seen later, simplifying the detection of changes in the ECG as surgery progresses (Perrino *et al.,* 1993; Heavner *et al.,* 1989b).

The monitor should have or provide output for a paper recorder. An audible signal for each QRS complex recognized is common and useful, especially if the volume level is adjustable. Competing ''beeps'' from multiple monitors, added to the noise generated by ventilators, electrosurgical units, and other equipment common to the operating room, can lend new meaning to the phrase ''driven to distraction.''

The standard screen sweep speed is 25 mm/sec. Many instruments permit faster and slower sweep speeds as well; fast sweep speeds (e.g., 50 mm/sec) are especially useful with small patients and high heart rates. Automated ECG analysis is available on some human instruments. This feature, if proven to be applicable over a wide range of species, would be a boon to the harried laboratory animal anesthetist. Finally, adjustable lower and upper heart rate alarms along with faulty lead alarms are found on most instruments.

In addition to the many ECG monitors made for human use, units intended for veterinary use also are available. Some are quite compact and are designed to register the high heart rates encountered with smaller animals.

Many of the problems encountered with using ECGs in anesthesia are traceable to the electrodes and lead wires. Proper placement is important, but good electrode contact and intact lead wires are essential. The electrodes can be disposable conductive pads used in humans, alligator clips, needles, or even steel sutures or staples. If patient size permits, disposable electrode patches are usually the least traumatic, and human pediatric patches may work well even with smaller animals. It is necessary to clip the hair in order to use these patches, and even with skin cleansing, adhesion may be a problem. In the author's experience, patch electrodes that are made for long-term or ICU monitoring seem to adhere best to animal skin, and the creative use of adhesive tape can usually overcome adhesion difficulties. Alligator clips are effective and reasonably atraumatic if the teeth are dulled or removed and the spring tension is appropriate. Small alligator clips are often most useful for rodents and other small laboratory species. Stainless steel clips will resist the corrosive effects of disinfectants, as well as the aggressive electrolyte pastes or solutions needed to ensure good electrical contact. Homemade or modified lead wires may not withstand the rigors of the operating room and even commercial leads eventually break. In the event of failure to recover an adequate signal, broken lead wires may be the culprit.

Silver metal strap-on plate electrodes, which have fallen out of use in human electrocardiography, are very useful for larger domestic animals. They are easily taped to the limbs, very conductive, and provide a generous contact area. These electrodes are often discarded by hospitals or are languishing in permanent storage and can frequently be had for the asking.

Surplus ECG monitors are commonly available from hospitals and clinics. As with all such used equipment the buyer should take care to verify that the equipment is safe, functioning, and that the correct operation is understood by the users. This will usually entail inspection by a biomedical technician. Equipment for which parts and service are not available is not a bargain, even if free.

3. Blood Pressure

Although electrocardiography monitors the electrical activity of the heart, measurement of blood pressure permits assessment of the mechanical activity. Blood pressure monitors may be divided into invasive and noninvasive instruments, and the noninvasive instruments may be further subdivided into those which detect mechanical pulsation and those which detect flow.

a. INVASIVE BLOOD PRESSURE MONITORS. Invasive pressure measurements remain the gold standard against which other methods are compared. A cannula is inserted into an artery or vein and is connected to a pressure transducer. Although it is possible to connect the cannula to a mechanical gauge or, in the case of a vein, to a simple fluid column, it is far more common to use an electromechanical transducer that converts the venous or arterial pressure into an electrical signal that is processed and displayed by the monitor. The features common to most clinical instruments are described below.

The monitor continuously displays the amplified waveform, providing useful qualitative information, and indicates the diastolic, systolic, and mean pressures. Mean blood pressure is determined electronically by integrating the area under the pressure curve (Gravenstein and Paulus, 1982b). Controls are provided for calibrating and zeroing the transducers and for adjusting the unit to the desired pressure range. Connections for a paper recorder or other data acquisition system are present as well.

Complex operations often require two or more arterial pressure measurements from different sites, as well as central venous or pulmonary artery pressure measurements. To meet this need, many monitors have multiple pressure amplifiers or are made to accept additional amplifiers as needed. As with ECG monitors, invasive blood pressure monitors are usually found as components of multifunction monitors.

Reusable or disposable transducers can be used. Reusable transducers typically have threaded plastic domes that attach to the body of the transducer. The dome has two Luer connectors: one for the blood pressure line leading from the patient and the other for a three-way stopcock or flushing syringe. Reusable transducers are cumbersome to sterilize. When asepsis is an issue, sterile plastic domes that have thin diaphragms separating the fluid pathway from the transducer surface are used. Permanent transducers are still found in research laboratories, where they find favor because of their economy (once purchased) and long-term stability, but in the human clinical setting they have largely been replaced by sterile, disposable transducers.

Disposable transducers are smaller than permanent ones. For clinical use, they are sterilely packaged, often with tubing sets and stopcocks. The transducer has a Luer connection to the line leading to the patient, often equipped with a three-way stopcock. A second port is usually equipped with a pinch valve between the transducer and another Luer connector. When attached to a pressurized bag of heparinized saline, the pinch valve permits a very slow continuous infusion through the transducer, effectively minimizing clots and blockage of the vascular cannula without interfering with recording. Disposable sets come with connections for one, two, or three transducers to be connected to a single pressurized bag of heparinized saline. Disposable transducers, if not excessively soiled, can be thoroughly rinsed with distilled water and dried by pulling air through the circuit, including the pinch clamp. The author has used ethylene oxide to sterilize disposable transducers cleaned in this fashion, without untoward consequences. The transducers are calibrated against a mercury manometer prior to each use. Resterilization is not recommended by any manufacturer.

The cannula is connected to the transducer by relatively stiff, noncompliant pressure tubing. Because air bubbles in the system will cause damping of the pressure wave, it is essential to purge the system of bubbles before connection to the patient and to avoid introducing bubbles during use. This is one reason to avoid unnecessary connections, stopcocks, etc. It is also important to keep the tubing lengths as short as feasible in order to minimize errors (Thys and Reich, 1993).

Once the transducers and cables have been prepared for use and the monitor turned on and allowed the specified warmup time, if any, the system should be calibrated and performance verified. Calibration should be carried out according to the directions of the manufacturer. Electronic calibration is often sufficient for clinical practice, but in more demanding circumstances, calibration can be verified using a mercury manometer (Gravenstein and Paulus, 1982b). This process is neither difficult nor especially time-consuming and should be part of the instrument routine for critical experimental surgery. After the patient has been connected, the transducer is adjusted to the level of the heart, if necessary, and rezeroed.

Transducers, cables, and blood pressure monitors are not universally interchangeable. It is important to make the correct connections with the correct cables. Transducer makers should be able to provide the correct cable for the monitor. This is rarely a problem when new equipment is purchased, but is a common problem in academic research settings where a wide variety of used, and sometimes obsolete, equipment remains in service. It becomes even more confusing when apparently similar models of equipment have different requirements. The manufacturer or a biomedical technician familiar with the equipment should be consulted if problems are encountered.

The use of peripheral or central catheters connected to transducers is the most common means of continuous, beat by beat pressure monitoring. However, catheters are available with miniature transducers at their tips. Because of the expense, they are not routinely used in anesthesia monitoring, but technological advances could make these catheters smaller and less expensive, permitting their routine use.

b. Noninvasive Blood Pressure Monitoring. The auscultatory technique of using a pressure cuff and stethoscope is not generally useful for measuring blood pressure in animals: the problem is not so much with the cuff as with the inability to hear with the stethoscope the sounds produced as blood flow is first interrupted and then resumed with cuff inflation and deflation. In the human operating room the classical method of sphygmomanometry, although retained as a manual backup, has been superseded in many cases by electronic noninvasive instruments. A number of such instruments, using a variety of detection methods, have been marketed, but only two will be discussed here: Doppler flow detection and oscillometric detection.

Both the Doppler and the oscillometric methods make use of blood pressure cuffs. Their correct use depends on selection of the appropriate cuff size. A cuff that is too narrow will overestimate arterial blood pressure whereas one that is too wide will underestimate it. For humans, the recommended cuff width is about 40% of the circumference of the arm, with a bladder length twice the width of the cuff; these values are usually recommended for animals as well (Perrino *et al.,* 1993; LeBlanc and Sawyer, 1993). Because the cuffs commonly used for animals are human pediatric cuffs, some compromise is inevitable and may contribute to inaccuracy.

c. Doppler Flow Detection. The ultrasonic Doppler flow probe can be used in combination with a pressure cuff and an aneroid manometer to estimate systolic blood pressure. The probe is placed over a peripheral artery, using ultrasonic coupling gel, and is positioned until a clear arterial pulse is heard.

The probe is then carefully fixed in position, taking care not to compress the artery. A simple way to do this is to use a Penrose drain and clamp, permitting easy adjustment of site and tension. A sphygmomanometer cuff with a width of about 40% of the diameter of the leg is placed proximal to the probe, and the cuff is inflated until the arterial pulse sound disappears. Cuff pressure is then slowly reduced until the arterial sound returns, corresponding to systolic pressure. This technique has been evaluated in dogs and cats and been found to be reasonably accurate in dogs at systolic pressure greater than 50 mm Hg (LeBlanc and Sawyer, 1993). In cats, the technique consistently underestimates systolic pressure compared to invasive monitoring by about 14 mm Hg (Grandy *et al.,* 1992). The technique just described can also be performed using pulse oximetry rather than Doppler flow as the pulse detector. The pulse oximeter should display the plethysmographic waveform. Two different techniques have been reported in human trials: in one case, the cuff was simply inflated to a level above the systolic pressure and then slowly deflated until the pulsatile plethysmographic waveform reappeared (Talke *et al.,* 1990). A variant of the technique involves noting the value at which pulsatile flow disappears with cuff inflation, then further inflating the cuff to 200 mm Hg and slowly deflating until the waveform reappears. The cuff pressures at disappearance and reappearance are averaged and taken to represent systolic pressure (Chalwa *et al.,* 1992). In human studies, the correlation between the pressures determined using the Doppler flow detector and the pulse oximeter were quite good. A study comparing the pulse oximeter method to direct pressure measurements in swine concluded that the technique "... may be useful for monitoring healthy animals undergoing low risk procedures under anesthesia, but lacks the accuracy required for critical or complicated cases" (Caulkett *et al.,* 1994).

d. Oscillometric Blood Pressure Monitoring. This method also uses a blood pressure cuff but, in this case, the instrument controls inflation and deflation while monitoring the minute air pressure changes in the cuff caused by arterial pulsation. Typically, systolic, diastolic, and mean arterial pressures are reported at selected intervals. Many noninvasive blood pressure modules are available as stand-alone instruments or as components of multifunction monitors intended for use with humans. A few instruments have been modified or designed for use with animals. Instruments designed for humans often perform poorly when used on animals. The buyer should carefully examine the performance of any oscillometric device to assure that it will perform as expected.

Current models of noninvasive oscillometric blood pressure monitors provide reasonable values in animals as small as cats, which are normotensive or only moderately hypotensive. They do not provide continuous readings and do not perform well at very low pressures or with very small animals. The systolic pressure provides the best correlation with direct pressure measurements (LeBlanc and Sawyer, 1993; Sawyer *et al.,* 1991). They are, however, useful monitors for larger laboratory animals, given their ease of use and acceptable accuracy in the range of blood pressures commonly expected. Further technical improvements may extend their usefulness. For cases that would not normally warrant the expense, effort, and potential injury associated with invasive monitoring, these monitors can provide an additional margin of safety.

B. Ventilation

When the patient is breathing spontaneously, the rate can be determined by observing the patient or the reservoir bag, and an estimate of volume can be made if the observer is sufficiently experienced and if the bag excursions are sufficiently large. In many cases, the patient is covered by drapes, the bag is not observed, and respiratory failure can occur without being noticed unless some other means of monitoring breathing is used.

Among the simplest monitors are the amplified esophageal stethoscopes previously described. Other simple devices consist of a thermistor placed between the circuit and the endotracheal tube that detects the changes in temperature as the patient inhales and exhales, sounding an audible alarm with each breath.

Animals having sufficiently large tidal volumes can be monitored using a respirometer, or spirometer. These devices are placed in the breathing circuit and measure tidal volume and minute volume. The older mechanical spirometers use precision clockwork mechanisms that depend on gas movement to move vanes or rotors, registering volume on the dial. Minute volume is determined using a stopwatch with some models whereas others have an integral timer. Newer, electronic respirometers offer alarms and digital readouts, but remain unable to sense the small tidal and minute volumes of many laboratory species.

Mechanical or manual ventilation can result in an increased risk of too much or too little pressure in the breathing circuit. Airway pressure can be monitored by observing the airway manometer, but airway pressure monitors are often used in conjunction with mechanical ventilation to warn of dangerous conditions. Generally, these monitors are mounted on the anesthetic machine or are incorporated into the ventilator. They detect low circuit pressures, warning of leaks or disconnects; sustained high pressures, as might occur with a ventilator malfunction, or scavenging tube occlusion; transient high pressures, which could be caused by endotracheal tube blockage or other causes; and subatmospheric pressure caused by a patient attempting to inhale independently of the ventilator, among other reasons. The monitor is connected to the breathing circuit by a tube. It is recommended that the connection be on the expiratory limb of the circuit, between the patient and the expiratory unidirectional valve (Dorsch and Dorsch, 1994m). Correctly maintained and used, these monitors contribute materially to the safety of mechanical ventilation during anesthesia.

1. Oxygenation

Oxygenation of the anesthetized patient can be assessed invasively by means of blood gas determinations and pulmonary artery oximetry, and noninvasively by means of transcutaneous oxygen sensors and pulse oximetry. Blood gas determinations and pulse oximetry are the two methods most commonly used.

a. BLOOD GASES. Arterial blood gas determinations have until recently depended on the anaerobic collection of an arterial blood sample and analysis by a blood gas machine. Although the information gained from a properly collected sample analyzed by a properly calibrated machine is frequently invaluable as a means of monitoring the respiratory and metabolic status of a patient, the blood gas analyzer itself is usually considered a laboratory instrument rather than a monitor. Blood gas analyzers use specific electrodes to determine the pH, partial pressure of oxygen (pO_2), and partial pressure of carbon dioxide (pCO_2) under precisely controlled conditions. In addition to pH, pO_2, and pCO_2, many instruments calculate additional parameters based on the directly determined values and the preset value for hemoglobin. Although the operation and maintenance of modern blood gas machines are much simpler and more convenient than in the past, the instruments still require a trained technician for calibration, cleaning, reagent servicing, and troubleshooting and the results are necessarily intermittent in nature. Two recent developments may change this situation.

The first development is the appearance of a new type of portable instrument that uses a disposable, single-use sample analysis module. Unlike conventional analyzers, these devices require little training to use and maintain, and can easily be brought to the patient. They are not intended to replace the more sophisticated laboratory instruments, but rather to extend the availability of vital information beyond the reach of the laboratory. Although the unit cost of a blood gas determination with these instruments may be higher than that of a conventional machine, the cost of obtaining and maintaining a traditional machine is avoided. This may be a reasonable trade-off for those facilities that only occasionally need blood gas values.

The second development is the appearance of blood–gas catheters, which promise to deliver continuous blood–gas information. One such instrument relies on a special catheter that uses a novel optical technology to detect oxygen, carbon dioxide, and pH. Although both the catheters and the monitor itself are technologically sophisticated and currently expensive, they may be the forerunners of routine, real-time blood–gas monitoring. The availability of affordable continuous blood–gas status would be a genuine boon in the management of many difficult patient support problems.

b. PULSE OXIMETRY. Two measures of the degree to which hemoglobin is saturated with oxygen are used in the clinical setting. *Fractional oxygen saturation* is the ratio of oxyhemoglobin to all hemoglobins in the blood, including dyshemoglobins such as methemoglobin and carboxyhemoglobin. Fractional saturation is determined by a co-oximeter, a laboratory instrument that uses multiple wavelength spectrophotometry to distinguish the various species of hemoglobin present. *Functional saturation* is the ratio of oxyhemoglobin to the sum of oxy- and reduced hemoglobin, and does not include the dyshemoglobins (Tremper and Barker, 1989). It can be measured *in vitro,* with a co-oximeter, or *in vivo,* with a pulse oximeter.

Pulse oximeters determine functional saturation by measuring the absorption of two wavelengths of light passing through pulsating blood vessels. After correcting for the nonpulsating or baseline absorbance at each wavelength, the ratio of the two wavelengths is calculated and is empirically correlated with the functional oxygen saturation determined by a co-oximeter. The light is generated by two light-emitting diodes (LEDs) that are switched on and off hundreds of time per second and is detected by photoelectric cells, usually located opposite the LEDs. The rapid cycling of the LEDs allows precise identification of the pulsations in the vessels. If both LEDs are off during a portion of the cycle, the influence of extraneous light can be measured and corrected for as well, although extraneous light remains a common problem. Because pulse oximeters use only two wavelengths of light, they distinguish only two forms of hemoglobin. The algorithm used to identify pulses and calculate the SpO_2 is determined for each instrument and probe type (Tremper and Barker, 1993).

Pulse oximeters are available as free-standing instruments and are common components of multifunction monitors. The simplest instruments usually display the SpO_2 and sound a tone with each pulse detected. By convention the pitch of the tone is related to the saturation level, so that a falling pitch indicates a drop in saturation. Many instruments also display some indication of signal strength, often a moving bar, and some instruments have a plethysmographic display of the pulse waveform. Both indicators are useful. The signal strength indicator warns of a deteriorating signal-to-noise ratio, as with very weak or disappearing pulses, or poor sensor positioning. The waveform or plethysmographic display traces the change in volume of the blood vessels with pulsatile flow and is useful because it aids in detecting motion artifact and provides graphical as well as auditory and numerical feedback to the anesthetist (Tremper and Barker, 1993). The plethysmographic display makes possible the indirect measurement of blood pressure mentioned earlier. A pulse oximeter with plethysmographic display capability is shown in Fig. 21.

Pulse oximeter sensors are not usually interchangeable among instruments from different manufacturers. An exception to this rule is that makers of multifunction monitors often obtain their pulse oximeter modules from one of a smaller number of original equipment manufacturers. In this instance, apparently different instruments may share the same sensors. A variety of sensors are available, depending on the instrument, including numerous disposable and reusable transmittance and reflectance sensors. Most transmittance sensors require that the LEDs and photocells be positioned directly opposite each other. Reflectance sensors depend on light passing through the tissue to underlying bone and being reflected back to photocells located adjacent to the LEDs. Pulse oximeters that use fiber-optic bundles to carry the light to and from the patient sensor are made for use in the hostile electromagnetic environment of the magnetic resonance imaging (MRI) suite.

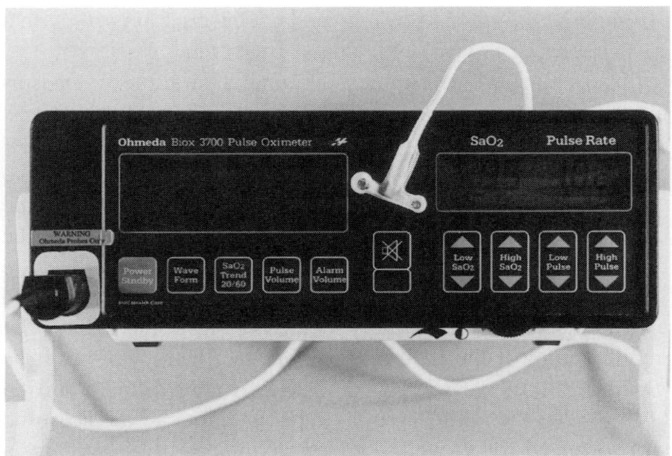

Fig. 21. A pulse oximeter designed for human patients. During operation, the left screen displays signal quality and the plethysmographic waveform, and the right screen is a numerical display of SpO_2 and pulse rate. The probe and part of the probe cable are seen between the display screens. Note the alarm controls for saturation and pulse rate settings on the right lower panel.

The preferred sites for sensor application vary with the animal, the limitations imposed by patient positioning and operative procedure, and the specific equipment being used. Perhaps the most popular site has been the tongue, in those cases when it is easily accessible and convenient. However, toes, tails, cheeks, mandibular mucosa, ears, flank folds, and other sites borne of desperation and ingenuity have all been successfully used at one time or another in a variety of domestic and laboratory animals (Huss *et al.,* 1995; Watney *et al.,* 1993; Hendricks and King, 1993; Jantzen and Hennes, 1991). Probe application in smaller animals is more challenging; human ear sensors have been used on mouse thighs and reflectance probes have been placed on the ventral neck of the rat (Sidwell *et al.,* 1992; Vender *et al.,* 1995).

Between SpO_2 of 100 and 90%, the pO_2 ranges from about 90 to about 60 mm Hg. Below a SpO_2 of 90%, a small change in hemoglobin saturation corresponds to a relatively large change in the partial pressure of oxygen. Pulse oximeters will not register a fall in oxygen tension until the PaO_2 reaches the point where hemoglobin begins to desaturate. For this reason, pulse oximeters have been described as sentinels standing at the edge of the cliff. Although they do provide a warning of desaturation, they may not provide much advance warning (Tremper and Barker, 1989).

Pulse oximeters have limitations. Because the instrument reports only the ratio of hemoglobin saturation and not the amount of hemoglobin present, a SaO_2 of 100% is no guarantee of adequate oxygenation. Pulse oximeters may well detect pulses that are adequate to report saturation, but not adequate to provide oxygenation: they cannot be relied on as perfusion monitors. Also, because pulse oximeters operate on the assumption that only two species of hemoglobin are present, if significant amounts of methemoglobin or carboxyhemoglobin are circulating in the patient, the indicated SaO_2 will be erroneous. Pigments or dyes that interfere with light transmission also compromise accuracy. These include dyes that may be used intraoperatively such as methylene blue or indocyanine green (Tremper and Barker, 1993).

Any condition that interferes with adequate pulse strength will cause signal degradation and failure. Such conditions include cardiac causes of low pulse pressure, vasoconstriction due to cooling, hypovolemia or other causes, the presence of pressure cuffs or other restrictive bandages or restraints, and excessive pressure or tension on the probe itself. The last complication is, in the author's experience, a common problem when human ear or finger probes are applied to less firm tissues, such as the tongue: the signal will gradually degrade over time as the tissue is compressed until the area is exsanguinated and the sensor can no longer detect a usable signal, requiring the sensor to be repositioned. Another cause of inaccuracy is optical interference due to stray light. For this reason, probes are best covered by an opaque barrier to prevent interference from ambient light. Pulse oximeters are very motion sensitive, a trait which should be of greater interest in the intensive care setting than in the operating room. Finally, almost all instruments made for use with humans have upper pulse display limits of 225–250 beats per minute and are unable to report the high heart rates commonly encountered in smaller animals. In some cases, the instruments will reject the high heart rate as artifact and fail to report any value at all.

As previously stated, the performance of pulse oximeters depends on the sensor design and characteristics, and to a very great extent on the operating algorithm used. Manufacturers continually refine and update the algorithms, as well as the hardware, making it impossible to extrapolate from the performance of one make or model of instrument to another. Vendors should be willing to provide instruments and probes for evaluation, and the buyer should examine the performance of the instrument in the setting in which it will be used and should contact other users of the same brand and model for their advice and experience.

Until recently, veterinary experience with pulse oximeters has been gained with instruments designed for humans. In many cases, these have performed satisfactorily for a wide range of species. However, the restrictions imposed by algorithms and sensors developed for humans have limited their usefulness with many smaller laboratory animals. Veterinary pulse oximeters that address some of these problems are now entering the market. One such instrument is seen in Fig. 22.

Pulse oximetry is simple, noninvasive, and, by detecting hypoxic episodes that are often otherwise unnoticed, contributes greatly to patient welfare. It is also a powerful teaching tool. If the instrument is consistently used throughout the course of anesthesia and recovery, the standard of anesthesia care cannot fail to improve.

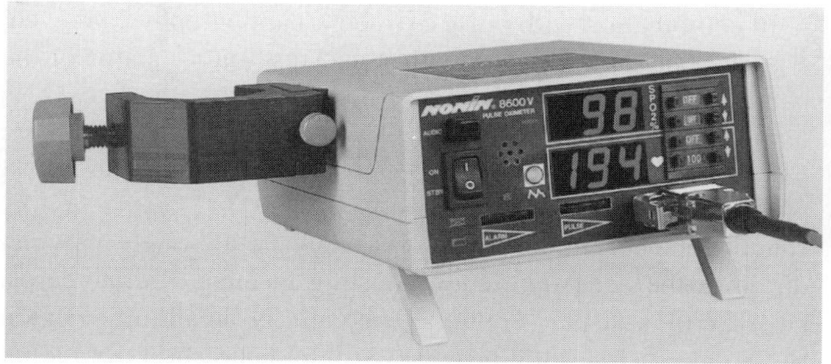

Fig. 22. A veterinary pulse oximeter. This unit has digital displays for SpO_2 and pulse rate. The instrument will register pulse rates up to 450 beats per minute. A signal quality indicator and adjustable alarms for saturation and pulse rate are also present. [Courtesy of Nonin Medical, Inc.]

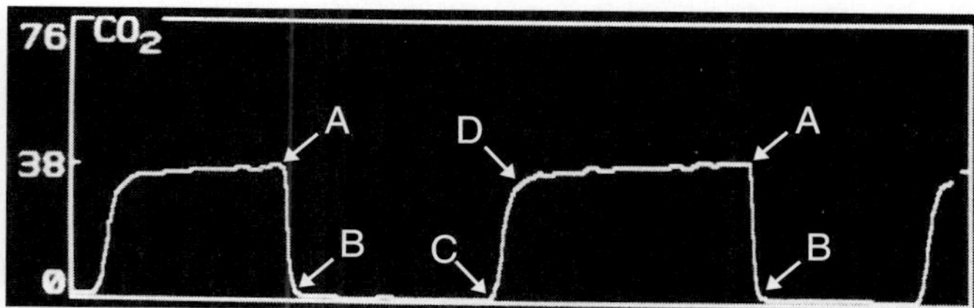

Fig. 23. Capnographic waveform. A screen photograph taken from a sidestream monitor displays the typical phases of a capnographic waveform over a period of two breaths: A–B, inspiratory downstroke; B–C, inspiratory baseline; C–D, expiratory upstroke; and D–A, expiratory plateau. The scale indicates carbon dioxide partial pressure in mm Hg.

2. Carbon Dioxide: Capnography

Carbon dioxide is produced by tissue metabolism and is carried by blood to the lungs where it is discharged into exhaled gas. The concentration of carbon dioxide in respiratory gas is measured because the concentration in exhaled gas depends on adequate ventilation, blood circulation, and tissue metabolism, and the concentration in inhaled gas reflects the function of the anesthetic machine breathing system. The continuous analysis and graphical display of carbon dioxide concentration is called capnography. An instrument that provides only numerical data, with no waveform, is called a capnometer. Capnography is the most common and useful form of carbon dioxide monitoring in anesthesia.

A typical capnographic waveform, or capnogram, is depicted in Fig. 23. The inspiratory baseline is followed by an abrupt expiratory upstroke, an expiratory plateau, and an inspiratory downstroke. The concentration of carbon dioxide is usually at its highest at the end of the expiratory plateau, just before the onset of inspiration. The peak value is identified as the *end-tidal* (et) concentration of CO_2, usually expressed as $PetCO_2$, the partial pressure of CO_2 in the expired gas in mm Hg. Under normal circumstances with adequate ventilation, the $PetCO_2$ is close to $PaCO_2$, the partial pressure of CO_2 in arterial blood. The capnogram also contains information about the inspired concentration of CO_2.

Any condition that alters the production of carbon dioxide, its delivery to or through the lungs, or its transport or absorption in the anesthesia circuit can result in an altered capnogram. In the most obvious case, the appearance of a normal capnographic waveform is strong evidence of a correctly positioned endotracheal tube, whereas the absence of such a waveform in the presence of respiratory efforts would indicate a malpositioned or blocked tube. A rise in inspired CO_2 can indicate rebreathing, which may occur with Bain circuits if the fresh gas flow rate is low, but also can occur in combination with elevated end-tidal CO_2 if the fresh gas delivery tube becomes dislodged. In a circle breathing system, malfunctions of the unidirectional valves can also result in elevated inspired and end-tidal CO_2. Efforts to breath independently of the ventilator are quickly manifested in an abnormal capnogram (Dorsch and Dorsch, 1994l; Berman and Pyles, 1988).

Carbon dioxide in respiratory gases can be detected and quantified by several methods, including infrared analysis, Raman spectroscopy, and mass spectroscopy. Many of these methods can also be used to detect other components of the anesthetic gas mixture, including volatile anesthetic agents. Mass spectroscopy is usually installed in hospitals as a central analytical device serving many operating rooms and is unlikely to be routinely used as a monitor in the research setting. Infrared analysis and Raman analysis are more likely to be used in the research setting. Infrared analysis makes use of the strong absorption by carbon dioxide of infrared light at specific wavelengths. Raman spectroscopy depends on the analysis of the frequency shift of scattered light produced by shining a laser through the gas sample (Dorsch and Dorsch, 1994l).

Infrared analyzers use either sidestream or mainstream sampling methods. Sidestream analyzers draw a constant flow of gas from the breathing circuit through a flexible tube to the optical analyzer. The sample gas withdrawal rate varies among instruments. The sample is usually taken from the area of the endotracheal tube connection or from the mask connection, but the sample can also be collected from nasal prongs. The optical system and pump are remote from the patient. A sidestream analyzer connection is seen in Fig. 24. Mainstream analyzers interpose the optical path into the breathing circuit, typically at the connector of the endotracheal tube. Each design has advantages.

Sidestream analyzers have lightweight connections and add little to the bulk and weight of the circuit. Depending on the capabilities of the instrument, the gas withdrawn can be analyzed for components other than carbon dioxide. However, sidestream analyzers are subject to blockage and interference due to moisture and mucus in the tubing and are somewhat slower in their response time than mainstream analyzers. The withdrawn gas must be scavenged into the waste gas system or returned to the breathing circuit. Finally, sidestream analyzers

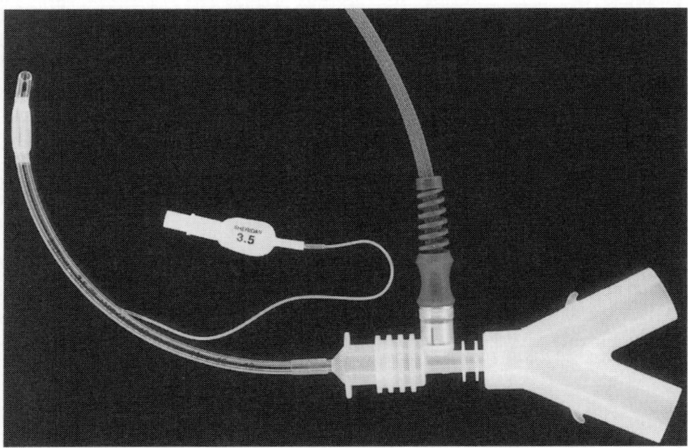

Fig. 24. An example of a sidestream connection between the endotracheal tube and the breathing circuit Y connector. In this case a filter and special tubing are used to conduct withdrawn gas to the analyzer. Some instruments use a simple plastic connecting line to a water trap mounted on the analyzer.

may present problems in obtaining adequate readings and waveforms with the Mapleson breathing systems because of contamination of the expired gas by fresh gases. With smaller patients and high respiratory rates, it is sometimes difficult to get a suitable waveform and reading with a sidestream analyzer. In this case, the sample tube is sometimes connected to a catheter which runs through the endotracheal tube to a site near the tip of the tube or to a needle inserted into the endotracheal tube immediately below the connector (Rich *et al.,* 1990; Halpern and Bissonette, 1994; Badgwell, 1992). Special endotracheal tubes can be purchased that have a separate sampling channel for this purpose. In both cases, problems with sample line blockage due to moisture and mucus are more likely.

Mainstream analyzers have rapid response times, are less affected by moisture, and do not present scavenging concerns. However, they usually have heated optical pathways and add weight and electrical connections, as well as some equipment dead space, to the circuit. Modern mainstream analyzers have more compact and lighter connections. However, mainstream analyzers, at least for now, can only detect carbon dioxide (Badgwell, 1992; Dorsch and Dorsch, 1994l).

Rats, mice, and other small laboratory animals have high respiratory rates and small tidal volumes that exceed the abilities of conventional analyzers, but at least one instrument is marketed for experimental physiology that is claimed to perform well with rats. For most larger animals, careful selection from the available instruments designed for use with human patients should result in satisfactory performance. Although dedicated capnographic monitors are made, they are often included as one component of multifunction monitors. Both sidestream and mainstream instruments have advantages and advocates, so the user is left to determine which best suits his needs and circumstances.

In combination with pulse oximetry, capnography can reveal a great deal about the circulatory and respiratory status of the patient. However, it should be emphasized that the $PetCO_2$ is *not* necessarily the same as $PaCO_2$. Although similar values are reassuring, substantial differences can and do occur and reflect physiologic abnormalities (Swedlow, 1993). These monitors cannot replace arterial blood gas analysis, although they can diminish the need for frequent testing in many routine cases. Because they display continuous breath-by-breath information to the anesthetist, they provide early and specific information about the function of the anesthetic machine and breathing circuit. Given the noninvasive nature of capnography and the value of the information gained, it must be considered among the most important and useful of the current monitoring technologies.

C. Anesthetic Agent Analysis

Many of the techniques described earlier for the analysis of carbon dioxide have been applied to the detection and quantitation of inhalant anesthetic agents. In the case of Raman spectroscopy and mass spectroscopy, other gases such as oxygen and nitrogen can be identified and their concentrations measured. In addition to these instruments, there are small dedicated agent analyzers that use piezoelectric crystals to determine the anesthetic agent concentration. The ability to measure volatile anesthetic agent concentration in the breathing circuit is of some interest when one considers that, regardless of the accuracy of vaporizer output, the concentration of agent the patient is breathing is rarely the same as the vaporizer setting. The effects of agent uptake by the patient, varying fresh gas flow rates, and dilution and delivery through the breathing circuit make the determination of inhaled concentration in the absence of an agent analyzer an educated guess.

Most instruments use sidestream sampling of the airway gas, similar to the sidestream capnography instruments previously described. Indeed, in many cases, the same instrument provides values for carbon dioxide and anesthetic agents. For agent analysis, they usually report numerical data for inspired and expired agent concentrations and do not display a real-time waveform, although many instruments can accumulate and display a graphical trend over several hours of anesthesia. Some instruments require that the user preidentify the specific agent to be determined, whereas others have the ability to recognize the agent or even mixtures of agents automatically. Many agent analyzers report the inspired and expired concentrations for nitrous oxide as well as for the halogenated anesthetic agents. The piezoelectric analyzers must be set in advance for the correct agent and can determine the concentration only for halogenated agents (Dorsch and Dorsch, 1994l).

Although agent analysis is obviously not essential for the safe conduct of inhalation anesthesia, it is very useful and adds another level of safety and convenience to the process. With an agent analyzer, maintaining a safe and effective concentration of volatile agent in the breathing circuit, and in the patient, is much easier. Knowing the expired agent concentration, especially when neuromuscular blocking agents are used, does not guarantee adequate anesthesia in itself, but it is a valuable and objective component of the overall assessment. As with pulse oximetry and capnography, the value of agent analysis as a teaching tool is considerable. It is much easier to train personnel in safe inhalation techniques when they can actually see the effects of vaporizer and fresh gas flows, as well as assisted ventilation, on the concentration of anesthetic agents.

As stated earlier, most agent analyzers are included as components of multifunction monitors, although single purpose stand-alone instruments are available. They usually include both high- and low-level alarms. As these instruments come out of service in the hospital operating room to be replaced by newer monitors, they will become more common in the research setting. Monitors of this sort become especially useful as longer and more complex surgical procedures are undertaken.

D. Oxygen Monitors

Continuous in-line oxygen monitoring is a routine practice in the human operating room, but is less commonly done in veterinary or experimental surgery. Again, a variety of analytical methods are used. Although infrared instruments cannot measure oxygen level, both mass spectrometry or Raman spectroscopy can. Single function oxygen analyzers usually use electrochemical means to detect and measure oxygen, either using a galvanic (fuel) cell or a polarographic electrode (Dorsch and Dorsch, 1994l). These small instruments are relatively inexpensive and easy to use. Many newer human anesthetic machines have oxygen-sensing ports built into the breathing circuit, and many instruments have sensors that directly connect to the ports. Others provide a connector with 22-mm fittings for the breathing circuit. Although these devices may seem unnecessary, especially if pure oxygen is used as the carrier gas, they are of particular value when low flow or closed circuit techniques are used. With these techniques, the concentration of oxygen in the circuit may be difficult to judge, and it is possible to deliver dangerously low oxygen levels.

E. Temperature

Animals undergoing surgery are first anesthetized, which decreases their ability to thermoregulate, the hair is removed and the skin is washed, cooling them further, and they are given cold, dry anesthetic gases to breathe, which carry away more body heat. Finally, an incision is made and a major body cavity may be entered and kept open for a prolonged time. As a result, surgical patients get cold. Although hyperthermia is an occasional problem, the principal reason for monitoring body temperature is to determine how cold the patient is and to assess the effect of efforts to combat unintentional hypothermia.

Operating room thermometers, whether free-standing instruments or components of other monitors, generate continuous readings. In most instances, the temperature is reported in degrees Celsius or Fahrenheit with an accuracy of 0.1°C. They usually use thermistor or thermocouple sensors. Other means of thermometry, such as infrared instruments and liquid crystal patches, can also be used, but suffer in terms of accuracy and flexibility in comparison with conventional instruments (Sessler, 1993).

In animals, rectal and esophageal sites are the most commonly used. In open chest cases and in smaller animals the esophageal site may be subject to cooling. Rectal temperature changes more slowly than esophageal temperature, is susceptible to error from operative exposure, vigorous peritoneal lavage, etc., and may not reflect ''core'' temperature because of fecal accumulation or during lower abdominal procedures. Skin temperature is generally lower than core temperature, but the difference between skin and core body temperature is determined by a number of variables and may vary in an unpredictable manner (Bissonnette, 1992). Although tympanic membrane temperature is taken to be a good indicator of brain temperature, it is complicated by the difficulty in correctly positioning the probe in contact with the tympanic membrane.

A wide variety of sizes and types of reusable and disposable temperature probes are available, including skin sensors and sensors small enough for rectal temperatures in mice. Disposable esophageal stethoscopes are marketed that have built-in temperature probes designed to connect to proprietary thermometers. Free-standing operating room thermometers frequently have two channels, as do some components of multifunction monitors. These instruments are relatively inexpensive and durable. The purchaser should be sure that if special probes are required, they are easily available at an acceptable cost. Otherwise, an instrument that will accept probes with more generic requirements might be a better investment.

F. Multifunction Monitors

Monitors can be purchased with virtually any combination of the individual components so far discussed. With new or used instruments, the purchaser should take the time to clearly understand the appropriate use and limitations of each component. To the degree that the individual components meet the needs of the users, multifunction monitors offer considerable advantages in terms of consolidating and formatting the desired information. They are more compact than a collection of individual instruments and simplify the arranging and connecting of the monitoring station. As hospitals replace equipment with newer instruments, older monitors become available at reduced cost or even free. Although older multifunction monitors may be very suitable choices for laboratory animal use, some caution is necessary. The instruments should be verified to be in good working order by a biomedical technician familiar with them. As with anesthesia machines, it is important that continued parts and technical support be available. Malfunctioning monitors, especially if the failure is not recognized, are not merely useless, they are dangerous.

A recent laudable trend among many monitoring instrument makers is an attempt to produce equipment that will communicate with other instruments, even from a different manufacturer. If the attempt is realized, it will become easier to configure monitors to display their output on one or two screens and to integrate and consolidate monitoring tasks. Many current monitors, such as pulse oximeters and capnographs, can store and display data collected over many hours as trends. Further development and refinement of monitors, with emphasis on ease of interconnection, data acquisition, manipulation, and display, should lead to a much safer and more ''intelligent'' anesthesia environment in which the anesthetist can concentrate on the patient rather than the equipment.

X. PATIENT SUPPORT EQUIPMENT

A. Heat

As noted earlier, anesthetized animals tend to lose more heat than they produce. Attempts to respond to this problem have resulted in a variety of products and practices. An obvious method of retarding heat loss is to raise the temperature in the operating room. Although uncomfortable for the operating room team, maintaining a high ambient room temperature can significantly benefit the patient and may be essential in some instances (Bissonnette, 1992). Because the high room temperatures needed are generally unacceptable to operating room personnel, more local means of thermal support are commonly used.

Among the simplest of measures to reduce the loss of body heat is adequate insulation between the patient and the operating table. This may be little more than a folded towel or blanket. Foam mattresses provide both insulation and the padding which is especially important with heavier animals. Although surgical drapes can provide some insulation against heat loss, they are usually insufficient unless other measures are taken. ''Space blankets,'' plastic film coated with a metallic reflective surface, may provide some advantage, but in one study were no more effective when used alone than towels (Haskins, 1981). Although passive measures such as these are simple and helpful, they cannot reverse the loss of heat entirely nor can they replace lost body heat. More aggressive measures are needed for that.

Alternatively, hot water bottles are easily employed and can be very effective with smaller animals for relatively short-term use. To avoid burns, bottles should not be placed in contact with the patient if the temperature exceeds 42°C and should be replaced when the temperature falls below the measured core temperature of the animal to prevent heat loss from the patient to the bottle (Haskins, 1987). Prewarmed intravenous fluid bags can serve effectively as warm water beds for rodents. One step beyond that are plastic bags or pouches filled with saturated solutions of sodium acetate. A small metal Bendix spring is ''clicked'' to activate crystallization with the release of latent heat. As the process continues, the clear liquid contents become a white opaque solid. These heating bags can be reactivated by slowly heating them to redissolve the sodium acetate. They are useful for small and larger animals for about 1 hr. They can also be folded over fluid lines and syringes to keep the contents warm.

Heating pads and blankets are often used for both small and large animals. Electric heating pads have gained a poor reputation due to their propensity to produce burns and even death from hyperthermia. Electrically heated panels have been marketed that appear to be reasonably effective in warming the veterinary operating table and minimizing conductive heat loss. Operating tables for small rodents and other very small animals also use electrically heated panels or plates, sometimes connected to a rectal thermometer through a switching device that can cycle the heating elements to keep body temperature within a prescribed range. Hyperthermia and burns can occur with electrically heated pads because they continue to produce heat regardless of skin temperature. This may be a problem even when a rectal thermometer is used as a component of an automatic control system because of the differences that occur between skin and rectal temperatures. If the skin is poorly perfused because of depth of anesthesia, vasoconstriction, or pressure, the local heat level may produce severe burns without raising the core temperature. If electric heating pads are heavily padded, if an air gap is maintained between the animal and the blanket, and if the temperature of the skin is monitored and kept below 42°C, they can be used safely (Haskins, 1987). In practice, these conditions are rarely met.

Warm water circulating pads partly address the problems associated with traditional electric heating pads. The water is warmed and circulated through the pad by a remote pump. Because the water is constantly moving and is at a controlled temperature, excess heat is removed and not increased. The pumps and pads most commonly seen are small units designed for the local application of heat in human patients. These small units may be purchased with connections for reusable or disposable pads. The reusable pads are made of heavy plastic and are somewhat expensive, whereas the disposable pads are lighter and are usually sold in cases of 6 or 12 for little more than the cost of a reusable pad. Water circulating pads are prone to damage from punctures and tears. Although the reusable pads can, in theory, be repaired using kits sold for the purpose, the disposable units must be discarded when punctured. In most cases, failure and repair of a reusable pad signal the onset of an aggravating cycle of leaks and repairs. Although disposable pads cannot be repaired, they can be washed and reused until damaged and are relatively inexpensive to replace. Some heating pumps can be equipped with adapters to supply two heating pads, permitting application of a warm pad both above and below the patient. Although these small circulating pads are very useful for smaller animals, such as rodents and rabbits, they are often inadequate for larger animals.

Circulating water blankets designed for human operating room use are considerably bigger and much more efficient for larger animals that the small units intended for bedside use. These blankets and their controllers have a much greater capacity, larger pumps, more sophisticated controls, and can be quite effective in maintaining body temperature, even in major surgeries in larger animals such as sheep, goats, and swine. Some models can provide cooling as well as heating. They are concomitantly more expensive than the smaller units.

Another approach to providing heat is by means of radiant heat, using either commercial infant warmers or infrared or even incandescent light bulbs. These can also be connected via control units to thermometers placed near the patient. Although radiant heaters are effective, they can excessively dry tissues. Skin temperature should be measured.

Convection heating with forced warm air blankets is a more recent mode of thermal support that has quickly gained an important place in human operating rooms. These devices force warm air through numerous small holes in special disposable plastic blankets. The blankets are made to adhere to the skin along one edge to channel the air flow and prevent air from being blown into the incision. Warm air blankets are very effective and can rapidly warm human patients. They are not, however, designed for use with animals and may require considerable modification before they can be adapted for animal use.

Heated humidifiers are used in the breathing circuit to limit heat loss, tracheal injury, and drying of mucus caused by cold, dry anesthetic gases. They are very effective in maintaining circuit humidity and limiting airway damage, but less so in terms of heating efficiency (Raffe and Martin, 1983; Bissonnette, 1992). Heated humidifiers are very useful, especially for prolonged cases, but for most applications a less expensive and simpler approach is the use of heat and moisture exchangers. These single-use devices have a large surface area usually made of a membrane or foam coated with a water-retaining chemical. They are interposed in the breathing circuit at or very near the endotracheal tube connector. As air is expired, water and heat are retained, to be picked up and returned to the airway upon inspiration. The degree to which they add dead space to the circuit, as well as some resistance to breathing, depends on the specific exchanger used, but should be taken into account, especially in small patients. They are, however, simple to use and relatively inexpensive.

Another approach to maintaining patient temperature is the use of warmed intravenous fluids, although the degree to which this is effective obviously depends on the volume of fluid infused. Warmed fluid bags can be wrapped in insulated bag jackets, to slow cooling, and insulated covers used for the fluid line. Instead of heating the fluid bag, the administration line can be heated. A number of intravenous line heaters are available. Many require the use of proprietary disposable heat exchangers, such as plastic coils, which fit into electrically heated blocks. Although most of them provide some sort of protection against overheating, they usually operate at a fixed temperature. However, unless the distance between the heating unit and the patient is quite short, fluid can cool considerably by the time it reaches the intravenous catheter. In order to avoid this problem, some devices use an electrical heating element that runs the length of the administration line to keep the fluid warm up to the point of entry into the vein. An interesting variation of this concept is the use of a pump to circulate warm water around a special administration line. This device uses a disposable three-channel administration set in which the intravenous fluid flows through the center channel while the warming water flows to and from the pump via the remaining two channels. In this way, the fluid is kept warm up to the point of patient connection. The circuit is long enough to provide adequate warming even at fairly high flow rates.

Steps taken to reduce heat loss and maintain normal body temperature are rewarded by a less stressful, and often faster, recovery. The technological sophistication of thermal support need not be high if the anesthesia provider is willing to invest energy and ingenuity in the effort to keep patients warm. Careful planning to minimize induction and surgical preparation time; appropriate, but not excessive, hair removal; careful skin preparation to avoid soaking and further cooling the animal; and rapid transfer to the operating room are all helpful. Table padding, simple warm water circulating pads, and the use of low gas flow rates are all relatively inexpensive and useful steps in the operating room. Certainly, rodents receiving postoperative fluids should not be further chilled by the use of cold fluids. More demanding and longer procedures may require correspondingly more aggressive support. In the process of planning surgeries and purchasing sophisticated anesthetic machines and monitors, the importance of effective thermal support should not be underestimated.

XI. ANCILLARY EQUIPMENT

A. Endotracheal Tubes

Endotracheal tubes are used to secure the airway, reduce anatomical dead space, and facilitate the administration of oxygen and anesthetic gases. They are available in a great variety of sizes, shapes, materials, and styles. Endotracheal tubes designed for human use have a standard 15-mm connector to the breathing circuit. Large animal tubes have connectors designed to fit the larger Y connectors used with 50-mm-diameter breathing circuits. Tubes designed for use in humans are usable for most laboratory animals, except for the smallest and largest.

Conventionally, endotracheal tubes are sized by the internal diameter in millimeters marked on the tube. Also printed is the length in centimeters and, if the tube is smaller than size 6.0, the outside diameter. Tube length increases with tube diameter (Dorsch and Dorsch, 1994k).

The materials most commonly used in constructing endotracheal tubes are rubber, polyvinyl chloride (PVC), and silicone. Rubber was once very common, but is less often used today. Although red rubber tubes are reusable and can be repaired within limits, they are opaque, heavy, and prone to kinking. With age, rubber tubes may become sticky and lose their elasticity. The most common material used is PVC, which is lighter than rubber or silicone, transparent, and offers a good ratio of internal diameter to wall thickness. PVC tubes can be washed and reused with caution, but are not easily sterilized. Silicone tubes are translucent, flexible, and can be easily sterilized in the autoclave. They are expensive compared to rubber or PVC, but can be repaired by the manufacturer and reused indefinitely. Veterinary tubes are available in rubber, PVC, and silicone.

Some tubes have an opening near the distal end, called a Murphy eye, intended to provide an alternate path for respiration should the tip of the tube become blocked with secretions. Tubes with an opening are called Murphy tubes, and those with no opening are designated as Magill-type tubes.

Endotracheal tubes may be cuffed or uncuffed. The cuff is intended to seal the trachea to prevent the aspiration of oral secretions or refluxed stomach contents, facilitate positive pressure ventilation, and minimize environmental contamination with anesthetic gases. Two types of cuff design are used. High-pressure, low-volume cuffs have a small tracheal contact area and require small volumes of gas at relatively high inflation pressures. These cuffs appear to be smooth and nearly continuous with the external surface of the endotracheal tube before they are inflated. In contrast, low-pressure, high-volume cuffs require larger amounts of gas to form a larger seal with the trachea at lower inflation pressures. The redundant cuff material of the low-pressure, high-volume design forms a wrinkled surface of cuff material in the undistended state. The low-pressure, high-volume design is said to provide a greater area of tracheal seal with less chance of pressure necrosis. High-volume cuffs may be somewhat more difficult to insert. High-pressure, low-volume cuffs have a smaller area of contact, but exert proportionally more pressure on the tracheal wall for a small increase in added volume. They may be somewhat easier to insert, especially in small patients (Dorsch and Dorsch, 1994k).

Cuffed tubes have a channel leading from the cuff to an inflation line and pilot balloon. The pilot balloon can indicate whether the cuff is inflated, but does not provide a good indication of inflation pressure. The inflation port may be either a one-way valve and Luer connector or simply an open port with a built-in stopper. Cuffed endotracheal tubes leak. The two most common causes of leakage are a torn or punctured cuff and failure of the inflation valve. However, damage to the pilot line is common if clamps are used to occlude the line, and even leaks in the inflation channel in the body of the tube occur if patients chew on the tube. Damage can also occur in the cleaning and storage process. There are enough reasons for cuff failure, even in new tubes, to warrant routine testing just before use.

An endotracheal tube can reduce anatomical dead space but, unless it is the correct length, will add to equipment dead space. Tubes should be cut so that they extend only far enough beyond the mouth to facilitate connection to the breathing circuit. The required length can be estimated by placing it next to the head of the animal at the level of the mouth and advancing the tip along the neck to the desired depth of insertion. The excess tube is then cut and the connector is firmly reattached. Care should be taken not to cut the pilot line or compress the molded pilot channel in the body of the tube when the connector is replaced.

Numerous modified and specialty tubes are made for nasal intubations, endobronchial intubations, tracheostomies, laser laryngeal surgery, and many other specialty applications. These are well-described and discussed at length elsewhere (Dorsch and Dorsch, 1994k). Only a few specialty tubes that are potentially of more general interest for laboratory animals will be discussed here.

Spiral reinforced tubes are useful in cases where there is a particular risk of kinking the endotracheal tube, such as when the head must be severely flexed or when the anesthetist may be unable to reach the head of the patient during the case, as in some ophthalmic, neurosurgical, and oral procedures. These tubes resist kinking by virtue of the embedded wire or nylon coil. For that reason, however, they cannot easily by trimmed to the appropriate length. The silicone tube seen in Fig. 25 is wire reinforced. Silicone wire-reinforced tubes are available in both standard and short lengths, and in sizes small enough to accommodate rabbits and most primates. They are reusable, repairable, and easily sterilized by autoclave, offsetting the initial relatively high cost.

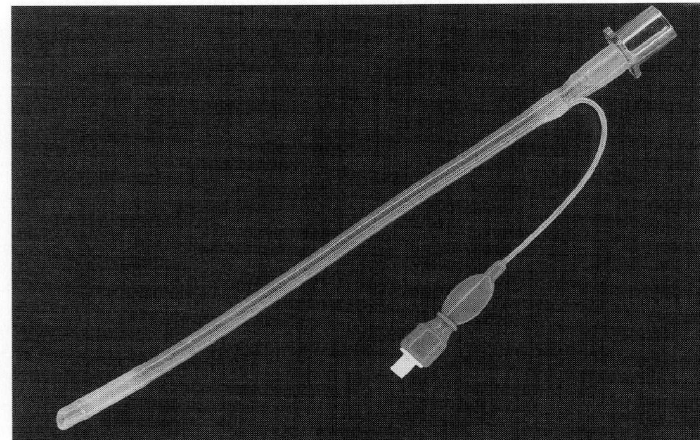

Fig. 25. A wire-reinforced cuffed silicone endotracheal tube. In addition to resistance to kinking, this reusable tube is autoclavable.

The smallest commercially available endotracheal tubes are Cole tubes, seen in Fig. 26. Cole tubes have a reduced distal end opening into a larger diameter body. These tubes were designed for pediatric human patients and, in the author's experience, the relatively short narrow intratracheal portion makes them more prone to displacement than conventional designs. Using the shoulder of the tube as a wedge to form a gas seal can result in laryngeal damage (Dorsch and Dorsch, 1994k). They are available in sizes as small as 1.5 mm at the patient end. Cole tubes are commonly recommended for very small patients.

Tubes with an extra channel for distal gas sampling are mentioned in the discussion of capnography. An example is seen in Fig. 27.

The very small animals encountered in laboratory animal practices may require custom fabrication of endotracheal tubes. Rats, for example, are often intubated using 16- or 14-gauge intravenous catheters. These can be further modified by cutting the end to form a bevel and smoothing the tip to reduce trauma. If desired, a Murphy eye can be cut in the catheter as well. Another method of minimizing trauma from the relatively stiff catheter is to trim it and add a small length of softer silicone tubing to the distal end (DeLeonardis *et al.,* 1995). A small length of rubber or plastic tubing can be slipped over the catheter to form a stop, preventing overinsertion and helping to secure a seal at the larynx. An example of a catheter modified to serve as a rat endotracheal tube is seen in Fig. 28. The Luer taper connector of an intravenous catheter will fit a standard 2.5-mm endotracheal tube connector.

Larger tubes can be made from PVC neonatal feeding tubes. The proximal end of these tubes is flared to accept a syringe and will also accept a small endotracheal tube connector. The catheter can be cut to the desired length and the tip smoothed to minimize trauma to the laryngeal and tracheal tissues. An endotracheal tube made from a 5 French feeding tube is shown in Fig. 29.

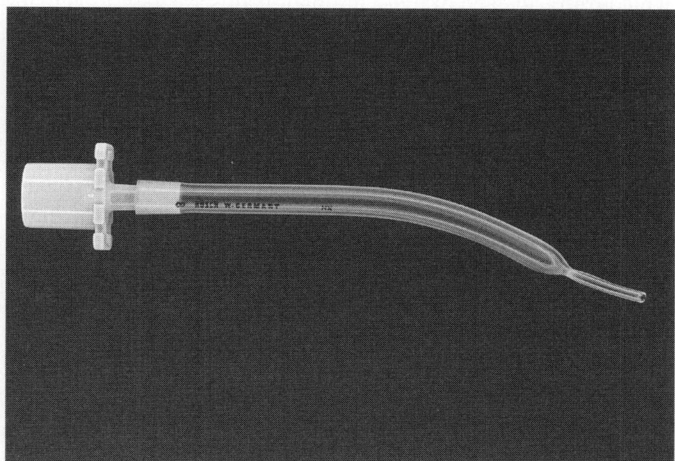

Fig. 26. A Cole tube. Note the sharply reduced distal end.

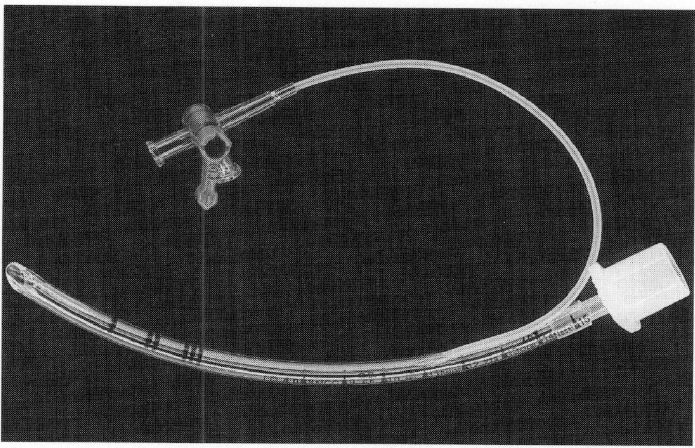

Fig. 27. Gas sampling endotracheal tube. This example lacks a cuff. A gas sampling channel leads from the distal end of the endotracheal tube to the line terminating in a three-way stopcock. A sidestream analyzer sampling line is connected to the stopcock.

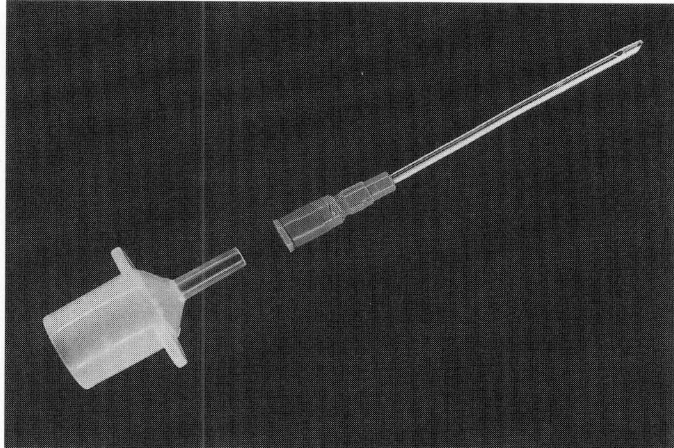

Fig. 28. A 14-gauge intravenous catheter modified to serve as a rat endotracheal tube. The tip has been beveled and smoothed, and a Murphy eye added. A small endotracheal tube connector fits the Luer taper catheter to provide a conventional 15-mm ID breathing circuit connection.

Endotracheal tube connectors can be metal or plastic. Most disposable tubes and some reusable tubes are sold with plastic connectors. The machine end of the connector is a standard 15-mm taper. Because the tube portion varies to fit the tube, they are usually identified by the tube size. Silicone tubes are often supplied with clear autoclavable polysulfone plastic connectors. In addition to conventional straight connectors, special low-volume connectors are available for very small tubes (Fig. 30). Right-angle adapters are made with or without sample

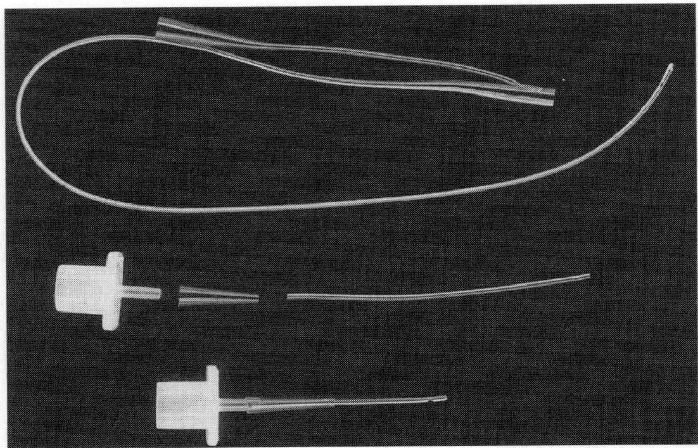

Fig. 29. An extemporaneous endotracheal tube constructed from a 5 French feeding tube. The feeding tube cap is trimmed to provide a transition sleeve between the conventional 15-mm endotracheal tube connector and the distal, 5 French segment of the tube. The transition sleeve permits the narrow distal tube to be advanced into the 15-mm connector, further reducing the dead space. Such tubes may meet an immediate, urgent need but are not reliable for routine use.

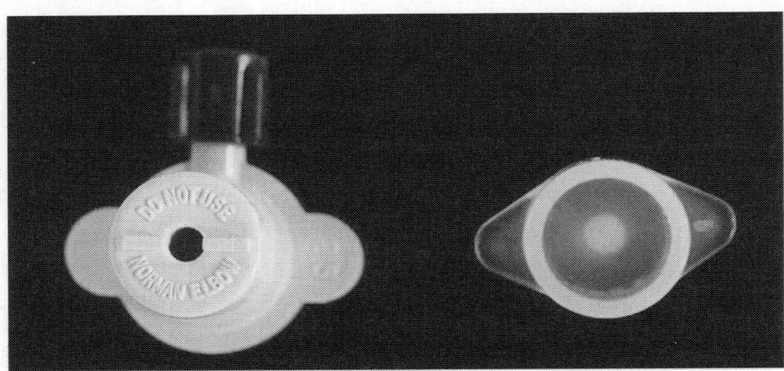

Fig. 30. The 15-mm breathing circuit end of two endotracheal tube connectors is shown. On the left is a low-volume, reduced dead space connector. On the right is a conventional connector. Note the difference in internal volume. The low-volume connector also has a gas sampling port, seen toward the top.

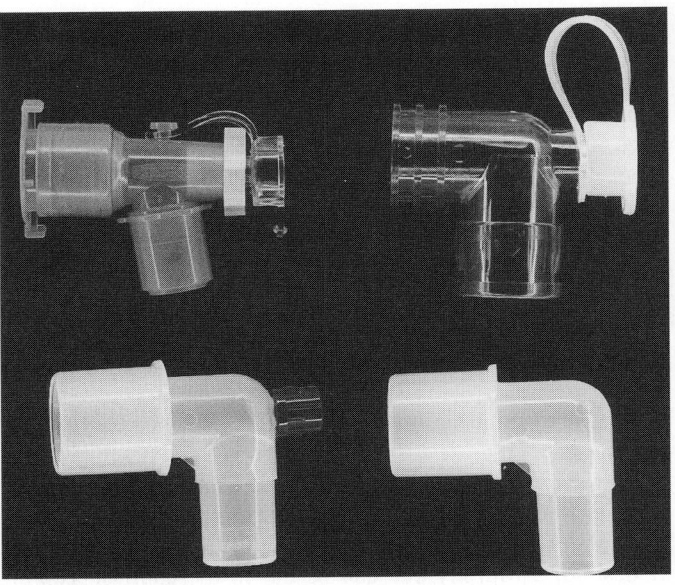

Fig. 31. Right-angle endotracheal tube adaptors or elbows. (Top right) Elbow with endoscope port. (Top left) Swiveling elbow with endoscope port. (Bottom right) Conventional elbow. (Bottom left) Elbow with gas sampling port. The sampling port is fitted with a sealing cap.

ports, as are swivel adapters with special ports for bronchoscopes. Examples of right-angle and swivel connectors are seen in Fig. 31. The connectors on new disposable tubes may be loosely mounted; they should be checked and tightened before use to avoid accidental disconnection from the anesthesia circuit.

A great variety of specialty endotracheal tubes, such as tubes for single lung or independent lung ventilation, laser surgery, special connectors, adapters, and sample ports, are commercially available. Veterinary and human anesthesiologists will be familiar with the sources and appropriate use of these devices and should be consulted for help in obtaining and using them.

B. Stylets, Intubation Guides, and Tube Changers

For the purposes of the following discussion, three types of intubation aids are identified on the basis of their intended use. *Stylets* are used to stiffen and shape the endotracheal tube to ease insertion. *Intubation guides* are passed into the trachea and the endotracheal tube is slipped over them and into place. *Tube exchangers,* as the name suggests, are first passed through an existing endotracheal tube, which is then withdrawn and replaced with a new tube. This choice of nomenclature is somewhat arbitrary: each of these intubation aids are identified in the literature by a confusing array of similar and, in some cases, identical names.

It is occasionally helpful to use a stylet to form the endotracheal tube to a desired curvature or to stiffen a very flexible tube. Although a large number of homemade stylets, ranging from coat hangers, copper wire, and even wooden dowels, have been used and recommended, for most purposes commercially available stylets are more convenient and safer. They are usually made of plastic-coated malleable wire. Some have a rubber or plastic stop that fits the endotracheal tube connector and serves to limit the advance of the stylet whereas others are simply bent sharply at the connector end of the tube to prevent further advance. Unless designed to be used in the trachea, it is commonly recommended that the stylet not extend beyond the tip of the endotracheal tube in order to avoid damage to delicate laryngeal and tracheal tissues. Examples of commercially available stylets are seen in Fig. 32.

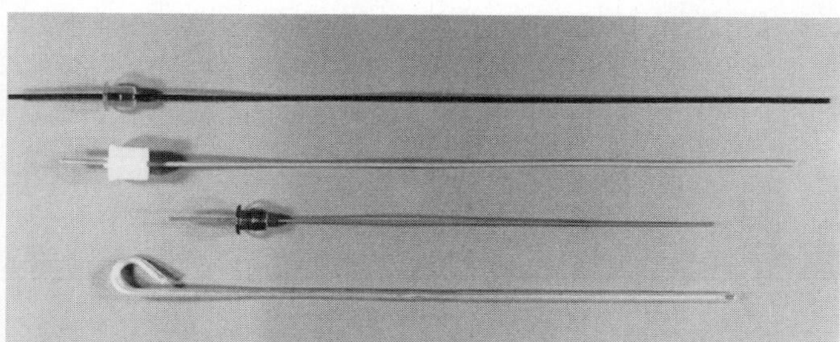

Fig. 32. Commercial endotracheal tube stylets. The top three are smooth, plastic-coated malleable wires with stops that can be adjusted to prevent the stylet from extending beyond the tip of the endotracheal tube. The white stop on the second stylet has a locking mechanism that will not allow the stylet to slip forward once adjusted. The bottom stylet is a heavier malleable rod encased in clear tubing.

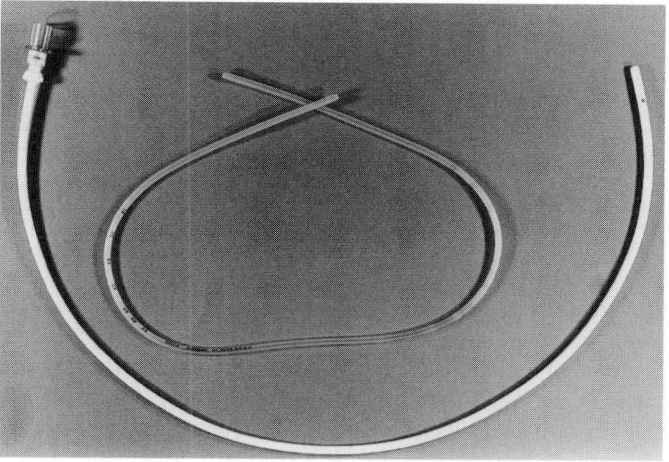

Fig. 33. Endotracheal tube exchangers. The inner, clear plastic exchanger is used for sizes 6.0 to 8.5 endotracheal tubes. The larger opaque exchanger has a detachable 15-mm fitting, seen at the top left, to permit delivery of breathing gases during the exchange procedure.

One technique for guiding an endotracheal tube into the trachea under conditions of limited vision is to first introduce a small-diameter intubation guide into the trachea and to then slip the endotracheal tube over the guide, which is then removed. This technique is sometimes used with very small animals, such as rats, where direct placement of the endotracheal tube can be difficult because the tube, as it advances, obscures the view of the laryngeal opening. Intubation guides should be stiff enough to permit easy manipulation, but soft enough to prevent laryngeal and tracheal trauma. For rats, a spring guide wire, such as is used for the Seldinger technique of vascular cannulation, is convenient and atraumatic. For larger animals, such as small rabbits, a PVC feeding tube has been used.

Intubation guides become tube exchangers if used to replace an existing endotracheal tube. Commercial tube exchangers, made for adult or pediatric human use, are convenient for larger endotracheal tubes. Some tube exchangers have adapters to permit insufflation with oxygen, or even jet ventilation, during the procedure (Dorsch and Dorsch, 1994k). Examples of tube exchangers are seen in Fig. 33.

C. Laryngoscopes

Laryngoscopes are used to obtain an adequate view of the larynx to facilitate endotracheal intubation. Although the most commonly used laryngoscopes are rigid designs, flexible fiberoptic stylet laryngoscopes and fiberoptic endoscopes are also used. For very small animals, modified or unmodified veterinary otoscopes are often used.

The most familiar design for rigid laryngoscopes is a cylindrical handle, housing one or more batteries, and a detachable blade. The blades have a standard hook-on connection to fit the handles and are, with some exceptions, interchangeable among handles. Handles are available in several sizes and styles, ranging from small penlight styles to heavy stubby handles designed for use with humans having limited oral access. The smaller handles are lightweight and convenient for smaller blades, which tend to be overbalanced by the larger standard handles. By the same token, larger blades are more easily controlled when an appropriately large handle is used. Although most blades form a right angle with the handle when they are in the operating position, special adapters and handles are made which allow the blade angle to be varied. Blades are usually made of stainless steel or chromed brass, but disposable plastic blades are also available. Disposable one-piece laryngoscopes are also made in a limited variety of patterns (Dorsch and Dorsch, 1994j).

The blade may use a light bulb or a fiber-optic bundle to provide illumination. In the latter case, the bulb is located in the handle or in the base of the blade. Standard 2.5-V incandescent lamps or the brighter 3.5-V halogen bulbs are used. Although blades and handles using the same style and voltage of illumination are usually interchangeable, fiber-optic blades may require specific handles. For most purposes, conventional blades, using distal lamps, are adequate, and the advantage of easy interchangeability among handles is compelling. Although the blades are somewhat expensive, conventional handles are not. Given the tendency for batteries to become exhausted at critical moments, it is economical and prudent to have at least one extra handle and bulb available.

There is a bewildering array of blades. Special blades have been designed for almost every conceivable variant of human anatomy and operator preference. In addition, a few manufacturers have modified human blades for veterinary use. Blades are made in straight and curved patterns, with and without a flange. Some blades are available with the light mounted on the left or on the right side. The flange, although advantageous in some animals, may be a hindrance in others. It can often be removed or reduced by a cooperative machinist. The variety of available blades can be confusing at first sight, but offers the possibility of selecting a blade that is closely tailored to the preferences and needs of the user. The types and sizes of blades needed depend on the species and size of animals encountered. A busy surgical facility may well accumulate a considerable blade collection. Under ordinary circumstances, however, only a few types of blades are needed. The point has been made that the skill of the operator is of more importance than the type of blade used (Dorsch and Dorsch, 1994j).

The most common straight blade is the Miller blade. It comes in a range of sizes and can be purchased with right- or left-mounted lamps. Although the Miller blade does have a flange, a modified version can be purchased without it. Other straight designs are the Phillips blade, which has a low flange, and the Alberts or Michaels blades, which have no flange. The Robertshaw blade has a flange that decreases in size near the tip and has been used in nonhuman primates such as cynomolgus monkeys. A modified Wisconsin blade, suitable for larger animals such as goats and larger swine, has been made specifically for veterinary use. Straight blades are usually placed over the epiglottis. Some straight laryngoscope blades are shown in Fig. 34.

Macintosh blades are curved and the tip is blunt, intended to be placed under the epiglottis in humans and lifted to provide a view of the vocal cords. Numerous modifications of the Macintosh blade have been made by altering the degree of curvature, the height of the flange, the placement of the light, etc. The Choi blade has a wide, flat double-angle design with no flange, which the author has found useful for domestic swine in the 12- to 20-kg weight range. Macintosh and Choi blades are shown in Fig. 35.

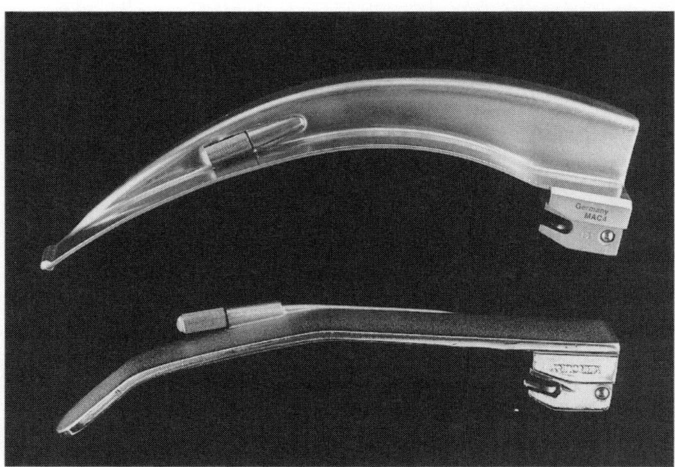

Fig. 35. Curved larygoscope blades. The Macintosh blade is shown at the top and a Choi double-angle blade at the bottom.

Conventional laryngoscope blades have been modified for use in smaller animals, such as rats and guinea pigs, by narrowing the tip (Turner *et al.,* 1992; Blouin and Cormier, 1987). Another approach is to use an otoscope cone. This method, however, requires the tube or stylet to be passed through the cone, which interferes somewhat with visualization, and requires the tube or stylet to be held in place while the otoscope is withdrawn over it. The modification of Tran and Lawson addresses this difficulty, and a modified cone is shown in Fig. 36 (Tran and Lawson, 1986).

In addition to conventional blades, fiber-optic stylets and endoscopes can also be used to accomplish intubation. The fiber-

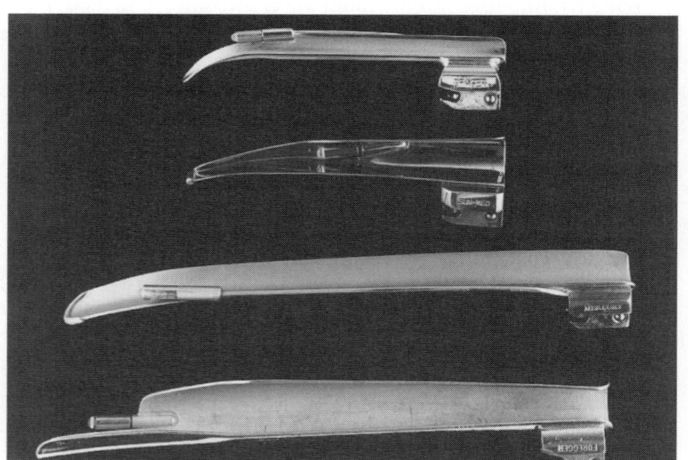

Fig. 34. Straight larygoscope blades. Shown from the top are Phillips, Robertshaw, modified Miller, and Wisconsin blades. The Miller blade has been modified by removing the flange and placing the light on the right side. This light placement is convenient for left-handed human anesthetists and for right-handed veterinary anesthetists whose patients are in sternal recumbancy.

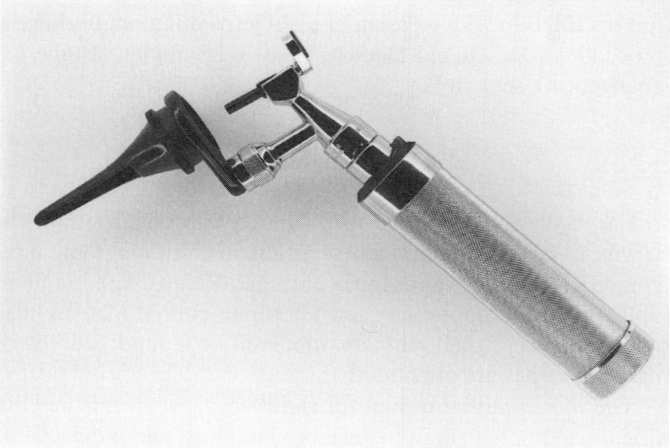

Fig. 36. An otoscope with a cone modified for use as a laryngoscope for rats. The modification of Tran and Lawson (1986) consists of removing a section of the distal half of the cone so that the endotracheal tube can be placed from the side while maintaining a clear view of the larynx.

optic stylets have a handle and viewing port attached to a fiber-optic bundle forming a stylet. The malleable stylet provides illumination and permits a view through the distal tip. The stylet is placed through the lumen of the endotracheal tube, the tube is guided into place under direct vision, and the stylet is withdrawn. A similar technique is used with a flexible fiber-optic endoscope (Dorsch and Dorsch, 1994j).

Laryngoscopes and blades should be cleaned and disinfected between uses. They can be gas sterilized, but are not usually autoclavable. In circumstances that require sterility, disposable laryngoscopes should be used, if feasible. The handles should be checked to ensure that the batteries are good and that the blade-locking mechanism is intact and secure. The blades should be inspected to assure that the bulbs work and are tight and that the locking mechanisms, small spring-loaded steel balls that fit into corresponding niches in the handle and keep the blade fixed in position during use, are intact. Handles and blades are subject to normal wear and tear: busy surgical facilities should budget for replacement on a regular basis.

References

Andrews, J. J. (1990). Inhaled anesthetic delivery systems. *In* ''Anesthesia'' (R. Miller, ed.), 3rd ed. pp. 171–223. Churchill-Livingstone, New York.

Badgwell, J. M. (1992). Respiratory gas monitoring in the pediatric patient. *In* ''International Anesthesiology Clinics: Anesthesia Equipment for Infants and Children'' (J. Pullerits and R. S. Holzman, eds.), Vol. 30, No. 2, pp. 131–146. Little, Brown, Boston.

Banner, M. J., and Lampotang, S. (1992). Mechanical ventilators—Fundamentals. *In* ''Handbook of Mechanical Ventilatory Support'' (A. Perel and M. C. Stock, eds.), pp. 7–30. Williams & Wilkins, Baltimore, Maryland.

Baxter, G. M., Adams, J. E., and Johnson, J. J. (1991). Severe hypercarbia resulting from inspiratory valve malfunction in two anesthetized horses. *J. Am. Vet. Med. Assoc.* **198**(1), 123–125.

Bednarski, R. M. (1993). Anesthetic breathing systems. *In* ''Anesthetic Equipment'' (R. M. Bednarski, ed.), Semin. Vet. Med. Surg. (Small Anim.), Vol. 8, No. 2, pp. 82–89. Saunders, Philadelphia.

Benumof, J. L., Shroff, P. K., and Skerman, J. H. (1992). Transtracheal ventilation. *In* ''Clinical procedures in Anesthesia and Intensive Care'' (J. L. Benumof, ed.), pp. 195–209. Lippincott, Philadelphia.

Berman, L. S., and Pyles, St. T. (1988). Capnographic detection of anaesthesia circle valve malfunctions. *Can. J. Anaesth.* **35**(5), 473–475.

Berryman, E. R., and Sharp, K. R. (1995). A device to maintain gaseous anesthesia in several rodents while using one anesthesia machine. *Abstr. Sci. Pap.: Platform Poster Presentations, 1995 AAALAS Natl. Meet.*, p. 19.

Bissonnette, B. (1992). Temperature monitoring in pediatric anesthesia. *In* ''International Anesthesiology Clinics: Anesthesia Equipment for Infants and Children'' (J. Pullerits and R. S. Holzman, eds.), Vol. 30, No. 2, pp. 63–76. Little, Brown, Boston.

Blouin, A., and Cormier, Y. (1987). Endotracheal intubation in guinea pigs by direct laryngoscopy. *Lab Anim. Sci.* **37**(2), 244–245.

Caulkett, N. A., Duke, T., and Bailey, J. V. (1994). A comparison of systolic blood pressure measurement obtained using a pulse oximeter, and direct systolic measurement in anesthetized sows. *Can. J. Vet. Res.* **58**, 144–147.

Chalwa, R., Kumarvel, V., Girdhar, K. K., Sethi, A. K., Indrayan, A., and Bhattacharya, A. (1992). Can pulse oximetry be used to measure systolic blood pressure? *Anesth. Analg.* **74**, 196–200.

Compressed Gas Association, Inc. (1989). ''Characteristics and Safe Handling of Medical Gases,'' 7th ed., Publ. P-2. Compressed Gas Association, Inc., Arlington, Virginia.

Dardai, E., and Heavner, J. E. (1987). Respiratory and cardiovascular effects of halothane, isoflurane and enflurane delivered via a Jackson-Rees breathing system in temperature controlled and uncontrolled rats. *Methods Findings Exp. Clin. Pharmacol.* **9**, 717–720.

Davis, P. D., Parbrook, G. D., and Kenny, G. N. (1995a). Volume and flow measurement. *In* ''Basic Physics and Measurement in Anaesthesia,'' 4th ed., pp. 29–43. Butterworth-Heinemann, Oxford.

Davis, P. D., Parbrook, G. D., and Kenny, G. N. (1995b). Vaporizers. *In* ''Basic Physics and Measurement in Anaesthesia,'' 4th ed., pp. 134–145. Butterworth-Heinemann, Oxford.

Davis, P. D., Parbrook, G. D., and Kenny, G. N. (1995c). Breathing and scavenging systems. *In* ''Basic Physics and Measurement in Anaesthesia,'' 4th ed., pp. 281–300. Butterworth-Heinemann, Oxford.

DeLeonardis, J. R., Clevenger, R., and Hoyt, R. F. (1995). Approaches in rodent intubation and endotracheal tube design. *Abstr. Sci. Pap., 46th AALAS Nat. Meet.*, p. 19.

Dorsch, J. A., and Dorsch, S. E. (1984a). The anesthesia machine. *In* ''Understanding Anesthesia Equipment: Construction, Care and Complications,'' 2nd ed., pp. 38–76. Williams & Wilkins, Baltimore, Maryland.

Dorsch, J. A., and Dorsch, S. E. (1984b). Vaporizers. *In* ''Understanding Anesthesia Equipment: Construction, Care and Complications,'' 2nd ed., pp. 77–135. Williams & Wilkins, Baltimore, Maryland.

Dorsch, J. A., and Dorsch, S. E. (1984c). The breathing system. II. The Mapleson Systems. *In* ''Understanding Anesthesia Equipment: Construction, Care and Complications,'' 2nd ed., pp. 182–196. Williams & Wilkins, Baltimore, Maryland.

Dorsch, J. A., and Dorsch, S. E. (1994a). ''Understanding Anesthesia Equipment: Construction, Care and Complications,'' 3rd ed. Williams & Wilkins, Baltimore, Maryland.

Dorsch, J. A., and Dorsch, S. E. (1994b). Medical gas cylinders and containers. *In* ''Understanding Anesthesia Equipment: Construction, Care and Complications,'' 3rd ed., pp. 1–23. Williams & Wilkins, Baltimore, Maryland.

Dorsch, J. A., and Dorsch, S. E. (1994c). The anesthesia machine. *In* ''Understanding Anesthesia Equipment: Construction, Care and Complications,'' 3rd ed., pp. 51–90. Williams & Wilkins, Baltimore, Maryland.

Dorsch, J. A., and Dorsch, S. E. (1994d). Vaporizers. *In* ''Understanding Anesthesia Equipment: Construction, Care and Complications,'' 3rd ed., pp. 91–147. Williams & Wilkins, Baltimore, Maryland.

Dorsch, J. A., and Dorsch, S. E. (1994e). The breathing system: General principles, common components, and classifications. *In* ''Understanding Anesthesia Equipment: Construction, Care and Complications,'' 3rd ed., pp. 149–165. Williams & Wilkins, Baltimore, Maryland.

Dorsch, J. A., and Dorsch, S. E. (1994f). The Mapleson breathing systems. *In* ''Understanding Anesthesia Equipment: Construction, Care and Complications,'' 3rd ed., pp. 167–189. Williams & Wilkins, Baltimore, Maryland.

Dorsch, J. A., and Dorsch, S. E. (1994g). The circle absorption system.. *In* ''Understanding Anesthesia Equipment: Construction, Care and Complications,'' 3rd ed., pp. 191–224. Williams & Wilkins, Baltimore, Maryland.

Dorsch, J. A., and Dorsch, S. E. (1994h). Anesthesia ventilators. *In* ''Understanding Anesthesia Equipment: Construction, Care and Complications,'' 3rd ed., pp. 255–279. Williams & Wilkins, Baltimore, Maryland.

Dorsch, J. A., and Dorsch, S. E. (1994i). Controlling trace gas levels. *In* ''Understanding Anesthesia Equipment: Construction, Care and Complications,'' 3rd ed., pp. 281–324. Williams & Wilkins, Baltimore, Maryland.

Dorsch, J. A., and Dorsch, S. E. (1994j). Laryngoscopes. *In* ''Understanding Anesthesia Equipment: Construction, Care and Complications,'' 3rd ed., pp. 399–437. Williams & Wilkins, Baltimore, Maryland.

Dorsch, J. A., and Dorsch, S. E. (1994k). Tracheal tubes. *In* ''Understanding Anesthesia Equipment: Construction, Care and Complications,'' 3rd ed., pp. 439–545. Williams & Wilkins, Baltimore, Maryland.

Dorsch, J. A., and Dorsch, S. E. (1994l). Gas monitoring. *In* ''Understanding Anesthesia Equipment: Construction, Care and Complications,'' 3rd ed., pp. 547–607. Williams & Wilkins, Baltimore, Maryland.

Dorsch, J. A., and Dorsch, S. E. (1994m). Vigilance aids and monitors. *In* ''Understanding Anesthesia Equipment: Construction, Care and Complications,'' 3rd ed., pp. 609–692. Williams & Wilkins, Baltimore, Maryland.

Dorsch, J. A., and Dorsch, S. E. (1994n). Checkout, management, and legal issues. *In* ''Understanding Anesthesia Equipment: Construction, Care and Complications,'' 3rd ed., pp. 693–718. Williams & Wilkins, Baltimore, Maryland.

Ehrenworth, J., and Eisenkraft, J., eds. (1993). ''Anesthesia Equipment: Principles and Practice.'' Mosby, St. Louis, Missouri.

Eichorn, J. H. (1993). Anesthesia equipment: Checkout and quality assurance. *In* ''Anesthesia Equipment: Principles and Applications'' (J. Ehrenworth and J. Eisenkraft, eds.), pp. 473–491. Mosby, St. Louis, Missouri.

Eisenkraft, J. B. (1993a). The anesthesia machine. *In* ''Anesthesia Equipment: Principles and Applications'' (J. Ehrenworth and J. Eisenkraft, eds.), pp. 27–56. Mosby, St. Louis, Missouri.

Eisenkraft, J. B. (1993b). Anesthesia vaporizers. *In* ''Anesthesia Equipment: Principles and Applications'' (J. Ehrenworth and J. Eisenkraft, eds.), pp. 57–88. Mosby, St. Louis, Missouri.

Flecknell, P. A., Cruz, I. J., Liles, J., and Whelan, G. (1994). Induction of anaesthesia with halothane or isoflurane in the rabbit. *Proc. Int. Congr. Vet. Anesth., 5th,* Guelph, Ontario, Abstr., p. 101.

Glen, J. B., Cliff, G. S., and Jamieson, A. (1980). Evaluation of a scavenging system for use with inhalation anesthesia techniques in rats. *Lab. Anim.* **14**(3), 207–211.

Grandy, J. L., Dunlop, C. I., Hodgson, D. S., Curtis, C. R., and Chapman, P. L. (1992). Evaluation of the Doppler ultrasonic method of measuring systolic blood pressure in cats. *Am. J. Vet. Res.* **53,** 1166–1169.

Gravenstein, J. S., and Paulus, D. A. (1982a). Perspectives on monitoring. *In* ''Monitoring Practice in Clinical Anesthesia,'' pp. 3–10. Lippincott, Philadelphia.

Gravenstein, J. S., and Paulus, D. A. (1982b). Arterial pressure. *In* ''Monitoring Practice in Clinical Anesthesia,'' pp. 37–88. Lippincott, Philadelphia.

Gravenstein, J. S., Paulus, D. A., and Hayes, T. J. (1989). Dead space. *In* ''Capnography in Clinical Practice,'' pp. 31–41. Butterworth, Boston.

Halpern, L., and Bissonette, B. (1994). The most proximal and accurate site for sampling end-tidal CO_2 in infants. *Can. Jour. Anaesth.* **41**(10), 984–990.

Hartsfield, S. M. (1987). Machines and breathing systems for administration of inhalation anesthetics. *In* ''Principles and Practice of Veterinary Anesthesia'' (C. Short, ed.), pp. 395–418. Williams & Wilkins, Baltimore, Maryland.

Hartsfield, S. M. (1994). Practical problems with veterinary anesthesia machines. *Proc. Int. Congr. Vet. Anesth., 5th,* Guelph, Ontario, p. 26.

Haskins, S. C. (1981). Hypothermia and its prevention during general anesthesia in cats. *Am. J. Vet. Res.* **42**(5), 856–861.

Haskins, S. C. (1987). Monitoring the anesthetized patient. *In* ''Principles and Practice of Veterinary Anesthesia'' (C. Short, ed.), pp. 455–477. Williams & Wilkins, Baltimore, Maryland.

Heavner, J. E., Flinders, C., McMahon, D. J., Branigen, T., and Bagdell, J. M. (1989a). Medical gases. *In* ''Technical Manual of Anesthesiology: An Introduction,'' pp. 130–155. Raven Press, New York.

Heavner, J. E., Flinders, C., McMahon, D. J., Branigen, T., and Bagdell, J. M. (1989b). Circulatory monitoring. *In* ''Technical Manual of Anesthesiology: An Introduction,'' pp. 68–101. Raven Press, New York.

Hendricks, J. C., and King, L. G. (1993). Practicality, usefulness, and limits of pulse oximetry in critical small animal patients. *Vet. Emerg. Crit. Care* **3**(1), 5–12.

Holzman, R. S. (1992). Intravenous infusion equipment. *In* Anesthesia Clinics: ''Anesthesia Equipment for Infants and Children.'' International (J. Pullerts and R. S. Holzman, eds.), Vol. 30, No. 3, pp. 35–50. Little, Brown, Boston.

Hubble, J. A. E. (1993). The anesthetic machine. *In* ''Anesthetic Equipment'' (R. M. Bednarski, ed.), Semin. Vet. Med. Surg. (Small Anim.), Vol. 8, No. 2, pp. 68–69. Saunders, Philadelphia.

Huss, B. T., Anderson, M. A., Branson, K. R., Wagner-Mann, C. C., and Mann, F. A. (1995). Evaluation of pulse oximeter probes and probe placement in healthy dogs. *J. Am. Anim. Hosp. Assoc.* 31(1), 9–14.

Jantzen, J.-P. A. H., and Hennes, H. J. (1991). Pigtail oximetry (letter). *Anesthesiology* **75,** 707–708.

Klide, A. M. (1992). Vaporizer position. *Vet. Clin. North Am.: Small Anim. Pract.* **22**(2), 389–391.

LeBlanc, P. H., and Sawyer, D. C. (1993). Electronic monitoring equipment. *In* ''Anesthetic Equipment'' (R. M. Bednarski, ed.), Semin. Vet. Med. Surg. (Small Anim.), Vol. 8, No. 2, pp. 119–126. Saunders, Philadelphia.

Levy, D. E., Zweis, A., and Duffy, T. E. (1980). A mask for delivery of inhalation gases to small laboratory animals. *Lab. Anim. Sci.* **30**(5), 868–870.

Ludders, J. W. (1993). Vaporizers used in veterinary anesthesia. *In* ''Anesthesia Equipment'' (R. M. Bednarski, ed.), Semin. Vet. Med. Surg. (Small Anim.), Vol. 8, No. 2, pp. 72–81. Saunders, Philadelphia.

Manley, S. V., and McDonell, W. N. (1979). Clinical evaluation of the Bain breathing circuit in small animal anesthesia. *J. Am. Anim. Hosp. Assoc.* **15,** 67–72.

Mason, D. E. (1993). Anesthesia machine checkout and troubleshooting. *In* ''Anesthesia Equipment'' (R. M. Bednarski, ed.), Semin. Vet. Med. Surg. (Small Anim.), Vol. 8, No. 2, pp. 104–108. Saunders, Philadelphia.

McMurphy, R. M., Hodgson, D. S., and Cribb, P. H. (1995). Modification of a nonrebreathing circuit adaptor to prevent barotrauma in anesthetized patients. *Vet. Surg.* **25,** 352–355.

Norris, M. L. (1981). Portable anaesthetic apparatus designed to produce and maintain anaesthesia by methoxyflurane inhalation in the Mongolian gerbil (*Meriones unguiculatus*). *Lab. Anim.* **15,** 153–155.

Perrino, A. C., Feldman, J. M., and Barash, P. G. (1993). Noninvasive cardiovascular monitoring. *In* ''Monitoring in Anesthesia'' (L. J. Saidman and N. T. Smith, eds.), pp. 101–143. Butterworth-Heinemann, Boston.

Pyles, S. T., Berman, L. S., and Modell, J. H. (1984). Expiratory valve dysfunction in a semi-closed circle anesthesia circuit—Verification by analysis of carbon dioxide waveform. *Anesth. Analg.* **63**(5), 536–537.

Raffe, M. R., and Martin, F. B. (1983). Effect of inspired air heat and humidification on anesthetic-induced hypothermia in dogs. *Am. J. Vet. Res.* **44**(3), 455–458.

Rich, G. F., Sullivan, M. P., and Adams, J. M. (1990). Is distal sampling of end-tidal CO_2 necessary in small subjects? *Anesthesiology* **73,** 265–268.

Sawyer, D. C. (1982a). The anesthetic period: maintenance. *In* ''The Practice of Small Animal Anesthesia'' (D. Piermattei, ed.), Major Probl. Vet. Med., Vol. 1, pp. 121–184. Saunders, Philadelphia.

Sawyer, D. C. (1982b). The induction period. *In* ''The Practice of Small Animal Anesthesia'' (D. Piermattei, ed.), Major Probl. Vet. Med., Vol. 1, pp. 13–43. Saunders, Philadelphia.

Sawyer, D. C., Brown, M., Striler, E. L., Durham, R. A., Langham, M. A., and Rech, R. H. (1991). Comparison of direct and indirect blood pressure measurement in anesthetized dogs. *Lab. Anim. Sci.* **41**(2), 134–138.

Schreiber, P. J. (1972). Ventilators. *In* ''Anaesthesia Equipment, Performance, Classification and Safety,'' pp. 138–182. Springer-Verlag, Berlin.

Sessler, D. A. (1993). Monitoring body temperature. *In* ''Monitoring in Anesthesia'' (L. J. Saidman and N. T. Smith, eds.), pp. 315–337. Butterworth-Heinemann, Boston.

Sidwell, R. W., Huffman, J. H., Gilbert, J., Moscon, B., Pedersen, G., Burger, R., and Warren, R. P. (1992). Utilization of pulse oximetry for the study of the inhibitory effects of antiviral agents on influenza virus in mice. *Antimicrob. Agents Chemother.* **36**(2), 473–476.

Smith, J. A. (1993). Waste anesthetic gas scavenging. *In* ''Anesthesia Equipment'' (R. M. Bednarski, ed.), Semin. Vet. Med. Surg. (Small Anim.), Vol. 8, No. 2, pp. 90–103. Saunders, Philadelphia.

Soma, L. R., and Klide, A. M. (1968). Techniques and equipment for inhalation anesthesia in small animals. *J. Am. Vet. Med. Assoc.* **152**(7), 957–972.

Steffey, E. P., Woliner, M. J., and Howland, D. (1983). Accuracy of isoflurane delivery by halothane-specific vaporizers. *Am. J. Vet. Res.* **144**(6), 1072–1078.

Swedlow, D. B. (1993). Respiratory gas monitoring. *In* ''Monitoring in Anesthesia'' (L. J. Saidman and N. T. Smith, eds.), pp. 27–50. Butterworth-Heinemann, Boston.

Talke, P., Nichols, R. J., and Traber, D. L. (1990). Does measurement of systolic blood pressure with a pulse oximeter correlate with conventional methods? *J. Clin. Monit.* **6**(1), 5–9.

Thys, D. M., and Reich, D. L. (1993). Invasive cardiovascular monitoring. *In* ''Monitoring in Anesthesia'' (L. J. Saidman and N. T. Smith, eds.), pp. 51–101. Butterworth-Heinemann, Boston.

Tran, D. Q., and Lawson, D. (1986). Endotracheal intubation and manual ventilation of the rat. *Lab. Anim. Sci.* **36**(5), 540–541.

Tremper, K. K., and Barker, S. J. (1989). Pulse oximetry. *Anesthesiology* **70**(1), 98–108.

Tremper, K. K., and Barker, S. J. (1993). Oxygenation and blood gases. *In* ''Monitoring in Anesthesia'' (L. J. Saidman and N. T. Smith, eds.), pp. 1–25. Butterworth-Heinemann, Boston.

Turner, M. A., Thomas, P., and Sheridan, D. J. (1992). An improved method for direct laryngeal intubation in the guinea pig. *Lab. Anim.* **26,** 25–28.

Vender, J. R., Hand, C. M., Sedor, D., Tabor, S. L., and Black, P. (1995). Oxygen saturation monitoring in experimental surgery: A comparison of pulse oximetry and arterial blood gas measurement. *Lab. Anim. Sci.* **45**(2), 211–215.

Ventrone, R., Baan, E., and Coggins, C. R. E. (1982). Novel inhalation device for the simultaneous anesthesia of several laboratory rodents. *Lab. Anim.* **16,** 231–233.

Watney, G. C., Norman, W. M., Schumacher, J. P., and Beck, E. (1993). Accuracy of a reflectance pulse oximeter in anesthetized horses. *Am. J. Vet. Res.* **54**(4), 497–501.

Chapter 8

Design of Surgical Suites and Postsurgical Care Units

William J. White and Joanne R. Blum

I. INTRODUCTION

Many terms are used to refer to the areas in which surgery on animals is conducted and supported. For the purposes of this discussion, the following terminology will be used.

Surgical facility or complex. All of the areas used to support and conduct surgery including, but not limited to, the surgical suite, postoperative care area, and associated support areas such as radiographic facilities, pharmacy, diagnostic laboratories, and locker rooms.

Surgical suite. Those areas most directly associated with the conduct of all surgical procedures, namely, the instrument/material preparation area, animal preparation area, surgeon preparation area, and the operating suite.

Operating suite. One or more operating rooms within a single area in which surgical procedures are performed.

Operating room. A single room in which surgery is conducted.

A well-designed and properly managed surgical facility is an integral part of most modern biomedical research facilities. The

ANESTHESIA AND ANALGESIA IN LABORATORY ANIMALS

ISBN 0-12-417570-8

environment created by this facility is important in fostering proper surgical techniques and appropriate care of animals undergoing surgery as part of research protocols or for routine veterinary care procedures.

Development of a surgical facility requires advanced planning and attention to detail during design and construction. Few published references or guidelines exist to assist in the architectural design and construction of surgical suites or postsurgical care units within biomedical research facilities [American Animal Hospital Association (AAHA), 1991–1992; American Institute of Architects Committee on Architecture for Health, 1987; American Society of Heating, Refrigerating and Air Conditioning Engineers (ASHRAE), 1977, 1987; Institute of Laboratory Animal Resources (ILAR), 1978; McSheehy, 1976; Ruys, 1991]. Use of guidelines established for animal housing areas, experience, and review of existing human and animal surgical suites and postsurgical areas are frequently utilized to assist in design of these facilities.

Appropriate facilities for surgery and postsurgical care must be provided in institutions conducting research programs involving surgical procedures. There are a number of regulatory, accreditation, and funding agencies that address the need for these facilities in their policies, guidelines, and/or regulations [Association for Assessment and Accreditation of Laboratory Animal Care (AAALAC) Activities Report, 1990; Code of Federal Regulations (CFR), 1992; ILAR, 1996]. Sound professional judgment, guided by the general guidelines promulgated by these agencies, should be used to define the overall surgical program and environment at each institution.

The Animal Welfare Act is a federal law administered by the United States Department of Agriculture (USDA) that requires aseptic technique, dedicated facilities, and provision of postsurgical care for all nonrodent species undergoing major surgical procedures. The National Research Council's Institute of Laboratory Animal Resources (ILAR) ''Guide for the Care and Use of Laboratory Animals'' (the ''Guide,'' ILAR, 1996) provides general guidelines on animal care and use programs including recommendations for physical plant construction. The ''Guide'' recommends that major survival surgery on nonrodent mammalian species should be done in facilities that provide functional areas for surgical support including animal preparation, surgeon preparation, operating room(s), and postsurgical care. The use of aseptic technique and sterile instruments as well as certain items of surgical attire may also be necessary for survival surgery on rodents. Institutions receiving NIH funding or seeking accreditation from the Association for Assessment and Accreditation of Laboratory Animal Care (AAALAC) should comply with these guidelines.

It is essential that an individual familiar with applicable rules, regulations, and policies be involved in planning the construction of the surgical facility to avoid compliance problems. Use of architectural and construction firms with previous experience in building and/or renovation of animal surgical facilities and associated support areas in biomedical research institutions can facilitate the planning and design process but does not assure success. Planning, design, and construction are dynamic processes requiring ongoing dialogue with those charged with developing final architectural and mechanical plans for the project, as well as those involved in construction.

Architects and engineers are often experienced in designing facilities for humans and apply similar specifications to animal facilities. Some features of human surgical suites, such as conductive flooring, may not be necessary in a research animal surgical facility. Moreover, energy loads produced by personnel and patients, as well as equipment, may be quite different. Considerable savings can be realized if those designing the facilities are appraised of these differences.

Adherence to relevant federal, state, and local building codes should be the primary responsibility of the architects and engineers associated with the project, with oversight from the responsible individual or group at the institution who will manage these areas once they become operational. Building codes address fire, plumbing, electrical, seismic, energy and energy conservation, handicapped access, health and safety, and/or environmental safety requirements, as well as mandate certain design and construction features.

Once general design criteria have been established and well before the facility design has been completed, it is often beneficial to visit surgical facilities at institutions with similar research programs to compare them with the proposed design and to assess the operational effectiveness of various design concepts. It is important to remember, however, that facility design and program concepts that are effective at one institution may not be applicable or easily transferable to another institution.

II. DESIGN CONSIDERATIONS

While some regulations and a few general standards have been established by various professional organizations and governmental agencies, each institution must establish its own design criteria for the animal surgical facility based on individualized needs and resources. It is important that institutional policies governing function and management of surgical facilities be clearly defined so that the subsequent design, planning, and construction of facilities support the original goals of the surgical program.

Design criteria for surgical facilities are based on the current or anticipated operational procedures used to manage the surgical facility as well as the nature and number of surgical procedures to be conducted within the facility. In essence, by carefully delineating the operational or management program, the institution delineates its policies for operation of the surgical facility. This in turn allows specific architectural and mechanical features to be incorporated in the facility that will enable it to function according to these operational policies.

It is extremely important that all aspects of the operation of the facility be considered early in the planning process and decisions made as to what standard operating procedures will be employed, what functions or practices will be allowed, and how the facility will be operated and maintained. Clearly, issues such as the use of the facilities for teaching, the range of sizes of animals to be used in the facility, who will manage the facility, the use of the facility for nonsurvival procedures, use of hazardous agents in the facility, and other similar management and procedural issues must all be considered at this stage of the design process. The input of investigators who will use the facility is also very important at this stage since it is their needs that must be met by the facility. If an operational plan is not clearly established or if the institution is not capable or willing to operate the surgical facility according to established procedures, the design criteria may result in the development of an unnecessarily complex or unworkable surgical facility that is not cost effective to operate and may not meet national guidelines or regulations.

A. Location of Surgical Facility

Depending on architectural and mechanical constraints, location of a surgical facility should be based on its functional relationship to other components of the animal resource facility, traffic flow patterns, and the characteristics of its use.

In order for surgical facilities to be operated effectively, easy access and a logical flow of people, animals, and materials between the surgical facility, research laboratories, and animal husbandry areas is essential. Prior to surgery and following recovery from surgery, movement of animals between existing housing rooms and the surgical facility will be necessary. Proximity to diagnostic laboratories or special analytical equipment might be important if samples critical to health care management or investigative needs are to be collected from animals during surgery or postsurgical recovery. Similarly, support areas such as research laboratories, material and equipment storage areas, and personnel areas including locker rooms and professional offices should bear a logical relationship to the surgical suite. Failure to take these factors into account can facilitate inappropriate use of the surgical facility and failure to adhere to required operational policies.

1. Decentralized versus Centralized Facilities

Location of the surgical complex will also depend on a clear delineation of responsibility for operation of the facility. To be effectively controlled and to ensure adherence to mandated federal and institutional procedures, the surgical suite and the postsurgical recovery room should be positioned such that they provide easy access for those designated to manage and provide professional oversight to these areas. If facilities are placed in remote locations or in areas managed by other divisions of the institution, the level of management and veterinary medical oversight will often diminish as a function of the distance from the remainder of the animal facility and offices of the veterinary staff.

The concept of centralized versus decentralized facilities must be addressed with regard to surgical and postsurgical facility location. Dedicated decentralized facilities are more resource intense, requiring separate space as well as duplication of equipment and personnel in order to assure proper operation. They do provide scheduling convenience to those individuals or departments that operate them. Decentralized surgical facilities might be necessary due to the need for proximity to laboratories or specialized research equipment that are an integral part of certain research projects but cannot be accommodated within the confines of a centralized surgical facility. Decentralized facilities are only successful if the institution can afford to support separate facilities while still assuring compliance with standards of sanitation, adherence to approved research protocols, and programs of perioperative veterinary medical care.

A gain in efficiency can be achieved in centralized facilities through careful scheduling and cost savings realized by sharing of equipment and the use of experienced personnel whose primary responsibility is limited to the surgical and/or postsurgical care of animals. Centralized postsurgical care units provide significant advantages to individual researchers with regard to cost savings for monitoring equipment, conservation of space, and ease of veterinary oversight to assure adherence to requirements for the care of postoperative animal patients in a centralized location. The additional security provided by locating the surgical facility within centralized animal facilities must also be considered.

2. Traffic Flow Patterns

The location of the surgical suite in relation to adjacent functional areas should be carefully studied. Factors such as noise, security, sanitation, and convenience dictate that components of the surgical suite should ideally be located adjacent to each other in a dedicated, low traffic hallway that can be closed to all but authorized personnel.

Noise is known to affect various physiological parameters of animals (Peterson, 1980), interfere with animals under certain anesthetics (such as combinations of fentanyl and droperidol) where the animals remain sensitive to audiogenic stimuli (Booth, 1977; Wixson *et al.,* 1987), and can also prove disruptive to the surgical team. Therefore, the location of components of the surgical complex adjacent to a cage-washing facility, loading dock, heavily trafficked corridor, or other noise-generating activities can be unwise unless noise abatement considerations are addressed. Similarly, disruption of air movement, possible aerosolization of contaminated materials, and the potential for entry of vermin are concerns that also discourage locating the surgical suite close to areas that might compromise the maintenance of a controlled, sanitary environment.

Surgical facilities must be designed to maintain a constant environment by controlling the flow patterns of materials and personnel through the facility as well as maintaining stable and clean physical conditions.

It is also important to control the physical environment of the surgical facility through adequate illumination and appropriate heating, ventilation, and air conditioning (HVAC) systems. Heat loss from anesthetized animals may influence the metabolism of anesthetic agents as well as other compounds administered during surgery or recovery (Lumb and Jones, 1984). Low relative humidity encourages fluid loss from exposed tissue surfaces, and poorly filtered, improperly diffused air could contribute to unacceptably high concentrations of bacteria in the environment surrounding the patient.

The issue of security in the prevention of equipment theft as well as protection of personnel and animals from unauthorized intrusions should also be considered. Although it may be possible to lock entry doors to an operating suite or postsurgical care unit when these areas are not in use, it is inconvenient and frequently undesirable to prevent ready access to these areas when surgical procedures are being performed or during the recovery of animals. All doors to critical areas of the surgical facility should be capable of being locked, yet provided with appropriate hardware that will allow rapid egress in case of emergencies. For example, a card key entry system is one means of providing controlled access to the entire surgical complex or selected rooms within the complex.

Another consideration in the traffic flow pattern of a surgical facility is the movement of dirty cages and equipment from the surgery room, postoperative care room, and other areas of the surgical complex to cage wash areas. Similarly, the transfer of sanitized cages from the cage wash area to the different components of the surgical facility should be patterned so as to minimize any disruption caused by noise or other factors associated with the movement of equipment and cages.

A number of mechanical considerations can also affect the location of surgical facilities. These facilities are mechanically intensive and often require sophisticated air filtration as well as HVAC systems and specialized electrical and plumbing services. During the planning process it is desirable to group areas having intensive mechanical requirements such as the surgical complex with other animal facility areas that have similar requirements. By doing so, the cost of extending services over great distances is reduced and economy in terms of system maintenance can also be obtained.

In those buildings requiring the use of elevators to access the surgical facility, location of the facility should bear a logical and convenient relation to the elevators without compromising security, sanitation standards, or other traffic flow considerations.

Design of the surgical facility must be done within the context of whether the facilities will be part of new construction or whether an existing area will be remodeled to accommodate these facilities. If an area is to be remodeled in order to construct a surgical facility, significant limitations may be placed on the flow patterns of people and equipment as well as on the size and shape of facilities that can be constructed. Often, the cost for demolition, relocating support columns, or replacement of existing surfaces may significantly reduce available funds for the project and impact overall planning and design. Despite these considerations, renovation may be the only option. For this reason, it is particularly important that personnel involved in the daily operation of the surgical facilities carefully review proposed plans in light of the activities to be conducted. This critical review will determine if compromises required to locate the surgical suite within existing structures will jeopardize efficiency and effectiveness of the operation of the surgical facility.

In contrast, new construction allows a much freer hand in planning and design but may encourage the tendency to overbuild. This tendency is usually due to poorly defined limits in the planning stage on the size and complexity of the facility. If a reasonable construction plan is to be developed it is important to carefully assess the real needs of the surgical program.

B. Dedicated versus Multiuse Facilities

A dedicated surgical facility is sometimes necessary in order to support specialized equipment associated with surgical and recovery procedures for specific studies. The use of anesthetic machines and gas scavenging units, extracorporeal bypass equipment, radiographic equipment, autoclaves, and other types of equipment with special mechanical, electrical, or shielding requirements is best accommodated in facilities that allow adequate space in a controlled environment for proper and safe operation of this equipment. Use of such equipment in less controlled environments can jeopardize not only the animals on which the equipment is being used, but also personnel in the vicinity of the equipment or those operating the equipment. A controlled and dedicated environment for such surgery and recovery provides the most efficient use of space, personnel, and equipment for conducting major survival surgical procedures.

Operational needs of a program might allow components of the surgical facility to be used for multiple purposes, such as use of the animal preparation area as a procedure room for minor experimental manipulations. A distinct separation of the functions, with appropriate sanitation between procedures, needs to be undertaken to assure that the compromise of surgical patients does not occur in these instances.

In some situations, anesthetized animals might have to be transported to a human patient area that might have been designed to also accommodate animals. Building dedicated animal facilities containing specialized equipment such as surgical lasers, magnetic resonance imaging devices, or radiotherapy equipment is not always economically feasible, and surgical facilities for animals might not be located in close proximity to this equipment. In such cases, concerns regarding the use of animals in human treatment areas, as well as the movement of animals through general public access areas, will need to be addressed. If it is anticipated that such situations will arise frequently, routes of travel from existing animal surgical facilities and holding areas to the specialized equipment should be considered when planning the location of the surgical facility.

C. Species of Animals

When designing a surgical facility, it is necessary to determine what species of animals will be used in the facility. One must consider whether the procedures will be conducted solely on rodents, small animals (e.g., rabbits, cats, dogs), large animals (e.g., pigs, sheep, horses), or a mixture of species.

If only rodent surgery is planned, the facility can be much smaller in size and less complex in nature. It is considered acceptable under current guidelines to conduct survival surgery on rodents in a laboratory space that is not solely dedicated to surgical procedures but is appropriately managed during surgeries to prevent contamination from any other activities occurring in that room. If a dedicated rodent surgical facility is planned, it will often have provisions for operating microscopes and will be oriented toward the surgeon being seated with the surgery being conducted around the perimeter of the room rather than in the center. Rodent surgical suites may not be supplied with as complex an assortment of medical gases, illumination sources, or other equipment as those surgical facilities that service nonrodent species. Postanesthetic recovery for rodents is often conducted in the same room as the surgical procedure. Availability of appropriate support equipment, such as external heat sources, and monitoring by trained personnel are still necessary to assure recovery for these animals.

Operating rooms for nonrodent species usually are larger than those for rodents in order to accommodate centrally located operating tables and complex support equipment along the periphery of the room. The procedures conducted on these animals are often more diverse than in rodents due to their larger body size and ease of scale, especially for procedures and devices intended for eventual human application. The need for specialized support equipment for the postsurgical recovery period becomes more complex as the size of the animal and/or complexity of the surgical procedures increases. Incubators or specialized enclosed chambers that can deliver an oxygen-enriched environment in climatically controlled conditions are frequently utilized to support small animals during recovery from surgery.

Large animal surgical facilities are designed to accommodate domestic farm animals weighing several hundred pounds or more. Commonly, procedures conducted on domestic farm animals require the use of specialized restraint devices, hydraulic lifts and tables, and a considerable amount of monitoring and support equipment to meet the specialized needs of animals of this size. Postsurgical facilities for large animals range from holding pens or stalls to specially padded recovery rooms constructed to prevent injury to animals as they emerge through the excitatory stages of anesthetic recovery.

For some small and most large animal surgical procedures, a team approach may be used with 2 to more than 12 people being engaged in the surgical procedure directly or in support roles within the operating suite. For this reason, the operating room will often need to be large enough to accommodate a surgical team as well as ancillary surgical and experimental equipment.

D. Utilization Objectives

It is essential to characterize the anticipated total caseload, as well as the frequency and duration of use for the surgical suite and support areas, in order to estimate appropriate space and personnel needs. Both the size and the complexity of any surgical suite or support area will be dictated by the nature and number of procedures to be conducted. Although it is risky to plan around current research programs since they can change, the scope of the institution's present research mission can often serve as a guide in gauging the level and nature of future research activity. If the procedures will be relatively minor, the complexity of the suite and its size might be limited, allowing some construction and operational cost savings. However, complex surgical procedures on both small and large animals will probably require more extensive support functions, thereby increasing construction costs.

If the surgical facility is large enough in size, multiple cases can be conducted simultaneously within the same facility provided that adequate consideration is given to the nature of the procedures. In situations where multiple surgical procedures and recovery of cases are occurring simultaneously, adequate space must be maintained such that cross contamination and/or interference with simultaneously occurring procedures and functions does not occur. Sometimes, multiple surgical procedures can be accommodated in a more orderly fashion by including several small operating rooms as a component of a larger surgical operating suite.

Inevitably the issue of conducting survival versus nonsurvival procedures within the same surgical suite becomes a consideration during the planning process. In many cases, it is not necessary for an animal to survive a surgical procedure in order to obtain the required research data. Cost savings can be realized if the facility is intended for only nonsurvival procedures, thereby obviating any need for a postsurgical care unit; however, this situation seldom occurs. It is also not always necessary to build separate surgical facilities for nonsurvival procedures as long as contamination control is exercised during the conduct of the procedure and adequate sanitation practices are utilized prior to the initiation of any survival surgery.

E. Special Considerations

In considering the development of plans for both surgical suites and postsurgical care areas, it is important to understand the nature of the environment that can be created within these areas. It is a common misconception that all surfaces and materials within the surgery suite are sterile and that personnel, patients, and equipment are somehow sterilized prior to entry into the room. In reality, only those surfaces that will come in direct contact with the surgical site will either be sterilized or covered with a sterilized coating. Other surfaces, including the skin of the patient, operating room equipment, and room surfaces, should be clean and in some cases disinfected, but will certainly not be sterile. During the course of an operative procedure, the patient can release contaminants (e.g., gastrointestinal contents, urine) onto the surgical field or into the operating room. Breaks can occur in aseptic techniques or in the sterile barrier surrounding the surgeon (e.g., holes in surgical gloves, fluids soaking through sterilized gowns). These breaks in sterility can only be overcome by good management procedures and cannot be eliminated solely by physical facilities.

It is important to design a surgical suite that can be easily cleaned and disinfected. The amount of airborne contamination can be minimized by filtering incoming air and carefully considering the location of air supply and exhaust vents as well as the room air exchange rate and air distribution patterns (Hughes and Reynolds, 1995). An effort should be made to separate either by distance or by physical barriers (i.e., walls) the operating suite/room from areas housing animals, areas used to prepare the hands and arms of the surgeon, and areas used for instrument preparation. All of these separations are relative, and their magnitude or presence does not negate the possibility of successfully conducting aseptic surgical procedures in areas in which such separations do not exist. Such separations do, however, facilitate the maintenance of asepsis.

In designing an operating suite, it must be remembered that absolute sterility will never be achieved and that those measures taken to maximize the ability to clean and disinfect an area must be balanced with the flow of materials and people through the area and the ease with which the area can be used. Designing cumbersome, complex facilities only encourages personnel to circumvent intended flow patterns and procedures.

Health and safety concerns regarding the use of certain inhalation anesthetic agents surfaced in the early 1970s when a number of studies indicated that exposure to certain anesthetic agents can be hazardous (Chang and Katz, 1976; Corbett, 1973; Manley and McDonell, 1980a; Stoyka and Havashi, 1977). Proper scavenging of waste anesthetic gases in those components of the surgical facility where these agents may be used continues to be a concern that should be addressed during the design phase of the facility [Manley and McDonell, 1980b; National Institute of Occupational Safety and Health (NIOSH), 1975, 1977].

III. PLANNING OF SURGICAL FACILITY

Many considerations go into determining the appropriate number and size of rooms that comprise a surgical complex. Careful assessment of the past, present, and projected surgical caseload in the institution should be evaluated. While past and present activity can be easily determined from records, future activity is often hard to assess. Approximations of future activity can be made from discussions with program or department heads, examination of pending grant or project proposals, and discussions with those investigators currently using surgical procedures in their research. Clearly, it is necessary to be realistic when making growth projections.

It is always good practice to develop plans for an orderly expansion of the facilities should a greatly increased level of activity develop. Such planning for expansion should not be restricted to how facilities will be oriented in the building. It should also provide for expansion of all components of the facility that are required to maintain the surgical program in proportion to possible increases in the future activity of the program. Hence, it is inappropriate to plan for a future expansion of twice the size of the proposed surgical facility if only the space for operating rooms is doubled with no concomitant increase of preoperative, postoperative, or instrument support space. In planning for orderly expansion, it is possible to design for flexibility and thereby construct surgical facilities in a multiphase process that allows the initial construction of a more modest facility that meets existing needs.

As discussed previously, evaluation of the complexity of the surgical procedures to be conducted and the amount of support equipment needed must be part of the planning phase. Similarly, the maximum size of the species to be used in the facility must be determined. A facility should not be overdesigned or overequipped for remote research possibilities.

Determining the number of investigators who will likely require the use of the various components of the surgical facility is essential to the planning process. Separate from the volume of surgery being done, one must address the number of different projects to be conducted daily within the space. If a relatively small number of projects generate almost all of the surgical volume, it is likely that the space for support equipment and the number of procedures being performed at any one time will be relatively small compared to a facility where there are many users conducting dissimilar surgical procedures. Under these circumstances, the amount of support equipment maintained in the surgical suite, as well as the need to accommodate conflicting schedules, will require a somewhat larger facility than might be anticipated from surgical volume alone. In some cases, these problems can be partially overcome by subdividing a large operating room into a series of smaller operating rooms, thereby allowing support equipment and monitoring devices to remain in place between cases on succeeding days and allowing dissimilar procedures to be conducted simultaneously. The disadvantage of this method is that it greatly reduces the flexibility of the operating suite by subdividing the space into smaller areas.

A. Components of Surgical Facility and Their Functions

1. Material Preparation Area

The material preparation area is used to clean, prepare, and sterilize instruments and equipment for operative procedures. Often, the area will contain cabinets, shelving, and countertops used to store instruments and supplies. A large sink with a spray arm will facilitate the cleaning of instruments and the preparation of disinfectant solutions. If the volume of surgery warrants, ultrasonic instrument cleaners that efficiently and effectively clean large numbers of instruments are often used in this area.

Such units usually require a separate hot water source and a standing drain pipe.

Commonly, the material preparation area is equipped with one or more autoclaves used to sterilize instruments and supplies. At least one autoclave should be capable of short sterilization cycles. The number and size of autoclaves will depend on the type and volume of surgery to be conducted. Due to the significant cost of large autoclaves as well as their size, plumbing, and electrical requirements, some materials used in the surgical suites (especially dry goods and liquids) can be sterilized in larger, multiuse autoclaves located elsewhere in the institution. It is important to consider the utilities required for the operation of autoclaves during the planning process as the use of in-house steam can greatly reduce the purchase and operational costs of this equipment. In addition, ethylene oxide sterilization presents special requirements which must be addressed. The installation of ethylene oxide equipment in animal surgical facilities is rarely cost effective.

To reduce possible cross contamination, the material preparation area should have one entrance from a common access corridor for purposes of supply and a separate access/entrance to the operating suite for the delivery of sterilized materials. Both entrances/accesses should be equipped with doors or pass through windows that separate the material preparation room from other functional areas.

Because the material preparation area is often quite small and might contain many pieces of equipment, it is important to be cognizant of door swings when planning this area in order to avoid injury to personnel and/or damage to equipment. It is also important to define fully the types of equipment that will be used in this area so that provision can be made for the appropriate number and types of electrical outlets and plumbing requirements.

2. Animal Preparation Area

Surgical facilities should have a dedicated area for the preparation of animals for surgery. This area ideally should be close to animal holding areas and in most cases should be immediately adjacent to the operating suite. The preparation area is used for administering preanesthetics and anesthetics as well as any necessary clipping of hair and cleaning of the operative site. It should be equipped with a table and sufficient lighting to allow complete visualization of the patient. Auxiliary lighting in the form of fixed or mobile surgical lamps is also commonly used in the animal preparation area to provide sufficient illumination to perform procedures such as intravenous catheter placement. A large sink with spray arm for the preparation of solutions used to scrub those areas of the patient directly involved in the operation is also recommended. Provision should be made for electric clippers and can include a cord reel or some other wall or ceiling-mounted device to house the clippers or clipper motor. Portable or central vacuum cleaner systems can be used for the collection of animal hair that is clipped from the surgical sites. Scales for obtaining accurate body weights should be located in this room or some nearby location. Depending on the size of the animal, a variety of options are available, including a scale built into the examination table used for preparing the animal. Sufficient space should be provided to allow equipment such as inhalation anesthetic machines, intravenous fluid stands, and monitoring equipment to be located within the room. If inhalation anesthetics are to be used, a system for the scavenging of waste anesthetic gases must be provided as should appropriate utilities such as oxygen and vacuum. Commonly, mobile or wall-mounted cabinets are used to house supplies and small pieces of equipment associated with the preparation of the animal for surgery. Many anesthetics and other preoperative medications are controlled substances and must be inventoried and kept in locked containers in accordance with the standards of the Drug Enforcement Agency (CFR, 1991). A suitable wall-mounted, double-locked cabinet or cabinet-mounted safe can be used for this purpose.

3. Surgical Personnel Preparation Area

There are three principal components in the process of preparation of personnel to conduct surgery. The first is an initial clothing change, usually done in a locker room, from street clothes to dedicated surgical attire. The surgical attire or scrub suits should be clean but are not sterilized. A shower facility is generally not required as a shower is not mandatory before changing into surgical attire. A multipurpose locker room will often suffice for this clothing change and temporary storage of personal items and can save valuable floor space in the design of a surgical facility. Ideally, the locker room should be located within a reasonable distance from the surgical suite, but does not have to be a dedicated component of the surgical facility.

Having donned surgical attire, the surgeon proceeds directly to a surgeon preparation area within the surgical suite. This surgeon preparation or scrub area, similar to the animal preparation area, should ideally be physically separate from the operating room. Provision should be made in the surgeon preparation area for mobile or fixed shelving to house containers of shoe covers, caps, masks, and prepackaged sterilized scrub brushes. Foot-, knee-, or electric eye-activated scrub sinks are the major equipment located here. Scrub sinks should be mounted at a sufficient height to prevent the splashing of surgical attire yet low enough to prevent contact of the surgeon's arms and hands with the sink basin. Typically, a height of 40 inches from the floor as measured at the top of the basin is adequate but is dependent on sink design. Once the surgeons have completed a preoperative scrub of their hands and arms, they cannot touch door handles or other surfaces prior to donning sterile gowns and gloves. Hence, it is important to consider door handle hardware and the direction of door swings from the surgeon preparation area into the operating room.

The final step in the preparation process of the surgeon is the donning of a sterilized gown and gloves. This activity is usually performed in the operating room and therefore does not require a separate area; however, adequate unobstructed space for this activity as well as space to hold gowning supplies should be available in the operating room.

4. Operating Room or Suite

The operating room is the heart of the surgical facility. It may be composed of a single room or multiple rooms. Ideally, it should contain little or no fixed equipment with all wall, ceiling, and floor surfaces being accessible for frequent cleaning and disinfection. Generally, the ceiling height should be at least 10 feet as many fixtures such as cord reels, surgical lighting, and medical gas outlets are ceiling mounted requiring considerable overhead space. However, rodent or small animal surgical facilities can be maintained using 8-foot ceilings if careful consideration is given to how services will be provided to operating tables located within the room.

Positioning of services within the operating room is an important design consideration. The location of all surgical tables or stations needs to be carefully planned so that wall- or ceiling-mounted receptacles for electricity, medical gases, and other services can be readily accessible to the surgical tables. All service outlets should be located such that the length of connecting cable or lines between the outlet and the device using the service is minimized and will not pose a hazard to operating room personnel. There are many commercially available, ceiling-mounted surgical columns that provide all services in a single unit that eliminates the need for any electrical or anesthetic machine lines or tubes trailing from the operating table across the floor to the walls. These units do limit the flexibility in locating surgical tables following construction.

By convention, electrical and medical gas outlets that are located on the walls are placed at least 60 inches off the floor surface to avoid the accumulation of dust and debris within the receptacles and to provide for the possibility of using flammable agents. Use of flammable and explosive agents in most surgical programs has been eliminated due to safety concerns and to the availability of nonflammable, nonexplosive alternatives.

Although conductive flooring is seldom used in the current construction of operating rooms, it is not uncommon to place ground fault circuits within the operating room and to provide separate grounding capabilities for specialized monitoring equipment. Ground fault panels are often mounted within the operating suite, and the location of the outlets served should be based on planned locations for the operating tables.

If track lighting is to be used to illuminate the surgical field, location of the tracks must be carefully planned. Track lighting should be situated so as not to interfere with regular room lighting fixtures, air diffusers, other ceiling-mounted equipment, or door swings. Often, more than one track will be required in large operating suites to accommodate all possible operating table locations. Track lighting usually requires extensive bracing in the ceiling due to the weight of the equipment and this should be considered in relation to the other service components that will be located within the ceiling space. If fixed surgical lighting is to be used, the placement of the lighting pedestals should be such that a maximum usable radius allowing easy positioning of the lighting head over a variety of surgical table locations is obtained. Obviously, the size of the room will influence the choice of fixed or track lighting.

A final consideration for the operating room is communication should additional personnel need to be summoned in the event of an emergency. An emergency notification system such as a simple audible alarm that alerts other personnel in the facility, an intercom system or telephone that allows voice messages to be transmitted and/or received, or even a management solution such as the provision of additional personnel in the operating room are all means of addressing this concern.

5. Postoperative Care Unit or Recovery Room

Any institution conducting survival surgical procedures on animals needs to provide for the postoperative care of these patients. Postoperative care units, also referred to as recovery rooms, provide the required physical environment to support the needs and monitor the vital functions of animals recovering from anesthesia and surgical procedures. Recovering animals typically stay in postoperative care units until they have recovered from the effects of the anesthetics, have reestablished stable physiological parameters, and have regained normal ambulatory and protective behaviors. The length of time an animal will need to remain in the recovery room will vary with the surgical manipulation performed as well as with the individual animal. Some programs may benefit from an intermediate care area where monitoring such as ECGs and blood gases can be done on recovered animals under less intense conditions.

Postoperative care units can be composed of a single room or of multiple rooms or cubicles located off of a central observation room. In programs with a low surgical volume, the animal preparation area can double as a postoperative recovery area providing there is no conflict between the two functions (e.g., preparation of one patient preventing adequate monitoring or care of a postoperative patient). Similarly, low surgical volume and few postoperative patients might allow an animal housing room to be converted temporarily into a postoperative care unit. However, caution should be exercised because many animal holding rooms might not be adequately equipped to address the monitoring and support of animals recovering from operative procedures. Recovering animals should not be placed in a housing situation where other animals could interfere with their recovery or well being.

Monitoring and support functions conducted in postoperative care units often have electrical and mechanical requirements similar to those of both the animal preparation and the operating room facilities. Minimally, oxygen should be available, either in the form of portable gas tanks or through a centralized system. Equipment such as electrocardiograph machines, circulating hot water blankets, intravenous infusion pumps, and other support equipment such as incubators or respirators require sufficient electrical outlets and possible provision for emergency power. Service outlets are typically located at least 60 inches from the floor or are provided by ceiling-mounted reels. The positioning of service outlets in the room is likewise an important design consideration that will be largely dictated by the location of caging, life support, and monitoring equipment as well as the number of patients using the facility. Whereas an operating room often is occupied by one patient, a postoperative care unit may provide services to multiple patients of different species recovering from a variety of procedures. Concern for contamination between multiple animals and/or species can be addressed by utilizing containment caging, building multiple containment cubicles or rooms to house each animal species (Ruys, 1988; White *et al.,* 1983), or using separate rooms for the recovery of different species.

Provision for computer terminals or access to similar types of research-related equipment may need to be considered in the electrical, mechanical, and other service requirements of a postoperative care area.

The lighting requirements of a postoperative care unit are similar to those of the animal preparation area—adequate illumination to visualize the patient and the provision of some form of auxiliary lighting for close examination of the surgical site or conducting manipulations such as catheter placement. Medical gas outlets provided must be compatible with those of the operating room and other areas to which the animals will be transported. Sinks are frequently found in or just outside of this area to allow for proper cleaning of hands prior to and following the examination of postoperative patients.

The placement of floor drains in the postoperative area is a common consideration. The inclusion of floor drains for postoperative care units servicing large animals might be advisable to facilitate cleaning. Depending on the caging type and patient load, floor drains are not necessary in most postsurgical care areas and may be undesirable from the standpoint of odor control and bacterial harborage. If design plans call for the placement of drains in the room, self-flushing drains should be considered.

Patient examinations are facilitated by use of an examination table. Some wall-mounted examination tables can be folded against the wall when not in use to conserve space.

The type and amount of caging required in the postoperative care unit are set both by the needs of the recovering patients and by the species of animals being brought to the facility. Portable caging provides flexibility as well as ease of sanitation and is preferred over fixed caging. Specialized intensive care caging units that provide temperature-controlled, oxygen-rich environments or aerosolized medication delivery might also be appropriate for certain programs. Standard caging used for normal housing of animals that is provided with supplemental heat sources and other support equipment can also be utilized. This alternative could provide cost savings and will allow sanitation to be easily accomplished in the animal facility's cagewash equipment. While private veterinary practices frequently use fixed caging for postoperative patients, it is difficult to adequately sanitize and reduces the flexibility of the rooms in which it is installed, making its use cumbersome in the biomedical research setting.

Storage of supplies in or near the postoperative care unit is necessary. Although there are many ways this task can be accomplished, one common solution for the storage of bandaging and other dry goods is the use of portable supply racks that are regularly sanitized. A small refrigerator for medications requiring refrigeration should be placed in the postoperative recovery room or an adjacent area. Postoperative patients frequently require analgesics, many of which are controlled substances. These medications must be kept within easy access of the postoperative care unit in a secured container in accordance with Drug Enforcement Agency standards (CFR, 1991).

6. Miscellaneous Support Facilities

A variety of specialized support areas might be located in or near the surgical suite. These facilities should be designed with the necessary lighting, HVAC system, and provision for medical gases to support surgical patients and minor surgical procedures. Radiology rooms or specially designed laboratories that house multiused equipment, such as magnetic resonance imagers, commonly incorporate these features.

Studies in human hospitals have documented the spread of contamination from simple housekeeping practices (Westwood *et al.,* 1971). It is, therefore, advisable to consider a dedicated janitorial supply space to house cleaning equipment for the surgical suite.

Unless disposable materials or a commercial laundry service is used, a facility for laundering surgical attire and associated dry goods must be included in the proposed resource.

Support areas such as a pharmacy, administrative support including offices or record-keeping areas, and research and diagnostic laboratory space should be readily accessible to the surgical suite yet distinctly separate from it. Planning for these support areas will depend on many of the same factors associated with planning for the construction of the surgical suite. Their existence, size, and location will all be dictated by the nature of the institution and its research programs.

Unassigned or equipment storage space should be located in the surgical facility or within a reasonable distance of the facility. Equipment that is only used for certain cases will need to be stored in a clean environment within easy access of the surgical suite. By providing such storage space, equipment will be available for use in the postsurgical care unit and operating room as well as other components of the surgical facility and will not be inappropriately stored in the operating room.

IV. CONSTRUCTION OF SURGICAL FACILITY

The final stage of planning requires characterization of the types of materials and finishes that will be used in construction of the surgical facility. Attention to details is required during this phase of the project to prevent errors that would make daily management of the facility difficult.

A. Walls

Where possible, walls should be of masonry construction. Such surfaces are resistant to physical damage, are capable of being adequately sealed, and are water resistant. One disadvantage of masonry walls is that the location of service outlets must be carefully planned in advance of construction in order to avoid future wall mounting of conduits and service boxes that may pose cleaning problems. A common alternative to masonry construction is the use of gypsum board or laminated wall boards. If properly treated with sealants that are durable and resistant to chemicals, these surfaces are economical and can provide a suitable alternative to masonry construction. The principal disadvantage of nonmasonry walls lies in the fact that once the outer surface has been penetrated by physical or chemical means, the inner surfaces are often highly water absorbent and rapid deterioration during frequent cleaning and disinfection can occur.

The walls of the facility should be free of cracks, unsealed utility penetrations, or imperfect junctions between doors, ceilings, floors, and corners. Seams or expansion joints should be sealed with high grade caulking materials. The surface finish applied to the walls should be capable of withstanding scrubbing with detergents and disinfectants. Those areas with great potential for damage from the impact of equipment should be protected using surface-mounted bumpers or by the use of bullnose curbing designed to keep equipment a safe distance from the walls.

Many wall surface finishes are available, with non-water-based epoxy sealants and paints being the most popular. Other types of paints and finishes can be used and many, such as some polyurethane and acrylic coatings, are capable of enduring the frequent cleaning that occurs in this type of facility. A choice of color for the surface is certainly a matter of individual taste. Lighter color wall finishes are more commonly used as they provide contrast to dirt or debris that may adhere to them. Lighter-colored finishes also reflect and diffuse light, providing more uniform illumination within the area. Pastel colors are usually chosen over pure monochromatic colors.

B. Ceilings

A ceiling height of 10 feet or greater is frequently recommended for the operating room in order to allow for overhead services. All other components of the surgical complex can usually be constructed with an 8-foot ceiling height. Suspended ceilings are undesirable unless they are constructed of impervious materials and are free from imperfect junctions. Most suspended ceiling materials do not meet these requirements as they often are designed for acoustical properties and do not have a smooth impervious surface. Ceilings constructed of fireproof plaster board that has been sealed and finished with a water-resistant coating are the most appropriate and are recommended. Ceilings formed by the concrete floor above or roof members are satisfactory in a surgical facility provided that they are smooth and sealed. In such instances, exposed conduits for overhead lighting or services cannot be avoided. Although this is not desirable, it is a manageable situation if these exposed services are kept to a minimum and if sanitation practices are effective in addressing this cleaning challenge. In all cases, ceiling materials should be capable of withstanding periodic scrubbing with aqueous solutions of disinfectants and detergents.

C. Floors

Floors within the surgical facility should be smooth, moisture-proof, nonabsorbent, skid-resistant, and resistant to the effects of detergents and disinfectants. Floors should be hard yet shatter resistant and able to resist pitting or cracking when heavy pieces of equipment are placed on them for long periods of time. Although seamless floors are preferred in a surgical suite, a few preplanned expansion joints might be appropriate, especially in large surgical suites. In the absence of explosive anesthetics, most modern operating rooms do not require conductive flooring. To be effective, conductive flooring uses quartz epoxy aggregate and requires adequate grounding, contact strips for all equipment, and the use of conductive shoe covers. This greatly increases construction and operating costs and is seldom necessary.

A variety of flooring materials have proved satisfactory for use in surgery, postsurgery, and other support areas. These include epoxy aggregates, rubber-based aggregates with special hardeners, primed surface-bonded aggregates applied to smooth surfaces, high durability sheet vinyls, and, in some special applications, sealed concrete. In all cases, adequate surface preparation is essential as is a proper moisture barrier. The use of door sills should be avoided wherever possible in order to minimize impediments to the passage of equipment and to facilitate cleaning. It is not uncommon to color code flooring as a reminder of the boundaries of the various functional components of the surgical suite.

With the possible exception of some postsurgical care unit applications, floors should be flat, with no slope and no floor drains. Depending on the species of animals in the postsurgical care units, it might be desirable to consider floor drains, with floors sloped to the drains, in order to facilitate cleaning practices. The location of any floor drains, as well as whether the drain will be self-flushing, and the degree of slope in the floor should be carefully considered in light of the types and amount of equipment and caging that will be located in this room.

D. Heating, Ventilation, and Air Conditioning

The HVAC requirements for the surgical suite and support areas are similar in most respects to animal holding areas. Heat loads will generally be lower than most animal holding rooms due to the low occupancy and the small number of heat-generating structures. Maintaining a temperature range of 65° to 82°F with a normal operating range of 72° ± 2°F should be within the capabilities of the system. It is highly desirable to have separate room temperature controls for the operating room and the postoperative care unit to accommodate the special needs of the occupants. Maintenance of relative humidity in the range of 30 to 70% is acceptable, although many systems are capable of a somewhat narrower range.

It is common practice to supply a large volume of air that has been appropriately filtered to the surgical suite for the purpose of diluting out contaminants. Usual air exchange rates vary from 10 to 20 complete changes per hour with rates of 12 to 15 fresh air exchanges per hour being the most commonly encountered. There is no good evidence to suggest that the energy expenditure associated with these higher air exchange rates is justified based on reduced levels of contamination (Besch, 1980). While never established, it has been assumed that high air exchange rates coupled with inappropriate diffusion might even encourage the distribution of dust, thereby enhancing the fomite transmission of bacteria.

Supply air to the surgical suite is usually highly filtered. In addition to prefilters, high efficiency particulate air (HEPA) filters may be used (Dyment, 1976). It should be noted, however, that no studies are available that link increased levels of postoperative bacterial complications to bacterial concentrations in the incoming air supply. Exhaust air is seldom filtered unless the recycling of air is contemplated. The use of recycled air is generally discouraged because of the difficulty in assuring the removal of escaped anesthetic agents or odor-causing compounds generated through sources such as the use of electronic equipment or electrocautery.

The type and placement of air diffusers depend on room configuration. Supply air diffusers are typically located in the ceiling but should not be located directly over operating tables or be obstructed by surgical track lighting fixtures. Generally, exhaust air outlets are placed low in the room along the side walls either opposite to the supply air diffusers if the diffusers are located along a wall or on the same side wall. Depending on the room and equipment configuration, trough and slot diffuser systems have also been used effectively.

In the past, laminar air flow and mass air displacement systems have been advocated for operating rooms and some postoperative care units (McGarrity and Coriell, 1973; McGarrity *et al.*, 1969; Sanderson and Bentley, 1976). These systems move large volumes of air through ceiling plenums in a downward vertical direction. They usually require substantial recycling of air if they are to be operated in a cost-efficient fashion. Laminar air flow systems suffer from the property that upon striking an object, the flow of air is no longer laminar and can result in eddying of up to three times the diameter of the object encountered (Whyte and Shaw, 1974). On a cost-benefit basis, the value of such systems in animal surgical facilities is questionable.

Although some studies indicate that a static air pressure differential between rooms is insufficient to completely prevent the transmission of airborne materials (Sullivan and Songer, 1966), it is nonetheless common practice to specify high static air pressures in operating rooms and other components of the surgical facility in comparison to corridors or other adjacent facilities. The concept of maintaining a higher static air pressure in areas that are designated as clean as compared to adjacent areas that are less clean (hence considered dirty) in theory should reduce the risk of airborne contaminants entering the operating room, thereby reducing the risk of contamination to patients. The real benefits of the presence and magnitude of such pressure differentials has never been established.

E. Electrical Services

Electrical requirements of the surgical suite and postoperative care unit are a function of the anticipated types of both fixed and portable equipment used in these areas. Ground fault circuitry is generally advisable in areas where water usage is likely to occur, including the material preparation area, animal preparation area, postoperative care unit, and the surgeon's preparation area. Sufficient 120-V, grounded receptacles should be placed around the perimeter of the rooms at 6- to 8-foot intervals at a height of 60 inches off the floor to provide easy access to electrically powered equipment. Enough circuits and amperage should be available to address peak needs, including the use of portable support equipment and research-related equipment. All wall-mounted electrical outlets should have moisture-resistant covers to prevent dust accumulation and to guard against shock hazard during the cleaning process. As a general rule, if any room in the surgical suite is more than 12 feet wide or 20 feet long, the use of overhead cord reels should be considered in order to minimize the hazards associated with long electrical cords stretching across the room. Provision for 220/240-V receptacles on a separate circuit with sufficient amperage to operate surgical lasers, portable radiographic machines, and other similar equipment is often warranted. Inclusion of 220/240 circuits is dependent on the nature of investigations to be conducted, but the need for equipment with this type of electrical requirement is becoming more commonplace.

Circuit breaker panels serving the surgical suite and the postoperative care unit should be located in close proximity to these areas. Operating and recovery room personnel should be familiar with the location of these panels as should maintenance personnel. Equipment used in the conduct of experiments as well as monitoring and support devices can occasionally cause circuit overloads resulting in the tripping of a breaker. Valuable time can be lost and patient care compromised if the circuit breaker panel is not easily accessible. It is prudent to provide emergency power to lighting as well as electrical outlets needed for life-support equipment within the surgical suite and postsurgical care area. All emergency power outlets should be clearly identified by the use of color-coded receptacles.

General purpose lighting is usually provided by covered fluorescent fixtures. Commonly, sufficient lighting fixtures are provided to allow an illumination intensity of 100 to 125 foot candles at a point 5 feet off the floor. The light should be well diffused and the fixtures should be amenable to cleaning. Flush-mounted ceiling light fixtures are often used because dust and other possible contaminants do not accumulate on as many surfaces as with exposed, hanging fixtures. However, with regular cleaning, hanging light fixtures can be used effectively.

Lighting for the postsurgical care unit poses several unique concerns. By nature of its primary function of monitoring and care for postoperative animals, activity in this area may occur on an around-the-clock basis such that timed light cycles normally found in animal housing rooms are impractical. However, numerous studies have indicated that disruption of the normal diurnal photoperiod of an animal can alter various biochemical, hormonal, and physiologic parameters (Bellhorn, 1980; Brainard, 1989; Lindsey *et al.,* 1978). It is essential that personnel monitor postoperative patients on a regular and frequent basis to ensure their uneventful recovery. For this reason, light cycles and intensity are usually not controlled by automatic timers but by manual switches.

Supplemental lighting for illumination of the surgical field and examination tables in the animal preparation or postsurgical areas is generally required. Three types of surgical lighting might meet these needs: fixed, track, and portable. Most lighting of both the fixed and the track variety can be adjusted either by the surgeon using handles that can be autoclaved or by operating room personnel. Fixed lighting is mounted by means of a center post on the ceiling and rotates on a radius arm around that center post. Fixed lighting is usually less expensive than track lighting but often places limitations on the location of surgical tables within the operating room.

With track lighting the center post can be moved along a horizontal track, greatly extending the range of a single light to many locations within the room. It offers the advantage of expansion of the system by adding more light fixtures at a later date to accommodate a greater number of procedures. The cost of track lighting is often significantly greater than fixed lighting.

Portable surgical or examination lights, as the name implies, are placed adjacent to the surgical or procedure tables using a center post mounted on casters. These lights are not always adjustable by the surgeon but may need to be manipulated by other operating room personnel not in sterile attire. Supplemental lighting using a fixture worn on the head of the surgeon and connected by a flexible fiber-optic cable to a high intensity light source mounted near the surgical table may also be used for certain types of surgery. This form of lighting is only supplemental to other permanently mounted surgical lighting, although it can be used as a sole surgical light source in some rodent surgical procedures.

F. Medical Gases

One of the most difficult choices in the design and construction of surgical suites and postsurgical care units is the selection of medical gases and the location of associated outlets. The operating room, animal preparation area, and the postsurgical area should be provided with vacuum, oxygen, anesthetic gas scavenging, and perhaps nitrogen or medical grade compressed air depending on the anticipated use of these areas. Outlets for these services should be placed along walls at locations similar to those for electrical outlets. Overhead hose reels for these services or surgical columns can also be used.

The decision to include nitrogen, nitrous oxide, or medical compressed air outlets must be weighed carefully. Nitrous oxide may be used to varying degrees in certain anesthetic protocols. Systems supporting its supply on a central basis usually cannot be economically justified in an animal facility. Similarly, the addition of nitrogen gas or medical grade compressed air for the operation of certain gas-driven equipment such as drills and pumps can only be justified if sufficient surgical procedures requiring these types of devices will be done to warrant the provision of a central system. Usually these gases can more economically be provided by using tanks of compressed gas on portable tank stands.

When planning surgical suites and postsurgical care units, one must decide whether to use portable tanks, fixed manifold tank stations, or central liquefied gas stations to provide the needed medical gases. Compressed gases come in a variety of tank sizes which, with proper regulators and support stands, provide a safe and economical means of delivering medical gases. Unfortunately, the use of free-standing tanks in safety stands requires ready access to auxiliary tanks during the operative procedure and a certain amount of room must be dedicated to holding tanks both in storage and in the room during use. The high gas flow rates required by some pieces of equipment might make the use of tank gas systems impractical, especially where large animals are concerned. In addition, concerns regarding the sanitation of tanks might need to be addressed if they are maintained within the operating room.

A tank manifold system provides a greater reserve gas volume as compared to individual tanks and is a less costly alternative to a centralized liquefied system. A tank manifold allows many large tanks of gas to be connected by a switch over mechanism to a central piping system into the various rooms within the surgical facility requiring medical gases. Remote display equipment in these rooms can be used to determine the status of the gas tank system and often provides both audible and visual warning when the system needs replenishing. Manifold systems should ideally be located within easy access of the surgical facility but can be located remotely even in an outdoor location. Long runs of piping between the manifold and the outlets, as well as the distribution within rooms, often adds expense similar to that of a liquefied central system due to the precision required in assembling this piping system as well as the necessary periodic testing and inspection.

A centralized liquid gas system provides the ultimate in reliability and overall economy of operation but with high initial installation costs. Generally, such a system is not warranted when the animal surgical facility is the only user of the system.

A centralized medical vacuum system can usually be provided by piping to remote vacuum equipment located in mechanical spaces within the building. Similarly, medical compressed air can also be provided by locating compressor equipment in mechanical spaces; however, sufficient separation is required between vacuum pumps, compressed air equipment, and the auxiliary electrical generators to ensure that contaminating gases will not be brought into the medical compressed air system. The system should also be fitted with a bacterial as well as a course prefilter and should use a greaseless compressor without oil.

Anesthetic gas scavenging is best accomplished with a separate exhaust system that directly vents to the outside of the building away from any air intake vents. Systems that dump waste anesthetic gas into exhaust ducts or into exhaust systems that vent into enclosed areas should be avoided since the potential for recirculation into human occupancy spaces or into recycled air exists.

G. Plumbing

A variety of plumbing fixtures will be required in the surgical facility. As previously mentioned, scrub sinks and preparation sinks for instruments, animals, and personnel are required. Because of the heavy use and potential for exposure to corrosive disinfectant and detergent solutions, sinks should be made of heavy-gauge stainless steel or of laboratory grade chemically resistant materials. The use of low-flow spray wands for instrument and animal preparation areas, as well as the provision of an eye wash station and shower for safety purposes, is recommended.

Some equipment, such as certain models of extracorporeal bypass pumps and ultrasonic cleaners, requires water outlets and standing drain pipes for their operation. If any of these devices are likely to be used, the appropriate plumbing fixtures need to be provided.

Floor drains have little application in most areas of the surgical facility, with the possible aforementioned exception of some postsurgical care units. In an operating room, the presence of a floor drain serves as an impediment to the movement of equipment and is an area for the accumulation of debris, growth of bacteria, and escape of noxious gases. The cleaning procedures within the operating room rarely entail pressure spraying of large volumes of water and hence any spills can be handled by the usual cleaning utensils or by a dedicated wet/dry vacuum with a HEPA-filtered exhaust.

As previously mentioned, to be economically operated, autoclaves often require piped in steam and usually require a floor drain for condensate removal. If equipment for electrical steam generation is incorporated in the autoclave, a water supply will also be required. If the autoclave is mounted into the wall, it is important that ready access be available to the plumbing as well as other operating parts of the autoclave through an access closet that is separated from the surgical suite.

H. Miscellaneous Considerations

Throughout the surgical facility, a variety of different types of casework to hold equipment and supplies may be required. The bulk of this casework should be mobile and easily removed for cleaning. The construction of these items can vary from stainless steel or laminates to rubberized or epoxy-coated materials. Within areas where surgery is to be conducted and where animals are prepared for surgery or are monitored postoperatively, the amount of fixed casework should be kept to a minimum and constructed of materials that are easily cleaned and adequately sealed. The use of open racks or shelving is preferable in some situations because it minimizes the tendency to sequester seldom used or outdated materials behind closed doors.

Doors to the surgical suite as well as the associated support areas should be constructed of materials that withstand the application of disinfectants and detergents. They should fit tightly within their frames and both the doors and the frames should be completely sealed to prevent harborage of insects or vermin. Metal-clad doors with viewing windows are often used; however, they must be provided with ventilation spaces at the tops or bottoms of the doors in order to allow the removal of condensation. Metal doors can be prone to rust and can provide insect harborage. Plastic laminate doors can be completely sealed and are durable. Wooden doors could also be acceptable provided they are appropriately sealed with a durable finish and are protected with strike plates. Drop handles or recessed handles are preferred to standard door fixtures because they resist mechanical damage. An observation window should be placed in each door to allow viewing of the room and its occupants without opening of the door. Door swings should be carefully considered, especially in the case of doors separating the surgeon preparation (scrub) area from the operating room. In such cases, doors should swing into areas of increasing cleanliness to minimize the chance of contaminating personnel that have completed a surgical scrub.

For security reasons, doors to all critical areas of the surgery suite should be capable of being locked, yet provide for immediate egress in case of emergencies. Doors to the different components of the facility will undoubtedly vary in size but must be large enough to allow movement of the equipment likely to be used in the areas served by them. Doors that are at least

42 inches wide and 84 inches high are generally recommended for principal access to the surgical facility. In the case of a facility designed for large animal use, double doors up to 6 feet in total width might be required. These doors should be equipped with large strike plates constructed of metal or plastic. Double doors seldom provide as tight a closure as single doors and will require that one of the two doors be manually fixed at the top and bottom by latches in order to have a secure locking mechanism. Entrance to support areas such as material preparation is most commonly served by single doors that are 42 inches wide. Wherever possible, self-sealing drop sweeps should be included and all doors should be self-closing. The use of automatic opening doors for the postsurgical care unit can be used if cost is not a factor and if the facility is located where traffic will not cause frequent inadvertent opening of the door. Doors to the postoperative care unit should be able to accommodate whatever portable caging, support, or monitoring equipment that might be needed in this area.

In some animal surgical facilities, teaching is a component of their operation. For this reason, it is sometimes necessary to make provision for authorized personnel not directly involved in the surgery to observe the procedure. This can be done by admitting personnel into the operating room in appropriate attire. In some cases, it may be necessary to provide for ceiling-suspended mirrors to allow for observation of the surgical procedure or alternatively to use suspended closed circuit television for this purpose. Required cabling for this type of service is usually run in surface-mounted conduits to remote locations.

In equipping the surgical suite and postsurgical care unit, a variety of moveable, general purpose items are often required. These might include radiographic viewers, clocks, operating room chairs, and surgical or examination tables. The type and complexity of these items are dictated by the nature of the procedures to be accomplished and available resources. These items must be capable of undergoing regular cleaning and disinfection. Other types of specialized surgical equipment such as intraoperative radiographic equipment, surgical lasers, and microsurgery equipment may be used in the operating or postsurgical rooms. When there is reasonable indication that such equipment will be used on a regular basis, special provision for space and mechanical or electrical support should be made.

V. CONCLUSION

This chapter has presented an overview of many design, planning, and construction issues that need to be addressed when developing surgical, postsurgical, and ancillary support areas in a biomedical research setting. This is a very dynamic process that must be individualized for each institution undertaking it. The process requires a clear definition of the surgical program to be conducted in the proposed facility as well as familiarity with existing guidelines and regulations governing surgery in research animals. Clearly, in any review such as this, it is not possible to address all of the potential factors that may affect the research animal surgery program of an institution or to foresee all situations from an architectural, mechanical, or operational basis that may have an impact on the design process. The lack of specific, scientifically based information regarding the interactions of animals with their environment prevents a rigorous and standardized approach to many of the issues covered in this chapter. Studies on lighting, air flow, and other parameters as they relate to surgery on animals need to be conducted, presented, and evaluated in peer-reviewed literature if a more definitive set of recommendations are to be developed.

REFERENCES

American Association for Accreditation of Laboratory Animal Care (AAALAC) Activities Report (1990). ''Accreditation Positions: Survival Surgery Facilities,'' Vol. 16, p. 9. AAALAC, Bethesda, Maryland.

American Animal Hospital Association (AAHA) (1991–1992). ''Standards for AAHA Hospitals.'' AAHA, Denver, Colorado.

American Institute of Architects Committee on Architecture for Health (1987). Guidelines for construction and equipment of hospital and medical facilities. *In* ''Guidelines,'' pp. 25–27. American Institute of Architects Press, Washington, D.C.

American Society of Heating, Refrigerating and Air Conditioning Engineers (ASHRAE) (1977). Physiologic principles, comfort and health. *In* ''Fundamentals Volume, ASHRAE Handbook and Product Directory,'' pp. 6.1–6.21. ASHRAE, Atlanta, Georgia.

American Society of Heating, Refrigerating and Air Conditioning Engineers (ASHRAE) (1987). ''HVAC Systems and Applications.'' ASHRAE, Atlanta, Georgia.

Bellhorn, R. W. (1980). Lighting in the animal environment. *Lab. Anim. Sci.* **30,** 440–448.

Besch, E. L. (1980). Environmental quality within animal facilities. *lab. Anim. Sci.* **30**(2, Part II), 385–406.

Booth, N. H. (1977). Neuroleptanalgesics, narcotic analgesics and analgesic antagonists. *In* ''Veterinary Pharmacology and Therapeutics'' (L. M. Jones, N. H. Booth, and L. E. McDonald, eds.), 4th ed., pp. 318–321. Iowa State Univ. Press, Ames.

Brainard, G. C. (1989). Illumination of laboratory animal quarters: Participation of light irradiance and wavelength in the regulation of the neuroendocrine system. *In* ''Science and Animals: Addressing Contemporary Issues'' (H. N. Guttman, J. A. Mench, and R. C. Simmonds, eds.), Sect. 3, pp. 69–74. Scientists Center for Animal Welfare, Bethesda, Maryland.

Chang, L. W., and Katz, J. (1976). Pathological effects of chronic halothane inhalation. An overview. *Anesthesiology* **45,** 640–653.

Code of Federal Regulation (CFR) (1991). Title 21, Food and Drugs, Part 1300 to END. Office of the Federal Register, Washington, D.C.

Code of Federal Regulations (CFR) (1992). Title 9, Animals and Animal Products, Subchapter A, Animal Welfare, Parts 1, 2, and 3. Office of the Federal Register, Washington, D.C.

Corbett, T. H. (1973). Incidence of cancer among Michigan nurse anesthetists. *Anesthesiology* **38,** 260–263.

Dyment, J. (1976). Air filtration. *In* ''Control of the Animal House Environment'' (T. McSheehy, ed.), Lab. Anim. Handb., pp. 209–246. Laboratory Animals Ltd., London.

Hughes, H. C., and Reynolds, S. (1995). The use of computational fluid dynamics for modeling airflow design in a kennel facility. *Contemp. Top.* **34,** 49–53.

Institute of Laboratory Animal Resources (ILAR) (1978). ''Laboratory Animal Housing.'' National Academy of Sciences, Washington, D.C.

Institute of Laboratory Animal Resources (ILAR) Commission on Life Sciences National Research Council (1996). ''Guide for the Care and Use of Laboratory Animals,'' National Academy of Sciences, Washington, D.C.

Lindsey, J. R., Conner, M. W., and Baker, H. J. (1978). Physical, chemical and microbial factors affecting biological response. *In* ''Laboratory Animal Housing,'' pp. 31–43. National Academy of Sciences, Washington, D.C.

Lumb, W. V., and Jones, E. W. (1984). ''Veterinary Anesthesia,'' 2nd ed., p. 693. Lea & Febiger, Philadelphia.

Manley, S. V., and McDonell, W. N. (1980a). Anesthetic pollution and disease. *I. Am. Vet. Med. Assoc.* **176,** 515–518.

Manley, S. V., and McDonell, W. N. (1980b). Recommendations for reduction of anesthetic gas pollution. *I. Am. Vet. Med. Assoc.* **176,** 519–524.

McGarrity, G. J., and Coriell, L. L. (1973). Mass airflow cabinet for control of airborne infection of laboratory rodents. *Appl. Microbiol.* **26**(2), 167–172.

McGarrity, G. J., Coriell, L. L., Schaedler, R. W., Mandle, R. J., and Greene, A. E. (1969). Medical applications of dust-free rooms. III. Use in an animal care laboratory. *Appl. Microbiol.* **18**(2), 142–146.

McSheehy, T., ed. (1976). ''Control of the Animal House Environment.'' Laboratory Animals Ltd., London.

National Institute of Occupational Safety and Health (NIOSH) (1975). ''Development and Evaluation of Methods for the Elimination of Waste Anesthetic Gases and Vapors in Hospitals,'' DHEW Publ. No. (NIOSH) 75–137. NIOSH, Washington, D.C.

National Institute of Occupational Safety and Health (NIOSH) and Centers for Disease Control (CDC) (1977). ''Criteria for a Recommended Standard Occupational Exposure to Waste Anesthetic Gases and Vapors,'' DHEW Publ. No. (NIOSH) 77–140. NIOSH/CDC, Washington, D.C.

Peterson, E. A. (1980). Noise and laboratory animals. *Lab. Anim. Sci.* **30,** 422–439.

Ruys, T. (1988). Isolation cubicles: Space and cost analysis. *Lab. Anim.* **17,** 25–30.

Ruys, T., ed. (1991). ''Handbook of Facilities Planning.'' Vol. 2. Van Nostrand-Reinhold, New York.

Sanderson, M. C., and Bentley, G. (1976). Assessment of wound contamination during surgery: A preliminary report comparing verticle laminar flow and conventional theater systems. *Br. J. Surg.* **63,** 431–432.

Stoyka, W. W., and Havashi, G. (1977). The effects of repeated ^{14}C halothane exposure in mice. *Can. Anesth. Soc. J.* **24,** 243–251.

Sullivan, J. F., and Songer, J. R. (1966). Role of differential air pressure zones in the control of aerosols in a large animal isolation facility. *Appl. Microbiol.* **14,** 674–678.

Westwood, J. C. N., Mitchell, M. A., and Legace, S. (1971). Hospital sanitation: The massive bacterial contamination of the wet mop. *Appl. Microbiol.* **21,** 693–697.

White, W. J., Hughes, H. C., Singh, S. B., and Lang, C. M. (1983). Evaluation of a cubicle containment system in preventing gaseous and particulate airborne cross-contamination. *Lab. Anim. Sci.* **33,** 571–576.

Whyte, W., and Shaw, B. H. (1974). The effect of obstructions and thermals in laminar air flow systems. *I. Hyg.* **72,** 415–423.

Wixson, S. K., White, W. J., Hughes, H. C., Jr., Lang, C. M., and Marchall, W. K. (1987). A comparison of pentobarbital, fentanyl-droperidol, ketamine-xylazine and ketamine-diazepam anesthesia in adult male rats. *Lab. Anim. Sci.* **37,** 726–730.

Chapter 9

Anesthesia and Analgesia in Rodents

Sally K. Wixson and Kathleen L. Smiler

I. INTRODUCTION

Rodents have enjoyed prominence as research subjects for at least a century. Despite this popularity, recommended regimens for anesthesia of these species have, amazingly, often been limited to a handful of injectable agents, with novel anesthetic drugs and delivery techniques applied to rodents after they have been popularized in humans and other ''larger'' laboratory animal species.

ANESTHESIA AND ANALGESIA IN LABORATORY ANIMALS

ISBN 0-12-417570-8

The goal of this chapter is to present a comprehensive review of the literature regarding methods to provide anesthesia as well as analgesia to laboratory rodents. Because rodents have historically been the mainstay of biomedical research and testing, many ''antiquated'' anesthetic regimens have been popularized in the past and may, in fact, still be utilized in some studies and research environments. For this reason, past as well as current drugs and regimens to provide a stable anesthetic plane in various rodent species will be critically reviewed. This chapter focuses primarily on the predominant rodent species used in research (i.e., mice and rats); however, discussions will also extend to guinea pigs, hamsters, and gerbils. Although it is becoming increasingly evident that different strains of rats may have varying responses to anesthetics, the majority of studies reported in this chapter have been done in Sprague–Dawley-derived rats. Therefore, all references to rats refer to Sprague–Dawley-derived unless otherwise stated.

The use of mice in experimental surgery or in studies requiring the use of anesthetics over long periods has been very restricted, as the extent and number of operative procedures possible are limited by the greater surgical skill required and the miniaturization of equipment made obligatory by the small size of the animal. As a result, experience with the various forms of anesthesia and knowledge of their consequences for this species are limited. Whenever possible, material presented is based on data gathered from scientific publications; however, many regimens for anesthetic and analgesic management of rodents are based on clinical experience or historical empirical use.

An analgesic agent(s) or a combination of agents that will obtund pain perception in rodents, particularly regarding postoperative pain, without causing adverse clinical and physiological effects that potentially may invalidate data obtained from the research study have not been defined for many rodent species. Because it is clearly difficult to objectively assess pain and/or distress in rodents, recommendations for analgesic management of these species have lagged much further behind guidelines suggested for intra- and postoperative analgesic therapy of humans and larger animals, such as dogs and cats. For these reasons, analgesic management will also remain a dynamic field of study as we seek newer and better analgesic agents for obtunding pain in rodents used in biomedical research and testing.

Controversial topics regarding appropriate anesthetic and analgesic techniques for use in rodents, such as the use of hypothermia as an anesthetic regimen for neonatal animals, preparation of stroke models, and the questionable need for sedatives and analgesics during orbital bleeding and tail transection for DNA analysis, are addressed at the conclusion of the chapter. These techniques are also critically evaluated. By using this format, the authors have attempted to make this chapter the comprehensive standard for investigators, veterinarians, and Institutional Animal Care and Use Committees (IACUC's), on whom the ultimate decisions rest regarding the most appropriate anesthetic and/or analgesic method for use in research studies using rodent species.

The Appendix to this chapter presents a comprehensive listing of anesthetics, tranquilizers, and analgesics for use in various rodents. Dosages, routes of injection, and duration of effects are presented.

II. UNIQUE FEATURES OF RODENT ANESTHESIA AND ANALGESIA

Rodent anesthesia offers unique challenges that are seldom encountered in any other veterinary setting. Due to the common practice of providing anesthesia for multiple patients simultaneously, a ''herd'' anesthesia approach is necessary.

Additional features that require consideration in planning an anesthetic protocol include dosing in milliliters per kilogram from premixed dilutions or combinations of drugs; poorly accessible peripheral vessels for IV induction; ''drug bias'' by certain scientific disciplines; difficulty in extrapolating dosages from humans and other animals to rodents; multiple species and strains that show differing sensitivity, efficacy, and duration of action with various anesthetics; a relative lack of scientific studies focusing on anesthetic efficacy in hamsters, gerbils, and guinea pigs; infrequent monitoring of physiologic parameters in rodents; the use of anesthetic agents in rodents that have no corollary in veterinary or human medicine, thus physiologic, toxicologic, and pathologic effects have not been determined; the difficulty in judging anesthetic and analgesic depth and muscle relaxation in rodent species; choice of anesthetic drugs based on expense, available equipment, and skill of the anesthetist; and requirements for aseptic surgical procedures and/or body size influence the ability to administer and monitor anesthesia.

III. PREOPERATIVE PATIENT EVALUATION AND CARE

Rodents require appropriate preoperative evaluation and care when planning experimental manipulations that require anesthesia. Colony history should be carefully reviewed to determine the source and health status of the animals. The age of the animals and the length of time they have been housed in the present facility should be determined, and the most current information available from the serological monitoring of sentinel animals or other available health surveillance reviewed. Newly arrived animals should be quarantined and allowed to rehydrate and acclimate after shipping for at least 3 days before use (Landi *et al.,* 1982). Body weights for the group should be determined when selecting dosages for a particular strain, sex, age, or species. Biochemical screening may be indicated for older or compromised animals. Preanesthetic fasting should generally be limited to 12 hr in light of the high metabolic rate and rapid dehydration and blood glucose detriments that occur rapidly in animals of small body mass. Water should not be restricted. Animals scheduled for anesthesia may be individually weighed and their weights recorded on the cage card or ID tag prior to induction.

IV. METHODS OF ANESTHETIC DELIVERY/ EQUIPMENT

There are basically three methods of anesthetic delivery to rodents: parenteral, inhalation, or a combination of both. ''Balanced anesthesia'' is a relatively novel method for producing anesthesia in humans that is becoming popular to use in animals. This format utilizes combinations of drugs, each in a sufficient amount to produce a desired effect(s), yet keeping its undesirable or unnecessary effects to a minimum.

A. Parenteral Anesthesia

The parenteral administration of anesthetic agents refers to intraperitoneal, intramuscular, and intravenous routes of administration.

1. Techniques for Administering Injectable Agents

The five critical factors to be considered when proposing parenteral drug delivery to rodents are (1) the drug volume to be delivered, (2) the site(s) for drug delivery, (3) the irritant properties, if any, of the anesthetic compound, (4) the method of administration, and (5) the rate of drug absorption. Parenteral anesthetics can be delivered by single bolus injection, intermittent injection to effect, and by continuous infusion, the latter method requiring some form of syringe infusion pump. Because the volume of drug to be injected is often quite small in rodents, it is preferable to dilute most agents. To further decrease the irritation at each site caused by the injection of irritating solutions, required dosages may be divided and smaller volumes injected at multiple sites.

a. Technique of Intraperitoneal Drug Injection. The IP route of drug injection is probably the most popular parenteral method of drug delivery to rodents because (1) minimal skills are required and those needed are easily mastered; (2) peripheral veins for IV injection are not easily accessible in several species of rodents (e.g., hamsters, guinea pigs); (3) IP injection puts the volume of a potentially irritating drug into a well-perfused space with a large surface area, which allows better physiologic buffering of each drug; and (4) many published regimens for anesthesia of rodents have been formulated for IP drug delivery.

Accepted practice for IP drug administration to rodents requires positioning of the animal so that it is held in a head-down position with the needle inserted into the lower left abdominal quadrant. In reality, this is not based on controlled studies but rather on historical dogma. Errors in IP drug injection can be minimized by both fasting the animal for 4–8 hr prior to drug injection and using a 20- to 22-gauge needle instead of a smaller (e.g., 25- or 26-gauge) needle that may not penetrate the subcutaneous tissue, fat, and abdominal wall and thus never deliver the anesthetic to the large peritoneal surface for absorption. One should also be aware that peritonitis may follow IP drug administration, particularly when irritating solutions are administered or bacterial contamination is introduced.

b. Technique of Intramuscular Drug Injection. Intramuscular drug injections are usually delivered into either the caudal thigh muscle mass or the epaxial muscles along the spine.

c. Technique of Intravenous Drug Injection. In mice, rats, and gerbils, IV anesthetic or analgesic drug injections are most often made into the lateral tail vein, particularly after this vessel has been vasodilated using warm water, counterirritants, a heating lamp, or a heated restrainer. With smaller rodents (e.g., mice), a 27-gauge needle is required for IV administration, whereas a larger gauge needle may be adequate for IV access in a larger rat. Other vessels, such as the dorsal metatarsal vein and, in particular, the sublingual vein, are mainly of theoretical interest and require marked restraint of the animal such that they are rarely used for the induction of anesthesia. The latter vessels would, however, be useful for postoperative analgesic drug delivery at the conclusion of surgery, before the animal has regained consciousness. When IV access is desired for long-term drug infusion in rats, a thin, flexible catheter can be inserted into the tail vein through a small bore needle and connected to a suitable perfusion system.

Intravenous anesthetic delivery to a guinea pig or hamster is rarely performed. A technique has been described for catheterization of the lateral saphenous vein in an anesthetized guinea pig (Nau and Schunck, 1993). After placing a tourniquet above the hock, a 22-gauge over-the-needle catheter is inserted 2–3 mm distal to the protruding vessel. Using this technique, the cannula remains patent for IV injections for up to 5 hr, although blood collection may not always be possible. Additionally, small volumes of drugs can be delivered into the dorsal metatarsal vein or into a catheterized femoral or jugular vein in these species.

d. Techniques to Minimize Irritancy/Tissue Response to Anesthetics. Various degrees of tissue damage almost inevitably result from IV or IM drug injections of several irritating anesthetics, e.g., ketamine and Innovar-Vet (Gaertner *et al.,* 1987; Smiler *et at.,* 1990). In the case of small laboratory animals, this irritation may be clinically evident. In some cases, however, sufficient tissue damage can occur which leads to lameness, self-mutilation of a limb, or cutaneous ulceration, thus undeniably causing some magnitude of pain and/or distress to the animal. Techniques that may be useful to minimize tissue irritation from IM drug delivery include deep rather than superficial IM delivery of the drug; use of small gauge needles;

delivery of multiple small injections into several IM sites rather than injection of a large volume at one site; and dilution of the drug prior to injection with physiological saline. Overall, the IM route of anesthetic delivery to small rodents (e.g., mice) is plagued with complications and inaccuracies, thus its use is discouraged in these species.

e. TECHNIQUE FOR CONTINUOUS INTRAVENOUS INFUSION. Anesthetics (e.g., propofol, alphaxolone–alphadolone) can be delivered by continuous IV infusion to rodents. Anesthetics are delivered by this method following indwelling catheterization as previously described. Drug administration requires a syringe infusion pump or similar equipment calibrated to continually deliver the relatively small drug volumes required for use in rodents.

B. Inhalation Anesthesia

Inhalation anesthesia involves the delivery of volatile anesthetic agents to the patient via the respiratory tract. From a practical standpoint, anesthesia of rodents often involves the induction of several animals simultaneously or in sequential fashion to permit surgical manipulations on a group of animals. This format is often more amenable to injectable anesthetics, which can be rapidly administered, rather than investing the greater time required for chamber, mask, or endotracheal tube induction using inhalant agents. Greater technical competence is also needed to deliver inhalant agents, particularly when endotracheal intubation of these small subjects is required. Even when skilled personnel are available, problems often encountered with the intubation of small rodents include trauma to oropharyngeal structures by insertion of the laryngoscope, local irritation or laceration of the trachea, and possible subsequent subcutaneous emphysema. Furthermore, in contrast to the minimal cost incurred in parenteral anesthetic delivery (i.e., drug + syringe + needle), the use of volatile agents is accompanied by the attendant costs of vaporizers, oxygen, customized endotracheal tubes and laryngoscopes, face masks, anesthesia chambers, and delivery circuits. Finally, rodent anesthetic and surgical procedures have frequently been performed outside of dedicated surgical areas (e.g., laboratories), which may not have scavenging equipment available for operator safety.

Despite these disadvantages, investigators and veterinarians are increasingly learning that inhalation anesthesia *can* be reliably and safely delivered to rodents with moderate temporal efficiency and cost effectiveness. Major advantages of inhalant techniques are (1) increased operator control over depth and duration of anesthesia, thus greater survivability, and (2) ability to choose agents (e.g., isoflurane) that require minimal metabolism, biotransformation, or excretion and are thus less likely to be a significant variable in the research project. Because equipment purchased and/or fabricated to deliver volatile agents to rodents can be reused, costs incurred are quickly recovered.

1. Methods of Delivery of Inhalant Agents to Rodents

The open drop (bell jar) does not permit control over anesthetic and presents high exposure for personnel. In the nonrebreathing system, the exhaled anesthetic mixture is released into the atmosphere. Methods of delivery include endotracheal intubation, use of an anesthesia chamber, or delivery using a face (head) mask. In the rebreathing system, the exhaled mixture flows through soda lime which removes all the carbon dioxide. This system uses less anesthetic and less oxygen. Due to the small tidal volume of rodents, this method is rarely practical or useful in these species.

Ancillary equipment useful for administering inhalant anesthetics to rodents includes an anesthesia chamber, an oxygen source, a vaporizer, various sized masks, endotracheal tubes (can be fabricated from over-the-needle catheters), a fiber-optic light source, a mouth speculum or customized laryngoscope, and miniaturized breathing circuits.

a. DELIVERY OF INHALANT ANESTHETIC GASES USING VARIOUS TYPES OF ANESTHESIA CHAMBERS AND FACE MASKS. The easiest method of delivering volatile anesthetic agents to rodents is by using an anesthesia chamber. ''Chambers'' can be fabricated simply and inexpensively by using a large covered glass container (bell jar). The liquid anesthetic is volatilized by placing it on cotton balls or gauze squares in the bottom of the jar. Prior to placing the rodent in this crude chamber, it is advisable to cover the anesthetic-impregnated cotton with a woven mesh grid to prevent local irritation of the feet of the animal by direct contact with the liquid anesthetic. The animal is then placed within the jar and visually observed for cessation of movement and recumbency, thereby signifying the onset of anesthesia. Because this system of induction involves no calibrated vaporizer, the anesthetic concentration within the jar cannot be controlled and lethal concentrations of certain anesthetics can rapidly accumulate. For this reason, the bell jar or open drop method is usually reserved for use with methoxyflurane, which reaches a maximum concentration of approximately 3% after full volatilization at room temperature in contrast to levels of approximately 30% inhalant gas which can be reached upon volatilization of halothane or isoflurane with this method. However, high potent volatile anesthetics can be used in closed containers when the volume of the container is known (refer to Chapter 2). Because this method contains no provisions for scavenging excess gases, it should be used within a fume hood or otherwise ventilated area to minimize inhalation of methoxyflurane by personnel. An additional disadvantage of the bell jar technique is that no apparatus is present to maintain anesthesia once the animal is removed from the jar, thus this method only provides anesthesia for very short-term procedures, e.g., orbital bleeding, subcutaneous tumor implantation, and tattooing.

A more sophisticated method for delivering volatile agents to rodents involves using an anesthesia chamber alone or in combination with a face mask appropriately sized for rodents. Anesthesia chambers are constructed with a fresh gas inlet as well as an outlet for exhaled and excess gases. Ideally, the rodent is placed within the chamber for induction, then removed from the chamber with anesthesia maintained by delivery through a face or head mask. Both chamber and mask delivery incorporate a calibrated vaporizer into the circuit for control of the concentration of anesthetic gas delivered to the patient. Because oxygen flow is required to volatilize the liquid anesthetic placed within the vaporizer, oxygen is also delivered to the patient and helps maintain the blood oxygen saturation. Because fairly high fresh gas flows are required for either chamber or mask delivery, adequate scavenging of waste anesthetic gases is necessary to avoid exposure to personnel. Anesthesia chambers are commercially available or can be fabricated from plexiglass or other clear materials or by modification of an aquarium tank. Face or head masks can be purchased or can be easily made from a funnel or proximal end of a 20-, 50-, or 60-cm^3 syringe barrel, depending on the size of the rodent. Several references for the construction of rodent anesthesia chambers or circuits are included at the conclusion of the chapter (Dardai and Heavner, 1987; Franz and Dixon, 1988; Glen *et al.,* 1980; Levy *et al.,* 1980; Mulder and Hauser, 1984; Norris, 1981; Norris and Miles, 1982; Querldo and Davis, 1985; Tarin and Sturdee, 1972).

b. Intubation of Rats, Guinea Pigs, and Hamsters. As with larger species, endotracheal intubation is most easily accomplished in rodents by direct visualization of the pharyngeal and laryngeal structures. Although customized rodent laryngoscopes are not commercially available, there are reports in the literature describing the equipment needed and techniques for fabricating laryngoscopes suitable for use in rats, hamsters, and guinea pigs. Even though there are no publications describing endotracheal intubation of the gerbil, it seems likely that techniques and equipment suggested for use in similarly sized rodents would be applicable.

Intravenous over-the-needle catheters (sizes 14–20, depending on the size of the rodent) are usually suitable for the intubation of rats, hamsters, and guinea pigs [(size 14 (60 mm) for guinea pigs, 16 (55 mm) for hamsters, and sizes 14–20 for various sizes of rats)]. When using over-the-needle intravenous catheters as endotracheal tubes, it is advisable to file down the pointed stylet prior to insertion in order to decrease the likelihood of oropharyngeal and/or tracheal trauma. Additionally, some modification of the catheter itself is advisable to minimize tracheal mucosal damage that may occur during prolonged surgical procedures. DeLeonardis *et al.* (1995) describe the modification of various sized angiocaths by adding silastic medical grade tubing glued to the tip of the catheter to minimize tracheal mucosal trauma. This modified catheter is then inserted using a flexible-tipped (thus atraumatic) catheter guidewire.

Guinea pigs, in particular, are a challenge to intubate due to their large tongue, which hides the larynx, and their small epiglottis, which is angled upward and positioned anteriorly. Intubation of many species of rodents can be accomplished by direct visualization using a Wisconsin pediatric laryngoscope size 0 blade attached to a pediatric laryngoscope handle and light source. When using this equipment in guinea pigs, it is advisable to bilaterally file the anterior aspect of the blade to facilitate its insertion into the narrow mouth of this species. This modification is not needed for laryngoscope insertion into the short and wide mouth of hamsters.

An alternative to intubation of rodents using a purpose-made laryngoscope is to position the animal on its back (i.e., the supine position), reflect the maxilla with an elastic band or adhesive tape attached to the incisors, insert an oral speculum (e.g., a dental mirror or spatula curved at a 145° angle or an otoscope fitted with a small, short cone tip), and transilluminate the pharynx by illuminating the ventral neck with a powerful light source (e.g., a 50-W electric light bulb). Commercially available fiber-optic penlights can also be used for this purpose. When placed on the ventral neck, the light source penetrates the tissues underlying the larynx and illuminates the oropharynx, thereby permitting direct visualization of the larynx. According to a publication by Cambron *et al.* (1995), thoracic transillumination causes the vocal cords of rats to appear as a luminous triangle through which an endotracheal tube can be readily inserted using appropriate supine positioning and direct visualization with a mouth speculum.

When using an intravenous catheter as an endotracheal tube, some modification of the Luer fitting is needed to provide connection to the appropriate anesthesia circuit. Connectors and tubing components of the anesthesia circuit should be miniaturized so that the smallest possible volume of dead space is present within the anesthesia circuit. The technique of endotracheal intubation in rodents and the equipment necessary are presented in Alpert *et al.* (1982), Gilroy (1981), Kujime and Natelson (1981), Pena and Cabrera (1980), and Sedgwick and John (1980).

V. INHALANT ANESTHETICS

Methods for delivery and equipment used for inhalation anesthesia of rodents have been discussed previously (refer to Section IV). A discussion of each specific agent is as follows.

A. Ether

A volatile anesthetic agent commonly used to anesthetize laboratory rodents was ethyl ether (diethyl ether, ether, ethyl oxide), although recommendations by institutional safety committees and the "Guide for the Care and Use of Laboratory Animals" have led to the replacement of ether by less hazardous inhalant anesthetics (Institute of Laboratory Animal Resources, 1996). Ether has the advantages of low cost, 5–10 min of anesthesia is suitable for minor procedures following the use of simple induction methods such as a bell jar or open drop method, and minimal physiological effects on some parameters (Flecknell, 1987[a]). Slow induction allows inexperienced anesthetists a wider margin of safety.

The disadvantages of ether are of much greater concern to biomedical scientists and are discussed in detail in Chapter 2. As specifically applies to rodents, the induction of anesthesia with ether is unpleasant and existing respiratory disease will be exacerbated by its use, resulting in high mortality. In some research applications (e.g., retro-orbital bleeding, tail transection for DNA analysis), substitution of methoxyflurane or brief exposure to carbon dioxide (CO_2) with oxygen may provide superior anesthesia and safety (Clifford, 1984). Carcasses of rodents anesthetized or euthanitized with ether should be well ventilated prior to sealing them in air tight bags for disposal and/or incineration. Any storage of such carcasses should take place in explosion-proof refrigerators and freezers. Atropine should be administered at 0.04 mg/kg SC (mice) or to 0.10 mg/kg (rats) prior to induction (Clifford, 1984; Poole, 1987). Mice lose corneal reflexes prior to the pedal reflex and use of ophthalmic ointment may be appropriate to protect the cornea (Clifford, 1984). Differences in susceptibility to ether were reported in six strains of inbred mice (Kobayashi, 1985). In order of sensitivity, the C57Bl/6 were found most resistant to ether, as measured in lethal effects at various concentrations, followed by the ICR, DBA/2, BALB/c, and C3H/He as most sensitive. Ether anesthesia does not alter hematocrit, white blood cell, red blood cell, or differential values in mice (Cunliffe-Beamer, 1983). In practical use, the volume of a bell jar used to anesthetize the rat should be at least four times the volume of the rat (Poole, 1987). Physiological parameters that may be affected by ether anesthesia include inhibition of hypothalamic function and effects on the liver (Muir and Hubbell, 1989), such as increased microsomal enzyme activity and effects on blood glucose levels (Fowler *et al.,* 1980).

Ether use has been reported in hamsters, although atropine is recommended as a preanesthetic and hamsters require very careful monitoring. Equipment to revive animals with depressed respiration such as oxygen and a mask insufflator should be available. Doxapram may be administered by placing one drop on the sublingual mucosa (Silverman, 1987) to stimulate respiration. Use of ether in guinea pigs is considered unsafe as they tend to hold their breath and salivate profusely (Hoar, 1969). Ether is considered the least satisfactory inhalant in the gerbil due to rapid induction; many emergencies and fatalities can be anticipated (Poole, 1987).

B. Nitrous Oxide

The use of nitrous oxide as the sole anesthetic agent in rodents is discouraged because it is not a complete anesthetic (DiFazio *et al.,* 1972; Flecknell, 1987b) and it is a health hazard to humans. The public health hazards of its use are discussed in Chapter 2.

Nitrous oxide has been frequently used as anesthesia for creating stroke models in rats and gerbils (Baughman *et al.,* 1988; Mahmoudi *et al.,* 1989). In creating stroke models, halothane and nitrous oxide are used for induction with maintenance using neuromuscular-blocking agents. Prior to vessel occlusion to create the ischemia, halothane volatile anesthesia is discontinued, and animals are intentionally maintained on only nitrous oxide analgesia at 50–80%. Anesthesia of nitrous oxide under these conditions may not adhere to humane principles of animal experimentation (Mahmoudi *et al.,* 1989).

C. Carbon Dioxide

A need for an acceptable replacement for ether or injectable anesthetics in rodents has led to the use of carbon dioxide, with oxygen, to produce brief periods of anesthesia (Fenwick and Blackshaw, 1989; Fowler *et al.,* 1980; Taber and Irwin, 1969; Urbanski and Kelley, 1991). Carbon dioxide sedation or anesthesia has been used most frequently for blood collection, particularly from the retro-orbital sinus. Advantages are that carbon dioxide is readily available, inexpensive, safe for personnel, and produces rapid recovery.

Carbon dioxide has been recommended for delivery in a 50:50 ratio with oxygen, using a charged induction chamber (Urbanski and Kelley, 1991). In this study, induction occurred within 1 min. After the animals were removed from the chamber, they regain consciousness rapidly unless exposure was prolonged. It is not known how long carbon dioxide/oxygen anesthesia can be safely maintained, although recommendations from 2 to 15 min (Fenwick and Blackshaw, 1989) have been made. In comparison to ether, CO_2 induction time is longer, but recovery is more rapid (Fowler *et al.,* 1980). Exposure to 100% carbon dioxide is not reliable or appropriate for anesthesia where recovery is planned.

Danneman (1994) explored the practicability and humane acceptability of CO_2 use for anesthesia as well as euthanasia in rats. In this study, anesthetic induction using 50–100% CO_2 + oxygen produced induction times of 16 and 2 min, respectively. Thus, lower CO_2 concentrations had a much longer induction time, as determined by loss of the toe pinch response. Most importantly, lower (50%) CO_2 concentrations led to frequent and severe adverse effects, such as nasal bleeding, excessive salivation, seizures, and death. In terms of its humane acceptability, 50–100% CO_2 were rated by humans from unpleasant to overtly painful, respectively. In summary, results of this study suggest that low (50%) CO_2 concentrations produce an excessive time of induction, with frequent severe and possibly fatal adverse effects with moderate distress and discomfort for the animals. In contrast, high (100%) CO_2 concentrations produce rapid anesthesia with fewer adverse effects but raise serious ethical and moral concerns. The authors of this study suggest that if CO_2 must be used, 70% + oxygen is the optimal concentration, based on practicality and humane acceptability. Accordingly, there is no consensus on the usefulness and humane acceptability of carbon dioxide anesthesia.

Physiologically, the effects of carbon dioxide anesthesia on blood parameters are well documented. Carbon dioxide anesthesia does not alter corticosterone, luteinizing hormone (LH), follicle stimulating hormone (FSH), and prolactin (Urbanski and Kelley, 1991). Carbon dioxide anesthesia has also been shown not to alter the total protein, alanine aminotransferase (ALT), urea, glucose, and prothrombin time, but glucose was elevated (Fowler *et al.,* 1980). Fenwick and Blackshaw (1989) summarized the increases in tidal and minute volume of respiration, decreased intestinal motility, arterial pH, and percent oxyhemoglobin saturation during carbon dioxide inhalation. In this study, cardiac arrhythmias and decreased heart rate were inversely proportional to the concentration of CO_2 and brain intracellular sodium and potassium concentrations were also decreased.

D. Methoxyflurane (Metophane, Penthrane)

Methoxyflurane may be administered to rodents by an open drop or induction chamber technique, although use of a calibrated vaporizer with a nonrebreathing circuit is preferable. Scavenging of anesthetic waste gases is important for personnel safety.

Methoxyflurane is highly soluble in blood and tissues such that induction and recovery are prolonged, making it particularly useful for procedures in neonates where monitoring the depth of anesthesia is difficult or during procedures on multiple rodent patients. A degree of analgesia during recovery from anesthesia may be an additional attribute of methoxyflurane. Induction time has been reported to require 4 to 20 min in various rodent species at concentrations of 3 to 4%. Using this technique, animals can be successfully maintained for prolonged periods: hamsters for up to 2 hr (Schreiber and Nettesheim, 1972) and guinea pigs for 4 to 6 hr (Clifford, 1984). In a study by Gardner *et al.* (1995), methoxyflurane delivered by the open drop method provided the shortest (0.9 min) and most repeatable induction time as compared to Telazol, tribromoethanol, ketamine/acetylpromazine, and pentobarbital anesthesia in ICR mice.

Physiological effects of methoxyflurane include less salivation than is produced by ether and cardiovascular and respiratory depression that is less than is produced by halothane. Hypotension can result with prolonged anesthesia. Methoxyflurane is exhaled by the lung, but greater than 40% is metabolized and eventually excreted in the urine. A diabetes insipidus-like syndrome has been observed in F344 rats (Clifford, 1984). Murray and Fleming (1972) reported renal damage from fluoride ion-induced renal damage. In rats, methoxyflurane has been shown to stimulate microsomal enzymes, inhibiting metabolism of P450-dependent substrates, and would be expected to have long-lasting effects well into the postanesthetic period (Rice and Fish, 1987). Wharton *et al.* (1980) reported that chronic exposure of pregnant mice to a 0.2% v/v concentration of methoxyflurane caused decreased fetal weight, decreased ossification, and delayed renal maturation of Swiss ICR mice. Additional references in the literature discussing the use of methoxyflurane in specific species include Tarin and Sturdee (1972) in mice, Farris and Snow (1987) in rats, and Clifford (1984) in guinea pigs. Norris (1981) considered that the use of methoxyflurane to produce anesthesia in gerbils was superior to pentobarbital, ether, and halothane.

E. Halothane (Fluothane)

Halothane is a volatile anesthetic gas that is nonflammable, nonexplosive, and nonirritating. Halothane has a much more rapid induction and recovery time than methoxyflurane, necessitating careful anesthetic monitoring to prevent overdose, and requires the use of a calibrated vaporizer.

In rats, microsomal enzymes are induced to a greater degree with halothane than with methoxyflurane (Rice and Fish, 1987), but may occur only after exposure for periods of greater than 30–60 min (Waynforth and Flecknell, 1992). Hepatotoxicity after halothane exposure occurs in some rodents. First reported by Hughes and Lang (1972), the guinea pig is an animal model for acute halothane-associated hepatotoxicity. Hepatocellular necrosis can develop even at low concentrations (Lind *et al.,* 1989, 1992). Markovic *et al.* (1993) reported inhibition of immune function [interferon stimulation of natural killer (NK) cell activity] in mice anesthetized with halothane or isoflurane.

The protective role of halothane and isoflurane when utilized in the rat model of forebrain ischemia in the rat was compared by Baughman *et al.* (1988) and Sano *et al.* (1993). Wistar rats maintained under 1% halothane with oxygen for 6 hr demonstrated altered arterial blood gas tensions (PCO_2 elevated above controls), but no significant change in acid base balance (Kaczmarczyk and Reinhardt, 1975).

Halothane is specifically a very useful anesthetic in guinea pigs, a species in which reliable anesthesia is very difficult to obtain. Franz and Dixon (1988) reported good results administering halothane with mask induction for 1–3 min at 5% concentration and maintenance at 2–3%. Recovery occurred in 10–15 min after 20–40 min of anesthesia. Use of halothane following preanesthesia with 30 mg/kg ketamine and 5 mg/kg xylazine was appropriate when the guinea pig was utilized as an otosurgical model (Wells *et al.,* 1986). Heavner (1989) recommends a Jackson–Rees breathing system developed for rats (Dardai and Heavner, 1987) as being a useful low-flow technique in the guinea pig. Seifen *et al.* (1989) determined the minimum alveolar concentration for halothane in the guinea pig to be 1.01 ± 0.03%. Use of halothane in hamsters is documented by Carvell and Stoward (1975), and the ED_{50} of halothane in gerbils is determined to be 1.32 volume %, which approximates the ED_{50} of the rat (Thielhart *et al.,* 1984).

F. Isoflurane (Forane)

Isoflurane is the most recent inhalant added to the anesthetic regimen available to veterinarians and it has rapidly become the anesthetic agent of choice for procedures requiring low risk and reliable rapid recovery of the individual patient (Burgman, 1991; Jenkins, 1992).

Although isoflurane appears to be the inhalant anesthetic of choice in some species of laboratory animals, only a few studies have been done to comparatively assess its suitability in rodents. Dardai and Heavner (1987) compared the respiratory and cardiovascular effects of isoflurane in Sprague–Dawley rats to those of halothane and enflurane, at 1.28 MAC, using a Jackson–Rees breathing system. Mean arterial pressure, heart rate, and respiratory rate and respiratory depression were the least under isoflurane anesthesia. Although studies have not been performed in rodents to assess the effects of prolonged delivery of isoflurane during mechanical ventilation, comparative studies on human pediatric patients did not show significant impairment of cardiovascular, hepatic, or renal function (Arnold *et al.,* 1993). Farris and Snow (1987) reported excellent anesthesia coupled with rapid postadministration recovery of reflexes with isoflurane use in Fisher 344 rats. Maekawa *et al.* (1986) examined local cerebral blood flow and glucose utilization during isoflurane anesthesia in the rat. In contrast to the other inhalants, isoflurane depresses metabolism to a greater degree, causes less increase in cerebral blood flow, and may offer a degree of protection to ischemic and hypoxic brains. In this study, isoflurane caused dose-related alterations in cerebral blood flow and metabolism.

Studies utilizing spontaneously hypertensive (SHR) and Wistar–Kyoto (WKY) rats found that the MAC for isoflurane was significantly lower in hypertensive strains as compared to normotensive Sprague–Dawley rats, suggesting that anesthetic effects may not be constant within a species (Cole *et al.,* 1990).

In humans, isoflurane has been reported to cause transient postoperative immunosuppression, which also occurs in mice (Markovic *et al.,* 1993). Animals were exposed to isoflurane or halothane prior to treatment with interferon and natural killer cell activity was assessed using *in vivo* and *in vitro* techniques. Isoflurane inhibited subsequent interferon-induced NK cell stimulation by greater than 90% as compared to halothane at 67%, respectively, and evidence of significant inhibition was observed 11 days after anesthesia. This study suggests that isoflurane should not be used for surgeries directly preceding immunologic research studies in mice.

In guinea pigs, Seifen *et al.* (1989) reported that the MAC for isoflurane (1.15 ± 0.05 volume %) is significantly lower than that reported in other rodent species. Similar studies comparing values for halothane and enflurane in guinea pigs (1.01 ± 0.03 and 2.17 ± 0.04 volume % for halothane and enflurane, respectively) were similar to those reported for other rodents. This suggests that isoflurane is a more potent general anesthetic in guinea pigs, possibly a benefit in this difficult to anesthetize species.

Author's Preference	
Agent:	Isoflurane
Route:	Inhalation
Species:	All rodent species
Comments:	Requires minimal metabolism or excretion

G. Sevoflurane

Sevoflurane is a new anesthetic, which has been introduced in Japan and is now marketed in the United States. Sevoflurane is now available in the U.S. It offers a significantly greater precision and control of maintenance of anesthesia and potentially a more rapid induction and recovery from anesthesia than the other inhalants. Sevoflurane has a blood solubility one-half to one-third of that shown by isoflurane, approaching that of nitrous oxide (Eger, 1994). Sevoflurane is also less pungent than isoflurane and causes a rapid induction that is not accompanied by struggling. This characteristic, coupled with its favorable kinetics for elimination and recovery, may make it a useful anesthetic for use in rodents. Studies have shown that rats given sevoflurane regained motor coordination in one-half the time of rats given isoflurane (Eger, 1994). In humans, this rapid awakening is associated with a more rapid return to the capacity to sustain an open airway and maintain spontaneous ventilation.

Sevoflurane is degraded by soda lime or Baralyme in a temperature-dependent manner (Eger, 1994). This breakdown is not significant to affect the course of anesthesia or be of economic importance. Of greater concern, however, is evidence showing that the breakdown product (an olefin called compound A) is lethal in rodents at high concentrations (400 ppm). Studies have shown that olefin production is flow dependent and averages 7.6 ppm (high fresh gas flows) to up to 61 ppm (with low fresh gas flows), thus it is now theorized that sevoflurane production of olefin may not approach lethal levels in clinical use (Eger, 1994; Kandel *et al.,* 1995). Further studies investigating the effect of a 1-hr exposure of Wistar rats to increasing concentrations of compound A plus oxygen showed that the threshold concentration for producing nephrotoxicity (i.e., the concentration which produces minimal toxicity) exceeded the maximum concentrations (particularly those at low inflow rates) reported in clinical practice by a factor of 2–3 (Kandel *et al.,* 1995).

The metabolic and physiologic properties of sevoflurane have been reviewed by Eger (1994). Metabolically, sevoflurane is degraded by cytochrome P450 2E1 to hexafluoro-2-propanol, which is excreted as the glucuronide conjugate. The peak excretion of the hexafluoro-2-propanol glucuronide occurs within 12 hr of cessation of anesthesia with an excretion half-life of 55 hr. Aside from the previously discussed olefin compound, which is produced by *in vitro* degradation and may cause nephrotoxicity at elevated levels in rats, sevoflurane shows a lack of toxic effects. Studies in either normal or enzyme-induced rats did not show hepatic or renal injury.

Physiologically, sevoflurane resembles isoflurane in that it depresses ventilation and blood pressure in a dose-dependent manner, but maintains heart rate. The stability of heart rate provided by sevoflurane is desirable because it neither increases myocardial oxygen comsumption nor decreases the time available for myocardial perfusion. Ventilatory depression occurs from a combination of central depression of medullary respiratory neurons and depression of diaphragmatic function and contractility. Hypotension occurs due to a decrease in total peripheral resistance. Sevoflurane preserves cardiac output at clinical doses, but excessive levels depress cardiac contractility and can produce cardiovascular collapse. Sevoflurane preserves splanchnic, including renal, blood flow. Cerebral vascular resistance and cerebral metabolic rate are decreased.

Because sevoflurane has been marketed only very recently in the United States, it is difficult to discuss the issue of cost of this new agent at this time and thus its feasibility for future use in rodents. Regardless of the final cost of the agent itself, capital investment costs will involve the purchase of new vaporizers and agent analyzers to monitor the safe use of this compound. The final decision on its future popularity in rodents will depend on these factors as well as further studies of its advantages over isoflurane as well as on the toxicity of its degradation product.

VI. PARENTERAL GENERAL ANESTHETICS

A. Barbiturates

Barbiturates have historically been the mainstay of anesthesia for laboratory rodents. The thiobarbiturates [thiopental, thiamylal, ethyl (1-methylpropyl) malonylthiourea salt, known as EMTU], in particular, are irritating perivascularly and are infrequently used in rodents due to difficulties in obtaining vascular access.

1. Thiopental

Recommendations for the use of thiopental in rats include 30–100 mg/kg of a 2.5% solution delivered IP (Kaczmarczyk and Reinhardt, 1975; Rao, 1990). These dosages have been shown to cause dose-dependent hypothermia, hypercarbia, acidosis, hypoxia, and hypoventilation. With the 100-mg/kg dosage, the duration of surgical anesthesia is up to 4 hr.

2. EMTU (Inactin)

EMTU (Inactin) has been used in rats at a dosage of 80–100 mg/kg IP (Buelke-Sam *et al.*, 1978). At this dosage, a suitable plane of surgical anesthesia was produced that lasted from 3–4 hr. Disadvantages of this regimen included marked hypercapnia and acidosis.

3. Methohexital (Brevital)

The methylated oxybarbiturate methohexital (Brevital) has also been used in rodents. Advantages of this drug include its short duration of action and rapid recovery, similar to inhalant agents. In the experience of one of the authors (SKW), methohexital (40 mg/kg IP) produces 15–20 min of profound restraint but analgesia insufficient for surgical manipulations in rats. When combined with pentobarbital (40 mg/kg methohexital IP followed 15 min later by 20 mg/kg pentobarbital IP), the combination yields variable sedation, poor analgesia, frequent respiratory depression that may be life-threatening, and a prolonged time to recovery, which usually lasts from 180 to 240 min.

Ferguson (1979) studied the effects of methohexital plus diazepam in hamsters. Using a mixture of 3 parts methohexital (1% solution) plus 1 part of 0.5% diazepam, an intraperitoneal dosage of 2–3 ml/kg produced sedation but insufficient anesthesia and analgesia for invasive procedures. Additional supplementation with methohexital to a total dosage of 4 ml/kg of the mixture produced ''good'' surgical anesthesia with no cyanosis. Higher dosages (5 ml/kg) produced a surgical plane of anesthesia but caused concomitant respiratory depression.

4. Pentobarbital (Nembutal)

The oxybarbiturate pentobarbital has been the most popular barbiturate used in all species of rodents. The wide use of pentobarbital stems from its generalized availability, modest cost, widely available database encompassing decades of use, rapid anesthetic onset, nonirritant nature, and ease of intraperitoneal injection to rodents of varying ages and body weights. Pentobarbital has been recommended for use in rats at dosages of 30–40 mg/kg IV (Aston, 1966; Hughes *et al.*, 1982) and 30–60 mg/kg IP (Rao, 1990; Seyde *et al.*, 1985; Wixson *et al.*, 1987a–d; Commissaris *et al.*, 1982; Routtenberg, 1968; Aston, 1966). At the higher dosage, anesthesia sufficient for mildly invasive surgical procedures is present. Because pentobarbital is a poor analgesic in rodents, dosages necessary to provide a surgical plane of anesthesia are accompanied by poor analgesia and a progressive decline in mean arterial blood pressure, heart rate, systolic, diastolic and pulse pressures, as well as acidosis, hypercarbia, and hypoxia (Wixson *et al.*, 1987a–d). In the authors' experience, rats that have eaten within 1 hr of injection have difficulty reaching a surgical plane of anesthesia. Alternatively, food deprivation enhances anesthesia and decreases the latency of onset and slightly lengthens the duration of anesthesia (Wixson *et al.*, 1987a).

Taber and Irwin (1969) reported on intravenous or intraperitoneal administration of pentobarbital to mice. When given an IP dosage of 40–70 mg/kg, the production of surgical anesthesia with a 20- to 30-min duration and less than 10% mortality was reported. In a study by Gardner *et al.* (1995), however, intraperitoneal administration of a 1:10 dilution of 60 mg/kg pentobarbital to ICR mice failed to produce loss of the interdigital toe pinch reflex, which is a commonly used criteria to identify a surgical anesthetic plane in mice. Pentobarbital has also been reported to have a slower onset (5–10 min) after IP administration to mice than for other species. Male mice have also been shown to be more sensitive to the anesthetic effects of this drug than are females (Taber and Irwin, 1969).

Lovell (1986, a,b,c) studied the effects of 60 mg/kg pentobarbital administered intraperitoneally to 23 inbred strains of mice. Significant differences in sleeping time occurred between the strains. Within strain differences were also present for age, sex, dose level, litter size, and fasting prior to anesthesia. In general, DBA mice slept the longest, whereas NZW slept the shortest. C57Bl/6 mice slept significantly longer than CBA, which in turn slept significantly longer than BALB/c. Mice homozygous for the obese (*ob/ob*) gene have shorter pentobarbital sleeping times than do their heterozygous (+/*ob*) litter mates.

Environmental variables that also affected the total sleeping time included diet, environmental temperature, and type of bedding material used. The pentobarbital sleeping time was significantly shorter for mice kept on unautoclaved softwood than on hardwood bedding. DBA/2 mice were an exception in that they slept much longer than other strains, regardless of the bedding material. Inbred strains were much more variable to the environment than were the F_1 hybrids (Lovell, 1986a,b,c). These studies suggest that pentobarbital, in particular, may show great clinical variability depending on the strain, age, and sex of the mice.

Recommendations for the use of pentobarbital in neonatal (1–4 days of age) mice include 5 mg/kg IP, delivered by a needle passed ventrally through the clavicle, subcutaneously past the ribs, and through the diaphragm into the peritoneal cavity. This somewhat circuitous route of injection avoids leakage of the drug directly through the thin abdominal wall.

Erhardt *et al.* (1984) compared the physiologic effects of pentobarbital (50 mg/kg IP) with ketamine–xylazine and etomidate in BALB/c mice. Results showed severe respiratory depression, adequate sedation but inadequate analgesia, and moderate circulatory depression. Excitation was also evident during induction, with a 30-min overall duration of anesthesia.

Pentobarbital has been used at dosages of 50–90 mg/kg IP in hamsters (Reid *et al.*, 1989; Skornik and Brain, 1990); however, a variable depth of anesthesia results. Physiologically, pentobarbital causes respiratory depression in hamsters similar in severity to that of rats and mice, with a decline in tidal volume, breathing frequency, minute ventilation, pulmonary resistance, and dynamic compliance (Skornik and Brain, 1990).

In gerbils, dosage recommendations for pentobarbital administration vary from 36 to 100 mg/kg IP (Edgehouse and Dorman, 1987; Flecknell *et al.*, 1983; Lightfoote and Molinari, 1978). Depending on the particular study, pentobarbital in gerbils produces sedation to anesthesia sufficient for short-term surgical procedures; however, similar to its adverse effects in other rodent species, profound respiratory depression and mortality accompany its use.

Author's Preference	
Agent:	Pentobarbital
Route:	IP
Dose:	30–40 mg/kg
Species:	Rats
Comments:	Preferable for terminal procedures of short to moderate duration

B. Dissociatives

1. Ketamine (Vetalar, Ketaset)

The most common dissociative anesthetic used in rodents is ketamine hydrochloride. Recommended dosages for ketamine in rats range between 10 and 175 mg/kg IM and are reported to produce effects from sedation to light surgical anesthesia (Green *et al.*, 1981). When used alone in rats and mice, ketamine produces minimal respiratory and cardiovascular depression, but causes muscle rigidity which results in incomplete anesthesia. In mice, the ketamine dosage ranges from 10 to 400 mg/kg IM (Green *et al.*, 1981). In rats and mice, the highest recommended dosages produce deep sedation but incomplete analgesia, such that some of the animals are not anesthetized sufficiently for major surgery. In guinea pigs, ketamine dosages of 20–200 mg/kg produce sedation with loss of the righting reflex, but poor analgesia and poor muscle relaxation (Green, 1975). In hamsters, dosages of ketamine up to 500 mg/kg do not provide sufficient analgesia for surgery (Curl and Peters, 1983).

Dosage recommendations for the use of ketamine in gerbils vary from 60 to 200 mg/kg IP or 100 mg/kg IM (Lightfoote and Molinari, 1978; Marcoux *et al.*, 1988; Flecknell *et al.*, 1983). In general, administration of ketamine without an accompanying tranquilizer causes sedation and incomplete mobility. Higher dosages (200 mg/kg IP) may produce short-duration (30 min) surgical anesthesia, but are accompanied by significant mortality.

2. Ketamine + Xylazine

To overcome the muscle rigidity and incomplete analgesia produced by ketamine, it is most frequently used in rodents with tranquilizers and sedatives. One of the most popular dissociative-tranquilizer combinations used in laboratory rodents is that of ketamine and the α_2-adrenergic agonist xylazine (Rompun). In rats, dosage recommendations for this combination include intramuscular or intraperitoneal 40–85 mg/kg ketamine + 5–21 mg/kg xylazine (Green *et al.*, 1981; Komulainen and Olson, 1991; Hsu *et al.*, 1986; Wixson *et al.*, 1987a–d). Lower dosages of this combination (40 mg/kg ketamine + 5 mg/kg xylazine IP) produce hyperacusia, good analgesia, minimal depression of core body temperature, and a light plane of anesthesia of fairly short duration (Wixson *et al.*, 1987a–d). Although moderate respiratory depression, hypotension, and hypothermia result, there is minimal mortality and a return to the preanesthetic values by recovery.

Dosages of 80–87 mg/kg ketamine + 10–13 mg/kg xylazine IP in rats produce excellent sedation, a rapid onset, and a 90- to 120-min duration of effect with the maximal effect lasting 15–30 min (Green *et al.,* 1981; Van Pelt, 1977; Wixson *et al.,* 1987a–d). Adverse effects of these dosages include acidosis, hypercarbia, hypoxia, hypotension, and hypothermia (Wixson *et al.,* 1987c,d). These adverse effects are of less severe intensity, however, than are caused by equianesthetic doses of pentobarbital and Innovar-Vet (Wixson *et al.,* 1987c,d). Overall, the ketamine–xylazine combination was judged to be superior to equianesthetic dosages of ketamine–diazepam, pentobarbital, and Innovar-Vet following a comparative study in Sprague–Dawley rats (Wixson *et al.,* 1987a–d).

Several authors have reported side effects in addition to those listed earlier for the ketamine–xylazine combination in rats. These effects include transient hyperglycemia, increased intraocular fluid glucose levels, polyuria, excessive salivation, and bradycardia (Hsu *et al.,* 1986; Wixson *et al.,* 1987a). Polyuria is theorized to result from a xylazine-induced decreased release of antidiuretic hormone and induction of hyperglycemia (Hsu *et al.,* 1986). These adverse effects may preclude use of this combination in weak, hypovolemic, or hypotensive animals unless supportive fluid therapy is administered.

Erhardt *et al.* (1984) reported on the effects of a combination of intramuscular 100 mg ketamine with 5 mg/kg xylazine in mice. After a calm, rapid (2–3 min) induction, an 80-min duration of surgical anesthesia is produced with an anesthetic interval of 110 min. Physiological changes caused by this combination include bradycardia and hypotension, with the latter being significant only during surgical levels of anesthesia. Excellent relaxation, sedation, and analgesia were present throughout the 80 min of deep anesthesia.

In guinea pigs, dosage recommendations for the ketamine–xylazine combination include 27–35 mg/kg ketamine + 0.2–10.0 mg/kg xylazine IM (Dawson and Scott-Conner, 1988; Brown *et al.,* 1989; Hart *et al.,* 1984; D'Alleinne, 1982). These dosages inconsistently produce an unreliable anesthetic depth in guinea pigs that may be sufficient for laparotomies, with a return to righting in 60–90 min (Dawson and Scott-Conner, 1988; Brown *et al.,* 1989; Hart *et al.,* 1984; D'Alleinne, 1982). Physiologic consequences of the ketamine–xylazine combination in guinea pigs include minimal respiratory depression and mild acute hypotension and bradycardia (Brown *et al.,* 1989; Hart *et al.,* 1984). This combination is also free from adverse hepatic or renal effects, as evidenced by normal values for blood urea nitrogen, creatinine, and serum glutamic–pyruvic transaminase throughout anesthesia and for 5 days thereafter (D'Alleinne, 1982). According to Dawson and Scott-Conner (1988), 27 mg/kg ketamine + 0.6 mg/kg xylazine IM supplemented with local infiltration of the incision site with 0.5 ml of 1% lidocaine + 1:200,000 epinephrine produces a surgical plane of anesthesia.

In hamsters, the ketamine–xylazine combination has been recommended at dosages ranging from 50 to 200 mg/kg ketamine + 10 mg/kg xylazine IP (Curl and Peters, 1983). With this combination, dosages of xylazine less than 10 mg/kg do not provide sufficient analgesia for surgery. Results of this study showed a dose-dependent anesthetic duration, ranging from 27 to 71 minutes. At the dosages just listed, this combination was also shown to provide safe and reliable anesthesia for use in pregnant hamsters, with no incidence of abortion or fetal abnormalities.

In gerbils, ketamine (50 mg/kg IM) has been combined with xylazine (2 mg/kg IM). Flecknell *et al.* (1983) reported that this combination produces sedation and immobility, but rarely produces a surgical plane of anesthesia. Sedation is further characterized by good skeletal muscle relaxation, but frequent "swimming or athetoid" movements.

Author's Preference	
Agent:	Ketamine + xylazine
Route:	IP
Dose:	40–60 mg/kg ketamine + 3–5-mg/kg xylazine; supplement with 1/3 of ketamine dose only
Species:	Rats
Comments:	Safe and reliable for procedures of short to moderate duration

a. KETAMINE–XYLAZINE REVERSAL. To overcome the adverse affects of ketamine–xylazine, Hsu *et al.* (1986) and Komulainen and Olson (1991) published reports on partial anesthetic reversal by the α_2-adrenergic antagonist, yohimbine, and tolazoline as well as the central nervous system stimulant 4-aminopyridine (4AP). Administration of yohimbine or tolazoline alone was superior to either 4AP alone or 4AP + yohimbine. Severe muscular tremors were produced by 4AP or 4AP + yohimbine in rats. At intraperitoneal dosages of 1.0–2.1 mg/kg yohimbine or 20 mg/kg tolazoline, a reversal of bradycardia, bradypnea, and polyuria occurred with a return to consciousness and righting present within 10 min of injection (Hsu *et al.,* 1986; Komulainen and Olson, 1991). Adverse effects were not caused by yohimbine unless excessively high dosages (20 mg/kg) were administered, with the latter causing acute respiratory arrest. None of the reversal agents listed earlier was able to ameliorate the pronounced hypothermia that is produced by ketamine–xylazine, thus thermal support is crucial (Komulainen and Olson, 1991).

3. Ketamine–Acetylpromazine

Ketamine has been frequently combined with acetylpromazine for use in rodents. Gardner *et al.* (1995) compared a combination of ketamine (44 mg/kg IP) and acetylpromazine (0.75 mg/kg IP) with other injectable and inhalant anesthetics in ICR mice. Results showed that none of the mice injected with this combination lost the interdigital toe pinch reflex, which suggests that the animals did not reach a plane of surgical anesthesia.

3. Ketamine–Diazepam

Ketamine is also frequently combined with diazepam for use in rats at dosages of 45–60 mg/kg ketamine + 2.5 mg/kg diazepam IM (Rao, 1990) or 40–80 mg/kg ketamine + 5–10 mg/kg diazepam IP (Wixson *et al.*, 1987a–d). Using these dosages, sedation to general anesthesia ensues, which is accompanied by a duration of 45–150 min, poor to fair analgesia, hyperacusia, and minor respiratory and cardiovascular depression (Wixson *et al.*, 1987a–d).

Flecknell *et al.* (1983) reported on use of the ketamine diazepam combination in gerbils. Intramuscular dosages of 50 mg/kg ketamine + intraperitoneal 5 mg/kg diazepam produce good muscle relaxation, sedation, and immobility, but lack a surgical plane of anesthesia.

4. Ketamine–Medetomidine

Medetomidine is a potent and selective α_2-adrenoceptor agonist with both sedative and analgesic effects that has been combined with ketamine in rats and guinea pigs (Nevalainen *et al.*, 1989). In Sprague–Dawley rats, dosages of intramuscular or subcutaneous 60–75 mg/kg ketamine + 0.25–0.5 mg/kg medetomidine produced a duration of effect of 20–180 min, prolonged immobility up to 300 min postinjection, and polyuria and hypothermia. Females are much more sensitive to the anesthetic effects of this combination than are males.

In guinea pigs, subcutaneous or intramuscular dosages of 40 mg/kg ketamine + subcutaneously injected 0.5 mg/kg medetomidine are suitable for immobilization but not for surgical anesthesia.

5. Tiletamine–Zolazepam (Telazol)

The dissociative anesthetic tiletamine, in combination with the benzodiazepine tranquilizer zolazepam (Telazol), has also been used in laboratory rodents. Dosages of 20–40 mg/kg IP produce anesthesia with variable analgesia in Sprague–Dawley and Fischer 344 rats (Silverman *et al.*, 1983; Ward *et al.*, 1974; Wilson *et al.*, 1992). The duration of anesthesia is dose dependent, lasting from 24 to 160 min, but most commonly providing a surgical duration of 30–60 min. With this combination, the corneal, pedal, and swallowing reflexes remain intact, such that these commonly used anesthetic parameters are not reliable for judging the depth of Telazol anesthesia. Physiologically, Telazol is cardiostimulatory in rats (Wilson *et al.*, 1992). Telazol has also been combined with xylazine or butorphanol for use in rats (Wilson *et al.*, 1993). When intraperitoneal 20–40 mg/kg Telazol is combined with 5–10 mg/kg xylazine, anesthesia results with a duration of 129–201 min, which is characterized by analgesia comparable to that of morphine, but with marked cardiovascular and minor respiratory depression (Wilson *et al.*, 1992, 1993). When 20–40 mg/kg Telazol is combined with 1.25–5.0 mg/kg butorphanol, an anesthetic duration of 59–140 min ensues. This combination is further characterized by transient hypotension and bradycardia, mild to severe dose-dependent respiratory depression, and analgesia comparable to that provided by morphine (Wilson *et al.*, 1992, 1993).

Telazol has also been used in mice, hamsters, and guinea pigs. A dosage of 80 mg/kg administered IP in BALB/c and CD-1 mice provides sedation but poor analgesia (Silverman *et al.*, 1983). Higher dosages produce severe respiratory depression and significant mortality, whereas lower dosages are not anesthetic. Telazol has also been combined with xylazine for use in ICR mice (Gardner *et al.*, 1995). Intramuscular dosages of 7.5 mg/kg Telazol plus 45 mg/kg xylazine and 45 mg/kg Telazol plus 7.5 mg/kg xylazine produced 36–99 (mean) min of adequate anesthetic depth (as determined by loss of the interdigital toe pinch reflex), respectively. Animals in the low-dose Telazol (7.5 mg/kg)/high-dose xylazine (45 mg/kg) slept for up to 145 min with no mortality, but they showed a sharp temperature decline with decreased respiration and immobility for up to 14 hr. Although this anesthetic combination requires attention to postoperative monitoring and patient support, it may prove valuable for studies requiring long-term restraint and sedation. The high-dose Telazol (45 mg/kg)/low-dose xylazine (7.5 mg/kg) produced adequate anesthesia useful for the duration of many experimental manipulations in mice (36 min). Approximately one-fifth of the animals in this group, however, did not lose the toe pinch response, suggesting variability and decreased predictability in anesthetic response.

Telazol dosages of 10–80 mg/kg IM or IP in guinea pigs provide sedation but insufficient analgesia and relaxation for surgery (Ward *et al.*, 1974). A dosage of 60 mg/kg Telazol plus local anesthesia may, however, be useful for minor surgeries and fetal injections (Ward *et al.*, 1974). One advantage of Telazol-induced sedation in guinea pigs is that a prolonged duration of effect ensues, which may be useful for procedures requiring prolonged restraint but with minimal painful manipulation, such as radioisotope scans.

Dosages of 50–80 mg/kg IP or IM in Syrian hamsters produce immobility, poor analgesia, and minor respiratory distress (Silverman *et al.*, 1983). Telazol has also been combined with xylazine for use in hamsters. Forsythe *et al.* (1992) reported that intraperitoneal 20–30 mg/kg Telazol + 10 mg/kg xylazine produces dose-dependent restraint to a surgical plane of anesthesia. This combination was further characterized by a smooth induction and recovery, a 10- to 40-min duration of surgical anesthesia, mild to moderate respiratory depression, freedom from nephrotoxicity, and a prolonged recovery period lasting up to 156 min. With this combination in hamsters, the pedal reflex is maintained throughout anesthesia and thus is not a reliable indicator of anesthetic depth.

Hrapkiewicz *et al.* (1989) reported on the effects of Telazol in gerbils. Dosages of 20–40 mg/kg IM produced sedation; however, 60 mg/kg is required to induce a surgical plane of anesthesia. At the latter dosage, a prolonged (4–6 hr) duration of anesthesia ensues, which is further characterized by occasional athetoid or involuntary movement and a rebound to somnolence and apparent return to anesthesia following initial recovery.

C. Neuroleptanalgesics

Neuroleptanalgesics are combinations of opioid agonists or mixed agonists/antagonists with tranquilizers. Although the opioid component can be reversed with suitable antagonists (e.g., naloxone), this is not commonly done in clinical practice in order to maximize the analgesia provided by the neuroleptanalgesic combination.

1. Fentanyl–Droperidol (Innovar-Vet)

Fentanyl–droperidol (Innover-Vet) is a neuroleptanalgesic composed of fentanyl, a narcotic that has 20 times greater analgesic potency than morphine, and the butyrophenone tranquilizer droperidol. There is both a veterinary (Innovar-Vet) and a human formulation (Innovar), with markedly differing drug concentrations in each formulation. This difference is important because this combination is most commonly dosed in rodents on a drug volume rather than a drug weight basis (i.e., ml/kg rather than mg/kg). Innovar-Vet contains 0.4 mg fentanyl citrate + 20 mg droperidol per milliliter whereas Innovar contains 0.05 mg fentanyl citrate + 2.5 mg droperidol per milliliter, thus the human drug has approximately one-tenth the potency per milliliter of the veterinary formulation. Innovar-Vet has been used for many years in mice, rats, guinea pigs, and hamsters. Its usefulness is compromised, however, by its irritant nature which may cause tissue necrosis and self-trauma to the digits following intramuscular injection.

In rats, Innovar-Vet has been recommended at a dosage of 0.09–0.26 ml/kg IM (Walden, 1978; Lewis and Jennings, 1972) or 0.2–0.6 ml/kg IP (Wixson *et al.*, 1987a–d). At these dosages, mild sedation to a surgical plane of anesthesia results. According to Walden (1978), Lewis and Jennings (1972), and Wixson *et al.* (1987a), 0.13–0.20 ml/kg is the minimally effective dosage to induce the immobilization of rats. At this dosage, the righting reflex is maintained and hyperacusia may be present (Wixson *et al.*, 1987b). Lewis found that this dosage range caused no gross tissue damage following IM administration. Anesthesia can also be prolonged by redosing with one-fourth to one-half of the initial IM dose of Innovar-Vet (Lewis and Jennings, 1972). Intraperitoneal administration has also been shown to cause no gross tissue damage or evidence of peritonitis (Wixson *et al.*, 1987a–d).

When used in mice, the commercial solution of Innovar-Vet should be diluted 1:10. Intramuscular dosages of 0.001–0.005 ml/g produce dose-dependent sedation to a surgical plane of anesthesia. The higher dosages require a 5- to 8-min onset, with an anesthetic duration of 30 min and full recovery in 1–2 hr (Lewis and Jennings, 1972). As previously discussed, however, the IM delivery of anesthetics to mice often leads to inaccurate dosing. For this reason, the authors do not advocate IM administration of Innovar-Vet to mice.

In guinea pigs, intramuscular dosages of 0.04–0.16 ml/kg produce dose-dependent sedation to anesthesia (Walden, 1978; Lewis and Jennings, 1972). The higher dosages produce an approximately 40-min duration of anesthesia with full recovery in 2–4 hr. Innovar-Vet has also been combined with pentobarbital for use in guinea pigs. Brown *et al.* (1989) showed that intramuscular 0.4 mg/kg Innovar-Vet plus intraperitoneal 15 mg/kg pentobarbital produced a 60- to 90-min duration of anesthesia that was characterized by mild hypotension and significant respiratory depression.

In hamsters, Innovar-Vet has been used at dosages ranging from 0.14 to 0.485 ml/100 g IP, which produce variable sedation and loss of the righting and pedal reflexes (Thayer *et al.*, 1972). This combination should not be administered IM to hamsters because it causes gross and histopathologic evidence of muscle necrosis at the injection site (Thayer *et al.*, 1972). In an attempt to produce a suitable short duration surgical anesthetic for use in hamsters, Ferguson (1979) combined fentanyl (0.03 mg/kg) + droperidol (24 mg/kg) + diazepam (5 mg/kg). Following IP administration, however, an inconsistent anesthetic depth ensued which was characterized by marked respiratory depression.

2. Fentanyl–Fluanisone (Hypnorm)

Hypnorm is a combination of the narcotic agonist fentanyl and the butyrophenone tranquilizer fluanisone, containing 0.30 mg fentanyl citrate and 10 mg fluanisone per milliliter. Because there is a lesser amount of fentanyl per milliliter and a different tranquilizer, Hypnorm and Innovar-Vet cannot be used interchangeably in rodents and the appropriate dosage must be used for each preparation. Although Hypnorm appears to be widely used in Europe, this compound is not currently commercially available in the United States and thus a DEA permit must be obtained to import it.

Hypnorm has been recommended for use in rats at intramuscular or intraperitoneal dosages of 0.2–1.5 ml/kg with optimal doses being 0.2–0.4 ml/kg (Green, 1975; Flecknell, 1987a; Royer and Wixson, 1992). At the optimal dosage, sedation to short duration anesthesia, results, which is accompanied by variable analgesia and some muscle rigidity. Higher dosages lead to labored respiration, cyanosis, and high mortality. Royer and Wixson (1992) compared equianesthetic dosages of Hypnorm and Innovar-Vet in Wistar rats and found that both drugs produced short-term (30 min) surgical anesthesia but profound sedation for 180 min. Innovar-Vet produced initial hypotension with severe, life-threatening hypoxia and hypercarbia which progresses to apnea at elevated dosages. Although Hypnorm also produced significant initial respiratory depression and hypotension, these aberrations were not life-threatening. Hypnorm also provided more profound analgesia for the duration of anesthesia. For these reasons, Hypnorm was judged to be the superior neuroleptanalgesic for use in rats.

Addition of the benzodiazepine tranquilizer midazolam (Versed, Hypnovel) to Hypnorm (using a mixture of 2 parts water for injection, 1 part Hypnorm, and 1 part midazolam) creates a reliable and longer-lasting surgical anesthetic with good muscle relaxation and analgesia (Flecknell, 1987a). Intraperitoneal dosages of 2.7–4.0 ml/kg of this mixture have been recommended for use in rats. The duration of anesthesia can also be prolonged by the supplemental intramuscular injection of 0.1 ml/kg Hypnorm, which may extend the duration of anesthesia for up to 6–8 hr (Flecknell and Mitchell, 1984). In rats, the premixed dilution of the Hypnorm–midazolam combination was shown to be stable for at least 2 months (Flecknell and Mitchell, 1984). When significant respiratory depression follows the use of Hypnorm alone or in combination with sedatives, it can be reversed with 0.1 mg/kg naloxone; however, this narcotic antagonist will reverse its analgesic as well as its respiratory depressant properties (Flecknell and Mitchell, 1984).

Author's Preference	
Agent:	Hypnorm + Midazolam
Route:	IP
Dose:	2.7–4.0 ml/kg of mixture explained in text
Species:	Rats
Comments:	Safe and reliable for short to moderate procedures

In guinea pigs, Hypnorm produces immobilization and restraint at intramuscular dosages of 0.5–3.0 ml/kg (Green, 1975; Flecknell, 1987a) with 1.0 ml/kg as the optimal dose (Flecknell, 1987a). Even at the highest dosage recommendation, Hypnorm does not provide potent muscle relaxation or abolish the righting reflex, but causes respiratory depression. An intraperitoneal injection of 1.0 ml/kg Hypnorm plus 2.5 mg/kg diazepam provides excellent anesthesia and muscle relaxation, but recovery can take up to 12 hr (Green, 1975). In pregnant guinea pigs, intraperitoneal 5 mg/kg diazepam plus intramuscular 0.2 ml/kg Hypnorm produces suitable short-term anesthesia, which can be prolonged by the administration of 0.05 ml/kg/hr Hypnorm (Carter, 1983). Physiologic complications of this regimen, however, include metabolic acidosis, hypoxia, and hypotension, such that "poor fetal condition" results. Intraperitoneal administration of 8.0 ml/kg of the Hypnorm–midazolam combination produces good surgical anesthesia and muscle relaxation with some respiratory depression but no cyanosis (Flecknell and Mitchell, 1984; Flecknell, 1987b).

Hypnorm should be diluted 1:10 prior to administration to hamsters, gerbils, and mice. In mice, an intraperitoneal dosage of 0.1–0.2 ml/30 g causes nervous twitching, paddling, and extreme hyperacusia (Green, 1975). When combined with intraperitoneal 5 mg/kg diazepam, Hypnorm induces a surgical plane of anesthesia with a 60- to 90-min duration of peak activity and 60–120 min of total sleep time (Green, 1975; Flecknell and Mitchell, 1984; Flecknell *et al.,* 1983). The time until complete recovery is up to 12 hr (Green, 1975). Although the respiratory rate is depressed, there is no cyanosis and minimal mortality. Flecknell and Mitchell (1984) reported on the effects of Hypnorm plus midazolam in mice. Dosages of 10.0–13.3 ml/kg IP produce good surgical anesthesia and muscle relaxation.

In hamsters, 1.0 ml/kg of the 1:10 Hypnorm dilution can be used to provide sedation that is not accompanied by the adverse effects that are present in mice; however, a surgical plane of anesthesia is not reached (Green, 1975). A mixture of Hypnorm (1.0–3.0 ml/kg IP) plus diazepam (5 mg/kg IP) results in excellent surgical anesthesia that is characterized by some respiratory depression, no cyanosis, minimal mortality, and a 5- to 6-hr recovery period (Green, 1975). Hypnorm can also be mixed with midazolam for administration to hamsters and gerbils (Flecknell and Mitchell, 1984; Flecknell *et al.,* 1983), using the Hypnorm–midazolam mixture detailed earlier. A dosage of 4.0 ml/kg IP produces surgical anesthesia accompanied by good muscle relaxation, some respiratory depression, but no cyanosis. Flecknell and Mitchell (1984) and Flecknell *et al.* (1983) have also reported that a combination of Hypnorm (0.3 ml/kg IM) plus diazepam (5 mg/kg IP) produces sedation but lack of a surgical anesthetic plane in *Meriones unguiculatus* but adequate surgical anesthesia in *M. libycus*. When the dosage is increased to 1.0 ml/kg Hypnorm + 5 mg/kg diazepam, 70% of *M. unguiculatus* are anesthetized, but there is attendant severe respiratory depression (Flecknell *et al.,* 1983). The addition of xylazine (5 mg/kg IP) to Hypnorm (0.3 ml/kg IM) produces sedation, but only approximately 30% of the gerbils reach a surgical plane of anesthesia.

3. Etorphine–Methotrimeprazine (Small Animal Immobilon)

This neuroleptanalgesic is a combination of the opioid etorphine (0.074 mg/ml) plus the phenothiazine tranquilizer methotrimeprazine (18 mg/ml). This anesthetic combination is not available in the United States. Etorphine is an opioid with analgesic properties 1000 times more potent than morphine. Methotrimeprazine has been added to improve sedation and counteracts the bradycardia induced by etorphine. In rats and mice, this combination does not provide suitable anesthesia for surgery but produces respiratory depression, bradycardia, and a relatively long-lasting duration of effect (Flecknell, 1987b; Whelan and Flecknell, 1994). When administered to mice (0.1–0.2 ml/mouse, SC), small animal Immobilon should be diluted 1:10 (Flecknell, 1987b).

To produce a deeper anesthetic plan, the etorphine–methotrimeprazine combination has also been combined with midazolam for use in Wistar rats and BALB/c and BKW mice (Whelan and Flecknell, 1994). Rats were subcutaneously administered 0.05 ml/100 g of a mixture composed of 1 ml Immobilon, 1 ml midazolam, and 2 ml sterile water for injection whereas mice received an intraperitoneal dosage of 0.1 ml/10 g of the mixture composed of 0.3 ml Immobilon, 1 ml midazolam, and 8.3 ml sterile water for injection.

Results in mice showed a prolonged anesthetic duration (mean loss of the righting reflex and pedal withdrawal reflex were 155 and 170 min, respectively) which was accompanied by moderate to severe hypoxia, hypercapnia, and acidosis. As compared to fentanyl/fluanisone plus midazolam in mice, etorphine/methotrimeprazine plus midazolam produced a twofold increase in the duration of surgical anesthesia, but an equivalent severe depression of respiratory parameters. Both neuroleptanalgesic combinations produced a prolonged recovery of up to 6 hr. Additionally, reversible cataracts occurred in some mice following recovery from the etorphine/methotrimeprazine/midazolam mixture. In summary, this combination produces a prolonged plane of surgical anesthesia in mice that is accompanied by moderate to severe respiratory depression; the latter is similar in magnitude to that produced by many other parenteral anesthetics in mice. Its use should be accompanied by respiratory support, including supplemental oxygen and, if feasible, positive pressure ventilation.

Etorphine/methotrimeprazine plus midazolam was judged to be a poor anesthetic combination in rats (Whelan and Flecknell, 1994). Although it produced a 4.7-min induction of anesthesia accompanied by what the authors define as "good surgical anesthesia," this combination produced persistent and severe hypercapnia, hypoxia, and acidosis in all experimental rats. For this reason, this combination was judged to be inferior to other available anesthetic methods in rats, and further trials to evalutate the duration of anesthesia and a comparison with other neuroleptanalgesics were not undertaken.

4. Carfentanyl–Etomidate

Carfentanyl is a highly potent, short-acting analgesic drug that induces excellent analgesia with little sedation and no muscle relaxation. As is true with many other opioids, it produces minimal circulatory compromise, but is a potent respiratory depressant. Etomidate produces sufficient sedation and muscle relaxation without any influence on circulation and has minor respiratory effects. Erhardt *et al.* (1984) investigated the anesthetic properties of this neuroleptanalgesic in mice. Results of this study showed strong excitation during induction, a 45-min duration of surgical anesthesia, and a prolonged recovery period that was accompanied by tonic–clonic spasms and trembling. Because of these adverse effects, the carfentanyl–etomidate combination should not be used in mice. There is no information available on its use in other rodents.

D. Miscellaneous Anesthetics

1. α-Chloralose

Field (1988) studied the effects of 45, 55, and 65 mg/kg α-chloralose in rats. Results showed that approximately 9% of the rats developed seizures, with one-third of the animals convulsing in the high-dose group. Other adverse effects of α-chloralose in rats included a prolonged onset, sternal rocking, catatonia, myoclonia, hyperacusia, mild hypothermia, and a thick, mucoid oral–nasal discharge. α-Chloralose produced a mild but progressive metabolic acidosis, mild hypertension, and tachycardia. This drug was a particularly poor analgesic in that even the highest dosage (65 mg/kg) did not provide analgesia sufficient for surgical procedures.

Based on this study, it appears that α-chloralose should not be used alone for surgical procedures in rats; however, further studies should be undertaken to assess its suitability when combined with analgesics such as morphine. Scientific justification is needed to support the use of chloralose to maintain sedation of animals that have already been instrumented using anesthetics with greater analgesic potency and/or other more classical anesthetic agents.

2. Alphaxolone–Alphadolone (Saffan, Althesin)

This anesthetic is composed of two steroids: alphaxolone (0.9%, w/v) and alphadolone acetate (0.3%, w/v). Similar to Hypnorm, this drug is not approved for use in the United States, but is commercially available in Europe and can be obtained for investigational use in the United States. When used in rodents, a 1:10 dilution should be made of the commercial preparation.

Green *et al.* (1978) studied the effects of this anesthetic in rats, mice, guinea pigs, and hamsters. In rats, alphaxolone–alphadolone is recommended at a dosage of 10–25 mg/kg IV or 25–30 mg/kg IP, which produces sedation to a light plane of anesthesia. Intravenous bolus delivery of 10–12 mg/kg produces surgical anesthesia of 5–12 min duration, with full recovery within 1 hr. Anesthesia can be prolonged for up to 8 hr by administrating 3–4 mg/kg by IV bolus every 15–20 min or 0.25–0.45 mg/kg/min by continuous IV infusion. By the latter method, stabile anesthesia is produced with no cardiovascular depression.

In mice, dosage recommendations include 60–150 mg/kg IM or 60–120 mg/kg IP (Green *et al.,* 1978). These dosages provide sedation to a light plane of surgical anesthesia that lasts for 45–90 min and are accompanied by hyperacusia, and fair to poor analgesia. No cumulative effects or tolerance is seen with repetitive doses, but muscle tremors and twitching are common during recovery. Additionally, the large volume of injection required makes IM dosage impractical in mice. Intravenous dosages of 12–50 mg/kg (optimal dose 14 mg/kg) provide 7–8 min of surgical anesthesia which are accompanied by respiratory depression but minimal mortality. Using the intravenous method of delivery, anesthesia can be prolonged by administration of 4–6 mg/kg every 15 min.

In guinea pigs, intramuscular dosages of 40 mg/kg IM produce sedation of a 40- to 90-min duration, which is characterized by poor analgesia with a recovery period of up to 3 hr (Green *et al.,* 1978). Intraperitoneal administration is unsatisfactory in this species because of its irritant nature. Intravenous injection of 16–20 mg/kg produces excellent surgical anesthesia and analgesia of a 10- to 12-min duration, with full recovery in 25–60 min (Green *et al.,* 1978). In contrast to other species, repetitive bolus administration of alphaxolone–alphadolone is contraindicated in guinea pigs because it causes respiratory distress and pulmonary edema.

Because alphaxolone–alphadolone does not produce an acceptable depth of anesthesia in guinea pigs, it has been combined with sedatives to permit surgical manipulations in this species. Brown *et al.* (1989) studied the effects of 45 mg/kg IM alphaxolone–alphadolone plus 5 mg/kg IP diazepam and assessed its suitability for survival surgical procedures in this species. Although a greater than 60-min duration of deep anesthesia was present, adverse effects of this combination included slight hypotension with severe respiratory depression, such that complete cessation of breathing occasionally occurred. For this reason, this combination should not be used in guinea pigs unless ventilatory support is available. Additionally, a large injection volume of the alphaxolone–alphadolone–diazepam combination is required, which may present difficulty due to the limited muscle mass available for IM injection in this species.

In hamsters, IM injection is also unsatisfactory due to the large injection volume. Intraperitoneal administration of 120–160 mg/kg produces surgical anesthesia with a peak effect of 40–60 min and full recovery within 2.0–2.5 hr (Green *et al.,* 1978). The optimal intravenous dosage is 14 mg/kg, which produces 6–9 min of surgical anesthesia with full recovery in 20–25 min. As with recommendations for alphaxolone–alphadolone in other species, 4–6 mg/kg can be intravenously administered every 15 min to prolong the duration of anesthesia.

Alphaxolone–alphadolone (120 mg/kg IP) has been tried in gerbils; however, an inconsistent plane of surgical anesthesia is produced that is accompanied by severe respiratory depression (Flecknell *et al.,* 1983).

3. Chloral Hydrate

Chloral hydrate, another hypnotic, has been a popular agent for use in rodents. In rats, dosage recommendations include 300–400 mg/kg IP or 400–600 mg/kg SC, using a 5% solution (Field *et al.,* 1993; Silverman and Muir, 1993). The most frequently used dosage in rats is 400 mg/kg IP followed by 40 mg/kg IP to maintain anesthesia. Using the latter recommendations, anesthesia ensues that lasts for approximately 2 hr (Silverman and Muir, 1993). Field *et al.* (1993) compared the effects of chloral hydrate, urethane, and α-chloralose in Sprague–Dawley rats. Results showed that a chloral hydrate dosage of 300–400 mg/kg IP produces surgical anesthesia with excellent analgesia. Unfortunately, because this dosage approaches the LD_{50} for this agent, dosages sufficient to provide surgical anesthesia and analgesia approach lethality. Adverse effects included severe respiratory, cardiovascular, and thermoregulatory depression that were characterized by acidosis, hypercarbia, hypoxia, hypotension, and bradycardia, as well as rapid and severe hypothermia. Although the intraperitoneal administration of chloral hydrate has classically been associated with gastric mucosal injury, peritonitis, and adynamic ileus, these effects may be related to the concentration of the solution used (Fleischman *et al.,* 1977; Ogino *et al.,* 1990; Silverman and Muir, 1993). Animals used in this study (using a 5% concentration of the agent) were followed for up to 14 days following drug injection with no apparent gross lesions, suggesting a lower concentration may not be as irritant.

Chloral hydrate is not commonly used in other species of rodents; however, a dosage of 800 mg/kg IP has been reported in guinea pigs (Silverman and Muir, 1993).

4. Propofol (Diprivan, Rapinovet)

Propofol is a ''novel'' hypnotic agent that is chemically classified as an alkyl phenol. In humans, advantages of propofol are its rapid onset after IV administration, short duration of action after bolus injection, lack of accumulation after repetitive injections, and rapid, tranquil recovery after the cessation of IV administration.

Glen (1980) and Glen and Hunter (1984) evaluated the effectiveness of propofol for anesthesia of rats and mice. Dosage recommendations in rats include 10–25 mg/kg IV for repetitive bolus administration or induction using 7.5–10.0 mg/kg IV followed by 44–55 mg/kg/hr for up to 1–2 hr of continuous IV infusion. This compound should be administered to mice at a dosage of 24 mg/kg for repetitive IV administration. At these dosages, the duration of immobilization is 5–15 min, the latter following repetitive boluses. As compared to its effects in humans, bolus administration of propofol to rodents causes 5–10 sec of apnea and hypotension following rapid IV administration, but a rapid recovery, some cumulative effects with repeat dosing, and no tissue damage following inadvertent perivascular infiltration. Continuous infusion, however, causes a minimal change in the heart rate and only a minor decline in the respiratory rate. In the experience of the primary author, propofol is useful for very short-term procedures in rodents (e.g., tattooing, blood sampling); however, its disadvantages include poor analgesia and the requirement for IV access, which practically preclude its use in guinea pigs, hamsters, and gerbils.

Brammer *et al.* (1993) compared propofol with other parenteral anesthetics regarding its onset, stress of induction, analgesia, depth of anesthesia, and respiratory and cardiovascular effects. Results showed that a combination of fentanyl–fluanisone induction followed by continuous propofol infusion was superior to pentobarbital, Inactin, fentanyl–fluanisone–midazolam, alphaxalone–alphadolone, or propofol used alone. Premedication with 0.5–1.0 ml/kg fentanyl–fluanisone IP, followed by a bolus dose of 1.0 ml IV propofol, with subsequent continuous infusion of 4–6 ml/kg/hr propofol produced stress-free induction, anesthesia, and analgesia sufficient for vascular catheterization, with a stable blood pressure, heart rate, and respiration throughout the 3-hr duration of the trials. Additionally, respiratory and cardiovascular parameters remained within published normal limits and there was 100% survivability of the study animals. Although this study was done under a nonsurvival format, the authors note that because propofol has been previously shown to cause a rapid recovery without sedation upon termination of the continuous infusion regimen, they predict that the fentanyl–fluanisone–propofol regimen may also be useful for survival studies.

5. Tribromoethanol (Avertin)

Tribromoethanol (TBE) is an anesthetic that has received renewed popularity for use in transgenic rodents to facilitate procedures such as embryo transfer, vasectomy, or distal tail amputation for Southern blot analysis. In mice, TBE has been recommended at a dosage of 0.2 ml/10 g (approximately equivalent to 250 mg/kg) IP, using a 1.2% w/v solution (Papaioannou and Fox, 1993). At this dosage, surgical anesthesia is produced that is characterized by rapid induction and recovery, a 16-min duration of surgical anesthesia, good skeletal muscle relaxation, a moderate degree of respiratory depression, and full recovery in 40–90 min. A less convincing study has been reported by Gardner *et al.* (1995) in which TBE was delivered intraperitoneally to ICR mice at a higher dosage (0.015–0.017 ml/g, approximately equivalent to 400 mg/kg, using a 2.5% solution). TBE was also prepared in this study according to previously published methods for storage to avoid the accumulation of toxic decomposition products. This latter study reported a 3-min onset, with a duration of anesthesia of 6.9 min. Five of the 12 mice used in the study, however, did not reach a surgical anesthetic plane, as judged by loss of the interdigital toe pinch.

The major disadvantages of TBE anesthesia include cardiovascular and respiratory depression at high dosages (e.g., 370 mg/kg) and, most importantly, high mortality due to toxic decontamination products (Nicol *et al.,* 1965; Tarin and Sturdee, 1972; Papaioannou and Fox, 1993). When TBE is exposed to light or is improperly stored, it decomposes into dibromoacetaldehyde and hydrobromic acid, which are potent gastrointestinal (GI) irritants, leading to fibrinous peritonitis, ileus, and fatalities. These adverse effects are particularly prominent in mice following the second exposure to TBE, regardless of the dosing interval (Papaioannou and Fox, 1993).

Papaioannou and Fox (1993) also retrospectively studied the efficiency, safety, and suitability of TBE (0.2 ml/10 g, using a 1.2% solution) for use in mice over a 2-year period. Results showed a rapid onset and recovery, an adequate plane of anesthesia, and lack of complications; the latter included morbidity or mortality from peritonitis or ileus. Proper storage of TBE at 4°C under dark conditions was crucial to avoid decomposition and subsequent mortality due to this anesthetic. When all of the just-described guidelines for preparation, dosage, and storage of TBE are strictly followed, these studies suggest that TBE appears to be acceptable for short duration survival surgical procedures in mice, although dosage adjustments may need to be made depending on the particular strain of mice being studied.

Norris and Turner (1983) reported on the efficacy of tribromoethanol for parenteral anesthesia of gerbils. Results showed that both the dose and the concentration influenced depth of anesthesia and survivability. Subcutaneous administration of 225–450 mg/kg of either a 1.25 or a 2.25% solution of tribromoethanol produced sedation, loss of the righting reflex, and no mortality, but lack of a surgical anesthetic plane. Intraperitoneal administration was required to produce surgical anesthesia; however, all dosages using the 2.25% solution produced a high percentage of visceral adhesions and death within a week of drug injection. Using the 1.25% solution, 250–325 mg/kg IP produced a surgical plane of anesthesia with a dose-dependent duration varying from 10 to 36 min. Higher dosages of the 1.25% solution also produced visceral adhesions and high mortality, thus suggesting that the duration of anesthesia cannot be safely prolonged by bolus supplementation.

Author's Preference	
Agent:	Tribromoethanol
Route:	IP
Dose:	250 mg/kg
Species:	Mice
Comments:	Must be carefully prepared and stored to avoid tissue reaction; under these conditions, reliable for short duration procedures

6. Urethane (Ethyl Carbamate)

Urethane has been popularized for producing long-lasting anesthesia in research animals with minimal cardiovascular and respiratory system depression. Recommended dosages for use in rats include 0.5–1.5 g/kg IP (Severs *et al.,* 1981; Field *et al.,* 1993). Field *et al.* (1993) studied the effects of urethane in Sprague–Dawley rats and showed that intraperitoneal dosages of 1.2 and 1.5 g/kg cause a prolonged period of deep surgical anesthesia, excellent muscle relaxation, and a total sleep time in excess of 24 hr. Anesthesia was further characterized by progressive acidosis, hypocapnia, hyperoxia, hypotension, and bradycardia. At similar dosages, Severs *et al.* (1981) showed that intraperitoneal urethane causes peritoneal fluid accumulation, hyperosmolality of body fluids, osmotic toxicity to the mesenteric vasculature, and increased plasma renin activity and aldosterone concentration. These effects indicate that urethane is not a physiologically inactive drug in rats, and its use is accompanied by profound respiratory and cardiovascular depression as well as a decline in renal function.

Urethane has been combined with α-chloralose for use in rats at intraperitoneal dosages of 250–400 mg/kg urethane + 35–40 mg/kg α-chloralose (Hughes *et al.,* 1982). With this regimen, urethane is injected 30 min prior to chloralose. The onset of anesthesia is prolonged (20–30 min) with a dose-dependent depth and anesthetic duration, the latter parameter increasing to 500 min using the higher dosages (Hughes *et al.,* 1982). Physiologic perturbations with this combination include minimal changes in respiration and arterial blood pressure over the approximately 6-hr anesthetic period.

Urethane has also been combined with α-chloralose for use in guinea pigs (Brown et al., 1989). In this study, animals were dosed with 0.8 ml/100 g of a combination of 1% w/v α-chloralose + 40% urethane. Results showed that the optimal drug formulation was composed of 2 parts chloralose: 1 part urethane, which produced deep surgical anesthesia with a prolonged duration of effect. Physiologically, this drug regimen produced minimal cardiovascular changes, but severe respiratory depression requiring artificial ventilation was noted. The authors concluded that this regimen would be suitable for long-term (several hours) nonsurvival experiments using mechanical ventilatory support.

In hamsters, urethane has been used at a dosage of 150 mg/100 g IP, which produces an inadequate plane of surgical anesthesia, but profound respiratory depression, acidosis, and up to a 6-hr duration of effect (Reid *et al.,* 1989). Reid *et al.* (1989) have also described a regimen for combining urethane (38 mg/100 g IP), α-chloralose (3.8 mg/100 g IP), and pentobarbital (2.6 mg/100 g IP) for use in hamsters. With this combination, a surgical plane of anesthesia ensues that is characterized by stable arterial blood pressure and ventilation. Anesthesia can be extended up to a 6-hr duration by giving supplemental dosages of urethane (13.5 mg/100 g) and chloralose (1.4 mg/100 g IP) every 2 hr as needed.

Perhaps the strongest disclaimer of urethane use is its proven status as a carcinogen and mutagen in rodents (Green, 1975). Due to all of the adverse effects listed earlier, use of urethane is strongly discouraged unless strict precautions are taken (e.g., gloves, face masks, mixing under a fume hood) to protect personnel. Because of its mutagenic and carcinogenic potential, the use of urethane should generally be limited to nonsurvival procedures.

VII. REGIONAL ANESTHESIA

Other than the administration of local anesthetics as a supplement to parenteral and inhalant delivery, regional anesthesia remains a research tool rather than a frequent method for clinical anesthesia of rodents. Limited reports are, however, available for the epidural and/or spinal administration of lidocaine to rodents. Eisele *et al.* (1993) reported on the technique for epidural administration of local anesthetics in guinea pigs. Using this model, a 27-gauge, 1.0- to 1.5-cm catheter is introduced caudally via the L3–L4 interspace and advanced to the epidural space. Epidural catheter placement can be confirmed using 0.12 ml/kg of 0.25% bupivicaine, which results in normal front leg function and responsiveness but impaired motor functions of the hind legs and cutaneous desensitization from the dorsal lumbar to midthoracic region.

Because regional drug delivery allows anesthetics to be delivered in much smaller quantities and potentially minimizes systemic toxicity and other adverse effects, this technology may be quite advantageous for incorporation into future research studies. To this end, methods need to be developed or existing materials and methods further refined for intraspinal and epidural drug delivery to various rodent species. Finally, regional delivery of analgesics is an active area of both clinical and investigative anesthesiology in humans for acute as well as intermittent and continuous chronic delivery of analgesic drugs. This technology would also enable us to potentially provide greater relief of pain to rodents used in biomedical research projects.

VIII. INTRAOPERATIVE ANESTHETIC MONITORING AND SUPPORT

Factors that influence whether or not rodents require intraoperative patient monitoring include the complexity and duration of the procedure as well as the particular anesthetic agent chosen. In some research situations, rodents are used for momentary surgical manipulations that do not require intraoperative monitoring and supportive care. In other circumstances, rodent surgery may be performed in a "herd format," in which individual patient monitoring is not practical or feasible. During more complex procedures, however, metabolic and respiratory acidosis, blood loss, hypothermia, and anesthetic overdose may contribute to cardiorespiratory collapse which can be fatal unless it is detected and corrected. In these circumstances, equipment and techniques are available for monitoring anesthesia and providing intraoperative supportive care to rodent patients.

Parameters used to monitor the physiological status of anesthetized rodents include monitoring of anesthetic depth; respiratory rate and pattern; arterial blood pH, blood gases, and oxygen saturation; pulse; arterial blood pressure; ECG; core and surface body temperature; and analysis of inhaled and gas for anesthetic agent and oxygen delivery.

Parameters used to assess the depth of anesthesia in rodents include recumbency and loss of purposeful movements (caution—"swimming" or purposeless limb movements often are retained in hamsters and gerbils, even at a deep plane of anesthesia); loss of reflexes (e.g., corneal, pedal, pinnae reflexes);

muscle relaxation; lack of vocalization; response to aversive stimulation (e.g., pinching the abdominal skin with a hemostat, pinching the toes or the tail); and alterations in respiration, cardiovascular function, and electroencephalograms.

In some instances, observations of chest wall movement to determine the respiratory rate and palpation of the apical pulse through the chest wall are sufficient assessments of cardiovascular and respiratory stability. More sophisticated methods are, however, available for the monitoring of anesthetized rodents when they are scientifically required. These methods include both invasive (i.e., using implanted arterial catheters connected in turn to transducers and recorders or using telemetry) and noninvasive (using tail cuffs connected to pneumatic pulse transducers, sphygmomanometers, and recorders) blood pressure measurement. With either invasive or noninvasive blood pressure measurement, data obtained are usually limited to the mean rather than both systolic and diastolic pressures. Commercially available portable monitors can also be used to noninvasively record the electrocardiogram and heart rate of anesthetized rodents. Arterial catheters can also be used to monitor hemodynamic changes and to provide arterial blood samples for blood gas measurements.

Pulse oximetry is being adapted for use in rodents as one means to assess respiratory function by modifying commercially available human neonatal digital as well as nasal sensors. The color of the eyes under reflected light can also be used as a crude measure of perfusion and oxygenation. To avoid aberrations in respiratory parameters, ventilation can be controlled using a commercially available rodent ventilator. Less sophisticated suggestions for manually assisting ventilation, particularly in mice, include pulling the tongue out with forceps to facilitate oropharyngeal air movement (Taber and Irwin, 1969).

IX. POSTOPERATIVE SUPPORTIVE CARE

It is not practical or feasible to provide the same intensity of supportive care to anesthetized rodents as is commonplace with that applied to larger animals. Furthermore, rodent surgery is unique in that "herd surgery" is often the research format, thus methods must be developed for the recovery of multiple animals simultaneously. Techniques that are available, however, for supportive care of rodents are briefly discussed in this section.

Because the ratio of body surface area to body mass is greater in rodents than in larger species, thermal support is often critical to the successful recovery of rodents from anesthesia. This hypothermia may, in turn, lead to a decline in both anesthetic metabolism and any urinary excretion of the anesthetic agent (Munson, 1970). Particularly with rats and mice, body heat may be dissipated from the tail, soles of the feet, and ears to the environment, with a resultant profound decline in the core and surface body temperature (Wixson *et al.,* 1987d).

Methods used to minimize heat loss to the environment during anesthesia of rodents include techniques such as increasing the ambient temperature of the operating room; placement of a thermal blanket (e.g., recirculating warm water blanket) or drape between the animal and heat-absorbing surfaces, such as the operating room table; the careful and controlled use of incandescent or heat lamps during surgery; minimization of organ exposure from body cavities during surgery; recovery of the animal on a warming blanket or within a temperature-supported cage or infant warmer; administration of warmed subcutaneous or intraperitoneal fluids intra- and/or postoperatively; recovery housing on bedding to provide thermal insulation; and recovery with cagemates (when permitted by the experimental protocol) to permit animals to huddle together and thus provide thermoregulation. In the experience of the authors, commercially available avian ICU cages are ideal for the group housing of rodents postoperatively. These cages can be supplied with oxygen, increased humidity, and an increased ambient temperature. These cages are relatively small in size, yet can be used for the recovery of multiple anesthetized rodents simultaneously.

Rodents have high energy requirements due to their small size and high metabolic rate, and often have minimal fat reservoirs that can be mobilized to supply needed energy. Nutritional support is critical on recovery to avoid hypoglycemia, particularly if the animal was fasted prior to induction. Nutritional support can be provided by simply providing a high-quality pelleted rodent diet as soon as the animal has recovered sufficiently to ambulate and eat. Fluid and nutritional support can also be administered by feeding gelatin or agar-based diets, ground moistened feed, or small amounts of peanut butter. Additionally, calories can be supplied to anesthetized rodents by injecting small volumes (3–12 ml, depending on the species of the animal) of a warmed 5% w/v dextrose solution subcutaneously.

Volume deficits can be corrected by the subcutaneous or intraperitoneal injection of warmed saline, lactated Ringer's solution, or other balanced replacement fluids (e.g., Normosol). Interosseous fluid administration remains an interesting possibility, particularly in small patients or long procedures and where small vessels are difficult to access and maintain patency. Although whole blood replacement is not usually done following surgical procedures or repetitive blood sampling in rodents, it has been shown that acute blood loss in rats causes respiratory alkalosis with an elevation in blood pH, a decline in blood carbon dioxide, and an increased red blood cell affinity of hemoglobin for oxygen (Miller *et al.,* 1976). If significant blood loss has occurred, blood transfusions can be administered, usually via the tail or jugular veins. Particularly with inbred and F_1 hybrid rodent strains and transfusions into naive animals, blood typing is usually not needed and transfusion reactions seldom occur.

Although the routine use of postoperative antibiotics is not recommended in lieu of strict adherence to aseptic surgical techniques, many investigators prefer to routinely apply topical antibiotic creams along surgical incisions or incorporate the use of parenteral antibiotics into the research protocol. Recommendations for parenteral antibiotic use include IM O.2 ml/mouse of a 300,000 U/ml penicillin suspension and 1 mg/mouse of IM terramycin dissolved in 5% glucose (Taber and Irwin, 1969); however, for the reasons previously presented, IM antibiotic administration may not be feasible in mice.

Special problems of neonatal rodents must be addressed for their postoperative care. They should be kept warmer than adult animals, with literature recommendations of 80–85°F ambient temperature (Taber and Irwin, 1969). Additionally, it is crucial to incorporate behavioral modification of the mother to prevent cannibalization of her offspring. Various methods have been suggested for this purpose, including rubbing soiled maternal bedding on the neonates and distracting the dam by placing shredded nesting material or cotton in the cage to preoccupy the mother with nest building rather than cannibalization (Taber and Irwin, 1969). If neonates are unable or not permitted to nurse, force feeding or IP injections of 5% dextrose (approximately 0.05–5.0 ml, depending on the species) may help maintain homeostasis.

Because rodents are frequently anesthetized with injectable agents that inhibit blinking (e.g., ketamine), ocular lubrication may be necessary to protect against corneal ulceration. The authors have found that C57Bl/6 mice, in particular, are sensitive to corneal ulceration after ketamine anesthesia if lubrication is not applied.

Catheters and implanted devices should be appropriately maintained (e.g., vascular catheters should be aseptically flushed with a diluted anticoagulant, such as heparin, on a regular basis). Exteriorized portals of implanted devices (e.g., skull caps) should be cleansed with a topical bacteriocidal disinfectant solution and the surrounding hair clipped on a regular basis. Caging may also be changed more frequently than normal to decrease local contact to debris and contaminants. Miniaturized bandages and splints should be removed frequently, replaced when needed, and the underlying tissue visually inspected for irritation, ulcers, or signs of infection. Materials used to discourage self-trauma to wounds (e.g., Elizabethan collars) should be sized such that they do not inhibit eating and drinking. Because self-grooming is an important component of the behavioral repertoire of a rodent, Elizabethan collars should be removed frequently. In the experience of the primary author, daily removal of Elizabethan collars is necessary for the survival of rats. Institutional investigator training programs should include training in postoperative care techniques needed to provide rodents with quality postoperative supportive care commensurate to the studies being done.

X. ACUTE AND CHRONIC ANALGESIC THERAPY

Ethical, legal, and scientific guidelines dictate the importance of appropriate analgesic therapy for rodents used in biomedical research. In the past, it was often argued that analgesic therapy was contraindicated for research animals because any drugs administered would impose additional variables over and above the hypothesis being studied. Studies in rodents, however, have shown that unrelieved pain, in itself, may profoundly alter the physiological responses in research animals. For example, Liebeskind (1991) has shown that pain and stress can inhibit immune function and metastatic tumor growth in rats. In humans, studies also indicate that effective pain management can decrease surgical morbidity and mortality (Liebeskind, 1991). Thus, it is both scientifically and ethically rational to design protocols to effectively relieve pain in research animals.

Identification and management of pain in rodents present unique problems. Because rodents are frequently group housed, alterations in behavior or mobility may be more difficult to discriminate than similar signs in a larger, individually housed animal. In addition, research requirements may dictate specialized caging (e.g., flexible film or semi-rigid isolators), which not only limit the observation of individual animals, but make physical examination and thus efforts to evaluate painful behavior difficult.

Before ethical and legal requirements to provide relief from pain and suffering can be fulfilled, logical and indisputable criteria to define and recognize pain in rodents must be established. Similar to humans, rodents show individual variability with regard to both their pain threshold and pain tolerance. As in humans, threshold intensity pain is perceived but can be tolerated, thus the animal often shows no overt signs of pain perception. In these cases, the need for and benefits of analgesic therapy are difficult to assess. When pain exceeds the intensity of the pain tolerance of the animal, however, it can no longer be accepted or masked and obvious signs of discomfort become evident. In these cases, analgesic therapy is clearly indicated and successful pain management is much easier to monitor.

A. Strategies for Recognizing and Alleviating Pain

1. In Individual Rodents: "Patient-Oriented Approach"

Detection of pain perception in individual rodents is based on subjective evaluation of behavioral and attitudinal changes as well as on an objective analysis of physiologic parameters. Behavioral signs indicative of pain perception may include a reluctance to move, abnormal posturing, teeth grinding, decreased appetite, and vocalization. Attitudinal changes indicative of pain perception include anxiety, apprehension, and hypersensitivity as well as depression or, uncommonly, overt aggression. Signs indicative of pain-induced distress that are particularly characteristic of rodents include polyphagia of bedding and digital irritation that may progress to autotomy. In the experience of the authors, recognition of attitudinal changes signifying pain in rodents depends on some preexisting knowledge of the temperament and behavior of each individual animal. In this regard, research staff and veterinary and animal husbandry technicians are crucial members of the veterinary care team, for these are the individuals that work with the animals on a daily basis, allowing assessment of temperamental changes in individual animals. A pain-scoring scheme has been developed (Morton and Griffith, 1984) for evaluating behavioral and attitudinal signs of pain perception in animals in which points are assigned to each parameter, such as posturing, and a cumulative score determined for each animal. Based on these point totals, decisions on analgesic therapy are made.

Physiologic signs of pain perception include fluctuations in blood pressure, heart rate, respiratory rate, and body temperature. Additional objective parameters used to infer pain perception include food and water consumption and body weight changes. These behavioral, attitudinal, and physiologic signs of pain perception are currently the best methods available to clinically detect individual rodents that are in some degree of pain as well as to assess the success/failure of ongoing analgesic therapy. In the experience of the authors, any one of the just-mentioned behavioral, attitudinal, and physiologic signs of pain perception may not be sufficient to designate an animal in need of analgesic therapy. Rather, a compilation of many of these parameters must be interpreted in the context of the overall condition of the animal and its departure from normal.

2. In Groups of Rodents: Herd Approach

In contrast to assessment and alleviation of pain in individual animals, rodents are frequently used in a "herd" concept of anesthesia and surgery. Instead of assessing individual animals for signs of discomfort and distress, investigators may choose to prospectively design an analgesic protocol that will make postoperative analgesic therapy available to each study participant. In these cases, investigators need to consider the pain intensity of a given research manipulation, often based on the discomfort level of that procedure in larger, companion animal species or in humans. This prediction is often based on the type of procedure and location of the lesion producing the noxious stimulus (e.g., spinal cord or peripheral nerves, producing neuropathic pain; thoracic or abdominal cavities, producing visceral pain; or peripheral skin, muscular, osseous, and subcutaneous tissue, producing somatic pain). Based on the type and location of pain, analgesics can be chosen to act selectively at central or peripheral sites. The pain-scoring scheme discussed earlier may also be used to develop a prospective plan for the postprocedural analgesic management of large numbers of rodents. In this format, procedures that have historically produced a high pain score (based on literature reviews or pilot study results) require a prospective plan for postoperative analgesic treatment of all study animals that will undergo this manipulation.

Many drugs currently marketed are reportedly efficacious for relief of pain in rodents. The major drugs available may be classified as opioid agonist, opioid agonist/antagonists, α_2-agonist, and anti-inflammatory agents. The latter group contains a diverse spectrum of compounds, including corticosteroids and nonsteroidal anti-inflammatory agents (NSAIDS). Other classes of drugs such as antihistamines and tricyclic antidepressants have been shown to possess some analgesic activity, but are, in general, clinically inferior to opioid agents. Additionally, local anesthetics are available and may be useful to obtund localized pain when systemic therapy is not needed or would be contraindicated in a research study.

B. Methods of Analgesic Drug Delivery to Rodents

Methods for delivery of analgesic agents to rodents are, at the present time, primarily limited to parenteral rather than oral delivery. With the exception of buprenorphine, which may maintain some of its effectiveness for up to 6–8 hr following parenteral administration, all other opioid agents require frequent bolus dosing to maintain adequate blood levels.

1. Oral Delivery of Analgesics to Rodents

It should be noted that putting any agent in water runs the risk of inaccurate dosing, lack of consumption due to palatability, and degradation of the agent over time due to hydrolysis.

Aspirin has been recommended at a dosage of 100–400 mg/kg PO in rodents (Hunskaar and Hole, 1987; Flecknell, 1984); however, these guidelines appear to be based on anecdotal reports rather than on a scientific comparison with other analgesics. Similarly, acetaminophen (Tylenol) has long been anecdotally reported to be an effective mild oral analgesic in rats and mice at an oral dosage of 110–305 mg/kg. In the experience of one of the authors (SKW), however, oral acetaminophen is both unpalatable to rats, regardless of the flavored formulation chosen, and ineffective because it does not show detectable analgesic potency. A study by Cooper *et al.* (1995) supported this observation. In the latter study, rats were given 0, 3.8, or 7.7 mg of acetaminophen per milliliter of drinking water. Results showed no difference in paw pressure latency (a test used in experimental analgesiometry) between the acetaminophen-treated and the control animals, with a decrease in water consumption in the treated group. Even at oral dosages three times in excess of the published parenteral dose, no detectable analgesia was present.

Walter *et al.* (1989) studied the efficacy of various oral NSAIDs to obtund *experimental* pain in mice (refer to Table A.VI in the appendix to this chapter for drugs and dosages). Unfortunately, these findings should be viewed with caution when attempting to apply them to *clinical* use because efficacy and adverse effects of chronic use were not studied in a format similar to that encountered in clinical practice. Although additional information is also available in the literature on the pharmacologic activity, drug metabolism, and toxicity of NSAIDs in rodents, dosages used in these studies are frequently well in excess of those used for analgesic therapy.

Limited information is available regarding the oral delivery of opioids to rats and mice. Oral formulations are available for the delivery of morphine, butorphanol, oxycodone, codeine, meperidine and pentazocine. Various diluents and solvents have also been used in an effort to create oral formations of the commercially available parenteral opioids. Although oral opioids are frequently used in humans, little is known about their oral effectiveness as analgesics in rodents. This void in the literature is most likely due to (a) oral opioids are primarily used to relieve chronic, lower intensity pain in humans, but research animals are frequently exposed to acute, higher intensity pain; (b) in general, because a marked first-pass metabolism rapidly degrades oral opioids, it is difficult to achieve efficacious blood and tissue drug levels, and (c) comparatively, oral opioids are quite expensive and thus may not be cost effective for ''herd'' analgesic therapy, the latter being a common format for analgesic therapy in rodents. Studies are also available suggesting that oral dosages of opioids, usually morphine, induce physiologic dependency and tolerance in mice and rats. Because dosages necessary to induce dependency and tolerance are often much higher than those necessary to induce analgesia, these studies may not be useful to provide guidelines for the relief of pain in rodent species.

There has been renewed interest in trying to develop effective oral opioid formulations to relieve chronic pain in laboratory rodents. Deeb *et al.* (1989) reported on the efficacy of oral administration of fentanyl or buprenorphine in rats. Dosages of 0.001 mg/ml fentanyl or 0.006 mg/ml buprenorphine were supplied *ad libitum* in the drinking water. Significant analgesia, as determined by measuring the reaction to the tail immersion test, was present with buprenorphine whereas no analgesic effect was shown by the orally delivered fentanyl. Unfortunately, this analgesia declined within 5 days. Minor side effects due to buprenorphine treatment included polydipsia, decreased grooming and investigative activities, and decreased food consumption without an accompanying decline in body weights. Clinical trials by one of the authors (SKW) using oral sustained-release morphine in the drinking water of rats have shown this formulation to be similarly ineffective to relieve experimental pain, even at dosages greatly exceeding those used clinically for SC or IM administration.

Buprenorphine mixed with flavored gelatin was advocated for postoperative analgesic therapy in some research facilities. Deeb and Frooman (1992) investigated the ability of oral buprenorphine (0.4 mg/kg) added to cherry gelatin to alleviate lameness in adjuvant-induced arthritic Lewis rats. Results indicated that buprenorphine enhanced the inflammatory response and that treated rats showed enhanced rather than ameliorated footpad swelling and rear-limb lameness. Thus, it appears that detectable analgesia is not afforded by the oral drug delivery to rodents.

2. Drugs and Methods Available for Bolus Parenteral Delivery of Analgesics to Rodents

Not all of the opioids, NSAID, and tricyclic antidepressants have been shown to be efficacious for bolus parenteral use in rodents. Reported dosages may be based on clinical experience, extrapolation from dosages used in humans, or founded on serendipitous discovery rather than on a truly scientific evaluation in a significant number of animals and in the various rodent species. Further investigation is needed to clearly identify those agents that are appropriate for the relief of pain in rodents.

Numerous drugs have been reported to be efficacious for bolus analgesic therapy in rodents and are presented below.

a. Acetylsalicylic Acid (Aspirin) and Sodium Salicylate. The salicylate analgesics have been shown to relieve only mild to moderate pain in humans in contrast to their poor efficacy in relieving deep-seated visceral pain or sharp, intense pain. Aspirin has been recommended at a dosage of 20 mg/kg SC in rodents (Flecknell, 1984); however, the authors are not aware of studies comparing its potency with that of other analgesics, such as opioids.

b. Xylazine hydrochloride (Rompun). Xylazine is a α_2-adrenergic agonist and, as such, possesses sedative, analgesic, and muscle relaxant properties. Xylazine is frequently used in combination with parenteral anesthetics, such as ketamine, to provide supplementary analgesia and muscle relaxation. The use of xylazine as a postoperative analgesic at dosages of 5–12 mg/kg SC in rats provides short duration, dose-dependent sedation and less potent analgesia, particularly for somatic pain, than is supplied by the opioid analgesics (Wixson *et al.,* 1991).

c. Flunixin Meglumine (Banamine). Flunixin meglumine is a nonnarcotic, nonsteroidal analgesic agent with anti-inflammatory and antipyretic activity. Flunixin is only available as a parenteral formulation for IV or IM administration and causes significant irritation when delivered subcutaneously. Its efficacy appears to vary with the type of pain and species in that it has been shown to provide moderate to potent analgesia for visceral pain in equines (Kalpravidh *et al.,* 1984), but may be poorly analgesic for somatic pain of a noninflammatory origin in many other species.

In the experience of the authors, this drug has minimal analgesic efficacy in rodents (Wixson *et al.,* 1991). This observation is in accord with a study by Liles and Flecknell (1994) in which flunixin was compared with buprenorphine and carprofen as a postoperative analgesic in rats. Results of this study showed that there were no significant differences between flunixin and saline-treated controls in regard to parameters such as postoperative weight loss, food and water consumption, and locomotor activity, which are all indices presumed to reflect overall depression of the animal due to postoperative pain.

d. Phenylbutazone (Butazolidin). Phenylbutazone is a synthetic nonhormonal, anti-inflammatory, antipyretic compound. This drug produces analgesia to pain of inflammatory origin in larger species, e.g., canines and equines. Anecdotal reports of its reputed efficacy in rodents should be viewed with caution as a controlled evaluation of its potency and adverse effects has not been undertaken.

e. IBUPROFEN (ADVIL, NUPRIN, MOTRIN). Ibuprofen is an NSAID drug that is efficacious in multiple species of animals and humans for relief of pain, particularly of inflammatory origin. A controlled study has not been undertaken to evaluate its efficacy and toxicity in rodents.

f. TRICYCLIC ANTIDEPRESSANTS. Clinically, tricyclic antidepressants (TCAs) have been used for many years to treat a variety of painful conditions in humans (Godefroy *et al.,* 1986). It is theorized that the analgesic effect of TCAs is due to either a blockade of serotonergic reuptake or a generalized antidepressant effect. Most of the putative mechanisms for their analgesic effect(s) are based on the results of experimental studies in rodents. These studies have focused mainly on the chronic administration of tricyclic antidepressants to rats, with and without accompanying opioid agonist analgesics, in an effort to investigate mechanisms for the long-term relief of pain of neuropathic origin in humans.

Results of these studies may offer valuable insights to laboratory animal veterinarians challenged with obtunding chronic pain in rodents. According to several reports, the tricyclic antidepressants amitriptyline (at dosages of 0.5–2.0 mg/kg IV, 1–10 mg/kg IP, or SC every 3–12 hr in rats and 1.2–5 mg/kg SC or IP in mice) and imipramine (at dosages of 10 mg/kg IP or SC every 12–24 hr in rats and 2.3 mg/kg every 12–24 hr in mice) decreased behavioral signs of pain perception, including digital irritation and autotomy, for up to 2 weeks with no tolerance or overt adverse effects (Abad *et al.,* 1989; Ardid and Guibaud, 1992; Butler *et al.,* 1985; DeFelipe *et al.,* 1986; Godefroy *et al.,* 1986). Although not as potent as morphine, chronically administered TCAs have been shown to be up to 70 times more effective than aspirin in experimental analgesiometry in mice (Spiegel *et al.,* 1983). Although there are no published reports evaluating TCAs for the relief of nonneuropathic pain in laboratory rodents, in the experience of one of the authors (SKW), amitriptyline (Elavil, 1.0 mg/kg every 12 hr) combined with methadone (1.0 mg/kg every 12 hr, as needed) can be useful in relieving pain associated with sciatic nerve injury in rats for up to 3 weeks postoperatively. Methadone was chosen as the narcotic component of this regimen due to its low addictive potential in humans (Urban *et al.,* 1986). This regimen effectively minimized self-trauma such that autotomy did not occur in any study animal. No overt adverse effects or drug tolerance were noted.

g. MORPHINE SULFATE. Morphine is the classic opioid agonist analgesic. It has been used in rodents for many years and can be relied on to provide potent, short duration (2–4 hr) analgesia as well as some sedation and respiratory depression in rats at a dosage of 10 mg/kg SC or IM (Jenkins, 1987; Flecknell, 1984; Flecknell and Liles, 1991; Wixson *et al.,* 1991). In the mouse, morphine has been shown to relieve experimental pain at a dosage of 0.98 mg/kg SC (Schmidt *et al.,* 1985).

Author's Preference	
Agent:	Morphine
Route:	SC
Dose:	10 mg/kg SC
Species:	Rats
Comments:	Provides excellent short duration (2–4 hr) analgesia

h. MEPERIDINE (DEMEROL, PETHIDINE). Meperidine is an opioid agonist analgesic that is less potent than morphine and, at pharmacologic dosages, is often accompanied by sedation and some degree of respiratory depression. Meperidine has been recommended for relief of pain at a dosage of 20 mg/kg SC or IM in rodents every 2–3 hr (Jenkins, 1987; Flecknell, 1984, 1991; Lightowler and Wilder Smith, 1963; Collier *et al.,* 1961). Additional studies in mice (Schmidt *et al.,* 1985) suggest that a lower dosage of 3.6 mg/kg SC may show some analgesic efficacy.

i. FENTANYL. Fentanyl is a synthetic opioid agonist that is estimated to be 80 times more potent than morphine (Short, 1987). Although fentanyl is used quite frequently in humans for intra- and postoperative analgesic therapy, its adverse effects (respiratory depression of a magnitude requiring positive pressure ventilation) and most effective method of delivery (patient-controlled analgesia using IV infusion ''on-demand'' pumps) make it impractical for many studies using rodents. Although plasma fentanyl concentrations can be measured and correlated with the desired dosage needed to achieve pain relief in humans, this information is also lacking at the present time in rodents. For these reasons, clinical dosages for the sole use of fentanyl are not available for rodents. Fentanyl does, however, enjoy great popularity for use in rodents as the narcotic component of two neuroleptanalgesics, Innovar-Vet and Hypnorm. These agents are discussed earlier in this chapter.

j. SUFENTANIL, ALFENTANIL, AND CARFENTANIL. All of these drugs are fentanyl congeners that have been developed as potent analgesics. Sufentanil and alfentanil are widely available and are currently quite popular for preoperative and intraoperative sedation and analgesia in humans, particularly when delivered by sustained intravenous infusion. Carfentanil is available for veterinary use and is most frequently used for the sedation of wildlife and exotics. Aside from their great analgesic potency, additional advantages of this group of compounds include high lipid solubility with a subsequent rapid onset following IV administration and comparatively stable hemodynamics, particularly following delivery by sustained infusion (Short, 1987).

Although the comparative potencies of these drugs have not been studied in rodents, bolus dosages in rats have shown analgesic potencies in excess of that of morphine of 4500 times with sufentanil, 73 times with alfentanil, and 10,000 times for carfentanil, respectively (Short, 1987). The range of duration for these drugs varies from ultrashort (alfentanil) and short (sufentanil) to intermediate/long (carfentanil). Dosages are not available for clinical use of these analgesics in rodents.

Following bolus administration, in particular, all of these agents seriously depress respiration such that mechanical ventilation is often required.

k. OXYMORPHONE HYDROCHLORIDE (NUMORPHAN). Oxymorphone is a semisynthetic opioid agonist. The relative potency has not been thoroughly evaluated in all rodent species, but it is likely to be more potent than morphine based on its action in other species. In rats, oxymorphone (0.22–0.33 mg/kg SC) provides potent analgesia similar in magnitude to that of morphine with a similar duration of action but with a greater incidence of respiratory depression (Wixson *et al.,* 1991). In mice, a dosage of 0.051 mg/kg SC has effectively obtunded experimental pain (Schmidt *et al.,* 1985).

Author's Preference	
Agent:	Oxymorphone
Route:	SC
Dose:	0.22–0.33 mg/kg SC
Species:	Rats
Comments:	Potent short duration analgesia

l. PENTAZOCINE LACTATE (TALWIN). Pentazocine is classified as an opioid with mixed actions. In this regard, it binds to the μ receptor as a competitive antagonist or exerts only limited action, e.g., is a partial agonist at the μ, κ, and σ receptors (Jaffee and Martin, 1985). Because pentazocine is a weak partial agonist at μ receptors, its analgesic potency is estimated to be one-half that of morphine (Short, 1987). Similarly to morphine, pentazocine may cause sedation, respiratory depression, and constipation in humans with prolonged use. These adverse effects have not been investigated in rodents. Pentazocine is available as a parenteral solution (pentazocine lactate, Talwin), but has shown relatively short-term analgesia at dosages of 1–21 mg/kg SC in mice (Flecknell, 1984; Schmidt *et al.,* 1985) and 8–10 mg/kg SC in rats (Collier *et al.,* 1961; Flecknell, 1984; Pircio *et al.,* 1976; Wixson *et al.,* 1991).

m. NALBUPHINE (NUBAIN). Nalbuphine is an opioid mixed agonist/antagonist that is structurally related to oxymorphone and naloxone. Its analgesic effects are mediated by agonist activity at κ receptors. Unlike morphine but similar to the other mixed agonist/antagonist analgesics, nalbuphine exhibits a ceiling effect such that further increases in dosage produce no further analgesia but may escalate its adverse effects. Nalbuphine has been recommended for use in rodents as a component of balanced anesthesia at a dosage of 40 μg/kg bolus injection, followed by a 100-μg/kg/min constant intravenous infusion (DiFazio *et al.,* 1981). According to DiFazio *et al.* (1981), continuous IV administration of nalbuphine is required in the rat because the half-life of nalbuphine in rats, in contrast to the 5-hr half-life in humans, is only 12 min. For postoperative analgesia, nalbuphine has been recommended at a dosage of 1 mg/kg SC administered every 4 hr for six total injections (Flecknell and Liles, 1991). According to Flecknell and Liles (1991), administration of nalbuphine following this dosage regimen in rats will prevent the depressant effects of surgery, perhaps due to a stimulatory effect rather than a specific analgesic action. In mice, nalbuphine has shown moderate analgesic activity versus experimental pain at a dosage of 1.2 mg/kg SC (Schmidt *et al.,* 1985). Because nalbuphine shows marked antagonist as well as agonist activity, it should be used with caution in animals initially receiving a potent opioid agonist so that any remaining analgesic activity of the agonist is not mitigated.

n. BUTORPHANOL TARTRATE (STADOL, TORBUGESIC, TORBUTROL). Butorphanol is an opioid mixed agonist/antagonist analgesic. Its pharmacology and analgesic efficacy have been extensively reviewed (Jaffee and Martin, 1985; Short, 1987; Pircio *et al.,* 1976). Butorphanol has been shown to be 3 to 5 times more potent than morphine in humans and up to 30 times more potent than morphine in rats. Butorphanol is only available in the following parenteral formulations: (i) butorphanol (Stadol), the human preparation, which is supplied at a concentration of 2 mg/ml; (ii) Torbugesic, the veterinary formulation, which is available at a concentration of 10 mg/ml; and (iii) Torbutrol, also a veterinary formulation, at a concentration of 0.5 mg/ml. Prior to administering this drug, care should be taken to note whether the human or veterinary formulation is being used and the animal should be appropriately dosed on a body weight basis. Butorphanol has been shown to provide moderate analgesia for 2–4 hr in rodents at dosages of 0.05–5.4 mg/kg SC in mice (Flecknell, 1984; Schmidt *et al.,* 1985) or 0.04–23.3 mg/kg SC in rats, with 10–12 mg/kg as the optimal dose (Pircio *et al.,* 1976; Wixson *et al.,* 1991).

o. BUPRENORPHINE (BUPRENEX). In rats, buprenorphine dosages of 0.05–1.0 mg/kg SC have been used to obtund pain (Cowan *et al.,* 1977a,b; Flecknell, 1984; Wixson *et al.,* 1991; Liles and Flecknell, 1993). Although most authors agree that buprenorphine provides a duration of effect of 6–12 hr in rats, there is some disagreement as to the analgesic potency of this drug in rodents. Studies have compared buprenorphine to several pure opioid agonists for the relief of acute, moderately intensive pain in rats (Wixson *et al.,* 1991). Even though the results of these studies showed that a 1.0-mg/kg subcutaneous bolus of buprenorphine provides up to 8 hr of analgesia, substantially greater relief from pain was shown by morphine and oxymorphone. Higher dosages of buprenorphine show a "ceiling effect" in that they do not lead to increasing analgesia and may, in fact, demonstrate lesser relief from the painful stimuli used in these studies. Thus, bolus administration of buprenorphine does not appear to provide the same magnitude of strong, long-lasting analgesia in rats that has been documented in humans. Further studies by one of the authors (SKW) have shown that repetitive dosing with high levels (exceeding 1.0 mg/kg) of buprenorphine may lead to hematuria and GI bleeding in rats, which can be fatal. For this reason, limiting the use of this agent to situations in which the animal requires mild, but longer duration, analgesic therapy may be prudent. Dosages in excess of 1.0 mg/kg by the SC or IM route of administration provide no greater analgesia and may precipitate urinary or GI bleeding. Recommendations for continuous IV infusion of buprenorphine to rats are discussed later in this chapter.

Liles and Flecknell (1993, 1994) studied the use of low-dose buprenophine as a postoperative analgesic in rats. Results showed that 0.05 mg/kg buprenorphine SC immediately following surgery was superior to 0.7 ml bupivacaine infiltrated into the wound margins to mitigate the decrease in food and water consumption, body weight, and locomotor activity that occurred in control animals upon recovery from laparotomies or bile duct ligation.

Limited guidelines are available for the use of buprenorphine in mice. Dosages of 0.008–0.029 mg/kg SC and 2.4 mg/kg IP have been recommended (Cowan *et al.,* 1977a,b; Schmidt *et al.,* 1985). Pieper *et al* (1995), however, studied excessive locomotor stimulation caused by postoperative buprenorphine administration to *scid/scid* mice. In this study, mice anesthetized with tribromoethanol were given 0.1 mg/kg buprenorphine on regaining the righting reflex postoperatively. Within several hours, all mice exhibited constant, rapid movements and Straub tail effect; the latter is indicative of buprenorphine stimulation of opiate receptors in the animals. The authors hypothesized that the alcohol-based tribromoethanol anesthetic may have potentiated the motor stimulation caused by buprenorphine, indicating these two drugs, in combination, may produce detrimental effects, such as decreased or delayed wound healing, in *scid/scid* mice.

3. Opioid Antagonists

Only minor structural changes between the opioid agonist and the antagonist drug families result in profound differences in activity as well as their analgesic and physiologic effects. Pure narcotic antagonists, in their own right, exert minimal effects at clinical dosages unless their use is preceded by the administration of opioid agonist drugs. In the latter circumstance, many of the adverse physiologic effects as well as the beneficial analgesic properties of the opioid agonist may be reversed.

In rodents, naloxone (Narcan) has been the primary opioid antagonist used clinically. This drug is a competitive antagonist at μ, σ, κ, and δ opioid receptors (Jaffee and Martin, 1985). Naloxone has been used in rats at a dosage of 0.005–0.1 mg/kg IV to reverse the respiratory depressant effects of opioid agonists (Flecknell, 1991).

4. Methods for Sustained Parenteral Delivery of Analgesics to Rodents

Although mechanisms are available for sustained drug delivery from implanted reservoirs in humans and other species of animals (e.g., subcutaneously implanted infusion pumps, slow-release pellets, osmotic pumps), these modalities show limited efficacy for chronic analgesic therapy in rodents. Preliminary studies have explored the use of osmotic minipumps to continuously deliver a buprenorphine dosage equivalent to that which rats would receive if given 0.05 or 0.10 mg/kg subcutaneously twice daily for 2 weeks (B. A. Deeb, personal communication). Results showed a decreased response to experimental painful stimuli in treated verses control rats. [*Note:* In the experience of one of the authors of this chapter (SKW), however, available commercial preparations of opioids usually lack sufficient concentration to deliver dosages of opioids by osmotic pump which are clinically necessary to obtund pain in rodents.]

Regimens have been described for continuous IV infusion of morphine to rats. In a study by Kissin *et al.* (1991), continuous IV delivery of 4 mg/kg/hr morphine produced maximal analgesia to experimental pain, but acute tolerance to the analgesic effect developed within 8 hr. Conversely, the brain morphine concentration continued to increase despite the decline in analgesic potency, suggesting the development of acute pharmacodynamic tolerance. Buprenorphine has also been administered by continuous intravenous infusion to relieve postoperative pain in a rat model of acute pancreatitis (Leese *et al.,* 1988; Wereszczynska-Siemiatkowska *et al.,* 1987). Following an intramuscular loading dose of 0.6 mg/kg at the conclusion of surgery, continuous IV infusion of 15 mg/kg/hr produced a significant decline in the response to toe and tail pinch and less pain on abdominal palpation in treated animals. Analgesic therapy was continued for up to 72 hr postoperatively, with no evidence of overt adverse physiologic and behavioral effects or tolerance to the analgesic effects of the drug. The optimal doses chosen for continuous IV analgesic administration to rats were 10 times the suggested human dose, which is consistent with the surprisingly high subcutaneous bolus dosage regimen discussed earlier in this chapter.

Additional problems with the earlier-cited methods for chronic parenteral drug delivery to rodents include technical expertise required to implant the depot reservoir and the excessive size of currently available infusion or osmotic pumps as well as the substantial costs that are incurred when using these delivery methods in multiple animals. Further research is needed both to overcome these problems and to develop alternative methodologies for sustained parenteral delivery of analgesics to various species of rodents.

Transdermal and transmucosal analgesic drug deliveries, particularly using fentanyl, have also become popular for sustained analgesic drug delivery to humans. The authors are not aware of attempts for delivery by these methods to rodents; however, this technology offers promise for use in future research studies.

XI. SPECIAL ANESTHETIC CONSIDERATIONS

A. Anesthetic Techniques for Use on Pregnant and Neonatal Rodents

There have been very few controlled studies investigating the reproductive safety of anesthetics in rodents using objective measures such as fetal resorptions, neonatal mortality, or congenital anomalies following surgery on pregnant animals. There is also a paucity of information available on the efficacy and safety of various anesthetic techniques for use in neonatal rodents. Clearly, additional studies need to be done to establish the best anesthetic techniques for use in pregnant and neonatal rodents. Anesthetic experience with these special applications is summarized below; however, the reader is advised to read each of these references in their entirety and evaluate the methodology used and the appropriateness of the conclusions presented within.

1. Anesthetic Techniques for Use in Pregnant Rodents

Injectable ketamine combinations are most commonly used for anesthesia of pregnant animals. Stickrod (1979) reported that a dosage of 90 mg/kg ketamine plus 10 mg/kg xylazine administered on days 17 and 19 of gestation produced surgical anesthesia in pregnant rats with no overt adverse effects on gestation or the fetuses. Maternal injection with ketamine (25–100 mg/kg IP) prior to day 16 of gestation has been shown to cause degeneration in the heart, liver, and kidneys of neonatal rat pups (Kochhar *et al.,* 1986). Although there was no effect on litter size or survivability, the magnitude of these degenerative effects was dependent on the dosage and duration of treatment. Diazepam is often combined with ketamine for rodent anesthesia; however, several studies in rats have implicated prenatal diazepam as the causative agent in behavioral and developmental as well as late postnatal teratogenic effects, including lower immunoglobulin levels, infections, arteriosclerosis, and neoplasia (Frieder *et al.,* 1984; Livezey *et al.,* 1986a,b). For these reasons, the authors suggest that diazepam should not be chosen as a component of anesthesia for pregnant rats. Barbiturates have also been implicated as embryotoxic agents in rats (Gunberg, 1982).

Currently available data suggest that inhalant agents may also have embryotoxic effects. Although no teratogenic effects due to isoflurane exposure have been observed in rabbits and rats, increased frequencies of cleft palate, skeletal variations, and fetal growth retardation have occurred in pregnant mice exposed to light anesthetic doses of isoflurane (Mazze *et al.,* 1985). Although studies have not been done in rodents, halothane significantly decreases pregnancy rate following embryo transfer in humans (Matt *et al.,* 1991). For these reasons, inhalant anesthetics should be used with caution for embryonic manipulations required for the maintenance of transgenic rodent colonies. Alternatively, the authors are unaware of any evidence contraindicating the use of inhalant agents for use in near-term pregnant or neonatal animals. However, because isoflurane crosses the placental barrier rapidly, investigators designing studies incorporating its use in the experimental design of pregnant animals should be aware of its potential effects. Reports in human medicine evaluated the use of isoflurane and halothane in obstetrical procedures (Abboud *et al.,* 1989a,b; Andrews *et al.,* 1992). The studies concluded that halogenated inhalants were suitable adjuncts as analgesia for vaginal delivery and cesarean section, but that mothers were at risk for increased maternal blood loss secondary to depressed uterine muscle contractility, as compared to blood loss during regional anesthesia.

2. Anesthetic Techniques for Use in Neonatal Rodents

Surgery on neonatal rodents is particularly challenging. Developing reliable and safe anesthetic methods in neonates is difficult because only a paucity of information exists on drug metabolism, excretion, and biotransformation; sophisticated techniques are not described to monitor; and equipment is not commercially available to utilize inhalant methods in neonatal animals. Equally frustrating is the frequently unpredictable response of neonatal animals to many drugs, e.g., the much greater induction time required for CO_2 immobilization in neonates. In order to provide standardization of anesthetic dosages in neonatal rodents, dosages should be calculated on the basis of body surface area rather than body weight. This format takes into account the proportionately greater body surface area of neonates that may require larger anesthetic dosages to attain a response similar to that which is seen in mature animals. Even if anesthesia is successful, surgical manipulations or chemical disinfectants can incite maternal cannibalism, thereby further complicating assessment of whether the anesthetic method could have produced viable animals. Finally, there have been virtually no standardized criteria established for the identification of pain perception in neonatal rodents, thus there are no guidelines accepted for analgesic management of these animals.

a. HYPOTHERMIA. Evidence shows that hypothermia is an appropriate form of analgesia/anesthesia. In humans, neural conduction velocity in peripheral nerves is reduced 75% when tissue is cooled to 20°C and is completely blocked at 9°C. Rats subjected to a cold water swim exhibit an analgesic potency similar to a 10-mg/kg dose of morphine, and the analgesic effect can remain up to 30 min following removal from hypothermia. Human infants anesthetized by hypothermia show no postoperative effects on somatic growth, psychomotor functioning, or intellectual development. Hypothermia has useful clinical applications that are utilized in neurosurgery (decreased brain volume and CSF pressure, and increased resistance of the nervous system to anoxia) and in open heart surgery (induces cardiac arrest, provides a bloodless surgical field, and avoids pulmonary complications by bypass).

Infant rodents have unique features that make them suitable for anesthesia utilizing hypothermia. Newborn rats and mice are poikilothermic and adult thermoregulatory capabilities do not develop until the third week of life. Because of their small body mass, rapid core cooling can be achieved by surface cooling. Newborns are more resistant to arrest of the blood supply to the brain, and infant rodents can tolerate extended periods of 1°C body temperature without known negative effects, whereas adult rats cooled to below 24°C develop potent retrograde amnesia lasting for several days. Evidence shows that pups suffer no profound physiologic/developmental damage from hypothermia anesthesia, as shortly after awakening, rat pups will crawl to the dam and commence suckling. Behavioral tests performed 1 day following hypothermia show no sign of decreased learning or respiratory distress (Phifer and Terry, 1986). Hypothermia is inexpensive, requires only moderate technical skill, and is safe for use by personnel. In a recent evaluation of anesthetics for use in 1–3 day old rat pups, hypothermia was shown to produce anesthesia superior to that produced by ketamine, pentobarbital and fentanyl-droperidol (Danneman and Mandrell, 1996). In this study, hypothermia produced consistent and safe anesthesia for minor surgical procedures with rapid induction and recovery times, minimal mortality and no evidence of maternal rejection. Older rat pups (12–18 days of age) were safely anesthetized using a combination of methoxyflurane, hypothermia, and maintenance on a cold pack during surgery (Phifer and Terry, 1986).

Disadvantages associated with hypothermia include increased risk of ventricular fibrillation, tissue hypoxia, and metabolic acidosis after rewarming. In surgical procedures that require lengthy anesthesia, it can be difficult to maintain a hypothermic state without additional equipment, such as a cooling platform, and there may be controversy considering the degree of analgesia achieved (Park *et al.,* 1992).

Several acceptable techniques used to induce hypothermia in neonates include placing pups on crushed ice or in paper-lined test tubes packed in crushed or dry ice (this method requires up to 15 min to reach torpor). If pups are placed in a cold room, up to 45 min is required to reach hypothermia. Rapid chilling can be achieved by submerging the pup to the cervical area in water and crushed ice. A 3- to 4-min exposure produces torpor that lasts up to 10 min.

Recovery from anesthesia can be rapid, but rewarming techniques should not use aggressive heating (heating pads or lamps) to avoid tissue damage. An incubator at 33°C for 20–30 min is appropriate.

b. PARENTERAL AND INHALANT TECHNIQUES. Park *et al.* (1992) reviewed available methods to induce anesthesia in neonatal rodents. In this study, halothane (5% in 100% oxygen) and injectable Innovar-Vet (0.02 ml/pup SC using a dilution of 0.5 ml in 9.5 ml saline) were contrasted. Both anesthetic methods require a prolonged induction (approximately 6.5 min) and produced acceptable anesthetic depth for approximately 30 min with approximately 95% survivability. Despite oxygen supplementation, halothane produced greater respiratory depression than did Innovar-Vet, with an approximately 15- to 40-min recovery period. Naloxone (Narcan, 0.02 ml/pup) was administered as a reversal of Innovar-Vet anesthesia, with an average of 3.5 min until return of the pedal reflex, increased respiratory rate, and vigorous motor activity. Pups were necropsied 7 days following recovery, and the only developmental anomalies noted were microphthalmia and retinal folding in Innovar-Vet anesthetized pups; however, the authors fail to note the incidence of these anomalies in pups not anesthetized. Renarcotization from recycling of the anesthetic frequently occurs following Innovar anesthesia in humans; however, this effect was not noted in neonates used in this study. Methods are also presented in Park *et al.* (1992) to condition both pregnant and postparturient females to minimize maternal rejection and cannibalism. Danneman and Mandrell recently reported that methoxyflurane provided effective and safe anesthesia of up to 30 minutes duration in neonatal (1–3 days of age) rat pups (Danneman and Mandrell, 1996).

B. Ultra-Short-Term Anesthesia: Useful Techniques for Orbital Bleeding, Ascites Tumor Taps, Tumor Injection, and DNA Analysis

Many commonly performed experimental manipulations using rodents require momentary restraint for procedures that humans deem to be uncomfortable or perhaps painful to the animals. These procedures include orbital bleeding, tail transection for DNA analysis, injection of core samples of transplantable tumors, and ascites taps. Arguably, the amount of time necessary to administer an inhalant or parenteral anesthetic agent may be equal to or exceed the time necessary to accomplish the experimental manipulation. Furthermore, investigators are often reticent to consent to the risk of anesthesia to obtund the possibility of a few seconds of distress or pain, particularly when these procedures have not been performed with anesthesia in the past.

Each Institutional Animal Care and Use Committee must develop standards for whether or not anesthesia will be required and, if so, list those agents deemed appropriate for these momentary procedures in rodents. In the experience of the authors, inhalant agents, such as CO_2, methoxyflurane, and isoflurane, permit these manipulations to be completed in a safe, humanely acceptable manner. Alternatively, short-acting parenteral agents, such as sedatives, neuroleptanalgesics with appropriate narcotic reversal, and ultra-short-acting barbiturates, also provide acceptable restraint and indifference to surroundings to enable completion of these brief procedures. Specific techniques used to administer each of these anesthetic agents are discussed earlier in this chapter.

ACKNOWLEDGMENTS

The authors gratefully acknowledge the contributions of Theresa Bosley, Dawn Mousseau, Charlotte Gonzales, Neal Storey, and Mark Rasmussen to this chapter.

APPENDIX

Comprehensive listings of anesthetics, tranquilizers, and analgesics are presented in Tables A.I through A.VI.

TABLE A.I

ANESTHETICS AND TRANQUILIZERS USED IN MICE

Drug	Dosage	Route of injection	Duration of anesthesia	Reference
Barbiturates				
Thiopental (Penthothal)	25–50 mg/kg	IV	10 min	Taber and Irwin (1960), Glen (1980),
	50 mg/kg	IP	Unproven	Rank and Jensen (1989), White and Field (1987)
Hexobarbital	100 mg/kg	IP	Unproven	Taber and Irwin (1969), White and Field (1987)
EMTU (Inactin)	80 mg/kg	IP	Unproven	Buelke-Sam *et al.* (1978)
Methohexitone	8–16 mg/kg	IV	2 min	Glen *et al.* (1980)
Pentobarbital	30–40 mg/kg (sedation)	IP	10–300 min	
(Nembutal)	50–90 mg/kg (anesthesia) dilute in saline to give a volume of 0.1 ml/10 g body weight	IP		Lovell (1986a,b,c) Gardner *et al.* (1995)
	50–60 mg/kg	IV		Rank and Jensen (1989)
Dissociatives				
Ketamine	80–100 mg/kg	IM	Unproven	Green *et al.* (1981)
(Vetalar, Ketanest)	100 mg/kg	IP	Unproven	White and Field (1987)
	50 mg/kg	IV	Unproven	White and Field (1987)
Ketamine +	44 mg/kg	IP	Sedation only	Gardner *et al.* (1995)
acetylpromazine (Acepromazine)	0.75 mg/kg	IP		
Ketamine +	100 mg/kg	IM	Unproven	Clifford (1984)
acetylpromazine (Acepromazine) +	2.5 mg/kg	IM		Flecknell (1987b)
xylazine	2.5 mg/kg	IM		
Ketamine + diazepam	200 mg/kg	IM	Unproven	
(Valium)	5 mg/kg	IP		
Ketamine +	100–200 mg/kg	IM		Erhardt *et al.* (1984)
xylazine	5–16 mg/kg	IP	60–100 min (anesthetic depth varies from sedation to anesthesia)	Mulder and Mulder (1979) Clifford (1984)
Tiletamine + zolazepam (Telazol)	80–100 mg/kg (restraint)	IP		Silverman *et al.* (1983)
Tiletamine + zolazepam (Telazol)	7.5–45 mg/kg	IM	36 min	Gardner *et al.* (1995)
+ xylazine	7.5–45 mg/kg refer to reference for methods for combining Telazol + xylazine	IM	99–145 min	
Neuroleptanalgesics				
Fentanyl + droperidol (Innovar-Vet)	0.001–0.01 ml/g (using a 10% solution of the commercial Innovar-Vet solution)	IM	Unproven	Walden (1978) Lewis and Jennings (1972)
Fentanyl + droperidol (Innovar-Vet) + diazepam	using a 1/10 saline dilution of Innovar-Vet, 0.1 ml/30 g after injecting	IP	Unproven	
	5 mg/kg diazepam	IP		
Fentanyl + fluanisone (Hypnorm)	0.01 ml/30 g	IP	Sedation only	Flecknell and Mitchell (1984), Flecknell (1987b)
Fentanyl + fluanisone (Hypnorm) +	0.1–0.2 ml/30 g of a 1/10 dilution of Hypnorm	IP, SC	20–40 min	Flecknell and Mitchell (1984), Flecknell (1987b) Green (1975)
diazepam	5 mg/kg	IP, SC		
Fentanyl + fluanisone (Hypnorm) + midazolam	10–13.3 ml/kg	IP	20–40 min	Flecknell and Mitchell (1984), Flecknell (1987b)
Carfentanyl +	0.003 mg/kg	IM	Unproven	Erhardt *et al.* (1984)
etomidate	15 mg/kg	IM		

TABLE A.I (*continued*)

ANESTHETICS AND TRANQUILIZERS USED IN MICE

Drug	Dosage	Route of injection	Duration of anesthesia	Reference
Other				
α-Chloralose (5% concentration)	114 mg/kg	IP	Unproven	White and Field (1987)
Alphaxolone–alphadolone	7–25 mg/kg	IV bolus	7–8 min (14 mg/kg dosage)	Green *et al.* (1978)
(Saffan, Althesin)	(14 mg/kg is optimal dose) maintain with 4–6 mg/kg every 15 min	IV bolus	10 min (25 mg/kg dosage)	Rank and Jensen (1989), Glen (1980)
Chloral hydrate	60–90 mg/kg	IP	Unproven	Green *et al.* (1978)
	370–400 mg/kg	IP		White and Field (1987)
Etomidate	23.7–33.0 mg/kg	IP	20 min	Gomwalk and Healing (1981)
(Hypnomidate)	11.2 mg/kg	IV	20 min	Gomwalk and Healing (1981)
Propofol (Rapinovet, Diprivan)	12–26 mg/kg	IV	5.6–6.9 min	Glen (1980)
Tribromoethanol (Avertin) (1.2% solution)	125–250 mg/kg	IP	30–45 min	Taber and Irwin (1969) Papaioannou and Fox (1993) Gardner *et al.* (1995)
Inhalants				
Carbon dioxide	50–70% mixed with oxygen; vaporized dry ice	Inhalant		Urbanski and Kelley (1991) Taber and Irwin (1969)
Ether	ambient vaporization of ether-soaked gauze			Taber and Irwin (1969) Tarin and Sturdee (1972)
Methoxyflurane	To effect (0.5–3%)			Tarin and Sturdee (1972) Gardner *et al.* (1995)
Halothane	1–4%			Tarin and Sturdee (1972) Markovic *et al.* (1993)
Isoflurane	1–4%			Markovic *et al.* (1993)

TABLE A.II

ANESTHETICS AND TRANQUILIZERS USED IN RATS

Drug	Dosage	Route of injection	Duration of anesthesia	Reference
Sedatives/tranquilizers				
Droperidol	0.5–2.0 mg/kg (sedation)	SC	Sedation only	Quinn *et al.* (1993)
Diazepam	5.0–15.0 mg/kg (sedation)	SC	Sedation only	Quinn *et al.* (1993)
Barbiturates				
Thiopental	20–40 mg/kg	IV	5–10 min	Rao (1990), White and Field (1987)
(Pentothal) (1.25% w/v solution)	40 mg/kg	IP		White and Field (1987)
Hexobarbital	100 mg/kg	IP	Unproven	Kaczmarczyk and Reinhardt (1975)
EMTU (Inactin)	80–100 mg/kg	IP	Unproven	Buelke-Sam *et al.* (1978) White and Field (1987)
Methohexital	10–15 mg/kg	IV	5–10 min	Yatsu *et al.* (1972)
(1% solution)	40 mg/kg	IP	15–20 min	S. K. Wixson (personal experience)
Methohexital + pentobarbital	40 mg/kg (induction) 20 mg/kg (maintenance)	IP IP	2 hr	S. K. Wixson (personal experience)
Pentobarbital (Nembutal)	40–60 mg/kg	IP	80–95 min	Rao (1990), Wixson *et al.* (1987a–d), Commissaris *et al.* (1982), Routtenberg (1968), Aston (1966), Hughes *et al.* (1982), Field *et al.* (1993)
	60–100 mg/kg	IM	Unproven	Seyde *et al.* (1985)
	induce with 50 mg/kg, then 500 μg/kg/min	IP Continuous IV infusion		
Dissociatives				
Ketamine (Vetalar, Ketanest)	50–100 mg/kg (sedation)	IM	Unproven	Green *et al.* (1981)

TABLE A.II *(continued)*

ANESTHETICS AND TRANQUILIZERS USED IN RATS

Drug	Dosage	Route of injection	Duration of anesthesia	Reference
Ketamine + acetylpromazine (Acepromazine)	75–80 mg/kg 2.5 mg/kg	IM IM	Unproven	Flecknell (1987b)
Ketamine + diazepam (Valium)	45–60 mg/kg 5–10 mg/kg or 2.5 mg/kg	IP IP IM	45–60 min	Rao (1990), Wixson *et al.* (1987a–d)
Ketamine + medetomidine	60–75 mg/kg 0.25–0.5 mg/kg	IP SC	Unproven	Nevalainen *et al.* (1989)
Ketamine + xylazine (Rompun)	40–87 mg/kg 5–13 mg/kg	IP, IM IP, IM	45–90 min	Green *et al.* (1981), Hsu *et al.* (1986), Van Pelt (1977), Wixson *et al.* (1987a–d), Smiler *et al.* (1990)
Tiletamine + zolazepam (Telazol)	20–40 mg/kg 20 mg/kg	IP IM	30–60 min	Silverman *et al.* (1983), Ward *et al.* (1974)
Tiletamine + zolazepam (Telazol) + Xylazine	20–40 mg/kg 5–10 mg/kg	IP IP	129–201 min	Wilson *et al.* (1993)
Tiletamine + zolazepam (Telazol) + butorphanol	20–40 mg/kg 1.25–5 mg/kg	IP IP	59–140 min	Wilson *et al.* (1992)
Neuroleptanalgesics				
Fentanyl + droperidol (Innovar-Vet)	0.02–0.06 ml/100 g 0.3 ml/kg	IP or IM		Wixson *et al.* (1987a–d) Lewis and Jennings (1972)
Fentanyl + fluanisone (Hypnorm)	0.4–0.5 ml/kg	IM or IP	20–90 min	Flecknell and Mitchell (1984), Green (1975), Royer and Wixson (1992)
Fentanyl + fluanisone (Hypnorm) + midazolam (Versed) (2 parts water for injection, 1 part Hypnorm and 1 part midazolam, the latter at 5 mg/ml concentration)	0.20–0.22 ml/kg	IP	60–90 min	Flecknell and Mitchell (1984)
Other				
α-Chloralose (5% w/v concentration)	31–65 mg/kg	IP	Unproven	White and Field (1987)
α-Chloralose + urethane	50–60 mg/kg 500–800 mg/kg (administer urethane 20–30 min prior to α-chloralose)	IP IP	500 min	Seyde *et al.* (1985), Hughes *et al.* (1982)
Alphaxolone + alphadolone (Saffan, Althesin)	10–12 mg/kg (supplement with 3–4 mg every 15–20 min)	IV bolus	5–12 min	Green *et al.* (1978) Child *et al.* (1971)
	0.25–0.45 mg/kg/min	Continuous IV infusion	Up to 8 hr	Green *et al.* (1978)
Chloral hydrate (5% concentration)	300–450 mg/kg	IP	60–136 min	Silverman and Muir (1993) Field *et al.* (1993)
Propofol (Rapinovet, Diprivan)	7.5–10.0 mg/kg (induction) 44–55 mg/kg/hr	IV Continuous IV infusion	8–11 min	Glen (1980) Adam *et al.* (1980)
	Sedate with 0.5–1.0 ml/kg fentanyl/fluanisone, induce with 0.1 ml propofol, maintain with 4–6 ml/kg/hr propofol	IP IV Continuous IV infusion	3 hr	Brammer *et al.* (1993)
Tribromoethanol (Avertin) (0.25% w/v concentration)	300 mg/kg	IP	Unproven	
Urethane (50% w/v concentration)	1000–1500 mg/kg	IP	Up to 24 hr	Field *et al.* (1993), Severs *et al.* (1981)

TABLE A.II *(continued)*

ANESTHETICS AND TRANQUILIZERS USED IN RATS

Drug	Dosage	Route of injection	Duration of anesthesia	Reference
Inhalants				
Carbon dioxide	50–80% CO_2 mixed with 20–50% oxygen	Inhalant		Fenwick and Blackshaw (1989)
Ether[a]	Ambient vaporization of ether-soaked gauze	Inhalant		Fowler *et al.* (1980)
Nitrous oxide	70%, in combination with halothane and oxygen	Inhalant		Mahmoudi *et al.* (1989)
Methoxyflurane	1–3%	Inhalant		
Halothane	1–3%	Inhalant		Dardai and Heavner (1987) Flecknell and Liles (1991) Mahmoudi *et al.* (1989)
Isoflurane	1–5%	Inhalant		Cole *et al.* (1990), Zucker (1991), Maekawa *et al.* (1986), Dardai and Heavner (1987)

[a]Ether should be used with extreme caution due to its explosive nature.

TABLE A.III

ANESTHETICS AND TRANQUILIZERS USED IN GUINEA PIGS

Drug	Dosage	Route of injection	Duration of anesthesia	Reference
Sedatives/tranquilizers				
Diazepam	2.5–5.0 mg/kg	IP	Sedation	Green (1975)
Xylazine	5–40 mg/kg	IP	Mild sedation	Green (1975)
Barbiturates				
Pentobarbital	15–40 mg/kg	IP	60 min	Frisk *et al.* (1982)
(Nembutal)	30 mg/kg	IV	60 min	White and Field (1987), Clifford (1984)
Pentobarbital +	20–25 mg/kg	IM		Carter (1983)
diazepam (Valium)	1–8 mg/kg	IM		
Dissociatives				
Ketamine	40–200 mg/kg (40 mg/kg is optimal dose)	IM		Green *et al.* (1981), Frisk *et al.* (1982)
Ketamine[a] +	33–44 mg/kg	IM	Sedation	Latt and Ecobichon (1984)
acetylpromazine (Acepromazine)	0.1–1.6 mg/kg or 125 mg/kg ketamine 5 mg/kg acepromazine	IM	Anesthesia	
Ketamine +	60–100 mg/kg	IM		Gilroy and Varga (1980)
diazepam	5–8 mg/kg	IM		
Ketamine +	125–150 mg/kg	IM	35 min	Tena *et al.* (1982)
droperidol	1 mg/kg			
Ketamine +	40 mg/kg	IM or IP	Unproven	Nevalainen *et al.* (1989)
medetomidine	0.5 mg/kg	SC		
Ketamine +	30–44 mg/kg	IM	77 min	Whur (1987), Gilroy and Varga (1980)
xylazine	0.1–5.0 mg/kg or	IM	(sedation)	Brown *et al.* (1989), Frisk *et al.* (1982)
	27 mg/kg ketamine + 0.6 mg/kg xylazine (combined with incisional infiltration with 1% lidocaine + 1:200,000 epinephrine), or	IM IM	60 min (sedation to general anesthesia)	Dawson and Scott-Conner (1988)
	40–100 mg/kg ketamine 4–5 mg/kg xylazine (optimal dose is 87 mg/kg ketamine + 13 mg/kg xylazine)	IM SC or IM	60 min (general anesthesia)	Wells *et al.* (1986), Brown *et al.* (1989), Reuter and Blaze (1987)

TABLE A.III *(continued)*

Anesthetics and Tranquilizers Used in Guinea Pigs

Drug	Dosage	Route of injection	Duration of anesthesia	Reference
Tiletamine + zolazepam (Telazol)	10–80 mg/kg (sedation)	IM or IP	Unproven	Ward *et al.* (1974)
Neuroleptanalgesics				
Fentanyl + fluanisone (Hypnorm)	0.5–1.5 ml/kg	IM	25–30 min (sedation)	Green (1975)
Fentanyl + fluanisone (Hypnorm) + diazepam	0.2–1.5 ml/kg	IM	60–70 min	Green (1975), Carter (1983)
	2.5–5 ml/kg			
	2.5– mg/kg	IP, IM	Up to 3.5 hr (with supplementation)	Whur (1987), Mulvey (1987)
	prolong with Hypnorm, 0.05 ml/kg	IM		Carter (1983)
Fentanyl + fluanisone (Hypnorm) + midazolam (Versed)	8.0 ml/kg of a mixture of 1 part Hypnorm, 2 parts water + 1 part midazolam (5 mg/ml concentration)	IP	Unproven	Flecknell and Mitchell (1984), Flecknell (1987a)
	supplement with 0.1–0.3 ml/kg	IP		
Fentanyl[a] + droperidol (Innovar-Vet)	0.44–0.88 ml/kg	IM	Unproven	Walden (1978) Rubright and Thayer (1970) Leash *et al.* (1973)
Fentanyl + droperidol (Innovar-Vet) + diazepam	1 mg/kg	IM	Unproven	Flecknell (1987a)
	2.5 mg/kg	IP		
Fentanyl + droperidol (Innovar-Vet) + pentobarbital	0.4 mg/kg	IM	60 min	Brown *et al.* (1989)
	15 mg/kg (pentobarbital is given 20 min prior to Innovar-Vet)	IP	Unproven	
Fentanyl (Sublimaze) + diazepam	0.32 mg/kg	IM	60–90 min	Brown *et al.* (1989)
	5 mg/kg (diazepam given 20 min prior to fentanyl)	IP		
Other				
Alphaxolone–alphadolone[b] (Saffan, Althesin)	40–45 mg/kg	IP or IM	40–90 min (sedation)	Green (1975) Brown *et al.* (1989)
	or 16–20 mg/kg	IV bolus	10–12 min (anesthesia)	Green *et al.* (1978)
Alphaxolone–alphadolone (Saffan, Althesin) + diazepam	45 mg/kg	IM	60 min	Brown *et al.* (1989)
	5 mg/kg (diazepam is injected 20 min prior to alphaxolone–alphadolone)	IP		
α-Chloralose (1%) + urethane (40%), in a 7:1 mixture	0.8 ml/100 g	IP	Greater than 120 min	Brown *et al.* (1989)
Chloral hydrate	400 mg/kg	IP	Unproven	Valenstein (1961) White and Field (1987)
Inhalants				
Methoxyflurane	to effect (0.5–3%)	Inhalant		Wells *et al.* (1986)
Halothane	1–5%	Inhalant		Franz and Dixon (1988), Seifen *et al.* (1989) Dardai and Heavner (1987), Cook *et al.* (1987)
Isoflurane	1–5%	Inhalant		Seifen *et al.* (1989), Dardai and Heavner (1987)

[a]IM injection of ketamine + acepromazine or fentanyl–droperidol may lead to self-mutilation in guinea pigs.

[b]Attempts to prolong the duration of alphaxolone–alphadolone anesthesia in guinea pigs by repetitive bolus administration may lead to pulmonary edema.

TABLE A.IV

ANESTHETICS AND TRANQUILIZERS USED IN HAMSTERS

Drug	Dosage	Route of injection	Duration of anesthesia	Reference
Barbiturates				
Pentobarbital (Nembutal)	70–90 mg/kg	IP	60–75 min	Skornik and Brain (1990) Reid *et al.* (1989), White and Field (1987)
Methohexitone + diazepam (combined in a mixture containing 7.5 mg/ml methohexitone + 1.25 mg/ml diazepam)	2–4 ml/kg	IP	15–30 min	Ferguson (1979) White and Field (1987)
EMTU (Inactin)	200 mg/kg	IP	Unproven	White and Field (1987)
Dissociatives				
Ketamine	40–80 mg/kg	IP	Sedation	Green *et al.* (1981)
	200 mg/kg	IP	Anesthesia	White and Field (1987)
	100 mg/kg	IM	Sedation	White and Field (1987)
Ketamine +	150 mg/kg	IM		
acetylpromazine	5 mg/kg	IM		
Ketamine[a] +	80–100 mg/kg	IP		Green *et al.* (1981), Gaertner *et al.* (1987)
xylazine	7–10 mg/kg	IP		Curl and Peters (1983)
Tiletamine + zolazepam (Telazol)	20–40 mg/kg (sedation)	IP or IM	Unproven	Silverman *et al.* (1983)
	50–80 mg/kg (anesthesia)	IP		
Tiletamine + zolazepam	20–30 mg/kg	IP	10–30 min	Forsythe *et al.* (1992)
(Telazol) + xylazine	10 mg/kg			
Neuroleptanalgesics				
Fentanyl[b] + droperidol (Innovar-Vet)	0.15 ml/100 g	IP	Unproven	Thayer *et al.* (1972)
Fentanyl + droperidol	1 ml/kg	IP	Unproven	
(Innovar-Vet) + diazepam	5 ml/kg	IP		
Fentanyl + fluanisone (Hypnorm)	1 ml/kg of a 1/10 dilution of Hypnorm	IM or IP	Sedation	Green (1975), Flecknell and Mitchell (1984)
Fentanyl + fluanisone	1.0 ml/kg (of a 1/10 dilution of Hypnorm)	IP	55–60 min	Green (1975), Flecknell and Mitchell (1984), Flecknell (1987b)
(Hypnorm) + diazepam	5 mg/kg	IP	Unproven	
Fentanyl + fluanisone (Hypnorm) + midazolam	4.0 ml/kg (of the Hypnorm–midazolam mixture)	IP	Unproven	Flecknell and Mitchell (1984), Flecknell (1987b)
Other				
α-Chloralose (1%, w/v)	8–10 mg/100 g	IP	Hypnosis only	Reid *et al.* (1989)
Alphaxolone–alphadolone (Saffan, Althesin)	120–160 mg/kg or	IP	40–60 min	Green *et al.* (1978)
	14 mg/kg induction,	IV bolus	6–9 min	Green *et al.* (1978)
	maintain with 4–6 mg/kg every 15 min	IV bolus		
Urethane (50%, w/v)	150 mg/100 g	IP	6 hr	Reid *et al.* (1989)
Urethane (50%) +	38 mg/100 g	IP	1.5–2 hr	Reid *et al.* (1989)
α-Chloralose (10%) +	3.8 mg/100 g	IP		
pentobarbital	2.6 mg/100 g supplement with 13.5 mg/100 g urethane and up to 6 hr 1.4 mg/100 g chloralose	IP		
Inhalants				
Methoxyflurane	To effect (1–3%)	Inhalant		Flecknell (1987b)
Halothane	1–4%	Inhalant		Flecknell (1987b)
Isoflurane	1–4%	Inhalant		Flecknell (1987b)

[a]Intramuscular ketamine + xylazine leads to irritation, lameness, and self-mutilation in hamsters.
[b]Innovar-Vet is contraindicated in hamsters because it induces central nervous system abnormalities.

TABLE A.V

ANESTHETICS AND TRANQUILIZERS USED IN GERBILS

Drug	Dosage	Route of injection	Duration of anesthesia	Reference
Barbiturates				
Pentobarbital (Nembutal)	36–100 mg/kg	IP	50–60 min	Flecknell (1987b) Lightfoote and Molinari (1978) Edgehouse and Dorman (1987)
Dissociatives				
Ketamine	44–100 mg/kg	IM, IP	30 min (mild sedation)	Flecknell (1987b) Lightfoote and Molinari (1978)
	150–200 mg/kg	IP	15 min (anesthesia)	Marcoux *et al.* (1988)
Ketamine +	75 mg/kg	IM	74 min (sedation)	Flecknell (1987b)
Acetylpromazine (Acepromazine)	3 mg/kg	IM		
Ketamine +	50 mg/kg	IM	51 min	Flecknell (1987b)
diazepam	5 mg/kg	IP	(sedation)	
Ketamine +	50 mg/kg	IM	41 min	Flecknell (1987b)
xylazine	2 mg/kg	IM	(sedation)	
Tiletamine + zolazepam	20–40 mg/kg (sedation)	IM	Unproven	Silverman *et al.* (1983)
(Telazol)	60 mg/kg (anesthesia)	IM		
Neuroleptanalgesics				
Fentanyl + fluanisone (Hypnorm)	0.5–1 ml/kg	IM or IP	Unproven	Flecknell *et al.* (1983), Flecknell (1987b)
Fentanyl + fluanisone (Hypnorm) +	1 ml/kg	IM or IP	21 min	Flecknell *et al.* (1983)
diazepam	5 mg/kg	IP		
Fentanyl + fluanisone (Hypnorm) +	0.6 ml/kg	IM	Sedation	Flecknell *et al.* (1983)
xylazine	5 mg/kg	IP		
Fentanyl + fluanisone (Hypnorm) + midazolam (Versed)	8.0 ml/kg (of the Hypnorm–midazolam mixture)	IP	Unproven	Flecknell (1987b)
Fentanyl +	0.05 mg/kg	SC	72 min	Flecknell *et al.* (1983)
metomidate	50 mg/kg	SC		
Other				
Alphaxolone–alphadolone (Saffan, Althesin)	80–120 mg/kg	IP	75 min	Flecknell *et al.* (1983)
Tribromoethanol (Avertin) (1.25% solution)	225–325 mg/kg	IP	10–35 min (anesthesia)	Norris and Turner (1983)
(2.25% solution)	225–450 mg/kg	SC	20–63 min (sedation)	Norris and Turner (1983)
Inhalants				
Methoxyflurane	To effect (1.3%)	Inhalant		Norris (1981)
Halothane	1–4%	Inhalant		
Isoflurane	1–4%	Inhalant		

TABLE A.VI

ANALGESICS USED IN LABORATORY RODENTS

Drug	Dosage in Rats and Mice	Dosage in Guinea pig	Duration of effect	Reference
Acetaminophen	110–305 mg/kg PO, IP	Unproven	Unproven	Flecknell (1984)
Alfentanil	0.05–1.5 mg/kg IP (mouse)	Unproven	15 min	Walter *et al.* (1989) Dingwall and Pleuvry (1986)
Aspirin	20 mg/kg SC, 100–120 mg/kg PO	20 mg/kg SC	Unproven	Flecknell (1984) Walter *et al.* (1989)
Buprenorphine	2.5 mg/kg SC, IP (mouse) 0.5–1.0 mg/kg SC 0.25–1.6 mg/kg IP (rat)	0.5–0.8 mg/kg SC	6–12 hr	Flecknell (1984, 1987b), Cowan *et al.* (1977a,b), Wixson *et al.* (1991), Liles and Flecknell (1993)
Butorphanol	5.6 mg/kg PO (mouse) 0.05–5.4 mg/kg SC (mouse) 0.04–23.3 mg/kg SC, 2.1 mg/kg PO (rat)	Unproven	2–4 hr	Flecknell (1984), Wixson *et al.* (1991)
Codeine	6.4–20 mg/kg SC, 12.5–25 mg/kg IP (mouse) 60–90 mg/kg PO (mouse) 60 mg/kg SC (rat) 30–50 mg/kg IP (rat)	25–40 mg/kg SC	4 hr	Schmidt *et al.* (1985), Nilsen (1961), Flecknell (1984), Lightowler and Wilder Smith (1963)
Diclofenac	14–100 mg/kg PO (mouse)	Unproven	Unproven	Walter *et al.* (1989)
Fentanyl	0.032 mg/kg SC (mouse ED$_{50}$)	Unproven	Unproven	Schmidt *et al.* (1985)
	0.0125–1.0 mg/kg IP (mouse)	Unproven	15 min	Dingwall and Pleuvry (1986)
Ibuprofen	7.5 mg/kg PO (mouse)	Unproven	Unproven	Flecknell (1984) Walter *et al.* (1989)
Indomethacin	1.5–10 mg/kg PO (mouse)	Unproven	Unproven	Walter *et al.* (1989)
Ketorolac	0.7–10 mg/kg PO (mouse)	Unproven	Unproven	Walter *et al.* (1989)
Meperidine	12.5–25 mg/kg IP, 20 mg/kg SC or IM (rat) 3–12 mg/kg IP, SC (mouse)	20 mg/kg SC or IM, 25–45 mg/kg IP	2–3 hr	Flecknell (1984), Lightowler and Wilder Smith (1963), Collier *et al.* (1961), Schmidt *et al.* (1985), Nilsen (1961)
Methadone	1.6–2.0 mg/kg SC (mouse) 1.0 mg/kg IP, 3.75 mg/kg SC (rat)	3.6 mg/kg SC	Unproven	Flecknell (1984), Piercey and Schroeder (1980), Collier *et al.* (1961), Paalzow (1969)
Morphine	1–4 mg/kg IV (rat) 2–10 mg/kg SC, 5–24 mg/kg IP (rat) 2–4 mg/kg IP (mouse ED$_{50}$)	5–12 mg/kg SC or IM	2–4 hr	Cowan *et al.* (1977a,b), DiFazio *et al.* (1981) Flecknell (1984), Paalzow (1969), Wixson *et al.* (1991), Nilsen (1961)
Naproxen	57–350 mg/kg PO (mouse)	Unproven	Unproven	Walter *et al.* (1989)
Nalbuphine	40 µg/kg bolus followed by 100 µg/kg/min continuous IV infusion, 1 mg/kg SC	Unproven Unproven	Unproven 4–6 hr	DiFazio *et al.* (1981) Flecknell (1991)
Oxymorphone	0.22–0.33 mg/kg SC (rat), 0.051 mg/kg SC (mouse ED$_{50}$)	Unproven	4 hr	Wixson *et al.* (1991) Schmidt *et al.* (1985)
Piroxicam	3.4–20 mg/kg PO (mouse)	Unproven	Unproven	Walter *et al.* (1989)
Pentazocine	1–21 mg/kg SC (mouse) 59–60 mg/kg PO (mouse ED$_{50}$) 8–10 mg/kg SC, 25–75 mg/kg IP (rat)	Unproven	3–4 hr	Flecknell (1984), Collier *et al.* (1961), Pircio *et al.* (1976), Wixson *et al.* (1991), Schmidt *et al.* (1985), Cowan *et al.* (1977a, b)
Phenylbutazone	31–250 mg/kg IP (mouse)	Unproven	Unproven	Nilsen (1961)
Sufentanil	0.0023 mg/kg SC (mouse ED$_{50}$)	Unproven	Unproven	Schmidt *et al.* (1985)
Xylazine	5–12 mg/kg SC (rat)	Unproven	2 hr	Wixson *et al.* (1991)

[a]Effective analgesic dose in 50% of experimental animals.

REFERENCES

Abad, F., Feria, M., and Boada, J. (1989). Chronic amitriptyline decreases autotomy following dorsal rhizotomy in rats. *Neurosci. Lett.* **99,** 187–190.

Abboud, T. K., D'Onofrio, L., Reyes, A., Mosaad, P., Zhu, J., Mantilla, M., Gangolly, J., Crowell, D., Cheung, M., Afrasiabi, A., Khoo, N., Davidson, J., Steffens, Z., and Zaki, N. (1989a). Isoflurane or halothane for cesarean section: Comparative maternal and neonatal effects. *Acta Anaesthesiol. Scand.* **33,** 578–581.

Abboud, T. K., Gangolly, J., Mosaad, P., and Crowell, D. (1989b). Isoflurane in obstetrics. *Anesth. Analg.* **68,** 388–391.

Adam, H. K., Glen, J. B., and Hoyle, P. A. (1980). Pharmacokinetics in laboratory animals of ICI 35868, a new I.V. anaesthetic agent. *Br. J. Anaesth.* **52,** 743–746.

Alpert, M., Goldstein, D., and Triner, L. (1982). Technique of endotracheal intubation in rats. *Lab. Anim. Sci.* **32**(1), 78–79.

Andrews, W. W., Ramin, S. M., Maberry, M. C., Shearer, V., Black, S., and Wallace, D. H. (1992). Effect of type of anesthesia on blood loss at elective repeat cesarean section. *Am. J. Perinatol.* **9**(3), 197–200.

Ardid, D., and Guibaud, G. (1992). Antinociceptive effects of acute and chronic injections of tricyclic antidepressant drugs in a new model of mononeuropathy in rats. *Pain* **49,** 279–287.

Arnold, J. H., Truog, R. D., and Rice, S. A. (1993). Prolonged administration of isoflurane to pediatric patients during mechanical ventilation. *Anesth. Analg.* **76,** 520–526.

Aston, R. (1966). Acute tolerance indices for pentobarbital in male and female rats. *J. Pharmacol. Exp. Ther.* **152,** 350–353.

Baughman, V. L., Hoffman, W. E., Miletich, D. J., Albrecht, R. F., and Thomas, C. (1988). Neurologic outcome in rats following incomplete cerebral ischemia during halothane, isoflurane, or N_2O. *Anesthesiology* **69**(2), 192–198.

Baughman, V. L., Hoffman, W. E., and Thomas, C. (1990). Comparison of methohexital and isoflurane on neurologic outcome and histopathology following incomplete ischemia in rats. *Anesthesiology* **72**(1), 85–94.

Brammer, A., West, C. D., and Allen, S. L. (1993). A comparison of propofol with other injectable anesthetics in a rat model for measuring cardiovascular parameters. *Lab. Anim.* **27,** 250–257.

Brown, J. N., Thorne, P. R., and Nuttall, A. L. (1989). Blood pressure and other physiological responses in awake and anesthetized guinea pigs. *Lab. Anim. Sci.* **39**(2), 142–148.

Buelke-Sam, J., Holson, J. F., Bazare, J. J., and Young, J. F. (1978). Comparative stability of physiological parameters during sustained anesthesia in rats. *Lab. Anim. Sci.* **28**(2), 157–162.

Burgman, P. M. (1991). Restraint techniques and anesthetic recommendations for rabbits, rodents, and ferrets. *Small Exotic Anim. Med.* **1**(2), 73–78.

Butler, S. H., Weil-Fugazza, J., Godefroy, F., and Besson, J. M. (1985). Reduction of arthritis and pain behavior following chronic administration of amitriptyline or imipramine in rats with adjuvant-induced arthritis. *Pain* **23,** 159–175.

Cambron, H., Latulippe, J. F., Nguyen, T., and Cartier, R. (1995). Orotracheal intubation of rats by transillumination, *Lab. Anim. Sci.* **45**(3), 303–304.

Carter, A. M. (1983). Acid-base status of pregnant guinea pigs during neuroleptanalgesia with diazepam and fentanyl-fluanisone. *Lab. Anim.* **17,** 114–116.

Carvell, J. E., and Stoward, P. J. (1975). Halothane anesthesia of normal and dystrophic hamsters. *Lab. Anim.* **9,** 345–352.

Child, K. J., Currie, J. P., and Davis, B. (1971). The pharmacological properties in animals of CT 1341 a new steroid anaesthetic agent. *Br. J. Anaesth.* **43,** 2–13.

Clifford, D. H. (1984). Preanesthesia, anesthesia, analgesia, and euthanasia. *In* "Laboratory Animal Medicine" (J. G. Fox, B. J. Cohen, and F. M. Loew, eds.), pp. 527–562. Academic Press, New York.

Cole, D. J., Marcantonio, S., and Drummond, J. C. (1990). Anesthetic requirement of isoflurane is reduced in spontaneously hypertensive and Wistar-Kyoto rats. *Lab. Anim. Sci.* **40**(5), 506–509.

Collier, H. O. J., Warner, B. T., and Skerry, R. (1961). Multiple toe-pinch method for testing analgesic drugs. *Br. J. Pharmacol.* **17,** 28–40.

Commissaris, R. L., Semeyn, D. R., and Reich, H. (1982). Dispositional without functional tolerance to the hypothermic effects of pentobarbital in the rat. *J. Pharmacol. Exp. Ther.* **220**(3), 536–539.

Cooke, S. W. (1987). Anaesthesia of guinea pigs and reversing pneumothorax. *Vet. Rec.* **120**(13), 309–310.

Cooper, D. M., DeLong, D., and Gillett, C. S. Analgesic efficacy of acetaminophen elixir added to the drinking water of rats. *Contemp. Top.* **34**(4), 61.

Cowan, A., Lewis, J. W., and Macfarlane, I. R. (1977a). Agonist and antagonist properties of buprenorphine, a new antinociceptive agent. **Br. J. Pharmacol. 60**(4), 537–545.

Cowan, A., Doxey, J. C., and Harry, E. J. R. (1977b). The animal pharmacology of buprenorphine, an oripavine analgesic agent. *Br. J. Pharmacol.* **60,** 547–554.

Cunliffe-Beamer, T. L. (1983). Biomethodology and surgical techniques. *In* "The Mouse in Biomedical Research" (H. L. Foster, J. D. Small, and J. G. Fox, eds.), Vol. 3, pp. 413–428. Academic Press, New York.

Curl, J. L., and Peters, L. J. (1983). Ketamine hydrochloride and xylazine hydrochloride anaesthesia in the golden hamster *(Mesocricetus auratus). Lab. Anim.* **17,** 290–293.

D'Alleinne, C. P. (1982). Evaluation of ketamine/xylazine anesthesia in the guinea pig: Toxicological parameters. *Vet. Hum. Toxicol.* **24,** 410–412.

Danneman, P. L. (1994). Humane and practical implications by rodent anesthesia and euthanasia using mixtures of carbon dioxide and oxygen. *Contemp. Top. Lab. Anim. Sci.* **33**(4).

Dannemann, P. L., and Mandrell, T. D. (1996). Personal communication.

Dardai, E., and Heavner, J. E. (1987). Respiratory and cardiovascular effects of halothane, isoflurane and enflurane delivered via a Jackson-Rees breathing system in temperature controlled and uncontrolled rats. *Methods Find. Exp. Clin. Pharmacol.* **9**(11), 717–720.

Dawson, D. L., and Scott-Conner, C. (1988). Adjunctive use of local anesthetic infiltration during guinea pig laparotomy. *Lab. Anim.* **17**(1), 35–36.

Deeb, B. J., and Frooman, R. (1992). Effect of buprenorphine on lameness and immune response in rats. *Contemp. Top. Lab. Anim. Sci.* **31**(4), 13.

Deeb, B. J., Eyman, P., Hutton, M. L., and Abbott, L. (1989). Efficacy of synthetic opioid analgesics administered in drinking water of rats. *Lab. Anim. Sci.* **39**(5), 473.

DeFelipe, M. D. C., DeCeballos, M. L., and Fuentes, J. A. (1986). Hypoalgesia induced by antidepressants in mice: A case for opioids and serotonin. *Eur. J. Pharmacol.* **125,** 193–199.

DeLeonardis, J. R., Clevenger, R., and Hoyt, R. F. (1995). Approaches to rodent intubation and endotracheal tube design. *Contemp. Top. Lab. Anim. Sci.* **34**(4), 60.

DiFazio, C. A., Brown, R. E., Ball, C. G., Heckel, C. G., and Kennedy, S. S. (1972). Additive effects of anesthetics and theories of anesthesia. *Anesthesiology* **36,** 57–63.

DiFazio, C. A., Moscicki, J. C., and Magruder, M. R. (1981). Anesthetic potency of nalbuphine and interaction with morphine in rats. *Anesth. Analg.* **60,** 629–633.

Dingwall, A. E., and Pleuvry, B. J. (1986). A comparison of the effects of fentanyl and alfentanil on the sleeping times of the intravenous induction agents. *J. Pharm. Pharmacol.* **38,** 323–325.

Edgehouse, N. L., and Dorman, R. V. (1987). Ischemia-induced development of cerebral edema in awake and anesthetized gerbils. *Neurochem. Pathol.* **7,** 169–179.

Eger, E. I. (1994). New inhaled anesthetics. *Anesthesiology* **80**(4), 906–922.

Eisele, P., Kaaekuahiwi, M., Caufield, D., and Golub. (1993). Epidural catheter placement for testing of obstetrical analgesics in female guinea pigs. *Contemp. Top.* **32**(4), 9–10.

Erhardt, W., Hebestedt, A., and Aschenbrenner, G. A. (1984). Comparative study with various anesthetics in mice (pentobarbitone, ketamine-xylazine, carfentanyl-etomidate). *Res. Exp. Med.* **184,** 159–169.

Farris, H. E., and Snow, W. S. (1987). Comparison of isoflurane, halothane, or methoxyflurane with combinations of ketamine hydrochloride and acepromazine for anesthesia of the rat. *Lab. Anim. Sci.* **37**(4), 520.

Fenwick, D. C., and Blackshaw, J. K. (1989). Carbon dioxide as a short-term restraint anaesthetic in rats with subclinical respiratory disease. *Lab Anim.* **23,** 220–228.

Ferguson, J. W. (1979). Anaesthesia in the hamster using a combination of methohexitone and diazepam. *Lab. Anim.* **13,** 305–308.

Field, K. J. (1988). The anesthetic effects of alpha chloralose, chloral hydrate, pentobarbital, and urethane on the respiratory system, cardiovascular system, core temperature, and noxious stimulus perception in adult male rats. Pennsylvania State University Thesis in Laboratory Animal Medicine, pp. iii–104.

Field, K. J., White, W. J., and Lang, C. M. (1993). Anaesthetic effects of chloral hydrate, pentobarbitone and urethane in adult male rats. *Lab. Anim.* **27,** 258–269.

Flecknell, P. A. (1984). The relief of pain in laboratory animals. *Lab. Anim.* **18,** 147–160.

Flecknell, P. A. (1987a). Anaesthesia of guinea pigs. *Vet. Rec.* **120**(7), 167.

Flecknell, P. A. (1987b). "Laboratory Animal Anaesthesia," pp. 90–98, 133–137. Academic Press, London.

Flecknell, P. A. (1991). Post-operative analgesia in rabbits and rodents. *Lab. Anim.* **20**(9), 34–37.

Flecknell, P. A., and Liles, J. H. (1991). The effects of surgical procedures, halothane anaesthesia and nalbuphine on locomotor activity and food and water consumption in rats. *Lab. Anim.* **25,** 50–60.

Flecknell, P. A., and Mitchell, M. (1984). Midazolam and fentanyl-fluanisone: Assessment of anaesthetic effects in laboratory rodents and rabbits. *Lab. Anim.* **18,** 143–146.

Flecknell, P. A., John, M., Mitchell, M., and Shurey, C. (1983). Injectable anaesthetic techniques in 2 species of gerbil *(Meriones libycus* and *Meriones unguiculatus). Lab. Anim.* **17,** 118–122.

Fleischman, R., McCracken, D., and Forbes, W. (1977). Adynamic ileus in the rat induced by chloral hydrate. *Lab. Anim. Sci.* **27,** 238–243.

Forsythe, D. B., Payton, A. J., and Dixon, D. (1992). Evaluation of Telazol®-xylazine as an anesthetic combination for use in Syrian hamsters. *Lab. Anim. Sci.* **42**(5), 497–502.

Fowler, J. S. L., Brown, J. S., and Flower, E. W. (1980). Comparison between ether and carbon dioxide anesthesia for removal of small blood samples from rats. *Lab. Anim.* **14,** 275–278.

Franz, D. R., and Dixon, R. S. (1988). A mask system for halothane anesthesia of guinea pigs. *Lab. Anim. Sci.* **38**(6), 743–744.

Frieder, B., Epstein, S., and Grimm, V. E. (1984). The effects of exposure to diazepam during various stages of gestation or during lactation on the development and behavior of rat pups. *Psychopharmacology* **83,** 51–55.

Frisk, C. S., Hermann, M. D., and Senta, K. E. (1982). Guinea pig anesthesia using various combinations and concentrations of ketamine, xylazine and/or acepromazine. *Lab. Anim. Sci.* **32**(4), 434.

Gaertner, D. J., Boschert, K. R., and Schoeb, T. R. (1987). Muscle necrosis in Syrian hamsters resulting from intramuscular injections of ketamine and xylazine. *Lab. Anim. Sci.* **37**(1), 80–83.

Gardner, D. J., Davis, J. A., Weina, P. J., and Theune, B. (1995). Comparison of tribromoethanol, ketamine/acetylpromazine, Telazol/xylazine, pentobarbital and methoxyflurane anesthesia in HSD:ICR mice. *Lab. Anim. Sci.* **46**(2), 199–204.

Gilroy, B. A. (1981). Endotracheal intubation of rabbits and rodents. *J. Am. Vet. Med. Assoc.* **179**(11), 1295.

Gilroy, B. A., and Varga, J. S. (1980). Ketamine-diazepam and ketamine-xylazine combinations in guinea pigs VM/SAC, *Vet. Med. Small Anim. Clin.* **75**(3), 508–509.

Glen, J. B. (1980). Animal studies of the anaesthetic activity of ICI 35 868. *Br. J. Anaesth.* **52,** 731–741.

Glen, J. B., and Hunter, S. C. (1984). Pharmacology of an emulsion formulation of ICI 35868. *Br. J. Anaesth.* **56,** 617–625.

Glen, J. B., Cliff, G. S., and Jamieson, A. (1980). Evaluation of a scavenging system for use with inhalation anaesthesia techniques in rats. *Lab. Anim.* **14**(3), 207–211.

Godefroy, F., Butler, S. H., Weil-Fugazza, J., and Besson, J. M. (1986). Do acute or chronic tricyclic antidepressants modify morphine antinociception in arthritic rats. *Pain* **25,** 233–244.

Gomwalk, N. E., and Healing, T. D. (1981). Etomidate: A valuable anesthetic for mice. *Lab. Anim.* **15,** 151–152.

Green, C. J. (1975). Neuroleptanalgesic drug combinations in the anesthetic management of small laboratory animals. *Lab. Anim.* **9,** 161–178.

Green, C. J., Halsey, M. J., and Precious, S. (1978). Alphaxolone-alphadolone anaesthesia in laboratory animals. *Lab. Anim.* **12,** 85–89.

Green, C. J., Knight, J., Precious, S., and Simpkin, S. (1981). Ketamine alone and combined with diazepam or xylazine in laboratory animals: A 10 year experience. *Lab. Anim.* **15,** 163–170.

Gunberg, D. L. (1982). Embryotoxic effects of pentobarbital (Nembutal®). *JODA* **24.**

Hart, M. V., Rowles, J. R., and Hohimer, R. (1984). Hemodynamics in the guinea pig after anesthetization with ketamine/xylazine *Am. J. Vet. Res.* **45**(1), 2328–2330.

Heavner, J. E. (1989). Correspondence. *Lab. Anim. Sci.* **39**(4), 293.

Hoar, R. M. (1969). Anesthesia in the guinea pig. *Fed. Proc. Fed. Am. Soc. Exp. Biol.* **28**(4), 1517–1521.

Hrapkiewicz, K. L., Stein, S., and Smiler, K. L. (1989). A new anesthetic agent for use in the gerbil. *Lab. Anim. Sci.* **39**(4), 338–341.

Hsu, W. H., Bellin, S. I., and Dellmon, H. D. (1986). Xylazine-ketamine induced anesthesia in rats and its antagonism by yohimbine. *J. Am. Vet. Med. Assoc.* **189**(9), 1040–1043.

Hughes, E. W., Martin-Body, R. L., Sarelius, I. H., and Sinclair, J. D. (1982). Effects of urethane-chloralose anaesthesia on respiration in the rat. *Clin. Exp. Pharmacol. Physiol.* **9,** 119–127.

Hughes, H. C., and Lang, C. M. (1972). Hepatic necrosis produced by repeated administration of halothane to guinea pigs. *Anesthesiology* **36,** 466–471.

Hunskaar, S., and Hole, K. (1987). The formalin test in mice: Dissociation between inflammatory and non-inflammatory pain. *Pain* **30,** 103–114.

Institute of Laboratory Animal Resources (ILAR) Committee to Revise the Guide for the Care and Use of Laboratory Animals, National Research Council (1996). "Guide for the Care and Use of Laboratory Animals." National Academy Press, Washington, D.C.

Jaffee, J. H., and Martin, W. R. (1985). Opioid analgesics and antagonists. *In* "The Pharmacological Basis of Therapeutics" (A. G. Gilman, L. S. Goodman, T. W. Rall, and F. Murad, eds.), 7th ed., pp. 491–531. Macmillan, New York.

Jenkins, J. R. (1992). Husbandry and common diseases of the chinchilla. *J. Small Exotic Anim. Med.* **2**(1), 15–17.

Jenkins, W. L. (1987). Pharmacologic aspects of analgesic drugs in animals: An overview. *J. Am. Vet. Med. Assoc.* **199**(10), 1231–1240.

Kaczmarczyk, G., and Reinhardt, H. W. (1975). Arterial blood gas tensions and acid-base status of Wistar rats during thiopental and halothane anesthesia. *Lab. Anim. Sci.* **25**(2), 184–189.

Kalpravidh, M., Lumb, W. V., Wright, M., and Heath, R. B. (1984). Effects of butorphanol, flunixin, levorphanol, morphine and xylazine in ponies. *Am. J. Vet. Res.* **45,** 217–223.

Kandel, L., Laster, M. J., Eger, E. I., Kerschmann, R. L., and Martin, J. (1995). Nephrotoxicity in rats undergoing a one-hour exposure to compound A. *Anesth. Analg.* **81,** 559–563.

Kissin, I., Brown, P. T., Robinson, C. A., and Bradley, E. (1991). Acute tolerance in morphine analgesia: continuous infusion and single injection in rats. *Anesthesiology* **74,** 166–171.

Kobayashi, K. (1985). Strain differences of ether susceptibility in mice. *Exp. Anim.* **34**(4), 379–385.

Kochhar, M. M., Aykac, I., and Davidson, P. (1986). Teratologic effects of d, 1-2-(0-chlorophenyl)-2-(Methylamino) cycloticxanone hydrochloride (Ketamine hydrochloride) in rats. *Res. Commun. Chem. Pathol. Pharmacol.* **54**(3), 413–416.

Komulainen, A., and Olson, M. E. (1991). Antagonism of ketamine-xylazine anesthesia in rats by administration of yohimbine, tolazoline, or 4-aminopyridine. *Am. J. Vet. Res.* **52**(4), 585–588.

Kujime, K., and Natelson, B. H. (1981). A method for endotracheal intubation of guinea pigs *(Cavia porcellus). Lab. Anim. Sci.* **31**(6), 715–716.

Landi, M. S., Kreider, J. W., and Lang, C. M. (1982). Effects of shipping on the immune function in mice. *Am. J. Vet. Res.* **43**(9), 1654–1657.

Latt, R. H., and Ecobichon, D. J. (1984). Self-mutilation in guinea pigs following the intramuscular injection of ketamine-acepromazine. *Lab. Anim. Sci.* **34**(5), 516.

Leash, A. M., Beyer, R. D., and Wilber, R. G. (1973). Self-mutilation following Innovar-Vet® injection in the guinea pig. *Lab. Anim. Sci.* **23**(5), 720–721.

Leese, T., Husken, P. A., and Morton, D. B. (1988). Buprenorphine analgesia in a rat model of acute pancreatitis. *Surg. Res. Commun.* **3,** 53–60.

Levy, D. E., Zwies, A., and Duffy, T. E. (1980). A mask for delivery of inhalation gases to small laboratory animals. *Lab. Anim. Sci.* **30**(5), 868–870.

Lewis, G. E., and Jennings, P. B. (1972). Effective sedation of laboratory animals using Innovar-Vet®. *Lab. Anim. Sci.* **22**(3), 430–432.

Liebeskind, J. C. (1991). Pain can kill. *Pain* **44,** 3–4.

Lightfoote, W. E., and Molinari, G. F. (1978). Comparison of ketamine and pentobarbital anesthesia in the Mongolian gerbil. *Am. J. Vet. Res.* **39**(6), 1061–1063.

Lightowler, J. E., and Wilder Smith, A. E. (1963). β-(N-pyrolidyl)-butyroanilid (W.S. 10), its sub-acute toxicity, influence on morphine analgesia in the rat and morphine analgesia and respiratory depression in the rabbit. *Arch. Int. Pharmacodyn. Ther.* **144**(1–2), 97–111.

Liles, J. H., and Flecknell, P. A. (1993). The influence of buprenorphine or bupivacaine on the post-operative effects of laparotomy and bile-duct ligation in rats. *Lab. Anim.* **27,** 374–380.

Liles, J. H., and Flecknell, P. A. (1994). A comparison of the effects of buprenorphine, carprofen and flunixin following laparotomy in rats. *J. Vet. Pharmacol. Ther.* **17,** 284–290.

Lind, R. C., Gandolphi, A. J., and Hall, P. de la M. (1989). Age and gender influence halothane associated hepatotoxicity in strain 13 guinea pigs. *Anesthesiology* **71,** 878–884.

Lind, R. C., Gandolphi, A. J., and Hall, P. de la M. (1992). Subanesthetic halothane is hepatotoxic in the guinea pig. *Anesth. Analg.* **74,** 559–563.

Livezey, G. T., Marczynski, T. J., and Isaac, L. (1986a). Prenatal diazepam chronic anxiety and deficits in brain receptors in mature rat progeny. *Neurobehav. Toxicol. Teratol.* **8,** 425–432.

Livezey, G. T., Marczynski, T. J., and McGrew, E. A. (1986b). Prenatal exposure to diazepam: Late postnatal teratogenic effect. *Neurobehav. Toxicol. Teratol.* **8,** 433–440.

Lovell, D. P. (1986a). Variation in pentobarbitone sleeping time in mice. 1. Strain and sex differences. *Lab. Anim.* **20,** 85–90.

Lovell, D. P. (1986b). Variation in pentobarbitone sleeping time in mice. 2. Variable affecting test results. *Lab. Anim.* **20,** 91–96.

Lovell, D. P. (1986c). Variation in barbiturate sleeping time in mice. 3. Strain X environment interactions. *Lab. Anim.* **20,** 307–312.

Maekawa, T., Tommasine, C., and Shapiro, H. M. (1986). Local cerebral blood flow and glucose utilization during isoflurane anesthesia in the rat. *Anesthesiology* **65,** 144–151.

Mahmoudi, N. W., Cole, D. J., and Shapiro, H. M. (1989). Insufficient anesthetic potency of nitrous oxide in the rat. *Anesthesiology* **70**(2), 345–349.

Marcoux, F. W., Goodrich, J. E., and Dominick, M. A. (1988). Ketamine prevents ischemic neuronal injury. *Brain Res.* **452,** 329–335.

Markovic, S. N., Knight, P. R., and Murasko, D. M. (1993). Inhibition of interferon stimulation of natural killer cell activity in mice anesthetized with halothane or isoflurane. *Anesthesiology* **78,** 700–706.

Matt, D. W., Steingold, K. A., Dastvan, C. M., James, C. A., and Dunwiddle, W. (1991). Effects of sera from patients given various anesthetics on preimplantation mouse embryo development *in vitro*. *J. In Vitro Fertil. Embryo Transfer* **8**(4), 191–197.

Mazze, R. I., Wilson, A. I., Rice, S. A., and Badenn, J. M. (1985). Fetal development in mice exposed to isoflurane. *Teratology* **32,** 339–345.

Miller, M. E., Rörth, M., and Stohlman, F. (1976). The effects of acute bleeding on acid-base balance, erythropoietin (Ep) production and *in vivo* P_{50} in the rat. *Br. J. Haematol.* **33,** 379–385.

Morton, P. B., and Griffith, P. H. M. (1984). Guidelines on the recognition of pain, distress and discomfort in experimental animals and on hypothesis for assessment. *Vet. Rec.* **166,** 431–436.

Muir, W. W., and Hubbell, J. A. E. (1989). Oxygen toxicity, anesthetic toxicity, waste gas scavenging, and anesthetic drug interactions. *In* "Handbook of Veterinary Anesthesia," pp. 127–136. Mosby, St. Louis.

Mulder, J. B., and Hauser, J. J. (1984). A closed anesthetic system for small laboratory animals. *Lab. Anim. Sci.* **34**(1), 77–78.

Mulder, K. J., and Mulder, J. B. (1979). Ketamine and xylazine anesthesia in the mouse. VM/SAC, *Vet. Med. Small Anim. Clin.*, p. 569.

Mulvey, S. (1987). Anaesthesia of guinea pigs. *Vet. Rec.* **120**(13), 309–310.

Munson, E. S. (1970). Effect of hypothermia on anesthetic requirements in rats. *Lab. Anim. Care* **20**(6), 1109–1113.

Murray, W. J., and Fleming, P. J. (1972). Defluorination of methoxyflurane during anesthesia. *Anesthesiology* **37**(6), 620–625.

Nau, R., and Schunck, O. (1993). Cannulation of the lateral saphenous vein—a rapid method to gain access to the venous circulation in anaesthetized guinea pigs. *Lab. Anim.* **27,** 23–25.

Nevalainen, T., Pyhälä, L., and Hanna-Maija, V. (1989). Evaluation of anaesthetic potency of medetomidine-ketamine combination in rats, guinea pigs and rabbits. *Acta Vet. Scand.* **85,** 139–143.

Nicol, T., Vernon-Roberts, B., and Quantock, D. C. (1965). Protective effect of oestrogen against the toxic decomposition products of tribromoethanol. *Nature (London)* **208,** 1098-1099.

Nilsen, P. L. (1961). Studies on algesimetry by electrical stimulation of the mouse tail. *Acta Pharmacol. Toxicol.* **18,** 10–22.

Norris, L. (1981). Portable anaesthetic apparatus designed to induce and maintain surgical anaesthesia by methoxyflurane inhalation in the Mongolian gerbil *(Meriones unguiculatus)*. *Lab. Anim.* **15**(2), 153–155.

Norris, M. L., and Miles, P. (1982). An improved, portable machine designed to induce and maintain surgical anaesthesia in small laboratory rodents. *Lab. Anim.* **16**(3), 227–230.

Norris, M. L., and Turner, W. D. (1983). An evaluation of tribromoethanol (TBE) as an anesthetic agent in the Mongolian gerbil *(Meriones unguiculatus)*. *Lab. Anim.* **17,** 324–329.

Ogino, K., Hobara, T., and Kobayashi, H. (1990). Gastric mucosal injury induced by chloral hydrate. *Toxicol. Lett.* **52,** 129–133.

Paalzow, L. (1969). An electrical method for estimation of analgesic activity in mice. *Acta Pharm. Suec.* **6,** 207–226.

Papaioannou, V. E., and Fox, J. G. (1993). Use and efficacy of tribromoethanol anesthesia in the mouse. *Lab. Anim. Sci.* **43**(2), 189–192.

Park, C. M., Clegg, K. E., Harvey-Clark, C. J., and Hollenberg, M. J. (1992). Improved techniques for successful neonatal rat surgery. *Lab. Anim. Sci.* **42**(5), 508–513.

Pena, H., and Cabrera, C. (1980). Improved endotracheal intubation technique in the rat. *Lab. Anim. Sci.* **30**(4, part I), 712–713.

Phifer, C. B., and Terry, L. M. (1986). Use of hypothermia for general anesthesia in preweanling rodents. *Physiol. Behav.* **38,** 887–890.

Pieper, J. O., Malatesta, P. F., and DeTolla, L. J. (1995). Locomotor stimulation in mice after surgery and buprenorphine administration. *Contemp. Top. Lab. Anim. Sci.* **34**(4), 61–62.

Piercey, M. F., and Schroeder, L. A. (1980). Aquantitative analgesic assay in the rabbit based on the response to tooth pup stimulation. *Arch. Int. Pharmacodyn. Ther.* **248,** 294–304.

Pircio, A. W., Gylys, J. A., and Cavanagh, R. L. (1976). The pharmacology of butorphanol, a 3, 14-dihydroxymorphinan narcotic antagonist analgesic. *Arch. Int. Pharmacodyn. Ther.* **220,** 231–257.

Poole, T. B., ed. (1987). "The UFAW Handbook on the Care and Management of Laboratory Animals," 6th ed. Longman Scientific and Technical, Essex, U.K.

Querldo, D., and Davis, B. L. (1985). A non-rebreathing, temperature-compensated device for administration of halothane to small laboratory animals. *S. Afr. J. Sci.* **81**(11), 688–690.

Quinn, R. H., Dannemen, P. J., and Dysko, R. C. (1993). Efficacy of sedating rats with diazepam or droperidol prior to noninvasive manipulation. *Contemp. Top. Lab. Anim. Sci.* **32**(4), 25.

Rank, J., and Jensen, A. G. (1989). The value of anaesthetic steroids alphaxolone-alphadolone in pregnant mice. *Scand. J. Lab. Anim. Sci.* **16**(3), 115–117.

Rao, S. B. (1990). Comparative study on the effects of pentobarbitone sodium, thiopentone sodium and ketamine-diazepam on body temperature in rats. *Indian Vet. J.* **67,** 339–342.

Reid, W. D., Davies, C., Pare, P. D., and Pardy, R. L. (1989). An effective combination of anaesthetics for 6-hour experimentation in the golden Syrian hamster. *Lab. Anim.* **23,** 156–162.

Reuter, R., and Blaze, C. (1987). Anaesthesia of guinea pigs. *Vet. Rec.* **121**(9), 207.

Rice, S. A., and Fish, K. J. (1987). Anesthetic considerations for drug metabolism studies in laboratory animals. *Lab. Anim. Sci.* **37**(4), 520.

Routtenberg, A. (1968). Pentobarbital anesthesia of albino rats. *J. Exp. Anal. Behav.* **11,** 52.

Royer, C. T., and Wixson, S. K. (1992). A comparison of fentanyl-fluanisone (Hypnorm®) with fentanyl-droperidol (Innovar-Vet®) for anesthesia and analgesia in laboratory rats. *J. Invest. Surg.* **5,** 282.

Rubright, W. C., and Thayer, C. B. (1970). The use of Innovar-vet® as a surgical anesthetic for the guinea pig. *Lab. Anim. Care* **20**(5), 989–991.

Sano, T., Patel, P. M., Drummond, J. C., and Cole, D. J. (1993). A comparison of the cerebral protective effects of etomoidate, thiopental, and isoflurane in a model of forebrain ischemia in the rat. *Anesth. Analg.* **76,** 990–997.

Schmidt, W. K., Tam, S. W., and Shotzberger, G. S. (1985). Nalbuphine. *Drug Alcohol Depend.* **14,** 339–362.

Schreiber, H., and Nettesheim, P. (1972). A new method for pulmonary cytology in rats and hamsters. *Cancer Res.* **32,** 737–745.

Sedgwick, C., and John, S. (1980). Techniques for endotracheal intubation and inhalation anesthesia for laboratory animals. *Calif. Vet.* **34**(3), 27–33.

Seifen, A. B., Kennedy, R. H., Bray, J. P., and Seifen, E. (1989). Estimation of minimum alveolar concentration (MAC) for halothane, enflurane and isoflurane in spontaneously breathing guinea pigs. *Lab. Anim. Sci.* **39**(6), 579–581.

Severs, W. B., Keil, L. C., Klase, P. A., and Deen, K. C. (1981). Urethane anaesthesia in rats. *Pharmacology* **22,** 209–226.

Seyde, W. C., McGowan, L., and Land, N. (1985). Effects of anesthetics on regional hemo-dynamics in normovolemic and hemorrhaged rats. *Am. J. Physiol.* **249**(1–Pt2), H164–H173.

Short, C. E. (1987). Pain, analgesics, and related medications. *In* "Principles and Practice of Veterinary Anesthesia" (C. E. Short, ed.), pp. 28–46. Williams & Wilkins, Baltimore, Maryland.

Silverman, J. (1987). Biomethodology. *In* "Laboratory Hamsters" (G. L. Van Hoosier and C. W. MacPherson, eds.), pp. 77–82. Academic Press, Orlando, Florida.

Silverman, J., and Muir, W. W. (1993). A review of laboratory animal anesthesia with chloral hydrate and chloralose. *Lab. Anim. Sci.* **43**(3), 210–216.

Silverman, J., Huhndorf, M., and Balk, M. (1983). Evaluation of a combination of tiletamine and zolazepam as an anesthetic for laboratory rodents. *Lab. Anim. Sci.* **33**(5), 457–460.

Skornik, W. A., and Brain, J. D. (1990). Breathing and lung mechanics in hamsters: Effect of pentobarbital anesthesia. *J. Appl. Physiol.* **68**(6), 2536–2541.

Smiler, K. L., Stein, S., and Hrapkiewicz, K. L. (1990). Tissue response to intramuscular and intraperitoneal injections of ketamine and xylazine in rats. *Lab. Anim. Sci.* **40,** 60–64.

Spiegel, K., Kalb, R., and Pasternak, G. W. (1983). Analgesic activity of tricyclic antidepressants. *Ann Neurol.* **13**(4), 462–465.

Stickrod, G. (1979). Ketamine/xylazine anesthesia in the pregnant rat. *J. Am. Vet. Med. Assoc.* **175**(9), 952–953.

Taber, R., and Irwin, S. (1969). Anesthesia in the mouse. *Fed. Am. Soc. Exp. Biol.* **28**(4), 1528–1532.

Tarin, D., and Sturdee, A. (1972). Surgical anaesthesia of mice: Evaluation of tribromoethanol, ether, halothane and methoxyflurane and development of a reliable technique. *Lab. Anim.* **6,** 79–84.

Tena, E., Ramirez, C., and Calderon, J. (1982). Anesthesia in suckling guinea pigs using a combination of ketamine hydrochloride and droperidol. *Lab. Anim. Sci.* **32**(4), 434.

Thayer, C. B., Lowe, S., and Rubright, E. C. (1972). Clinical evaluation of a combination of droperidol and fentanyl as an anesthetic for the rat and hamster. *J. Am. Vet. Med. Assoc.* **161,** 665–668.

Thielhart, L. M., Reitan, J. A., and Martucci, R. W. (1984). Determination of halothane ED-50 in Mongolian gerbils *(Meriones unguiculatus). Lab. Anim. Sci.* **34**(1), 75–76.

Urban, B. J., France, R. D., Steinberger, K., Scott, D. L., and Maltbie, A. A. (1986). Long-term management of phantom limb pain. *Pain* **24,** 191–196.

Urbanski, H. F., and Kelley, S. T. (1991). Sedation by exposure to a gaseous carbon dioxide-oxygen mixture: Application to studies involving small laboratory animal species. *Lab. Anim. Sci.* **41**(1), 80–82.

Valenstein, E. (1961). A note on anesthetizing rats and guinea pigs. *J. Exp. Anal. Behav.* **4,** 6.

Van Pelt, L. F. (1977). Ketamine and xylazine for surgical anesthesia in rats. *J. Am. Vet. Med. Assoc.* **171,** 842–844.

Walden, N. B. (1978). Effective sedation of rabbits, guinea pigs, rats and mice with a mixture of fentanyl and droperidol. *Aust. Vet. J.* **54,** 538–540.

Walter, T., Chau, T. T., and Weichman, B. M. (1989). Effects of analgesics on bradykinin-induced writhing in mice presensitized with PGE_2. *Agents Actions* **27**(3–4), 375–377.

Ward, G. S., Johnson, D. O., and Roberts, C. R. (1974). The use of CI 744 as an anesthetic for laboratory animals. *Lab. Anim. Sci.* **24**(5), 737–742.

Waynforth, H. B., and Flecknell, P. A. (1992). Anesthesia and postoperative care. *In* "Experimental and Surgical Technique in the Rat" 2nd ed., pp. 101–163. Academic Press, London.

Wells, J. R., Gernon, W. H., Ward, G., Davis, K., and Hays, L. L. (1986). Otosurgical model in the guinea pig *(Cavia porcellus). Otolaryngol. Head Neck Surg.* **95**(4), 450–457.

Wereszczynska-Siemiatkowska, U., Nebendahl, K., and Pohl, U. (1987). Influence of burenorphine on acute experimental pancreatitis. *Res. Exp. Med.* **187,** 211–216.

Wharton, R. S., Sievenpiper, T. S., and Mazze, R. I. (1980). Development of toxicity of methoxyflurane in mice. *Anesth. Analg.* **59,** 421–425.

Whelan, G., and Flecknell, P. A. (1994). The use of etorphine/methotrimeprazine and midazolam as an anesthetic technique in laboratory rats and mice. *Lab. Anim.* **28,** 70–77.

White, W. J., and Field, K. J. (1987). Anesthesia and surgery of laboratory animals. *Vet. Clin. North Am.: Small Anim. Pract.* **17**(5), 989–1017.

Whur, P. (1987). Anaesthesia of guinea pigs. *Vet. Rec.* **120**(5), 119.

Wilson, R. P., Zagon, I. S., and Larach, D. R. (1992). Antinociceptive properties of tiletamine-zolazepam improved by addition of xylazine or butorphanol. *Pharmacol. Biochem. Behav.* **43,** 1129–1133.

Wilson, R. P., Zagon, I. S., and Larach, D. R. (1993). Cardiovascular and respiratory effects of tiletamine-zolazepam. *Pharmacol. Biochem. Behav.* **44,** 1–8.

Wixson, S. K., White, W. J., and Hughes, H. C. (1987a). A comparison of pentobarbital, fentanyl-droperidol, ketamine-xylazine and ketamine-diazepam anesthesia in adult male rats. *Lab. Anim. Sci.* **37**(6), 726–730.

Wixson, S. K., White, W. J., and Hughes, H. C. (1987b). The effects of pentobarbital, fentanyl-droperidol, ketamine-xylazine and ketamine-diazepam on noxious stimulus perception in adult male rats. *Lab. Anim. Sci.* **37**(6), 731–735.

Wixson, S. K., White, W. J., and Hughes, H. C. (1987c). The effects of pentobarbital, fentanyl-droperidol, ketamine-xylazine and ketamine-diazepam on arterial blood pH, blood gases, mean arterial blood pressure and heart rate in adult male rats. *Lab. Anim. Sci.* **37**(6), 736–742.

Wixson, S. K., White, W. J., and Hughes, H. C. (1987d). The effects of pentobarbital, fentanyl-droperidol, ketamine-xylazine and ketamine-diazepam on core and surface body temperature regulation in adult male rats. *Lab. Anim. Sci.* **37**(6), 743–749.

Wixson, S. K., Hobbs, B. A., Karchner, L. E., and Lawrence, W. (1991). Determination of appropriate analgesics for relief of pain in laboratory rats. *J. Invest. Surg.* **4,** 359.

Yatsu, F. M., Diamond, I., Graziano, C., and Lindquist, P. (1972). Experimental brain ischemia: Protection from irreversible damage with a rapid-acting barbiturate (methohexital). *Stroke* **3,** 726–732.

Zucker, J. (1991). Central cholinergic depression reduces MAC for isoflurane in rats. *Anesth. Analg.* **72,** 790–795.

Chapter 10

Anesthesia and Analgesia in Rabbits

Neil S. Lipman, Robert P. Marini, and Paul A. Flecknell

I. INTRODUCTION

Rabbits are routinely utilized as research subjects for *in vivo* studies. Their popularity is a result of availability, cost, size, ease of handling, vascular access, and a variety of specific anatomical, physiological, and biochemical features. Their use requires the routine administration of pharmacologic agents for restraint, vascular access, surgery, and pain relief. The rabbit is

ANESTHESIA AND ANALGESIA IN LABORATORY ANIMALS

ISBN 0-12-417570-8

often considered to be one of the most difficult laboratory animals to anesthetize. Numerous literature citations attribute this difficulty to individual variation in response to anesthetic agents, enhanced sensitivity of the brain stem to respiratory depression, and anatomic features making endotracheal intubation difficult (Chen and Bohner, 1968; Gardner, 1964; Gay, 1963; Green, 1982b; Murdock, 1969). Most of these characteristics, which only apply to the use of pentobarbital or to rabbits that were compromised with pneumonia, have unfairly stigmatized the rabbit, which can be easily and safely anesthetized for a wide variety of procedures. In fact, rabbits are routinely used as an animal model to investigate and define the physiologic effects of new anesthetics and analgesics (Bernards and Artru, 1991; Blake *et al.,* 1991; Chakrabarty *et al.,* 1991; Dhasmana *et al.,* 1984; Dorward *et al.,* 1987; Drummond *et al.,* 1987; Harris and Best, 1979; Hess *et al.,* 1971, 1981; Horita *et al.,* 1983; Hovav and Weinstock, 1987; Lockhart *et al.,* 1991; Maggi *et al.,* 1984; McGrath and MacKenzie, 1977; Mills *et al.,* 1987; Morita *et al.,* 1987; Patel and Mutch, 1990; Ruta and Mutch, 1989; Sainz *et al.,* 1987; Warren and Ledingham, 1978; Wyler and Weisser, 1972).

This chapter reviews the topic of anesthesia and analgesia as it pertains to the rabbit. Biomethodology, pre-, intra-, and postoperative considerations, as well as special anesthetic procedures, are also discussed. The chapter provides a general overview of the subject, but the reader should seek more detailed information from the references provided throughout. Because the science of anesthesia is constantly changing, the reader should also consult pertinent journals for ''state of the art'' information. Tables A.I–A.III in the Appendix at the end of the chapter can be consulted for quick reference to the agents and doses used for sedation, tranquilization, anesthesia, and analgesia in rabbits.

II. PREOPERATIVE CONSIDERATIONS

A. Choice of Subject

The selection of the experimental subject is as important as the choice of the appropriate anesthetic agent. Rabbits may be infected with a variety of agents that can adversely impact anesthesia. As an example, *Pasteurella multocida* causes a variety of clinical syndromes, including pneumonia. Rabbits may appear clinically normal despite significant respiratory pathology. These rabbits may fail to respire spontaneously once anesthetized as the animal has adapted to hypercapnia resulting from pulmonary insufficiency (Sedgwick, 1986). The careful selection, purchase, and maintenance of specific pathogen-free (SPF) rabbits will eliminate the risks associated with using animals that are compromised by pathogens.

Breed selection is frequently mandated by the size of the desired subject, specific anatomical or physiological features, and availability. The New Zealand White and Dutch Belted are the most frequently used breeds for research in the United States. These outbred stocks are maintained by vendors in closed colonies. Genetic differences do exist in members of the same stock obtained from different vendors. The choice of sex may also be a consideration. Cyclic hormonal changes occur in does during the estrus cycle which may have subtle effects that can alter research or anesthetic responses (Lawson, 1985). Diurnal and seasonal effects may also contribute to response (Deimling and Schnell, 1980; Hastings and Menaker, 1976; Jori *et al.,* 1971; Radzialowski and Bousquet, 1968; Scheving *et al.,* 1968).

Time should be allotted when planning and scheduling experimental procedures to allow for stabilization following shipment and acclimation to a new environment. Stress from shipment and handling may result in increased levels of circulating catecholamines which may adversely affect the anesthetic course. Ideally, rabbits should be quarantined for a minimum of 72 hr, habituated to handling, and evaluated clinically before being subjected to anesthesia.

Animals with specialized anatomic, physiologic, or biochemical features may be required in studies utilizing anesthesia. These specialized characteristics may occur spontaneously or be experimentally induced. Examples include the Watanabe, a model of familial hypercholesteremia and the adriamycin-induced model of cardiomyopathy (Arnolda *et al.,* 1985; Goldstein *et al.,* 1983). Rabbits with these features should be evaluated preoperatively to anticipate special factors that may impact anesthesia. Steps may be required pre- and intraoperatively to ensure a stable course of anesthesia. Careful consideration should be given to the selection of the anesthetic agent. For example, xylazine has been shown to increase blood glucose levels so it should be avoided in the diabetic animal (Salonen, 1992); agents that reduce or minimally impact intracranial pressure should be selected for neurosurgical procedures; and consideration of the effect of anesthetic agents on intraocular pressure is important in ophthalmologic procedures.

B. Patient Evaluation and Preparation

Rabbits should be subject, at a minimum, to a physical examination prior to inclusion in an anesthetic protocol. The chest should be auscultated carefully for evidence of pulmonary or cardiovascular disease and rectal temperature should be determined. Baseline laboratory evaluation including a complete blood count and serum biochemistry may be useful in animals that have been subject to prior manipulation. Rabbits with pasteurellosis are not good anesthetic candidates. Thoracic radiography is advisable in rabbits from colonies with enzootic pasteurellosis to identify latent pulmonary consolidation and abcessation. Specialized studies may require additional preoperative diagnostics.

There is no consensus on withdrawal of food from rabbits preoperatively (Flecknell, 1987; Kaplan and Timmons, 1979). The rabbit is unable to vomit and its stomach does not empty despite 5 days of food withdrawal (Murdock, 1969). It has been theorized that a 12-hr fast yields more consistent anesthesia and that the decreased stomach volume may aid respiration, as rabbits breathe primarily by diaphragmatic excursion. Blood glucose levels remain stable for up to 96 hr without eating (Kozma *et al.,* 1974). However, rabbits less than 3 kg develop metabolic acidosis and a significant decline in blood glucose levels as a result of food deprivation (Bonath *et al.,* 1982). Rabbits less than 2 kg are not able to compensate for the acidosis. It is recommended that rabbits less than 3 kg not be fasted for longer than 12 hr (Bonath *et al.,* 1982). Generally there is no need to remove water more than several hours before surgery.

C. Premedication

1. Parasympatholytics

Parasympatholytics may be employed to prevent bradycardia from vagal reflex and to reduce salivary and bronchial secretions that can occlude the airway. Rabbits may produce atropinesterase (AtrE) which degrades atropine into inactive products. The presence of AtrE is inherited and is found in varying frequencies (up to 50%) in the sera and/or tissues of some rabbits in a variety of breeds (Ecobichon and Comeau, 1974). The enzyme may be present in liver and brain without being detectable in the sera (Margolis and Fiegelson, 1963). AtrE activity in serum has been shown to increase with age from 3 until 10 weeks at which time the maximal activity is reached (van Zutphen, 1972). At least three different phenotypes have been described with low, intermediate, and high levels of AtrE activity (Ecobichon and Comeau, 1974). The enzyme hydrolyzes other amino alcohol esters and appears to be related to or identical to the enzyme cocainesterase which hydrolyzes cocaine (van Zutphen, 1972). *In vitro* and *in vivo* screening tests have been described for AtrE detection (Ecobichon and Comeau, 1974; Linn and Liebenberg, 1979; Tucker and Beattie, 1983; van Zutphen, 1972). A rapid screening test (*in vivo*) in which the presence of the pupillary light reflex is evaluated 45 min after receiving atropine sulfate (0.5 mg/kg SC) has compared favorably to qualitative *in vitro* tests (Linn and Liebenberg, 1979). The presence of AtrE has led some authors to recommend preoperative doses of atropine as high as 1–2 mg/kg (Hall and Clarke, 1991) and redosing as frequently as every 10–15 min (Sedgwick, 1986). Glycopyrrolate, a quaternary ammonium with powerful antisialagogue activity with a more rapid onset than atropine, can be utilized in rabbits with AtrE. Glycopyrrolate was shown to be an effective anticholinergic agent in rabbits when administered at a dose of 0.1 mg/kg IM (Olson *et al.,* 1994). Glycopyrrolate produced a significantly elevated heart rate that lasted 60 min when administered alone to rabbits free of AtrE as determined by the *in vivo* screening test, whereas doses of atropine as high as 2 mg/kg did not. Glycopyrrolate also prevented the bradycardia associated with the administration of ketamine/xylazine.

2. Tranquilizers and Sedatives

Tranquilizers/sedatives are frequently used as premedicants for anesthesia or alone for noninvasive procedures such as phlebotomy. Their use frequently reduces the dose of anesthetic and may counteract or limit the adverse effects of some agents.

The phenothiazine tranquilizers (acepromazine, chlorpromazine, and propiopromazine) have been utilized alone or in combination with a variety of agents. Acepromazine has been extensively used in the rabbit (Dolowy and Hesse, 1959; Freeman *et al.,* 1972; Lipman *et al.,* 1990; Ludders *et al.,* 1987; McCormick and Ashworth, 1971; Moore *et al.,* 1987). Peripheral vasodilation, a result of its blockade of adrenoceptors, and sedation make it extremely useful for phlebotomy. It is used alone at doses of 0.25–0.75 mg/kg IV, IM, or SC (N. S. Lipman, unpublished observations) and in combination with ketamine–xylazine at 1 mg/kg IM for blood collection (Ludders *et al.,* 1987). Acepromazine significantly increased the duration and depth of anesthesia achieved with ketamine–xylazine and urethane (Lipman *et al.,* 1990; Moore *et al.,* 1987). Acepromazine dosage ranging from 0.75 to 10 mg/kg IM have been recommended when used in combination with other agents (Dolowy and Hesse, 1959; Freeman *et al.,* 1972; Lipman *et al.,* 1990; Ludders *et al.,* 1987; McCormick and Ashworth, 1971; Moore *et al.,* 1987; Sedgwick, 1986). In the authors' experience, doses of 0.75 to 2 mg/kg are recommended when acepromazine is utilized in combination with other agents. Chlorpromazine, at doses of 25–100 mg/kg IM, reduces the amount of pentobarbital required to induce anesthesia while partially overcoming the respiratory depression caused by the barbiturate (Dolowy and Hesse, 1959). Chlorpromazine has been shown to cause severe muscle necrosis following IM injection; intravenous administration (7.5 mg/kg) has therefore been advocated (Bree *et al.,* 1971). Propiopromazine administered with paraldehyde has been used safely as a preanesthetic cocktail prior to the administration of Equi-Thesin (Hodesson *et al.,* 1965).

The benzodiazepine tranquilizers (diazepam, midazolam, and zolazepam) have either been used alone or combined with other agents to produce general anesthesia. Diazepam has been used at doses of 1–2 mg/kg IM or IV (Flecknell *et al.,* 1983). Higher dosages have been cited (5–10 mg/kg IM) in the literature (Green *et al.,* 1981). When diazepam (5 mg/kg IM) is utilized with ketamine (30 mg/kg IM), the combination produces good sedation and muscle relaxation but insufficient analgesia to carry out surgical procedures (Green *et al.,* 1981). Midazolam, which is more potent than diazepam and is water soluble, has been successfully utilized for short-term anesthesia at 2 mg/kg IP when combined 5 min later with 0.3 ml/kg IM of fentanyl–fluanisone and at 0.5 mg/kg IM combined with medetomidine (0.25 mg/kg IM) followed in 5 min by propofol (2 mg/kg IV) (Flecknell and Mitchell, 1984; Ko *et al.,* 1992). The use of midazolam (1 mg/kg) has also been reported in combination with xylazine–alfentanil (1 mg/kg; 0.1 mg/kg) for intravenous anesthesia in the rabbit (Borkowski *et al.,* 1990). However, muscle rigidity often accompanied by seizures was reported with this combination.

The α_2-adrenergic agonists (xylazine, medetomidine, and detomidine) have been employed alone as tranquilizers or in combination with other agents for anesthesia. The pharmacology of the

α_2-adrenoceptor agonists has been reviewed (Bertolet and Hughes, 1980). Xylazine administered to rabbits at 3–9 mg/kg IV produces sedation, excellent muscle relaxation, and, at the high dose, analgesia; however, when used alone it is not sufficient to provide surgical anesthesia and is reported to cause hyperacusia (Green, 1975). Moderate depression of cardiopulmonary parameters is observed following IM administration at 5 mg/kg (Sanford and Colby, 1980). Xylazine has been used in combination with a variety of agents to induce and maintain surgical anesthesia (Borkowski *et al.,* 1990; Green, 1975; Green *et al.,* 1981; Lipman *et al.,* 1990; Marini *et al.,* 1992; Popilskis *et al.,* 1991; Sanford and Colby, 1980; Sedgwick, 1986). The α_2-antagonist yohimbine (0.2 mg/kg IV) was shown to reverse the effects of anesthesia in ketamine–xylazine-anesthetized rabbits (Lipman *et al.,* 1987).

Medetomidine, the most selective α_2-agonist available for use in veterinary medicine, is useful as a sedative when administered to rabbits at 0.25 mg/kg IM (Ko *et al.,* 1992). Endotracheal intubation should not be attempted when the agent is used alone (Ko *et al.,* 1992). Medetomidine in combination with midazolam/propofol, propofol, and ketamine is useful for rabbit anesthesia of short to moderate duration (Ko *et al.,* 1992; Nevalainen *et al.,* 1989.) The effects of medetomidine are successfully reversed following the administration of 1 mg/kg IM of the α_2-antagonist atipamezole (Nevalainen *et al.,* 1989). Detomidine, another α_2-adrenergic agonist, has also been evaluated in rabbits (Hurley *et al.,* 1994). Detomidine administered at 150 μg/kg IM provided tranquilization only. When combined at 150 and 300 μg/kg IM with ketamine and 150 μg/kg IM with ketamine–diazepam, anesthesia was attained only with the high dose of detomidine or when diazepam was included in the regimen. Anorexia and myocardial necrosis were noted in many of the rabbits receiving the high detomidine dose. The authors of the study concluded that the use of detomidine does not provide any advantages over other established parenteral anesthetic regimens and that detomidine should be avoided until the etiology of the myocardial necrosis is determined.

3. Opioids

The use of opioid drugs pre- or intraoperatively may reduce or eliminate the need for postoperative analgesia. Experimental evidence in rodents suggests that opioids used preoperatively may block afferent impulses from arriving within the central nervous system (CNS) and prevent sensitization to pain that may develop from CNS nociceptive stimulation, thereby reducing postoperative pain and analgesic requirements (Coderre *et al.,* 1993). Classically, narcotic analgesics have been employed alone for pre- and intraoperative analgesia or with sedatives for chemical restraint preoperatively. A more detailed discussion of opioid analgesics is provided later in the chapter.

4. Antimicrobials

Parenteral antimicrobial administration may be useful for specific procedures, e.g., surgical manipulation of the bowel. Antimicrobials should be administered pre- or intraoperatively so that sufficient levels are present in blood and/or tissue at the time of the expected insult. Particular antibiotics such as amoxicillin, ampicillin, cephalothin, clindamycin, and lincomycin may precipitate potentially fatal clostridial enterotoxemias (Carman and Evans, 1984). Postoperative administration of cholestyramine may be useful in reducing mortality associated with antibiotic administration (Lipman *et al.,* 1992). Certain antibiotics may also interact adversely with paralytics and/or inhalation agents.

III. METHODS OF DELIVERY OR ADMINISTRATION

A. Injection Sites

The intramuscular route is commonly used for the administration of a variety of parenteral agents used for tranquilization and anesthesia. As in other species, injections should be made into the body of large muscles, avoiding vessels and nerves. IM injections can be made into either the anterior or the posterior aspect of the thigh or the lumbar epaxial musculature. Hodesson *et al.* (1965) recommend administering drugs into the vastus lateralis and rectus femoris muscles of the anterior thigh at right angles to the femur (Fig. 1). Injections into the caudal thigh,

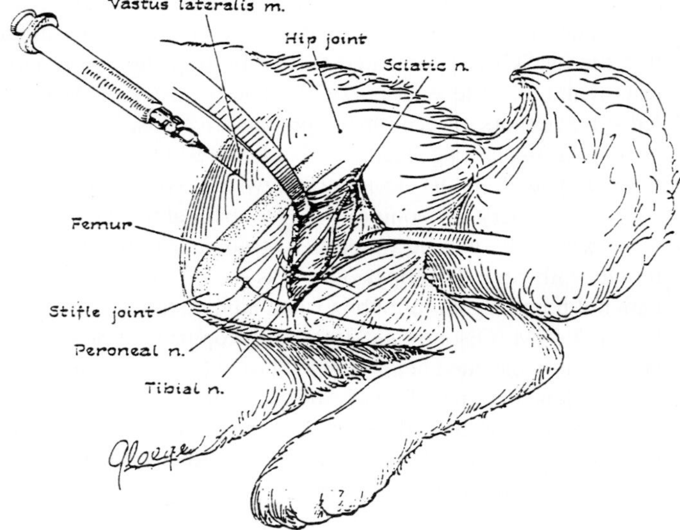

Fig. 1. Major nerves and blood vessels of the thigh. Reprinted with permission from Hodesson, S. *et al.* (1965). Anesthesia of the rabbit with Equi-Thesin® following the administration of preanesthetics. *Lab. Anim. Care* **15,** 336–337.

specifically the biceps femoris, semimembranosis, semitendinosus, and the adductor magnus, can be performed safely if the needle is directed away from the sciatic nerve and its branches. The needle should be inserted into the lateral aspect of the thigh at right angles to the femur or directed caudally away from the femur. Despite appropriate techniques, some drugs, notably chlorpromazine and ketamine–xylazine, have been associated with pathology following IM administration (Beyers *et al.,* 1991; Bree *et al.,* 1971). Myonecrosis, vasculitis, and axonal degeneration with attendant self-trauma and amputation have been reported. Injection volumes should be limited to less than 1.5 ml IM for adult New Zealand White and 1 ml per site for Dutch Belted rabbits, respectively.

The subcutaneous administration of anesthetics, although uncommon, may be useful with particular agents. Postoperative fluid administration by the SC route is routine. The cervical region should be avoided as rabbits are routinely handled by the scruff of the neck. Substances that are irritating should be avoided or diluted prior to SC injection.

Vascular access for intravenous administration of drugs is readily attainable. The lateral (marginal) auricular veins are preferred and easily accessed. Injection and catheterization techniques have been described that can also be used for constant IV or intermittent bolus infusion techniques (Green, 1982b; Grice, 1964; Melich, 1990; Paulsen and Valentine, 1984). The cephalic and recurrent tarsal veins can also be utilized (Gardner, 1964; Green, 1982b). The volume and rate of IV infusion must be monitored in order to prevent volume overload which may lead to pulmonary edema. Air bubbles should be voided from the material to be administered to prevent air emboli. The application of lidocaine–prilocaine cream to the ear before venipuncture has been recommended to reduce pain, especially when training inexperienced staff (Flecknell *et al.,* 1990). The central auricular artery is easily cannulated for arterial blood sampling for blood gas determination and is routinely used for the collection of large volumes of blood (Fick and Schalm, 1986; Ludders *et al.,* 1987; Marini *et al.,* 1992; Smith *et al.,* 1988; Stickrod *et al.,* 1981). Drugs with vasodilatory action, such as acepromazine or droperidol, can be administered to aid in blood collection. Vasospasms can be prevented by applying 2% nitroglycerin ointment to the artery (Smith *et al.,* 1988). Adequate postphlebotomy hemostasis is critical, especially following arterial puncture.

The intraperitoneal administration of anesthetics is primarily of historical interest. Although IP injections can be performed safely, the risk of puncturing a hollow viscus, the difficulty of restraint for injections, the possibility of irritation, and inconsistent absorption limit their consideration in light of alternatives.

Although used infrequently, techniques have been described for the intrathecal administration of local anesthetics for spinal anesthesia in the rabbit. Techniques for both single injection and chronic administration through a surgically implanted cannula have been described (Adams *et al.,* 1974; Bieter *et al.,* 1936; Hughes *et al.,* 1993; Kero *et al.,* 1981; Langerman *et al.,* 1990). Spinal anesthesia has been recommended for studies on the fetus in which depressant anesthetic effects are to be avoided (Kero *et al.,* 1981).

B. Tracheal Access

The anatomy of the oropharynx of the rabbit makes endotracheal intubation more difficult than in most species. The epiglottis and larynx are difficult to visualize because the oral cavity is long and narrow (Fig. 2). The epiglottis is large, U-shaped, soft, and flexible. Just beyond the base of the epiglottis is a deep sagittal niche, bordered on both sagittal recesses by the hamuli epiglottici (Fig. 3). This structure lifts the local epithelial layer and can be damaged easily by rough or incorrect endotracheal intubation (Schuyt and Leene, 1977). The diameter

Fig. 2. Dorsal view of the tongue, larynx, and proximal trachea from an adult New Zealand White rabbit. Note the long tongue and notched U-shaped epiglottis.

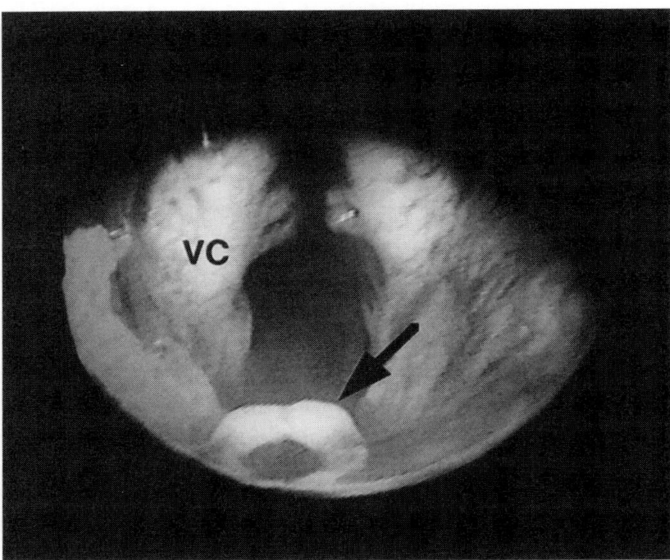

Fig. 3. View of laryngeal structures in an anesthetized rabbit utilizing a 3.5-mm fiber-optic bronchoscope. The tip of the bronchoscope is depressing the epiglottis. The hamuli epiglottici are visible (arrow) as are the vocal cords (VC). Photograph courtesy of P. Padrid.

of the aditus laryngicus is smaller than the diameter of the tracheal lumen (Fig. 4) and dictates the size of the endotracheal tube (Schuyt *et al.,* 1978). The vocal cords are situated extremely craniad and run obliquely in a dorso–ventral direction (Schuyt and Leene, 1977).

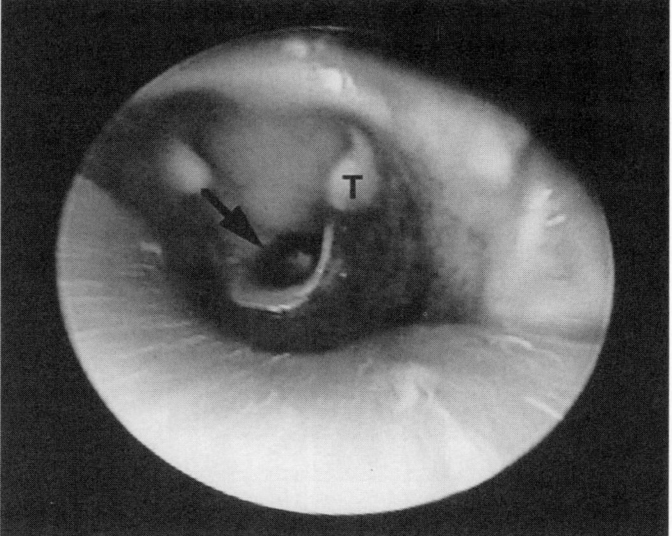

Fig. 4. View of the oropharynx and larynx in an anesthetized rabbit utilizing a 3.5-mm fiber-optic bronchoscope. The laryngyscope blade is visible in the foreground. Note the tonsils (T) and the aditus laryngicus (arrow).

Numerous descriptions of techniques and devices are useful in aiding endotracheal tube placement (Alexander and Clark, 1980; Bertolet and Hughes, 1980; Davis and Malinin, 1974; Fick and Schalm, 1987; Hoge *et al.,* 1969; Howard *et al.,* 1990; Kumar *et al.,* 1993; Lindquist, 1972; Macrae and Guerreiro, 1989; Schuyt and Leene, 1977; Schuyt *et al.,* 1978). Parenteral or volatile anesthetics are administered prior to intubation. Endotracheal intubation can be performed with or without direct visualization of the larynx in the prone or supine position. For successful intubation, the mouth, larynx, and trachea must be brought into linear alignment. With the rabbit in an outstretched prone position the head of the rabbit is tipped and extended until it is upright at a right angle to the rest of its body (Alexander and Clark, 1980). If intubated in the supine position, the neck is hyperextended by placing a rolled towel under the cervical region (Schuyt and Leene, 1977) or by fashioning a holder from Styrofoam (Davis and Malinin, 1974). Uncuffed (Cole or straight) or cuffed endotracheal tubes 2.5–3.5 mm in diameter are used with or without an intubation stylet. The pressure of the inflated cuff against the epithelium has been reported to interfere with blood flow and possibly cause necrosis (Nordin *et al.,* 1977). Wire-reinforced endotracheal tubes should be used for procedures where kinking the tube is possible, such as stereotaxic surgery. A pediatric laryngoscope (size 0–1 Wisconsin or neonatal or size 1 Miller blade) may be used dependent on the size of the rabbit (Lindquist, 1972; Macrae and Guerreiro, 1989). The larynx may be sprayed with a topical anesthetic (without epinephrine) or anesthetic lubricant applied to the tube to reduce laryngeal irritability. Extreme caution should be used when spraying the larynx with any of the topical anesthetics as overdosage is possible. Benzocaine topical anesthetic spray has been shown to cause methemoglobinemia, which may confound experimental results, in rabbits following a 2-sec application (Davis *et al.,* 1993). Following visualization of the larynx the tube is inserted gently on inspiration so that the tube passes through the aditus laryngicus when the vocal cords are abducted. If placement of the endotracheal tube in the supraglottic region severely limits vision, the endotracheal tube can be placed over a 5 or 6 French nylon intravenous catheter so that approximately 50 mm of the catheter protrudes from the tube's lip (Macrae and Guerreiro, 1989). The catheter can be advanced into the larynx under direct visualization before advancing the tube.

When intubation is performed blindly the tube is placed in the supraglottic region and is inserted into the larynx during inspiration using the respiratory pattern and rate, tube condensation, or listening for breath sounds with or without a respiratory monitor to coordinate the advancement (Hoge *et al.,* 1969; Howard *et al.,* 1990). Proper placement of the tube can be verified, as in other species, by palpation of the trachea and esophagus, by capnometry, by listening for breath sounds, or by observing air flow at the distal end of the tube by looking for displacement of plucked fur or condensation on a dental mirror.

Intubation has also been accomplished by inserting percutaneously a 19-gauge, 20-cm catheter into the trachea of an anesthetized rabbit through a guide needle, advancing the catheter retrograde through the larynx into the mouth, and finally advancing the endotracheal tube over the catheter which serves as a guide (Bertolet and Hughes, 1980). Endotracheal tubes should be fixed in place with gauze or tape to avoid accidental extubation.

Tracheostomies may be appropriate for nonsurvival surgical preparations requiring tracheal access. They are not generally recommended for survival surgical procedures because of postoperative considerations. Techniques have been described for the administration of gas anesthetics by mask (Betteridge, 1973; Kent, 1971; Sherrard, 1966). However, the use of these methods should be limited as controlled or assisted ventilation cannot be provided and waste gas pollution may lead to personnel exposure.

The authors' recommended regimen for intubation of the rabbit is ketamine (35 mg/kg IM) and xylazine (5 mg/kg IM). Saffan, methohexital, thiamylal, thiopentone, pentobarbital, medetomidine–propofol, medetomidine–xylazine–propofol, and propanidid have all been recommended for anesthetic induction prior to endotracheal intubation and administration of inhalants in the rabbit (Green, 1982b; Ko *et al.,* 1992). Rabbits should be premedicated with sedatives if mask or chamber induction is used as they are inclined to struggle violently when exposed to the inhalant. Modified cat restraint bags have been used to mask induce previously sedated animals (Lindquist, 1972). Breath holding is common in rabbits; as with other species, gradually increasing the concentration of inhalant may enhance patient acceptance.

IV. PARENTERAL TECHNIQUES

For practical purposes, anesthetics used in the rabbit can be divided into the broad categories of injectable and inhalational agents. Injectable techniques have a long history of use in the rabbit. Ease of administration, predictability, reasonable efficacy, and avoidance of the technical demands of administering inhalants are features that have popularized the use of these agents in the rabbit. At the same time, however, untoward physiologic effects, inability to control anesthetic depth, and prolonged recovery attendant with their use have prompted the search for newer agents or novel combinations of agents. The literature abounds with descriptions of the use of various injectables; some are of historical interest, others are of limited availability, and still others have become valuable agents in the armamentarium of the anesthetist. In this section, injectable agents will be discussed based on functional or chemical similarities.

A. Barbiturates

The oxybarbiturate agent pentobarbital has been alternately lauded (Conn and Langer, 1978; Jacobs and Krohn, 1975; Koch and Dwyer, 1988; Marston *et al.,* 1965; Zhou *et al.,* 1981) and disparaged (Borkowski *et al.,* 1990; Flecknell *et al.,* 1983; Green, 1975; Peeters *et al.,* 1988) as an anesthetic agent in the rabbit. It is now widely held that the dose at which surgical anesthesia occurs is extremely close to the dose at which apnea occurs. This apnea may be reversed by utilizing the Hering–Breuer reflex. Most commonly, chest compression can be provided by manually compressing or by placing an elastic band around the thorax. Anecdotally, success in the use of this agent requires considerable finesse. Although pentobarbital is customarily given intravenously, the intramuscular (Krogh, 1975) and intrahepatic routes of administration have also been reported (Jacobs and Krohn, 1975). These latter routes are discouraged because of variability in effect, the potential for tissue irritation, and the ease of intravenous administration in the rabbit. Dosages range from 25 to 60 mg/kg IV. A typical schedule of administration is to inject some fraction of the calculated dose (generally one-third) over a specified time period, followed by slow infusion of the remainder. Incremental dosages of 3–5 mg/kg IV can be used to prolong anesthesia (Green, 1982b). Pentobarbital may be diluted with saline to enhance the precision of dose administration and, consequently, its safety (Booth and McDonald, 1988). Continuous intravenous infusions of 7.5 and 20 mg/kg/hr after an initial pentobarbital bolus of 50 mg/kg have also been reported (Todd *et al.,* 1994).

The hallmark physiologic effect of pentobarbital is respiratory depression. A number of studies using a variety of doses and infusion schedules describe severe respiratory depression or arrest (Borkowski *et al.,* 1990; Flecknell *et al.,* 1983; Peeters *et al.,* 1988). Respiratory acidosis, hypoxemia, hypercarbia, and depression of respiratory rate are all characteristic features of this drug (Borkowski *et al.,* 1990; Flecknell *et al.,* 1983). In contrast, several studies have demonstrated minimal blood gas alterations when using pentobarbital at doses that did not eliminate the pedal reflex (Korner *et al.,* 1968; Morita *et al.,* 1987). The additional dosage required to eliminate the response to noxious stimuli may induce prolonged apnea.

The features of the effects of pentobarbital on the cardiovascular system include preservation of (Borkowski *et al.,* 1990) or increased heart rate (Korner *et al.,* 1968; Morita *et al.,* 1987), decreased arterial blood pressure (Borkowski *et al.,* 1990; Korner *et al.,* 1968; Morita *et al.,* 1987), peripheral vasodilation (Korner *et al.,* 1968; Morita *et al.,* 1987), decreased cardiac output (Korner *et al.,* 1968), and depression of the vasoconstrictor response to hemorrhage (Warren and Ledingham, 1978). Pentobarbital maintained cardiovascular parameters better than either ketamine–xylazine or midazolam–xylazine–alfentanil in one study (Borkowski *et al.,* 1990). The tachycardia of pentobarbital anesthesia has been shown to be attributable to baroreflex compensation (Morita *et al.,* 1987).

Pentobarbital has been used in combination with a variety of agents to reduce the dosage required for anesthesia, to improve safety, and to enhance efficacy (Dolowy and Hesse, 1959; Green, 1975, 1982b; Hobbs *et al.,* 1991; Krogh, 1975; Olson *et al.,* 1987; Schutten and Van Horn, 1977). The following combinations have been used: (1) Ketamine (50 mg/kg) and pentobarbital (30 mg/kg) administered in separate syringes intramuscularly caused deep anesthesia of a 60- to 80-min duration (Krogh, 1975). (2) Chlorpromazine premedication (2 mg/kg IM) followed by pentobarbital (20–30 mg/kg IV) has been reported to be a safe anesthetic combination for the rabbit (Green, 1975). (3) A range of chlorpromazine doses (25–100 mg/kg IM) was used prior to pentobarbital (20 mg/kg IV) to produce anesthesia, but pedal ''withdrawal reflexes were often present'' (Dolowy and Hesse, 1959). (4) Five percent guaifenesin in 5% dextrose (200 mg/kg IV) was followed by 20 mg/kg pentobarbital (20 mg/kg IV) to produce anesthesia in six rabbits (Olson *et al.*, 1987). Tachycardia with reductions in respiratory rate and mean arterial blood pressure characterized this combination; pedal reflex returned in all rabbits by 30 min after drug administration (Olson *et al.*, 1987). (5) Xylazine (5 mg/kg SC) followed in 10 min by 11.8 to 28.4 mg/kg pentobarbital IV produced 37.5 min of anesthesia, although one of eight rabbits died (Hobbs *et al.*, 1991).

Pentobarbital has been used in the rabbit as a component of Equi-Thesin (Hodesson *et al.*, 1965). Rabbits were premedicated with paraldehyde (0.3 ml/kg IM) and diazepam (7.5 mg/kg IM) and then administered varying doses of Equi-Thesin intrarectally (1–3 ml/kg). Results were variable with regard to safety and the depth and duration of anesthesia produced. Equi-Thesin was also administered intravenously to effect after paraldehyde (0.3 ml/kg) and diazepam (10 mg/kg IM). Effective dosages of Equi-Thesin ranged from 0.83 to 2.35 ml/kg; anesthesia of varying durations was achieved in 10 of 11 rabbits with one fatality. Similar results were attained with Equi-Thesin used after premedication with propiopromazine (5 mg/kg IM) and paraldehyde (0.3 ml/kg IM). Equi-Thesin is no longer commercially available but can be prepared by combining pentobarbital, magnesium sulfate, and chloral hydrate.

Several miscellaneous features of pentobarbital anesthesia deserve mention. When pentobarbital was used in a coronary artery ligation study, less myocardial damage was observed when compared to either halothane or α-chloralose (Chakrabarty *et al.*, 1991), underscoring the impact of choice of anesthetic on research results. Pentobarbital caused reductions in plasma potassium concentrations that were associated with increases in plasma renin activity and aldosterone concentration (Robson *et al.*, 1981). Pentobarbital appears to exhibit tolerance (tachyphylaxis) in the rabbit; equivalent anesthetic effects required doubling the pentobarbital doses when attempts were made to anesthetize animals more than once weekly (Pandeya and Lemon, 1965). Finally, opioids appear to modify the duration of action of pentobarbital by a central cholinergic mechanism (Horita and Carino, 1978; Horita *et al.*, 1983).

Other barbiturates that have been used in the rabbit include the thiobarbiturates thiamylal, thiopental, ethylmalonylthiourea (EMTU), and the methylated oxybarbiturate, methohexital (Gardner, 1964; Green, 1975, 1982b; Hobbs *et al.*, 1991; Holland, 1973; Lockhart *et al.*, 1991; MacLeod, 1977; Murdock, 1969; Sedgwick, 1986; Stunkard and Miller, 1974). These are outlined below.

Dosages reported for thiamylal are 31 mg/kg of a 1% solution, 29 mg/kg of a 2% solution (Gardner, 1964), and 15 mg/kg IV (Sedgwick, 1986). Thiopentone dosages recommended in the literature include 30 mg/kg (one-half given rapidly IV with the rest administered slowly over 60 sec) (Green, 1982b), 6 mg/kg slowly IV to effect prior to endotracheal intubation (Green, 1982b), 15 mg/kg (Sedgwick, 1986), 50 mg/kg of a 1.25% solution (Holland, 1973), and 50 mg/kg of a 2.5% solution (Lumb and Jones, 1984). These dosages vary widely and may be influenced by other factors such as breed or conditioning. Whatever dose is chosen, safety is enhanced by using the more dilute solutions and injecting agents slowly to effect. The duration of action for these agents is usually 5–10 min (Murdock, 1969).

The use of EMTU has been reported in conjunction with xylazine or ketamine (Hobbs *et al.*, 1991). Xylazine (5 mg/kg SC) or ketamine (35 mg/kg IM) was followed in 10 min by EMTU administered intravenously to effect (12.5–47.6 mg/kg following xylazine; 25–54 mg/kg following ketamine). The combinations caused 34 and 19 min of anesthesia, respectively. Anesthesia was not consistently induced with ketamine–EMTU whereas xylazine–EMTU provided ''the most consistent and safest anesthetic episode'' of those tested (Hobbs *et al.*, 1991).

Methohexitone has been recommended for use as an induction agent prior to endotracheal intubation (10 mg/kg of a 1% solution rapidly IV) Green, 1982b). Two to 5 min of light anesthesia are provided; increments of 4–6 mg/kg IV may prolong anesthesia for short procedures. Methohexital (5 mg/kg) administered IV has also been recommended for use in the rabbit (Antal, 1985).

B. Dissociatives

The dissociative agent that has been used most widely in rabbit anesthesia is ketamine. It is typically administered in combination with other agents as it does not provide adequate muscle relaxation or analgesia for surgical anesthesia when used alone (Green, 1982b; Green *et al.*, 1991; White and Holmes, 1976). As a sole agent, ketamine should be restricted to minimally invasive procedures; both intramuscular and intravenous routes have been used. Intramuscular doses ranging from 20 to 60 mg/kg have been reported (Green, 1975, 1982b), although doses greater than 50 mg/kg probably provide little additional restraint. Endotracheal intubation has been reported with ketamine alone (15–20 mg/kg); 10% ketamine diluted 1:2 with saline and injected at a rate of 100 mg in 10–15 sec was used for this purpose (Lindquist, 1972). In the authors' experience, the combination ketamine–xylazine is preferred for intubation.

The agent most commonly used in combination with ketamine is the α_2-agonist xylazine. Ketamine–xylazine doses ranging from 22 mg/kg ketamine and 2.5 mg/kg xylazine to 50 mg/kg ketamine and 10 mg/kg xylazine have been reported (Beyers *et al.*, 1991; Rich *et al.*, 1990; White and Holmes, 1976). The dose that appears most frequently in the literature is 35 mg/kg ketamine and 5 mg/kg xylazine. The combination at this dose provided mean withdrawal reflex loss durations of 46.5 (Hobbs *et al.*, 1991), 32 (Marini *et al.*, 1992), and 57 min (Lipman *et al.*, 1990). Considerable variation in the response to ketamine–xylazine exists among rabbits (Marini *et al.*, 1992; Peeters *et al.*, 1988). Incremental doses of one-third the original dose IM can be used to prolong anesthesia. In the authors' experience, the combination of ketamine (50 mg/kg) and xylazine (10 mg/kg) provides generally effective anesthesia for procedures of moderate surgical stimulus intensity (e.g., carotid endarterectomy and subcutaneous implantation). More intense surgical stimuli associated with intraabdominal and intrathoracic procedures are accompanied by undesirable muscle fasciculation and autonomic response (e.g., increased heart rate, arterial blood pressure, and respiratory rate).

A myriad of physiologic effects occurs with intramuscularly administered ketamine/xylazine. Depressions of the respiratory rate ranging from 40 to 77% of preanesthetic baseline values have been reported (Hobbs *et al.*, 1991; Lipman *et al.*, 1990; Marini *et al.*, 1992; Peeters *et al.*, 1988; Popilskis *et al.*, 1991; Sanford and Colby, 1980; Wyatt *et al.*, 1989). Associated blood gas changes included hypoxemia (43–50% reduction) and CO_2 retention (25–50% increase) (Marini *et al.*, 1992; Peeters *et al.*, 1988; Popilskis *et al.*, 1991; Wyatt *et al.*, 1989). Arterial blood pH either remained unchanged (Marini *et al.*, 1992; Popilskis *et al.*, 1991; Wyatt *et al.*, 1989) or increased marginally due to metabolic alkalosis with partial respiratory compensation (Peeters *et al.*, 1988). Cardiovascular changes were manifest as decreases in heart rate (from insignificant change to 35% reduction) and hypotension (20–35% reduction). The maximal alterations produced by the ketamine–xylazine combination paralleled those seen when either drug was administered alone; ketamine preservation of the heart rate was the only exception (Sanford and Colby, 1980).

Authors' Preference

Agent:	Ketamine–xylazine
Route:	IM injection
Dose:	35–50 mg/kg ketamine 5–10 mg/kg xylazine
Comments:	Perineural injection may lead to self-trauma; adequate for arterial cutdowns and other somatic procedures. Consistency may be improved and depth enhanced by the addition of butorphanol (0.1 mg/kg IM) or acepromazine (0.75 mg/kg IM). Incremental doses of one-third the original dose of ketamine–xylazine may be used to prolong anesthesia

Several investigators have described the intravenous use of ketamine, either alone or in combination with xylazine (Borkowski *et al.*, 1990; Dhasmana *et al.*, 1984; Lindquist, 1972; McGrath *et al.*, 1975; Wyatt *et al.*, 1989). Borkowski *et al.* (1990) evaluated ketamine (25 mg/kg) and xylazine (5 mg/kg); the agents were mixed in the same syringe (1:1,v:v) and one-third the calculated dose was administered into the marginal ear vein over 1 min. The remainder was administered over the next 4 min. The physiologic changes measured were similar to those described previously for the intramuscular administration of this combination with the exception that apnea occurred in two of seven rabbits. In another study, a single intramuscular dose of ketamine (35 mg/kg) and xylazine (5 mg/kg) was followed with a 4-hr, constant rate intravenous infusion of the same agents at dose levels of 1 mg/min ketamine and 0.1 mg/min xylazine (Wyatt *et al.*, 1989). Blood pressure progressively declined throughout the duration of this regimen from an initial reduction of 21% from baseline to a reduction of 49%. Arterial O_2 tension initially declined to 45% of baseline but increased progressively as the infusion continued. The authors concluded that the regimen was a safe and effective method of producing light anesthesia in the rabbit but cautioned that hypotension and hypoxemia might be unacceptable experimental variables.

Other ketamine combinations that have been reported include ketamine–pentobarbital (Krogh, 1975; Schutten and Van Horn, 1977), ketamine–EMTU (Hobbs *et al.*, 1991), ketamine–guaifenesin (Olson *et al.*, 1987), ketamine–chloral hydrate (Chen and Bohner, 1968; Hobbs *et al.*, 1991), ketamine–medetomidine (Nevalainen *et al.*, 1989), ketamine–xylazine–acepromazine (Lipman *et al.*, 1990), ketamine–detomidine (Hurley *et al.*, 1994), ketamine–diazepam (Green, 1982b; Sedgwick, 1986), ketamine–detomidine–diazepam (Hurley *et al.*, 1994), and ketamine–xylazine–butorphanol (Marini *et al.*, 1992).

Ketamine–xylazine–acepromazine increased the duration of anesthesia (as measured by absence of pedal reflex) when compared to ketamine–xylazine but did so at the expense of moderate increases in hypotension and hypothermia. Reported doses include ketamine (35 mg/kg), xylazine (5 mg/kg), acepromazine (0.75 mg/kg) and ketamine (40 mg/kg), xylazine (3 mg/kg), and acepromazine (1 mg/kg) (Hobbs *et al.*, 1991; Lipman *et al.*, 1990; Ludders *et al.*, 1987). The addition of butorphanol to ketamine–xylazine also increased the duration of reflex loss but with less physiologic embarrassment than that attending the addition of acepromazine (Marini *et al.*, 1992). Ketamine (50 mg/kg IV) and 5% guaifenesin in 5% dextrose (200 mg/kg IV) were reported to provide safe and effective anesthesia for 30 min with preservation of heart rate and arterial blood pressure (Olson *et al.*, 1987). The respiratory rate decreased by approximately 50% from baseline.

Chen and Bohner (1968) reported that a combination of ketamine (20 mg/kg IM) and 10% chloral hydrate (250 mg/kg) injected IV over a 2- to 3-min period after ketamine-induced catalepsy had occurred produced 1–1.5 hr of anesthesia without dramatic changes in arterial blood pressure or respiratory rate. In contrast, Hobbs found that the identical regimen produced only 20 min of anesthesia and a 67% reduction in respiratory rate from baseline (Hobbs *et al.*, 1991).

Telazol is a combination agent consisting of the dissociative agent, tiletamine, and the benzodiazepine, zolazepam. Administered alone it does not provide sufficient analgesia for surgery but is useful for restraint and immobilization (Brammer *et al.,* 1991; Doerning *et al.,* 1992; Ward *et al.,* 1974). The agent is also nephrotoxic, as manifest by azotemia and urinary casts. Intranasal administration of Telazol (10 mg/kg) produced a mean duration of righting reflex loss of 44 min while eliminating the pedal reflex in only one of four rabbits (Robertson and Eberhart, 1994). Additional features of intranasal Telazol include good muscle relaxation, increased heart rate, and decreased respiratory rate. Serum creatinine and BUN were not measured. The addition of xylazine (5 mg/kg IM) to Telazol (15 mg/kg IM) produced surgical anesthesia lasting approximately 70 min (Popilskis *et al.,* 1991). In contrast, Doerning *et al.* (1992) reported that as little as 7.5 mg/kg tiletamine was sufficient to produce mild nephrosis manifest as ''scattered dilated tubules and occasional cellular casts.'' Azotemia did not occur at this dose. Tiletamine use is best avoided in the rabbit because of attendant nephrotoxicity.

C. Neuroleptanalgesia–Neuroleptanesthesia

The most widely investigated neuroleptanalgesic combinations used in the rabbit are fentanyl–droperidol (Innovar-Vet) and fentanyl–fluanisone (Hypnorm).

Innovar-Vet, available in the United States, contains fentanyl (0.4 mg/ml) and droperidol (20 mg/ml). It has been recommended for use as a sole agent for anesthesia (0.22 ml/kg) (Strack and Kaplan, 1968), for sedation during attempts at group housing (Love and Hammond, 1991), for sedation prior to cardiac puncture (0.17 ml/kg) (Lewis and Jennings, 1972) or prior to arteriocentesis of the central auricular artery (0.15, 0.17, 0.19, 0.30 ml/kg) (Sartick *et al.,* 1979); (0.125 ml/kg) (Tillman and Norman, 1983), and for stable neuroleptanalgesia using constant intravenous infusion (255 μl/min of a 0.05-mg/ml fentanyl, 0.13-mg/ml droperidol solution in 5% dextrose) after a single IM dose (0.125 ml/kg) (Guerreiro and Page, 1987). Fentanyl–droperidol administered intranasally at a dosage of 0.3 mg/kg produced bradycardia, apnea, and 50% mortality (Robertson and Eberhart, 1994). The droperidol component of Innovar-Vet has a longer duration of action than fentanyl; accordingly, animals may remain sedated well after the peak of neuroleptanalgesia has passed. Atropine or glycopyrrolate should be considered as a premedicant to avoid the bradycardia associated with the use of fentanyl or any of its congeners.

When Innovar-Vet (0.15 ml/kg IM) was used in combination with either diazepam (2 mg/kg IV) or the α_2-agonist detomidine (20 μg/kg IV), neither combination was found to improve the consistency or reliability of fentanyl–droperidol (Marini *et al.,* 1993). Physiologic changes associated with Innovar-Vet included decreases in heart rate (35%), arterial blood pressure (11%), respiratory rate (85%), and PaO_2 (46%). Blood pressure was the parameter most severely affected by the addition of diazepam or detomidine. The authors concluded that fentanyl–droperidol use at the dose selected be restricted to providing restraint for physical examination and minimally invasive diagnostic procedures.

Hypnorm is available in Europe but must be imported as an investigational drug in the United States. The commercial preparation contains 0.3 mg/ml fentanyl and 10 mg/ml fluanisone. At doses of 0.3–0.5 ml/kg, Hypnorm has been used for superficial surgeries and as a sedative prior to the administration of inhalational agents (Green, 1975, 1982b). Respiratory depression, bradycardia, and poor muscle relaxation are characteristics of this agent (Green, 1975). The addition of diazepam or midazolam to Hypnorm results in good surgical anesthesia of moderate duration (Flecknell *et al.,* 1983; Flecknell and Mitchell, 1984). Premedication with diazepam (1 mg/kg IV or 2 mg/kg IP), followed 5 min later by Hypnorm (0.2 ml/kg IM), produced 45–60 min of anesthesia and respiratory rate depression of 30–60% (Flecknell *et al.,* 1983). Similarly, midazolam (2 mg/kg IP) has been used with Hypnorm (0.3 ml/kg IM) to produce surgical anesthesia of a 100-min mean duration with a respiratory rate depression of 29–62% (Flecknell and Mitchell, 1984). Increments of 0.05–0.1 ml/kg Hypnorm IM every 20–30 min and 1 mg/kg diazepam IV every 2–4 hr have been used to prolong anesthesia. Hypnorm (0.3 ml/kg SC) and diazepam (5 mg/kg SC) were used after atropine premedication (1 mg/kg SC) to provide anesthesia for orthopedic procedures in 565 rabbits (Mero *et al.,* 1987). The subcutaneous route was used so that drug absorption might be delayed and the likelihood of respiratory depression reduced. The mortality rate for use of the combination in this fashion was 1.4%.

Authors' Preference

Agent:	Fentanyl–fluanisone–midazolam
Route:	Fentanyl–fluanisone IM Midazolam IV
Dose:	0.3 ml/kg fentanyl–fluanisone 0.5–2 mg/kg midazolam to effect
Comments:	Fentanyl–fluanisone is administered as a premedication. Midazolam is then administered to the sedated rabbit to produce surgical anesthesia. Fentanyl–fluanisone is available in Europe

A constant rate infusion technique has been described in which a 1:10 dilution of Hypnorm is administered intravenously at a rate of 1–3 ml/kg/hr (Flecknell, 1987). Although most reports extol the virtues of the Hypnorm–benzodiazepine combinations, Peeters *et al.* (1988) reported that only three of six rabbits were adequately anesthetized with Hypnorm (0.5 ml/kg IM) and diazepam (2 mg/kg IM). It should be noted, however, that forelimb withdrawal was used as the criterion for anesthesia in that study instead of the more commonly used parameter of hindlimb withdrawal. These responses are lost at different times in the rabbit.

Naloxone (at doses of 0.005, 0.01, and 0.1 mg/kg IV), doxapram (5 mg/kg IV), and various mixed agonist/antagonist opioids can reverse the respiratory depression caused by Hypnorm

(Flecknell *et al.,* 1989). The use of mixed agonist/antagonist opioids following neuroleptanalgesia has been termed anesthesié analgesique sequentielle, and permits restoration of physiologic variables to preanesthetic values while providing postoperative pain relief (DeCastro and Viars, 1968). The mixed agonist/antagonists were notable in that despite reversal of sedation and respiratory depression, analgesia persisted for varying periods after administration. Ranked by the duration of analgesia provided after reversal of fentanyl, meptazinol < pentazocine < butorphanol < nalbuphine < buprenorphine. Buprenorphine (0.3 mg/kg IV) reversed the Hypnorm associated depression of respiratory rate, normalized hypoxemia and arterial CO_2 retention, and provided 420 min of analgesia.

A combination containing etorphine (0.074 mg/ml) and methotrimeprazine (18 mg/ml) (Small Animal Immobilon) was used at dose rates of 0.05 and 0.025 ml/kg IM in the rabbit (Flecknell *et al.,* 1983). There was insufficient analgesia at the lower dose but combining this dose with diazepam (1 mg/kg IV) produced surgical anesthesia in six of nine rabbits. Severe respiratory depression was associated with the higher dose of this agent.

Midazolam–xylazine–alfentanyl has been evaluated as an intravenous neuroleptanesthetic for rabbits (Borkowski *et al.,* 1990). Midazolam (1 mg/kg) was administered over 15 to 30 sec; xylazine (1 mg/kg) was administered over 10 to 15 sec. Alfentanyl (0.1 mg/kg) was given incrementally over a 5-min period; infusion was terminated for 2 min if muscle rigidity or opisthotonus occurred. Additional increments of one-half the calculated dose were administered every 15 min. Features of this regimen included good analgesia, incomplete or variable muscle relaxation, apnea in one of seven rabbits during induction, and muscle rigidity with seizure activity during redose in three of seven rabbits. Physiologic changes included a marked depression of respiratory rate (90%), hypoxemia (60%), CO_2 retention, and decreased pH. Heart rate and mean arterial blood pressure were also depressed (30 and 35%, respectively, from baseline). The usefulness of this combination appears to be limited by these adverse effects.

D. Miscellaneous Agents

The steroidal combination alphaxalone–alphadolone (12 mg/ml total steroids, Saffan or Althesin) was evaluated in rabbits and was found to produce useful, light to medium depth anesthesia of 8–10 min duration when used at a dose of 12 mg/kg IV (Green *et al.,* 1978). Full recovery took 2–2.5 hr; incremental doses every 20 min could be used to prolong anesthesia for up to 8 hr. The agent has been recommended for superficial surgery or for endotracheal intubation; the latter is accomplished by injecting one-half the calculated dose rapidly with the remainder given slowly to effect (Green, 1982b). Analgesia may be inconsistent with this agent despite reliable muscle relaxation.

A bolus injection (1 mg/kg IV), followed by constant rate infusion of alphaxalone–alphadolone (0.1 mg/kg/min IV) in chronically instrumented rabbits, was used to determine the hemodynamic and reflex responses to this agent in comparison with propofol (Blake *et al.,* 1988). The regimen used was sufficient to produce loss of righting and palpebral reflex but did not eliminate the response to noxious stimuli. Mean arterial blood pressure, cardiac output, blood gas tensions, and total peripheral resistance were unchanged whereas heart rate increased approximately 20%. At low doses, alphaxalone–alphadolone preserves peripheral vasoconstrictor reflexes and may therefore be used in pharmacologic studies that require integrity of these reflexes in a lightly anesthetized animal.

The alkyl phenol agent propofol, reviewed by Sebel and Lowdon (1989), has been used to produce deep sedation during constant intravenous infusion in the rabbit (Adam *et al.,* 1980; Blake *et al.,* 1988; Glen, 1980). Glen (1980) used doses of 5–15 mg/kg of 2% propofol in 10% Cremophor EL and found only light anesthesia of short duration with "little reflex depression." A mean "utilization rate," obtained by the quotient of total dose administered and time from induction to recovery (head lift), was determined to be 1.55 mg/kg/min for the rabbit. This infusion rate provided light anesthesia without increasing recovery time. Blake *et al.* (1988) found that a bolus dose of 1.5 mg/kg followed by a constant rate infusion of 0.2, 0.4, and 0.6 mg/kg/min produced either sedation (characterized by loss of righting and palpebral reflex) at the two lower infusion rates or anesthesia at the higher infusion rate. Blood gas tensions were preserved at all infusion rates and there were no changes in mean arterial pressure at the lower infusion rates. Cardiac output and heart rate were increased from baseline at the two lower infusion rates; total peripheral resistance was decreased. At the highest infusion rate, cardiac output and mean arterial pressure were decreased while total peripheral resistance remained depressed. Propofol caused a dose-related reduction in the range and gain of the baroreceptor reflex. The pharmacokinetics of single bolus propofol (5 mg/kg) in the rabbit have also been described (Cockshott *et al.,* 1992).

The Cremaphor EL diluent used in these studies was associated with hypersensitivity in humans and has now been replaced by an oil-in-water emulsion that presumably yields a less potent propofol formulation (Aeschbacher and Webb, 1993a). Using this formulation at an injection rate of 20 mg/kg/min, an ED_{50} for induction was determined to be 6.44 mg/kg. Induction and recovery were smooth and rapid, with rabbits regaining the righting reflex 2–8.5 min after induction. Further studies using propofol induction (mean 7.3 mg/kg) and long-term intravenous infusion (mean infusion rate 0.876 mg/kg/min) produced light levels of anesthesia characterized by persistence of the pinna, pedal, and palpebral reflexes. Hypoxemia, hypertension, and prolonged recovery were other features of this regimen (Aeschbacher and Webb, 1993b). Propofol as a sole agent is apparently suitable only for induction and minimally invasive diagnostic procedures.

Ko *et al.* (1992) studied propofol induction after medetomidine–atropine or medetomidine–midazolam–atropine premedication. Rabbits were administered medetomidine (0.25 mg/kg IM) and atropine (0.5 mg/kg IM) or medetomidine (0.25 mg/kg IM), midazolam (0.5 mg/kg IM), and atropine (0.5 mg/kg IM) 5 min prior to propofol (4 mg/kg IV for the former group; 2 mg/kg IV for the latter). Each premedicant combination produced loss of the righting reflex but persistence of pedal, ear pinch, palpebral, and corneal reflex. Propofol administration caused a further loss of palpebral, ear pinch, and corneal reflex, but the pedal withdrawal reflex was preserved. The addition of midazolam prolonged the loss of ear pinch reflex from a mean of 25 min to a mean of 37 min. No changes were seen from the preanesthetic baseline in heart rate, respiratory rate, mean arterial pressure, or end-tidal CO_2. These combinations provided sufficient anesthesia to achieve endotracheal intubation and may therefore be useful for induction or short-term anesthesia.

Urethane (ethyl carbamate) has been used for anesthesia in rabbits both alone (Bree and Cohen, 1965; Murai and Ogura, 1978) and in combination with other agents (Moore *et al.*, 1987; Wyler, 1974). The popularity of this agent can be attributed to its long duration of action (5–6 hr) and excellent muscle relaxation (Maggi *et al.*, 1984). Disadvantages of urethane include carcinogenicity, slow recovery, hypotension, a reduced response of vascular smooth muscle to norepinephrine, a wide range of endocrine effects, tissue slough at extravasation, venodilation of the vessel used for injection, hemolysis, and a transient reduction of hematocrit (Bree and Cohen, 1965; Maggi *et al.*, 1984). Its carcinogenic properties and the attendant hazards to personnel require that its use be regulated by institutional safety personnel. Doses of 1.5 g/kg (20% urethane slowly IV to effect) (Bree and Cohen, 1965) and 1–1.6 g/kg IP have been recommended (Green, 1982b). These recommendations notwithstanding, intraperitoneal doses of 1.5 and 1.75 g/kg of a 20% urethane solution were associated with postoperative death in one study (Bree and Cohen, 1965).

Urethane combinations include chloralose–urethane (Dorward *et al.*, 1987; Korner *et al.*, 1968; Warren and Ledingham, 1978; Wyler, 1974) and urethane–acepromazine (Moore *et al.*, 1987). Chloralose–urethane has been widely used for nonsurvival procedures requiring long-term anesthesia. Physiologic features of this combination include respiratory rate depression (65%), tachycardia, increased total peripheral resistance, increased venous blood pressure, decreased total body blood flow, and unchanged arterial blood pressure (Wyler, 1974). It is widely held that chloralose–urethane preserves or enhances baroreceptor reflexes, making it popular in studies where preservation of these reflexes is desirable (Sebel and Lowdon, 1989; Warren and Ledingham, 1978). Dosage recommendations include (1) 800 mg/kg of 25% urethane IV with 40–60 mg/kg of 1% chloralose IV initially followed by 3–4 ml/hr of 1% chloralose (Dorward *et al.*, 1987) and (2) 400 mg/kg of 25% urethane IV with 60 mg/kg of 1% chloralose IV initially followed by 1–3 ml of 1% chloralose every 30–50 min (Jenkins, 1987).

Urethane–acepromazine has been evaluated in the rabbit as a long-term anesthetic (Moore *et al.*, 1987). Light anesthesia was produced by premedication with acepromazine (1 mg/0.46 kg IM), followed in 15 min by 1 g/kg urethane IV. The urethane was injected over a period of 5–10 min followed by a saline flush. Loss of consciousness persisted for over 12 hr. The authors described the level of anesthesia as comprising loss of responsiveness to superficial, but not to deep visceral pain. Heart and respiratory rates decreased over the first hour but then stabilized. Administration of a higher dose of urethane (1.3 g/kg IV) after acepromazine premedication resulted in deep anesthesia appropriate for abdominal and thoracic procedures. Pedal reflex was still absent 18 hr after administration with this dose. Physiologic changes paralleled those measured with the lower dose regimen.

V. INHALATION ANESTHESIA

Although requiring special equipment and training, the use of inhalants provides excellent reliability, efficacy, and anesthetic depth. These features are especially important in rabbit anesthesia where many injectable combinations that are adequate for procedures involving superficial structures are inadequate for more invasive manipulations. Reduction in recovery time is an additional benefit of inhalant anesthesia.

Once the rabbit is intubated, inhalants may be administered using various circuits, and animals may be ventilated or be allowed to breathe spontaneously. The Rees modified T-piece and Bain circuits are traditionally considered the most appropriate circuits for animals comprising the range of weights of most rabbits; Magill (Peeters *et al.*, 1988) and circle circuits (Nelson *et al.*, 1990) have also been used. The range of fresh gas flows reported in the maintenance of spontaneously breathing animals is 1 to 3 liters/min (Green, 1982b; Kent, 1971; Kumar *et al.*, 1993; Peeters *et al.*, 1988). Alternatively, one may choose to administer a volume of fresh gas that is two to three times the minute respiratory volume of the animal (Flecknell, 1987). Twice the minute respiratory volume of 100% O_2 delivered via the Bain circuit is the method used for inhalant administration in our surgical laboratory (R. P. Marini, personal experience).

Artificial ventilation has most commonly been used in the rabbit in the setting of acute preparations involving paralytics. Ventilatory parameters during intermittent positive pressure ventilation can be determined by estimates of tidal volume and empiric settings of ventilatory rate, by use of pressure cycling to terminate inspiration at a chosen airway pressure, or by guidance of ventilatory settings by blood gas parameters. Tidal volume and respiratory rate settings of 10 ml/kg, 45 breaths per minute (Chakrabarty *et al.*, 1991; Ishibe *et al.*, 1993), 15 ml/kg, 20 breaths per minute (Drummond, 1985; Drummond *et al.*, 1987), 8 ml/kg, 20 breaths per minute (Aeschbacher and Webb, 1993a,b), and 15 ml/kg, 30 breaths per minute (Kumar *et al.*, 1993; Patel and Mutch, 1990) have been reported. The newer ventilation modalities, high frequency jet ventilation, continuous positive pressure ventilation with positive end expiratory pressure, oscillatory ventilation, and negative impedance ventilation have been described in the rabbit, often as models for ventilation in human beings, but have not replaced traditional methods (Baum *et al.,* 1989; Lebowitz, 1990; Mook *et al.,* 1984).

Methoxyflurane and ether, once widely used for anesthesia in the rabbit, have been largely supplanted by isoflurane and halothane. Reports on the use of desflurane, sevoflurane, and enflurane exist, but these agents do not appear to be in common usage (Ishibe *et al.,* 1993; Doorley *et al.,* 1988; Scheller *et al.,* 1988; Drummond, 1985; Lockhart *et al.,* 1991; Stadnicka *et al.,* 1993; Tashiro *et al.,* 1986). The inhalant of choice in many surgical laboratories is isoflurane. Advantages of isoflurane include cardiac safety, rapid induction and recovery due to low blood solubility, minimal hepatic transformation, and reduced viscerotoxicity. Disadvantages are cost, potential increase in breath holding at the start of exposure, reduced respiratory anesthetic index, and hypotension. Physiologic effects of 1.3 MAC isoflurane anesthesia include 21% reduction in mean arterial pressure, 18% increase in heart rate, 21% reduction in renal blood flow, and preservation of hepatic blood flow, cardiac output, respiratory rate, and $PaCO_2$ (Blake *et al.,* 1991). The cardiac output in rabbits with adriamycin-induced heart failure was unchanged by isoflurane but was decreased (20%) by halothane (Blake *et al.,* 1991), a reflection of the high cardiac anesthetic index of isoflurane (the quotient of the dose of an agent required to produce cardiac arrest and the MAC). Cerebral uptake and elimination of isoflurane were slower than that of desflurane but faster than that of halothane (Lockhart *et al.,* 1991). The MAC value for isoflurane in rabbits is $2.05 \pm 0.18\%$ (Drummond, 1985). In contrast, the MAC for halothane is $1.39 \pm 0.32\%$ and that for enflurane is $2.86 \pm 0.18\%$. Anesthetists accustomed to the use of halothane will notice that maintenance concentrations of isoflurane are higher for a given procedure or surgical stimulus.

Authors' Preference	
Agent:	Ketamine–xylazine followed by isoflurane
Route:	Ketamine–xylazine (IM) Isoflurane (inhalation)
Dose:	Induction 35 mg/kg ketamine 5 mg/kg xylazine Maintenance 1–4% isoflurane
Comments:	Our anesthetic regimen of choice for invasive procedures such as celiotomy or thoracotomy

Halothane is relatively inexpensive and provides safe and effective anesthesia in rabbits. Physiologic effects of halothane anesthesia include reductions in cardiac output, arterial blood pressure, PaO_2, and respiratory rate and elevation in venous blood pressure, plasma renin, $PaCO_2$, and heart rate (Sartick *et al.,* 1979; Wyler, 1974; Wyler and Weisser, 1972). Total peripheral resistance remains unchanged or is mildly elevated, suggesting that the mechanism of hypotension and reduced cardiac output is myocardial depression (Wyler and Weisser, 1972). Reduction in the organ blood flow occurs in proportion to the reduction in cardiac output with the preservation of flow to brain, adrenals, testes, and muscle in one study (Wyler and Weisser, 1972) and brain, stomach, appendix, arterial hepatic flow, and muscle in another (Wyler, 1974). Sartik *et al.* (1979) reported that cardiac output and arterial blood pressure returned to preanesthetic baseline values 15 min after the termination of halothane anesthesia. Plasma renin concentrations returned to baseline by 210 min after anesthesia.

The use of nitrous oxide as a carrier gas during induction has been advocated in the past, but concern over operating theater pollution in today's setting may limit the use of this agent. Nitrous oxide may be used as a carrier gas or as a background agent in long-term injectable techniques (Green, 1982b). When compared to pentobarbital, ketamine–xylazine, and fentanyl–fluanisone–diazepam, halothane in O_2–nitrous oxide (1:2) was found to provide the most reliable anesthesia (Peeters *et al.,* 1988). Halothane–nitrous oxide anesthesia was associated with a reduction in blood pressure (37.5%) and an increase in heart rate (24%); rabbits recovered from 50 min of general anesthesia in 25 min. Care should be taken to administer 100% O_2 without nitrous oxide for 5–10 min at the end of a procedure to avoid diffusion hypoxia. Investigators performing intracranial procedures should be aware that nitrous oxide will increase cerebral blood flow to varying degrees depending on background anesthesia (Drummond *et al.,* 1987).

VI. REGIONAL ANESTHESIA

A. Local or Regional Anesthesia

Local or regional techniques may be used in combination with inhalant or injectable anesthetics as they are in other species (Lockhart *et al.,* 1991; Madsen *et al.,* 1993; Raman *et al.,* 1989). Although the issue is still being debated, studies in human beings suggest that the preoperative infiltration of incision sites with local anesthetics may affect the course of postoperative pain, causing a decrease in the number of patients requiring analgesics or increasing the time to first request for analgesics (Coderre *et al.,* 1993). Lidocaine and bupivacaine are commonly used in our surgical laboratory for the local infiltration of incision sites. Specific nerve block techniques and sites have not been described for use in the rabbit.

B. Spinal Anesthesia

Spinal anesthesia has not been widely used clinically in the rabbit, but reports promoting its use as a model for evaluating the pharmacology and toxicology of spinal anesthesia or analgesia have appeared in the literature (Adams *et al.,* 1974; Crawford *et al.,* 1993; Durant and Yaksh, 1986; Hughes *et al.,* 1993; Jensen *et al.,* 1992; Langerman *et al.,* 1990; Madsen *et al.,* 1993). Both the epidural (Madsen *et al.,* 1993) and the subarachnoid (Durant and Yaksh, 1986; Jensen *et al.,* 1992; Langerman *et al.,* 1990) spaces have been cannulated to allow chronic administration of local anesthetics. Both procedures required surgical exposure of the lumbar spinal column and incision of the ligamentum flavum. In one study, the duration of motor blockade from injected anesthetics was dose dependent; the relative pharmacologic activity of the agents tested was amethocaine (duration, approximately 170 min) > bupivacaine (duration, approximately 100 min) > lidocaine (duration, approximately 27 min) > procaine (duration, approximately 20 min) (Langerman *et al.,* 1990).

Kero *et al.* (1981) recommended spinal anesthesia for hysterotomy, using 20 mg (0.5 ml) of mepivacaine injected "through the lumbar spinal interspaces at a level 1 to 2 interspaces higher than the iliac crest." Rabbits were held so that their spines were arched; 0.6 × 25-mm needles were used, the points being directed in a caudoventral direction. Motor function was eliminated for 55 min using this technique. The same procedure performed two to three lumbar interspaces craniad produced permanent paralysis. Hughes *et al.* (1993) described a technique for the epidural administration of local anesthesia in conscious rabbits. Twenty-gauge, 1.5-cm, short-beveled spinal needles were passed through the lumbosacral space with the needle oriented perpendicular to the skin. Duration of sensory loss, loss of weight bearing, and paresis were determined for 2% lidocaine with and without 1:200,000 epinephrine and 0.5% bupivacaine used at a volume of 0.2 ml/kg. These studies notwithstanding, it is our opinion that rabbits should be heavily sedated prior to spinal anesthesia so that the potential for movement during administration and subsequent nerve damage are reduced.

VII. SPECIAL ANESTHETIC CONSIDERATIONS

A. Hypnosis

Although fascinating to the animal scientist, hypnosis, or the immobility response, is limited in its utility by variability in response and duration among rabbits (Danneman *et al.,* 1988). Many techniques have been described to induce this phenomenon; maintaining rabbits in lateral recumbency in a darkened room for 15 sec (Klemm, 1965), traction on the head and neck while in dorsal recumbency (Danneman *et al.,* 1988), forced dorsal recumbency while stroking the abdomen of the animal and incantation (Rapson and Jones, 1964), and restraint of the animal in dorsal or ventral recumbency with the limbs immobilized while exerting gentle traction on the head (Gruber and Amato, 1970). Features of hypnosis include miosis, increased depth of respiration, analgesia, reduced respiratory and heart rate, and decreased blood pressure. Danneman *et al.* (1988) showed that naloxone had little effect on hypnosis and that some rabbits showed an autonomic response to noxious stimuli while showing classical signs of the immobility response. The authors concluded that hypnosis was not sufficiently reliable or efficacious to be considered a replacement for analgesics or anesthetics.

B. Anesthesia during Ophthalmic Procedures

Important criteria for anesthesia during ophthalmic surgery include adequate depth, relaxation of the extraocular muscles, and control of intraocular pressure (Ludders, 1993). Procedure length, ability to increase anesthetic depth with increments of the anesthetic, site of manipulation within the eye, and research goals will also determine the choice of anesthetic. Topical anesthetic agents like 0.3% proparacaine are sometimes used as adjuncts to anesthesia in procedures involving the cornea (Sherrard, 1966). Atropine or glycopyrrolate is used in ophthalmic procedures to avoid the oculocardiac reflex.

Anesthetics may affect retinal fine structure, intraocular pressure, and the volume of intraocular air (Antal, 1985; Boucher and Meyers, 1983; Johnson *et al.,* 1973; Schutten and Van Horn, 1977). Ketamine and ketamine–pentobarbital have been shown to affect peak increases in intraocular pressure by 2.2 and 7.1 mm, respectively. The increase was unaffected by atropine premedication and lasted for 10 min (Schutten and Van Horn, 1977). Nitrous oxide use may increase intraocular pressure and has been shown to promote the absorption of air from eyes that have received total air–fluid exchange following lensectomy and vitrectomy (Boucher and Meyers, 1983). In contrast, intravenous methohexital anesthesia was associated with a significant decrease in intraocular pressure, but produced hypoxic alterations of retinal photoreceptors (Antal, 1985). These changes persisted for the duration of the study (48 hr); effects on vision were not evaluated. Urethane alone and a combination of 0.5% halothane–nitrous oxide, suxamethonium and phencyclidine anesthesia, were found to preserve retinal ultrastructure whereas 0.5–2.5% halothane–nitrous oxide caused variable degrees of damage to photoreceptors and retinal pigment epithelium. The latter change was duration dependent; anesthesia of 5–6 hr duration was associated with the injury whereas anesthesia of a 15-min duration was not (Johnson *et al.,* 1973).

C. Anesthesia for Fetal Surgery

Reports on anesthetics used in fetal rabbit surgery include barbiturates, ether, and halothane (Cowen and Laurenson, 1959; Nelson *et al.,* 1990; Thomasson and Ravitch, 1969). The former two agents were characterized by high fetal and/or maternal loss whereas the latter was used in over 4000 fetal operations with less than 5% maternal mortality and postoperative fetal survival close to 90% (Nelson *et al.,* 1990). In that study, does were premedicated with ketamine (20 mg/kg) and acepromazine (1 mg/kg); one-half the dose was administered IM and the remainder was administered IV 20 min later. Halothane was administered via face mask and circle circuit with CO_2 absorption and a vaporizer setting of 1–1.5% halothane in 1 liter/min flow of O_2. Uterine relaxation during halothane anesthesia was considered by the authors to be an important determinant of success.

D. Long-Term Anesthetic Preparations

Many regimens have been described for the maintenance of anesthesia for extended periods of time during nonsurvival procedures or under special experimental circumstances such as extracorporeal circulation. Although they are presented briefly here, readers should consult the original papers for details to determine if a particular regimen is appropriate in their setting. For those regimens using paralytics, guidelines advocated by the NIH Workshop on "Preparation and Maintenance of Higher Mammals During Neuroscience Experiments" (National Institutes of Health, 1991) and Chapter 4 of this volume should be used to provide assurance that adequate depth of anesthesia is produced.

Green (1982b) reported three methods of providing basal narcosis-light anesthesia, all of which involved intubation and IPPV with oxygen–nitrous oxide (1:1). They are listed in the order of his preference: (1) Induction: propanidid (no longer available) (10 mg/kg IV), maintenance: α-chloralose IV (60–80 mg/kg); (2) induction: Hypnorm (0.3 ml/kg IM) and inhalation of halothane in oxygen–nitrous oxide (1:1), maintenance: Hypnorm (0.15 ml/kg IM) at 30- to 40-min intervals and pentobarbital (3–5 mg/kg IV) as needed every 45–60 min; and (3) induction: alphaxalone–alphadolone (8 mg/kg IV), maintenance: continuous IV infusion alphaxalone–alphadolone (6 mg/kg/hr) starting 15 min after the initial dose.

The rabbit has become an important animal model for studying the physiology of cerebral blood flow and cerebral anesthetic uptake and elimination (Drummond *et al.,* 1987; Hindman *et al.,* 1990; Lockhart *et al.,* 1991; Mills *et al.,* 1987; Patel and Mutch, 1990). These investigations have provided innovative regimens for anesthesia of long duration and include (1) induction and initial instrumentation: IV methohexital, 70% nitrous oxide, local infiltration of 1% lidocaine, maintenance: pancuronium (2 mg/hr IV); 70% nitrous oxide; methohexital (3 mg/kg/hr IV) (Lockhart *et al.,* 1991); (2) induction and initial instrumentation: ketamine (35 mg/kg IM) and xylazine (5 mg/kg IM), maintenance: 60% nitrous oxide, ketamine (10 mg/kg/hr IV), pancuronium as needed (Mills *et al.,* 1987); (3) induction and initial instrumentation: chamber induction with 4% halothane or 6% isoflurane in oxygen; pancuronium (2 mg IV); halothane or isoflurane (0.75–1.0 MAC) and 70% nitrous oxide. This study evaluated the effects of background anesthesiaon cortical blood flow. Maintenance anesthesia was therefore provided by the regimens under study. Wound margins were infiltrated with 0.25% bupivacaine and the following treatments were administered: 0.5 and 1.0 MAC isoflurane with 70% nitrous oxide or 70% N_2; 0.5 and 1.0 MAC halothane with 70% nitrous oxide or 70% N_2; 0.5 MAC isoflurane or halothane and 10 mg/kg morphine sulfate, administered as a slow IV bolus (2 mg/kg/hr) with 70% nitrous oxide or 70% N_2 (Drummond *et al.,* 1987). (4) Induction and initial instrumentation: Chamber induction 5% halothane; succinylcholine 2 mg/kg IV; 1.5% halothane with oxygen/nitrous oxide (1:2); succinylcholine (3 mg/kg/hr), maintenance: IV fentanyl (loading dose 100 μg/kg; infusion 2.5 μg/kg/min) and diazepam (loading dose 2 mg/kg, infusion) (50 μg/kg/min); pancuronium (0.1 mg/kg) (Hindman *et al.,* 1990).

VIII. INTRAOPERATIVE SUPPORT AND MONITORING

Constant vigilance is the anesthetist's watchword; one person who is trained in appropriate decision making should be assigned the task of monitoring anesthesia. Patient monitoring can be considered to be composed of three areas: reflexes, cardiopulmonary parameters, and body temperature. These three areas should be evaluated with regard to presurgical baseline and response to surgical stimuli or intraoperative events.

A. Reflexes

Traditional reflexes used in the monitoring of rabbit anesthesia include righting, palpebral, corneal, pedal withdrawal, and pinna reflex. It is widely held that the pinna reflex (ear movement in response to a compressive force) is the most accurate measure of depth of anesthesia, followed by the pedal withdrawal, corneal, and palpebral reflexes, in that order (Borkowski *et al.,* 1990). The corneal reflex may be preserved until very deep levels of anesthesia are achieved. Care should be exercised, however, as the presence or absence of these reflexes with different anesthetics in the rabbit defies generalization.

Muscle tone, jaw tone, vocalization, and gross purposeful movement in response to surgical stimulus may also be used as indices of anesthetic depth.

B. Body Temperature

Hypothermia can alter the metabolic clearance of injectable agents and decreases MAC of inhalants. Rectal temperature can be measured intermittently or continuously using available instrumentation. Esophageal temperature probes may provide less variability than rectal probes and provide continuous monitoring. Feedback control of body temperature can be achieved through commercially available devices and should be considered for procedures involving the prolonged exposure of body cavities or extended durations of anesthesia. Rabbits should be protected from cold surfaces during anesthesia by table drapes, foam pads, or similar devices. Supplemental heating can be provided with circulating hot water blankets, warmed irrigation fluids, judiciously placed heat lamps, and intravenous fluids heated by fluid warmers.

C. Cardiopulmonary Parameters

In the absence of hemorrhage, hypothermia, hyperthermia, drug effects, and vasovagal or oculocardiac reflexes, changes in respiratory and cardiovascular parameter values from preanesthetic baseline may comprise an autonomic response to a surgical stimulus. The value of ablation of these responses and their relationship to anesthetic depth has been debated (Kulli and Koch, 1991; Weinger and Koob, 1990) and is beyond the scope of this chapter. Perhaps it is sufficient in the setting of experimental anesthesia and surgery to say that ablation of the autonomic response provides compelling evidence that there is adequate depth of anesthesia whereas absence of ablation does not necessarily indicate that anesthetic depth is inadequate. The reader is referred to the discussion of this issue by Stanski (1990).

The respiratory rate can be assessed by the direct observation of chest wall or abdominal wall movements, use of respiratory monitors that adapt to the endotracheal tube, esophageal stethoscopes, pediatric pneumotachometers, and chest wall plethysmographs. Cerebrospinal fluid movement may also be used to determine the respiratory rate during intracranial procedures. Alternatively, the respiratory rate may be set by the anesthetist during mechanical ventilation of the lungs. The assessment of breathing pattern, depth of respiration, and respiratory effort may be made by direct observation or through use of pneumotachometers with circuit integrators that measure or calculate respiratory flow, tidal volume, and minute ventilation (Guerreiro and Page, 1987; Rich *et al.,* 1990). Information on the adequacy of ventilation can be derived through observation of mucous membrane color, blood gas analysis, end-tidal CO_2 determination, and pulse oximetry. The rabbit is well suited for routine blood gas evaluation because of the availability of peripheral sites for arterial cannulation or puncture. Pediatric pulse oximeters may be used to provide information on arterial O_2 saturation (Aeschbacher and Webb, 1993a,b; Robertson and Eberhart, 1994). The end-tidal CO_2 measurement has been specifically evaluated during spontaneous and controlled ventilation in the rabbit with specific regard to the site of sampling (Rich *et al.,* 1990). End-tidal CO_2 determined at the pulmonary tip of the endotracheal tube and that determined 12 cm from the pulmonary tip differed by less than 1 mm Hg, independent of the mode of ventilation or fresh gas flows. The end-tidal CO_2 was also found to be a reasonable indicator of $PaCO_2$; the values determined for each differ by only 2.9 and 3.6 mm Hg at the distal tip and 12-cm mark, respectively.

Cardiovascular parameters that may be routinely monitored in the rabbit include mucous membrane color and capillary refill time, heart rate, arterial blood pressure, and pulse rate and character. Heart rate may be monitored continuously by an esophageal stethoscope or EKG monitor and intermittently by direct auscultation wtih a stethoscope or palpation of the maximal apex beat through the chest wall. Arterial blood pressure may be measured directly through arterial cannulation or indirectly through the use of appropriately sized limb bands (Ko *et al.,* 1992). Pulse rate and character may be evaluated by palpation of the saphenous artery, carotid artery, or femoral artery or through the use of pressure pulse waveforms available with some direct pressure monitoring instruments. Visual determination of pulse pressure and character can be used when appropriate vascular beds are exposed during surgical procedures (e.g., laparotomy). Central venous pressure can be used as a direct measure of preload and as an indirect measure of myocardial function but is not widely reported in the literature.

Cardiovascular support during surgery includes the maintenance of body temperature, administration of fluids, maintenance of adequate ventilation, and use of agents to alter blood pH, heart rate, rhythm, and blood pressure. In the experience of the authors, routine procedures of short to moderate duration incorporate an intravenous fluid administration rate of at least 10 ml/kg/hr. Administration rates reported during neurosurgical procedures are 4 ml/kg/hr (Hindman *et al.,* 1990; Lockhart *et al.,* 1991; Mills *et al.,* 1987). Although drug doses used for cardiovascular support are usually extrapolated from those used in other species, dosages of vasopressors specific for the rabbit may be found in Ruta and Mutch (1989), Patel and Mutch (1990), and Dorward *et al.* (1987).

IX. POSTOPERATIVE CONSIDERATIONS

''Every anesthetic is a form of physiologic trespass, and the anesthetist must always consider the effects his drugs will produce on body homeostasis'' (Holland, 1973). Cardiopulmonary function should continue to be monitored during recovery into the postoperative period. Hypostatic pulmonary congestion should be avoided by altering the position of the rabbit from left to right lateral recumbency every 15 min during anesthetic recovery.

As in other species, thermal support is critical during recovery. This is especially important in the rabbit because of their relatively small size, high surface area/body weight ratio, and the hypothermic effects of most anesthetics. Rectal temperature should be monitored regularly during recovery. Thermal support consisting of circulating hot water blankets, hot water bottles, infant incubators, ICU cages, warming lamps, or blankets should be used to maintain body temperature. Rectal temperature should continue to be monitored for 7–10 days postsurgery to aid in the detection of infection.

Nutritional and fluid support may be necessary postoperatively. Isotonic fluids should be administered, especially if there is significant fluid loss during surgery. To minimize hypothermia, fluids can be warmed before administration using a fluid warmer, microwave oven, or incubator. The maintenance fluid requirement of the rabbit is reported to be 100 ml/kg/day (Hillyer, 1992). Fluids can be administered subcutaneously, intravenously, or intraperitoneally.

Rabbits recovering from anesthesia may be lightly wrapped in a blanket or towel (burrito style) with the legs flexed against the body to prevent thrashing and damage to the spinal column. Postoperative rabbits with accessible sutures may be fitted with an Elizabethan or cervical collar for the first postoperative week to prevent inadvertent suture removal.

Rabbits are prone to hypoglycemia because of high metabolic rates and, in neonates, limited fat reserves. They should be allowed access to a nutritious pelleted diet as soon after surgery as feasible. Anorexic animals can be given 5% dextrose solution SC, 50% dextrose orally, parenteral nutritional support through a stomach or nasoesophageal tube; offered palatable supplements such as hay, alfalfa cubes, or dandelions; and/or administered high-calorie nutritional supplements such as Nutrical PO (EVSCO Pharmaceuticals, Buena, NJ) (Hillyer, 1992; Reed *et al.,* 1987). Persistent untreated anorexia causes fat mobilization and often leads to irreversible fatal hepatic lipidosis. The oral administration of probiotics such as *Lactobacillus* culture or a slurry of feces or cecal contents from healthy rabbits has been promulgated to return the gastrointestinal flora to normal following gastrointestinal surgery or the administration of flora-altering antibiotics (Gillett *et al.,* 1983; Reed, 1990).

Antibiotics may be necessary following select surgical procedures. They should not be used indiscriminately or as a substitute for aseptic technique. Clostridial enterotoxemia has been associated with the administration of a wide variety of antibiotics (Carman and Evans, 1984). Surgical sites should be evaluated daily for dehiscence, infection, and fluid accumulation. Nonabsorbable suture materials or staples should be removed as soon as wound healing permits to avoid their serving as a nidus of infection.

Analgesics should be used routinely postoperatively to reduce or eliminate pain or discomfort. Agents should be selected based on the intensity and nature of the stimulus. The following section should be consulted for detailed information.

X. ANALGESIA

A. Pain Assessment

Recognition and accurate assessment of pain intensity, location, and cause are essential if effective therapy is to be instituted. As with other animal species, our understanding of pain expression and pain-related behavior in rabbits is limited. Nevertheless, some attempt to assess pain should be made.

In common with several other species, rabbits experiencing pain may reduce their food and water consumption and as a consequence may lose weight. Failure to groom may result in the coat becoming ruffled and unkempt because of the buildup of shed hair that would normally be removed by grooming. Behavioral changes associated with pain often include reduced activity and almost complete immobility in animals in severe pain. Guarding behavior may be seen, although frequently this is manifested as an escape reaction or as a generalized tensing of muscles with a rigid feel to the trunk and limbs. Because this behavior also occurs in rabbits that are frightened or apprehensive, it is important to obtain an accurate history of any experimental procedures and their possible impact on the animal. It is also helpful to discuss the normal behavior of the animal with the regular caretaker of the animal. Because the response of many rabbits to an unfamiliar handler is to remain immobile, assessment of the behavior of the animal may be best carried out using a video camera or via an observation panel to view the animal in its cage or pen.

Before initiating a procedure that is likely to cause pain, e.g., a surgical operation, it is important to assess and record the normal behavior, growth rate, and food and water consumption of the rabbit. Observations made following the procedure can then be made, and the influence of analgesic treatment can then be assessed. Careful observation of the responses to a particular research procedure can be used to formulate a pain scoring system specifically tailored for that procedure. This approach is likely to be considerably more successful than the use of a generalized scoring system.

B. Selection of Analgesics

A range of compounds are available to provide pain relief in the rabbit. However, because this species is less widely used than rodents in the development of new analgesics, relatively limited pharmacokinetic and pharmacodynamic data are available. Some data concerning the use of analgesics in rabbits are available, but these are primarily results from analgesiometric tests. Several different methods of analgesiometry have been used in rabbits. Thermal stimuli have been applied to the muzzle (Zhou *et al.,* 1981), ear (McCallister *et al.,* 1986; Piercey and Schroeder, 1980; Wynn *et al.,* 1984), or skin (Flecknell and Liles, 1990) using a radiant heat source. Electrical stimulation of the ears (Ayhan *et al.,* 1983) or of the tooth pulp (Mattila and Saarnivaara, 1968) has also been used. Use of pressure on the hindlimb was described by Lightowler and Smith (1963). In all of these tests, the noxious stimulus is applied until the rabbit makes a defined response, e.g., a skin twitch when radiant heat is applied to the skin or a chewing response to electrical stimulation of the tooth pulp. Changes in the response threshold after administration of an analgesic are used to determine the potency of the agent. These analgesiometric tests are most useful for assessing the action of opioids while they are relatively ineffective for nonsteroidal anti-inflammatory drugs (NSAIDS) (Taber, 1974). Tests equivalent to the inflamed paw pressure test, used in rodents to assess the efficacy of NSAIDs, do not appear to have been developed in the rabbit.

When basing recommendations on data obtained using analgesiometry, it is important to appreciate that the severity of the painful stimuli used may vary considerably and may also be qualitatively different from postsurgical pain. The potency of different analgesics assessed using analgesiometry correlates well with their potency in humans, but the doses used to control clinical pain may vary considerably. At present, it can only be noted that dose rates established using analgesiometry provide a safe starting point for clinical use of the agent. Adjustment of the doses used requires the development of reliable methods of postoperative pain assessment. When no data concerning particular analgesics are available for the rabbit, extrapolations have been made from data obtained in other species, or recommendations have been made based simply on clinical experience. The source of each recommendation for analgesics use is made clear in the following section.

Opioid analgesics can be used both to provide intraoperative analgesia in rabbits and to provide postoperative pain relief. The efficacy of a range of agents has been assessed using analgesiometry (McCallister *et al.,* 1986; Piercey and Schroeder, 1980; Wynn *et al.,* 1984) (Table A.II in the Appendix), and these data can be used to formulate dose rates for the treatment of clinical pain (Table A.III in the Appendix). The mixed agonist/antagonist opioids butorphanol, pentazocine, and nalbuphine, and the partial agonist buprenorphine cause either no respiratory depression or only a minor depression of respiratory rate, which is unlikely to be of clinical significance (Flecknell and Liles, 1990). Buprenorphine administered at a dose of 0.03 mg/kg IV was shown to produce over 10 hr of analgesia by using thermal stimulus analgesiometry (Flecknell and Liles, 1990). Its onset of action is 30 min so it should be administered intraoperatively or prior to recovery from anesthesia when needed for postoperative pain relief. Buprenorphine is the authors' choice of analgesic for the control of postoperative pain of moderate intensity. Fentanyl and other μ-agonists can cause marked respiratory depression and depress the responses of the respiratory center to CO_2 (Brown *et al.,* 1980; Hunter *et al.,* 1968; Pleuvry and Rees, 1969). The clinical significance of some of this data is difficult to interpret since the rabbits studied are likely to have had a marked tachypnea induced by the stress associated with handling and restraint. The pharmacokinetics of fentanyl have been investigated in detail in the rabbit (Hess *et al.,* 1971, 1972). A dose of 20 μg/kg IV produced an analgesic effect for approximately 30 min. Pharmacokinetic data published by these authors enable the calculation of continuous intravenous infusion regimens of fentanyl in the rabbit.

Authors' Preference

Agent:	Buprenorphine
Route:	SC or IV
Dose:	0.01–0.05 mg/kg
Comments:	An opioid analgesic (partial μ-agonist) useful for treating postoperative pain of moderate intensity

Morphine has been shown to produce hypertension and hyperglycemia in conscious rabbits (May *et al.,* 1988), as a consequence of increased sympathetic activity, and moderate histamine release. Meperidine (Pethidine) (4–16 mg/kg) and pentazocine (2–8 mg/kg) have been reported to cause convulsions in rabbits after IV administration (Hunter *et al.,* 1968), although in a subsequent report it is unclear whether true convulsions were produced (Hunter, 1968). Doses of 1–5 mg/kg of pentazocine IV (Flecknell and Liles, 1990) and pethidine (5–10 mg/kg SC) (P. A. Flecknell, clinical experience) did not cause any visible undesirable reactions. Pentazocine at these dose rates produced analgesia assessed using thermal analgesiometry. Administration of the potent μ-opioids fentanyl, alfentanil, and sufentanil can all produce muscle rigidity unless suitable sedatives and/or tranquilizers are administered simultaneously (Borkowski *et al.,* 1990; P. A. Flecknell, unpublished observations). Even after the administration of sedatives, these undesirable side effects may still be encountered (Marini *et al.,* 1993). It is therefore recommended that these potent μ-opioids are administered only as adjuncts to general anesthetics or in well-characterized neuroleptanalgesic combinations. Morphine has been reported to produce sedation in rabbits (May *et al.,* 1988), as have butorphanol, pentazocine, and nalbuphine (Flecknell and Liles, 1990). Detailed studies of the effects of opioids on behavior in rabbits have not been undertaken, however, so their nonspecific effects on pain scoring remain to be established.

The efficacy of orally administered morphine and codeine has been investigated by Leaders and Keasling (1962), using a tooth pulp stimulation model. In comparison to subcutaneous administration, approximately a 10-fold increase in oral dosage was required to have an equivalent analgesic effect. Despite their poor bioavailability, oral codeine or morphine might provide a useful means of providing pain relief in rabbits if palatability issues could be overcome.

There appear to be no published clinical trials of the use of opioids to control postoperative pain in rabbits. Clinical experience in our laboratory suggests that a variety of opioids can be useful (see Table A.II), but detailed studies using pain scoring techniques are urgently required.

The use of nonsteroidal anti-inflammatory drugs in laboratory species has been reviewed by Liles and Flecknell (1992), and only limited data concerning the use of these agents in the rabbit appear to have been published.

Piroxicam (0.1 and 0.2 mg/kg PO every 8 hr) and flunixin (1.1 mg/kg IM) were shown to be effective in reducing limb swelling in an experimental fracture model in rabbits (More *et al.,* 1989), but no attempt was made to assess the analgesic effects of these compounds. Of particular clinical relevance, however, was the finding that both flunixin and piroxicam reduced tissue swelling without affecting the strength of the healed fracture. No animals were reported to develop adverse reactions during the 3-week treatment period in this study, although these authors state that in a pilot study, aspirin was found to be poorly tolerated by rabbits. Unfortunately, no details of the dose rate used or the problems encountered were described.

Aspirin has been used clinically in rabbits, but no controlled trials have been carried out to monitor its toxicity or efficacy in this species. A dose of 10 mg/kg SC had a detectable analgesic effect in a tooth pulp assay (Ayhan *et al.,* 1983), and doses over 300 mg/kg IV were lethal (Piercey and Schroeder, 1980). Ketoprofen, administered for 8 days (3 mg/kg IV), produced no signs of renal toxicity, but affected renal prostaglandin synthesis (Perrin *et al.,* 1990). Aspirin (10 mg/kg SC) was effective in an electrical stimulation model (Ayhan *et al.,* 1983), but was relatively ineffective, even in high doses, in other pain models (Murai and Ogura, 1978; Piercey and Schroeder, 1980). It is important to note that NSAIDs generally are relatively ineffective in standard analgesiometric tests, and data from tests that more effectively predict the clinical efficacy of NSAIDs (e.g., the inflamed paw pressure test) appear to be unavailable for the rabbit. As with the opioids, there appear to be no published clinical trials of NSAIDs in rabbits, and the suggested dose rates in Table A.II require validation using a pain scoring system.

Acetaminophen with and without codeine has been anecdotally reported to have a mild analgesic effect of short duration when administered orally to rabbits (Wixson, 1994). Palatability is markedly enhanced by using grape- or cherry-flavored acetaminophen or a fruit-flavored acetaminophen–codeine suspension. The recommended dose of either product is 1 ml drug/100 ml water (Wixson, 1994).

Local anesthetics such as lidocaine and bupivacaine can be used to provide local anesthesia and pain relief in rabbits either by local infiltration of surgical wounds or by spinal anesthesia. Topical ophthalmic anesthetics can be used to produce anesthesia of the cornea in rabbits.

As with other species, corticosteroids can be used to reduce tissue inflammation and so reduce pain. Dose rates must be obtained by extrapolation from other species.

The α_2-agonists, xylazine and medetomidine, have been extensively used as adjuncts to ketamine anesthesia in rabbits. Their utility as analgesics is limited as they are relatively short-acting and produce profound sedation.

APPENDIX

This appendix lists doses and dose rates for anticholinergics, sedatives, tranquilizers, anesthetics, and analgesics used for rabbits.

TABLE A.I.

SUGGESTED DOSES OF ANTICHOLINERGICS, SEDATIVES, TRANQUILIZERS, AND ANESTHETICS FOR CLINICAL USE IN THE RABBIT[a]

Drug	Dosage	Route of administration	Reference
Anticholinergics			
Atropine	0.04–2.0 mg/kg (0.5 mg/kg commonly recommended)	IM, SC	Hall and Clarke (1991)
Glycopyrrolate	0.1 mg/kg	IM, SC	Olson *et al.* (1994)
Sedatives/tranquilizers			
Diazepam	5–10 mg/kg	IM	Sedgwick (1986); Green *et al.* (1981)
	1–2 mg/kg	IM, IV	Flecknell *et al.* (1983)
Midazolam	2 mg/kg	IP, IV	Flecknell and Mitchell (1984)
Acepromazine	0.75–10.0 mg/kg (0.75–1.0 mg/kg is most frequently used)	IM	McCormick and Ashworth (1971); Freeman *et al.* (1972)
Chlorpromazine hydrochloride	25–100 mg/kg	IM	Bivin and Timmons (1974); Dolowy and Hesse (1959)
Xylazine	3–9 mg/kg	IM, IV	Sanford and Colby (1980); Green (1975)
Medetomidine	0.25 mg/kg	IM	Ko *et al.* (1992)
	6 mg/kg	IV (1.25% sol slowly to allow endotracheal intubation)	Green (1982b)
Barbiturates			
Thiopental	15–30 mg/kg	IV (1% sol) GTE[b]	Clifford (1984); Sedgwick (1986)
	50 mg/kg	IV (2.5% sol) GTE	Lumb and Jones (1984)
Thiamylal sodium	15 mg/kg	IV (1% sol) GTE	Clifford (1984); Sedgwick (1986)
	29 mg/kg	IV (2% sol) GTE	Gardner (1964)

TABLE A.I. *(continued)*

SUGGESTED DOSES OF ANTICHOLINERGICS, SEDATIVES, TRANQUILIZERS, AND ANESTHETICS FOR CLINICAL USE IN THE RABBIT[a]

Drug	Dosage	Route of administration	Reference
EMTU	47.5 mg/kg	IV to effect	Hobbs *et al.* (1991)
Methohexital	5–10 mg/kg	IV (1% sol) GTE[b]	Green (1975); Antal (1985)
Pentobarbital	20–60 mg/kg	IV	Marston *et al.* (1965); Koch and Dwyer (1988); Jacobs and Krohn (1975); Zhou *et al.* (1981); Conn and Langer (1978); Flecknell *et al.* (1983); Borkowski *et al.* (1990); Green (1975); Peeters *et al.* (1988)
	30 mg/kg	IM	Krogh (1975)
		Intrahepatic	Jacobs and Krohn (1975)
Pentobarbital +	20 mg/kg	IV	Olson *et al.* (1987)
guaifenesin	200 mg/kg	IV	
Pentobarbital +	20–30 mg/kg	IV	Green (1975)
chlorpromazine	2 mg/kg	IM prior to pentobarbital	
Pentobarbital +	11.8–28.4 mg/kg	IV	Hobbs *et al.* (1991)
xylazine	5 mg/kg	SC followed in 10 min by pentobarbital	
Pentobarbital +	30 mg/kg	IV	Raman *et al.* (1989)
ketamine +	10 mg/kg	IM administration 10 min following pentobarbital	
1% Lidocaine hydrochloride	Local infiltration of surgical incision	SC	
Dissociatives			
Ketamine	20–60 mg/kg	IM	Clifford (1984); Sedgwick (1986); Green (1975, 1982b)
Ketamine +	10 mg/kg	IV	Flecknell (1987)
xylazine	3 mg/kg	IV	
Ketamine +	22–50 mg/kg	IM	Clifford (1984); Sanford and Colby (1980); Sedgwick (1986); Koch and Dwyer (1988); Peeters *et al.* (1988); Lipman *et al.* (1990); Popilskis *et al.* (1991); Beyers *et al.* (1991); White and Holmes (1976); Rich *et al.* (1990); Lipman *et al.* (1987)
xylazine	2.5–10 mg/kg		
(reverse with yohimbine)	0.2 mg/kg	IV	
Acepromazine +	0.75–1.0 mg/kg	SC	Hobbs *et al.* (1991)
ketamine +	35–40 mg/kg	IM	Lipman *et al.* (1990); Ludders *et al.* (1987)
xylazine (pre-op with atropine, 0.04 mg/kg IM)	3–5 mg/kg		
Ketamine +	75 mg/kg	IM	Clifford (1984)
acepromazine	5 mg/kg	IM (given 30 min prior to ketamine)	
Ketamine +	60–80 mg/kg	IM	Sedgwick (1986)
diazepam	5–10 mg/kg	IM (given 30 min prior to ketamine)	
Ketamine +	35 mg/kg	IM	Hobbs *et al.* (1991)
EMTU *(Inactin)*	25–55 mg/kg	IV to effect	
Ketamine +	50 mg/kg	IM	Krogh (1975)
pentobarbital	30 mg/kg	IM	
Ketamine +	20 mg/kg	IV	Chen and Bohner (1968)
chloral hydrate	250 mg/kg	IV	Hobbs *et al.* (1991)
Ketamine +	25 mg/kg	IM	Nevalainen *et al.* (1989)
medetomidine	0.5 mg/kg	SC	
Tiletamine[c] + zolazepam	32–64 mg/kg	IM	Brammer *et al.* (1991)
Tiletamine[c] + zolazepam +	15 mg/kg	IM (inject all simultaneously but use separate syringes)	Popilskis *et al.* (1991)
xylazine	5 mg/kg		
Neuroleptanalgesics			
Fentanyl–droperidol	0.125 mg/kg	SC	Tillman and Norman (1983)
	0.15–0.44 ml/kg (0.22 ml/kg is optimal dose)	IM	Bivin and Timmons (1974); Walden (1978); Strack and Kaplan (1968); Lewis and Jennings (1972)
Fentanyl–fluanisone	0.2–0.6 ml/kg	IM, SC	Green (1975); Flecknell (1987); Flecknell *et al.* (1989); Alberius *et al.* (1989)

TABLE A.I. *(continued)*

SUGGESTED DOSES OF ANTICHOLINERGICS, SEDATIVES, TRANQUILIZERS, AND ANESTHETICS FOR CLINICAL USE IN THE RABBIT[a]

Drug	Dosage	Route of administration	Reference
Diazepam +	1.5–5 mg/kg	IM, IV, IP	Green (1975)
Fentanyl–fluanisone	0.2–0.5 ml/kg (administer diazepam 5 min prior to fentanyl–fluanisone)	IM, SC	Flecknell (1987); Flecknell *et al.* (1983); Mero *et al.* (1987)
Midazolam +	2 mg/kg	IP, IV	Flecknell and Mitchell (1984)
fentanyl–fluanisone	0.3 ml/kg (administer midazolam 5 min prior to fentanyl–fluanisone)	IM	Flecknell (1987)
Etorphine + methotrimeprazine	0.025–0.05 ml/kg	IM	Flecknell (1987); Flecknell *et al.* (1983)
Diazepam +	1.0 mg/kg	IV, IP	Flecknell (1987); Flecknell *et al.* (1983)
etorphine + methotrimeprazine	0.25 ml/kg	IM	
Other			
Alphaxalone–alphadolone	6–20 mg/kg (optimal dose = 12 mg/kg)	IV	Green (1975); Green *et al.* (1978)
Chloral hydrate	250 mg/kg	IV	Harvey and Walberg (1987)
Chloral hydrate + magnesium sulfate + Pentobarbital + propylene glycol (Equi-Thesin)	0.5–3.0 ml/kg in increments of 0.5 ml	Per rectum	Hodesson *et al.* (1965) Bivin and Timmons (1974)
Equi-Thesin	To effect	IV	Hodesson *et al.* (1965); Bivin and Timmons (1974)
Ketamine +	20 mg/kg	IM	Bivin and Timmons (1974)
chloral hydrate	250 mg/kg	IV	Hobbs *et al.* (1991)
α-Chloralose	100 mg/kg	IV	Chakrabarty *et al.* (1991); Harvey and Walberg (1987)
Propofol	7.5–15 mg/kg	IV	Adam *et al.* (1980)
Guaifenesin	5% solution in 5% dextrose, given at a dosage of 200 mg/kg	IV	Olson *et al.* (1987)
Medetomidine +	0.25 mg/kg followed in 5	IM	Ko *et al.* (1992)
propofol	min by 4 mg/kg	IV	
Medetomidine +	0.25 mg/kg	IM	Ko *et al.* (1992)
midazolam +	0.5 mg/kg	IM	
propofol	2 mg/kg	IV	
Guaifenesin +	200 mg/kg	IV	Olson *et al.* (1987)
ketamine	50 mg/kg	IM	
α-Chloralose + urethane	32 mmol (10 g)/liter in saline at dose of 258 μmol (80 mg)/kg 400–500 mg/kg (5.61 mmol/kg) in 1 liter of saline (2.81 mol/liter)	IV (slowly)	Korner *et al.* (1968) Warren and Ledingham (1978)
Urethane	1–1.6 g/kg	IP	Bree and Cohen (1965)
	1.5 g/kg	IV	
Urethane +	1–1.30	IV	Moore *et al.* (1987)
acepromazine	1 mg/0.46 kg	IM	
Paraldehyde	1 ml/kg	IM, IP, orally	Hodesson *et al.* (1965); Pandeya and Lemon (1965); Green (1982a)
Intermittent bolus or continuous IV infusion regimens			
1. Sedate with fentanyl +	0.05 mg/kg	IM	Guerreiro and Page (1987)
droperidol	2.5 mg/kg		
Induce with infusion of	2.4 mg	IV	
fentanyl + droperidol	40 mg in 100 ml of 5% dextrose, rate of 15–20 drops (volume of 17 μl/drop) per min. Maintain with 15 drops/min		
2. Dilute with Hypnorm 1:10	Infuse 1–3 ml/kg/hr	IV	Flecknell (1987)

TABLE A.I. *(continued)*

Suggested Doses of Anticholinergics, Sedatives, Tranquilizers, and Anesthetics for Clinical Use in the Rabbit[a]

Drug	Dosage	Route of administration	Reference
3. Midazolam +	1 mg/kg	IV	Borkowski *et al.* (1990)
xylazine +	1 mg/kg		
alfentanil	0.1 mg/kg		
4. α-Chloralose +	60 mg/kg	IV	Jenkins (1987)
urethane	400 mg/kg, followed by 1–3 ml 1% α-chloralose q 30–50 min		
5. α-Chloralose +	40–60 mg/kg	IV	Dorward *et al.* (1987)
urethane	800 mg/kg, followed by 3–4 ml/hr 1% α-chloralose		
6. Sedate with ketamine +	35 mg/kg	IM	Wyatt *et al.* (1989)
xylazine	5 mg/kg	IM	
Maintain with ketamine +	1 mg/min	Continuous	Wyatt *et al.* (1989)
xylazine	0.1 mg/min	IV infusion	
7. Ketamine +	25 mg/kg	IV	Borkowski *et al.* (1990)
xylazine	5 mg/kg over 1 min, remainder over 4 min	one-third bolus dose	
8. Pentobarbital	40 mg/kg one-third bolus dose initially over 1 min, remainder over 4 min	IV	Borkowski *et al.* (1990)
9. Sedate with propofol	1.5 mg/kg	IV bolus	Blake *et al.* (1983)
Maintenance–propofol	0.2–0.6 mg/kg/min.	Continuous IV Infusion	
10. Sedate with alphaxalone +	1 mg/kg	Maintain with infusion	Blake *et al.* (1988)
alphadolone	0.1 mg/kg/min	IV bolus Continuous IV	

[a]Modified from Wixson (1994).
[b]Given to effect.
[c]May cause severe nephrotoxicity.

TABLE A.II

Analgesic Dose Rates in Rabbits[a]

Drug	Test	Dose range	Route of administration	Reference
Opioids				
Alfentanil	Electrical	0.03 mg/kg	IV	Wynn *et al.* (1984)
Buprenorphine	Thermal	0.0075–0.3 mg/kg	IV	Flecknell and Liles (1990)
Butorphanol	Thermal	0.1–1.5 mg/kg	IV	Flecknell and Liles (1990)
Codeine	Electrical	35–70 mg/kg	SC	Leaders and Keasling (1962)
		280–560 mg/kg	Per os	
Fentanyl	Electrical	0.0074 mg/kg	IV	Wynn *et al.* (1984)
Methadone	Electrical	0.1–5.0 mg/kg	IV	Piercey and Schroeder (1980)
Morphine	Electrical	1.25–10 mg/kg	IV	Murai and Ogura (1978)
	Electrical	1.0–20.0 mg/kg	SC	Ayhan *et al.* (1983)
	Electrical	1.1 mg/kg	IV	Wynn *et al.* (1984)
Nalbuphine	Thermal	1.0–10.0 mg/kg	IV	Flecknell and Liles (1990)
Pentazocine	Thermal	1.0–5.0 mg/kg	IV	Flecknell and Liles (1990)
	Electrical	1–4 mg/kg	IV	Murai and Ogura (1978)
	Electrical	8.2 mg/kg	IV	Wynn *et al.* (1984)
NSAID				
Aspirin	Electrical	1–20 mg/kg	SC	Ayhan *et al.* (1983)
Flunixin	Reduction inflammation	1.1 mg/kg	IM	More *et al.* (1989)
Piroxicam	Reduction inflammation	0.1–0.2 mg/kg	Per os	More *et al.* (1989)
Miscellaneous				
Baclofen	Electrical	1.5–12 mg/kg	IV	Piercey and Schroeder (1980)
Mefenamic acid	Electrical	200 mg/kg	Per os	Murai and Ogura (1978)
Nefopam	Electrical	1.5–12 mg/kg	IV	Piercey and Schroeder (1980)

[a]Assessed using experimental analgesiometry or related animal models.

TABLE A.III

Suggested Dose Rates of Analgesics for Clinical Use in Rabbits

Analgesic	Dose rate	Reference
Acetaminophen (with or without codeine)	1 ml drug/100 ml drinking water	Wixson (1994)
Buprenorphine	0.01–0.05 mg/kg SC, IV 6–12 hourly	Flecknell and Liles (1990)
Butorphanol	0.1–0.5 mg/kg IV, 4 hourly	Flecknell and Liles (1990)
Flunixin	1.1 mg/kg IM, ? 12 hourly	More *et al.* (1989)
Meperidine	5–10 mg/kg SC, 2–3 hourly	P. A. Flecknell (clinical experience)
Methadone	1 mg/kg IV	Piercey and Schroeder (1980)
Morphine	2.5 mg/kg SC, 2–4 hourly	P. A. Flecknell (clinical experience), Murai and Ogura (1978)
Nalbuphine	1–2 mg/kg IV, 4–5 hourly	Flecknell and Liles (1990)
Pentazocine	5 mg/kg IV, 2–4 hourly	Flecknell and Liles (1990)
Piroxicam	0.2 mg/kg per os, 8 hourly	More *et al.* (1989)

REFERENCES

Adam, H. K., Glen, J. B., and Hoyle, P. A. (1980). Pharmacokinetics in laboratory animals of ICI 35 868, a new I. V. anaesthetic agent. *Br. J. Anaesth.* **52,** 743–746.

Adams, H. J., Mastri, A. R., Eischolzer, A. W., and Kilpatrick, G. (1974). Morphologic effects of intrathecal etidocaine and tetracaine on the rabbit spinal cord. *Anesth. Analg.* **53,** 904–908.

Aeschbacher, G., and Webb, A. I. (1993a). Propofol in rabbits. 1. Determination of an induction dose. *Lab. Anim. Sci.* **43,** 324–327.

Aeschbacher, G., and Webb, A. I. (1993b). Propofol in rabbits. 2. Long-term anesthesia. *Lab. Anim. Sci.* **43,** 328–335.

Alberius, P., Klinge, B., and Isaksson, S. (1989). Management of craniotomy in young rabbits. *Lab. Anim.* **23,** 70–73.

Alexander, D. J., and Clark, G. C. (1980). Simple method of oral endotracheal intubation in rabbits *(Oryctolagus cuniculus)*. *Lab. Anim. Sci.* **30,** 871–873.

Antal, M. (1985). Ultrastructural changes in the rabbit retina following methohexital anaesthesia. *Acta Morphol. Hung.* **33,** 13–19.

Arnolda, L., McGrath, B., Cocks, M., Sumithran, E., and Johnston, C. (1985). Adriamycin cardiomyopathy in the rabbit: An animal model of low output cardiac failure with activation of vasoconstrictor mechanisms. *Cardiovasc. Res.* **19,** 378–382.

Ayhan, I. H., Turker, R. K., and Melli, M. (1983). A new method for the rapid measurement of analgesic activity in rabbits. *Arch. Int. Pharmacodyn. Ther.* **262,** 215–220.

Baum, M., Putz, G., Mutz, N., Putinsen, C., Klima, G., and Benzer, H. (1989). Influence of high frequency ventilation at different end-expiratory lung volumes on the development of lung damage during lung lavage in rabbits. *Br. J. Anaesth.* **63,** 615–705.

Bernards, C. M., and Artru, A. A. (1991). Hexamethonium and midazolam terminate dysrhythmias and hypertension caused by intracerebroventricular bupivacaine in rabbits. *Anesthesiology* **74,** 89–96.

Bertolet, R. D., and Hughes, H. C. (1980). Endotracheal intubation: an easy way to establish a patent airway in rabbits. *Lab. Anim. Sci.* **30,** 227–230.

Betteridge, K. J. (1973). Simple and inexpensive apparatus for halothane anaesthesia in rabbits and other small animals. *Vet. Rec.* **93,** 398–399.

Beyers, T. M., Richardson, J. A., and Prince, M. D. (1991). Axonal degeneration and self-mutilation as a complication of the intramuscular use of ketamine and xylazine in rabbits. *Lab. Anim. Sci.* **41,** 519–520.

Bieter, R. N., Cunningham, R. W., Lenz, O., *et. al.* (1936). Threshold anesthetic and lethal concentrations of certain spinal anesthetics in the rabbit. *J. Pharmacol. Exp. Ther.* **57,** 221–244.

Bivin, W. S., and Timmons, E. H. (1974). Anesthesiology. *In* ''The Biology of the Laboratory Rabbit'' (S. M. Weisbroth, R. E. Flatt, and A. L. Kraus, eds.), pp. 77–83. Academic Press, New York.

Blake, D. W., Jover, B., and McGrath, B. P. (1988). Haemodynamic and heart rate reflex responses to propofol in the rabbit. *Br. J. Anaesth.* **61,** 194–199.

Blake, D. W., Way, D., Trigg, L., Langton, D., and McGrath, B. P. (1991). Cardiovascular effects of volatile anesthesia in rabbits: Influence of chronic heart failure and enapril treatment. *Anesth. Analg.* **73,** 441–448.

Bonath, K., Nolte, I., Schniewind, A., Sandmann, H., and Failing, K. (1982). [Food deprivation as precaution and aftercare measure for anaesthesia—the influence of fasting on the acid base status and glucose concentration in the blood of rabbits]. *Berl. Muench. Tieraerztl. Wochenschr.* **95,** 126–130.

Booth, N. H., and McDonald, L. E., eds. (1988). ''Veterinary Pharmacology and Therapeutics,'' 6th ed. Iowa State Univ. Press, Ames.

Borkowski, G. L., Danneman, P. J., Russell, G. B., and Lang, C. M. (1990). Evaluation of three intravenous anesthetic regimens in New Zealand rabbits. *Lab. Anim. Sci.* **40,** 270–276.

Boucher, M. C., and Meyers, E. (1983). Effects of nitrous oxide anesthesia on intraocular air volume. *Can. J. Ophthalmol.* **18,** 246–247.

Brammer, D. W., Doerning, B. J., Chrisp, C. E., and Rush, H. G. (1991). Anesthetic and nephrotoxic effects of Telazol® in New Zealand White rabbits. *Lab. Anim. Sci.* **41,** 432–435.

Bree, M. M., and Cohen, B. J. (1965). Effects of urethane anesthesia on blood and blood vessels in rabbits. *Lab. Anim. Care* **15,** 254–259.

Bree, M. M., Cohen, B. J., and Abrams, G. D. (1971). Injection lesions following intramuscular administration of chlorpromazine in rabbits. *J. Am. Vet. Med. Assoc.* **159,** 1598–1602.

Brown, J. H., Pleuvry, B. J., and Kay, B. (1980). Respiratory effects of a new opiate analgesic, R39209, in the rabbit: Comparison with fentanyl [R39209 = alfentanyl]. *Br. J. Anaesth.* **52,** 1104–1106.

Carman, R. J., and Evans, R. H. (1984). Experimental and spontaneous clostridial enteropathies of laboratory and free living lagomorphs. *Lab. Anim. Sci.* **34,** 443–452.

Chakrabarty, S., Thomas, P., and Sheridan, D. J. (1991). Arrhythmias, haemodynamic changes and extent of myocardial damage during coronary ligation in rabbits anaesthetized with halothane, alpha chloralose or pentobarbitone. *Int. J. Cardiol.* **31,** 9–14.

Chen, G., and Bohner, B. (1968). Surgical anesthesia in the rabbit with 2-(ethylamino)-2-(2-Thienyl) cyclohexanone - HC1 (CI-634) and chloral hydrate. *Am. J. Vet. Res.* **29,** 869–875.

Clifford, D. H. (1984). Preanesthesia, anesthesia, analgesia and euthanasia. *In* ''Laboratory Animal Medicine'' (J. G. Fox, B. J. Cohen, and F. M. Loew, eds.), pp. 543–545. Academic Press, Orlando, Florida.

Cockshott, T. D., Douglas, E. J., F, P. G., and Simons, P. J. (1992). The pharmacokinetics of propofol in laboratory animals. *Xenobiotica* **22,** 369–375.

Coderre, T. J., Katz, J., Vaccarino, A. L., and Melzack, R. (1993). Contribution of central neuroplasticity to pathological pain: Review of clinical and experimental evidence. *Pain* **52,** 259–285.

Conn, H., and Langer, R. (1978). Continuous long-term intra-arterial infusion in the unrestrained rabbit. *Lab. Anim. Sci.* **28,** 598–602.

Cowen, R. H., and Laurenson, R. D. (1959). Technique of operating upon the fetus of the rabbit. *Surgery* **45,** 321–323.

Crawford, M. E., Jensen, F. M., Toftdahl, D. B., and Madsen, J. B. (1993). Direct spinal effect of intrathecal and extradural midazolam on visceral noxious stimulation in rabbits. *Br. J. Anaesth.* **70,** 642–646.

Danneman, P. J., White, W. J., Marshall, W. K., and Lang, C. M. (1988). Evaluation of analgesia associated with the immobility response in laboratory rabbits. *Lab. Anim. Sci.* **38,** 51–56.

Davis, J. A., Greenfield, R. E., and Brewer, T. G. (1993). Benzocaine-induced methemoglobinemia attributed to topical application of the anesthetic in several laboratory animal species. *J. Am. Vet. Med. Assoc.* **54,** 1322–1326.

Davis, N. L., and Malinin, T. I. (1974). Rabbit intubation and halothane anesthesia. *Lab. Anim. Sci.* **24,** 617–621.

DeCastro, G., and Viars, P. (1968). Anesthesie analgesique sequentielle, ou A.A.S. *Arch. Med.* **23,** 170–176.

Deimling, M. J., and Schnell, R. C. (1980). Circadian rhythms in the biological response and disposition of ethanol in the mouse. *J. Pharmacol. Exp. Ther.* **213,** 1–8.

Dhasmana, K. M., Saxena, P. R., Prakash, O., and Van Der Zee, H. T. (1984). Study on the influence of ketamine on systemic and regional haemodynamics in conscious rabbits. *Arch. Int. Pharmacodyn. Ther.* **269,** 323–334.

Doerning, B. J., Brammer, D. W., Chrisp, C. E., and Rush, H. G. (1992). Nephrotoxicity of tiletamine in New Zealand White rabbits. *Lab. Anim. Sci.* **42,** 267–269.

Dolowy, W. C., and Hesse, A. L. (1959). Chlorpromazine premedication with pentobarbital anesthesia in the rabbit. *J. Am. Vet. Med. Assoc.* **134,** 183–184.

Doorley, B. M., Waters, S. J., Terrell, R. C. *et al.* (1988). MAC of 1–653 in beagle dogs and New Zealand white rabbits. *Anesthesiology* **69,** 89–91.

Dorward, P. K., Burke, S., Janig, W., and Cassell, J. (1987). Reflex responses to baroreceptor, chemoreceptor and nociceptor inputs in single renal sympathetic neurones in the rabbit and the effects of anaesthesia on them. *J. Auton. Nerv. Syst.* **18,** 39–54.

Drummond, J. C. (1985). MAC for halothane, enflurane, and isoflurane in the New Zealand White rabbit: And a test for the validity of MAC determinations. *Anesthesiology* **62,** 336–338.

Drummond, J. C., Scheller, M. S., and Todd, M. M. (1987). Effect of nitrous oxide on cortical cerebral blood flow during anesthesia with halothane and isoflurane, with and without morphine, in the rabbit. *Anesth. Analg.* **66,** 1083–1089.

Durant, P. A. C., and Yaksh, T. L. (1986). Epidural injection of bupivacaine, morphine, fentanyl, lofentanyl and DADL in chronically implanted rats: a pharmacologic and pathologic study. *Anesthesiology* **64,** 43–53.

Ecobichon, D. J., and Comeau, A. M. (1974). Genetic polymorphism of plasma carboxylesterases in the rabbit: Correlation with pharmacologic and toxicologic effects. *Toxicol. Appl. Pharmacol.* **27,** 28–40.

Fick, T. E., and Schalm, S. W. (1986). Procedure for arterial blood sampling in the rabbit. *Lab. Anim.* **20,** 138–139.

Fick, T. E., and Schalm, S. W. (1987). A simple technique for endotracheal intubation in rabbits. *Lab. Anim.* **21,** 265–266.

Flecknell, P. A. (1987). "Laboratory Animal Anaesthesia: An Introduction for Research Workers and Technicians." Academic Press, London.

Flecknell, P. A., and Mitchell, M. (1984). Midazolam and fentanyl-fluanisone: Assessment of anaesthetic effects in laboratory rodents and rabbits. *Lab. Anim.* **18,** 143–146.

Flecknell, P. A., and Liles, J. H. (1990). Assessment of the analgesic action of opioid agonist-antagonists in the rabbit. *J. Assoc. Vet. Anaesth.* **17,** 24–29.

Flecknell, P. A., John, M., Mitchell, M., Shurey, C., and Simpkin, S. 1983). Neuroleptanalgesia in the rabbit. *Lab. Anim.* **17,** 104–109.

Flecknell, P. A., Liles, J. H., and Wootton, R. (1989). Reversal of fentanyl/fluanisone neuroleptanalgesia in the rabbit using mixed agonist/antagonist opioids. *Lab. Anim.* **23,** 147–155.

Flecknell, P. A., Liles, J. H., and Williamson, H. A. (1990). The use of lignocaine-prilocaine local anaesthetic cream for pain-free venepuncture in laboratory animals. *Lab. Anim.* **24,** 142–146.

Freeman, M. J., Bailey, S. P., and Hodesson, S. (1972). Premedication, tracheal intubation, and methoxyflurane anesthesia in the rabbit. *Lab. Anim. Sci.* **22,** 576–580.

Gardner, A. F. (1964). Development of general anesthesia in the albino rabbit for surgical procedures. *Lab. Anim. Care* **14,** 214–225.

Gay, W. I., ed. (1963). "Methods of Animal Experimentation," Vol. 1. Academic Press, New York.

Gillett, N. A., Brooks, D. L., and Tillman, P. C. (1983). Medical and surgical management of gastric obstruction from a hairball in the rabbit. *J. Am. Vet. Med. Assoc.* **183,** 1176–1178.

Glen, J. B. (1980). Animal studies of the anaesthetic activity of ICI 35 868. *Br. J. Anaesth.* **52,** 731–741.

Goldstein, J. L., Kita, T., and Brown, M. S. (1983). Defective lipoprotein receptors and atherosclerosis. *N. Engl. J. Med.* **309,** 288–296.

Green, C. J. (1975). Neuroleptanalgesic drug combinations in the anaesthetic management of small laboratory animals. *Lab. Anim.* **9,** 161–178.

Green, C. J. (1982a). "Animal Anaesthesia." Laboratory Animals Ltd., London.

Green, C. J. (1982b). Orders lagomorpha, rodentia, insectivora and chiroptera. *In* "Animal Anesthesia, Laboratory Animal Handbook," Vol. 8, pp. 131–138. Laboratory Animals Ltd., London.

Green, C. J., Halsey, M. J., Precious, S., and Wardley-Smith, B. (1978). Alphaxolone-alphadolone anaesthesia in laboratory animals. *Lab. Anim.* **12,** 85–89.

Green, C. J., Knight, J., Precious, S., and Simpkin, S. (1981). Ketamine alone and combined with diazepam or xylazine in laboratory animals: A 10 year experience. *Lab. Anim.* **15,** 163–170.

Grice, H. C. (1964). Methods for obtaining blood and for intravenous injections in laboratory animals. *Lab. Anim. Care* **14,** 483–493.

Gruber, R. P., and Amato, J. J. (1970). Hypnosis for rabbit surgery. *Lab. Anim. Care* **20,** 741–742.

Guerreiro, D., and Page, C. P. (1987). Effect of neuroleptanalgesia on some cardiorespiratory variables in the rabbit. *Lab. Anim.* **21,** 205–209.

Hall, L. W., and Clarke, K. W. (1991). "Veterinary Anaesthesia," 9th ed. Baillière Tindall, London.

Harris, R. H., Jr., and Best, C. F. (1979). Ureteral catheter system for renal function studies in conscious rabbits. *Lab. Anim. Sci.* **29,** 781–784.

Harvey, R. C., and Walberg, J. (1987). Special considerations for anesthesia and analgesia in research animals. *In* "Principles and Practice of Veterinary Anesthesia" C. E. Short, ed.), pp. 381–382. Williams & Wilkins, Baltimore, Maryland.

Hastings, J. W., and Menaker, M. (1976). Physiological and biochemical aspects of circadian rhythms. *Fed. Proc., Fed. Am. Soc. Exp. Biol.* **35,** 2325–2357.

Hess, R., Herz, A., and Friedel, K. (1971). Pharmacokinetics of fentanyl in rabbits in view of the importance for limiting the effect. *J. Pharmacol. Exp. Ther.* **179,** 474–484.

Hess, R., Stiebler, G., and Herz, A. (1972). Pharmacokinetics of fentanyl in man and the rabbit. *Eur. J. Clin. Pharmacol.* **4,** 137–141.

Hess, R., Husmann, K., and Kettler, D. (1981). Blood levels of fentanyl during multiple injections and intravenous infusions of low and high doses: Approaching optimal conditions for "stress-free anaesthesia." *Methods Find. Exp. Clin. Pharmacol.* **3,** 107S–114S.

Hillyer, E. V. (1992). Approach to the anorexic rabbit. *J. Small Exotic Anim. Med.* **1,** 106–108.

Hindman, B. J., Funatsu, N., Cheng, D. C. H., Bolles, R., Todd, M. M., and Tinker, J. H. (1990). Differential effect of oncotic pressure on cerebral and extracerebral water content during cardiopulmonary bypass in rabbits. *Anesthesiology* **73,** 951–957.

Hobbs, B. A., Rolhall, T. G., Sprenkel, T. L., and Anthony, K. L. (1991). Comparisons of several combinations for anesthesia in rabbits. *Am. J. Vet. Res.* **52,** 669–674.

Hodesson, S., Rich, S. T., Washington, J. O., and Apt, L. (1965). Anesthesia of the rabbit with Equi-Thesin® following the administration of preanesthetics. *Lab. Anim. Care* **15,** 336–337.

Hoge, R. S., Hodesson, S., Snow, I. B., and Wood, A. I. (1969). Intubation technique and methoxyflurane administration in rabbits. *Lab. Anim. Care* **19,** 593–595.

Holland, A. J. C. (1973). Laboratory animal anaesthesia. *Can. Anaesth. Soc. J.* **20,** 693–705.

Horita, A., and Carino, M. A. (1978). Analeptic and antianaleptic effects of naloxone and naltrexone in rabbits. *Life Sci.* **23,** 1681–1686.

Horita, A., Carino, M. A., and Yamawaki, S. (1983). Morphine antagonizes pentobarbital-induced anesthesia. *Neuropharmacology* **22,** 1183–1186.

Hovav, E., and Weinstock, M. (1987). Temporal factors influencing the development of acute tolerance to opiates. *J. Pharmacol. Exp. Ther.* **242,** 251–256.

Howard, D. L., Wittry, J. P., and Blum, J. R. (1990). Use of an audible respiratory monitor to aid endotracheal intubation of the rabbit [abstract P22]. *Lab. Anim. Sci.* **40,** 562–563.

Hughes, P. J., Doherty, M. M., and Charman, W. N. (1993). A rabbit model for the evaluation of epidurally administered local anaesthetic agents. *Anaesth. Intensive Care* **21,** 298–303.

Hunter, A. R. (1968). The respiratory effects of pentazocine in rabbits. *Anesthesia* **23,** 338–342.

Hunter, A. R., Pleuvry, B. J., and Rees, J. M. H. (1968). The respiratory depressant effects of barbiturates and narcotic analgesics in the unanaesthetised rabbit. *Br. J. Anaesth.* **40,** 927–935.

Hurley, R. J., Marini, R. P., Avison, D. L., Murphy, J. C., Olin, J. M., and Lipman, N. S. (1994). Evaluation of detomidine anesthetic combinations in the rabbit. *Lab. Anim. Sci.* **44,** 472–478.

Ishibe, Y., Gui, X., Uno, H., Shiokawa, Y., Umeda, T., and Suekane, K. (1993). Effect of sevoflurane on hypoxic pulmonary vasoconstriction in the perfused rabbit lung. *Anesthesiology* **79,** 1348–1353.

Jacobs, R., and Krohn, D. L. (1975). Anesthesia in rabbits by intrahepatic sodium pentobarbital. *J. Surg. Res.* **19,** 115–119.

Jenkins, W. L. (1987). Pharmacologic aspects of analgesic drugs in animals: An overview. *J. Am. Vet. Med. Assoc.* **191,** 1231–1240.

Jensen, F. M., Dahl, J. B., and Frigast, C. (1992). Direct spinal effect of intrathecal acetaminophen on visceral noxious stimulation in rabbits. *Acta Anaesthesiol. Scand.* **36,** 837–841.

Johnson, N. F., Strang, R., and Wilson, T. M. (1973). Further observations on the retinal fine structure after long-term anesthesia. *Exp. Eye Res.* **17,** 73–85.

Jori, A., DiSalle, E., and Santini, V. (1971). Daily rhythmic variation and liver drug metabolism in rats. *Biochem. Pharmacol.* **29,** 2965–2969.

Kaplan, H. M., and Timmons, E. H. (1979). "The Rabbit: A Model for the Principles of Mammalian Physiology and Surgery." Academic Press, New York.

Kent, G. M. (1971). General anesthesia in rabbits using methoxyflurane, nitrous oxide, and oxygen. *Lab. Anim. Sci.* **21,** 256–257.

Kero, P., Thomasson, B., and Soppi, A.-M. (1981). Spinal anaesthesia in the rabbit. *Lab. Anim.* **15,** 347–348.

Klemm, W. R. (1965). Drug potentiation of hypnotic restraint of rabbits, as indicated by behavior and brain electrical activity. *Lab. Anim. Care* **15,** 163–167.

Ko, J. C. H., Thurmon, J. C., Tranquilli, W. J., Benson, G. J., and Olson, W. A. (1992). Comparison of medetomidine-propofol and medetomidine-midazolam–propofol anesthesia in rabbits. *Lab. Anim. Sci.* **42,** 503–507.

Koch, K. L., and Dwyer, A. (1988). Effects of acetylsalicyclic acid on electromechanical activity of *in vivo* rabbit ileum. *Dig. Dis. Sci.* **33,** 962–968.

Korner, P. I., Uther, J. B., and White, S. W. (1968). Circulatory effects of chloralose–urethane and sodium pentobarbitone anaesthesia in the rabbit. *J. Physiol. (London)* **199,** 253–265.

Kozma, C., Macklin, W., Cummins, L. M., and Mauer, R. (1974). The anatomy, physiology, and the biochemistry of the rabbit. *In* "The Biology of the Laboratory Rabbit" (S. H. Weisbroth, R. E. Flatt, and A. L. Kraus, eds.), pp. 50–90. Academic Press, New York.

Krogh, E. (1975). Intramuscular rabbit anesthesia: Ketamine hydrochloride and mebumalnatrium (NFN): A safe and easy combination. *Acta Ophthalmol.* **53,** 367–368.

Kulli, J., and Koch, C. (1991). Does anesthesia cause loss of consciousness? *Trends Neurosci.* **14,** 6–10.

Kumar, R. A., Boyer, M. I., and Bowen, C. V. A. (1993). A reliable method of anesthesia for extensive surgery in small rabbits. *Lab. Anim. Sci.* **43,** 265–266.

Langerman, L., Chaimsky, G., Golomb, E., Tverskoy, M., and Kook, A. I. (1990). Rabbit model for evaluation of spinal anesthesia: Chronic cannulation of the subarachnoid space. *Anesth. Analg.* **71,** 529–535.

Lawson, D. M. (1985). Anesthetics and neuroendocrine function. *In* "CRC Handbook of Pharmacologic Methodologies for the Study of the Neuroendocrine System" (R. W. Steger and A. Johns, eds.), pp. 385–406. CRC Press, Boca Raton, Florida.

Leaders, F. E., and Keasling, H. H. (1962). Oral versus subcutaneous potency of codeine, morphine, levorphan, and anileridine as measured by rabbit toothpulp changes. *J. Pharm. Sci.* **51,** 46–49.

Lebowitz, H. (1990). Negative impedance ventilation as a novel mode of respiratory assist. MD Thesis, Harvard/MIT Division of Health Sciences and Technology, Cambridge, Massachusetts.

Lewis, G. E., Jr., and Jennings, P. B., Jr. (1972). Effective sedation of laboratory animals using Innovar-Vet. *Lab. Anim. Sci.* **22,** 430–432.

Lightowler, J., and Smith, A. E. W. (1963). B-(N-pyrolidyl)-butyroanilid(ws 10): Its sub-acute toxicity, influence on morphine analgesia in the rat and morphine analgesia and respiratory depression in the rabbit. *Arch. Int. Pharmacodyn. Ther.* **144,** 97–111.

Liles, J. H., and Flecknell, P. A. (1992). The use of non-steroidal anti-inflammatory drugs for the relief of pain in laboratory rodents and rabbits. *Lab. Anim.* **26,** 241–255.

Lindquist, P. A. (1972). Induction of methoxyflurane anesthesia in the rabbit after ketamine hydrochloride and endotracheal intubation. *Lab. Anim. Sci.* **22,** 898–899.

Linn, J. M., and Liebenberg, S. P. (1979). *In vivo* detection of rabbit atropinesterase. *Lab. Anim. Sci.* **29,** 335–337.

Lipman, N. S., Phillips, P. A., and Newcomer, C. E. (1987). Reversal of ketamine/xylazine anesthesia in the rabbit with yohimbine. *Lab. Anim. Sci.* **37,** 474–477.

Lipman, N. S., Marini, R. P., and Erdman, S. E. (1990). Comparison of ketamine/xylazine and ketamine/xylazine/acepromazine anesthesia in the rabbit. *Lab. Anim. Sci.* **40,** 395–398.

Lipman, N. S., Weischedel, A. K., Connors, M. J., Olsen, D. A., and Taylor, N. S. (1992). Utilization of cholestyramine resin as a preventative treatment for antibiotic (clindamycin) induced enterotoxaemia in the rabbit. *Lab. Anim* **26,** 1–8.

Lockhart, S. H., Cohen, Y., Yasuda, N., Freire, B., Taheri, S., Litt, L., and Eger, E. I., II (1991). Cerebral uptake and elimination of desflurane, isoflurane, and halothane from rabbit brain: An *in vivo* NMR study. *Anesthesiology* **74,** 575–580.

Love, J. A., and Hammond, K. (1991). Group housing rabbits. *Lab. Anim.* **20,** 37–43.

Ludders, J. W. (1993). Anesthesia for ophthalmic surgery. *In* "Textbook of Small Animal Surgery" (D. Slatter, ed.), 2nd ed., Vol. 2, pp. 2276–2278. Saunders, Philadelphia.

Ludders, J. W., Thomas, C. B., Sharp, P., and Sedgwick, C. J. (1987). Anesthetic technique for repeated collection of blood from New Zealand white rabbits. *Lab. Anim. Sci.* **37,** 803–805.

Lumb, W. V., and Jones, E. W. (1984). "Veterinary Anesthesia," 2nd ed., Lea & Febiger, Philadelphia.

MacLeod, B. (1977). Comparison of the effects of thiopentone and halothane anaesthesia on critical oxygen tension. *Can. Anaesth. Soc.* **24,** 609–614.

Macrae, D. J., and Guerreiro, D. (1989). Simple laryngoscopic technique for the endotracheal intubation of rabbits. *Lab. Anim.* **23,** 59–61.

Madsen, J., Jensen, F., Faber, T., and Bille-Hansen, V. (1993). Chronic catheterization of the epidural space in rabbits: A model for behavioural and histopathological studies—examination of meptazinol neurotoxicity. *Acta Anaesthesiol. Scand.* **37,** 307–313.

Maggi, C. A., Manzini, S., Parlani, M., and Meli, A. (1984). Analysis of the effects of urethane on cardiovascular responsiveness to catecholamines in terms of its interference with Ca^{++} mobilization from both intra and extracellular pools. *Experientia* **40,** 52–59.

Margolis, R., and Fiegelson, P. (1963). Atropinesterase, a genetically determined rabbit serum esterase. *J. Biol. Chem.* **238,** 2620–2627.

Marini, R. P., Avison, D. L., Corning, B. F., and Lipman, N. S. (1992). Ketamine/xylazine/butorphanol: A new anesthetic combination for rabbits. *Lab. Anim. Sci.* **42,** 57–62.

Marini, R. P., Hurley, R. J., Avison, D. L., and Lipman, N. S. (1993). Evaluation of three neuroleptanalgesic combinations in the rabbit. *Lab. Anim. Sci.* **43,** 338–345.

Marston, J. H., Rand, G., and Chang, M. C. (1965). Care, handling and anesthesia of the snowshoe hare (*Lepus americanus*). *Lab. Anim. Care* **15,** 325–328.

Mattila, M., and Saarnivaara, L. (1968). Modification by antihistaminic drugs of the morphine analgesia in rabbits. *Ann. Med. Exp. Fenn.* **46,** 72–77.

May, C. N., Ham, I. W., Heslop, K. E., A. S. F., and Mathias, C. J. (1988). Intravenous morphine causes hypertension, hyperglycaemia and increases sympatho-adrenal outflow in conscious rabbits. *Clin. Sci.* **75,** 71–77.

McCallister, L. W., Lipton, J. M., Giesecke, A. H., Jr., and Clark, W. G. (1986). Rabbit ear-withdrawal test: A new analgesiometric procedure. *Pharmacol., Biochem. Behav.* **25,** 481–482.

McCormick, M. J., and Ashworth, M. A. (1971). Acepromazine and methoxyflurane anesthesia of immature New Zealand White rabbits. *Lab. Anim. Sci.* **21,** 220–223.

McGrath, J. C., and MacKenzie, J. E. (1977). Effects of intravenous anaesthetics on the cardiovascular system of the rabbit. *Br. J. Pharmacol.* **61,** 199–212.

McGrath, J. C., MacKenzie, J. E., and Millar, R. A. (1975). Circulatory responses to ketamine: dependence on respiratory pattern and background anaesthesia in the rabbit. *Br. J. Anaesth.* **47,** 1149–1156.

Melich, D. (1990). Method for chronic intravenous infusion of the rabbit via the marginal ear vein. *Lab. Anim. Sci.* **40,** 327–328.

Mero, M., Makela, A., Vainionpaa, S., Vihtonen, K., and Rokkanen, P. (1987). Use of neuroleptanaesthesia for experimental orthopaedic surgery in the rabbit. *Acta Vet. Scand.* **28,** 251–252.

Mills, P., Sessler, D. I., Moseley, M., Chew, W., Pereira, B., James, T. L., and Litt, L. (1987). *In vivo* 19F nuclear magnetic resonance study of isoflurane elimination from the rabbit brain. *Anesthesiology* **67,** 169–173.

Mook, P. H., Proctor, H. J., Zee, H. V. D., Ennema, J. J., and Wildevuur, C. R. H. (1984). High-frequency ventilation in rabbits with respiratory insufficiency. *J. Surg. Res.* **36,** 614–619.

Moore, L. R. C., Chang, S.-F., and Greenstein, E. T. (1987). Urethane–acepromazine: A novel method of administering parenteral anesthesia in the rabbit. *Methods Find. Exp. Clin. Pharmacol.* **9,** 711–715.

More, R. C., Kody, M. H., Kabo, J. M., Dorey, F. J., and Meals, R. A. (1989). The effects of two non-steroidal antiinflammatory drugs on limb swelling, joint stiffness, and bone torsional strength following fracture in a rabbit model. *Clin. Orthop.* **247,** 306–312.

Morita, H., Nishida, Y., Uemura, N., and Hosomi, H. (1987). Effect of pentobarbital anesthesia on renal sympathetic nerve activity in the rabbit. *J. Auton. Nerv. Syst.* **20,** 57–64.

Murai, S., and Ogura, Y. (1978). Effects of several analgesics on the number of vocalization discharges in the rabbit. *Jpn. J. Pharmacol.* **28,** 636–639.

Murdock, H. R., Jr. (1969). Anesthesia in the rabbit. *Fed. Proc., Fed. Am. Soc. Exp. Biol.* **28,** 1510–1516.

National Institutes of Health (NIH) (1991). *''Preparation and Maintenance of Higher Mammals during Neuroscience Experiments,''* Publi. No. 91-3207. NIH, Washington, D.C.

Nelson, J. M., Krummel, T. M., Haynes, J. H., Flood, L. C., Sauer, L., Flake, A. W., and Harrison, M. R. (1990). Operative techniques in the fetal rabbit. *J. Invest. Surg.* **3,** 393–398.

Nevalainen, T., Pyhälä, L., Voipio, H., and Virtanen, R. (1989). Evaluation of anaesthetic potency of medetomidine-ketamine combination in rats, guinea-pigs and rabbits. *Acta Vet. Scand.* **85,** 139–143.

Nordin, U., Lindhalm, C. E., and Wolgast, M. (1977). Bloodflow in the rabbit tracheal mucosa under normal conditions and under the influence of tracheal intubation. *Acta Anaesthesiol. Scand.* **21,** 81–94.

Olson, M. E., McCabe, K., and Walker, R. L. (1987). Guaifenesin alone or in combination with ketamine or sodium pentobarbital as an anesthetic in rabbits. *Can. J. Vet. Res.* **51,** 383–386.

Olson, M. E., Vizzutti, D., Morck, D. W., and Cox, A. K. (1994). The parasympatholytic effects of atropine sulfate and glycopyrrolate in rats and rabbits. *Can. J. Vet. Res.* **57,** 254–258.

Pandeya, N. K., and Lemon, H. M. (1965). Paraldehyde: An anesthesia for recovery experiments in albino rabbits. *Lab. Anim. Care* **15,** 304–306.

Patel, P. M., and Mutch, W. A. C. (1990). Cerebral pressure-flow relationship during 1.0 MAC isoflurane anesthesia in the rabbit: the effect of different vasopressors. *Anesthesiology* **72,** 118–124.

Paulsen, R., and Valentine, J. L. (1984). Procedure for intravenous administration of drugs and repeated arterial blood sampling in the rabbit. *Lab. Anim.,* **13,** 34–36.

Peeters, M. E., Gil, D., Teske, E., Eyzenbach, V., van der Brom, W. E., Lumeij, J. T., and de Vries, H. W. (1988). Four methods for general anaesthesia in the rabbit: A comparative study. *Lab. Anim.* **22,** 355–360.

Perrin, A., Milano, G., Thyss, A., Cambon, P., and Schneider, M. (1990). Biochemical and pharmacological consequences of the interaction between methotrexate and ketoprofen in the rabbit. *Br. J. Cancer* **62,** 736–741.

Piercey, M. F., and Schroeder, L. A. (1980). A quantitative analgesic assay in the rabbit based on the response to tooth pulp stimulation. *Arch. Int. Pharmacodyn. Ther.* **248,** 294–304.

Pleuvry, B. J., and Rees, J. M. H. (1969). The effects of etorphine and of morphine on respiration, blood carbon dioxide tension, and carbon dioxide sensitivity in the conscious rabbit. *J. Pharmacol. Pharmacodyn. Ther.* **21,** 814–825.

Popilskis, S. J., Oz, M. C., Gorman, P., Florestal, A., and Kohn, D. F. (1991). Comparison of xylazine with tiletamine-zolazepam (Telazol®) and xylazine–ketamine anesthesia in rabbits. *Lab. Anim. Sci.* **41,** 51–53.

Radzialowski, F. M., and Bousquet, W. F. (1968). Daily rhythmic variation in hepatic drug metabolism in the rat and mouse. *J. Pharmacol. Exp. Ther.* **163,** 229–238.

Raman, J., Montano, S. R., and Lord, R. S. A. (1989). Assisted local anaesthesia in rabbits. *Microsurgery* **10,** 75–76.

Rapson, W. S., and Jones, T. C. (1964). Restraint of rabbits by hypnosis. *Lab. Anim. Care* **14,** 131–133.

Reed, R. P., Marcello, J., Welzel, G., and Schechter, A. (1987). Gavaging a rabbit with a rubber catheter. *Lab. Anim.* **16**(6), 51–52.

Reed, T. E. (1990). Did you know? Probiotics. *Domestic Rabbits,* pp. 28–29.

Rich, G. F., Sullivan, M. P., and Adams, J. M. (1990). Is distal sampling of end-tidal CO_2 necessary in small subjects? *Anesthesiology* **73,** 265–268.

Robertson, S. A., and Eberhart, S. (1994). Efficacy of the intranasal route for administration of anesthetic agents to adult rabbits. *Lab. Anim. Sci.* **44,** 159–165.

Robson, W. L., Bayliss, C. E., Feldman, R., Goldstein, M. B., Chen, C.-B., Richardson, R. M. A., Stinebaugh, B. J., Tam, S.-C., and Halperin, M. L. (1981). Evaluation of the effect of pentobarbital anaesthesia on the plasma potassium concentration in the rabbit and the dog. *Can. Anaesth. Soc. J.* **28,** 210–216.

Ruta, T. S., and Mutch, W. A. C. (1989). Regional cerebral blood flow following hemorrhage during isoflurane anesthesia in the rabbit: Comparison of techniques to support blood pressure. *Anesthesiology* **70,** 978–983.

Sainz, M., Martinez, F., Ciges, M., de Carlos, R., and de la Cruz, T. (1987). Brainstem and middle latency auditory evoked responses in rabbits with halothane anaesthesia. *Acta Oto-Laryngol.* **103,** 613–619.

Salonen, J. S. (1992). Chemistry and pharmacokinetics of the α_2-adrenoreceptor agonists. *In* ''Animal Pain'' (C. E. Short and A. Van Poznak, eds.), pp. 191–200. Churchill-Livingstone, New York.

Sanford, T. D., and Colby, E. D. (1980). Effect of xylazine and ketamine on blood pressure, heart rate and respiratory rate in rabbits. *Lab. Anim. Sci.* **30,** 519–523.

Sartick, M., Eldridge, M. L., Johnson, J. A., Kurz, K. D., Fowler, W. L., Jr., and Payne, C. G. (1979). Recovery rate of the cardiovascular system in rabbits following short-term halothane anesthesia. *Lab. Anim. Sci.* **29,** 186–190.

Scheller, M. S., Saidman, I. J., and Partridge, B. L. (1988). MAC of sevoflurane in humans and the New Zealand white rabbit. *Can. J. Anaesth.* **35,** 153–156.

Scheving, L. E., Vedral, D. F., and Pauly, J. E. (1968). A circadian susceptibility rhythm in rats to pentobarbital sodium. *Anat. Rec.* **160,** 741–749.

Schutten, W. H., and Van Horn, D. L. (1977). Effects of ketamine sedation and ketamine-pentobarbital anesthesia upon the intraocular pressure of the rabbit. *Invest. Ophthalmol. Visual Sci.* **16,** 531–534.

Schuyt, H. C., and Leene, W. (1977). Improved method of tracheal intubation in the rabbit. *Lab. Anim. Sci.* **27,** 690–693.

Schuyt, H. C., Meeder, P., and Leene, W. (1978). Bit to immobilize the endotracheal tube in the intubated rabbit. *Lab. Anim. Sci.* **28,** 470–471.

Sebel, P. S., and Lowdon, J. D. (1989). Propofol: A new intravenous anesthetic. *Anesthesiology* **71,** 250–277.

Sedgwick, C. J. (1986). Anesthesia for rabbits. *Vet. Clin. North Am.: Food Anim. Pract.* **2,** 731–736.

Sherrard, E. (1966). Anesthesia of rabbits [letter]. *Vet. Rec.* **253,** 254.

Smith, P. A., Prieskorn, D. M., Knutsen, C. A., and Ensminger, W. D. (1988). Method for frequent blood sampling in rabbits. *Lab. Anim. Sci.* **38,** 623–625.

Stadnicka, A., Flynn, N. M., Bosnjak, Z. J., and Kampine, J. P. (1993). Enflurane, halothane, and isoflurane attenuate contractile responses to exogenous and endogenous norepinephrine in isolated small mesenteric veins of the rabbit. *Anesthesiology* **78,** 326–334.

Stanski, D. R. (1990). Monitoring depth of anesthesia. *In* ''Anesthesia'' (R. D. Miller, ed.), 3rd ed., pp. 1101–1029. Churchill-Livingstone, New York.

Stickrod, G., Ebaugh, T., and Garnett, C. (1981). Use of a mini-peristaltic pump for collection of blood from rabbits. *Lab. Anim. Sci.* **31,** 87–88.

Strack, L. E., and Kaplan, H. M. (1968). Fentanyl and droperidol for surgical anesthesia of rabbits. *J. Am. Vet. Med. Assoc.* **153,** 822–825.

Stunkard, J. A., and Miller, J. C. (1974). Outline guide to general anesthesia in exotic species. VM/SAC, *Vet. Med. Small Anim. Clin.* 1181–1186.

Taber, R. I. (1974). Predictive value of analgesic assays in mice and rats. *Adv. Biochem. Psychopharmacol.* **21,** 191–211.

Tashiro, C., Muranishi, R., Gomyo, I., Mashimo, T., Tomi, K., and Yoshiya, I. (1986). Electroretinogram as a possible monitor of anesthetic depth. *Arch. Clin. Exp. Ophthalmol.* **224,** 473–476.

Thomasson, B., and Ravitch, M. (1969). Fetal surgery in the rabbit. *Surgery* **66,** 1092–1102.

Tillman, P., and Norman, C. (1983). Droperidol-fentanyl as an aid to blood collection in rabbits. *Lab. Anim. Sci.* **33,** 181–182.

Todd, M. M., Wu, B., Warner, D. S., and Maktabi, M. (1994). The dose-related effects of nitric oxide synthase inhibition on cerebral blood flow during isoflurane and pentobarbital anesthesia. *Anesthesiology* **80,** 1128–1136.

Tucker, F. S., and Beattie, R. J. (1983). Qualitative microtest for atropine esterase. *Lab. Anim. Sci.* **33,** 268–269.

van Zutphen, L. F. M. (1972). Qualitative and quantitative detection of atropinesterase and cocainesterase in two breeds of rabbits. *Enzymologia* **42,** 201–218.

Walden, N. B. (1978). Effective sedation of rabbits, guinea pigs, rats and mice with a mixture of fentanyl and droperidol. *Aust. Vet. J.* **54,** 538–540.

Ward, G. S., Johnsen, D. O., and Roberts, C. R. (1974). Use of CI 744 as an anesthetic for laboratory animals. *Lab. Anim. Sci.* **24,** 737–742.

Warren, D. J., and Ledingham, J. G. G. (1978). Renal vascular response to haemorrhage in the rabbit after pentobarbitone, chloralose-urethane and ether anaesthesia. *Clin. Sci. Mol. Med.* **54,** 489–494.

Weinger, M. B., and Koob, G. F. (1990). What constitutes adequate anesthesia in animals? In neonates? [letter]. *Anesthesiology* **72,** 767–768.

White, G. L., and Holmes, D. D. (1976). Comparison of ketamine and the combination ketamine-xylazine for effective surgical anesthesia in the rabbit. *Lab. Anim. Sci.* **26,** 804–806.

Wixson, S. K. (1994). Anesthesia and analgesia for rabbits. *In* ''The Biology of the Laboratory Rabbit'' (P. J. Manning), 2nd ed., pp. 87–109. Academic Press, San Diego, California.

Wyatt, J. D., Scott, R. A. W., and Richardson, M. E. (1989). Effects of prolonged ketamine-xylazine intravenous infusion on arterial blood pH, blood gases, mean arterial blood pressure, heart and respiratory rates, rectal temperature and reflexes in the rabbit. *Lab. Anim. Sci.* **39,** 411–416.

Wyler, F. (1974). Effect of general anesthesia on distribution of cardiac output and organ blood flow in the rabbit: Halothane and chloralose-urethane. *J. Surg. Res.* **17,** 381–386.

Wyler, F., and Weisser, K. (1972). Effect of halothane anaesthesia on distribution of cardiac output and organ blood flow in the rabbit. *Br. J. Anaesth.* **44,** 551–555.

Wynn, R. L., El'Baghdady, Y. M., Ford, R. D., Thut, P. D., and Rudo, F. G. (1984). Rabbit tooth-pulp assay to determine ED_{50} values and duration of action of analgesics. *J. Pharmacol. Methods* **11,** 109–117.

Zhou, Z., Du, M., Wu, W., Jiang, Y., and Han, J. (1981). Effects of intracerebral microinjection of naloxone on acupuncture- and morphine-analgesia in the rabbit. *Sci. Sin. (Engl. Ed.)* **24,** 1166–1178.

Chapter 11

Anesthesia and Analgesia in Nonhuman Primates

Sulli J. Popilskis and Dennis F. Kohn

ANESTHESIA AND ANALGESIA IN LABORATORY ANIMALS

ISBN 0-12-417570-8

I. Introduction

Nonhuman primates are important models for a wide variety of biomedical and behavioral research because of their close phylogenetic relationship to humans. They are frequently used in surgical protocols of organ transplantation, neuroscience, cardiovascular disease, and fetal–maternal studies. Competent clinical management of nonhuman primates has helped to assure the most effective use of these valuable animals. Advances in methods of anesthesia and analgesia have played an important role in ensuring their humane treatment. Of the approximate 190 species of prosimians, New World and Old World monkeys, and apes, the majority of nonhuman primates used in biomedical research are the macaques (*Macaca* spp.), baboon (*Papio* spp.), squirrel monkey (*Saimiri sciureus*), patas monkey (*Erythrocebus patas*), marmosets and tamarins (*Callitrichidae),* and chimpanzee (*Pan troglodytes*) (see Table I).

When selecting methods for anesthesia and restraint, the diversity of the order Primata must be considered. The wide range in body size and weight of nonhuman primates plays an important role in selecting an appropriate anesthetic, the route of administration, and dosage of the drug (Sainsbury *et al.,* 1989). Extrapolation of anesthetic or analgesic doses from one primate species to another should be done with caution because of differences in the responses of some species to certain agents. For example, the anesthetic dose of alphaxolone–alphadolone for macaques can exceed the lethal dose for the adult *S. sciureus* (Logdberg, 1988). Other criteria within the same species, such as age and gender of the animal, should be considered when determining appropriate anesthesia or analgesia management. The decreased ability of the neonate to metabolize drugs and the immaturity of the blood–brain barrier are examples of factors that could reduce the safe and effective dose of some anesthetics. The purpose of this chapter is to provide the reader with a review of the literature on techniques, methodologies, and agents that have been reported in commonly used nonhuman primates, along with those that have been found to be effective by the authors. In some instances in which there is little information available in the literature and for which the authors have had insufficient experience, non-peer-reviewed information has been obtained from knowledgeable clinicians by personal communication. The overall goal of this chapter is to provide veterinarians and investigators with information that will help ensure that nonhuman primates are provided with optimal anesthesia and analgesia during their use as research subjects.

TABLE I

NOMENCLATURE OF NONHUMAN PRIMATES

Scientific name	Common name
Callithrix jacchus geoffroyi	White-fronted marmoset
C. jacchus jacchus	Common marmoset
Cebus albifrons	White-fronted capuchin
Cercopithecus aethiops	African Green monkey
Erythrocebus patas	Patas monkey
Galago spp.	Bush babies
Gorilla gorilla	Gorilla
Macaca arctoides	Stump-tailed macaque
M. cyclopis	Formosan-rock macaque
M. fascicularis	Cynomolgus
M. mulatta	Rhesus
M. nemestrina	Pig-tailed macaque
Pan troglodytes	Chimpanzee
Papio papio	Baboon
P. hamadryas	Sacred baboon
P. ursinus	Chacma baboon
Saguinus spp.	Tamarin spp.
Saimiri sciureus	Squirrel monkey

A. Public Health Concerns

In any discussion of nonhuman primate use, the topic of zoonotic disease must be considered. Because of their phylogenetic relationship to humans, many infectious agents can be transmitted from nonhuman primates to humans [e.g., *Mycobacterium tuberculosis, Shigella* spp., *Salmonella* spp., hepatitis B virus in apes, Cercopithecine herpesvirus 1, and *Filovirus* (Adams *et al.,* 1995a; Holmes *et al.,* 1995)]. Conversely, infectious agents can be transmitted from humans to nonhuman primates. All personnel involved in providing anesthesia, analgesia, and perioperative care of nonhuman primates should be fully aware of means to prevent possible exposure to zoonotic diseases, should be provided with protective clothing, gloves and masks, and should be enrolled in an appropriate occupational health program [National Research Council (NRC), 1997; Adams *et al.,* 1995b].

II. PREOPERATIVE EVALUATION

Preoperative assessment includes history of previous use, physical examination, pertinent laboratory data, and the influence of the current experimental protocol on anesthetic management. Despite the limitations associated with performing a thorough physical examination in the awake nonhuman primate, important signs of illness that can be readily identified are unusual posture or behavior, anorexia, and abnormal urine or feces. Obtaining body weight and temperature, and auscultation for heart rhythm and bilateral lung sounds during the preoperative period will provide additional information on the physical status of the animal. Routine laboratory tests (e.g., complete blood counts, blood chemistries) should be performed while nonhuman primates are in quarantine and periodically thereafter as part of a preventive medicine program (Southers and Ford, 1995). Variations in testing within an institution may reflect specific needs associated with the use of nonhuman primates. It is usually not necessary to perform preoperative clinical laboratory testing in animals that are in good physical condition. However, if a complex surgical procedure is to be done, additional laboratory data may be necessary. For example, baseline hematology would be desirable for nonhuman primates undergoing extracorporeal bypass and other surgical procedures that may produce vascular volume deficiencies.

Concomitant natural and experimental disease may influence the selection of an appropriate anesthesic. Nonhuman primates with induced hypertension may be less tolerant to repeated doses of ketamine and might be better managed with volatile anesthetics such as isoflurane and halothane. Splenectomized or immunosuppressed *Macaca fascicularis* monkeys are at added risk for developing anemia due to naturally occurring malarial infections. Anemia, in turn, may result in some physiologic changes (e.g., increase in heart rate and cardiac output) to meet a greater demand for tissue oxygenation. Anesthetics that reduce a previously elevated cardiac output could interfere with tissue oxygen delivery. Anemia may also reduce the solubility of volatile anesthetics and, consequently, accelerate the rate at which the alveolar concentration can be increased or decreased.

Preoperative fasting is an accepted practice for nonhuman primates undergoing surgical procedures. Although the optimal fasting time in nonhuman primates has not been established, it is conventional practice to fast primates for at least 12 hr in order to decrease the risk of pulmonary aspiration. Exceptions to this are the *Callitrichidae* (marmosets and tamarins) and other small species that generally are fasted only 6 to 8 hr to help avoid perioperative hypoglycemia. Human (Schreiner *et al.,* 1990) and nonhuman primate (Popilskis *et al.,* 1992) studies suggest that prolonged withholding of clear fluids neither reduces gastric volume nor increases pH, and does not decrease the risk of aspiration of gastric contents when compared to 3 hr of withholding fluids. It is probable that withholding water for only 3 hr may reduce the likelihood of hypotension, especially in juvenile primates; however, clearly this recommendation cannot be applied to solid foods.

In situations requiring emergency surgery or in pregnant animals in which gastric emptying is delayed, inclusion of H_2 antagonists may reduce the risk of aspiration pneumonia by blocking the histamine-induced secretion of gastric fluid. The addition of H_2 antagonists such as cimetidine (10 mg/kg) or ranitidine (1.5 mg/kg) 30–40 min before induction was shown to protect against aspiration pneumonia in *Papio* spp. (Popilskis *et al.,* 1992). Ranitidine is preferred because it does not effect cytochrome P450 (Somogyi and Gugler, 1982), and accordingly does not influence the hepatic detoxification of drugs.

III. ANESTHETIC DELIVERY

Caudal thigh muscles are most commonly used for intramuscular injections. For repeated intramuscular injections, it is advisable to alternate the leg to minimize the possibility of muscle or nerve irritation by drugs with low pH such as ketamine (Davy *et al.,* 1987). Using "squeeze-back" cages, subcutaneous injections can be easily done in the lateral thigh or dorsolateral sites in the back. In large enclosures in which animals cannot be restrained for IM or SC injections, agents such as ketamine may be delivered by darts propelled from blow pipes or air guns (Fowler, 1986; Klein and Murray, 1995). This method for delivery of agents is frequently needed for the chemical restraint of *P. troglodytes* and adult male *Papio* spp. In adult *P. troglodytes,* D. Cohn (personal communication, 1995) prefers to use a modified hand-pump air pistol instead of a CO_2-propelled gun because the latter is more likely to result in unnecessary physical trauma due to the high velocity of the dart. If a blow gun is used, care must be taken to ensure that an animal does not physically redirect the pipe in the direction of another animal or human.

Chemical restraint is usually needed for intravenous injections or blood withdrawal. Cephalic or saphenous veins can be used both for venipuncture and for the administration of drugs or fluids. The authors usually prefer the cephalic vein, which is visible and readily accessible, originating on the radial surface of the arm and achieving prominence just above the elbow. In juvenile animals, visibility and access can be enhanced by the use of a tourniquet or by direct application of a warm towel to the site. The femoral vein is commonly used for the withdrawal of relatively large blood volumes. An indwelling catheter or needle (20- or 22-gauge) is inserted at the femoral triangle just medial to the femoral artery which can be identified by pulsation. The femoral artery can be used for blood collection or direct intraarterial blood pressure monitoring. Whenever the femoral artery is being used, direct pressure must be applied to prevent a hematoma which may be of clinical importance in an animal with natural or experimental coagulopathies (Loeb *et al.,* 1976).

Chronic intravascular catheterization of nonhuman primates has been described (Barnstein *et al,* 1966; DaRif and Rush, 1983; Scalese *et al.,* 1990). The placement of indwelling vascular catheters allows multiple blood sampling, administration of various therapeutic agents, and monitoring of cardiovascular parameters in a conscious animal. Thus, indwelling catheters reduce the number of animals used and eliminate repeated venipuncture. The internal and external jugular vein and the femoral vein and artery are the most common vascular access sites. Polyethylene and medical grade silicon rubber tubing are materials commonly used for indwelling catheters. Tethering systems, consisting of a jacket, main swivel, and pump assembly, are commonly used to protect catheters from damage induced by the animal and as a means to enable movement of the animal in its cage. Major problems associated with indwelling catheter systems are loss of the catheter patency and infection. The risk of infection is related, in part, to the catheter exit site in the skin. Totally implanted subcutaneous vascular access ports reduce the risk of infection and stress associated with multiple cutaneous venous access techniques (Schmutzler *et al.,* 1988).

IV. ANTICHOLINERGIC DRUGS

Anticholinergics are used to diminish salivary and bronchial secretions in nonhuman primates during anesthesia and to prevent reflex bradycardia. These actions make atropine or glycopyrrolate beneficial adjuncts to ketamine–xylazine and fentanyl–droperidol use. Bradycardia is of great significance in pediatric nonhuman primates because their cardiac output is heart rate dependent. The addition of atropine (0.02–0.05 mg/kg) will effectively reduce xylazine- and fentanyl-induced bradycardia. When an antisiallagogue effect is desired, glycopyrrolate (0.005–0.01 mg/kg) is preferable to atropine because it is twice as potent and has a longer duration of action. Nevertheless, inclusion of an anticholinergic as part of premedication is not always necessary or indicated. Noteworthy is that atropine possesses arrhythmogenic properties and may predispose nonhuman primates to ventricular tachycardia and bigeminal patterns (Sedgwick, 1986). Because of this, atropine premedication as part of protocols for cardiac surgery is usually avoided.

V. PARENTERAL ANESTHETICS

A. Dissociative

1. Phencyclidine

Phencyclidine was the prototype agent of the 1-arylcyclohexylamine dissociative anesthetics used in nonhuman primates (Chen and Weston, 1960; Spalding and Heymann, 1962). It proved to be quite useful as an agent for chemical restraint in a number of species at doses varying from 0.5 to 3.0 mg/kg IM. Prior to its availability, few drugs were efficacious for restraining nonhuman primates, and restraint was commonly done manually with the use of protective gloves, nets, and transfer boxes. Intramuscular administation of phencyclidine induces unconsciousness within several minutes. Animals characteristically remain sedated for 1 to 3 hr depending on the dose administered and differences associated with a particular animal or species. Phencyclidine became a drug of abuse in humans because of hallucinatory effects and is no longer manufactured.

2. Ketamine

Ketamine is an analog of phencyclidine and has been widely used for chemical restraint and as an anesthetic agent in nonhuman primates since the 1970s. Ketamine produces a cataleptoid state involving unconsciousness and somatic analgesia in nonhuman primates. Ketamine has a wide margin of safety in nonhuman primates, and it has been used in dosages varying from 5 to 20 mg/kg IM in many species as an agent for restraint and induction for subsequent administration of other injectable or gaseous anesthetics. Induction is usually achieved within 5 min after intramuscular ketamine administration, and a single bolus will provide chemical restraint for 15 to 30 min that is sufficient for tuberculin testing, clinical examination, or minor procedures. Complete recovery occurs within 1 to 2 hr, depending on the dosage used. Nonhuman primates retain pharyngeal, laryngeal, and palpebral reflexes after ketamine administration. Laryngospasm is not uncommon during intubation after ketamine administration due to pharyngeal and laryngeal reflexes, and irritation of the vocal cords by oropharyngeal secretions. Atropine, at a dose of 0.02–0.05 mg/kg IM, can be given to limit the salivation that usually occurs after the administration of ketamine. Poor muscular relaxation and, in some instances, tonic–clonic movements or psychotomimetic emergence reactions are associated with ketamine administration in nonhuman primates (Green *et al.,* 1981; Sainsbury *et al.,* 1989; Jacobs *et al.,* 1993). For this reason, if a procedure requires complete immobilization or if a surgical procedure is to be done, another agent is administered in combination with ketamine.

The effects of ketamine on hematologic and serum biochemical values in nonhuman primates have been examined. As with other species of animals and humans, sympathetic stimulation due to stress and excitement affects a number of indices such as leukocyte and lymphocyte counts (Ives and Dack, 1956). Accordingly, differences in some indices are related to whether a nonhuman primate has been given ketamine or physically restrained with or without conditioning. Loomis *et al.* (1980) compared hemograms of *M. mulatta* that were either ketamine treated or physically restrained for venipuncture and found that leukocyte and lymphocyte counts and total plasma protein levels were reduced in the ketamine-treated animals. Bennett *et al.* (1992) reported similar findings with significant decreases in erythrocyte, lymphocyte, and total leukocyte counts, hemoglobin, hematocrit, and serum concentrations of glucose, total protein, cholestrol, and albumin.

The effect that ketamine has on the cardiovascular system and on selected plasma hormone levels in *M. fascicularis* has been studied (Castro *et. al.,* 1981). The animals in this study were well acclimated to restraining chairs to negate effects that could be due to animal handling rather than to ketamine. Ketamine produced no significant changes in mean arterial blood pressure, plasma insulin, glucose, or cortisol concentrations. After insulin challenge, plasma glucose concentrations, plasma ACTH (adrenocorticotropin), growth hormone, and cortisol responses were similar in both ketamine-treated and control monkeys which suggests that ketamine does not alter the magnitude of many endocrine responses. Brady and Koritnik (1985) reported similar findings for glucose tolerance testing in *Cercopithecus aethiops.*

Yohimbine has not been commonly used as a reversal agent in nonhuman primates. Although yohimbine acts by a competitive blockade of presynaptic α_2-adrenoreceptors, it has been reported to partially reverse ketamine in cats (Hatch *et al.,* 1983). However, in adult *M. mulatta,* yohimbine (0.5 mg/kg IV) did not decrease the length of ketamine-induced anesthesia (Lynch and Line, 1985).

Ketamine combined with other agents has been shown to be effective for anesthesia in many species of nonhuman primates. The following combinations are the most frequently reported.

3. Ketamine–Xylazine

Combining xylazine with ketamine provides muscular relaxation and analgesia sufficient for minor surgical procedures. Ketamine (7 mg/kg) and xylazine (0.6 mg/kg) given to *M. mulatta,* varying in weight from 1 to 11 kg, provided adequate anesthesia for cisternal and lumbar spinal puncture, insertion of urinary catheters, tattooing, and digit amputations (Banknieder *et al.,* 1978). The effects of various ratios of ketamine and xylazine in *M. mulatta* have been tested (Naccarato and Hunter, 1979). In that study, anesthesia was defined only as a loss of response to a pinprick stimulus. Therefore, the dosages used cannot be correlated with their efficacy for surgical anesthesia, only for their ability to produce chemical restaint. Among the dosage combinations used, ketamine (10 mg/kg) and xylazine (0.25 mg/kg) produced a mean anesthesia time of 45 min; increasing the xylazine dosage to 2 mg/kg extended the anesthesia duration to a mean of 138 min.

Ketamine (15–20 mg/kg) and xylazine (1 mg/kg) given IM to juvenile *P. troglodytes* provided anesthesia with sufficient analgesia for performing minor procedures such as plasmapheresis and percutaneous liver biopsies. Laryngeal and pharyngeal reflexes remained intact and endotracheal intubation was not possible. The average time of anesthesia was 25 min with recovery in about 2 hr (April *et al.,* 1982).

Reutlinger *et al.* (1980) studied the effects that ketamine and xylazine have on cardiovascular and pulmonary values in *M. mulatta.* Four groups of six adult males were given ketamine (7 mg/kg), xylazine (0.6 mg/kg), a ketamine (7 mg/kg) and xylazine (0.6 mg/kg) combination or saline. No significant differences were found among the three drug regimens and saline control animals with regard to mean respiratory rate, acid–base status, or arterial blood gases. However, xylazine or xylazine combined with ketamine produced statistically significant decreases in mean arterial blood pressure and heart rate compared to animals given ketamine alone. Xylazine-induced bradycardia and hypotension are common in most species and are associated with a decreased outflow of sympathetic nervous system impulses to the periphery. The significant decrease in blood pressure and heart rate encountered in xylazine and xylazine–ketamine groups indicates that xylazine overrides the stimulatory effects of ketamine.

4. Ketamine–Medetomidine

Ketamine–medetomidine has been successfully used to anesthetize *P. troglodytes* (Lewis, 1993; Jalanka and Roeken, 1990). In these reports, medetomidine (30–60 μg/kg) IM and ketamine (2–6 mg/kg) IM produced a rapid induction, prolonged and stable immobilization, excellent relaxation, and calm recovery. The duration of anesthesia was about 60 min and was sufficient to perform dental extraction and the subcutaneous placement of implants. Advantages of the ketamine–medetomidine combination over ketamine alone were better muscle relaxation and a lengthened period of anesthesia. Another advantage of combining medetomidine with ketamine is that medetomidine can be reversed with atipamezole, a specific α_2-antagonist. Chimpanzees given atipamezole at five times the medetomidine dose recovered within 15 to 20 min whereas those not receiving the antagonist recovered in 1–2 hr.

In *S. sciureus,* 100 μg/kg of medetomidine given either IM or SC facilitates mask induction with isoflurane (T. Morris, personal communication, 1995). Medetomidine can be reversed after isoflurane anesthesia is discontinued with 200 μg atipamezole to produce a very rapid recovery.

S. Capuano (personal communication, 1995) has found that medetomidine administered to *M. mulatta* at dosages of 50, 100, 150, or 200 μg/kg. IV induced extreme stupor with excellent muscle relaxation. However, this sedation was often punctuated by unexpected periods of arousal and aggressiveness, and the degree of analgesia was inconsistent, thus making medetomidine alone of questionable value in *M. mulatta.* The combination of ketamine (5mg/kg IM) and medetomidine (100 μg/kg IM), however, produced excellent sedation and muscle relaxation in *Cercopithecus aethiops* and *Papio* species (S. Capuano, personal communications, 1995).

5. Ketamine–Diazepam

The addition of diazepam to ketamine eliminates excessive involuntary movements and helps maintain adequate immobilization. The combination of ketamine (10 mg/kg) and diazepam (0.2 to 0.35 mg/kg) given IM has been reported to produce effective restraint in an adult male *Papio* species during a dental study that required muscle relaxation of the mouth and head (Woolfson *et al.,* 1980). In *S. sciureus,* ketamine (15–20 mg/kg) and diazepam (1 mg/kg) given IM provide light to moderate anesthesia (T. Morris, personal communication, 1995). J. Wyatt (personal communication, 1995) premedicates *Callithrix jacchus* with ketamine (15 mg/kg) and diazepam (1.0 mg/kg), given subcutaneously. Intubation must be done within several minutes after the onset of sedation due to the short duration of maximal sedation.

However, it is important to note that IM injection of diazepam may be painful with unreliable absorption and that multiple doses may lead to prolonged recovery due to its long elimination half-life.

6. Ketamine–Midazolam

Midazolam offers certain advantages over diazepam. Compared to diazepam, it is better absorbed after IM injection, provides more effective anxiolytic sedation, and has a shorter elimination half-life (Jacobs *et al.*, 1993). Jacobs *et al.* (1993) demonstrated that after an initial bolus of ketamine (15 mg/kg IM), an IV injection of midazolam (0.05–0.09 mg for animals less than 1 kg and 0.05–0.15 mg for animals over 1kg) followed by a ketamine infusion (12 mg/kg/hr) provided complete immobilization in young *M. mulatta* and *C. aethiops* during positron emission tomography. No noticeable side effects were noticed and the animals were returned to their cages within 30 min after the discontinuation of anesthesia. *Cercopithecus aethiops* exhibited age-dependent sensitivity to midazolam, with older animals requiring lower dosages. In *M. mulatta,* the effective dose of midazolam decreased until 4 months of age and then gradually increased. This may be associated with higher metabolic rates in infants and juvenile animals (Sedgwick, 1986).

7. Tiletamine–Zolazepam (Telazol)

Telazol is a 1:1 combination of the dissociative anesthetic tiletamine and the benzodiazepine tranquilizer zolazepam. It has been reported to be a useful agent for chemical restraint and as an anesthetic for minor surgical procedures in nonhuman primates (Kaufman and Hahnenberger, 1975; Cohen and Bree, 1978; Booker *et al.,* 1982). In juvenile *M. mulatta,* the minimum dosage for restraint is 1.5 mg/kg IM and 3.0 mg/kg for anesthesia sufficient for minor surgical procedures. At these dosages, Telazol provides a rapid onset and smooth recovery, and with the exception of depressed myocardial contractility, it has minimal cardiopulmonary effects (Booker *et al.,* 1982). At 4 to 6 mg/kg IM, chemical restraint is achieved for about 45 to 60 min in *M. mulatta, M. nemestrina, M. arctoides,* and *P. papio,* whereas the duration is about 100 min in *E. patas* (Cohen and Bree, 1978). T. Morris (personal communication, 1995) uses Telazol at a dosage of 10 mg/kg IM to produce light to moderate anesthesia in *S. sciureus.* At a dosage of 5 mg/kg IM, Telazol provides 15 min of anesthesia adequate for minor invasive procedures in *C. jacchus* (J. Wyatt, personal communication, 1995).

Telazol given to *P. troglodytes* at 3–5 mg/kg IM provides sufficient chemical restraint to perform physical examinations, sample collection, and minor procedures. An advantage of Telazol over ketamine is that its greater potency allows for administration of smaller volumes. This is particularly helpful when a dart is used for delivery of the agent to apes. Also, the occurrence of cyclohexylamine-induced seizures is nearly eliminated in comparison to ketamine (B. Swenson, personal communication, 1995).

B. Alphaxolone–Alphadolone

Alphaxolone–alphadolone acetate, a combination of two pregnanedione derivatives, is available in Europe from Glaxo Ltd. as Saffan. Currently, it is not available commercially in the United States. One milliliter of Saffan contains 12 mg of steroid (9 mg alphaxolone and 3 mg alphadolone acetate). It has been reported as an effective anesthetic in several species of nonhuman primates. Phillips and Grist (1975) reported the use of Saffan in *Callithrix jacchus* of various ages at dosages of 15 to 19.5 mg/kg. Surgical anesthesia was induced within 10 min after IM administration of Saffan and was sufficient for abdominal procedures of 1 hr in length. Full recovery occurred within 2 to 3 hr of induction.

Saffan has been used successfully in *S. sciureus* undergoing hysterotomy, bone tissue excision or implantation, and magnetic resonance imaging procedures (Logdberg, 1988). A single IM dosage of 11.5 to 15.5 mg/kg of alphaxolone–alphadolone provided up to 1 hr of surgical anesthesia with good muscle relaxation in adult and juvenile animals. An additional dosage of 3 to 5 mg/kg extended anesthesia for another hour. Other advantages of Saffan anesthesia include rapid induction and uneventful recovery. However, in *S. sciureus,* respiratory depression and hypothermia accompany Saffan anesthesia (Logdberg, 1988).

In larger primate species (e.g., *M. fascicularis*), at a dose similar to that used in New World primates, Saffan produces only sedation. This agent was found to provide effective surgical anesthesia in *M. fascicularis* at a dosage of 120 mg/kg IM given as a single bolus; alternatively, an IM injection of Saffan at 18 mg/kg, followed by 6- to 12-mg/kg dosages given IV as needed, provided satisfactory surgical anesthesia (Box and Ellis, 1973).

Surgical anesthesia can be provided by a continuous intravenous infusion of Saffan. This method of anesthesia has been reported in *Erythrocebus* and *Papio* species (Dhiri, 1985; Cookson and Mills, 1983). Up to 4 hr of anesthesia was provided by a continuous IV infusion of Saffan at a dosage of 0.14 mg/kg/min in *E. patas.* In these animals, the heart and respiratory rates dropped by as much as 62 and 40%, respectively, in animals breathing air and by 25% in those breathing oxygen (Dhiri, 1985). The use of alphaxolone–alphadolone in *P. hamadryas* via a continuous IV infusion was used for a minor surgical procedure in which anesthesia was maintained for up to 6 hr without cumulative effects (Cookson and Mills, 1983). The anesthetic infusion was given at a rate of 0.2 to 0.25 mg/kg/min in animals given an initial bolus of 4 mg/kg ketamine IM. It was found to be safe and reliable with a smooth and rapid induction. Muscular incoordination and twitching occurred during recovery, but was controlled by chlorpromazine (2.5 mg/kg IM) given twice, 1 hr apart.

C. Propofol

Propofol provides a smooth induction with adequate muscle relaxation sufficient for procedures of short duration. Because rapid clearance of propofol contributes to a relatively fast awakening, repeated boluses of 2–5 mg/kg IV can be administered to extend the duration of anesthesia without delaying recovery. A noticeable side effect is the occurrence of apnea following an induction dose. A slow induction will minimize this respiratory effect of the agent.

In *M. fascicularis,* the duration of anesthesia varied from about 5 to 40 min with propofol dosages of 2.5 to 10 mg/kg IV (Sainsbury *et al.,* 1991). At 2.5 mg/kg IV, given repetitively as needed, propofol provided adequate anesthesia for laparoscopy with good muscle relaxation and minimal effects on cardiovascular and respiratory functions. A dosage of 10 mg/kg produced moderate surgical anesthesia for 20 to 40 min. However, apnea for 1 to 2 min occurred at dosages exceeding 2.5 mg/kg IV. Recovery was uneventful and animals were standing 20 to 57 min after propofol administration.

In the authors' experience, propofol given at a dosage of 2–5 mg/kg provides smooth induction with adequate muscle relaxation in *M. Mulatta, M. fascicularis,* and *Papio* species sufficient for procedures of short duration (e.g., catheter placement, wound suturing, radiography) or as an induction agent for inhalational anesthesia. The authors have also used a 0.4- to 0.6-mg/kg/min propofol infusion in *Papio* spp. to provide reliable sedation during 60- to 90-min magnetic resonance imaging (MRI) studies after induction with a bolus of propofol (2.5–3 mg/kg IV). During the MRI, the baboons were maintained on oxygen at 2–3 liters/min. Recovery occurred within 15 to 30 min after propofol was discontinued.

An effective dosage in *P. troglodytes* is 1–2 mg/kg as an initial bolus followed by a propofol infusion at 0.13–0.2 mg/kg/min. This provides anesthesia for procedures of short duration and a recovery that is smooth and rapid (B. Swenson, personal communication, 1995).

Bennett *et al.* (1995) described postoperative bacterial and fungal infections in humans due to the extrinsic contamination of propofol associated with lapses in aseptic technique by anesthesia personnel. Unlike other intravenous agents, propofol is formulated in an emulsion of soybean oil, glycerol, and egg lecithin which readily supports bacterial growth. Because refrigeration is not recommended by the manufacturer, multiuse of propofol vials or syringes between patients often will allow sufficient time and a suitable environment for microbial replication. Accordingly, it is important that sterile techniques be practiced when this agent is used in nonhuman primates.

D. Barbiturates

1. Pentobarbital

Prior to the general use of inhalant anesthetics in veterinary medicine, pentobarbital was almost exclusively used to anesthetize nonhuman primates. The major difficulties with pentobarbital are a severe respiratory depression at high dosages necessary for adequate analgesia, an inability to modulate the depth of anesthesia well, and a very long period of recovery from anesthesia.

The usual dosage in nonhuman primates for pentobarbital, without the use of other agents, is 20–30 mg/kg IV; however, there may be considerable variation between animals. Usually, animals have been chemically restrained with ketamine prior to the administration of pentobarbital, and this may reduce by about one-third to one-half the needed dosage of pentobarbital to effect surgical anesthesia. Pentobarbital anesthesia is induced by delivering approximately one-half the calculated dosage as a bolus and then administrating additional amounts to effect. The duration of surgical anesthesia will vary between 30 and 60 min. Despite the advantages of inhalant anesthetics, there are applications for which pentobarbital may be useful. For example, because pentobarbital induces minimal changes in cerebrospinal fluid pressure and decreases cerebral blood flow and metabolic rate, it is useful in some neurosurgical procedures (Branston *et al.,* 1979; Dormehl *et al.,* 1992). This agent may also be appropriate for long-term, nonsurvival procedures that do not require deep anesthesia with concomitant analgesia.

The effect of pentobarbital anesthesia on respiratory and heart rates and on body temperature was studied in adult *M. fascicularis* that were anesthetized for 3 to 9 hr during neurosurgical studies (Zola-Morgan and Micheletti, 1986). After ketamine induction, the animals were given an initial bolus of pentobarbital (11 mg/kg IV). This was followed by additional periodic boluses to effect throughout the procedures. There was a wide range of effective initial pentobarbital dosages among the animals, and it was noted that the mean dosage was only about one-half of the usually described pentobarbital dosage. The mean recovery period was more than 3 hr. The authors of this report suggest that *M. fascicularis* may be more sensitive to pentobarbital than *M. mulatta*; however, we have not observed significant differences in pentobarbital dosages between these macaque species.

2. Thiopental

This ultra-short-acting barbiturate is commonly used to facilitate intubation prior to the induction of anesthesia with inhalation agents at 10–15 mg/kg IV with one-half given as a bolus and the rest to effect; the dosage is less (5–7 mg/kg) if the animal has received ketamine.

Slow infusion of thiopental at 15–17 mg/kg/hr has been reported to provide satisfactory chemical restraint and stable physiological baseline values for 90 min in *P. ursinus* with animals recovering within 20 min after discontinuation of the infusion (Goosen *et al.,* 1984). This rate of infusion resulted in a lower heart rate, a slower respiration, and a moderate decrease in body temperature. Of interest is that the mean arterial blood pressure did not differ from the values obtained from awake animals. The administration of thiopental typically produces a modest reduction in blood pressure due to peripheral vasodilation, reflecting barbiturate-induced depression of the vasomotor center and a decreased sympathetic nervous system outflow (Stoelting and Miller, 1994a). The unaffected blood pressures observed in these baboons may be partially explained by the fact that they were given a slow infusion rather than being induced with a bolus of thiopental.

E. Opioids

Opioids have been used in nonhuman primates to reduce the requirement for inhalational anesthetic and to enhance intraoperative analgesia during balanced anesthesia. In humans, their use as a sole anesthetic remains somewhat controversial based on the reports of awareness during high-dose narcotic anesthesia. The main advantage of high-dose opioid anesthesia is that it does not seriously impair cardiovascular function, and it effectively suppresses sympathetic nervous system responses to surgical stress (Sofianos *et al.,* 1985). Among the opioids available, morphine and oxymorphone are commonly used in nonhuman primates for postoperative analgesia, and fentanyl is most often used as the analgesic component for balanced anesthesia.

In the authors' experience, fentanyl is a valuable supplement to inhalational anesthesia for cardiovascular procedures in Old World primates. It can be administered either as an intravenous bolus (5 to 10 μg/kg) or as a continuous infusion (10 to 25 μg/kg/hr) in combination with a low inspired concentration of isoflurane with minimal effects on heart rate and blood pressure.

The reduction in blood pressure and bradycardia associated with opioid infusion is usually dose dependent. In *M. mulatta* (Nussmeier *et al.,* 1991), fentanyl produced minimal changes in mean arterial pressure, cardiac output, pulmonary and central venous pressures, and systemic vascular resistance in dosages not exceeding 16 μg/kg IV. A 30% decrease in mean arterial blood pressure was evident only after a fentanyl bolus of 64 μg/kg IV.

Increasing the doses of morphine up to 1 mg/kg IV in *S. sciureus* resulted in a maximum decrease in mean arterial blood pressure of 15 mm Hg from baseline values (Dourish *et al.,* 1990).

As with humans, respiratory depression is detectable in nonhuman primates after relatively small doses of narcotic analgesics. Morphine, given at dosages of 0.125 to 1.0 mg/kg IV to *S. sciureus,* resulted in a dose-dependent reduction in respiratory rate, arterial pO_2, and elevation in arterial pCO_2 (Dourish *et al.,* 1990).

A significant decrease in respiratory rates was observed in *M. mulatta* after a 2-μg/kg fentanyl dosage, whereas $PaCO_2$ increased significantly after 4 μg/kg and apnea occurred after a 64-μg/kg IV fentanyl bolus (Nussmeier *et al.,* 1991). Reversal of opioid-induced respiratory depression is achieved with naloxone (0.01 mg/kg IV). Naloxone should be administered in nonhuman primates before extubation to counteract any elevation of $PaCO_2$ induced by opioids.

The anesthetic, analgesic, and respiratory effects of one of the most potent opioids, carfentanyl, were evaluated in *M. mulatta* (Port *et al.,* 1984). At 0.3 μg/kg IV, carfentanyl produced a moderate degree of sedation and analgesia with mild respiratory depression. Anesthesia and severe respiratory depression requiring reversal occurred at dosages of 0.6 to 1.0 μg/kg.

Neuroleptanalgesic combinations such as fentanyl–droperidol (Innovar-Vet) at dosages of 0.1 to 0.3 ml/kg IM and fentanyl–fluanisone (Hypnorm) at a dosage of 0.3 ml/kg IM have been used in nonhuman primates for performing minor procedures (Green, 1982; Martin *et al.,* 1972). D. Elmore (personal communication, 1995) uses Innovar-Vet (0.1 ml/kg) IM in macaques for minor procedures. Fentanyl–fluanisone and fentanyl–droperidol at 0.3 ml/kg given IM or SC provide sedation in *S. sciureus* (T. Morris, personal communication, 1995). Good analgesic effects have been reported with their use, and specific narcotic antagonists (i.e., naloxone, nalorphine) are effective if needed. Unlike dogs, nonhuman primates do not become transiently aroused by auditory stimuli after the administration of Innovar-Vet. The major disadvantages of neuroleptanalgesic combinations are respiratory depression, bradycardia, and relatively poor muscle relaxation (Green, 1981). Accordingly, atropine should be included with neuroleptanalgesics to prevent opioid-induced bradycardia.

VI. MUSCLE RELAXANTS

Muscle relaxants are not anesthetics or analgesics and should only be used in fully anesthetized animals. Used as an adjunct in adequately anesthetized nonhuman primates, they can facilitate mechanical ventilation and reduce skeletal muscle tone during surgery. The authors use pancuronium (0.08–0.1 mg/kg IV) and vecuronium (0.04–0.06 mg/kg IV) to produce muscle relaxation during maintenance of inhalational anesthesia. Pancuronium is a long-acting muscle relaxant. It produces a moderate elevation in heart rate and blood pressure; these effects have been found to be desirable in hypovolemic and bradycardic nonhuman primates. Vecuronium, however, has no hemodynamic effects and is a viable alternative to pancuronium in cardiovascular procedures.

The interaction of ketamine with commonly used nondepolarizing neuromuscular relaxants has been evaluated in adult *M. cyclopis* monkeys anesthetized with 0.5 to 1% halothane (Tsai and Lee, 1989). Intravenous infusion of *d*-tubocurarine, atracurium, vecuronium, or pancuronium was done to provide a steady depression of the thumb twitch for 15 min at 50% of baseline. The effect of ketamine on thumb twitch was done at three dosage levels: 2, 5, and 10 mg/kg IV. In a dose-dependent manner, except for pancuronium at 2 mg/kg, ketamine significantly increased the 50% depression of the thumb twitch values for each of the four muscle relaxants. The maximum potentiation induced by ketamine occurred within 5–15 min, and the average duration of effect associated with pancuronium was 1 hr; shorter durations occurred with the other three agents. Ketamine, in the absence of the four neuromuscular relaxants, had no neuromuscular blocking effects. It has also been shown that ketamine potentiates the Phase I and Phase II neuromuscular blocks of succinylcholine in *M. cyclopis* monkeys (Tsai *et al.,* 1989).

VII. INHALATIONAL ANESTHESIA

A. Induction

The induction of anesthesia in most nonhuman primate species is usually initiated with ketamine (5–10 mg/kg IM) to provide chemical restraint and intravenous access. Thiopental is commonly used as an induction agent for inhalational anesthesia. This is usually given at a dosage of 3–5 mg/kg IV as a bolus, followed by additional amounts to effect. As an alternative to thiopental, the authors have found that propofol given at 2–5 mg/kg IV as a bolus followed by additional amounts to effect provides a very satisfactory and smooth induction in macaques and *Papio* spp. Once induced, the animal can be easily intubated and the inhalational agent administered.

However, intubation of small species and pediatric animals may pose a certain challenge. *Callithrix jacchus* can be intubated using a modified No. 8 French feeding catheter with an endotracheal tube adaptor end (J. Wyatt, personal communication, 1995). To pass the tube, the neck of the animal is hyperextended and digital pressure is applied on the outside of the larynx to aid in opening the larynx using a No. 1 Miller blade. In *Galago* spp., a 3.0-mm endotracheal tube, either cuffed or uncuffed, can be inserted with the aid of a laryngoscope (T. Blankenship, personal communication, 1995). Intubation of pediatric *Macaca* and *Papio* spp. is discussed in Section X,C.

B. Nitrous Oxide

Unlike the potent volatile anesthetics, nitrous oxide has a relatively high minimum alveolar concentration (MAC), 200% in macaques versus 105% in humans, preventing its use as a complete surgical anesthetic. Nitrous oxide is sometimes used in combination with other volatile anesthetics such as isoflurane and halothane because it allows for a lower inhaled concentration of either agent. Tinker *et al.* (1977) demonstrated that the MAC of halothane in *M. fascicularis* was reduced from the mean of 1.15 to 0.75% when halothane was supplemented with 30% nitrous oxide. Similarly, the MAC concentration of enflurane (1.84%) was decreased to 1.46% when supplemented with 30% nitrous oxide. The principal advantage of nitrous oxide as an adjunct to other potent volatile anesthetics is a less pronounced circulatory depression, which may be seen with the sole administration of isoflurane or halothane. The minimal cardiovascular effects produced by nitrous oxide can be attributed to its stimulatory effects on the sympathetic nervous system, characterized by slight elevations in heart rate and systemic and pulmonary arterial pressures. In humans, this sympathomimetic effect of nitrous oxide is most evident when it is used together with halothane and to a lesser degree with isoflurane (Stoelting and Miller, 1994b).

The authors do not routinely use nitrous oxide in nonhuman primates; however, when it is employed, 50–70% inhaled nitrous oxide and 30–50% oxygen are used with the remaining anesthetic requirement being provided by either 0.8–1.25% isoflurane or 0.5–1.0% halothane in *Papio* spp. and macaques. McCully *et al.* (1990) reported a similar anesthetic regimen in *M. mulatta* using 0.5 to 1.0% halothane plus a 1:1 ratio of nitrous oxide to oxygen at a total flow rate of 4 liters/min. Nitrous oxide produces a second gas effect when administered concomitantly with volatile anesthetics and as such accelerates induction. Induction in nonhuman primates can be augmented by allowing the animal to spontaneously breathe 1.5–3.0% halothane supplemented with 50–60% inspired nitrous oxide. The drawback to this technique is the increased risk of personnel being exposed to waste anesthetic gases. Because of the rapid elimination of large volumes of nitrous oxide from the blood into alveoli with a consequent diminution in alveolar oxygen concentrations, diffusion hypoxia may occur if an animal is allowed to breathe air at the conclusion of anesthesia. The authors administer 100% oxygen for at least 5 min at the end of anesthetic administration to prevent this occurrence.

C. Halothane

Halothane has been widely used in nonhuman primates for several decades and has proven to be, in most applications, a relatively safe and effective inhalational agent. The MAC of halothane in various species of primates has been determined (Tinker *et al.,* 1977). There seems to be some variance in MAC of halothane even between closely related species of nonhuman primates. For example, in *M. arctoides,* the MAC of halothane is 0.89% (Steffey *et al.,* 1974a) but in *M. fascicularis* it is 1.15% (Tinker *et al.,* 1977).

Cardiovascular effects of halothane have been well documented in nonhuman primates. The noticeable effect of halothane on the cardiovascular system is a decline in systemic blood pressure and heart rate proportional to the depth of anesthesia. Ritzman *et al.* (1976) reported that increasing halothane from 1 to 1.5 MAC resulted both in a significant decrease in blood pressure and in a reduction of heart rate in *M. mulatta* (Table II). In a study comparing hemodynamic effects in *M. arctoides* during spontaneous and controlled ventilation at different MAC of halothane, Steffey *et al.* (1974b) confirmed and extended these findings. In that study, cardiovascular function progressively decreased as inhaled concentrations of halothane were increased with both types of ventilation. Mean arterial

TABLE II

Effect of Halothane MAC on Cardiovascular Parameters in *M. mulatta*[a]

MAC[b]	CVP[c]	MAP[d]	HR[e]
0	1.84 ± 0.8	98.5 ± 6.3	195 ± 10.7
1	3.13 ± 1.4	55.9 ± 5.8	152.5 ± 10.7
1.5	4.19 ± 1.6	26.1 ± 6.6	117 ± 10.2

[a]Modified from Ritzman *et al.* (1976).
[b]Mean alveolar concentration.
[c]Central venous pressure.
[d]Mean arterial pressure.
[e]Heart rate.

pressure declined 40–50% and cardiac output was reduced 20–60% when halothane was increased from 1 to 2 MAC. This dose-dependent cardiovascular depression seems to be greater in macaques than in other nonhuman primate species and is due to the direct effect of halothane on myocardial contractility (Steffey *et al.,* 1974b). Substitution of nitrous oxide for equipotent levels of halothane tends to blunt this response.

Halothane sensitizes the heart to the dysrhythmogenic effects of epinephrine. The dysrhythmogenic effect of halothane is not dose dependent and there is no association between initiation of dysrhythmia and type or duration of surgical manipulations. Care should be taken to avoid hypoventilation which may result in hypercarbia and increase the occurrence of premature ventricular contractions (Sanders *et al.,* 1991).

Halothane usually produces rapid and shallow breathing. Because the increase in the rate of breathing is countered by a greater reduction in tidal volume, alveolar ventilation decreases and the arterial carbon dioxide concentration is elevated. A progressive respiratory acidosis occurs which may lead to higher serum potassium levels (Goosen *et al.,* 1984). Providing ventilatory support and lowering inspired concentration of halothane by supplementing it with nitrous oxide will counteract these effects.

Halothane causes a dose-dependent increase in cerebral blood flow (CBF) and cerebral vasodilation. In *M. nemestrina,* the autoregulation of cerebral blood flow begins to fail at a 1.0% end-tidal halothane concentration with complete loss occurring at 2.0% (Morita *et al.,* 1977). The administration of 2.0% halothane was associated with a maximum increase in cerebral blood flow by 97% from the baseline value. These authors noted that impaired CBF autoregulation renders the brain highly sensitive to changes in cerebral perfusion pressure. Because this may also have a potentially adverse effect on intracranial pressure, halothane should be used with caution for neurosurgical procedures.

Eng *et al.* (1975) investigated the effects of halothane anesthesia on maternal and fetal hemodynamics in pregnant *M. mulatta.* As expected at 1.5% halothane, the maternal blood pressure and cardiac output were reduced with a consequential reduction in uterine blood flow. A decrease in uterine blood flow and the direct effect of halothane-induced depression on the fetal cardiovascular system resulted in a 20% reduction in both fetal blood pressure and heart rate. There were pronounced fetal acidosis and hypoxia in some fetuses. Similar results were reported in pregnant *M. mulatta* in which serious hypotension frequently occurred (Sanders *et al.,* 1991). These data, as well as the authors' personal observations with macaques and *Papio* spp., suggest that systemic blood pressure monitoring should be instituted during experimental procedures in pregnant nonhuman primates. Careful attention to the depth of anesthesia, positioning of the animal, and intraoperative fluid management will minimize maternal and fetal cardiovascular depression.

Renal perfusion and glomerular filtration may be decreased secondary to a decline in blood pressure and cardiac output produced by halothane. This may affect radiorenography studies due to the delay in excretion of various isotopes and contrast materials (Dormehl *et al.,* 1984).

The controversy concerning the role of halothane in the production of hepatic necrosis was investigated in *E. patas* (Fleming and Bearcroft, 1966). They were able to show that despite multiple and frequent, albeit relatively short-duration exposures, halothane was not a specific hepatotoxin in this species.

D. Isoflurane

Isoflurane is a widely used inhalational agent in nonhuman primates and is the anesthetic of choice of the authors for most applications. An advantage of isoflurane over halothane is that it is minimally metabolized due to its chemical stability and low solubility, and accordingly, it is exhaled essentially unchanged (see Chapter 2). This characteristic makes it a particularly safe agent if the nonhuman primate has hepatic or renal deficits, or if the research objective may be compromised by an agent that has potentially toxic metabolites. Cook and Clarke (1985) used isoflurane in a *Gorilla gorilla* that had cholestatic jaundice and found that all hepatic function data remained stable throughout the period of anesthesia.

Isoflurane, unlike halothane, does not sensitize the myocardium to the arrhythmogenic properties of circulating catecholamines. Heart rhythm remains stable with isoflurane, and based on the authors' experience, ectopic cardiac beats are very uncommon in normocapnic nonhuman primates. Isoflurane produces minimal depression of cardiac output, but there is a dose-dependent decrease in blood pressure due to reductions in systemic vascular resistance. The hypotension is especially pronounced if isoflurane is mask-induced at inspired concentrations of 3–4% or when animals are maintained at or above 2%. Inclusion of fentanyl attenuates the decrease in blood pressure associated with isoflurane. In young animals, isoflurane-induced hypotension most likely reflects unrecognized hypovolemia.

Isoflurane has a MAC of 1.28% in *M. fascicularis* (Tinker *et al.,* 1977). In the authors' experience, maintaining nonhuman primates at about 1.3 MAC (1.6–1.75%) provides satisfactory anesthesia for most surgical applications. Alternatively, balanced anesthesia consisting of 0.5–0.75% isoflurane and fentanyl titrated to achieve an appropriate depth of anesthesia may be used in nonhuman primates. This inhalational and opioid combination is useful when cardiovascular stability is needed for a procedure. Although the authors are unaware of controlled studies done to determine the effects of various concentrations of nitrous oxide on the MAC of isoflurane in nonhuman primates, we would anticipate that the MAC of isoflurane would be about 1.0% with 60–70% nitrous oxide.

In *Galago* spp., ketamine (10–15 mg/kg IM) may be used for sedation to allow for placement of a mask for the delivery of isoflurane, initially at 3.0% and then reduced to about 1.5% for maintenance (T. Blankenship, personal communication, 1995).

In *C. jacchus,* 1 to 3% isoflurane at 3 liters/min with oxygen can be used for maintenance anesthesia after induction with ketamine (15 mg/kg) and diazepam (1.0 mg/kg) given subcutaneously (J. Wyatt, personal communication, 1995).

E. Methoxyflurane

Although methoxyflurane is a potent inhalational anesthetic and provides good muscle relaxation and analgesia, it is seldom used in nonhuman primates due to its low vapor pressure at room temperatures and its high solubility. These characteristics significantly increase induction and recovery times relative to those obtainable from either isoflurane or halothane. Hughes and Lang (1972) reported that induction and recovery times were significantly longer with methoxyflurane than with halothane in two species of macaques and *Cebus albifrons.* The authors also reported the occurrence of vomiting with methoxyflurane in *M. arctoides.* Another consideration that limits the use of this agent is its potential nephrotoxicity so exposure time should be limited.

F. Enflurane

This inhalational agent, based on the authors' personal knowledge and the lack of references in the literature, appears to be used infrequently in nonhuman primates. Enflurane, like halothane, produces dose-dependent depression of myocardial contractility. This results in a decrease of arterial blood pressure and cardiac output, and an increase in central venous pressure. At equal MAC levels, both enflurane and halothane produced similar effects on measured cardiovascular parameters in *M. mulatta* (Ritzman *et al.,* 1976). However, in comparison to halothane, enflurane is less likely to sensitize the myocardium to the effects of epinephrine and to elicit premature ventricular contractions. In humans, enflurane has been shown to produce electroencephalographic patterns resembling epileptiform activity (Liu, 1992). This effect is exacerbated by high anesthetic concentrations and hypocapnia. Although tonic–clonic convulsions have been observed in humans with enflurane anesthesia, no adverse neurological sequelae have been reported in the postanesthesia period (Liu, 1992).

VIII. INTRAOPERATIVE MONITORING

Intraoperative monitoring provides the means to assess physiological function during anesthesia and to ascertain the proper functioning of anesthetic equipment. It allows prompt recognition of adverse reactions and improves the effectiveness of therapeutic interventions. However, the use of monitoring equipment is intended to enhance but not substitute for the awareness on the part of the anesthesiologist.

A. Cardiovascular

In larger nonhuman primates insertion of an esophageal stethoscope during the intraoperative period facilitates the early detection of changes in heart rate and rhythm, and provides information on adequacy of ventilation. Usually, heart rhythm and rate are best evaluated by displaying the electrocardiogram of the animal on an oscilloscope. Lead II is commonly used for the detection of cardiac dysrhythmia because it produces an easily recognizable P wave and depicts the relationship of atrial to ventricular depolarization. The authors find that an intravenous bolus of lidocaine (1.0–2.0 mg/kg) followed by lidocaine infusion (20–50 μg/kg/min) is helpful in suppressing recurring premature ventricular contractions. Regardless of the type of cardiac dysrhythmia, the first priority must be directed toward correcting the underlying cause.

Indirect arterial blood pressure can be measured by placing a pediatric cuff on the lower arm or leg of the nonhuman primate. Utilizing an automated device allows for measurements as frequently as every minute which frees personnel for other responsibilities.

Direct arterial pressure monitoring is achieved by percutaneous cannulation or cutdown of the femoral artery in smaller nonhuman primates. To cannulate the femoral artery, an 18-gauge catheter is inserted in animals over 10 kg, whereas a 20-gauge Teflon catheter is used for smaller macaques. To avoid accidental displacement of the arterial catheter during the procedure, the femoral artery is routinely catheterized using the Seldinger technique. The catheter-over-the-needle (Angiocath) is used for percutaneous cannulation of the femoral artery. The skin at the cannulation site is pierced with the stylet of the catheter to facilitate advancement of the catheter and prevent a skin plug from occluding the stylet. After obtaining pulsatile blood flow from the cannulation site, a stylet is removed and a guide wire is placed through the catheter and into the vessel. The catheter used for monitoring is then threaded over the wire into the artery. Placement of a three-way stopcock between the catheter and the arterial tubing allows for frequent blood sampling to determine arterial blood gases. Direct arterial pressure monitoring is usually reserved for animals with cardiac and respiratory dysfunctions and for complex surgical procedures such as extracorporeal bypass that may produce vascular volume deficiencies.

Assessment of central venous pressure is useful in the management of fluid replacement. The external jugular vein is easily accessible and therefore is the most common site for cannulation. A 20-gauge polyurethane indwelling catheter is well suited for medium and larger size macaques and *Papio* spp. Following initial cannulation, a J wire is inserted through the catheter and advanced into the central vein. It is used as a guide for the placement of a longer venous catheter. Cannulation of the internal jugular vein usually requires a cut down.

Measurements of cardiac output and calculation of systemic and pulmonary vascular resistance can be obtained with a pulmonary artery catheter (Swan–Ganz). The Swan–Ganz catheter is essential for assessment of cardiac function, preload volume status, and responses to therapeutic interventions during major cardiovascular surgery. Placement of the pulmonary artery catheter into the right internal jugular vein represents the most direct approach to the right heart. Catheterization via the femoral vein is also commonly performed, but usually requires fluoroscopic guidance. A 5-French pediatric Swan–Ganz catheter can be successfully inserted into adult macaques (4–8 kg), whereas a 7.5-French pulmonary catheter is used in larger primates (e.g., 15- to 20-kg baboons). Udelsman *et al.* (1984) reported that *M. fascicularis* had cardiac output, systemic, and pulmonary vascular resistance values (adjusted for basal surface area) similar to those of humans. With few exceptions (e.g., lower pulmonary arterial pressure in *Papio* spp.), a close correlation of hemodynamic data between *Papio* spp. and humans has been observed.

B. Respiratory

Assessment of oxygenation and ventilation during the administration of anesthetics is essential. The color of the mucous membranes is the simplest method to ascertain adequacy of oxygenation. Quantitative pulmonary evaluation can be achieved by obtaining arterial blood for blood gas analysis or noninvasively with monitors such as pulse oximetry, end-tidal CO_2 monitors, or mass spectrometry. Pulse oximetry provides continuous measurement of arterial oxygen saturation. The authors use an infant digit oxygen transducer that can be attached to the tongue or ear of a nonhuman primate. Because the tongue site seems to be less influenced by intraoperative conditions such as hypothermia or hypotension, the authors find it preferable to an ear site.

It is well recognized that the measurement of expired CO_2 is the single most effective method of determining airway patency and adequacy of ventilation. End-tidal CO_2 monitors sample gas from the airway of an animal, usually at the site where the breathing circuit is connected to the endotracheal tube. In addition to displaying CO_2 levels, it also generates a CO_2 waveform called a capnogram. The shape of the capnogram provides useful information on a variety of conditions such as obstruction of the endotracheal tube and rebreathing of expired gases.

Airway pressure created by assisted or controlled ventilation is measured by a pressure manometer within the airway circuit of the anesthesia machine. Peak inspiratory pressures of 15–30 cm H_2O are usually sufficient to expand the lungs. High peak inspiratory pressures are usually indicative of a kinked endotracheal tube, endobronchial intubation, or an insufficiently opened ''pop-off'' valve. Any of these conditions may lead to barotrauma and eventually result in pneumothorax or subcutaneous emphysema.

C. Depth of Anesthesia

Anesthetic depth is usually assessed by monitoring a variety of parameters. Loss of palpebral and corneal reflexes, degree of muscle relaxation, rate and depth of breathing, and lack of somatic response to surgical stimuli are most commonly used to ascertain adequacy of anesthesia. If muscle relaxants are used during the maintenance of anesthesia, monitoring autonomic responses to surgical events must be done to assess the depth of anesthesia. Increases in heart rate and systolic blood pressure of 20% or more over baseline values can be interpreted as indications of an inadequate depth of anesthesia. Lacrimation and attempts to breathe out of synchrony with a ventilator might also indicate inadequate anesthesia.

D. Body Temperature

It is not uncommon for a nonhuman primate to lose a significant amount of body heat during anesthesia and surgery. Very young and small animals are particularly at risk of becoming hypothermic. Severe hypothermia can result in a decrease of cardiac output and bradycardia. Hypothermia may also delay awakening and cause postoperative shivering with a concomitant increase in oxygen consumption.

Intraoperative monitoring of body temperature is routinely performed by inserting a temperature probe into the esophagus or rectum and connecting it to the monitor. The drawback of rectal temperature is that it does not accurately reflect blood or ''core'' temperature when changes in temperature are very rapid. Deep hypothermia induced during cardiopulmonary bypass surgery is an example that calls for monitoring core temperature by an esophageal probe inserted into the lower third of the esophagus. A tympanic probe placed in the external auditory canal will reflect brain temperature by measuring the temperature of blood perfusing the brain and is useful during cerebral cooling procedures.

E. Urinary Output

Urinary output is a direct indicator of renal function and serves as a useful guide of intravascular volume status. In female macaques and *Papio* spp., bladder catheterization is usually performed by inserting a soft rubber Foley catheter through the urethral meatus, located between two papillary folds posterior to the clitoris, into the bladder and then connecting it into a urinary drainage bag.

IX. INTRAOPERATIVE SUPPORT

Infusion of crystalloid solutions during surgery helps to maintain normovolemia necessary for adequate tissue perfusion. Intraoperative fluid and blood loss replacement can usually be predicted by the type of procedure and duration of surgery. For minor procedures in nonhuman primates, administration of isotonic electrolyte solutions such as lactated Ringer's solution at the rate of 5–10 ml/kg/hr is sufficient to maintain normal fluid composition. During abdominal surgery, infusion of lactated Ringer's solution at the rate of 15–20 ml/kg/hr is needed to compensate for third space loss. Because of their small intravascular volumes, young and small primates are at increased risk of some electrolyte disturbances. As a guideline for the maintenance of fluids for pediatric macaques and *Papio* species, the authors routinely administer 2.5% dextrose in 0.45% saline at the rate of 5 ml/kg/hr.

A urinary catheter should be used to evaluate the adequacy of fluid therapy whenever surgery is prolonged or if significant changes in intravascular volume are anticipated. Minor to moderate blood loss can be replaced with crystalloid solutions given in the amounts equal to about three times the amount of blood loss. Monitoring blood pressure and urinary output allows one to determine whether intravascular fluid volume is being adequately replaced. Significant intraoperative blood loss warrants serial determination of the hematocrit of a nonhuman primate. If the hematocrit falls below 20%, blood is usually administered as part of the replacement solution. There is some disagreement on the necessity of cross-matching blood between the same species of nonhuman primates prior to transfusion (Socha *et al.,* 1984). Initial blood transfusions in macaques and *Papio* spp. do not normally cause any of the symptoms associated with an acute transfusion reaction because preformed isoantibodies to erythrocyte antigens are absent (Socha *et al.,* 1982). Autologous blood transfusions can be used in studies where clinically inapparent, but immunologically significant, incompatibilities may occur (e.g., organ transplantation). In such instances, 10–12% of the nonhuman primate's blood is withdrawn 10–14 days before surgery, and the blood is placed in a blood collection bag (with citrate–phosphate–dextrose as anticoagulant) that is stored at 1–6° C until the day of surgery.

Anesthesia in conjunction with surgery produces a complex and unstable physiological status. Understanding the effects of various inotropes and vasopressors is important for treating alterations of physiological functions during the intraoperative period. The authors consider phenylephrine as the drug of choice in treating isoflurane-induced hypotension. A 1- to 2-μg/kg bolus of phenylephrine, followed by an infusion of 0.5–1.0 μg/kg/min, will elevate blood pressure by increasing systemic vascular resistance. For macaques and *Papio* spp., when hypotension is accompanied by bradycardia, a 1.5- to 2.5-mg intravenous bolus of ephedrine, repeated if needed, will increase blood pressure and improve cardiac output. Cardiac dysrhythmias have not been observed in nonhuman primates anesthetized with isoflurane and treated with ephedrine. During cardiovascular collapse characterized by decreased cardiac output and markedly reduced blood pressure, dopamine and norepinephrine infusions can be used. Dopamine has the advantage over other sympathomimetic drugs due to its ability to maintain perfusion pressure and cardiac output without reducing renal blood flow.

X. SPECIAL ANESTHETIC CONSIDERATIONS

A. Cardiac; Cardiopulmonary Bypass

Our anesthetic management is directed toward preserving hemodynamic stability throughout the perioperative period. The authors separate anesthesia for cardiac transplantation into five periods: preincision, prebypass, cardiopulmonary bypass, postbypass, and immediate postsurgical. The overall goal is to adjust the depth of anesthesia for each period to help avoid wide fluctuations in blood pressure and heart rate. Because hemodynamic complications can occur during cardiac anesthesia, inotropes and vasopressors are part of the intraoperative management to stabilize circulatory function.

The preincision period is one of minimal surgical stimulation. For this reason, after ketamine (5 mg/kg, IM) and induction with either thiopental (5–7 mg/kg) or propofol (2–4 mg/kg) intravenously, macaques and *Papio* spp. are maintained on a low inspired concentration of isoflurane. Isoflurane mask induction is avoided due to pronounced vasodilatory effects of inhalational anesthetic. The authors routinely place two large-bore peripheral intravenous lines to maintain adequate intravascular volume and to administer emergency drugs. Animals are often volume-depleted from overnight fasting and may require fluid boluses. A urinary catheter is placed to monitor the effectiveness of fluid therapy. After percutaneous placement of a femoral arterial line, animals are given the muscle relaxant vecuronium (0.04 mg/kg IV) and, presurgically, are maintained on 0.8–1.2% isoflurane. This concentration of isoflurane is sufficient to produce loss of consciousness and avoid hemodynamic instability. It has been found that maintaining macaques and *Papio* spp. on isoflurane concentrations higher than 1.5–1.75%, in the absence of surgical stimuli, often leads to a reduction in blood pressure.

Because the period from incision to initiation of cardiopulmonary bypass (prebypass period) is characterized by intense surgical stimulation, 10 to 15 min prior to surgical incision fentanyl (5–10 μg/kg), followed by fentanyl infusion (10–25 μg/kg/hr), is typically administered while the nonhuman primate is maintained on 0.5–0.75% isoflurane. This combination produces minimal cardiac depression while maintaining the ability to change anesthetic concentration rapidly. Skin incision, sternotomy followed by sternal retraction, and aortic dissection are frequently associated with the greatest surgical stimulation and the animal should be observed closely for any signs of inadequate depth of anesthesia. The authors routinely use pancuronium (0.1 mg/kg) or vecuronium (0.05 mg/kg) intravenously to facilitate mechanical ventilation and reduce skeletal muscle tone. Because this abolishes somatic reflexes, we rely on autonomic responses to assess the depth of anesthesia. Any increase in blood pressure and heart rate greater than 20% from baseline values is treated as an indication of inadequate depth of anesthesia.

The hypotension that occurs during maintenance of anesthesia is treated with the intravenous infusion of fluids and the use of vasopressors. A bolus of phenylephrine (1–2 μg/kg) followed by phenylephrine infusion (0.5–1.0 μg/kg/min) is the drug of choice for intraoperative hypotension. Bradycardia and hypotension are occasionally observed during sternotomy and cardiac manipulation, and are best managed with incremental intravenous dosages of 2.5 mg/kg ephedrine. Reduction of blood pressure after sternotomy usually is a result of a decrease in venous return due to the equilibration of negative intrathoracic pressure with atmospheric pressure. The use of nitrous oxide is avoided because of its tendency to enlarge air bubbles that may form during cannulation and bypass. Prior to initiation of the cardiopulmonary bypass, 300 units/kg of heparin is given intravenously to prevent clot formation. The adequacy of anticoagulation is confirmed by maintaining the activated clotting time (ACT) above 400 sec.

After beginning cardiopulmonary bypass, the heart is arrested and cooled. This is a critical period and is often associated with an initial decline in mean arterial pressure, presumably reflecting hemodilution that results from the infusion of priming solutions. Any decrease in mean perfusion pressure below 40 mm Hg can be effectively treated with phenylephrine. Although anesthetic requirements are usually reduced during hypothermic cardiopulmonary bypass, blood pressure often begins to rise, reflecting activation of the sympathetic nervous system. Repeated intravenous fentanyl (1–2 μg/kg) and midazolam (0.2–0.4 mg/kg) boluses can be used to maintain perfusion pressures of 50–60 mm Hg. Midazolam is given slowly to avoid hypotension that may occur when it is given in combination with fentanyl.

Repeated blood gases are done to evaluate flow and perfusion. Low mixed venous O_2 saturation accompanied by metabolic acidosis in the absence of hypoxemia may indicate inadequate flow rates. After completion of surgical anastomoses and rewarming, efforts are directed toward restoring electrical activity to the heart. The authors routinely administer 1–2 mg/kg of lidocaine as an intravenous bolus followed by defibrillation. Lungs are gently inflated to facilitate pulmonary blood return into the left heart and to expel residual air emboli. Ventricular volume and contractility can be evaluated visually, and cardiac output, central venous, pulmonary arterial, and left atrial pressures are obtained with the aid of a Swan–Ganz catheter. Nonhuman primates with good ventricular function and filling pressures tend to rapidly develop good cardiac output and systemic blood pressures which allows separation from bypass. Inadequate cardiac output should prompt a search for explanations. Kinked grafts, air embolism, and myocardial ischemia due to inadequate myocardial protection may require surgical correction and pharmacological support. A dopamine infusion is usually required (3–5 μg/kg) to augment cardiac contractility and improve renal blood flow. Urinary output serves as a guide to the adequacy of renal perfusion, with an output of 1 ml/kg/hr being a reasonable expectation. Continuous intravenous infusions of isoproterenol (0.01–0.03 μg/kg/min), dobutamine (2–10 μg/kg/min), and norepinephrine (0.05–0.2 μg/kg/min) are often needed to maintain optimal cardiac output.

During the postbypass period, bleeding is controlled with a dose of protamine to reverse heparin activity. Protamine at the ratio of 1 mg to 80 units of heparin is routinely administered. The protamine must be administered slowly to avoid pronounced hypotension. If aortic cannula are still in place, the blood remaining in a bypass reservoir can be given directly or can be collected with a cell-saver device and infused intravenously. The authors try to achieve a postbypass hematocrit of at least 22–25%.

Following bypass, nonhuman primates are maintained on a low inspired concentration of isoflurane and supplemented with the infusion of fentanyl at 5–7 μg/kg/hr. In our experience, transplanted nonhuman primates do not tolerate surgical levels of isoflurane well.

Diuresis with furosemide (1–2 mg/kg IV) is usually required to reduce water and sodium retention, especially after hypothermic cardiopulmonary bypass. Lung ventilation with a positive end expiratory pressure of 5–10 cm is indicated to avoid atelectasis and pulmonary edema. Repeated arterial blood samples every 15–20 min are useful in recognizing and correcting any disturbances in the acid–base balance.

Before closure of the sternotomy incision, intercostal injections of 0.25% bupivicaine (1 mg/kg) are given to produce prolonged sensory analgesia. A chest tube is placed to evacuate intrathoracic air and fluids.

Postsurgically, nonhuman primates are maintained on a low inspired concentration of isoflurane (0.2–0.4%) for at least 2–3 hr to minimize movement while monitoring drainage from the chest tube and to achieve hemodynamic stability. In the authors' experience, a 1- to 4-μg/kg/min nitroprusside infusion may be used to control an excessively elevated blood pressure that could induce a cerebrovascular accident. Arterial blood gases are measured until the level of consciousness of the animal necessitates removal of the arterial catheter. Extubation is done only if the animal is hemodynamically stable and can maintain arterial oxygenation above 60 mm Hg and adequate alveolar ventilation (pCO_2 35–45 mm Hg) on room air. Before extubation, the oropharynx and trachea are thoroughly suctioned to prevent the obstruction of respiratory passages with secretions.

B. Obstetric Anesthesia

Physiological changes occurring during pregnancy can directly influence the anesthesia management for obstetric surgery. Increased plasma volume leading to dilutional anemia, increased alveolar ventilation, prolonged gastric emptying, and decreased requirements for inhaled anesthetics are the most common physiological alterations. After ketamine administration, thiopental (3–5 mg/kg) or propofol (2–4 mg/kg) IV can be used for induction. For maintenance of anesthesia in *Papio* spp., 0.8–1.25% isoflurane or 0.5–1.0% halothane supplemented with 50–60% inspired nitrous oxide is used. The advantage of including nitrous oxide with other volatile anesthetics is that there is a less pronounced circulatory depression. Eng *et al.* (1975) have shown that maintaining pregnant macaques at around 1.5% halothane reduced maternal blood pressure and cardiac output with a consequential reduction in uterine blood flow. This, in turn, decreased fetal perfusion, resulting in fetal acidosis and hypoxia. Similar findings were reported in *M. mulatta* by Sanders *et al.* (1991). These authors noted that maintaining the animals on an end-tidal halothane concentration of 1.1 ± 0.1% resulted in hypotension, particularly in animals not given fluids throughout the anesthetic period.

Because maternal hypotension is the most common complication encountered during anesthesia, blood pressure should be routinely monitored. Indirect blood pressure is monitored with the proper pediatric size cuff placed around the area of the tibial or radial artery. In some cases, even though a blood pressure reading may be normal with the cuff around the arm, an animal in a supine position may have decreased uterine and placental blood flow due to aortocaval compression by the gravid uterus. Tilting the pregnant animal to the left by elevating the right sacral area helps prevent supine hypotension. Maternal hypotension from any cause requires immediate volume restoration and, if needed, careful vasopressor therapy. Hydration can be maintained by intravenous infusion of lactated Ringer's solution at the rate of 10–20 ml/kg/hr. Preoperative placement of a urinary catheter allows for monitoring the effectiveness of fluid therapy. If hypotension persists, vasopressor therapy should be considered. Ephedrine, intravenously given in 1.25- to 2.5-mg increments, is the safest vasopressor for use during maternal hypotension. Because of its predominant β-adrenergic activity, ephedrine will maintain uterine arterial perfusion by increasing maternal cardiac output and restoring uterine blood flow without uterine arterial vasoconstriction (Levinson *et al.,* 1979).

Maternal normocapnea should be maintained during the intraoperative period. Extreme hyperventilation of the lungs should be avoided. In humans, hypocarbia has been shown to reduce uteroplacental blood flow (Stoelting and Miller, 1994a). Conversely, hypoventilation during halothane anesthesia has been associated with an increased incidence of premature ventricular contractions in *M. mulatta* (Sanders *et al.,* 1991). Sanders *et al.* (1991) resolved premature ventricular complexes by increasing alveolar ventilation and increasing the depth of anesthesia, measures that corrected probable hypercapnea, and reduced circulating catecholamines.

C. Pediatric Anesthesia

Successful pediatric anesthesia management requires an understanding of the physiologic and pharmacologic differences between adult and pediatric animals. Cardiac output is heart rate dependent in pediatric animals, as stroke volume is relatively fixed due to the noncompliant left ventricle in young nonhuman primates. The high heart rate makes ECG monitoring difficult with routine electrocardiograph units. Arterial blood pressure is lower than in adult nonhuman primates. Accurate measurement of blood pressure can be a problem if adult human blood pressure cuffs are used. The authors have used pediatric (size one) blood pressure cuffs for 1-kg nonhuman primates with intermittent success.

Therefore, based on physiologic differences, one of the main goals of preanesthesia and induction is the avoidance of heart rate reduction. This can be achieved by administering ketamine (5 mg/kg) and atropine (0.02 mg/kg) IM. Inhalational mask induction is also an accepted technique in pediatric anesthesia. But because neonatal primates are usually allowed to nurse until anesthesia time, assisted ventilation during mask induction should be avoided. Assisted ventilation may lead to the insuflation of the stomach with gases and place the neonate at risk of gastric distension, vomiting, and subsequent aspiration. For the same reason, the authors often induce neonatal primates with propofol (2–4 mg/kg) IV. The suckling reflex can be used as one of the indicators for depth of anesthesia in neonatal nonhuman primates. This is done by placing a gloved finger into the mouth of the neonate which should elicit sucking. The loss of this reflex serves as a reliable indicator of adequate depth of anesthesia.

Intubation of pediatric nonhuman primates can be done with the use of both uncuffed and cuffed endotracheal tubes. The authors use 2.5- to 3.5-mm-internal-diameter endotracheal tubes for neonatal and juvenile *Papio* and *Macaca* spp. Placing the animal in a sitting position with its head extended toward the anesthetist has been found to be useful during intubation. After the mouth is opened by depressing the mandible with a thumb, the straight laryngoscope blade is passed beneath the epiglottis. Applying slight pressure forward and downward on the blade depresses the epiglottis and facilitates exposure of the glottic opening. The endotracheal tube is advanced through the glottic opening with the distal end of the tube situated midway between the carina and vocal cords. The proper placement of the tube is confirmed by bilateral movement of the chest with manual compression of the reservoir bag or by the presence of bilateral lung sounds upon auscultation. The tube should be well secured to avoid accidental dislodging, especially during movement of the animal to the operating table. Postintubation laryngeal edema is unlikely if the size of the endotracheal tube in the trachea is such that a slightly audible air leak occurs around the tube when positive pressure of 15–20 mm Hg is applied.

Small tidal volumes in pediatric animals require the use of nonrebreathing circuits such as modified Jackson Rees and Bain systems. A fresh gas flow of 1.5–2.0 liter/min is adequate to prevent rebreathing during spontaneous ventilation. Because volatile anesthetics depress spontaneous ventilation, the authors use controlled or assisted ventilation for all but short procedures to prevent CO_2 accumulation.

ECG, indirect blood pressure, oxygen saturation, and body temperature should be routinely monitored in pediatric cases. The pulse oximetry transducer is best attached to the tongue or to the palm of the hand in the neonate. Pediatric animals have a larger surface area per kilogram of body weight than adults, resulting in an increased heat loss to the environment. This, together with an increased metabolic rate and lack of body fat, increases the potential for significant hypothermia in pediatric nonhuman primates. Using warming lights, thermal blankets, and warm intravenous fluids will minimize the severity of hypothermia.

The authors consider isoflurane to be the anesthetic of choice for pediatric animals. Even though it may increase the heart rate slightly, a stable rhythm is maintained. The hypotension that accompanies isoflurane anesthesia is not uncommon in pediatric macaques and *Papio* spp. The authors treat isoflurane-induced hypotension with phenylephrine, and in some cases it may be more effectively treated with a norepinephrine infusion (0.05–0.1 μg/kg/min).

As a guideline for maintenance fluids in pediatric primates, the authors use 4 ml/kg/hr of warm 5% dextrose or 2.5% dextrose in 0.45% NaCl. This will not correct third space intraoperative losses and blood loss. Because vasoconstrictive responses of neonates to hemorrhage are less than those of adults, even moderate blood loss will result in precipitous declines in blood pressure. Therefore, a crystaloid solution, equal to three times the blood loss, should be given immediately.

In general, neonatal *Papio* spp. and macaques recover more quickly from anesthesia and are less disturbed by minor complications than are adult nonhuman primates.

XI. POSTOPERATIVE CARE

Postoperative monitoring ensures prompt recognition of postsurgical complications and helps provide for an overall safe recovery from anesthesia. Physiological disorders most commonly encountered during postoperative recovery include pulmonary and circulatory complications, hypothermia, and pain.

A. Pulmonary Complications

It is not uncommon for nonhuman primates to vomit when they emerge from anesthesia. Extubation should be delayed until the animal regains the swallowing reflex or other signs of voluntary movement (e.g., head movement, chewing). If vomiting occurs after extubation, the animal should be placed in a prone position with its head lowered to avoid aspiration of the vomitus and the oropharynx suctioned.

Maintaining an arterial catheter until the nonhuman primate regains sufficient consciousness allows for the determination of arterial blood gases after the animal is spontaneously breathing room air. To avoid respiratory inadequacy during the postoperative period, the authors generally do not extubate the nonhuman primate if PaO_2 is less than 60 mm Hg and $PaCO_2$ is more than 50 mm Hg. *Papio* spp. and macaques initially seem to tolerate mild to moderate hypoxemia (PaO_2 55–60 mm Hg) relatively well; however, over a period of time this condition may progress into severe acidosis and circulatory depression.

Before extubation, the oropharynx and trachea are thoroughly suctioned to clear secretions and blood that may have accumulated during surgery. Suctioning of the trachea is particularly important for smaller nonhuman primates because respiratory passages can be easily obstructed with bronchial secretions. Attaching a soft rubber catheter to standard suction tubing allows for effective clearing of the trachea and pharynx.

B. Circulatory Complications

Animals may require careful clinical evaluation for inadequate surgical hemostasis. Sequential hematocrit determinations help provide a rapid and accurate evaluation of a bleeding animal. Animals receiving large amounts of heparin (e.g., during cardiopulmonary bypass) require evaluation of prothrombin time when a cardiopulmonary bypass is discontinued. Protamine is usually administered over a period of time to reverse heparin anticoagulation (see Section X,A on bypass).

Hypotension and tachycardia are common circulatory complications associated with the early postoperative period. Residual effects of anesthetics and inadequate replacement of blood lost during surgery are often causes for hypotension. Uncontrolled postoperative bleeding can result in hypovolemia and hypotension. Because bladder catheterization is a relatively simple technique, maintaining a urinary catheter in the early postoperative period helps in evaluating the adequacy of fluid and blood loss replacement.

Sinus tachycardia is a common occurrence in the postoperative period. Although specific treatment for tachycardia is not required, correction of the underlying cause (e.g., pain, arterial hypoxemia or hypovolemia) must be initiated.

C. Hypothermia

Postoperative hypothermia is frequently encountered during a surgical procedure. Exposure to a cool ambient temperature in the operating room and the use of unwarmed intravenous fluids may contribute to significant heat loss. Young animals are particularly at risk of developing severe hypothermia because of their large body surface area, which increases heat loss to the environment. Severe hypothermia may delay awakening by reducing metabolism and excretion of anesthetic drugs. It may also result in hemodynamic compromise manifested in a decrease of cardiac output and bradycardia. Hypothermia and associated shivering should be treated with warming lights or heating blankets to raise the body temperature. Space heaters may be used to preheat the recovery area.

D. Pain

Pain is a predictable response as the effect of anesthetics dissipates in the early postoperative period. Pain is an extraordinarily complex phenomenon, which is both difficult to assess (see Chapter 6) and equally difficult to treat in an effective and timely manner in uncooperative animals. The goal of postoperative pain management is, on the one hand, to decrease an animal's experience of a noxious stimuli and, on the other hand, to maintain normal physiological and cardiovascular stability. The drugs available to treat nonhuman primates can be broadly divided into two groups: nonsteroidal anti-inflammatory drugs (NSAID) and opioids. The selection of appropriate postoperative analgesics is partially dependent on the surgical site. Generally, thoracotomy, orthopedic, and upper abdominal surgeries are considered to be the most painful.

Aspirin is the best known compound in the NSAID group. Although the authors know of no studies that have explored an effective aspirin regimen in nonhuman primates, 325 mg is considered empirically by some to be an effective dose in Old World nonhuman primates (Rosenberg, 1991). It should be noted that aspirin is only a mild analgesic and as such may not be sufficient for most types of postsurgical pain. Moreover, oral formulations are usually impractical for use in the immediate postoperative period. For pain associated with minor surgeries, the authors routinely administer pediatric rectal aspirin suppositories (125 mg for a 5-kg primate) at the end of the procedure. The authors have also used the injectable NSAID, ketorolac (15–30 mg IM in macaques and *Papio* spp.), to provide analgesia for moderate postoperative pain. It offers certain advantages over opioids (e.g., lack of sedation and cardiopulmonary depression). Because ketorolac may prolong bleeding, it should not be used in nonhuman primates with acquired or natural coagulopathies.

The traditional management of pain usually consists of opioid administration given either intramuscularly or intravenously. The inclusion of opioids as part of an intraoperative anesthesia regimen may delay the need for immediate pain relief, but it should not be considered a substitute for further postoperative analgesia. Effective postoperative analgesia can be provided by 1–2 mg/kg morphine IM or SQ every 4 hr, 0.15 mg/kg (Old World primates) and 0.075 mg/kg (New World primates) oxymorphone IM every 4–6 hr, or 2–4 mg/kg meperidine IM every 2–4 hr. Oxymorphone is often considered to be the analgesic of choice in nonhuman primates because it provides effective postoperative pain relief without producing noticeable respiratory depression in both Old and New World monkeys (Rosenberg, 1991). Although morphine produces reliable postoperative analgesia, it causes a dose-dependent reduction in the respiratory rate and pO_2 and an increase in pCO_2 (Dourish *et al.,* 1990). It should be used with care, especially in New World primates.

In the authors' experience, meperidine (2 mg/kg IM) can safely and reliably produce postoperative analgesia in *M. mulatta* undergoing obstetric procedures. Nevertheless, some anectodal reports suggest that meperidine may cause sudden death in healthy nonhuman primates (Rosenberg, 1991).

For mild to moderate postoperative pain, oral codeine plus acetaminophen or other oral opioids are given at 4-hr intervals for 24 to 48 hr. Apes are reluctant to take hydrocodone, possibly because it causes nausea. Oxycodone appears to be more effective and better tolerated (B. Swenson, personal communication, 1995).

Opioid agonist–antagonists (buprenorphine and butorphanol) have been used to provide postoperative analgesia in nonhuman primates. Buprenorphine (0.01 mg/kg IM, IV) has been recommended for use in nonhuman primates (Flecknell, 1987). The long duration of effect (6–8 hr) and relative absence of respiratory depression make buprenorphine an attractive alternative to other analgesia regimens. It has been used to alleviate postoperative discomfort associated with laparotomy at a dosage of 0.01 mg/kg IV every 12 hr for a total of four doses in pregnant *M. mulatta* monkeys (Sanders *et al.,* 1991). Buprenorphine (0.015 mg/kg) provides effective analgesia for mild to moderate pain in *S. sciureus* (T. Morris, personal communication, 1995). In *Galago* spp., buprenorphine, at a dosage of 0.01 mg/kg IM or SC, provides satisfactory postoperative analgesia for up to 12 hr without excessive sedation (T. Blankenship, personal communication, 1995). In apes, a combination of buprenorphine (0.01–0.02 mg/kg) and flunixin meglumine (1 mg/kg IM or SQ) is used every 6 hr to provide postoperative pain relief (B. Swenson, personal communication, 1995). Butorphanol (0.02 mg/kg SC) has been administered for postoperative analgesia in *Sanguinus* spp. and *C. geoffroyi* (Wolff *et al.,* 1990). A potential disadvantage of all agonist–antagonist opioids is the ''ceiling effect'' when repeated doses do not induce further analgesia but instead produce marked respiratory depression.

In case of potential overdose with opiate analgesics, naloxone can be of invaluable assistance. When administered intravenously it effectively reverses respiratory depression induced by opioids. The usual dose of naloxone is 0.1–0.2 mg IV, repeated as needed. Naloxone given in the lower dosage range may reverse the respiratory depression without completely reversing analgesia. Because naloxone is a short-acting antagonist, a second dose may be necessary to avoid the return of respiratory depression.

Postoperative intramuscular administration of opiates results in variable absorption, especially in hypotensive and hypothermic animals, and this may reflect a considerable delay between administration of the drug and effective analgesia. The continuous intravenous infusion of opiates may be more effective, but it is impractical in uncooperative nonhuman primates. It also may carry a greater risk of respiratory depression, sedation, and hypotension, thus requiring close monitoring. The discovery of opiate receptors in the spinal cord has led to great interest in the use of opiates by the epidural route. The action of epidural opioids is mediated by opiate receptors located in the substantia gelatinosa in the dorsal horn of the spinal cord. Previous human studies have suggested that the administration of epidural opioids produces effective and safe postoperative analgesia with minimal sympathetic blockade and no sedation or respiratory depression (Shulman *et al.,* 1984; Rutberg *et al.,* 1984; Morgan, 1989).

Placement of an epidural catheter is a relatively simple technique in nonhuman primates such as macaques and *Papio* spp. The authors use an 18-gauge Tuohy epidural needle with a curved distal end at the lumbar 5–6 or 6–7 interspace. Once the epidural needle penetrates the intraspinous ligament, the characteristic ''pop'' sound is heard. A glass syringe with a freely movable plunger is attached and the needle is advanced into the epidural space. Entry into the epidural space is confirmed by the ease of depression of the plunger in the syringe. The nylon catheter is then inserted and the epidural needle is removed. The catheter is usually advanced 5–7 cm from the introduction site to ascertain the proper placement of the epidural catheter.

Morphine is one of the most commonly used opiate analgesics via the epidural route. One of the main advantages of epidural morphine is that it has low lipid solubility relative to other opioids, which makes systemic absorption less likely to occur from the epidural site. This suggests that more of the drug is available at the receptor sites in the spinal cord, resulting in prolonged analgesia. The authors have used epidural morphine (a single injection of 0.1 mg/kg) for pain relief after thoracotomy and hysterotomy in baboons with postoperative analgesia lasting up to 24 hr (Popilskis *et al.,* 1994).

Morphine (0.01 mg/kg) given intrathecally as a single dose has been reported to provide excellent postoperative analgesia for 18–24 hr in *P. troglodytes* with no concomitant clinically recognizable hypotension. One case of generalized pruritis was observed; in humans this has been attributed to possible histamine release (B. Swenson, personal communication, 1995).

XII. SUMMARY

In conclusion, Table III presents a summary of the drugs and dosages for nonhuman primates. Data are also presented on the route of administration and duration of effect.

TABLE III
Summary of Drugs and Dosages for Nonhuman Primates

Species and drugs	Dosage	Duration	Reference
Macaca			
Anticholinergics			
Atropine	0.02–0.05 mg/kg IM		
Glycopyrrolate	0.005–0.01 mg/kg IM		
Dissociatives			
Ketamine	5.0–20 mg/kg IM	15–30 min	
Ketamine +	7 mg/kg IM		Bankneider *et al.* (1978)
xylazine	0.6 mg/kg IM		
Ketamine +	10 mg/kg IM		Naccarato and Hunter (1979)
xylazine	0.25–2 mg/kg IM	45–138 min	
Ketamine +	15 mg/kg IM followed by		Jacobs *et al.* (1993)
midazolam	0.05–0.15 mg IV followed by ketamine infusion 12 mg/kg/hr		
Tiletamine + zolazepam	1.5–3.0 mg/kg IM		Booker *et al.* (1982)
(Telazol)	4.0–6.0 mg/kg IM	45–60 min	Cohen and Bree (1978)

TABLE III *(continued)*

SUMMARY OF DRUGS AND DOSAGES FOR NONHUMAN PRIMATES

Species and drugs	Dosage	Duration	Reference
Other injectable anesthetics			
Alphaxolone–alphadolone	120 mg/kg IM bolus		
(Saffan)	18 mg/kg IM followed by 6–12 mg/kg IV		Box and Ellis (1973)
Propofol	2.5–5.0 mg/kg IV bolus		Sainsbury *et al.* (1991)
	2.5–5.0 mg/kg IV followed by infusion 0.3–0.4 mg/kg/min		
Barbiturates			
Pentobarbitol	20–30 mg/kg IV	30–60 min	
	11 mg/kg IV preceded by ketamine 20 mg/kg IM		Zola-Morgan and Micheletti (1986)
Thiopental	5–7 mg/kg IV for induction		
Opioids			
Fentanyl	5–10 μg/kg IV bolus or as a continuous infusion at 10–25 μg/kg/hr in combination with low MAC isoflurane		
	16–64 μg/kg IV bolus		Nussmeier *et al.* (1991)
Neuroleptanalgesics			
Fentanyl–droperidol (Innovar-Vet)	0.1–0.3 ml/kg IM		Green (1982), D. Elmore (personal communication, 1995)
Fentanyl–fluanisone (Hypnorm)	0.3 ml/kg IM		Martin *et al.* (1972)
Muscle relaxants			
Pancuronium	0.08–0.1 mg/kg IV		
Vecuronium	0.04–0.06 mg/kg IV		
Inhalation anesthesia			
Nitrous oxide	Addition of 30% reduces MAC halothane from 1.15 to 0.75% and enflurane from 1.84 to 1.46%		Tinker *et al.* (1977)
Halothane	1 MAC = 0.89–1.15%		Tinker *et al.* (1977), Steffey *et al.* (1974a)
Isoflurane	1 MAC = 1.28%		Tinker *et al.* (1977)
Enflurane	1 MAC = 1.84%		Tinker *et al.* (1977)
Analgesics			
NSAIDs			
Aspirin	325 mg PO		
	125 mg/5 kg rectal suppositories		
Ketorolac	15–30 mg IM		
Opioids			
Morphine	1–2 mg/kg IM, SQ	4 hr	
Oxymorphone	0.15 mg/kg IM	4–6 hr	
Meperidine	2 mg/kg IM	4 hr	
Buprenorphine	0.01 mg/kg IM	6–8 hr	
Opioid antagonist			
Naloxone	0.1–0.2 mg as needed		
Papio			
Anticholinergics			
Atropine	0.02–0.05 mg/kg IM		
Glycopyrrolate	0.005–0.01 mg/kg IM		
Dissociatives			
Ketamine	5–10 mg/kg IM	15–30 min	
Ketamine + diazepam	10 mg/kg IM 0.2–0.35 mg/kg IM		Woolfson *et al.* (1980)
Ketamine + medetomidine	5 mg/kg IM 100 μg/kg IM		S. Capuano (personal communication, 1995)
Tiletamine + zolazepam (Telazol)	4.0–6.0 mg/kg IM	45–60 min	Cohen and Bree (1978)

TABLE III *(continued)*

Summary of Drugs and Dosages for Nonhuman Primates

Species and drugs	Dosage	Duration	Reference
Other injectable anesthetics			
Alphaxolone–alphadolone (Saffan)	0.2–0.25 mg/kg/min infusion preceded by ketamine 4 mg/kg IM	Up to 6 hr	Cookson and Mills (1983)
Propofol	2.0–4.0 mg/kg IV for induction, repeated as needed		
Barbiturates			
Thiopental	15–17 mg/kg/hr IV infusion		Goosen *et al.* (1984)
Opioids			
Fentanyl	5–10 μg/kg IV bolus or as a continuous infusion at 10–25 μg/kg/hr in combination with low MAC isoflurane		
Muscle relaxants			
Pancuronium	0.08–0.1 mg/kg IV		
Vecuronium	0.04–0.06 mg/kg IV		
Inhalation anesthesia			
Halothane	0.5–1.0% supplemented with 2:1 ratio of nitrous oxide to oxygen for anesthesia maintenance		
Isoflurane	1.5–2.0% on 100% oxygen		
	0.8–1.25% supplemented with 2:1 ratio of nitrous oxide to oxygen for anesthesia maintenance		
Analgesics			
NSAIDs			
Aspirin	325 mg PO		
	125 mg/5 kg rectal suppositories		
Ketarolac	15–30 mg IM		
Opioids			
Morphine	1–2 mg/kg IM, SQ	4 hr	
	0.15 mg/kg epidurally	Up to 24 hr	
Oxymorphone	0.15 mg/kg IM	4–6 hr	
Buprenorphine	0.01 mg/kg IM	6–8 hr	
Opioid antagonist			
Naloxone	0.1–0.2 mg as needed		
Saimiri squireus			
Dissociatives			
Ketamine	10–30 mg/kg IM	15–30 min	
Ketamine +	10–30 mg/kg IM		
xylazine	3.0 mg/kg IM	Up to 30 min	D. Elmore (personal communication, 1995)
Ketamine +	15–20 mg/kg IM		T. Morris (personal communication, 1995)
diazepam	1.0 mg/kg IM		
Tiletamine + zolazepam (Telazol)	10 mg/kg IM		T. Morris (personal communication, 1995)
Sedatives/tranquilizers			
Medetomidine	100 μg/kg IM or SQ		T. Morris (personal communication, 1995)
α_2-Antagonists			
Atipamezole	200 μg IV		T. Morris (personal communication, 1995)
Other injectable anesthetics			
Alphaxolone–alphadolone (Saffan)	11.5–15.5 mg/kg IM bolus	Up to 1 hr	Logdberg (1988)

TABLE III *(continued)*

SUMMARY OF DRUGS AND DOSAGES FOR NONHUMAN PRIMATES

Species and drugs	Dosage	Duration	Reference
Neuroleptanalgesics			
Fentanyl–droperidol (Innovar-Vet)	0.3 ml/kg IM or SQ		T. Morris (personal communication, 1995)
Fentanyl–fluanisone (Hypnorm)	0.3 ml/kg IM or SQ		T. Morris (personal communication, 1995)
Analgesics			
Opioids			
Morphine	1–2 mg/kg IM, SQ	4 hr	
Oxymorphone	0.075 mg/kg IM	4–6 hr	
Buprenorphine	0.015 mg/kg IM	6–8 hr	T. Morris (personal communication, 1995)
Opioid antagonist			
Naloxone	0.1–0.2 mg as needed		
Pan troglodytes			
Dissociatives			
Ketamine	5–20 mg/kg IM	15–30 min	
Ketamine +	15–20 mg/kg IM		
xylazine	1.0 mg/kg IM	Up to 25 min	April *et al.* (1982)
Tiletamine + zolazepam (Telazol)	3.0–5.0 mg/kg IM		
Sedatives/tranquilizers			
Medetomidine	30–60 μg/kg IM, preceded by 2.0–6.0 mg/kg ketamine IM	60 min	Lewis (1993), Jalanka and Roeken (1990)
α_2-Antagonists			
Atipamezole	150–300 μg/kg IV		Lewis (1993), Jalanka and Roeken (1990)
Other injectable anesthetics			
Propofol	1–2 mg/kg as initial bolus followed by infusion to effect		B. Swenson (personal communication, 1995)
Analgesics			
Opioids			
Morphine	0.01 mg/kg intrathecally	18–24 hr	
Buprenorphine +	0.01–0.02 mg/kg IM	6 hr	B. Swenson (personal communication, 1995)
flunixin meglumine	1.0 mg/kg IM or SQ		
Callithrix jacchus			
Anticholinergics			
Atropine	0.04 mg/kg SQ or IM		
Dissociatives			
Ketamine	15–20 mg/kg IM	15–30 min	
Ketamine +	15–22 mg/kg IM		
xylazine	1.0–1.5 mg/kg IM	Up to 30 min	D. Elmore (personal communication, 1995)
Ketamine +	15 mg/kg IM		Bankneider *et al.* (1978)
diazepam	1.0 mg/kg IM		
Tiletamine + zolazepam (Telazol)	5.0 mg/kg IM	15 min	J. Wyatt (personal communication, 1995)
Other injectable anesthetics			
Alphaxoline–alphadolone (Saffan)	15–19 mg/kg IM bolus	Up to 1 hr	Phillips and Grist (1975)
Inhalational anesthesia			
Isoflurane	1.0–3.0% for maintenance		
Analgesics			
Opioids			
Oxymorphone	0.075 mg/kg IM	4–6 hr	
Butorphanol	0.02 mg/kg SQ	6 hr	Wolff *et al.* (1990)
Opioid antagonist			
Naloxone	0.1–0.2 mg as needed		

REFERENCES

Adams, S. R., Muchmore, E., and Richardson, J. H. (1995a). Biosafety. Part B. Zoonoses, biohazards, and other health risks. *In* ''Nonhuman Primates in Biomedical Research: Biology and Management'' (B. T. Bennett, C. R. Abee, and R. Henrickson, eds.), pp. 391–412. Academic Press, San Diego, California.

Adams, S. R., Muchmore, E., and Richardson, J. H. (1995b). Biosafety. Part C. Model occupational health program for persons working with nonhuman primates. *In* ''Nonhuman Primates in Biomedical Research: Biology and Management'' (B. T. Bennett, C. R. Abee, and R. Henrickson, eds.), pp. 412–414. Academic Press, San Diego, California.

April, M., Tabor, E., and Gerety, R. J. (1982). Combination of ketamine and xylazine for effective anaesthesia of juvenile chimpanzees (*Pan troglodytes*). *Lab. Anim.* **16,** 116–117.

Banknieder, A. R., Phillips, J. M., Jackson, K. T., and Vinal, S. I., Jr. (1978). Comparison of ketamine with the combination of ketamine and xylazine for effective anesthesia in the rhesus monkey (*Macaca mulatta*). *Lab. Anim. Sci.* **28,** 742–745.

Barnstein, N. J., Gilfillan, R. S., Pace, N., and Rahlman, D. F. (1966). Chronic intravascular catheterization. A technique for implanting and maintaining arterial and venous catheters in laboratory primates. *J. Surg. Res.* **12,** 511–521.

Bennett, J. S., Gossett, K. A., McCarthy, M. P., and Simpson, E. D. (1992). Effects of ketamine hydrochloride on serum biochemical and hematologic variables in rhesus monkeys. (*Macaca mulatta*). *Vet. Clin. Pathol.* **21,** 15–18.

Bennett, S. N., McNeil, M. M., Bland, L. A., Arduino, M. J., Villarino, M. E., Perrotta, D. M., Burwen, D. R., Welbel, S. F., Pegues, D. A., Stroud, L., Zeitz, P. S., and Jarvis, W. R. (1995). Postoperative infections traced to contamination of an intravenous anesthetic, propofol. *N. Engl. J. Med.* **333,** 147–154.

Booker, J. L., Erickson, H. H., and Fitzpatrick, E. L. (1982). Cardiodynamics in the rhesus macaque during dissociative anesthesia. *Am. J. Vet. Res.* **43,** 671–676.

Box, P. G., and Ellis, K. R. (1973). Use of CT 1341 anaesthetic ('Saffan') in monkeys. *Lab. Anim.* **7,** 161–170.

Brady, A. G., and Koritnik, D. R. (1985). The effects of ketamine anesthesia on glucose clearance in African Green Monkeys. *J. Med. Primatol.* **14,** 99–107.

Branston, N. M., Hope, D. T., and Symon, L. (1979). Barbiturates in focal ischemia of primate cortex effects on blood flow distribution, evoked potentials and extracellular potassium. *Stroke* **10,** 647–653.

Castro, M. I., Rose, J., Green, W., Lehner, N., Peterson, D., and Taub, D. (1981). Ketamine-HCL as a suitable anesthetic for endocrine, metabolic, and cardiovascular studies in *Macaca fascicularis* monkeys. *Proc. Soc. Exp. Biol. Med.* **168,** 389–394.

Chen, G. M., and Weston, J. K. (1960). The analgesic and anesthetic effect of 1-(1-phenylcyclohexyl) piperidine HCl on the monkey. *Anesth. Analg.* **39,** 132–137.

Cohen, B. J., and Bree, M. M. (1978). Chemical and physical restraint of nonhuman primates. *J. Med. Primatol.* **7,** 193–201.

Cook, R. A., and Clarke, D. A. (1985). The use of isoflurane as a general anesthetic in the western lowland gorilla (*Gorilla g. gorilla*). *J. Zoo Anim. Med.* **16,** 122–124.

Cookson, J. H., and Mills, F. J. (1983). Continuous infusion anaesthesia in baboons with alphaxolone-alphadolone. *Lab. Anim.* **17,** 196–197.

DaRif, C. A., and Rush, H. G. (1983). Management of septicemia in rhesus monkeys with chronic indwelling venous catheters. *Lab. Anim. Sci.* **33,** 90–94.

Davy, C. W., Trennery, P. N., Edmunds, J. G., Altman, J. F. B., and Eichler, D. A. (1987). Local mycotoxicity of ketamine hydrochloride in the marmoset. *Lab. Anim.* **21,** 60–67.

Dhiri, A. K. (1985). Continuous intravenous infusion using Saffan (CT 1341) in the *Erythrocebus patas*. *Lab. Anim.* **19,** 159.

Dormehl, I. C., Jacobs, D. J., duPlessis, M., and Goosen, D. J. (1984). Standardization of anaesthesia for radiorenography with primates under laboratory conditions. *J. Med. Primatol.* **13,** 5–10.

Dormehl, I. C., Redelinghuys, F., Hugo, N., Oliver, D., and Pilloy, W. (1992). The baboon model under anesthesia for in vivo cerebral blood flow studies using single photon emission computed tomographic (SPECT) techniques. *J. Med. Primatol.* **21,** 270–274.

Dourish, C. T., O'Neill, M. F., Schaffer, L. W., Siegl, P. K. S., and Iversen, S. D. (1990). The cholecystokinin receptor antagonist devazepide enhances morphine-induced respiratory depression in the squirrel monkey. *J. Pharmacol. Exp. Ther.* **255,** 1158–1165.

Eng, M., Bonica, J. J., Alamatsu, T. J., Berges, P. U., Yuen, D., and Ueland, K. (1975). Maternal and fetal responses to halothane in pregnant monkeys. *Acta Anaesthesiol. Scand.* **19,** 154–158.

Flecknell, P. A. (1987). ''Laboratory Animal Anesthesia,'' p. 128. Academic Press, London.

Fleming, S. A., and Bearcroft, W. G. C. (1966). The effect of repeated administration of halothane on the livers of healthy monkeys. *Can. Anaesth. Soc. J.* **13,** 247–250.

Fowler, M. E. (1986). Restraint. *In* ''Zoo and Wild Animal Medicine'' (M. E. Fowler, ed.), 2nd ed., Chapter 6, pp. 44–47. Saunders, Philadelphia.

Goosen, D. J., Davies, J. H., Maree, M., and Dormehl, I. C. (1984). The influence of physical and chemical restraint on the physiology of the chacma baboon (*Papio ursinus*). *J. Med. Primatol.* **13,** 339–351.

Green, C. J. (1982). Order primata (nonhuman). *In* ''Animal Anaesthesia,'' Lab. Anim. Handb. 8, pp. 217–226. Laboratory Animals Ltd., London.

Green, C. J., Knight, J., Precious, S., and Simpkin, S. (1981). Ketamine alone and combined with diazepam or xylazine in laboratory animals: A 10 year experience. *Lab. Anim.* **15,** 163–170.

Hatch, R. C., Booth, N. H., Kitzman, J. V., Wallner, B. M., and Clark, J. D. (1983). Antagonism of ketamine anesthesia in cats by 4-aminopyridine and yohimbine. *Am. J. Vet. Res.* **44,** 417–423.

Holmes, G. P., Chapman, L. E., Stewart, J. A., Straus, S. E., Hilliard, J. K., Davenport, D. S., and the B Virus Working Group (1995). Guidelines for the prevention and treatment of B-virus infections in exposed persons. *Clin. Infect. Dis.* **20,** 421–439.

Hughes, H. C., Jr., and Lang, C. M. (1972). A comparison of halothane and methoxyflurane anesthesia in three species of nonhuman primates. *Lab. Anim. Sci.* **22,** 664–667.

Ives, M., and Dack, G. M. (1956). ''Alarm reaction'' and normal blood picture in *Macaca mulatta*. *J. Lab. Clin. Med.* **47,** 723–729.

Jacobs, B., Harris, G. C., Allada, V., Chugani, H. T., Pollack, D. B., and Raleigh, M. J. (1993). Midazolam as an effective intravenous adjuvant to prolonged ketamine sedation in young rhesus. (*Macaca mulatta*) and vervet (*Cercopithecus aethiops sabaeus*) monkeys: a preliminary report. *Am. J. Primatol.* **29,** 291–298.

Jalanka, H. H., and Roeken, B. O. (1990). The use of medetomidine, medetomidine-ketamine combinations, and atipamezole in nondomestic mammals: A review. *J. Zoo Wildl. Med.* **21,** 259–282.

Kaufman, P. L., and Hahnenberger, R. (1975). CI-744 anesthesia for ophthalmological examination and surgery in monkeys. *Invest. Opthalmol.* **14,** 788–791.

Klein, H. J., and Murray, K. A. (1995). Medical management. Part C. Restraint. *In* ''Nonhuman Primates in Biomedical Research: Biology and Management'' B. T. Bennett, C. R. Abee, and R. Henrickson, eds.), pp. 287–297. Academic Press, San Diego, California.

Levinson, G., Shnider, S. M., deLorimier, A. A., and Steffenson, J. L. (1974). Effects of maternal ventilation on uterine blood flow and fetal oxygenation and acid-base status. *Anesthesiology* **40,** 340–347.

Lewis, J. C. M. (1993). Medetomidine-ketamine anaesthesia in the chimpanzee (*Pan troglodytes*). *J. Vet. Anaesthesiol.* **20,** 18–20.

Liu, P. (1992). ''Principles and Procedures in Anesthesiology'' p. 169. Lippincott, Philadelphia.

Loeb, W. F., Cicmonec, J. L., and Wickum, M. (1976). A coagulopathy of the owl monkey (*Aotus trivirgatus*) associated with high antitrombin III activity. *Lab. Anim. Sci.* **26,** 1084–1086.

Logdberg, B. (1988). Alphaxolone-alphadolone for anesthesia in squirrel monkeys of different ages. *J. Med. Primatol.* **17,** 163–167.

Loomis, M. R., Henrickson, R. V., and Anderson, J. H. (1980). Effects of ketamine hydrochloride on the hemogram of rhesus monkeys (*Macaca mulatta*). *Lab. Anim. Sci.* **30,** 851–853.

Lynch, S., and Line, S. (1985). Failure of yohimbine to reverse ketamine anesthesia in rhesus monkeys. *Lab. Anim. Sci.* **35,** 417–418.

Martin, D. P., Darrow, C. C., Valerio, D. A., and Leiseca, S. A. (1972). Methods of anesthesia in nonhuman primates. *Lab. Anim. Sci.* **22,** 837–843.

McCully, C. L., Balis, F. M., Bacher, J., Phillips, J., and Poplack, D. G. (1990). A rhesus monkey model for continuous infusion of drugs into cerebrospinal fluid. *Lab. Anim. Sci.* **40,** 520–525.

Morgan, M. (1989). The rational use of intrathecal and extradural opioids. *Brit. J. Anaesthesiol.* **63,** 165–188.

Morita, H., Nemoto, E. M., Bleyaert, A. L., and Stezoski, S. W. (1977). Brain blood flow autoregulation and metabolism during halothane anesthesia in monkeys. *Am. J. Physiol.* **233,** H670–H676.

Naccarato, E. F., and Hunter, W. S. (1979). Anaesthetic effects of various ratios of ketamine and xylazine in rhesus monkeys (*Macaca mulatta*). *Lab. Anim.* **13,** 317–319.

National Research Council (NRC) (1996). ''Occupational Health and Safety in the Care and Use of Research Animals.'' A report of the Institute of Laboratory Animal Resources Committee on Occupational Safety and Health of Personnel in Research Animal Facilities. National Academy Press, Washington, D. C.

Nussmeier, N. A., Benthuysen, J. L., Steffey, E. P., Anderson, J. H., Carstens, E. E., Eisele, J. H., Jr., and Stanley, T. H. (1991). Cardiovascular, respiratory and analgesic effects of fentanyl in unanesthetized rhesus monkeys. *Anesth. Analg.* **72,** 221–226.

Phillips, I. R., and Grist, S. M. (1975). Clinical use of C 1341 anaesthetic ('Saffan') in marmosets (*Callithrix jacchus*). *Lab. Anim.* **9,** 57–60.

Popilskis, S., Danilo, P., Acosta, H., and Kohn, D. (1992). Is preoperative fasting necessary? *J. Med. Primatol.* **21,** 349–352.

Popilskis, S., Daniel, S., and Smiley, R. (1994). Effects of epidural versus intravenous morphine analgesia on postoperative catecholamine response in baboons. *Proc. Int. Cong. Vet. Anesth., 5th,* Guelph, Canada. Aug. 21–25.

Port, J. D., Stanley, T. H., Steffey, E. P., Pace, N. L., Henrickson, R., and McJames, S. W. (1984). Intravenous carfentanil in the dog and rhesus monkey. *Anesthesiology,* **61,** A378.

Reutlinger, R. A., Karl, A. A., Vinal, S. I., and Nieser, M. J. (1980). Effects of ketamine HCl-xylazine HCl combination on cardiovascular and pulmonary values of the rhesus macaque (*Macaca mulatta*). *Am. J. Res.* **41,** 1453–1457.

Ritzman, J. R., Erickson, H. H., and Miller, E. D. (1976). Cardiovascular effects of enflurane and halothane on the rhesus monkey. *Anesth. Analg.* **55,** 85–90.

Rosenberg, D. P. (1991). Nonhuman primate analgesia. *Lab. Anim.* **20,** 22–32.

Rutberg, H., Hakanson, E., Anderberg, B., Jorfeldt, L., Martensson, J., Schildt, B. (1984). Effects of the extradural administration of morphine or bupivacaine, on the endocrine response to upper abdominal surgery. *Brit. J. Anaesthesiol.* **56,** 233–238.

Sainsbury, A. W., Eaton, B. D., and Cooper, J. E. (1989). Restraint and anaesthesia of primates. *Vet. rec.* **125,** 640–643.

Sainsbury, A. W., Eaton, B. D., and Cooper, J. E. (1991). An investigation into the use of propofol (Rapinovet) in long-tailed macaques (*Macaca fascicularis*). *J. Vet. Anaesthesiol.* **18,** 38–41.

Sanders, E. A., Gleed, R. D., and Nathanielsz, P. W. (1991). Anesthetic management for instrumentation of the pregnant rhesus monkey. *J. Med. Primatol.* **20,** 223–228.

Scalese, R. J., DeForrest, J. M., Hammerstone, S., Parente, E., and Burkett, D. E. (1990). Long term vascular catheterization of the cynomolgus monkey. *Lab. Anim. Sci.* **40,** 530–532.

Schmutzler, R., van Uem, J. F. H. M., Elkind-Hirsch, K., Hahn, D., McGuire, J. L., Danforth, D. R., and Hodgen, G. D. (1988). Subcutaneous implantable infusion device for chronic intravenous pulsatile gonadotropin-releasing hormone treatment. *J. Med. Primatol.* **17,** 281–285.

Schreiner, M., Triebwasser, A., and Keon, T. (1990). Ingestion of liquids compared with preoperative fasting in pediatric outpatients. *Anesthesiology* **72,** 593–597.

Sedgwick, C. J. (1986). Scaling and anesthesia for primates. *In* ''Primates: The Road to Self-Sustaining Populations'' (K. Benirschke, ed.), pp. 815–822. Springer-Verlag, New York.

Schulman M., Sandler, A. N., Bradley, J. W., Young, P. S., and Brebner, J. (1984). Postthoracotomy pain and pulmonary function following epidural and systemic morphine. *Anesthesiology* **61,** 569–575.

Socha, W. W., Rowe, A. W., Lenny, L. L., Lasano, S. G., and Moor-Jankowski, J. (1982). Transfusion of incompatible blood in rhesus monkeys and baboons. *Lab. Anim. Sci.* **32,** 48–56.

Socha, W. W., Moor-Jankowski, J., and Ruffie, J. (1984). Blood groups of primates: present status, theoretical implications and practical applications: A review. *J. Med. Primatol.* **13,** 11–40.

Sofianos, E., Alevizou, F., Zissis, N., Kostaki, P., and Balamoutsos, N. (1985). Effects of high-dose fentanyl anesthesia, compared to halothane anesthesia. *Acta Anaesthesiol. Belg.* **36,** 89–96.

Somogyi, A., and Gugler, R. (1982). Drug interactions with cimetidine. *Clin. Pharmacokinet.* **7,** 23.

Southers, J. L., and Ford, E. W. (1995). Medical management. Part A. Preventive medicine. *In* ''Nonhuman Primates in Biomedical Research. Biology and Management'' (B. T. Bennett, C. R. Abee, and R. Hendrickson, eds.), pp. 257–270. Academic Press, San Diego, California.

Spalding, V. T., and Heymann, C. S. (1962). The value of phencyclidine in the anaesthesia of monkeys. *Vet. Rec.* **74,** 158.

Steffey, E. P., Gillespie, J. R., Berry, J. D., Eger, E. I., II, and Munson, E. S. (1974a). Anesthetic potency (MAC) of nitrous oxide in the dog, cat, and stump-tail monkey. *J. Appl. Physiol.* **36,** 530–532.

Steffey, E. P., Gillespie, J. R., Berry, J. D., Eger, E. I., II, and Rhode, E. A. (1974b). Cardiovascular effects of halothane in the stump-tailed macaque during spontaneous and controlled ventilation. *Am. J. Vet. Res.* **35,** 1315–1320.

Stoelting, R. K., and Miller, R. D. (1994a). *In* ''Basics of Anesthesia'' (R. Stoelting and R. Miller, eds.), Vol. 3, p. 60. Churchill-Livingstone, New York.

Stoelting, R. K., and Miller, R. D. (1994b). *In* ''Basics of Anesthesia'' (R. Stoelting and R. Miller, eds.), Vol. 3, p. 360. Churchill-Livingstone, New York.

Tinker, J. H., Sharbrough, F. W., and Michenfelder, J. D. (1977). Anterior shift of the dominant EEG rhythm during anesthesia in the java monkey: correlation with anesthetic potency. *Anesthesiology* **46,** 252–259.

Tsai, S. K., and Lee, C. (1989). Ketamine potentiates nondepolarizing neuromuscular relaxants in a primate. *Anesth. Analg.* **68,** 5–8.

Tsai, S. K., Lee, C., and Tran, B. (1989). Ketamine enhances phase I and phase II neuromuscular block of succinylcholine. *Can. J. Anaesth.* **36,** 120–123.

Udelsman, R., Bacher, J., Gallucci, W. T., Gold, P. W., Morin, M. L., Renquist, D., Lorieux, D. L., and Chrousos, G. P. (1984). Hemodynamic and endocrine parameters in the anesthetized cynomolgus monkey: A primate model. *J. Med. Primatol.* **13,** 327–338.

Wolff, P. L., Pond, J., and Meehan, T. (1990). Surgical removal of *Prosthenorchis elegans* from six species of *Callithricidae. Proc. Am. Assoc. Zoo Vet.,* pp. 95–97.

Woolfson, M. W., Foran, J. A., Freedman, H. M., Moore, P. A., Shulman, L. B., and Schnitman, P. A. (1980). Immobilization of baboons (*Papio anubis*) using ketamine and diazepam. *Lab. Anim. Sci.* **30,** 902–904.

Zola-Morgan, S., and Micheletti, C. (1986). Respiration rate, heart rate, and body temperature values in cynomolgus monkeys (*Macaca fascicularis*) during barbiturate anesthesia. *J. Med. Primatol.* **15,** 399–408.

Chapter 12

Anesthesia and Analgesia in Dogs, Cats, and Ferrets

Ralph C. Harvey, Robert R. Paddleford, Sulli J. Popilskis, and Sally K. Wixson

ANESTHESIA AND ANALGESIA IN LABORATORY ANIMALS

ISBN 0-12-417570-8

INTRODUCTION

The wealth of anesthetic options for the management of dogs, cats, and ferrets results not only from the long-standing popularity of these animals in research models, but also from the accumulation of clinical experience with each of these species as domestic pets. Customizing anesthetic techniques to be used for each research model or protocol is possible through the application of this information. The state or condition referred to as anesthesia, or particularly general anesthesia, includes the component conditions of unconsciousness, amnesia, lack of sensation to potentially noxious stimuli, and muscle relaxation. The characteristics of each of these constituent components of general anesthesia are affected by the choice of anesthetics and anesthetic techniques. Anesthetic protocols should include consideration of sedation; pharmacological restraint (occasionally appropriate as an alternative to general anesthesia); induction and maintenance of general anesthesia; perioperative physiological management or support; anesthetic monitoring and plans for responding to anticipated physiological changes; and, in chronic studies, the recovery from anesthesia and postoperative care.

The plan to either recover the animal from anesthesia or to euthanatize the animal while still anesthetized is one of the fundamental considerations in designing an anesthetic protocol for research purposes. For acute studies (those with no recovery from anesthesia) there may be several additional options available in the choice of pharmaceuticals. Drugs that may not afford suitable recovery (e.g., chloralose) may be quite appropriate for research protocols in which recovery is not planned. Although the provision of operative analgesia is a fundamental requirement in any study, obviously there would be no need to consider agents for strictly postoperative analgesia if there is to be no recovery from the anesthetic episode. Euthanasia at the end of a study may be the most suitable option for humane or other reasons, and it may afford considerable latitude in the anesthetic management.

PART A. ANESTHESIA AND ANALGESIA IN DOGS AND CATS

Ralph C. Harvey, Robert R. Paddleford, and Sulli J. Popilskis

I. PREANESTHETIC EVALUATION

A. Historical Information and Vendor Health Profile

Review of documentation accompanying animals on receipt and review of daily records may similarly aid in the identification of abnormalities. Vendors of purpose-bred dogs and cats usually provide a health profile or the results of previous screening tests used in determining the status of specific pathogen-free animals. Vendor-supplied profiles may serve as useful medical history, but reexamination or repeat testing may provide warranted confirmation.

B. Physical Examination

Two benefits of preanesthetic evaluation are the establishment of baseline data and the identification of any preexisting physical or physiological abnormalities. Laboratory animal handlers and caretakers who have regular contact and familiarity with individual animals can provide insight into unique characteristics that may call for more detailed evaluation. A limited physical examination should cover the major body systems and include palpation of pulses, recording of the heart and respiratory rates, mucous membrane color and capillary refill time, ascultation of the heart and lungs, and evaluation of the attitude and neurologic function of the animal. Preanesthetic evaluation is also customized to the needs of each protocol and animal. Variations from normal direct further attention and may lead to additional testing.

C. Laboratory Data and Parasitology Screen

The extent of medical profiling needed varies with the source of the animals (i.e., random bred, purpose bred), intended use, and the standards set for each research program and/or the facility. Quarantine or perhaps a less restricted, but specified, holding time is suggested to verify health status. This holding period, prior to use of the animal, contributes to the pertinent medical history in such animals. Protection of the existing population of animals in a facility from contagious diseases often requires such screening prior to acceptance of incoming animals. Although vendors may provide screening tests, the tests may be repeated or the ''panel'' of tests may be expanded during holding or quarantine periods.

Many clinical programs have structured various panels or combinations of assays or tests to screen for significant deviations from referenced normal serum chemistries, hematological indexes, and parasitological tests. The appropriateness of such screening tests in clinically healthy or normal subjects is open to debate. Standards for inclusion of subjects into research protocols often include normal results of such screening tests. Although a variety of conditions in ''normal'' or ''healthy'' subjects can result in deviations from normal readings, these tests can successfully identify individual animals with subclinical dysfunction.

II. DELIVERY OF ANESTHETICS AND ADJUNCTIVE DRUGS

A. Venous Access Sites and Intravenous Catheter Placement

Peripheral venous access is relatively easy in dogs and cats, with cephalic and saphenous veins commonly used, and the jugular veins used for central vein access. In dogs, the lateral saphenous vein is used; whereas in cats, the medial saphenous vein is used. Gentle and minimal restraint is valuable in successful injection technique. A minimally stressful approach makes possible unassisted injections or blood sampling in cats (Johnson, 1981). Jugular vein catheters are less subject to positional problems but may be more often associated with significant thrombosis. Secure placement of an intravenous catheter for administration of anesthetics, and for physiological supportive measures, adds considerable safety in anesthetic management. Administration of potentially irritating anesthetics through a catheter decreases the risk of extrasavation.

B. Sites for Intramuscular and Subcutaneous Drug Delivery

Traditionally, intramuscular injections in dogs and cats have most often been delivered to the caudal muscles of the thigh, in the semimembranosus or semitendinosus muscles. Muscles of the caudal thigh are probably not the best choice for IM injections. Superior absorption has been documented with injection into either the cranial thigh muscles (the quadriceps) or the lumbar spinal epaxial muscles (Autefage *et al.,* 1990). Choice of sites for IM injection may be limited by local pathology, presence of implants, or other factors. When using the caudal thigh muscles, inadvertent injection of the drug into fascial planes posterior to the femur decreases drug absorption and allows for tracking along the fascial planes, potentially to the area of the sciatic nerve.

Infrequently, either septic or sterile abscessation results from an IM injection. It is unlikely that a brief application of alcohol decreases the risk. In those rare cases of abscess formation, it would be easier to establish good drainage if the injection has been made in the muscles of the limbs rather than the epaxial muscles.

The uptake of drugs following subcutaneous injection may be variable and influenced by the state of hydration and local perfusion. Subcutaneous injections seem to elicit less pain than do intramuscular injections and thereby can help to maintain a calmer animal. Injection sites often used include the interscapular region, the lateral thoracic, and the lumbar dorsal region.

C. Chemical Restraint

Tranquilizers, sedatives, or other central nervous system (CNS) depressants are occasionally used when chemical restraint is needed to perform minor procedures. Tranquilizers or sedatives administered alone are useful for calming and quieting excited animals. Acetylpromazine (Acepromazine) (0.05–0.1 mg/kg IV, IM, SC; maximum 3-mg dose) is commonly used to tranquilize dogs and cats. The α_2-adrenergic agonist sedatives, xylazine and medetomidine, may be used in dogs. Vainio *et al.* (1989) reported that xylazine (2.2 mg/kg IM) and medetomidine (0.03 mg/kg IM) had similar sedative and analgesic effects lasting for about 1 hr. Ketamine (10–20 mg/kg IM) can be used in cats for chemical restraint and as an anesthetic for minor procedures of 45 min or less. Innovar-Vet (fentanyl–droperidol) provides 30 min of heavy sedation and moderate analgesia in dogs at a dose of 0.05 ml/kg IM (Thurmon *et al.*, 1996).

D. Induction of Anesthesia with Inhalant Anesthetics

There are few techniques as frequently employed in veterinary anesthesia which are more dependent on good judgment, technical skill, and finesse than mask or chamber induction (Harvey, 1992). These techniques can be extremely valuable in selected research animal protocols but may be quite inappropriate in others. The avoidance of other anesthetics and their physiologic effects and the simplification of the protocol are often cited as justifications for inhalant induction of anesthesia in research settings. The volatile anesthetics halothane and isoflurane are suitable for inhalation induction of anesthesia using a chamber or mask. Methoxyflurane is not suitable due to the high solubility that results in an unacceptably long transition from awake to asleep with associated delirium.

A variety of suitable face masks are commercially available for dogs and cats. Masks are selected that match the animal in order to minimize the rebreathing of dead-space gas and to minimize the leakage of anesthetic gases. Poorly fitted masks not only unnecessarily expose personnel to anesthetic waste gases but can also allow for entrainment of room air as the patient inhales, thereby altering the delivery concentration of the anesthetic. For some breeds of dogs, and particularly neonates, the contours of the face may be best matched with custom prepared (or improvised) masks built from plastic bottles, syringe cases, or other materials sufficiently padded for comfort, fit, and avoidance of substantial gas leaks. Commercially available anesthetic induction chambers, specifically designed for cats and small dogs, facilitate inhalant inductions. Minimizing excess space in induction chambers helps limit the waste of anesthetic gases and reduces waste gas release.

Among the prerequisites for mask induction, the cooperation of the animal is essential to maintain the safety of the technique for both animal and anesthetist. Although debilitated and depressed animals often accept mask induction without preanesthetic tranquilizers or sedatives, healthy young animals may not cooperate as readily. It is only with great finesse that a mask induction will be practical without prior administration of CNS depressants. A struggling animal will not have a smooth induction or recovery. Excitement leads to catecholamine release and associate hemodynamic effects including increased risk of arrhythmias. Ventilatory insufficiency can occasionally result from related spasms of the larynx or thoracic muscles. Induction delirium may be accompanied by scratching, biting, vocalization, and a very traumatic and unpleasant experience. The lightly anesthetized animal can easily become dangerously excited in the process of inhalational induction of anesthesia. Prior tranquilization or sedation greatly facilitates inhalation induction and is generally indicated if acceptable within the requirements of the research protocol. During the transition from awake to asleep it is imperative that the level of extraneous stimulation be minimized. As the animal loses consciousness it will relax and ventilation will usually become smooth and regular. There is less objection to gentle restraint as consciousness fades. The sense of hearing can be the last sensation to succumb to anesthetic depression. Loud or abrupt noises during a mask or chamber induction often trigger exaggerated responses. Delirium with struggling, vocalization, pawing, expression of anal sacs, and physiologic stress responses can be initiated by untimely stimulation and particularly by relatively minor noises during inductions.

Two basic approaches have been used in mask inductions: (1) a gradual step-up in delivered concentration of the anesthetic gases, or (2) an immediate administration of high partial pressures of the anesthetic. The first method is generally indicated.

E. Techniques of Endotracheal Intubation

Airway control is a fundamental of safe anesthetic management. Securing the airway by means of placing an endotracheal tube is more easily accomplished in dogs, cats, and ferrets than in perhaps any other species (with the possible exception of birds). Because the pharyngeal anatomy of cats and ferrets is similar to that of young children, these species have been useful in training sessions for personnel learning intubation skills to apply in pediatric emergency care. The techniques for canine and feline laryngoscopy and endotracheal intubation have been described in detail (Sawyer, 1984). These are so important for successful anesthesia, regardless of the choice of anesthetics, that the fundamentals are repeated here. These principles were described as long ago as 1927 (Hardenbergh and Mann, 1927).

An adequate anesthetic depth is necessary for endotracheal intubation. This will not only help to maintain cutaneous integrity among personnel, but also decreases the risk of laryngospasm and excessive vagal response to pharyngeal instrumentation. Endotracheal intubation should be accomplished quickly and gently, and the techniques of intubation are fundamental in anesthetic training. Accomplished anesthetists should be able to perform this task in a wide variety of research animal species such as rabbits, rats, pigs, and sheep, which are substantially more difficult to intubate than dogs and cats. Anesthetic apparatus available should include endotracheal tubes of various sizes: cats, 3.0 to 5.0 mm; dogs, 5.0 to 12 mm ID. These size ranges are appropriate for most adult animals of these species. In general, cuffed Murphy endotracheal tubes are chosen, but uncuffed Murphy or Cole tubes may be preferred for smaller animals. Although these animals can usually be intubated without a laryngoscope by well-experienced persons, a functioning laryngoscope, with an appropriate size blade, is usually indicated to ensure accurate placement of an endotracheal tube. A stylet can be used to stiffen very flexible endotracheal tubes and facilitate difficult intubations. When a stylet is used inside an endotracheal tube, care must be taken to avoid causing additional trauma. Tubes can be lubricated with a small amount of sterile, water-soluble gel or with water. To minimize laryngospasm in cats, a small amount of lidocaine (0.1 ml) can be sprayed on the larynx and epiglottis. Benzocaine may be associated with methemoglobinemia and should be avoided (Harvey *et al.*, 1979). Dogs and cats are usually intubated while in sternal recumbency with the head elevated and the neck extended. This position reduces the risk of regurgitation and aspiration while securing the airway. Intubation may also be accomplished with the animal in either dorsal or lateral recumbency. Once placed in the trachea such that the tip is between the larynx and the thoracic inlet, the tube is secured using tape or gauze strips around the tube and around the maxilla, mandible, or behind the neck. The cuff on the endotracheal tube is carefully inflated with just enough air to prevent leaks when the anesthetic breathing bag is gently squeezed to a pressure of about 15 cm water. This ''minimum no-leak volume'' technique will suffice to allow for positive pressure ventilation, prevent aspiration, and will maintain circulation to the tracheal mucosa. Excessive cuff pressure may result in mucosal ischemia and necrosis or even tracheal rupture.

Alternative techniques for endotracheal intubation such as retrograde intubation and intubation over a previously placed cannula, guide wire, or stylet are usually not necessary for these species, but should be considered in difficult cases.

III. PREANESTHETIC MEDICATIONS

The dosage, route of administration, and effects of some preanesthetic drugs used for cats and dogs are presented in Table A.I.

A. Anticholinergics

The vagal parasympathetic nervous system tone increases with certain anesthetic and surgical procedures, including pharyngeal instrumentation (endotracheal intubation), visceral traction or manipulation, ocular procedures, and manipulation of the carotid sheath or vagus nerve. Vagal tone also increases with the administration of vagotonic drugs such as most opioids and α-$_2$-agonists. Anticholinergic medications are quite commonly administered to prevent bradycardia and to decrease salivary and respiratory secretions.

Atropine (0.02–0.04 mg/kg IM or SC) is more widely used than glycopyrrolate (Robinul-V) and will more substantially raise heart rate. Glycopyrrolate (0.02 mg/kg IM or SC) provides more control of salivation and may have a longer duration of action than atropine. There are some risks of tachyarrhythmias with the use of anticholinergics and there is a bronchodilating effect that increases dead-space ventilation. Anticholinergic therapy is optional in many anesthetic protocols.

B. Tranquilizers and Sedatives

Acepromazine is the most popular preanesthetic tranquilizer and is usually given SQ or IM at 0.05–0.1 mg/kg 15 min before

TABLE A.I

PREANESTHETIC DRUGS FOR USE IN DOGS AND CATS

Drug	Dose	Route of injection	Duration of action	Reference
Anticholinergics				
Atropine	0.02–0.04 mg/kg	IM, SC	30–60 min	Daunt (1994)
Glycopyrrolate (Robinul-V)	0.02 mg/kg	IM, SC	60–120 min	Daunt (1994)
Sedatives and tranquilizers				
Acetylpromazine (Acepromazine)	0.05–0.1 mg/kg (maximum 3-mg total dose)	IM, SC	4 hr	Daunt (1994)
Xylazine (Rompun)	0.1–0.5 mg/kg 0.4–0.9 mg/kg	IV (slowly) IM	30–60 min	Tranquilli and Benson (1992)

the administration of general anesthesia. Hypotension and prolonged tranquilization are frequent effects of acepromazine. The prolonged and smooth recovery can be advantageous in many situations but requires appropriate postoperative attention. Diazepam (Valium) (0.2–0.4 mg/kg IM or IV; maximum 10 mg) is a valuable alternative for more rapid recovery and is generally less hypotension. Midazolam (Versed) (0.2–0.4 mg/kg IM or IV), a water-soluble benzodiazepine, is a useful alternative to diazepam because it is better absorbed and less painful when administered IM.

Xylazine (Rompun, Anased, Xyla-Ject) (0.1–0.5 mg/kg IV, slowly; or 0.4–0.9 mg/kg IM) is an α-$_2$-adrenergic agonist sedative and a short-acting analgesic that is used either alone or in combinations with other drugs. Medetomidine, an almost pure α-$_2$-agonist, has been widely used as a sedative (0.04–0.08 mg/kg IM or IV) (Kramer *et al.,* 1996) and premedicant in dogs (0.01–0.04 mg/kg IM) and cats (0.01–0.08 mg/kg IM) in Europe (Tranquilli and Benson, 1992; Vaha-Vahe, 1989). α_2-Agonists can produce mild to marked vagal bradycardia and reduce cardiac output (Thurmon *et al.,* 1994). Although advanced second-degree heart block with rates less than 30 beats per minute may occur, less severe slowing of the heart rate is common. Pretreatment with atropine or another anticholinergic usually prevents xylazine-induced bradycardia but does not prevent the decrease in cardiac output (Dunkle *et al.,* 1986). Anticholinergics for prevention of the bradycardia are more successful than treatment with anticholinergics once bradycardia has occurred. The use of anticholinergics is, however, somewhat controversial because there is a resulting transient increase in myocardial work with a decrease in myocardial blood flow and oxygen supply. The cardiopulmonary effects of medetomidine and atropine combinations in dogs have been studied by Alibhai *et al.* (1996). They administered medetomidine (0.04 mg/kg IM) 30 min after atropine (0.03 mg/kg IM), and although atropine counteracted medetomidine-induced bradycardia, it caused prolonged and severe hypertension.

C. Opioids in Anesthesia

Opioids are often used in combination with other anesthetic drugs to provide control of surgical and postoperative pain. Among the most useful are morphine, meperidine (Demerol), oxymorphone, butorphanol (Stadol, Torbutrol, Torbugesic), and buprenorphine (Temgesic).

Morphine is quite inexpensive and provides sedation in dogs with a considerable degree and duration of analgesia at dosages of 1–2 mg/kg IM or SC. Meperidine (2–4 mg/kg IM or SC) is short-acting and tends to maintain heart rate without the addition of anticholinergics. Both morphine and meperidine can cause histamine release (especially in dogs), with subsequent hypotension, and therefore are usually not injected intravenously in dogs. Morphine is more likely to cause vomiting and decreased heart rate.

Oxymorphone, at a dosage of 0.1–0.2 mg/kg IV, IM, or SC, provides a longer duration of analgesia (3–6 hr) with little risk of histamine release. Oxymorphone (0.22 mg/kg IV) works well in combination with acepromazine (0.02 mg/kg IV) to produce sedation and analgesia. This combination, in comparison to butorphanol (0.22 mg/kg IV or IM) and acepromazine (0.02 mg/kg IV or IM), produced significantly better analgesia. Anticholinergics such as atropine may be given to prevent the bradycardia that can otherwise result after oxymorphone administration.

Among the agonist–antagonist opioids, butorphanol (0.1–0.5 mg/kg IM or IV) and buprenorphine (0.01–0.02 mg/kg in dogs, and 0.005–0.01 mg/kg in cats) are most frequently used. Both butorphanol and buprenorphine are useful in anesthesia to provide limited sedation and analgesia. Buprenorphine (0.0075–0.01 mg/kg) combined with acepromazine (0.03–0.05 mg/kg) when given as a single IV bolus results in better sedation than if either drug is used alone (Stepien *et al.,* 1995). Combina- with diazepam or midazolam may be particularly useful in situations where some residual analgesia or sedation is desired. The duration and the magnitude of effect can be considerably greater with buprenorphine than with butorphanol. The limited potency and side effects of butorphanol makes reversal rarely necessary. Buprenorphine has a very high affinity for opioid receptors. The significance is that it can be quite difficult to reverse the effects of buprenorphine. Although rarely significant, this has limited the anticipated usefulness of this relatively long-acting opioid for postoperative analgesia in research.

IV. INTRAVENOUS AGENTS FOR INDUCTION OR MAINTENANCE OF GENERAL ANESTHESIA

In many situations, injectable anesthetic techniques are preferred to inhalants in order to rapidly control the airway, to avoid the delays and delirium that can accompany mask or chamber induction techniques using inhalants, to avoid the significant decreases in blood pressure frequently associated with inhalants, and to avoid the increasingly appreciated potential problems that are attributed to the exposure of personnel to waste anesthetic gases. Table A.II lists some of the anesthetic drugs used for dogs and cats and includes data on dosages, route of administration, and effects.

A. Barbiturates

Despite its long history of use in research animal anesthesia, pentobarbital is now less frequently used as an anesthetic for survival procedures in the dog and cat. This agent is less desirable due to its narrow margin of safety, its poor analgesic properties, and the long and occasionally violent (vocalizing,

TABLE A.II

Anesthetic Drugs for Use in Dogs and Cats

Drug	Dose	Route of injection	Duration of action	Reference
Barbiturates				
Pentobarbital	20–30 mg/kg	IV	30–45 min	Green (1979)
Thiopental (Pentothal)	8–12 mg/kg	IV	15 min	Paddleford (1995)
Dissociatives				
Ketamine (cat)	10–20 mg/kg	IM	30–45 min	Green (1979)
Ketamine +	10 mg/kg	IV	20 min	Daunt (1994)
diazepam	0.5 mg/kg	IV		Haskins *et al.* (1986)
Ketamine +	10 mg/kg	IV		Jacobson and Hartsfield (1993)
midazolam	0.5 mg/kg	IV		
Ketamine +	7 mg/kg (cats)	IM	45 min	Young and Jones (1990)
medetomidine	0.08 mg/kg	IM		
Ketamine +	2–4 mg/kg	IV	10 min	Paddleford (1995)
acepromazine	0.1 mg/kg	IM, IV		
Telazol	6–12 mg/kg (dogs)	SC, IM	20–30 min	Paddleford (1995)
	9–12 mg/kg (cats)	SC, IM		
Other injectables				
Propofol	6 mg/kg	IV (induction)	15 min	Paddleford (1995)
(Diprivan)	0.2–0.4 mg/kg/min	IV (infusion)	Maintenance	
	1 mg/kg	IV (bolus, as needed)	Maintenance	Daunt (1994)
Etomidate (Amidate)	1.5–3 mg/kg	IV	5–10 min	Paddleford (1995)
Medetomidine +	15 μg/kg	IM		Ko *et al.* (1994)
etomidate +	0.5 mg/kg	IV (bolus)		
etomidate	50 μg/kg/min	IV (infusion)		
Medetomidine	0.01–0.08 mg/kg (cats)	IM	Dose dependent	Tranquilli and Benson (1992)
	0.01–0.08 mg/kg (dogs)	IM	Dose dependent	
Diazepam (Valium)	0.2–0.4 mg/kg (maximum 10 mg)	IM, IV	30–180 min	Paddleford (1995)
Midazolam	0.2–0.4 mg/kg	IM, IV	Less than 2 hr	Paddleford (1995)
Innovar-Vet (fentanyl, droperidol)	1 ml/10–50 kg (dogs)	IM, IV	30–60 min	Wilson (1992)
Opioids				
Butorphanol (Torbugesic)	0.1–0.5 mg/kg	IM, IV	Up to 4 hr	Paddleford (1995)
Buprenorphine	0.01–0.02 mg/kg (dogs)	IM, SC	6–8 hr	Flecknell (1985)
	0.005–0.01 mg/kg (cats)	IM, SC	6–8 hr	
Buprenorphine +	0.0075–0.01 mg/kg	IV		Stepien *et al.* (1995)
acepromazine	0.03–0.05 mg/kg	IV		
Morphine	0.25–0.5 mg/kg (dogs)	IM, SC	4 hr	Daunt (1994)
	0.1 mg/kg (cats)	IM, SC	4 hr	
Morphine	0.1–0.15 mg/kg (dogs)	Epidural	24 hr	Popilskis *et al.* (1993)
Meperidine (Demerol)	2–4 mg/kg	IM, SC	30–60 min	Daunt (1994)
Oxymorphone	0.1–0.2 mg/kg	IV, IM, SC	4 hr	Daunt (1994)
(Oxymorphone-PM)	0.05–0.2	IV, IM, SC	4 hr	
Oxymorphone +	0.22 mg/kg	IV		
acepromazine	0.02 mg/kg	IV		
Oxymorphone	0.1 mg/kg (dogs)	Epidural	10 hr	Popilskis *et al.* (1991)
Fentanyl (Sublimaze)	0.02–0.04 mg/kg	IV, IM, SC	30 min	Wilson (1992)
Fentanyl (Duragesic)	50 μg/hr	Transdermal	1–3 days	Kyles *et al.* (1996)
NSAIDs				
Aspirin	10–20 mg/kg	PO	4–8 hr (dogs)	Cribb (1992)
	10 mg/kg	PO	48 hr (cats)	
Ketorolac	30 mg, dogs <17 kg	IM	6 hr	
	60 mg, dogs >17 kg			

TABLE A.II *(continued)*

ANESTHETIC DRUGS FOR USE IN DOGS AND CATS

Drug	MAC	Clinical dose range	Duration of anesthesia	Reference
Inhalants				
Methoxyflurane	0.29% (dogs) 0.23% (cats)	0.2–2%	Until discontinued	Daunt (1994)
Halothane	0.87% (dogs) 1.19% (cats)	Up to 3%	Until discontinued	Daunt (1994)
Isoflurane (Aerrane)	1.3% (dogs) 1.6% (cats)	Up to 3%	Until discontinued	Daunt (1994)
Nitrous Oxide	200% (dogs) 250% (cats)	50–66% 50–66%	Until discontinued Until discontinued	Paddleford (1995)

involuntary motor activity) period of recovery of the animal. However, it remains as a common and useful anesthetic for nonsurvival procedures. The dosage in dogs and cats is 20–30 mg/kg IV, with about one-half given as a bolus and the remaining amount given to effect. This dosage will provide light surgical anesthesia for about 30 to 45 min (Green, 1979). For longer procedures, IV infusions at 3–5 mg/kg/hr or small intermittent boluses (generally 10–20% of the calculated dose) may be given to effect (Frederiksen *et al.,* 1983). Animals should be ventilated because the deep state of anesthesia needed to produce adequate analgesia will usually be accompanied by severe respiratory depression.

Thiobarbiturates are best used for induction or for very brief anesthesia rather than for maintenance of prolonged anesthesia by repeated injection of small dosages. The ultra-short-acting thiobarbiturates, thiamylal (formerly available as Anestal, Biotal, or Surital) or thiopental (Pentothal) (8–12 mg/kg IV), are powerful and useful anesthetics for the induction of anesthesia prior to use of inhalants or as an alternative to the mask or chamber techniques of inhalational induction. Thiamylal is no longer available. The "rapid sequence" induction of anesthesia by intravenous injection is highly desired when immediate control of the airway is needed. The margin of safety (ratio of ED_{50}:LD_{50} or therapeutic index) is limited for barbiturates in any animal. In high-risk situations, particular care is warranted in use of barbiturates. Methohexital (Brevital or Brevane) is an ultra-short-acting barbiturate, given at 4–8 mg/kg IV, that has been recommended for induction in sight hounds (e.g., greyhounds) that typically respond poorly to thiobarbiturates like thiopental.

Combination techniques using intravenous lidocaine and pentothal have been recommended for dogs as a means to reduce the dose of barbiturate needed and to minimize some of the cardiac side effects (Rawlings and Kolata, 1983).

B. Propofol

Propofol (diisopropylphenol, Diprivan or Rapinovet) is a nonbarbiturate, nondissociative injectable sedative–hypnotic. The characteristics of rapid elimination with full recovery have led to the use of propofol anesthesia in dogs and cats for (1) induction of anesthesia prior to use of an inhalant, (2) very brief procedures with rapid full recovery, and (3) maintenance of anesthesia by controlled intravenous infusion with rapid and full recovery. Constant rate infusions are usually administered with an electronic syringe pump for maintenance of anesthesia after induction by bolus intravenous injection (Hall and Chambers, 1987). In cats, Brearley *et al.* (1988) used 6.8 mg/kg for induction and an infusion rate of 0.51 mg/kg/min for maintenance. They noted a 30-sec period of apnea if the induction dosage was given as a single, rapid bolus, but none if one-half of the dose was given slowly. Pascoe (1992) noted that propofol is primarily metabolized in the liver by glucuronidation, and that because cats are deficient in glucuronide synthase, recovery from a propofol infusion may be prolonged. Andress *et al.* (1995) suggested that propofol administered on consecutive days to healthy cats induces oxidative injury to feline red blood cells reflected by the formation of excessive Heinz body formation which may result in clinical signs of illness.

The infusion rate for propofol has been reported to be 0.3 to 0.5 mg/kg/min in dogs (Pascoe, 1992). Similarly to cats, apnea commonly occurs after bolus administration of propofol; accordingly, ventilatory support should be available if needed.

During the induction, propofol decreases systemic blood pressure but myocardial contractility and heart rate are minimally affected. Hemodynamic side effects can be attenuated by reducing the induction dose. This makes propofol, in combination with opioids, an attractive alternative for dogs with compromised cardiovascular function. One of the authors (SP) routinely administers propofol (1–2 mg/kg) IV preceded by a slow bolus of fentanyl (7–10 μg/kg) and atropine (0.005–0.01 mg/kg) IV for induction of dogs with clinical symptoms of heart failure.

Propofol has only minimal analgesic properties and must be augmented with an opioid or an α_2-agonist if a painful procedure is to be performed. Medetomidine–propofol combinations have been reported in the dog. Premedication with medetomidine (30 μg/kg IM) and atropine (0.044 mg/kg IM) followed by a 2-mg/kg IV bolus of propofol and a propofol infusion at 165 μg/kg/min produced adequate short-term general anesthesia in dogs (Thurmon *et al.,* 1994). Similarly, Hall *et al.* (1994) obtained a light plane of anesthesia in dogs by premedicating with medetomidine (10 μg/kg IM) 15 min prior to administration of a 4- to 6-mg/kg bolus of propofol followed by infusion at 0.2 mg/kg/min for 120 min. Reversal of effects from medetomidine was achieved by administrating atipamezole (50 μg/kg IM) at the termination of infusing propofol.

C. Etomidate

Etomidate (Amidate) is an intravenous, ultra-short-acting, imidazole derivative anesthetic. The most prominent desirable feature of etomidate anesthesia is a relative preservation of cardiovascular and respiratory function. Ko *et al.* (1994) reported safe and adequate anesthesia in beagles premedicated with medetomidine (15 μg/kg IM) prior to a bolus of etomidate (0.5 mg/kg IV) followed by an infusion at 50 μg/kg/min. There is rapid recovery from either bolus injection or continuous infusion techniques and the therapeutic index of etomidate in dogs is quite favorable. Etomidate suppresses the adrenal cortical stress responses to anesthesia and surgery (Dodam *et al.,* 1990). Although this suppression may be desirable in some research, the relative risk or benefit of blocking the stress response is debatable and is quite dependent on the status of the animal and the physiologic demands presented. Pain on injection and relatively high costs have limited the use of propofol and etomidate in clinical veterinary patients.

D. Dissociatives

The dissociative anesthetics have a wide therapeutic index and tend to be relatively sparing of adverse hemodynamic effects. Cardiac output is often increased as a result of an increased heart rate. This may improve circulation but may do so at the expense of increased myocardial work. In the absence of underlying heart disease, the animal may have significantly improved hemodynamics relative to that which would accompany other anesthetic techniques. Elimination and recovery from the dissociatives require hepatic metabolism, more so in dogs than in cats. In patients with impaired liver function, conservative administration of the dissociatives, perhaps supplemented with the use of opioids or other CNS depressants, is recommended. Recovery from dissociative anesthetics can be accompanied by delirium. Haskins *et al.* (1985) found ketamine (Ketaset, Vetalar, Ketalar) at a dose of 10 mg/kg IV to be unsatisfactory for surgical purposes in dogs due to significant hypertonus and spontaneous movements. In cats, 10–20 mg/kg IM of ketamine provides 30–45 min of sedation and analgesia (Green, 1979). Ketamine alone in cats should be used only for diagnostic or minor surgical procedures.

Ketamine has been proven effective and safe in a wide variety of combinations with other injectable anesthetics. Perhaps the most useful for consideration is the combination of ketamine and diazepam, usually given by intravenous injection. Haskins *et al.* (1986) noted that this combination, ketamine (10 mg/kg) and diazepam (0.5 mg/kg), should be given as a single IV bolus because diazepam alone does not induce sedation and frequently results in excitement in dogs and cats. Hartsfield (1992) reported dosages for induction and intubation in dogs and cats to be 10 mg/kg of ketamine and 0.2–0.4 mg/kg of diazepam given IV, with one-third of the calculated dose as a bolus with the rest to effect. Anticholinergics are often, but not always, needed to control excessive salivation.

Ketamine (10 mg/kg IV) and midazolam (0.5 mg/kg IV) given to dogs as a bolus provide a smooth induction and chemical restraint of short duration. Cardiovascular effects were minimal except for a significant increase in heart rate. Side effects include salivation, hypertonia, and defecation (Jacobson and Hartsfield, 1993).

Recovery from dissociative anesthetics can be accompanied by delirium. Administrating low dosages of diazepam (0.1–0.2 mg/kg) is usually quite effective in smoothing an otherwise rough recovery from ketamine.

In cats, ketamine (7 mg/kg) and medetomidine (0.08 mg/kg) given as a single IM bolus produced good anesthesia and muscle relaxation for diagnostic and minor surgical procedures of less than 45 min in duration (Young and Jones, 1990).

The injectable anesthetic Telazol is a combination of the dissociative tiletamine and the benzodiazepine zolazepam. Telazol is most useful for substantial restraint or light general anesthesia rather than for surgical anesthesia of long duration. The onset of action is very rapid following intramuscular or subcutaneous injection. In dogs, there seems to be a considerable variability in time to full recovery from Telazol anesthesia. There is risk of significant respiratory depression at higher dosages.

Combinations of Telazol with ketamine and xylazine have been developed which offer advantages for general anesthesia. In cats, 500 mg of Telazol powder can be dissolved in 400 mg (4 ml) of ketamine and 100 mg (1 ml) of 10% xylazine. The resultant combination is given IM at 0.03 ml/kg. Induction is rapid, laryngeal relaxation is excellent, and surgical anesthesia lasts approximately 30 to 45 min. This combination works well for anesthetic induction followed by inhalant anesthetic maintenance. Recovery is smooth and more rapid than when Telazol is used alone (G. J. Benson, personal communication, 1996). In dogs, Telazol can be combined with α_2-adrenergic agonists and opioids to produce a rapid induction and surgical anesthesia of approximately 1 hr. Six to 8 mg/kg of Telazol is administered along with 0.5 mg/kg of xylazine or 15 μg/kg of medetomidine and 0.2 mg/kg butorphanol or 0.5 mg/kg morphine. An anticholinergic agent has been recommended to prevent vagal-mediated bradycardia. All drugs can be mixed in the same syringe and administered IM (G. J. Benson, personal communication, 1996).

E. Chloralose

Chloralose is frequently the anesthetic of choice for cardiovascular physiological studies because the hemodynamic profile in dogs given the agent is similar to that of conscious animals (Silverman and Muir, 1993). It is particularly useful for noninvasive, nonsurvival, cardiovascular procedures that require light anesthesia for 4 to 8 hr. It is considered to be an ineffective analgesic or anesthetic for major surgical procedures when used alone in dogs (Holzgrefe *et al.,* 1987). The pharmacologic effects and poor analgesic properties of chloralose are addressed in Chapter 1. Chloralose should always be used in combination with a thiobarbiturate, methohexital, tranquilizer, or opioid to eliminate the tonic convulsions that occur when chloralose is administered IV at dosages of 80–100 mg/kg. Lang *et al.* (1992) premedicated dogs with morphine (5 mg/kg SC) followed with chloralose (100 mg/kg IV) and maintained the animals on room air respiration with a Harvard volume respirator. Over a 3-hr period, they found that this combination produced a hemodynamic profile that more closely resembled that of conscious dogs than did two other regimens, pentobarbital or fentanyl.

F. Reversal or Antagonism of Injectable Anesthetics

Among the most notable improvements in injectable anesthesia is the availability of a specific reversal agent, yohimbine (Yobine), to antagonize the effects of the potent sedative/analgesic xylazine that has been used extensively in veterinary medicine for years. Although widely used in general practice, many specialists in veterinary anesthesiology have been reluctant to use xylazine in small animal patients because they recognize the variety of adverse side effects and the potential for complications with use of the drug.

Xylazine can produce mild to extreme bradycardia and reduce cardiac output. Advanced second-degree heart block with rates less than 30 beats per minute are noted with some frequency. Less severe slowing of the heart rate is common. Pretreatment with an anticholinergic usually prevents xylazine-induced bradycardia but does not prevent the decrease in cardiac output. Pretreatment with an anticholinergic may actually be detrimental to cardiovascular function (Dunkle *et al.,* 1986). Although controversial, prevention of the bradycardia is more successful than treatment with anticholinergics once it has occurred. The advent of yohimbine as a receptor-specific antagonist of xylazine (and other α_2-agonists) gives veterinarians a means of definitive, immediate, and effective treatment of xylazine-associated side effects and complications.

The reversibility of opioids (narcotics) has long been recognized as an advantage to their use in high-risk patients. Naloxone is the drug of choice for full reversal of any of the opioids used in small animals. Naloxone (0.005 to 0.02 mg/kg IV) or (0.04 mg/kg IM) reverses most of the effects induced by oxymorphone. A single dose of naloxone may permanently reverse the effects of a single dose of a narcotic; however, multiple dosages of oxymorphone may require repeated IM dosages of naloxone every 40–70 min (Copland *et al.,* 1989). The mixed-acting agonist–antagonist butorphanol can be used for partial reversal of the depressant effects of other opioids while preserving some of the analgesia. In a study by McCrackin *et al.* (1994), a 0.2-mg/kg bolus of butorphanol caused significant and sustained partial reversal of oxymorphone-induced respiratory depression and bradycardia, while preserving analgesia. The authors suggested, however, that it may be prudent to titrate butorphanol IV in 0.05-mg/kg increments rather than a 0.2-mg/kg bolus to avoid sudden arousal.

Flumazenil has become available for the specific reversal of benzodiazepine tranquilizers such as diazepam and midazolam. Although combinations of opioids and benzodiazepine tranquilizers can be excellent neuroleptanalgesics for geriatrics and other high-risk patients, the availability of antagonists for both the opioid and the tranquilizer adds even more safety to this injectable technique.

V. BALANCED ANESTHESIA

Tranquilizers, sedatives, opioids, or other CNS depressants are components of ''balanced anesthesia'' in which their use reduces dosages of injectable or inhalant anesthetics, thereby minimizing the more adverse effects of high dosages of any one drug. Pypendop *et al.* (1996) evaluated the hemodynamic effects of using either buprenorphine (10 μg/kg IV) or butorphanol (0.1 mg/kg IV) in combination with medetomidine (39–46 μg/kg IM) and midazolam (1 mg/kg IV) in dogs. Both combinations induced a comparable cardiovascular depression due to bradycardia; however, buprenorphine led to greater respiratory depression. The authors indicate that additional studies using different dosages, timing and routes of administration, and supplemental oxygen are needed to achieve a more acceptable cardiac index.

VI. OVERVIEW OF INHALATION ANESTHESIA

The safety that often has been associated with inhalant, as opposed to injectable, anesthetics is largely due to the provision of supplemental oxygen as the carrier gas for the volatile anesthetics. Endotracheal intubation and administration of supplementary oxygen can easily be incorporated into injectable anesthetic techniques and can add substantial safety. If anesthesia is deep enough to allow for placement of an endotracheal tube, then the animal is no longer able to protect its airway from the aspiration of regurgitated or foreign material. Although not all anesthetized animals will require supplementary oxygen administration, most animals should be considered for this level of support during anesthesia and surgery.

The ability to titrate anesthesia to the desired effect and to safely provide more prolonged anesthesia has popularized inhalant anesthesia. Halothane, methoxyflurane, and isoflurane have long histories of use in research animals. Halothane and isoflurane have the widest application for dogs and cats. Techniques for use of the newer inhalants, desflurane (Suprane) and sevoflurane (Ultane), have been described, but these anesthetics are not yet in common use for research protocols (Steffey, 1992).

Comparison of the costs of various anesthetics reveals that the expense per animal per unit of time under inhalant anesthesia is several times greater with isoflurane than with the other inhalants, most notably halothane (Bednarski *et al.,* 1984). Costs will vary dramatically, depending on many factors such as the use of preanesthetic CNS depressants, carrier gas flow rates, type of anesthetic equipment, and duration of procedures (Klide, 1992).

The utility of nitrous oxide is limited by its low potency. Except in humans, relatively little reduction in other anesthetics is achieved through the use of nitrous oxide. The attendant reduction in delivered oxygen concentration, the potential for diffusion hypoxia at recovery, and occasional difficulties due to diffusion of nitrous oxide into closed body cavities represent additional disadvantages (Klide and Haskins, 1992). Based on its very low solubility, however, nitrous oxide can be quite useful in facilitating the mask or chamber induction of anesthesia and for quick adjustments in the depth of anesthesia. In addition, substituting a portion of other inhalational anesthetics with nitrous oxide produces a sparing effect on cardiovascular function.

Isoflurane provides almost ideal characteristics of rapid induction and recovery with reduced potential toxicity and acceptable levels of cardiopulmonary depression. The vapor pressure of isoflurane is very close to that of halothane. The same calculations or guidelines can be used for the determination of vaporizer flow rates for either of these two anesthetics in Copper-Kettle (Foregger, Allentown, Pennsylvania) or Verni-Trol-type (Ohio Medical Products, Madison, Wisconsin) vaporizers. The similarity also means that most halothane-specific precision vaporizers can be used for the accurate delivery of isoflurane (Steffey, 1983).

Other inhalants do not provide the (appearance of) postanesthetic analgesia that has been attributed to methoxyflurane. It should be noted, however, that some research data cast doubt on the quality of residual analgesia with methoxyflurane anesthesia (Sawyer *et al.,* 1989).

VII. HEMODYNAMIC EFFECTS OF INHALATIONAL ANESTHESIA

The cardiopulmonary depressant effects of anesthetics are generally dose related. All of the inhalant anesthetics used in veterinary medicine cause a considerable decrease in ventilation and blood pressure. These effects are dose related in that the effects are more extreme at the higher dosages required for surgical anesthesia, particularly if other (injectable) CNS depressants are not used as a part of the anesthetic management and thereby reduce the dose requirement for the inhalants. The notable hypotension that can occur is due to a combination of dilatation of the vascular smooth muscle and a decrease in cardiac output. One of the most significant advantages of isoflurane anesthesia is the relative circulatory stability provided (Eger, 1988). At clinically useful concentrations, cardiac output remains near normal. As with halothane, there is dose-dependent peripheral vasodilatation that leads to a drop in blood pressure. Unlike halothane, isoflurane produces a less pronounced effect on both the myocardium and the pumping function of the heart at equipotent anesthetics concentrations. The maintenance of cardiac output, however, results in improved tissue perfusion under isoflurane anesthesia. Isoflurane does not sensitize the heart to catecholamine-induced arrhythmias (Joas and Stevens, 1971) and may be useful in managing ventricular arrhythmias during anesthesia (Harvey and Short, 1983).

One of the authors (SP) uses isoflurane exclusively for cardiac surgeries with few adverse effects observed. The most common hemodynamic effect associated with isoflurane is a moderate hypotension that is easily corrected with phenylephrine given as a 1- to 2-μg/kg bolus or continuous infusion to effect. To help preclude isoflurane-induced dose-dependent myocardial depression, a 2- to 10-μg/kg fentanyl bolus is administered if intraoperative management would have required a concentration of isoflurane above 2%.

VIII. METHODS OF INHALATIONAL INDUCTION OF ANESTHESIA

During inhalational induction of isoflurane anesthesia using a mask or chamber technique, the most obvious feature of isoflurane is evident. The very low solubility of this agent allows for a remarkably rapid induction and recovery. In clinical practice, the inhalational induction of anesthesia is popular for many high-risk animals, for exotic species or animals that are difficult to handle, and for use in "outpatients" to allow for recovery that is free of residual, injectable anesthetic-related depression. Isoflurane is by far the most suitable anesthetic for induction of anesthesia using a mask or chamber. The facility of chamber inductions with isoflurane has popularized this technique for use in cats, small dogs, and ferrets. Exposure of personnel to waste anesthetic gases is considerable during mask and chamber inductions. Attention to appropriate precautions, such as good room ventilation, helps minimize exposures.

Rapid recoveries from inhalant halothane or isoflurane anesthesia are usually smooth and free of excitement and delirium, particularly if a preanesthetic tranquilizer or sedative has been given. A brisk recovery from anesthesia is most readily apparent on the discontinuation of inhalant anesthesia using isoflurane and the newer inhalants when no injectable CNS depressants have been used. However, recovery from a purely inhalant technique is occasionally too rapid and can be accompanied by excitement or delirium. Administration of small dosages of a tranquilizer, such as acepromazine or diazepam, can be effective in avoiding or treating such problems with the emergence from anesthesia. Analgesics such as morphine, oxymorphone, or butorphanol are recommended in place of (or in addition to) tranquilizers if postoperative pain is anticipated.

IX. INTRAOPERATIVE MONITORING AND PHYSIOLOGICAL SUPPORT

The monitoring of anesthesia should be customized to meet the requirements of each research protocol just as should the choice of anesthetic drugs and supportive measures. Basic monitoring is directed at physical changes that may reflect the development of toxic side effects of anesthetics or may signal deterioration or inappropriate anesthetic depth. Priorities in monitoring address circulation, ventilation, oxygenation, and neurological status. The provisions of appropriate instrumentation and (particularly) personnel are necessary to meet the chosen objectives. The intensity and invasiveness of monitoring will vary with the protocol design, availability of resources, choice of anesthetic methods, physiologic status of the animal, duration of the procedure, and the anticipated physiologic challenges. Maintenance of an anesthetic record should be a part of monitoring in that it provides an ongoing record of trends and changes, and it represents a potentially valuable component of the research records.

The size of dogs, cats, and ferrets makes them quite suitable for the application of much of the monitoring equipment originally designed for anesthetized human patients, particularly that designed for human neonatal and pediatric use. With many research laboratories operating in association with clinical hospitals (human or veterinary), a variety of clinical equipment is often reasonably available for short-term use in research. Increasingly, vendors are supplying used and reconditioned clinical monitoring equipment such that economic barriers to obtaining adequate devices are minimal (see Chapter 7).

The assessment of tissue perfusion can be performed by obtaining blood pressure and heart rate. Observing recovering animal's mucous membranes together with measuring capillary refill time will provide general guidance to the condition of tissue perfusion and oxygenation. Heart rate and rhythm can be evaluated by either continuously recording electrocardiogram (ECG) or intermittent auscultation of the animal's chest. Arterial blood pressure can be estimated by placing an inflatable cuff around the extremity or by palpating the pulse. Although the pulse method is less accurate, it is easy to perform and can provide information on the pulse pressure or the difference between diastolic and systolic pressures. The most precise method to obtain arterial blood pressure is by intraarterial cannulation of the femoral or cranial tibial artery. This method also provides an access for samples of arterial blood gases, allowing one to assess the alveolar ventilation and oxygenation of the animal. The drawback of invasive monitoring is that it requires continuous attention of postsurgical unit personnel because of possible technical mishaps and danger of local bleeding into surrounding tissues.

X. POSTOPERATIVE CARE

The physiologic instability of the animal recovering from anesthesia and surgery requires close surveillance. General anesthesia not only can impair the adequacy of alveolar ventilation and tissue perfusion, but also can inhibit compensatory responses to surgical stress.

A. Circulatory Complications

Circulatory complications frequently encountered during postoperative care include hypotension and cardiac dysrhythmias. The most common causes of hypotension are hypovolemia and ventricular dysfunction. Hypovolemia encountered in the immediate postoperative period is usually a reflection of inadequately replaced intraoperative fluid or blood losses. Unrecognized postoperative bleeding should be considered if a progressive deterioration of blood pressure is noticed. Postanesthetic myocardial depression due to experimental cardiac manipulations may produce ventricular dysfunctions leading to reduced cardiac output and hypotension.

Treatment is determined by identifying the cause of hypotension. Providing the animal with a bolus of intravenous crystalloid solution (5–10 ml/kg) usually corrects hypovolemia. An increase in urinary output after a fluid challenge is also suggestive of the presence of hypovolemia rather than myocardial depression. If blood pressure improves only transiently after fluid treatment, consideration should be given to the possibility of continuing surgical bleeding. The use of vasopressors (phenylephrine) or inotropes (dopamine) may be necessary to increase arterial blood pressure until the underlying cause of hypotension is corrected.

The detection of cardiac dysrhythmias in the absence of direct cardiac manipulations is usually suggestive of hypoxemia or respiratory acidosis. A rapidly developing bradycardia in the face of deteriorating tissue perfusion may be highly indicative of severe hypoxemia. Sinus bradycardia may also reflect hypothermia and the excessive use of opiates (oxymorphone, fentanyl) during anesthesia. Tachycardia is commonly seen during immediate postoperative recovery. It is usually the result of pain, hypovolemia, and hypercarbia. Consecutive premature ventricular contractions (PVC) and particularly ventricular tachycardia are considered ominous signs and should be given prompt attention. A lidocaine bolus (1–1.5 mg/kg) followed by continuous infusion (10–50 μg/kg/min) is administered to suppress potentially lethal ventricular dysrhythmias. Hypoxemia and respiratory acidosis are corrected by establishing adequate oxygenation and ventilation. Pharmacologic interventions used to treat significant cardiac dysrhythmias include atropine to increase heart rate and β-blockers (esmolol, propranolol) to eliminate excessive sympathetic activity and tachycardia and to treat dysrhythmias (i.e., PVCs and ventricular tachycardia).

B. Respiratory Complications

Respiratory complications commonly occur in the immediate postoperative period. The residual effects of most anesthetic drugs produce a depression of ventilation evidenced by the excessive accumulation of carbon dioxide. A residual neuromuscular block can also contribute to hypoventilation and hypoxemia by interfering with the optimal activity of respiratory muscles. Inadequate pain relief, particularly after thoracic and upper abdominal surgeries, will affect the ability of the animal to take a deep breath, leading to hypoventilation. Alveolar hypoventilation is considered when $PaCO_2$ exceeds 45 mm Hg and is best diagnosed by the analysis of arterial blood gases.

Intrapulmonary shunting due to atelectasis, accumulation of secretions, or airway obstruction may lead to arterial hypoxemia. Clinical signs of arterial hypoxemia (PaO_2 below 60 mm Hg) include restlessness, tachycardia, tachypnea, and may later progress into bradycardia, cyanosis, and severe depression. Cyanosis may be absent in anemic animals and clinical signs are not specific for the detection of hypoxemia. The analysis of arterial blood gases should be performed to confirm diagnosis. Animals recovering from chest surgery may develop atelectasis because of air or fluid entrapment in the pleural space. Increased respiratory rates, reduced lung sounds, and cyanosis may lead to suspicion of pneumothorax. Diagnosis can be confirmed on chest X-rays. Placement of the chest tube before closure of the thoracotomy incision will help prevent the accumulation of air or fluids.

Airway obstruction most often occurs during the immediate postoperative period in animals that have not remained sufficiently conscious to protect their airway. Therefore, extubation should not be attempted until the animal regains a swallowing reflex. If necessary, the trachea and oropharynx should be suctioned before removal of the endotracheal tube to clear secretions from airway.

Treatment of respiratory complications should be directed at correcting the contributing factors. Adequate postoperative analgesia will alleviate splinting after thoracic or upper abdominal procedures and improve respiratory function (see Section XI on postoperative analgesia). Hypoxemia in the recovery room is treated with supplemental oxygen. Supplemental oxygen can be administered via a mask, nasal catheter, or cannula with minimal discomfort to the animal. Marked hypoventilation and hypoxemia should be corrected by reintubating the animal and providing controlled ventilation while concomitant measures are employed.

C. Hypothermia

Postoperative hypothermia is commonly encountered due to many factors, including vasodilatation, intraoperative heat loss from surgical wounds, and from the high flows of unhumidified gases in the breathing circuits. An exposure to a cold ambient temperature in the operating room and the use of unwarmed intravenous fluids will also contribute to significant heat loss. Young animals are particularly at risk of developing severe hypothermia because of a large body surface area, which is responsible for increased heat loss to the environment. Severe hypothermia can result in hemodynamic compromise manifested in a decrease of cardiac output and bradycardia. It may delay awakening by reducing metabolism and excretion of anesthetic drugs. Decreased platelet function and clotting factor activity can cause increased bleeding in the postoperative period. During awakening from general anesthesia, hypothermic animals will often begin shivering to increase heat production and raise body temperature. This postoperative shivering increases oxygen consumption and may be poorly tolerated by animals with existing cardiac and pulmonary impairment.

Although it is difficult to maintain normothermia during surgical procedures, an effort should be made to minimize intraoperative heat loss. The use of a circulating water-warming blanket or overhead heating lamp and covering the animal with surgical drapes and blankets will decrease heat loss through the skin. Passive airway humidification by a heat and moisture exchanger will help prevent heat loss via the respiratory system. Shivering in the postoperative period should be treated with warming lamps or heating blankets and supplemental oxygen provided to offset the marked increase in oxygen consumption. The rise in the body temperature of the animal should be monitored because as the temperature rises, venous capacitance increases and intravascular volume must be restored to avoid hypotension. Whenever active rewarming is attempted, care must be taken to ensure against inadvertent hyperthermia.

D. Renal Failure

Postoperative renal failure can be seen as a decrease in urine production to less than 0.5 ml/kg/hr. This may reflect reduced renal blood flow secondary to hypovolemia or decreased cardiac output. Intraoperative insult to the kidneys due to hypoxia, renal vascular injury (arterial, venous obstruction), or toxic drugs (aminoglycoside antibiotics) are other precipitating causes leading to acute renal failure.

Placement of an indwelling urinary catheter is important in the early recognition of acute renal failure. It also helps to rule out an obstruction in the lower urinary tract. Treatment should be directed toward restoring intravascular volume and providing adequate cardiac output. The use of diuretics such as furosemide, mannitol, or low-dose dopamine (1–2 μg/kg/min) as a specific renal vasodilator may be indicated to stimulate urine production. Fluid challenge with the rapid infusion of 5–10 ml/kg can be attempted to correct the volume deficit, but central venous pressure should be closely monitored.

E. Hemorrhage

Bleeding in the postoperative period may indicate inadequate surgical hemostasis or coagulation disorders (disseminated intravascular coagulation, excessive heparin dose). Blood should be collected to evaluate hemostatic profile (platelet count, bleeding time, prothrombin time, partial thromboplastin time, and fibrin split products). While awaiting laboratory results, a simple whole blood clotting time test can be performed. Repeated blood samples for hematocrit evaluation will help rule out a continuous blood loss.

XI. POSTOPERATIVE ANALGESIA

Pain is a predictable response as the effects of general anesthetics dissipate in the early postoperative period. Severity of pain may vary depending on the individual animal response and site of surgery with thoracotomy, upper abdominal, and orthopedic surgeries considered to be the most painful.

The goal of postoperative pain management is not only to decrease the animal's experience to noxious stimuli, but also to restore a more physiologic state to the components affected by pain. The release of catecholamines has been long associated with significant changes in cardiovascular homeostasis. Ensuing tachycardia and hypertension may increase myocardial oxygen consumption and lead to the further deterioration of myocardial function, particularly in animals with experimental or natural cardiovascular disease. Pain in the early postoperative period may also lead to marked respiratory impairment that results in atelectasis, decreased clearance of secretions, and arterial hypoxemia. These changes may have a profound effect on recovering animals with borderline pulmonary function, i.e., cardiogenic pulmonary edema, pneumonia, and pulmonary resection.

One of the major difficulties in assessing pain in animals is their inability to communicate pain. Vocalization, changes in ambulation, depression, and other signs associated with abnormal behavior are interpreted as signs associated with pain. Unfortunately, the aforementioned behavioral displays of pain may become obtunded during the initial stages of recovery from anesthesia. However, agitation or distress associated with recovery from anesthesia may mimic some of the signs of postsurgical pain. To overcome this uncertainty and to facilitate appropriate analgesic administration, quantification of pain may be necessary. Calculation of pain score (higher number correlates with the severity of pain) has been widely used in pediatric patients (Hannalah *et al.,* 1987; Krane *et al.,* 1987). Although subjective, numerical rating scales allow a greater degree of sensitivity. Objective methods of pain assessment rely on measurements of biochemical indices such as changes in plasma catecholamines and cortisol but these tend to be expensive and inaccurate.

The opioids are effective in alleviating postsurgical pain. The main advantage of opioids is that they produce a profound analgesia with mild sedation. Oxymorphone and morphine are opioid agonists most commonly used to alleviate postoperative pain. Their side effects may include cardiorespiratory depression, a decrease in intestinal motility, and release of histamine associated with the administration of morphine. A 0.25- to 0.5-mg/kg IM dosage of morphine produces analgesia for up to 6 hr. In cats, the dose of morphine is 0.1 mg/kg IM. Dosages at 0.5 mg/kg or higher in the cat frequently cause excitement and mania (Green, 1979). Although the duration of analgesia provided by oxymorphone (0.05–0.15 mg/kg IM, SC, or IV) is only 3–5 hr, it is often a preferred opioid analgesic for postoperative pain relief because it induces minimal respiratory depression and provides excellent analgesia.

Opioid agonist–antagonists, butorphanol and buprenorphine, provide moderate pain relief. In contrast to opioid agonists, the agonist–antagonists have limited analgesic properties related to a ''ceiling effect'' after which additional dosages do not induce further analgesic action. Butorphanol (0.05 mg/kg IV or 0.2–0.5 mg/kg SC) provides 3–4 hr of analgesia characterized by less sedation than produced by opioid agonists. Buprenorphine induces prolonged analgesia (6–12 hr) after a single intramuscular or subcutaneous injection. The recommended dose of buprenorphine is 0.01–0.03 mg/kg IM or SC.

Epidural administration of opioids is an effective and safe alternative to parenteral opioids. The action of epidural opioids is mediated by opiate receptors located in the central nervous system and in particular their existence in the substantia gelatinosa of the dorsal horn of the spinal cord. This method of producing a selective blockade of pain transmission has been shown to cause minimal depression of the sympathetic nervous system or interference with neuromuscular function. Morphine and oxymorphone have been administered by epidural injection in dogs to alleviate postoperative pain (Bonath and Saleh, 1985; Popilskis *et al.,* 1991). Although many factors (i.e., volume of injectate, molecular weight, and dose of opioids) are considered important, opioid solubility appears to be a key determinant of onset, duration, and effectiveness of epidural analgesia. Morphine is less lipid soluble than other opioids and, accordingly, it has a slow onset and a long duration. It has a limited uptake from the cerebrospinal fluid (CSF), allowing the drug to persist for a longer period of time with more of the drug available at the receptor sites in the spinal cord. This results in prolonged analgesia, and a single injection of epidural morphine (0.1–0.15 mg/kg) provides postoperative analgesia in dogs for up to 24 hr (Popilskis *et al.,* 1993). The distribution of morphine throughout the CSF also makes lumbar neuroaxial administration of morphine, via epidural needle or catheter, as effective as thoracic neuraxial administration after upper abdominal and thoracic surgeries. Keegan *et al.* (1995) used an epidural bolus of morphine (0.1 mg/kg) and xylazine (0.02 mg/kg) in dogs during isoflurane (1.5 MAC) anesthesia. They found that the morphine and xylazine combination was not associated with any significant cardiovascular changes and suggest that the combination may be of value as preemptive analgesics to provide improved postoperative analgesia. With opioids that are more lipid soluble (e.g., oxymorphone), the spinal level of neuroaxial administration produces a greater impact on efficacy. Increased lipid solubility requires that the administration of oxymorphone be closer to the target dermatome. Oxymorphone (0.1 mg/kg epidurally) has also been shown to be efficacious for prolonged (10 hr) postoperative analgesia after thoracotomy in dogs when administered at the thoracic epidural space via an indwelling epidural catheter (Popilskis *et al.,* 1991).

Other methods of opioid administration have been identified. The finding that opioid receptors are located on the peripheral terminal of primary afferent neurons has been clinically applied to effect analgesia after intraarticular administration in humans and dogs (Raja *et al.,* 1992; Day *et al.,* 1995). A 0.1-mg/kg dose of morphine (diluted in saline at 1 ml/10 kg) provided reliable pain relief for up to 6 hr in dogs following exploratory stifle arthrotomy (Day *et al.,* 1995).

A transdermal drug delivery system offers a method of administering a drug in a slow and continuous fashion, resulting in fairly constant plasma drug concentrations for a prolonged period (Holley and Van Steenis, 1988). In dogs, transdermal fentanyl (50 μg/hr) resulted in plasma concentrations that are generally considered to be analgesic from 24 to 72 hr after application of the patch (Kyles *et al.,* 1996).

The rationale of using nonsteroidal anti-inflammatory drugs (NSAIDs) is that soft tissue inflammation may be a potent factor in postoperative pain. The disadvantage of most NSAIDs is that they have to be administered orally and may not be suitable for the treatment of acute postoperative pain in a semiconscious animal. These agents are also not likely to possess sufficient analgesic properties and their use is limited to adjunctive therapy for acute states. In dogs, aspirin (10–20 mg/kg PO) and acetaminophen (10 mg/kg PO) can be used to control short-term (2–4 hr) moderate somatic pain. Cats are more susceptible than dogs to aspirin toxicity due to the lack of hepatic glucuronyltransferase, and the plasma half-life of aspirin in cats is 38 hr versus 8 hr in the dog (Cribb, 1992). The dosage for aspirin in cats is 10 mg/kg/48 hr PO.

More potent NSAIDs have become available to treat postoperative pain and various musculoskeletal conditions. (Schmitt and Guentert, 1990; Vasseur *et al.,* 1995). Their analgesic effect is enhanced if given before the onset of pain. Because the mechanism of action of NSAIDs does not involve opioid receptors, it offers the advantage of producing no respiratory depression and minimal sedation. Carprofen, five times as potent as ibuprofen, has been used orally (2.2–4.0 mg/kg) and intravenously (4 mg/kg) in dogs (Nolan and Reid, 1993; Lascelles *et al.,* 1994). Carprofen administered by either route provided effective and lasting analgesia, and was associated with a faster return to the normal conscious state. One of the authors (SP) has administered ketorolac in dogs (30 mg $<$ 17 kg and 60 mg $>$ 17 kg) to treat postoperative pain with similar results. Ketorolac inhibits platelet function for 24–48 hr and should be used with caution in animals with natural and acquired coagulopathies.

Local anesthetics placed in proximity to the injury site produce a transient loss of sensory, motor, and autonomic functions. Interruption of neural pathways in the immediate postoperative period will significantly reduce not only pain perception but also overall stress response. The duration of the postoperative pain relief is dependent on properties of the local anesthetic. Bupivicaine is more lipid soluble and is metabolized less rapidly than lidocaine, making it more potent and longer acting. Both local anesthetics are used for local infiltration to produce sensory anesthesia in injected areas without blocking specific nerves. Common nerve block techniques include the intercostal nerve block, lumbar epidural block, and intrapleural regional analgesia. Bupivicaine (0.25–0.5%) injected caudal to each rib before the closure of the incision provides pain relief after thoracic surgery. The administration of bupivicaine via an intrapleural catheter has been shown to ameliorate pain associated with thoracic or upper abdominal procedures. Placement of the chest tube after thoracotomy allows for the continuous or intermittent infusion of local anesthetic into the pleural space, producing block of adjacent intercostal nerves. In a study by Thompson and Johnson (1991), a dosage of intrapleural bupivicaine (1.5 mg/kg) provided adequate analgesia in dogs for 3–12 hr. This dose of bupivicaine did not result in undesirable hemodynamic and respiratory effects (Kushner *et al.,* 1995).

Providing adequate pain relief is one of the most important tasks in immediate postoperative care. A heightened awareness of this problem will help ensure that postsurgical unit personnel are aware of the variables involved. Improvements in the ability to quantify pain and the introduction of effective alternative analgesic methods should lead to considerable improvement in acute pain therapy. Preemptive analgesia and balanced analgesic methods offer such improvements.

XII. SPECIFIC CONSIDERATIONS

Attention should be directed to more exhaustive sources in preparing anesthetic protocols for specific conditions. The considerations listed here are intended to emphasize a few pertinent points.

A. Neonatal and Pediatric Animals

Neonates and relatively young animals are subject to hypothermia under anesthesia as a result of immature thermogenesis and their relatively great surface area to body mass ratio. There is a tendency for these animals to develop hypoglycemia for a variety of reasons, including the limited glycogen reserves and high metabolic rate. Neonates are dependent on maintaining heart rate in order to maintain cardiac output. It is therefore important to avoid bradycardia (Grandy and Dunlop, 1991). The metabolic elimination of anesthetics is limited in neonatal animals.

B. Cesarean Section

Because all drugs capable of crossing the blood–brain barrier (e.g., anesthetics) also cross the placental barrier, principles of neonatal anesthesia should be considered. Regional anesthetic techniques that avoid fetal depression include epidural anesthesia (suitable for dogs, cats, and ferrets) and use of a local anesthetic line block. Either one of these techniques can substantially reduce the depth of general anesthesia needed, thereby facilitating brisk recovery. Anesthetic drugs that can be reversed, particularly opioids, may be useful. Prompt recovery will often facilitate maternal care and will be advantageous unless the intent is to obtain barrier-derived offspring with no maternal contact. Adequate provision should be made for warming the offspring and providing supplemental oxygen. Equipment and supplies for other resuscitative measures should be available.

C. Geriatric Animals

The limited reserve of the geriatric animal for hepatic and renal metabolism and the elimination of drugs are important considerations in anesthetic management (Paddleford, 1995). Delayed recovery and prolonged side effects may reflect the persistence of anesthetics in the body. As an example, reduced hepatic metabolism, delayed recovery, and the adverse hypotensive effects of acepromazine make it a relatively poor choice for most anesthetic needs for geriatrics.

For geriatrics, it is useful to select anesthetics that (1) can be fully reversed (opioids, possibly xylazine, potentially diazepam), (2) can be totally eliminated by supported ventilation (isoflurane), (3) have neither intrinsic toxicity nor significant adverse effects should their effects persist (glycopyrrolate, diazepam, butorphanol), or (4) have very rapid metabolic elimination (propofol). In situations where inadequate metabolism or elimination is recognized, physiologic support including judicious fluid therapy, support of hemodynamics and body temperature, and ventilatory support should be provided as indicated.

D. Extracorporeal Bypass

Although swine and ruminants are more popular models, dogs have been used in cardiopulmonary bypass and extracorporeal membrane oxygenation procedures (Klement *et al.*, 1987). Species-dependent limitations and logistical concerns in planning and support for extracorporeal bypass warrant careful consideration (Fish *et al.*, 1994). Despite substantial problems, controlled cross circulation has been used as a method of cardiopulmonary bypass in dogs. The extracorporeal oxygenator is functionally replaced by a donor patient (Hunt, 1994; Kuntz *et al.*, 1996).

E. Prolonged Surgical Procedures

Any anesthetic/surgical procedure of longer than 3 or 4 hr may warrant special attention. Airway management to minimize desiccation, careful management of intravenous volume, avoidance of cumulative toxicity, monitoring of urine output, and maintenance of body temperature and hemodynamics are concerns in prolonged procedures.

F. Breed Considerations

Although many research dogs are beagles or hound dogs, there may be some significant breed considerations in laboratory dogs. Breed-specific differences in drug metabolism have been noted in ''sight hound'' dog breeds, including greyhounds, that are occasionally used in biomedical research. Inefficient hepatic biotransformation and altered distribution of injectable anesthetics in sight hounds make barbiturates less suitable for these dogs (Sams *et al.*, 1985). Additional canine breed sensitivities and unique features are occasionally encountered in clinical veterinary anesthesia. Anecdotal reports of various breed sensitivities abound.

G. Ocular and Intraocular Surgery

The two goals of anesthesia for ophthalmic surgery are (1) to avoid increases in intraocular pressure and (2) to support cardiovascular function (Ludders, 1993). The significant potential for oculocardiac reflex vagal stimulation suggests the use of anticholinergics in many cases. Neuromuscular blocking agents are commonly used during general anesthesia to immobilize the eye and to facilitate ocular surgery. It is imperative that paralyzed animals are adequately anesthetized. The use of paralytic agents also mandates controlled ventilation. Postoperative analgesics are warranted after intraocular procedures, and sedatives or tranquilizers can help provide a smooth and calm recovery.

H. Cardiac Dysfunction

The susceptibility of each of these species to cardiac disease and the suitability of each for modeling human cardiac dysfunction have led to their use in a variety of cardiovascular research. Cardiomyopathy in each species has been exploited as animal models of human disease and in research on the animal diseases themselves. There may be special considerations for autonomic blockade in studies of cardiovascular reflexes and cardiac conduction systems. Anesthetic management varies with the particular circulatory disease or model. The potential for anesthetic-mediated destabilization of compensatory mechanisms requires appropriate preparation and management (Hellyer, 1992).

REFERENCES

Alibhai, H. I. K., Clarke, K. W., Lee, Y. H., and Thompson, J. (1996). Cardiopulmonary effects of combinations of medetomidine hydrochloride and atropine sulphate in dogs. *Vet. Rec.* **138,** 11–13.

Andress, J. L., Day, T. K., and Day, D. G. (1995). Effects of consecutive day propofol anesthesia on feline red blood cells. *Vet. Surg.* **24,** 277–282.

Autefage, A., Fayolle, P., and Toutain, P. L. (1990). Distribution of material injected intramuscularly in dogs. *Am. J. Vet. Res.* **51,** 901–904.

Bednarski, R. M., Bednarski, L. S., and Muir, W. W., III (1984). Cost comparison of anesthetic regimens in the dog and cat. *J. Am. Vet. Med. Assoc.* **185,** 869–872.

Bonath, K. H., and Saleh, A. S. (1985). Long term pain treatment in the dog by peridural morphines. *Proc. Int. Congr. Vet. Anesthesiol. 2nd,* Santa Barbara, California, p. 161.

Brearley, J. C., Kellagher, R. E. B., and Hall, L. W. (1988). Propofol anesthesia in the cat. *J. Small Anim. Prac.* **29,** 315–322.

Copland, V. S., Haskins, S. V., and Patz, J. (1989). Naloxone reversal of oxymorphone effects in dogs. *Am. J. Vet. Res.* **50,** 1854–1858.

Cornick, J. L., and Hartsfield, S. M. (1992). Cardiopulmonary and behavioral effects of combinations of acepromazine/butorphanol and acepromazine/oxymorphone in dogs. *J. Am. Vet. Med. Assoc.* **200,** 1952–1956.

Cribb, P. H. (1992). Precautions when using antiprostaglandins. *Vet. Clin. North Amer.: Small Anim. Pract.* **22**(2), 370–374.

Daunt, D. A. (1994). Dogs and cats: Anesthesia and analgesia. *In* ''Research Animal Anesthesia, Analgesia and Surgery'' (A. C. Smith and M. M. Swindle, eds.), pp. 93–105. Scientists Center for Animal Welfare, Greenbelt, Maryland.

Day, T. K., Pepper, W. T., Tobias, T. D., Flynn, M. F., and Clarke, K. M. (1995). Comparison of intra-articular and epidural morphine for analgesia following stifle arthrotomy in dogs. *Vet. Surg.* **24,** 522–530.

Dodam, J. R., Kruse-Elliot, K. T., Aucoin, D. P., and Swanson, C. R. (1990). Duration of etomidate-induced adrenocortical suppression during surgery in dogs. *Am. J. Vet. Res.* **51,** 786–788.

Dunkle, N., Moise, N. S., Scarlett-Kranz, J., and Short, C. E. (1986). Cardiac performance in cats after administration of xylazine or xylazine and glycopyrrolate: Echocardiographic evaluations. *Am. J. Vet. Res.* **47**(10), 2212–2216.

Eger, E. I., II (1988). Circulation. *In* ''Isoflurane (Forane): A Compendium and Reference,'' 2nd ed., pp. 37–72. Anaquest, Inc., Madison, Wisconsin.

Fish, R., Hogan, M. C., Wright, J. A., and Walls, J. T. (1994). Planning and support for experimental cardiopulmonary bypass and extracorporeal membrane oxygenation studies. *In* ''Research Animal Anesthesia, Analgesia and Surgery'' (A. C. Smith and M. M. Swindle, eds.) pp. 147–157. Scientists Center for Animal Welfare, Greenbelt, Maryland.

Flecknell, P. A. (1985). The management of post-operative pain and distress in experimental animals. *Anim. Technol.* **36**(2), 97–103.

Frederiksen, M. C., Henthorn, T. K., Ruo, T. I., and Atkinson, A. J., Jr. (1983). Pharmacokinetics of pentobarbital in the dog. *J. Pharmacol. Exp. Ther.* **225,** 355–360.

Grandy, J. L., and Dunlop, C. I. (1991). Anesthesia of pups and kittens. *J. Am. Vet. Med. Assoc.* **198**(7), 1244–1249.

Green, C. J. (1979). Order Carnivora. *In* ''Animal Anaesthesia,'' pp. 199–215. Laboratory Animals Ltd., London.

Hall, L. W., and Chambers, J. P. (1987). A clinical trial of propofol infusion anesthesia in dogs. *J. Small Anim. Pract.* **28,** 623–637.

Hall, L. W., Lagerweij, E., Nolan, A. M., and Sear, J. W. (1994). Effect of medetomidine on the pharmacokinetics of propofol in dogs. *Am. J. Vet. Res.* **55,** 116–120.

Hannalah, R. S., Broadman, L. M., Belman, A. B., Abramovitz, M. D., and Epstein, B. S. (1987). Comparison of caudal and ilioinguinal/iliohypogastric nerve blocks for control of post-orchidopexy pain in pediatric ambulatory surgery. *Anesthesiology* **66,** 832–834.

Hardenbergh, J. G., and Mann, F. C. (1927). The auto-inhalation method of anesthesia in canine surgery. *J. Am. Vet. Med. Assoc.* **71,** 493–501.

Hartsfield, S. 1992). Advantages and guidelines for using ketamine for induction of anesthesia. *Vet. Clin. North Am.: Small Anim. Pract.* **22**(2), 266–267.

Harvey, J. W., Sameck, J. H., and Burgard, F. J. (1979). Benzocaine-induced methemoglobinemia in dogs. *J. Am. Vet. Med. Assoc.* **175,** 1171–1175.

Harvey, R. C. (1992). Precautions when using mask induction. *Vet. Clin. North Am.: Small Anim. Pract.* **22**(2), 310–311.

Harvey, R. C., and Short, C. E. (1983). The use of isoflurane for safe anesthesia in animals with traumatic myocarditis or other myocardial sensitivities. *Canine Pract.* **10,** 18–23.

Haskins, S. C., Farver, T. B., and Patz, J. D. (1985). Ketamine in dogs. *Am. J. Vet. Res.* **46,** 1855–1860.

Haskins, S. C., Farver, T. B., and Patz, J. D. (1986). Cardiovascular changes in dogs given diazepam and diazepam-ketamine. *Am. J. Vet. Res.* **47,** 795–798.

Hellyer, P. W. O. (1992). Anesthesia for patients with cardiopulmonary disease. *In* ''Current Veterinary Therapy. XI: Small Animal Practice''(R. W. Kirk and J. D. Bonagura, eds.), pp. 655–660. Saunders, Philadelphia.

Holley, F. O., and Van Steenis, C. (1988). Post-operative analgesia with fentanyl: Pharmacokinetics and pharmacodynamics of constant rate IV and transdermal delivery. *Br. J. Anaesth.* **60,** 608–613.

Holzgrefe, H. H., Everitt, J. M., and Wright, E. W. (1987). Alpha-chloralose as a canine anesthetic. *Lab. Anim. Sci.* **37,** 587–595.

Hunt, G. (1994). Cross-circulation for cardiopulmonary bypass. *Semin. Vet. Med. Surg.* **9,** 217–220.

Jacobson, J. D., and Hartsfield, S. M. (1993). Cardiorespiratory effects of intravenous bolus administration and infusion of ketamine-midazolam in dogs. *Am. J. Vet. Res.* **54,** 1710–1715.

Joas, T. A., and Stevens, W. C. (1971). Comparison of the arrhythmic dosages of epinephrine during Forane, halothane, and floroxene anesthesia in dogs. *Anesthesiology* **35,** 48–53.

Johnson, D. L. (1981). Venipuncture in unrestrained cats. *Lab. Anim.* **10**(7), 35–38.

Keegan, R. D., Greene, S. A., and Weil, A. B. (1995). Cardiovascular effects of epidurally administered morphine and a xylazine-morphine combination in isoflurane-anesthetized Dogs. *Am. J. Vet. Res.* **56,** 496–506.

Klement, P., del Nido, P. J., Mickleborough, L., MacKay, C., Klement, G., and Wilson, G. J. (1987). Technique and postoperative management for successful cardiopulmonary bypass and open-heart surgery in dogs. *J. Am. Vet. Med. Assoc.* **190,** 869–874.

Klide, A. M. (1992). The case for low gas flows. *Vet. Clin. North Am.: Small Anim. Pract.* **22**(2), 384–387.

Klide, A. M., and Haskins, S. C. (1992). Precautions when using nitrous oxide. *Vet. Clin. North Am.: Small Anim. Pract.* **22**(2), 314–316.

Ko, J. C. H., Thurmon, J. C., Benson, G. J., Tranquilli, W. J., Olson, W. A., and Vaha-Vahe, A. T. (1994). Hemodynamic and anesthetic effects of etomidate infusion in medetomidine-premedicated dogs. *Am. J. Vet. Res.* **55,** 842–846.

Kramer, S., Nolte, I., and Jochle, W. (1996). Clinical comparison of medetomidine with xylazine/l-methadone in dogs. *Vet. Rec.* **138,** 128–133.

Krane, E. J., Jacobson, L. E., Lynn, A. M., Parrot, C., and Tyler, D. C. (1987). Caudal morphine for postoperative analgesia in children: A comparison with caudal bupivicaine and intravenous morphine. *Anesth. Analg.* **66,** 647–653.

Kuntz, C. A., Johnston, S. A., Jacobson, J., Martin, R. A., Moon, M., Shires, P., and Lee, J. C. (1996). Controlled cross circulation in dogs: Effects on donor hemodynamics. *Vet. Surg.* **25,** 29–39.

Kushner, L. I., Trim, C. M., Madhusudhan, S., and Boyele, C. R. (1995). Evaluation of the hemodynamic effects of intrapleural bupivicaine in dogs. *Vet. Surg.* **24,** 180–187.

Kyles, A. E., Papich, M., and Hardie, E. M. (1996). Disposition of transdermally administered fentanyl in dogs. *Am. J. Vet. Res.* **57**(5), 715–719.

Lang, R. M., Marcus, R. H., Neumann, A., Janzen, D., Hansen, D., Fujii, A., and Borow, K. (1992). A time-course study of the effects of pentobarbital, fentanyl, and morphine-choralose on myocardial mechanics. *J. Appl. Physiol.* **73,** 143–150.

Lascelles, B. D., Butterworth, S. J., and Waterman, A. E. (1994). Postoperative analgesic and sedative effects of carprofen and petidine in dogs. *Vet. Rec.* **134**(8), 187–191.

Ludders, J. (1993). Anesthesia for ophthalmic surgery. *In* ''Textbook of Small Animal Surgery'' (D. Slatter, ed.), 2nd ed., pp. 2276–2278. Saunders, Philadelphia.

McCrackin, M. A., Harvey, R. C., Sackman, J. E., McLean, R. A., and Paddleford, R. R. (1994). Butorphanol tartrate for partial reversal of oxymorphone-induced postoperative respiratory depression in the dog. *Vet. Surg.* **23,** 67–74.

Nolan, A., and Reid, J. (1993). Comparison of the postoperative analgesic and sedative effects of carprofen and papavertum in the dog. *Vet. Rec.* **133**(10), 240–242.

Paddleford, R. R. (1995). Anesthesia. *In* ''Geriatrics and Gerontology of the Dog and Cat'' (R. T. Goldston and J. D. Hoskins, eds.), pp. 363–377. Saunders, Philadelphia.

Pascoe, P. J. (1992). The case for maintenance of general anesthesia with an injectable agent. *Vet. Clin. North Amer.: Small Anim. Pract.* **22**(2), 275–277.

Popilskis, S. J., Kohn, D. F., Sanchez, J. A., and Gorman, P. (1991). Epidural versus intramuscular oxymorphone analgesia after thoracotomy in dogs. *Vet. Surg.* **20,** 462–467.

Popilskis, S. J., Kohn, D. F., Laurent, L., and Danilo, P. (1993). Efficacy of epidural morphine versus intravenous morphine in alleviation of post-thoracotomy pain in dogs. *J. Vet. Anaesthesiol.* **20,** 22–26.

Pypendop, B., Serteyn, D., and Verstegen, J. (1996). Hemodynamic effects of medetomidine midazolam-butorphanol and medetomidine-midazolam-buprenorphine combinations and reversibility by atipamezole in dogs. *Am. J. Vet. Res.* **57,** 724–730.

Raja, S. N., Dickstein, R. E., and Johnson, C. A. (1992). Comparison of post-operative analgesic effects of intraarticular bupivicaine and morphine following arthroscopic knee surgery. *Anesthesiology* **77,** 1143–1147.

Rawlings, C. A., and Kolata, R. J. (1983). Cardiopulmonary effects of thiopental/lidocaine combination during anesthetic induction in the dog. *Am. J. Vet. Res.* **44**(1), 144–149.

Sams, R. A., Muir, W. W., Detra, R. L., and Robinson, E. P. (1985). Comparative pharmacokinetics and anesthetic effects of methohexital, pentobarbital, thiamylal, and thiopental in Greyhound dogs, and non-Greyhound, mixed-breed dogs. *Am J. Vet. Res.* **46**(8), 1677–1683.

Sawyer, D. C. (1984). Canine and feline endotracheal intubation and laryngoscopy. *Compend. Contin. Educ.* **6**(11), 973–984.

Sawyer, D. C., Rech, R., and Adams, T. (1989). Analgesic effects of meperidine and oxymorphone following methoxyflurane and halothane in the dog (abstr. 2). *Proc. Am. Coll. Vet. Anesthesiol.,* New Orleans, p. 8.

Schmitt, M., and Guentert, T. W. (1990). Biopharmaceutical evaluation of carprofen following Single intravenous, oral, and rectal doses in dogs. *Biopharm. Drug Dispos.* **11,** 585–594.

Silverman, J., and Muir, W. W. (1993). A review of laboratory animal anesthesia with chloral hydrate and chloralose. *Lab. Anim. Sci.* **43,** 210–216.

Steffey, E. P. (1983). Accuracy of isoflurane delivery by halothane-specific vaporizers. *Am. J. Vet. Res.* **44,** 1072–1078.

Steffey, E. P. (1992). Other new and potentially useful inhalational anesthetics. *Vet. Clin. North Am.: Small Anim. Pract.* **22**(2), 335–340.

Stepien, R. L., Bonagura, J. D., Bednarski, R. M., and Muir, W. W. (1995). Cardiorespiratory effects of acepromazine maleate and buprenorphine hydrochloride in clinically normal dogs. *Am. J. Vet. Res.* **56,** 78–84.

Thompson, S. E., and Johnson, J. M. (1991). Analgesia in dogs after intercostal thoracotomy: A comparison of morphine, selective intercostal nerve block, and interpleural regional analgesia with bupivicaine. *Vet. Surg.* **20**(1), 73–77.

Thurmon, J. C., Ko, J. C., Benson, G. J., Tranquilli, W. J., and Olson, W. A. (1994). Hemodynamic and analgesic effects of propol infusion in medetomidine-premedicated dogs. *Am. J. Vet. Res.* **55**(3), 363–367.

Thurmon, J. C., Tranquilli, W. J., and Benson, G. J. (1996). Perioperative pain and distress. *In* ''Veterinary Anesthesia'' 3rd ed., (W. V. Lumb and E. W. Jones, eds.), p. 56. Williams & Wilkins, Baltimore, Maryland.

Tranquilli, W. J., and Benson, G. J. (1992). Advantages and guidelines for using alpha-2 agonists as adjuvants. *Vet. Clin. North Am.: Small Anim. Pract.* **22**(2), 289–293.

Vaha-Vahe, T. (1989). Clinical evaluation of medetomidine, a novel sedative and analgesic drug for dogs and cats. *Acta Vet. Scand.* **30,** 267–273.

Vainio, O., Vaha-Vahe, T., and Palmer, L. (1989). Sedative and analgesic effects of medetomidine in dogs. *J. Vet. Pharmacol. Ther.* **12,** 225–231.

Vasseur, P. B., Johnson, A. L., Budsberg, S. C., Lincoln, J. D., Toombs, J. P., Whitehair, J. G., and Lentz, E. L. (1995). Randomized, controlled trial of the efficacy of carprofen, a nonsteroidal anti-inflammatory drug, in treatment of osteoarthritis in dogs. *J. Am. Vet. Med. Assoc.* **206**(6), 807–811.

Wilson, D. V. (1992). Advantages and guidelines for using opioid agonists for induction of anesthesia. *Vet. Clin. North Am.: Small Anim. Pract.* **22**(2), 269–272.

Young, L. E., and Jones, R. S. (1990). Clinical observations on medetomidine/ketamine anaesthesia and its antagonism by atipamezole in the cat. *J. Small Anim. Pract.* **31,** 221–224.

PART B. ANESTHESIA AND ANALGESIA IN FERRETS

Sally K. Wixson

I. INTRODUCTION

Ferrets have become popular research animals, particularly as a replacement for dogs and cats. Ferrets have been used as animal models for bacterial and viral diseases, reproduction, and gastrointestinal research. Ferrets have also become a popular model for simulating human infants for teaching intubation skills to physicians, nurses, and emergency medical personnel. Anesthetic techniques for ferrets are, in general, adapted from those which have proven to be successful in dogs and particularly in cats. Modifications of these techniques, as well as novel methods that have been proven to be successful in the experience of the authors, are summarized in the following sections.

II. UNIQUE FEATURES OF FERRET ANESTHESIA

Laboratory ferrets were originally domesticated from the wild polecat and, in general, cannot be ''trusted'' to have the same docile temperament and tractable nature as is often encountered with purpose-bred canines and felines. For this reason, sedation or general anesthesia is a frequent necessity in order to use them in research projects. Because of their small size and rapid metabolic rate, intraoperative and postoperative supportive techniques are often the same as many of those used for rabbits and large rodents rather than those used for larger-sized carnivores or omnivores.

III. PREOPERATIVE PATIENT EVALUATION AND CARE

Laboratory ferrets can be purchased from either purpose-bred or random-source facilities. As with other laboratory species, it is far superior to incorporate purpose-bred, specific pathogen-free animals into a research project than it is to use animals with a questionable health status. Ferrets should be purchased only from USDA-licensed vendors and should be housed in a closed indoor colony. As with dogs and cats, it is generally recommended that ferrets be quarantined, usually for 7 days or more prior to initiation of experimental studies. Depending on the type of study to be undertaken, quarantine screening measures may include physical examination, hematologic and biochemical screening, and fecal examination for ova and parasites.

In ferrets, blood samples for quarantine screening are most frequently obtained by cephalic, tarsal, or, less frequently, coccygeal venipuncture or by toenail clipping. Although cardiac puncture and orbital bleeding have been popular methods for blood withdrawal from sedated or anesthetized ferrets in the past, these methods are now used infrequently due to their attendant risks and the availability of other, easily accessible blood vessels for sampling. Depending on the nature of the study (e.g., cardiovascular project will require baseline ECGs), additional specified prestudy testing may be warranted.

For chronic studies, ovariohysterectomy is often performed on female ferrets during the quarantine period. This procedure eliminates bone marrow suppression and aplastic anemia, which may occur in cycling females during chronic research studies. Behavioral conditioning to frequent handling, restraint for blood sampling, bandage changing, and care of exteriorized catheters or prosthetic devices may also be needed. Technicians need to be aware of special handling techniques that must be applied to ferrets, e.g., ferrets cannot be restrained by stretching them out with traction on the legs, as is a common restraint technique with cats. Acclimation of both the staff and the research animal to these procedures should occur during the quarantine period. Quarantine procedures for ferrets may also include descenting, particularly if they will be subsequently used in chronic studies. Laboratory ferrets are also frequently vaccinated for feline and canine distemper during the quarantine period.

IV. METHODS OF ANESTHETIC DELIVERY AND EQUIPMENT

In general, methods for delivery of parenteral or inhalant anesthetics that are appropriate for use in cats, rabbits, or small dogs may be used on ferrets.

For inhalant anesthesia, neonatal or infant-size masks appropriate for rabbits can be used to cover the mouth and nose of the ferret. Alternatively, a large mask suitable for use in a large cat can be used with insertion of the entire head of the ferret into the mask. In this latter format, the gasket on the mask forms a seal around the neck of the animal to minimize waste gas contamination of the ambient air. Ferrets can also be induced using a plexiglass anesthesia chamber appropriate for use with cats; however, in the experience of the author, ferrets tend to become excitable when placed within an anesthetic chamber.

Depending on the size of the ferret, Cole or cuffed endotracheal tubes (size 3–4 mm internal diameter) can easily be placed to deliver inhalant anesthesia. Similarly to cats, ferrets may be susceptible to laryngospasm so it is beneficial to apply topical lidocaine to the vocal folds prior to attempts to intubate. Pediatric nonrebreathing anesthesia circuits (i.e., Bain, Mapleson, or Magill) and ventilators should be used to support inhalant anesthesia of ferrets.

Parenteral anesthetics can be delivered intramuscularly using a 25- to 27-gauge needle to inject into the semitendinosus/semimembranosus muscle mass. Anesthetics are infrequently delivered by the intraperitoneal route to ferrets, using a 0.55-mm-diameter, 2.5-cm-long needle. It is particularly important in ferrets not to position IP injections into the left upper abdominal quadrant because ferrets frequently have a disproportionately large spleen that may be lacerated during the injection. Anesthetics and sedatives can be delivered intravenously using a 24-gauge butterfly or over-the-needle catheter or 26-gauge needle on a tuberculin syringe. The most commonly used sites for IV anesthetic delivery are the anterior cephalic or recombinant tarsal veins. The coccygeal artery of the tail is less commonly used for vascular access but is appropriate for use in emergencies.

V. SEDATIVES AND TRANQUILIZERS

Due to their potentially fractious nature, ferrets are often sedated for handling to facilitate research manipulations. A listing of sedatives/tranquilizers appropriate for use in ferrets is presented in Table BI. Recommended regimens include acepromazine (0.2–0.5 mg/kg SC or IM), xylazine (1.0 mg/kg SC or IM), diazepam (1.0–2.0 mg/kg IM), and ketamine (10–20 mg/kg IM) (Fox, 1988; Bernard *et al.,* 1984; Janssens, 1978). In the experience of the author, 0.5 mg/kg acepromazine IM, using a dilute solution of 1 mg/ml acepromazine, is quite useful for ferret sedation.

VI. PARENTERAL ANESTHETICS

In the past, parenteral anesthetics were the mainstay of general anesthetic techniques used in ferrets. Recent requirements for dedicated operating room facilities for nonrodent species used in research now present investigators using ferrets with the opportunity to use inhalant methods and sophisticated scavenging technology. Concomitantly, the use of parenteral anesthetics in ferrets has declined. When using injectable anesthetics, as

TABLE B.I

ANESTHETICS AND TRANQUILIZERS FOR USE IN FERRETS

Drug	Dosage	Route of injection	Duration of anesthesia	Reference
Sedatives/tranquilizers				
Acetylpromazine (acepromazine)	0.2–0.5 mg/kg	IM, SC	Sedation	Ryland *et al.* (1983), Fox (1988)
Xylazine (Rompun)	1.0 mg/kg	IM, SC	Sedation	Ryland *et al.* (1983), Fox (1988)
Diazepam (Valium)	1.0–2.0 mg/kg	IM	Sedation	Fox (1988)
Barbiturates				
Pentobarbital (Nembutal)	30–50 mg/kg	IP	Up to 1 hr	Ryland *et al.* (1983), Andrews and Illman (1987), Fox (1988), Grundy (1990)
Dissociatives				
Ketamine	10–35 mg/kg	IM	30–60 min (sedation only)	Andrews and Illman (1987), Janssens (1978), Moreland and Glaser (1985)
Ketamine	20–35 mg/kg	IM	41 min	Moreland and Glaser (1985)
Diazepam	0.5–3.0 mg/kg	IM, SC	Sedation only	Andrews and Illman (1987)
Ketamine +	20–30 mg/kg	IM	41 min	Moreland and Glaser (1985), Ryland *et al.* (1983),
xylazine	1–4 mg/kg	IM	Light anesthesia	Burgmann (1991)
reverse with yohimbine	0.5 mg/kg	IM		Messina *et al.* (1988)
Tiletamine + zolazepam (Telazol)	12–22 mg/kg	IM	Sedation only	Payton and Pick (1989)
Telazol +	T(3 mg/kg) +	IM	30–45 min	Ko *et al.* (1996)
ketamine +	K(2.4 mg/kg) +	IM		
xylazine	X(0.6 mg/kg)	IM		
Neuroleptanalgesics				
Fentanyl + fluanisone	1 ml/2.5 kg or	SC	Sedation	Harper (1978)
(Hypnorm) +	0.5 ml/kg	IM	15–30 min	Andrews and Illman (1987)
xylazine or	1.0 mg/kg	IM	Sedation	
diazepam	2.0 mg/kg	IM		Andrews and Illman (1987)
reverse with				
naloxone	0.04–1.0 mg/kg	IV, IM, SC		
Other				
Alphaxolone + alphadolone (Saffan, Althesin)	6–8 mg/kg	IM	Preanesthetic sedation	Fox (1988)
	12–15 mg/kg	IM	15–30 min (light sedation)	Green (1978), Andrews and Illman (1987)
	15–20 mg/kg	IM	10–20 min (deep sedation)	Green (1978)
	10–12 mg/kg	IV	10–15 min (anesthesia)	Green (1978)
Urethane (50% w/v)	13–15 g/kg	IP	4–6 hr	Andrews and Illman (1987), Andrews *et al.* (1979, 1984), Grundy (1990)
		Route of administration		
Inhalants				
Nitrous oxide	50/50 with oxygen and 3% halothane	Inhalant	Until discontinued	Janssens (1978), Fox (1988)
Halothane	1–5%	Inhalant	Until discontinued	Murat and Housmans (1988), Housmans and Murat (1988a,b), Adams *et al.* (1979), Ryland *et al.* (1983), Andrews and Illman (1987)
Methoxyflurane ± N_2O/O_2 (1:1)	1–3%	Inhalant	Until discontinued	Ryland *et al.* (1983), Andrews and Illman (1987)
Isoflurane	1–5%	Inhalant	Until discontinued	Burgmann (1991), Fox (1988), Housmans and Murat (1988a,b), Murat and Housmans (1988)

with small rabbits and rodents, it is best to dilute the drugs to at least half-strength to aid absorption and accuracy of dosage and to minimize the tissue-damaging effects of some drugs.

A. Barbiturates

Similar to its popularity for use in other laboratory species, pentobarbital has been the barbiturate most frequently used to induce general anesthesia of ferrets. Intraperitoneal dosages of 30–100 mg/kg pentobarbital provide deep sedation to anesthesia with a 5- to 10-min onset and a 60- to 120-min duration of effect (Andrews and Illman, 1987; Ryland *et al.,* 1983; Grundy, 1990; Housmans and Murat, 1988a,b; Moreland and Glaser, 1985; Jellie, 1978; Fox, 1988). As has been shown in other species, pentobarbital provides minimal analgesia and causes marked respiratory depression with a prolonged recovery period of up to 1–2 hr (Andrews and Illman, 1987).

B. Dissociatives

1. Ketamine

Ketamine has been used both alone in ferrets and combined with sedatives/tranquilizers to produce a greater depth of anesthesia. Dosages of 60 mg/kg IM ketamine provide a rapid induction (2–3 min) and a 55-min loss of righting but provide only brief periods of analgesia. The sole use of ketamine as a general anesthetic/sedative in ferrets is characterized by profound muscle rigidity and paddling movements during recovery (Moreland and Glaser, 1985).

Ketamine combined with xylazine is a popular dissociative–tranquilizer combination used in ferrets. Recommended dosages include intramuscular or subcutaneous 1–4 mg/kg xylazine followed by 20–30 mg/kg ketamine (Ryland *et al.,* 1983; Burgmann, 1991; Moreland and Glaser, 1985). A further depth of anesthesia can be produced by supplementation with inhalants. Moreland and Glaser (1985) judged this combination to be the superior dissociative–tranquilizer mixture for use in ferrets because it produced a 69-min loss of righting, 47 minutes of excellent analgesia, and a smooth recovery. The only disadvantage noted from this combination was the presence of cardiac arrhythmias during the immediate postinjection period. An added advantage of the ketamine–xylazine combination is the ability to reverse the xylazine component with α_2-antagonists, such as yohimbine hydrochloride. Sylvina *et al.* (1990) showed that 0.5 mg/kg intramuscular yohimbine injected 15 min after the ketamine–xylazine combination decreased the duration of recumbency by approximately 50% as well as antagonized the bradycardia caused by this anesthetic combination.

The ketamine–diazepam combination has also been utilized in ferrets to provide sedation to light anesthesia. Intramuscular dosages of 20 mg/kg ketamine plus subcutaneous 0.5 mg/kg diazepam produce short-term sedation (Andrews and Illman, 1987). A higher dosage of 35 mg/kg ketamine plus 3 mg/kg diazepam IM produces 41 min of immobilization characterized by excellent muscle relaxation but poor analgesia and occasional cardiac irregularities (Moreland and Glaser, 1985). The ketamine–diazepam combination is also frequently characterized by paddling movements and sneezing on recovery. For these reasons, this combination is not suitable for surgical procedures requiring extensive manipulation of the animal and qualified personnel should be available to support the animal during recovery.

2. Tiletamine–Zolazepam (Telazol)

Telazol is an anesthetic composed of the dissociative agent tiletamine plus the benzodiazepine tranquilizer zolazepam. Based on its popularity in cats, in particular, this combination has been used alone and combined with ketamine plus xylazine in ferrets (Payton and Pick, 1989; Ko *et al.,* 1996). Results showed that a low dosage (12 mg/kg IM) of Telazol produced only sedation; however, a higher dosage of 22 mg/kg IM caused sedation, muscle relaxation, and variable analgesia adequate for minor, short-duration surgical procedures. At both dosages, Telazol caused abnormal respiration that was characterized by shallow, rapid breaths with the low dosage and an apneustic breathing pattern with the higher dosage. Additionally, ferrets anesthetized with either Telazol dosage manifested random, occasionally violent sneezing throughout the duration of anesthesia and a rough recovery characterized by opisthotonos and excessive paddling and swimming motions. Because the muscle relaxation provided by zolazepam lasted longer than the analgesia produced by tiletamine, recovery was prolonged and lasted from 117 min to 4 hr. Telazol in combination with ketamine and xylazine (3 mg/kg Telazol, 0.6 mg/kg xylazine, and 2.4 mg/kg ketamine IM) provided a rapid onset and approximately 30-min losses of the toe pinch, tail pinch, and skin pinch responses. This combination was further characterized by an approximately 45-min duration of tolerance to endotracheal intubation and smooth recovery. Heart rate and systolic blood pressure decreased following the administration of Telazol + xylazine + ketamine; however, hypotension and bradycardia did not occur. Because physiologically significant hypoventilation did occur with this combination, the authors suggest that oxygen insufflation should accompany this regimen in order to maintain normal oxygen saturation of hemoglobin.

C. Neuroleptanalgesics

The most popular neuroleptanalgesic used in the United States, fentanyl–droperidol (Innovar-Vet), has not been described for use in ferrets. As with other small laboratory species, fentanyl–fluanisone (Hypnorm) is the neuroleptanalgesic of choice for research ferrets used in Europe. Dosages of 0.5

ml/kg IM provide a 6- to 8-min onset with a 15- to 30-min duration of effect and rapid recovery within 1 hr following injection (Andrews and Illman, 1987). Recovery time can be further shortened by reversing the narcotic component of this neuroleptanalgesic by administrating the narcotic antagonists naloxone hydrochloride (at a dosage of 0.04–1.0 mg/kg IV, IM, or SC) or nalorphine hydrobromide (at a dosage of 2.0 mg/kg IM) (Andrews and Illman, 1987; Harper, 1978). Following reversal of anesthesia, investigators must be careful to monitor the animal during recovery because renarcotization, with attendant reanesthesia, has occurred in ferrets (Harper, 1978). Additional muscle relaxation and analgesia can be produced by the simultaneous administration of xylazine (1.0 mg/kg IM) or diazepam (2.0 mg/kg IM). For induction of anesthesia or sedation only, dosages of 0.3–0.4 ml/kg IM have been given, followed by supplementation with inhalant halothane and oxygen (Harper, 1978; Fox, 1988).

D. Alphaxolone–Alphadolone (Saffan, Althesin)

This combination steroid anesthetic, which is only available for commercial use outside the United States, has been frequently used in ferrets. Dosages of 6–8 mg/kg IM produce light sedation or serve as a useful preanesthetic (Fox, 1988). Higher intramuscular dosages of 15–20 mg/kg require a 3- to 4-min onset of effect and yield a 10- to 20-min peak effect of profound restraint to light surgical anesthesia (Green, 1982; Andrews and Illman, 1987). Intravenous dosages of 10–12 mg/kg produce a surgical plane of anesthesia that lasts for 10–15 min and is characterized by moderate muscle relaxation and good analgesia with a prolonged recovery of up to 1–2 hr. An added advantage of this combination anesthetic is that repetitive incremental dosages can be added by intramuscular or intravenous bolus or by continuous intravenous infusion (Green, 1982).

E. Urethane

Ferrets have been anesthetized for acute experiments by administrating intraperitoneal 1.3–1.5 g/kg urethane (Grundy, 1990; Andrews *et al.,* 1979; Andrews and Illman, 1987). Using this regimen, there is a 5- to 10-min onset of effect with a 4- to 6-hr duration of full surgical anesthesia.

VII. INHALANT ANESTHESIA

Until recently, inhalant anesthetics were infrequently used in ferrets. This philosophy probably was motivated by the necessity to use parenteral agents for any degree of animal manipulation as well as a perceived difficulty in intubating and maintaining ferrets on inhalant equipment. In this format, use of inhalants was restricted to mask delivery of usually halothane plus oxygen as supplements to extend the depth and duration of parenteral anesthetics (Janssens, 1978; Adams, 1979; Harper, 1978; Ryland *et al.,* 1983; Andrews and Illman, 1987; Fox, 1988).

The popularity of isoflurane as the inhalant anesthetic of choice for laboratory species coupled with increased availability and use of pediatric anesthetic circuits as well as regulatory pressure to perform all nonrodent research animal surgery in a dedicated operating suite have clearly made inhalant techniques an integral part of ferret anesthesia. In this format, inhalants are frequently used as a component of balanced anesthetic regimens. Murat and Housmans (1988) developed comparative minimal alveolar concentration values for halothane, enflurane, and isoflurane in ferrets. Results showed that MAC values at 37°C were 1.52 for isoflurane, 1.99 for enflurane, and 1.01 for halothane. These results indicate that the relative potencies of the halogenated anesthetics in ferrets are of the same order as those reported for large animals and for humans. Housmans and Murat (1988a,b) further investigated the cardiovascular effects of isoflurane in ferrets. Results showed that isoflurane is not as potent a negative inotrope as are halothane and enflurane in ferrets. For this reason, isoflurane may become the inhalant anesthetic of choice for use in cardiovascular research studies in this species.

VIII. REGIONAL ANESTHESIA

Further research needs to be undertaken to apply techniques of regional anesthesia to ferrets. Due to their sometimes fractious nature, it is doubtful that even minor surgical manipulations could be completed using this as the sole anesthetic method. Possible applications of regional anesthesia to ferrets include postoperative incisional anesthesia, local anesthesia of the area surrounding exteriorized or imbedded prosthetic devices, and, in combination with deep sedation, minor peripheral surgical procedures.

IX. MONITORING OF ANESTHESIA

Similar parameters and instrumentation used to monitor anesthesia in cats and rabbits can be applied to ferrets. Parameters used to monitor the physiological status of anesthetized ferrets include respiratory rate and pattern, arterial blood gases, pH and oxygen saturation, arterial pulse, arterial pulse pressures, central venous pressure, electrocardiogram (ECG), core and surface body temperature, analysis of inhaled and exhaled gas to estimate alveolar oxygen concentration, en-tidal carbon dioxide content, and anesthetic agent concentration, and urine output.

Parameters used to assess the depth of anesthesia in ferrets include recumbency and loss of purposeful movement, loss of reflexes, muscle relaxation, response to aversive stimulation (e.g., pinching the toes, tail, or abdomenal skin with a hemostat), and alterations in respiration and cardiovascular function in response to aversive stimulation.

Because of their rapid heart rate, it is difficult to accurately apply noninvasive blood pressure monitors to ferrets. Although there are no published reports in this species, several laboratories throughout the United States are now adapting human neonatal pulse oximetry cuffs for use in this species. Commercially available portable monitors can also be adapted to noninvasively record the ECG, heart, and respiratory rates of anesthetized ferrets. Similarly, several models of small animal or human neonatal ventilators are also appropriate for use in this species. Implanted venous and arterial catheters can be used to supply samples for biochemical and blood gas monitoring throughout anesthesia.

X. INTRAOPERATIVE AND POSTOPERATIVE SUPPORTIVE CARE

Technology developed for use in human neonates and infants has been used by the author to provide intraoperative and postoperative supportive care measures to ferrets used in biomedical research studies. Volume deficits can be corrected by subcutaneous, intraperitoneal, intravenous, or interosseous administration of warmed fluids. Pediatric infusion sets, automated syringe and bag infusion pumps, and IV infusion warmers allow precise volumes of warmed fluids to be delivered throughout and subsequent to surgical procedures. The author is unaware of publications discussing bood typing in ferrets, but successful transfusions have been experienced by administering suitable volumes of donor blood using pediatric transfusion equipment and standard methods. Commercially available avian or canine intensive care cages can be used to supply ambient heat, oxygen, and humidity during recovery and indefinitely during the postoperative period. Nutritional support can be provided by administering appropriate volumes of parenteral nutrients via indwelling catheters, by gavage using a 3½–55 French red rubber feeding tube, or, if the animal is able to eat, by supplying a high-quality, high-calorie diet (in the experience of the author, a diet created by mixing moistened mink chow with egg whites is palatable and provides a high caloric density). Because ferrets are "escape artists," it is difficult to keep bandages or jackets to protect prosthetic devices in place. Elizabethan collars that are marketed for use in cats or rabbits can be used on ferrets; however, the collar must be removed several times per day to permit the animal to groom, eat, and drink.

XI. ACUTE AND CHRONIC ANALGESIC THERAPY

There are virtually no published guidelines for the analgesic management of ferrets. Undoubtedly, standard regimens for relief of pain in research ferrets will become available as this species enjoys greater popularity in biomedical research. Fox (1988) presented dosages of aspirin (200 mg/kg PO), phenylbutazone (100 mg/kg PO), meperidine (4 mg/kg IM), and fentanyl–fluanisone (Hypnorm, 0.2 ml/kg IM); however, no further information is presented as to duration of effect, induction of narcotic drug tolerance, or types of clinical pain that have been effectively obtunded using each drug and dosage. In the experience of the author, the analgesic management of research ferrets can follow guidelines for pain control in cats. Oxymorphone hydrochloride (0.4–1.5 mg/kg SC, IM, or IV every 4–6 hr) is an effective analgesic in controlling acute postoperative pain. Following its use, animals show some degree of sedation, but respiratory depression is not overtly evident.

REFERENCES

Adams, C. E. (1979). Anaesthesia of the ranch mink (*Mustela vison*) and ferret (*Mustela putorius furo*). *Vet. Rec.* **105**, 492.

Andrews, P. L. R., and Illman, O. (1987). Chapter 27. The ferret. *In* "The UFAW Handbook on the Care and Management of Laboratory Animals" (T. B. Poole, ed.), pp. 436–455. Longman Scientific and Technical, Edinburgh.

Andrews, P. L. R., Bower, A. J., and Illman, O. (1979). Some aspects of the physiology and anatomy of the cardiovascular system of the ferret, *Mustela putorius furo*. *Lab. Anim.* **13**, 215–220.

Andrews, P. L. R., Dutia, M. B., and Harris, P. J. (1984). Angiotensin II does not inhibit vagally-induced bradycardia of gastric contractions in the anaesthetized ferret. *Br. J. Pharmacol.* **82**(4), 833–837.

Bernard, S. L., Gorham, J. R., and Ryland, L. M. (1984). Biology and diseases of ferrets. *In* "Laboratory Animal Medicine" (J. G. Fox, B. J. Cohen, and F. M. Loew, eds.), p. 386. Academic Press, Orlando, Florida.

Burgmann, P. M. (1991). Restraint techniques and anaesthetic recommendations for rabbits, rodents, and ferrets. *J. Small Exotic Anim. Med.* **1**(3), 73–78.

Fox, J. G. (1988). Anesthesia and surgery. *In* "Biology and Diseases of the Ferret," pp. 289–302. Lea & Febiger, Philadelphia.

Green, C. J. (1982). *In* "Animal Anaesthesia," Lab. Anim. Handb. 8, pp. 213–214. Laboratory Animals Ltd., London.

Grundy, D. (1990). The effect of surgical anaesthesia on antral motility in the ferret. *Expl. Physiol.* **75**, 701–708.

Harper, R. C. (1978). Anaesthetising ferrets. *Vet. Rec.* **102**(10), 224.

Housmans, P. R., and Murat, I. (1988a). Comparative effects of halothane, enflurane, and isoflurane at equipotent anesthetic concentrations on isolated ventricular myocardium of the ferret. I. Contractility. *Anesthesiology* **69**, 451–463.

Housmans, P. R., and Murat, I. (1988b). Comparative effects of halothane, enflurane, and isoflurane at equipotent anesthetic concentrations on isolated ventricular myocardium of the ferret. II. Relaxation. *Anesthesiology* **69**, 464–471.

Janssens, L. (1978). Anaesthetising ferrets. *Vet. Rec.* **102**(15), 350.

Jellie, D. A. (1978). Anaesthetising ferrets. *Vet. Rec.* **102**(17), 388.

Ko, J. C. H., Pablo, L. S., Bailey, J. E., and Heaton-Jones, T. G. (1996). Anesthetic effect of Telazol and combinations of ketamine-xylazine and Telazol-ketamine-xylazine in ferrets. *Contemp. Top. Lab. Anim. Sci.* **35**(2), 47.

Messina, J. E., Sylvina, T. J., Hotaling, L. C., Pecquet Goad, M. E., and Fox, J. G. (1988). A simple technique for chronic jugular catheterization in ferrets. *Lab. Anim. Sci.* **38**(1), 89–90.

Moreland, A. F., and Glaser, C. (1985). Evaluation of ketamine, ketamine-xylazine and ketamine-diazepam anesthesia in the ferret. *Lab. Anim. Sci.* **35**(3), 287–290.

Murat, I., and Housmans, P. R. (1988). Minimum alveolar concentrations (MAC) of halothane, enflurane, and isoflurane in ferrets. *Anesthesiology* **68**(5), 783–786.

Payton, A. J., and Pick, J. R., (1989). Evaluation of a combination of tiletamine and zolazepam as an anesthetic for ferrets. *Lab. Anim. Sci.* **39**(3), 243–246.

Ryland, L. M., Bernard, S. L., and Gorham, J. R. (1983). A clinical guide to the pet ferret. *Compend. Contin. Educ.* **5**(1), 25–32.

Sylvina, T. J., Berman, N. G., and Fox, J. G. (1990). Effects of yohimbine on bradycardia and duration of recumbency in ketamine/xylazine anesthetized ferrets. *Lab. Anim. Sci.* **40**(2), 178–182.

Chapter 13

Anesthesia and Analgesia in Ruminants

Colin I. Dunlop and Robert F. Hoyt, Jr.

ANESTHESIA AND ANALGESIA IN LABORATORY ANIMALS

ISBN 0-12-417570-8

I. INTRODUCTION

The purpose of this chapter is to review considerations and techniques for anesthesia, monitoring, and support of healthy small ruminants including sheep, goats, and calves used for biomedical research. Where appropriate, information will also be provided relating to anesthesia of adult cattle. This discussion will focus on sheep because of the volume of published anesthesia research in that species compared to goats and calves, and because the methodology is generally applicable to sheep, goats, and calves. There are numerous reviews of anesthetic considerations and techniques for ruminants from the perspective of clinical veterinary medicine (Gray and McDonell, 1986; Hubbell *et al.,* 1986; Thurmon, 1986; Thurmon and Benson, 1993; Trim, 1981). This chapter will describe anesthetic and analgesic techniques that vary from simple, regional anesthesia applied to a conscious or sedated animal to general anesthesia needed for more complex surgical procedures. In addition, practical monitoring techniques and commonly used monitoring equipment will be discussed.

Small ruminants have become widely used in biomedical research for a variety of reasons: (1) they are readily available; (2) they are generally quiet, docile, and easily restrained; (3) they rapidly acclimate to the restricted spaces available in a biomedical research laboratory setting; (4) they can be easily matched for age, weight, sex, breed, and stage of gestation; (5) their large blood volume allows for frequent collection of blood and other body fluid samples of adequate volume for analytical determinations; and (6) their body size allows for the adaptation of traditional small animal surgical, anesthesia, and monitoring techniques and equipment. In addition, because both their body size and many physiological parameters are comparable to human beings, they can be used as animal models for studies of drug pharmacodynamics, intrauterine manipulations of fetuses during pregnancy, cardiopulmonary responses of neonates, cardiovascular and orthopedic procedures, biomedical device and artificial internal organ development, osteoporosis, bone healing, organ transplantation, tissue rejection, and a variety of other specialized studies. Furthermore, critical care techniques used in human beings or dogs such as parenteral fluid administration and monitoring of cardiopulmonary function as well as the use of routine and emergency drugs and procedures are directly applicable to small ruminants.

Sheep and young cattle have historically been used as animal models in cardiopulmonary research because their heart size-to-body weight ratio as well as other thoracic cardiovascular structures are analogous in size to those structures in humans. This allows for the implantation of many bioprosthetic devices intended for human use directly into the animal model without modifications to adjust for anatomic size or variations. It is important to take into account the significant breed variation in ruminants with respect to mature body size, growth patterns, and conformation. These ultimately can have an impact not only in the suitability of a particular breed and species for research, but also in anesthetic management. Although most ruminants obtain sexual maturity on or before 1 year of age, they may not reach mature body weight or stop growing until several years of age. Sheep may require 4 to 5 years to reach maximum body size, which may be in excess of 90 kg, whereas cattle may take 2 to 3 years to reach maximum body size, which can exceed 700 kg in females and over 1000 kg in males. Most biomedical research facilities are not equipped nor have the appropriate facilities to accommodate animals of a mature body size in excess of 140 kg.

II. ANIMAL HEALTH CONSIDERATIONS

A. Zoonotic Diseases

The use of small ruminants in biomedical research poses a small but real risk of contact with zoonotic disease agents. With few exceptions, most ruminants used in biomedical research are obtained from agricultural sources rather than dedicated commercial sources as might be the case for rodents or other laboratory animal species. For this reason, the health status of individual animals as well as their potential exposure to zoonotic agents can be quite variable. There are, however, federal and state programs in place to detect and eradicate certain zoonotic agents, thereby reducing the risk of exposure to disease such as tuberculosis or brucellosis. Given the close contact required during surgical manipulation and administration of anesthesia in a biomedical research environment, anyone working with these animals should be aware of the potential risks involved. The agents of zoonotic concern that are most commonly encountered are *Coxiella burnetti, Brucella* spp., *Mycobacterium bovis* and related species, *Trichophyton mentagrophytes,* and contagious ecthyma virus.

Q fever is a zoonotic disease caused by the rickettsia *C. burnetti* which affects both domestic and wild ruminants [Bernard *et al.,* 1982; Pike, 1976; Centers for Disease Control/National Institutes of Health (CDC/NIH), 1993]. Sheep are the most common sources of infection for humans with pregnant ruminants posing the greatest likelihood of infection due to the high titer of the organism in fetal tissues and fluids. The presence of the organism has been documented serologically in many sheep flocks throughout the United States, but causes little, if any, clinical disease in sheep. The organism is most likely to be spread from infected animals at parturition, especially through contact with fetal membranes and amniotic fluids. The organism is extremely stable in the environment and may be transmitted by aerosol from fomites (e.g., dust), blood, or urine.

Humans infected with this organism are often asymptomatic when they contract the disease. Those who become ill usually have flu-like symptoms that are characterized by acute onset, high fever, severe headaches, chills, and generalized malaise. In some patients, these symptoms progress into chronic infections that are characterized either by atypical pneumonia or by gastrointestinal disturbances including vomiting and anorexia. Deaths are generally attributed to complications from hepatitis and endocarditis (Meiklejohn *et al.,* 1981, Spinelli *et al.,* 1981).

Serologic tests for the organism are available but their sensitivity and specificity are unclear. An interpretation of test results can be difficult, especially if decisions have to be made on individual animals rather than on a herd basis.

Several *Brucella* spp. (e.g., *B. abortus* and *B. melitensis*) infect ruminants and are capable of causing disease in humans. In humans they cause a febrile systemic disease without rash, often associated with fatigue and a long convalescence that is frequently associated with relapses following stress. Cattle and goats can be infected with species that will infect humans. Transmission occurs by contact with infected tissues or fluids, although airborne transmission is possible. The organism can penetrate intact skin or mucus membranes. The organism can be detected by serologic methods and human infection can be treated with antibiotics.

Several species of *Mycobacterium* have been associated with disease in ruminants. Infection in ruminants commonly resides in the lungs, although peripheral lymph node involvement is possible. In humans, pulmonary involvement as well as lymph node involvement is common. Intradermal testing of both ruminants and people is helpful for diagnosis and screening.

Contagious ecthyma is a disease of sheep, goats, and wild ungulates caused by a poxvirus. In animals the disease is characterized by papules on the skin, eye lids, ears, lips, and nostrils, but also may occur in other locations. In humans, infection is characterized by a singular maculopapular or pustular lesion at the site of inoculation which is most often on the hands, arms, or face. This can be associated with region lymphadenitis. Lesions in animals and humans tend to resolve themselves within 10 to 14 days conferring immunity.

Young cattle are commonly infected with the fungus *T. mentagrophytes* which is capable of causing dermatomycosis in humans. The most common location for infection in young ruminants is on the face, particularly around the eyes. The lesions in humans and animals are often circular or even angular in nature and are associated with scaling, hair loss, and occasional itching. Topical and systemic treatments are available.

B. Procurement

Ruminants used for research should be in good health and individually identified by tattoo, tag, transponder, or other permanent method. A record of each individual animal should accompany shipment and ideally include age (date of birth), sex, preshipment weight, vaccination and deworming history, and medical records of clinical tests and treatments. Movement of ruminants between states is controlled and requires examination by an accredited veterinarian as well as the appropriate documents, including a health certificate. Given the stress associated with the shipment of animals as well as the need to procure small ruminants from multiple agricultural sources, vaccination for common disease-causing agents well in advance of shipment is often advisable. Vaccines for *Pasteurella multocida, P. haemolytica,* contagious ecthyma, and parainfluenza III as well as bacterial toxoids for *Clostridium chauvoei, C. septicum, C. haemolyticum, C. novyi, C. tetani, C. perfrigens* types C and D are available and, depending on the species, may be appropriate components of a procurement specification. A period of quarantine during which clinical, as well as laboratory, determinations can be conducted and the animal allowed to stabilize to a new environment is particularly important. This decreases the likelihood of anesthetic-related complications due to infectious disease.

C. Physical Examination of Newly Acquired Animals

Because variations in body size, condition, and wool or hair growth, an accurate weight should be determined for all ruminants. Body temperature should be measured and a fecal sample analyzed for gastrointestinal parasites. Because respiratory disease is commonly observed in ruminants that may be stressed and housed in confined quarters, particular emphasis should be given to the respiratory examination. Should evidence of lung disease be detected, either the animal should be rejected from the study or further workup, including chest radiographs, should be undertaken to determine the extent and cause of the problem.

Due to the unknown health status of newly arrived animals it is often advisable to house them separately from animals existing in the facility for a quarantine period often lasting up to 14 days. During this period, the newly arrived animals should have one or more complete physical examinations and any necessary treatment, including the use of anthelminthics, administered as appropriate for the species. Animals that are obviously sick upon arrival may need to be rejected or a course of treatment initiated. It is important to remember that even if a sick animal is rejected, other animals in the group have been exposed and may have contracted the illness. Purchasing animals from closed herds may reduce the likelihood of such problems but will not eliminate it. Hematology and clinical chemistry determinations done during the quarantine period may vary considerably from those conducted on animals that have completed quarantine and are stabilized to their new environment.

D. Specific Considerations for Ruminant Anesthesia

1. Hypoventilation Caused by Bloat or Abdominal Distention

Restraint in either lateral or dorsal recumbency, with or without anesthesia, can prevent a ruminant from belching gas that accumulates in the forestomach. This problem does not occur when ruminants are positioned in sternal recumbency. Bloat during anesthesia can be minimized by withholding food for 24 to 48 hr before anesthesia, which reduces the rate of rumen fermentation. Administration of antidiarrheal boluses containing mixed sulfonamides has been used by some to reduce rumen fermentation rates to prevent bloating. Withholding food in young ruminants for more than 12 hr may not be required when such a regimen is used. Minimizing the duration of anesthesia or time in the lateral or dorsal postion will also help prevent bloat. When bloat does occur during anesthesia, it can be relieved by passing a stomach tube into the rumen. Depending on the surgical procedure and positioning of the animal, rumenal gas can also be evacuated by percutaneous insertion of a large-bore needle (e.g., 16-gauge) into the rumen and channeling away the gas through attached tubing. This procedure overcomes the positional difficulties associated with the insertion of a stomach tube and the potential for evacuating unintentionally large quantities of fluid and ingesta from the rumen through the stomach tube. Nitrous oxide should not be used in ruminant anesthesia because the difference in solubility between nitrogen (less soluble) and nitrous oxide results in increased volume and pressure in closed gas cavities such as the rumen. This increased pressure results in ventilatory depression and may cause postanesthetic abdominal colic.

Inhibition of ventilation is a serious problem that may be caused by rumenal bloat or simply the mass of the abdominal viscera. In lateral or dorsal recumbency the weight of the viscera is downward and forward, which puts pressure on the diaphragm. Anesthetized ruminants have rapid, shallow ventilation which, combined with the physical restriction caused by the abdominal contents, frequently results in marked hypoventilation (arterial carbon dioxide tension [$PaCO_2$] > 55 mm Hg for sheep and 70 mm Hg for cattle). In cases of extreme hypoventilation, the use of intermittent positive pressure ventilation (IPPV) should be considered. Abdominal pressure can also impede venous return to the heart, resulting in reduced cardiac output and pulmonary perfusion. Reduced lung–blood flow causes the ventilation/perfusion ratio to widen, which lowers the arterial oxygen tension (PaO_2). Even with inspiring concentrations of oxygen greater than 95% ($FiO_2 > 0.95$), ruminants tend to have a lower than expected PaO_2 and, in some cases, become hypoxemic (at a pH of 7.40, $PaO_2 < 70$ mm Hg for sheep and <60 mm Hg for cattle) (Bartels and Harms, 1959).

2. Regurgitation

Because ruminants normally eructate and have a large volume of ingesta in the forestomach, regurgitation during anesthesia is a common problem. Anesthetized animals also have depressed or absent swallowing reflexes, so that aspiration pneumonia can result. The rumen liquid contains small particles of feed material and microbial organisms. When it is aspirated, this fluid can reach small airways causing acute bronchial constriction, inability to ventilate, hypoxia, and death. In such cases if the animals do not die, they develop severe pneumonia, which is difficult to treat successfully. The solution to this problem is prevention, including placement of a cuffed orotracheal tube in ruminants anesthetized for long procedures.

Two types of regurgitation are seen during anesthesia (Thurmon and Benson, 1993): (a) Active regurgitation is characterized by an explosive discharge of ruminal contents which is observed at very light levels of anesthesia, particularly if the larynx is stimulated such as during intubation. The cause of this is inadequate anesthetic depth at the time of laryngeal stimulation. This can be prevented by examining the larynx either manually or with a laryngoscope before attempting intubation. If marked swallowing, coughing, or chewing occurs, a further dose of parenteral anesthetic should be administered prior to attempting intubation. If regurgitation occurs during intubation, first administer further parenteral anesthetic, then attempt to place the endotracheal tube. If unsuccessful, the endotracheal tube might be positioned in the esophagus and the cuff inflated to attempt to direct rumenal fluid past the larynx. Positioning the head (mouth) below the pharynx may facilitate fluid drainage. Periodic suctioning of the mouth can also be helpful. In the experience of one author (Dunlop), intubation of the esophagus to facilitate ruminal fluid flow past the larynx is of limited usefulness. Without airway intubation with a cuffed tube, such animals either die from airway obstruction or subsequently develop aspiration pneumonia. (b) Passive regurgitation is characterized by a continual stream of rumenal fluid draining from the mouth and is often seen at deeper levels of anesthesia (Blaze *et al.,* 1988a). Decreasing the depth of inhalation anesthesia may cause this response to cease. Therefore, the presence or absence of passive regurgitation may be a useful indicator of depth of anesthesia in ruminants. Anesthetic drugs may also influence the amount of passive regurgitation. Ketamine is less commonly associated with passive regurgitation compared to thiobarbiturates, xylazine, and inhalation anesthetics. It is not easy to predict which animals will passively regurgitate. However, this can be minimized by withholding food for 48 hr in adult, nonpregnant cattle and for 24 hr in sheep, and water for between 8 and 24 hr. This simply decreases the rumen volume. Because of the unpredictable nature of passive regurgitation, ruminants anesthetized for long procedures (greater than 15 to 20 min) or those positioned in dorsal recumbency should be intubated with cuffed endotracheal tubes. Once anesthetized, positioning the head below the body may facilitate drainage of this fluid from the mouth. Periodic suctioning of the mouth can also be helpful. In sheep and goats where regurgitation persists during anesthesia, passage of a thick-walled, 1- to 2-cm ID tube, such as a large equine stomach tube, via the esophagus into the rumen may enable drainage of some fluid directly from the rumen. However, this manipulation frequently does not resolve this problem, and fluid will continue to flow up the esophagus, outside of the stomach tube.

3. Salivation and Use of Anticholinergic Drugs

A similar problem to regurgitation is salivation. Ruminants produce large volumes of saliva and this continues to flow during anesthesia. To facilitate visualization of the larynx and, therefore, intubation, some authors advocate the administration of atropine (0.04 mg/kg IV). This generally decreases the volume of saliva, but may make it more viscous. The duration of action of atropine in ruminants is short (around 5 min). Also, the repeated use of anticholinergic drugs may cause rumenal stasis. In one author's practice (Dunlop), atropine is not routinely administered to sheep or cattle during anesthesia because the volume of saliva produced generally does not cause difficulties with intubation. Appropriate head positioning as outlined previously will help drain saliva from the mouth during anesthesia. If administration of an anticholinergic is required (e.g., to treat bradycardia or vagal stimulation), glycopyrrolate (Robinol) (0.022 mg/kg administered IM or SC) may provide a longer duration of effect than atropine, although it also has a longer onset of action.

4. Temperament and Restraint

Domestic ruminants, particularly sheep and goats, are usually docile but stoic, i.e., they usually do not resist physical restraint and manipulation, or even painful procedures that may cause considerable stress (Grandin, 1989; Randolph, 1994). Adult cattle in particular become unmanageable when markedly stressed and require appropriate facilities to permit safe handling. When physical restraint is employed to facilitate painful manipulations, local anesthetics or systemic analgesics should be used, generally in conjunction with tranquilizing drugs. Physical restraint should not be employed as an alternative for poor quality local, regional, or general anesthesia.

Struggling is minimized when good methods of restraint are employed, such as positioning sheep on their rump, goats in lateral recumbency on a padded surface, and adult cattle in a chute with a head-restraining device. Small calves (less than 300 pounds) can often be effectively restrained by backing them rump first into a corner and placing firm lateral pressure to push them against one wall while maintaining control of the animal's head. Sheep tolerate being restrained in lateral recumbency and will rest quietly without attempting to struggle if their feet are tied together. Wheelbarrows, modified to allow the head and neck to be in a natural position, provide an excellent, stress-free method of transporting such restrained sheep as well as allowing minor procedures such as venous catheterization to be performed at a comfortable working height. As an alternative to this, unanesthetized animals may be placed in a sling (e.g., Panipinto sling) similar to procedures used for the restraint of pigs. Like pigs, sheep adapt very quickly to slings and will remain relatively quite for long periods of time. This allows minor procedures such as radiography or bandage changing to be accomplished with minimal stress to the animal.

5. Postanesthetic Myopathy/Neuropathy

Anesthetized large ruminants, especially cattle, are susceptible to postanesthetic myopathies and neuropathies from poor positioning, padding, or hypoperfusion during anesthesia. All anesthetized ruminants should be positioned on a padded surface, particularly for long procedures. In cattle positioned in lateral recumbency, the upper limbs should be supported on platforms and sufficient padding used between the hard surface and animal to elevate the point of the down hip and shoulder above the hard surface. In calves it is also important to place padding around the lower portions of the legs, especially where any ropes or other tie-downs are used to position the legs.

III. PREPARATION FOR ANESTHESIA

Because anesthesia is required for a variety of procedures, a reasonable approach is as follows: (1) examine the physical status of the animal, including basic clinical pathology evaluation; (2) consider the specific problems and requirements of ruminants; (3) devise an anesthetic plan suitable for the procedure; (4) provide appropriate monitoring and support during anesthesia; and (5) provide appropriate analgesia, monitoring, and support during recovery.

A. Management for Bloat/Regurgitation

In order to minimize problems with bloat, abdominal distention, and regurgitation of rumenal fluid, feed can be withheld for at least 48 hr in adult, nonpregnant cattle and for 24 hr in sheep, goats, and calves. Water should be withheld for between 8 and 24 hr depending on the age and state of health, including pregnancy. The authors do not advocate withholding feed, water, or milk from young ruminants that are not weaned (e.g., less than 30 days of age).

B. Preanesthesia Examination

The preanesthetic examination should include determination of body weight and temperature, assessment of the adequacy of hydration and perfusion, evaluation of cardiac function (especially heart rate and rhythm), and ventilation. Normal values for healthy, conscious sheep are given in Table I. For animals involved in major surgical procedures where major blood loss or fluid therapy occur, the hematocrit, total plasma solids (or plasma protein), and, in some cases, a complete blood count and serum chemistry panel can be determined in order to serve as a baseline for perioperative determinations (Table I).

C. Venous Access

In conscious ruminants, the jugular vein is the most accessible vessel from which large volumes of blood can be obtained using a syringe and needle. Other sites commonly used for venous injection in sheep and goats include the central vein on the dorsal surface of the ear, the cephalic vein on the anterior aspect of the thoracic limb, and the saphenous vein located on the lateral aspect of the pelvic limb just above the tarsus. In adult cattle, the coccygeal vein on the ventral side of the tail and, in cows, the mammary vein can also be used.

TABLE I

CARDIOVASCULAR AND ARTERIAL BLOOD GAS PARAMETERS IN AWAKE ADULT SHEEP[a]

Parameter	Kuchel *et al.* (1990)	Pollack *et al.* (1990)	Runciman *et al.* (1984a)	Runciman *et al.* (1985)
Number of sheep	15	5 (pregnant)	10	6
Measurements/sheep	3	3	7	7
Weight (kg)	38 to 51 (range)	66 ± 8	32 to 45	Not given
Heart rate/min	100 ± 4	107 ± 9		117 to 121 (range)
Cardiac output (liter/min)	5.8 ± 0.23	8.7 ± 2.5	3.05 ± 0.96	4.45 ± 0.33
Mean arterial pressure (mm Hg)	98 ± 2.5	93 ± 5		92 to 100 (range)
Renal blood flow (liter/min)	0.95 ± 0.12		0.61 ± 0.21	0.66 ± 0.05
Hepatic blood flow (liter/min)	1.6 ± 0.28		1.13 ± 0.2	
Body temperature (°C)	39.2 ± 0.1		39.3 ± 0.5 ($n = 10$)	
PaO_2 (mm Hg)	101 ± 2	104 ± 7	95 ± 3	
$PaCO_2$ (mm Hg)	34 ± 1	37.0 ± 1.6	34 ± 2	
pHa	7.45 ± .01	7.42 ± 0.04	7.50 ± 0.02	
BUN (mmol/liter)			6 ± 2	
Creatinine (mmol/liter)			0.07 ± 0.02	
Albumin (g/liter)			28 ± 3	
Plasma protein (g/liter)			74 ± 5	
Hematocrit (%)			22 to 35 (range)	

[a]Mean ± SD.

D. Catheterization

To facilitate the administration of repeated doses of intravenous anesthetics or the continuous administration of intravenous fluids, it is best to use an indwelling, percutaneous catheter. Catheters can be placed in small ruminants restrained in a standing posture, usually by one individual, without causing undue stress to the animal or handler. However, such restrained animals can still move during the procedure. Better restraint for catheterization may be obtained by positioning sheep and goats in lateral recumbency or sheep on their rump.

1. Jugular Vein

To catheterize the jugular vein in adult sheep and goats or calves, a 16- or 18-gauge 14-cm polyethylene catheter (Angiocath) is commonly used. In adult cattle, a 12- or 14-gauge catheter is used. Following clipping, the insertion site is aseptically prepared (e.g., alternating povidone iodine and isopropyl alcohol washes). Then 1 to 2 ml of lidocaine can be infiltrated intradermally and subcutaneously to anesthetize the skin over the catheter entry site if an over-the-needle catheter (e.g., Angiocath) is to be used. Making a small incision in the skin with the tip of a scalpel at the catheter insertion site will prevent ripping of the catheter as it is thrust through the skin and will minimize the amount of discomfort associated with the procedure.

An aseptic technique needs to be used for catheter placement, especially in sheep used for transplantation studies. Contamination of the catheter by the operator is least likely when the operator wears sterile surgical gloves. The vein is occluded by one hand and, using the other hand, the catheter and stylet are advanced until a flashback of blood is seen. It is important to advance the catheter at least a further 0.5 cm before trying to slide it off the stylet. After removing the stylet, an injector cap or stopcock is then used to cap the open end of the catheter and is used to flush sterile heparinized saline (3 units heparin/ml) into the catheter dead space. A catheter should not be left in the jugular vein that is improperly capped because air can be aspirated via the vein into the heart and pulmonary arteries. To maintain the catheter in position, a drop of cyanomethacrylate glue can be applied between the catheter hub and skin. As an alternative, a piece of adhesive tape can be placed around the catheter just below the hub and the tape sutured to the skin.

2. Auricular Vein

In conscious sheep and goats, the central vein on the abaural surface of the ear is a useful site for catheter placement to facilitate anesthetic drug administration. The auricular vein is accessible, easily fixed in position, vascular injury results in only minor potential injury, and catheterization causes minimal stress to the animal. However, compared to jugular vein catheterization, this is technically more difficult for the operator. After clipping the lateral surface of the ear and aseptic skin preparation, either an 18- or a 20-gauge 5-cm catheter can be placed in the vein. It is usually easiest if an assistant occludes the vein at the base of the ear with one hand and restrains the head of the animal with the other hand. The vein can also be occluded by use of a tourniquet such as a rubber band placed around the base of the ear and held in place by a hemostat. During catheterization, it is important for the operator to then keep tension on the tip of the ear with the free hand. The catheter is inserted and advanced in a manner similar to that used for the jugular vein. After the catheter is positioned in the vein, capped, and flushed with heparinized saline, the insertion site should be dried using a 4 × 4 sponge and the catheter maintained in place by applying a drop of cyanoacrylate between the catheter hub and the skin.

3. Cephalic Vein

In sheep and goats, the cephalic vein can be visualized on the anterior aspect of the thoracic limb just below the elbow. Compared to dogs, where the cephalic vein is generally straight and relatively large, this vein in sheep and goats is relatively small and curves from the medial to lateral side of the limb. Because of the small vessel size, its curvature, and animal restraint considerations, this is a more difficult vessel to catheterize compared to dogs. Sheep can be positioned on their rump and usually sit quietly. Goats may be better restrained in lateral recumbency. Usually an 18- or 20-gauge 5-cm catheter can be placed in the vein. An assistant restrains the animal and holds the limb to be catheterized forward using the palm and fingers behind the elbow. The thumb can then be positioned over the anterior aspect of the elbow, which then occludes the vein. The operator stretches out the distal limb with one hand and uses the other hand to place the catheter in the manner described for the jugular vein. Cephalic vein catheters can be held in place using either cyanoacrylate or tape that encircles the limb.

4. Saphenous Vein

The technique for catheterizing the saphenous vein is similar to that for the cephalic vein. However, sheep and goats will need to be restrained in lateral recumbency by an assistant to facilitate catheterization. The major concern with using this vessel is maintaining the catheter in position, especially over a number of days. The vessel is located above the hock on the lateral surface of the pelvic limb so is subject to considerably more movement compared to a jugular, auricular, or cephalic vein catheter. The catheter is also in a position where it is subject to fecal contamination.

E. Arterial Access

Arterial catheterization is used to facilitate rapid, direct arterial blood pressure measurement in both conscious and anesthetized animals and for the collection of arterial blood samples for measurement of O_2 and CO_2 tensions, pH, and parenteral drug levels. In conscious sheep and goats, a 20-gauge 5-cm catheter can be placed in the central auricular artery on the abaural surface of the ear. The artery is located rostral to the central auricular vein. The technique for catheterizing this vessel is the same as that for catheterizing the auricular vein. A ''T-port'' connector can be used to cap the arterial catheter to distinguish it from the auricular vein catheter and thus minimize the chance of inadvertent arterial administration of anesthetic drugs.

During many types of experimental surgical procedures, direct access to larger arteries is required for arterial blood sampling and direct central arterial blood pressure measurement. The two most common peripheral vessels used for this purpose are the femoral and carotid arteries. Sheep have very small peripheral blood vessels relative to their body size. The femoral artery is located deep in the inguinal region whereas the carotid artery in ruminants is located deep within the neck, close to the trachea. For this reason, percutaneous catheterization of these vessels is difficult. Fortunately, both vessels can be rapidly accessed and catheterized through a surgical cut-down procedure. It should be noted that it is not uncommon for the arteries to go into spasm on manipulation. These vasospasms can be countered by the topical administration of 5–10 drops of papaverine or 1% lidocaine to the vessel. Once the catheter has been placed in the vessel, it is necessary to secure it with one or more ligatures on the ''heart side'' of the vessel in order to avoid slippage. It is also necessary to use a three-way stopcock to control flow from the catheter and to keep the catheter filled with an anticoagulant containing flush solution.

IV. LOCAL AND REGIONAL ANESTHESIA

A. Application

Local or regional anesthesia is the temporary loss of sensation to an area of the body without loss of consciousness. Local anesthesia is achieved by the infiltration of local anesthetic agents into the surgical site. Regional anesthesia is achieved by the perineural injection of local anesthetic at major nerves that may be distant from the intended surgical site. In clinical veterinary medicine, local or regional anesthesia is commonly used to facilitate surgical procedures on conscious ruminants because of economy and because ruminants usually tolerate physical restraint. However, effective local and particularly regional anesthesia require good anatomic knowledge, appropriate selection of local anesthetic drugs, skilled needle placement, and drug infiltration. In addition, the operator should be sufficiently skilled in the surgical or manipulative procedure to ensure that it is completed in a timely fashion and that the surgical field remains within the limits of the anesthetized region. Therefore, in a biomedical research setting where such knowledge, training, or skill may not be readily available, general anesthesia may be more appropriate. Two reference sources that provide detailed descriptions of commonly used regional anesthesia blocks for ruminants are Benson and Thurmon (1993) and Skarda (1986). Local anesthetic techniques for castration and dehorning will be described here.

B. Drug Selection

Selection of the drug to be used for local or regional anesthesia depends on the required speed of onset of analgesia, the duration of the procedure, and, to some extent, the site of drug administration. For practical purposes, commonly used local anesthetic drugs can be considered in two groups: (1) rapid onset (minutes) and short duration of action (60 to 90 min), including lidocaine (Xylocaine) and mepivacaine (Carbocaine); and (2) slower onset (15 to 20 min) and long duration of action (4 to 6 hr) such as bupivacaine (Marcaine). To increase the duration of action through peripheral vasoconstriction, epinephrine can be added to these agents, particularly to lidocaine used for infiltration, at a concentration between 5 and 20 μg/ml (1:50,000 to 1:200,000). Tissue inflammation can reduce the effectiveness (analgesic potency and onset of action) of local anesthetics. Local anesthetic drugs are basic salts dissolved in a slightly acidic water-based solution. In tissues, the solution is neutralized, liberating the base anesthetic, which is the form that penetrates cell membranes and blocks nerve conduction. Inflammation causes a reduction in tissue pH, decreasing the effectiveness of the anesthetic agent.

C. Toxicity

Excessive doses of local anesthetic drugs, particularly if inadvertently administered IV, can cause muscle twitching and seizures followed by depression, hypoventilation, tachycardia, and cardiovascular collapse. Treatment of such cases includes support of ventilation, administration of intravenous fluids, and vasopressors (Benson and Thurmon, 1993). Because local anesthetics are rapidly eliminated by sheep (elimination half-life is about 1 hr; Rutten *et al.,* 1989) and probably other ruminants, supportive therapy following signs of toxicity should only be required for a relatively short period. Rutten *et al.* (1989) reported seizures following the intravenous bolus administration of 3 to 7 mg/kg lidocaine and 1 to 2 mg/kg bupivacaine to adult sheep (45 kg average body weight). All sheep survived the toxic doses of lidocaine, but two of five sheep died within 2 min of the high dose of bupivacaine. Cardiovascular changes in sheep that were seizuring included an increased heart rate, cardiac output, systemic vascular resistance, and mean arterial pressure.

D. Techniques

1. Castration

Immature sheep, goats, and cattle are usually restrained in a lateral position for castration. Following aseptic preparation of the scrotal skin surface (refer to jugular vein catheterization/skin preparation), a testicle is grasped and the skin tensed. The incision line is infiltrated with 2 to 5 ml of 2% lidocaine, injecting anesthetic as the needle is advanced subcutaneously. Local anesthetic (1 to 3 mg/kg) is then injected into the body of each testicle using a 2.5- to 3.75-cm 20-gauge needle inserted below the tail of the epididymis and directed toward the ventral aspect of the scrotum (Skarda, 1986). Because local anesthetic is rapidly absorbed via the spermatic chord and lymph vessels into the blood, the testicle should be removed in a timely fashion.

2. Dehorning/Disbudding

Removing horn buds from immature ruminants is a much simpler procedure than dehorning conscious adult ruminants, especially cattle and goats. Appropriate physical restraint is required to enable dehorning of adult ruminants, particularly cattle, where a chute with a head gate is required. The regional anesthetic technique is similar for both immature and adult ruminants.

In immature cattle and sheep, analgesia of the horn and skin at its base is achieved by blocking the cornual nerve (Benson and Thurmon, 1993; Skarda, 1986). A 2.5-cm 22-gauge needle is inserted off the lateral edge of the palpable ridge of the frontal bone, about halfway between the lateral canthus of the eye and horn base. The nerve is located just below the skin surface on the ventral aspect of the bone ridge. In immature ruminants, usually 1 to 3 ml of 2% lidocaine (in young lambs, less than 1 to 2 mg/kg lidocaine for each horn bud) is injected between 0.5 and 1.0 cm deep. Failure to cause analgesia results from either a too deep an injection or inserting the needle too close to the base of the horn, after the nerve has divided.

In goats, both the cornual and the infratrochlear nerves must be anesthetized. The injection sites are located halfway between the lateral canthus of the eye and the lateral horn base (cornual nerve) and halfway between the medial canthus of the eye and the medial horn base (infratrochlear nerve) (Skarda, 1986). The cornual nerve is 1.0 to 1.5 cm deep, beneath the frontalis muscle. The infratrochlear nerve is palpable, just below the skin surface. Using a 22-gauge 2.5-cm needle, 0.5 to 1 ml of 2% lidocaine (in young kids, less than 1 to 2 mg/kg lidocaine for each horn bud) is injected at each site. The volume of lidocaine available for each horn can be increased by dilution with saline.

V. EPIDURAL OR SPINAL ANESTHESIA

A. Application

Epidural anesthesia usually refers to the deposition of local anesthetic drugs into the epidural space. Spinal anesthesia usually refers to the deposition of local anesthetic drugs into the cerebrospinal fluid (CSF) in the subarachnoid space. In biomedical research, this technique is most commonly used in sheep and goats, where needle placement at the lumbosacral junction is relatively easy and results in entrance into the subarachnoid space allowing anesthetics to be delivered into CSF fluid. In either immature or adult sheep and goats, and calves, lumbosacral epidural or spinal anesthesia, usually combined with sedation, can be used to facilitate surgical procedures of the pelvic limbs, perineal region, including urogenital surgery, and caudal abdominal surgery, including caesarean section. Administration of local anesthetics at the level of the lumbosacral junction results in loss of motor innervation of the pelvic limbs, therefore temporary paresis or paralysis. Lumbosacral administration of local anesthetics should not be used in adult cattle because they tend to struggle and therefore injure themselves if they are unable to support weight on their pelvic limbs. An excessively high spinal anesthetic in sheep or goats can result in thoracic spinal anesthesia, causing intercostal muscle dysfunction and respiratory depression. Hypotension or increased incisional bleeding may result from spinal anesthesia because anesthesia of the sympathetic nerves results in vasodilatation. Treatment of hypotension is discussed in Section XII on support of anesthetized ruminants.

B. Factors Affecting Extent of Spinal Anesthesia

Several factors have to be taken into consideration when administering spinal anesthesia. (1) Location of injection: Compared to spinal administration, epidural injections require a one-third increase in the local anesthetic drug dose; (2) the mass (volume × concentration) of drug administered: Increasing the mass increases the extent of anesthesia; (3) speed of administration: Increasing the speed or force of injection increases the extent of anesthesia; (4) size of the epidural space: Animals with long backs generally require more drug than those with shorter backs to achieve the same extent of anesthesia; accumulation of fat in the epidural space reduces its volume and increases the extent of the anesthesia; (5) volume of blood and lymphatic fluid in the epidural pace: Pregnancy causes engorgement of blood vessels so the dose of anesthetic should be reduced by one-third to achieve the same extent of anesthesia; and (6) animal position: Gravity generally causes local anesthetics to migrate downward, so the body area where analgesia is desired should be positioned most ventrally for the period of time required for drug onset (see later).

C. Drug Selection and Dose

The duration of anesthesia for practical purposes depends on the nature of the local anesthetic drug (see comments in Section II). Generally, epinephrine should not be mixed with local anesthetics used for spinal injection. Although there are specific products made for spinal administration, they are usually packaged in single (human) dose ampules and are relatively expensive. Multidose vials may be a suitable alternative providing asepsis can be maintained. Typically, 2% lidocaine can be used for procedures less than 90 min (onset of action is 2 to 3 min), and 0.75% bupivacaine for longer procedures lasting up to 4 to 6 hr (onset of action is 15 to 20 min). At these drug concentrations, the dose for spinal administration is between 1 ml/3.0 kg and 1 ml/4.5 kg. The higher dose results in a higher extent of anesthesia, but more likelihood of respiratory depression. For epidural administration, the dose would be increased by one-third.

D. Technique

In sheep and goats, epidural injections are technically simple. The animal is usually restrained in lateral recumbency on a table. A 10-cm-square area is clipped and the skin is aseptically prepared over the lumbosacral region, extending caudal from a line joining the anterior border of each ileum, palpable on each side of the lumbar spine. The operator should then wear sterile surgical gloves. Material that needs to be readily available prior to inserting the needle includes a sterile 5- to 7.5-cm 20-gauge spinal or intravenous needle and either a 6- or a 12-ml sterile syringe already filled with the calculated dose of anesthetic. These items can be placed on the sterile surface of the opened surgical glove packet. The injection site is the most palpable depression caudal to the line joining the anterior border of each ileum. Usually the junction between lumbar vertebrae 6 and 7 is palpable on this line, and the lumbosacral junction is 2 to 3 cm more caudal. The continuous dorsal spinous processes of the sacrum are then palpable further caudally. With the back of the animal resting at the edge of the table, the hand holding the needle can be stabilized by resting the wrist either on the back of the animal or on the table edge (Fig. 1). The needle is inserted through the skin over the center of the palpable depression, maintaining a plane parallel to the table surface and a second plane that is perpendicular to the spine of the animal. There is some resistance as the needle advances through the skin, subcutaneous tissue, and the lumbodorsal fascia. The resistance increases as the needle passes through the interspinous ligament, then suddenly ceases. If the stylet of a spinal needle is then removed, CSF will drip from the needle hub if the needle is in the subarachnoid space. If this does not occur, advancing or withdrawing the needle 1 to 2 mm will usually result in subarachnoid placement. In sheep and goats, the epidural space is very narrow and the needle passes through it usually without the operator perceiving its location. At this point, a right-handed operator would then stabilize the needle hub with the thumb and index fingers of the left hand, resting the left wrist on the back of the animal. The right hand would then connect the syringe to the needle hub and inject the local anesthetic at a rate of 1 ml/2 sec. The pressure required to depress the syringe plunger should be equivalent to injecting the anesthetic liquid from a syringe that is not connected to a needle. The needle is then withdrawn and the animal is positioned with the surgical site in the most ventral position for the onset time of the local anesthetic drug. After repositioning the animal for surgery, the adequacy of chest excursions and ventilatory effort should be observed.

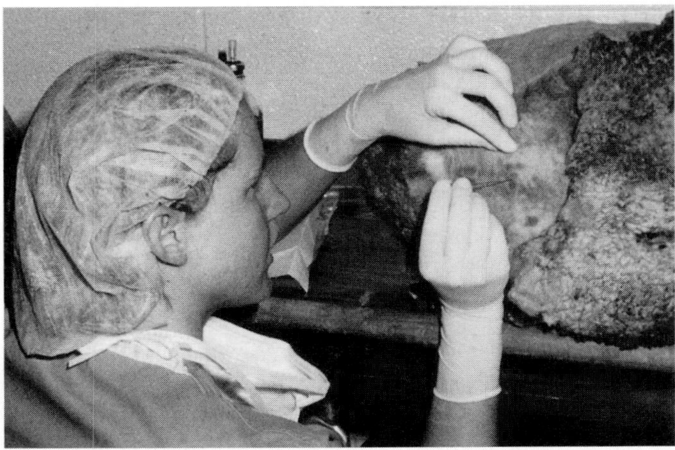

Fig. 1. Epidural needle placement. The injection site is the most palpable depression caudal to the line joining the anterior border of each ileum (left index finger). The back of the sheep is toward the edge of the table, and the hand holding the needle is stabilized by resting the wrist on the edge of the table.

VI. PHARMACOLOGIC CONSIDERATIONS FOR ANESTHESIA OF RUMINANTS

Factors that influence the speed of induction and recovery from sedation or anesthesia caused by parenteral drug administration include the dose per kilogram administered, the plasma albumen level (for highly protein-bound drugs), the extracellular fluid volume, the relative distribution of cardiac output, sites available for drug redistribution, body temperature, and the efficiency of hepatic metabolism and renal excretion. Ruminants may have thick hair or wool coats so body weight should be accurately determined and drug doses carefully calculated to improve the predictability of the response to sedative and anesthetic agents.

Ruminants have efficient capabilities for redistribution, metabolism, and elimination of anesthetic drugs, frequently resulting in short durations of action as compared to those in dogs. Because of the large mass of their abdominal viscera, which presumably receives a high proportion of the cardiac output, ruminants also have a large capacity to redistribute anesthetic drugs. Therefore, drugs such as thiopental, which depend on redistribution for anesthetic recovery, have a shorter duration of action compared to that in nonruminant species such as dogs. For drugs such as pentobarbital, where recovery depends on hepatic metabolism and renal excretion, the elimination half-life can be used to estimate the speed of recovery (in one half-life, 50% of the drug will be eliminated from the body; in two elimination half-lives, 75% will be removed, etc., Table II). For anesthetic drugs such as propofol, xylazine, and ketamine that have very high clearance rates, recovery from sedation or anesthesia depends on multiple factors, including drug redistribution and metabolism.

VII. Sedation and Premedication

Reasons for using sedative and analgesic drugs include minimizing physical resistance, animal stress, pain perception, blunting cardiopulmonary responses, and reducing the requirement for more potent general anesthetic agents. Considerations and reasons for preanesthetic administration of analgesic agents are discussed in Section XV on pain management. Domestic ruminants, particularly sheep and goats, are usually docile and do not resist or become overly stressed by physical restraint. Sheep and cattle do not exhibit excitement during recovery from general anesthesia. For these reasons, sheep and cattle are not routinely sedated prior to induction of general anesthesia. However, goats may become distressed during anesthesia recovery and may vocalize so the use of long-acting, mild sedatives might be desirable. The reaction of goats to even mild stimulation during the postoperative recovery period, especially following procedures such as thoracotomy, can be quite violent. Diazepam given at the rate of 5–10 mg IV has been used to blunt this reaction in the postoperative period. The infiltration of long-acting local anesthetics in the wound site of thoracotomies in goats has also been used to obtund this reaction. Placing goats in a sling postoperatively can be used to minimize this postanesthetic excitement.

All ruminants may hypoventilate and have difficulty maintaining sternal recumbency and laryngeal function if sedation is profound or if anesthesia recovery is prolonged. Because of their mild temperament, many local and regional anesthesia techniques have been developed to facilitate even major surgical procedures.

Desirable qualities of sedative and tranquilizing drugs include a rapid onset, a predictable duration of action and level of sedation, minimal undesirable cardiopulmonary effects, and the potential to limit the action of the drug. Drugs commonly used for sedation include acepromazine, xylazine, and, in cattle, pentobarbital or chloral hydrate. Newer drugs that more closely characterize the ''ideal'' sedative include detomidine, medetomidine, romifidine, and midazolam. Common doses for anesthetic drugs discussed in this section are given in Table III.

A. Phenothiazine Tranquilizers

Of the phenothiazine tranquilizers, acepromazine is most commonly used in veterinary medicine. When administered intravenously, it causes a slow onset (20 min) of mild sedation, usually without causing recumbency. It has a 2- to 4-hr duration of action that can be prolonged following IM administration. The long duration of action is useful for improving anesthesia recovery in goats, but will prolong recovery from general anesthesia in all ruminants (Thurmon and Benson, 1993). Acepromazine alone does not cause analgesia (Nolan *et al.,* 1987b), although it may potentiate analgesia caused by other agents. In excited or stressed cattle such as those undergoing caesarian section from dystocia, acepromazine may fail to cause adequate sedation. Sedation following acepromazine administration causes

TABLE II

PHARMACOKINETIC VARIABLES FOR INJECTABLE ANESTHETICS IN SHEEP[a]

Drug	Dose (mg/kg)	Number of sheep	Distribution half-life $t_{1/2}\ \alpha$ (min)	Volume of distribution: Central (liter/kg)	Volume of distribution: Steady state (liter/kg)	Clearance (ml/kg/min)	Elimination half-life (hr)	Reference
Alfentanil	1.15 to 1.5 (infusion)	5	11.6 ± 6.4	0.04 ± 0.02	0.7 ± 0.3	13 ± 3	1.1 ± 0.5	Ilkiw and Benthuysen (1991)
Thiopental	20 to 30 (infusion)	6	7.4 ± 3.0	0.04 ± 0.01	0.95 ± 0.14	3.0 ± 0.9	4.2 ± 1.8	Ilkiw *et al.* (1991)
Thiopental	20	6		0.24 ± 0.05	1.0 ± 0.2	3.8 ± 0.8	3.31 ± 1.1	Toutain *et al.* (1983)
Pentobarbital	14.3	7	4.8 ± 1.2	0.5 ± 0.08	1.0 ± 0.2	11 ± 3	1.2 ± 0.26	Short *et al.* (1985)
Propofol (and halothane)	6.0	7	0.73 ± 1.4	0.16 ± 0.001	1.15 ± 0.002	85 ± 59	9.4 ± 1.8 (min)	Handel *et al.* (1991)
Xylazine	1.0	6	1.89		2.74	83	23.1 (min)	Garcia-Villar *et al.* (1981)
Ketamine (and halothane)	11.6	8	8.0		2.0	250 for 15 min, then 67	40 (min)	Waterman and Livingston (1978)

[a]Mean ± SD.

TABLE III

DRUG DOSAGES USED FOR RUMINANTS

Compound	IV (mg/kg)	IM/SQ (mg/kg)
Sedative drugs		
Atropine	0.02–0.04	0.13
Acepromazine	0.02–0.05	0.05–0.2
Xylazine	0.02–0.15	0.05–0.3
Clonidine	0.015	
Detomidine	0.03	
Medetomidine	0.01	
Romifidine	0.05	
Pentobarbital	1.0–2.0	Usually not given IM
Chloral hydrate	50–70	Do not use IM/SQ
Diazepam	0.2–0.4	Usually not given IM
Midazolam	0.4–1.3	
Butorphanol	IV causes stimulation	0.5 (SQ)
General anesthetic agents		
Pentobarbital	20–30	
Thiopental	25	
Propofol	4–6	
Ketamine	10	20–30
Ketamine +	2.2–7.5	5–15
xylazine	0.1	0.1–0.2
Ketamine +	2.2–7.5	
diazepam	0.1–0.375	
Ketamine +	0.5	
medetomidine	0.02	
Atipamezole	0.02	0.06
Telazole	4.0	

minimal respiratory depression (Adetunji *et al.,* 1985; Hodgson *et al.,* 1991; Thurmon *et al.,* 1975). There is minimal cardiac depression, although mild hypotension occurs as a result of peripheral vasodilation, presumably via its blockade of peripheral α_1-adrenergic receptors (Adetunji *et al.,* 1985; Hodgson *et al.,* 1991). Peripheral vasodilation can cause the body temperature to decrease, which will also prolong recovery from general anesthesia. Acepromazine may be particularly useful for the sedation of pregnant ruminants as phenothiazines cause a minimal alteration of uterine blood flow and therefore maintain oxygenation of the fetus (Hodgson *et al.,* 1991; Pollack *et al.,* 1990). Promethazine administered to pregnant ewes with a hypoxic fetus did cause further fetal stress (Cottle *et al.,* 1983), which suggests that the cardiovascular response may differ for an already stressed fetus.

In summary, acepromazine causes mild sedation, minimal cardiovascular depression and, in pregnant ruminants, minimal depression of fetal oxygenation. Acepromazine does cause prolonged vasodilation which can lead to hypotension, especially in dehydrated ruminants, those losing blood during a surgical procedure, or animals that are administered other drugs that cause profound hypotension (e.g., IV propofol, epidural local anesthetics).

B. α_2-Adrenergic Agonists

Xylazine is the most widely used tranquilizing drug in ruminants. Other drugs in this class include clonidine, detomidine, medetomidine, and romifidine. α_2-Adrenergic agonists cause dose-dependent sedation that may be profound, frequently resulting in recumbency (Raptopoulos and Weaver, 1984). In addition, drugs in this class cause analgesia. However, the level of analgesia is not sufficient to prevent kicking, vocalizing, and struggling in response to a surgical incision, even in recumbent animals (Raptopoulos and Weaver, 1984). The main advantages of α_2-adrenergic agonists are their rapid onset, especially following IV administration, and profound sedation in normal animals. However, sedation, recumbency, and duration of affect are not predictable, especially in stressed animals. In the study reported by Raptopoulos and Weaver (1984) out of 84 trials in which steers were administered around 0.2 mg/kg xylazine IV, animals in 56% of cases became recumbent unaided and the recumbency time varied from 10 min to 3 hr (average 75 min). Because the duration of action of xylazine can be longer than the elimination half-life would predict (Table II), it suggests that the specific α_2-receptor interaction is important in determining α_2-adrenergic agonist action and duration (Daunt *et al.,* 1994).

α_2-Adrenergic agonists cause some undesirable side effects, and in the case of high doses of xylazine (0.22 mg/kg IM, 0.17 mg/kg IV), this has been reported to cause a 10% incidence of morbidity and mortality in 125 cattle that was mostly attributed to aspiration of ruminal contents, even though these cattle had been intubated during the period of profound sedation (Adetunji *et al.,* 1984). Cattle can be intubated without regurgitation using high doses of xylazine (0.22 mg/kg IM) with atropine being administered at the rate of 0.13 mg/kg IM 10 min before xylazine dosage (Floyd and Randle, 1989). Because an increased incidence of regurgitation can occur either during or following orotracheal intubation in healthy sheep given xylazine, routine sedation using this agent prior to general anesthesia is not recommended. This problem may be related to altered ruminal motility, generalized muscle relaxation, or poor oxygenation and ventilatory distress as reported by Raptopoulos *et al.* (1985).

Cardiovascular effects of α_2-adrenergic agonists include bradycardia, second degree A–V blocks, ventricular dysrhythmias, hypotension, and decreased cardiac output (Adetunji *et al.,* 1985; Celly *et al.,* 1993; Kariman *et al.,* 1994). Hypotension is probably related to the reduction in cardiac output, as α_2-adrenergic agonists cause increased peripheral vascular resistance. Although these drugs cause minimal ventilatory depression, there is a profound decrease in PaO_2 that can result in hypoxemia and persist for the duration of the sedative effect (Adetunji *et al.,* 1985; Celly *et al.,* 1993; Hodgson *et al.,* 1991; Kariman *et al.,* 1994; Raptopoulos *et al.,* 1985). In sheep, the administration of atropine (0.04 mg/kg IV) with xylazine resulted in less depression of PaO_2 (Raptopoulos *et al.,* 1985). However, hypoxemia can be abolished by prior administration of an α_2-adrenergic antagonist, suggesting that this response is mediated via α_2-receptors (Waterman *et al.,* 1987). Because xylazine depresses cardiac output and PaO_2, it causes a profound reduction in fetal oxygenation in cattle (Hodgson *et al.,* 1991). Eisenach *et al.* (1987) have demonstrated a similar response in pregnant sheep, suggesting that α_2-adrenergic agonists should be used with caution in pregnant ruminants. In third trimester pregnant cows, xylazine can cause premature parturition (Rosenberger *et al.,* 1969).

Other undesirable effects of α_2-adrenergic agonists include reduced ruminal motility and bloat (Hikasa *et al.,* 1988; Seifelnasr *et al.,* 1974), hyperglycemia (Hsu and Hummel, 1981), increased urine output (Thurmon *et al.,* 1978), and disturbed thermoregulation (Dehghani *et al.,* 1991; Fayed *et al.,* 1989). In thermoneutral conditions, body temperature and ventilation are normal, but when the environmental temperature is high, as might occur in barns or in fields, respiratory rate is depressed and hyperthermia occurs. Death can occur if xylazine-sedated animals are recumbent and not protected from the sun on hot days (C. I. Dunlop, personal experience).

An important advantage of α_2-adrenergic agonists is that their effects can be reversed by specific α_2-antagonists, including idazoxan (0.03 mg/kg IV; Thompson *et al.,* 1989), tolazoline (1.5 mg/kg IV; Bonath *et al.,* 1985), and atipamezole (0.02 mg/kg IV, 0.06 mg/kg IM Raekallio *et al.,* 1991; Sharma *et al.,* 1994). Therefore, ruminants could be sedated sufficiently to cause recumbency, a procedure carried out, then the sedation antagonized resulting in recovery to standing and restoration of homeostasis to near normal function. This overcomes many of the concerns of prolonged recovery from α_2-adrenergic agonist sedation in ruminants. Yohimbine has been tried as a reversal agent for xylazine in sheep with little success (R. F. Hoyt, Jr., personal experience).

In summary, α_2-adrenergic agonists cause dose-dependent sedation and analgesia with some undesirable effects. However, the level of sedation may enable a variety of procedures to be carried out without the need to administer further anesthetic drugs and this sedation can be antagonized. Xylazine sedation should be supplemented with effective analgesia such as from a local or regional nerve block if a surgical procedure is to be performed. α_2-Adrenergic agonists should be used with caution in pregnant animals, especially in the third trimester. If these drugs are used for premedication prior to administration of other general anesthetic drugs, the dose of the general anesthetic agents needs to be substantially reduced.

C. Benzodiazepines

Diazepam and midazolam have been used alone for muscle relaxation and mild sedation in calves, sheep, and goats. These drugs can cause excitement and ataxia. More usually, these drugs are used in combination with general anesthetics to provide muscle relaxation and to potentiate the anesthetic effects.

Diazepam administered IV to sheep causes an increase in systemic blood pressure and hypoventilation and a decrease in PaO_2 (Celly *et al.,* 1993). The change in PaO_2 was less than that seen following α_2-adrenergic agonists. Diazepam given at the rate of 0.25–0.3 mg/kg intravenously has been used successfully to sedate sheep prior to either pentobarbital or thiopental anesthesia (R. F. Hoyt, Jr., personal communication). In calves, diazepam IV caused a short duration of sedation (6 to 12 min) but a marked reduction in PaO_2, sufficient to result in hypoxemia (Mirakhur *et al.,* 1985).

Midazolam causes a rapid onset of sedation and recumbency in sheep (1.3 mg/kg) with an increase in heart and respiratory rates (Klein *et al.,* 1988). There was increased salivation, pupil constriction, and sheep stood after 25 min of recumbency. The main advantage of midazolam is the availability of a specific benzodiazepine antagonist, flumazenil. When administered IV (1.0-mg total dose) to recumbent, midazolam-sedated sheep, standing occurred immediately. These sheep "jumped to their feet, ran a few steps and began eating." Most sheep were mildly ataxic a few minutes later, but did not become recumbent. Midazolam probably has similar effects on oxygenation as diazepam.

D. Pentobarbital

Although most commonly used as a general anesthetic, pentobarbital is a useful sedative at lower doses. It causes reliable sedation with a more rapid onset than chloral hydrate, but still has a long duration of action. Valverde *et al.* (1989) reported that pentobarbital (2 mg/kg IV) in adult cows caused moderate sedation in 15 min, but the effect decreased by 60 min. Sedated cows were ataxic, but did remain standing. There were minimal changes in heart rate, blood pressure, PaO_2, or $PaCO_2$.

E. Chloral Hydrate

Chloral hydrate is a reliable sedative that has a slow onset (the active compound, trichloroethanol, is produced by liver metabolism) but a long duration of action. Its main use in biomedical research is for either long-term sedation or to permit the sedation of unmanageable cattle via oral administration (100 to 150 g in 8 to 10 liter of drinking water for adult cattle; Thurmon and Benson, 1993). It is very irritating if administered perivascularly so it should not be given IM or SQ. There is a wide range of intravenous doses, but sedation can be achieved after 5 to 7 g/100 kg IV. Chloral hydrate causes minimal cardiovascular effects in cattle, but depresses ventilation and PaO_2 (Adetunji *et al.,* 1985).

F. Opioids

Butorphanol has been used clinically in ruminants for sedation. Opioid drugs may not provide effective analgesia in ruminants and certainly have a shorter duration of action than is usually expected in dogs. Butorphanol administered to sheep at a dose of 0.1 to 0.2 mg/kg IV causes agitation, bleating, and chewing (Waterman *et al.,* 1991). At 0.4 mg/kg IV, butorphanol causes ataxia. However, a study by O'Hair *et al.* (1988) found that 0.5 mg/kg SQ butorphanol in sheep caused sedation and analgesia (in response to skin, toe, or nose pinch) from 15 to 120 min following administration. Clinically, butorphanol seems to decrease the response to nasopharyngeal stimulation (e.g., passage of an endoscope) in adult cattle, which supports the findings of O'Hair *et al.* (1988) in sheep. There is variability in the analgesic and sedative response to opiate drugs in ruminants, but if they are used, abnormal behavior is less commonly observed following SQ administration.

VIII. INDUCTION OF GENERAL ANESTHESIA

A. Pentobarbital

Pentobarbital is still widely used as a general anesthetic for small ruminants in biomedical research, particularly sheep. Because pentobarbital is rapidly eliminated by hepatic microsomal oxidative metabolism (Table II), it causes a relatively short duration of anesthesia in sheep (10 to 20 min following 20 to 30 mg/kg IV; Komar, 1991; Short *et al.,* 1985). Recovery to standing will take about 1 hr following a single dose. Prolonging anesthesia with repeated dosages will cause longer recoveries; at deeper levels of anesthesia, pentobarbital depresses ventilation. Komar (1991) reported that pentobarbital caused profuse salivation, slow respiration rates, and, in 50 out of 60 sheep, hemoglobinuria.

It has been suggested that pentobarbital alone is not appropriate for general anesthesia because barbiturates do not cause analgesia. Although the authors are not aware of specific studies in ruminants, pentobarbital does inhibit sympathetic nervous system activity (Baum *et al.,* 1985) and has been shown to alter nociception at subanesthetic levels in cats (Sandkuhler *et al.,* 1987). Therefore, it would seem that pentobarbital alone can provide acceptable general anesthesia for surgical procedures.

B. Thiopental

Thiopental is commonly used for the induction of general anesthesia in ruminants, either as a sole anesthetic agent or in combination with muscle relaxants because of its rapid onset, short duration of action, and rapid recovery to standing. In sheep, 25 mg/kg resulted in 15 min of anesthesia and sheep were standing in 30 to 45 min (Komar, 1991). Recovery from thiobarbiturates is via redistribution to muscle and abdominal viscera. Ruminants efficiently eliminate thiopental because of their large muscle mass (about 30% of the live weight in sheep, receiving about 10% of the cardiac output) and visceral mass (receiving about 40% of the cardiac output in sheep) (Hales *et al.,* 1976; Rae, 1962). Although fat is a large portion of the body mass of a ruminant (15% of the live weight of sheep), it is poorly perfused so the concentration of thiopental increases slowly (in sheep, from 30 min to 2 hr). Therefore, it has little to do with recovery to consciousness. Because goats can have relatively less muscle mass than sheep, recovery from thiopental anesthesia can be longer and less predictable, particularly if repeated doses are used.

Thiopental alone has been reported to result in profuse salivation and in a high incidence of regurgitation (22 out of 45 sheep, Komar, 1991), but this may be due to either attempting to intubate inadequately anesthetized animals (Thurmon and Benson, 1993) or excessive anesthetic depth. This issue has been discussed in the section on regurgitation. Intubation with a cuffed tube will minimize such problems. In sheep where anesthesia was maintained with thiopental alone, an increase in heart rate and systemic blood pressure and a slight decrease in cardiac output and hepatic and renal perfusion were observed (Runciman *et al.,* 1990). These changes in cardiovascular function persisted for several hours after anesthesia, which might be of importance in studies where tissue oxygenation or drug distribution was being evaluated.

Thiopental is an irritant if administered perivascularly (it should not be administered IM, IP, or SQ), particularly if used in concentrations greater than 5%. Therefore, it is advisable to administer it through an indwelling catheter. Anesthesia is induced by administering a bolus of the drug, usually one-third to one-half of the calculated dose, depending on the level of sedation of the animal. Animals should become recumbent in 30 sec. Thiopental alone is also useful, especially in sheep undergoing short procedures such as radiography, where 5 to 10 min of nonmovement is required.

C. Propofol

Propofol is a newer general anesthetic agent that has been used especially in small ruminants for the induction and maintenance of anesthesia. Propofol has similar usage characteristics to thiopental and is administered in a similar fashion. Its singular advantages are (1) that it is not an irritant if administered perivascularly and (2) that it is very rapidly eliminated from the plasma. It has the shortest elimination half-life of any parenteral anesthetic drug in sheep (Table II) and is similarly efficiently eliminated in goats (Nolan *et al.,* 1991). Because of its extremely rapid elimination, anesthesia can be maintained using a constant rate of infusion of propofol and still have a relatively rapid recovery. Infusion of propofol in sheep was associated with an increase in plasma triglyceride concentrations, which might be of concern for laboratory tests if blood samples were taken in the immediate anesthetic period. In addition, alterations in red blood cells have been seen in cats anesthetized with propofol on multiple occasions.

Bolus administration of propofol usually results in a brief period of apnea and transient hypotension. These effects are reduced by administering the drug more slowly (e.g., over 1 min). In sheep where anesthesia was subsequently maintained with a constant rate of propofol infusion (20 to 25 mg/min in adult 35-to 65-kg sheep), heart rate and blood pressure increase above resting awake values in a similar fashion to that observed with thiopental (Runciman *et al.*, 1990). However, as compared to thiopental, propofol did not depress cardiac output, although it did cause a slight reduction in hepatic and renal blood flow. These alterations returned to normal almost twice as fast following propofol as compared to thiopental. Because it is rapidly eliminated in adult ruminants, propofol might also be useful in sheep and goats undergoing cesarean section. In a report by Vesce and Lucisano (1991), propofol and halothane used for cesarean section in sheep resulted in neonatal depression, but newborn lambs also rapidly eliminated propofol and all animals survived. In this study, propofol rapidly crossed the placental barrier in both sheep and goats.

Propofol alone is probably not sufficient to prevent movement in response to surgical stimulation. This may indicate insufficient analgesia, requiring supplementation with either local anesthesia or other systemic analgesic or anesthetic drugs if surgery is performed.

D. Ketamine

Ketamine is a dissociative drug, used most commonly in combination with sedatives or tranquilizers to induce general anesthesia. Ketamine has a short duration of action due to its large volume of distribution (weak organic base) and its rapid metabolism (Table II). Although it is occasionally used as a sole anesthetic agent, ketamine causes some undesirable effects, including increased muscle tone, forelimb rigidity, salivation, and apneustic ventilation. One advantage of increased muscle tone is that regurgitation seems to be less common following ketamine as compared to thiopental. In addition, laryngeal function seems to be better maintained. Because increased muscle tone results in the eyes being open, the head should be padded and ointment should be instilled into the eyes and be covered to prevent damage from trauma or exposure to sunlight if outside. These undesirable affects of ketamine are reduced by administering xylazine, diazepam, or acepromazine.

In sheep (Thurmon *et al.*, 1973) and goats (Kumar *et al.*, 1983), 22 mg/kg IM ketamine causes recumbency in 2 to 3 min, about 10 min of anesthesia, but recovery to the point of standing, eating, and drinking takes about 100 min. Komar (1991) reported similar results following 20 mg/kg of ketamine IV in sheep, except that the anesthesia time was longer (25 min). Anesthetic recovery is faster if a lower dose of ketamine is administered IV. Ketamine causes profuse salivation and sympathetic stimulation which increases heart rate, cardiac output, and blood pressure. Mild hypoventilation, slow respiration rates, and a period of low PaO_2 sufficient to cause hypoxemia for 5 min after recumbency are also reported.

E. Commonly Used Ketamine Combinations

1. Ketamine–Xylazine

This combination is commonly administered to ruminants intramuscularly to provide between 1 and 1.5 hr of anesthesia (Blaze *et al.*, 1988b; Dehghani *et al.*, 1991; Rings and Muir, 1982). The onset of anesthesia is rapid following IM administration. However, recovery to standing following anesthesia with this combination (90–160 min) is the slowest of any single-dose parenteral technique. Anesthetic duration is around 30 min following IV administration and recovery to standing occurs in around 90 min (Coulson *et al.*, 1989; Komar, 1991). Because most small ruminants used in biomedical research are easily restrained for IV injections, the prolonged recovery limits the usefulness of IM ketamine and xylazine. However, this combination does have a place, particularly for anesthesia of unmanageable cattle, for smaller ruminants where inhalation anesthetic drugs are not available, or where IV injections cannot be given. In addition, it causes good muscle relaxation.

The cardiopulmonary responses to this combination of drugs include a slight decrease in heart rate, a larger decrease in systemic blood pressure, but little change in cardiac output (Coulson *et al.*, 1989; Kumar *et al.*, 1983; Rings and Muir, 1982). Apneustic breathing, mild hypoventilation, a low PaO_2, and frequently a period of hypoxemia are also reported (Blaze *et al.*, 1988b; Coulson *et al.*, 1989; Kumar *et al.*, 1983). Hypoxemia and acidemia seem to be primarily caused by hypoventilation (Blaze *et al.*, 1988b). Because hypoxemia does occur, the administration of supplemental oxygen such as via a face mask is advisable, particularly for pregnant and debilitated animals. This combination causes good muscle relaxation. Regurgitation is not common, but for long procedures in lateral or dorsal recumbency, intubation would be prudent and would also facilitate the administration of supplemental oxygen. Intravenous administration of lower doses (xylazine 0.1 mg/kg; ketamine 2.2 mg/kg) than those used by Coulson *et al.* (1989) results in induction of anesthesia but a shorter duration of action with lesser undesirable cardiopulmonary changes.

2. Ketamine–Diazepam

This combination, administered IV, is the most common method used to induce general anesthesia in healthy sheep used for research at Colorado State University. At a dose of 2.75 mg/kg ketamine and 0.2 mg/kg diazepam IV there is a rapid onset of anesthesia, sheep are easily intubated with few cases of regurgitation, there is little cardiopulmonary dysfunction, and small volumes of drugs are used and are not irritant if administered perivascularly. Anesthetic recovery is rapid (15 to 20 min), but of sufficient duration to allow initiation of inhalation anesthesia for longer procedures.

Coulson *et al.* (1991) reported that higher doses (7.5 mg/kg ketamine and 0.375 mg/kg diazepam) caused effective anesthesia lasting 15 min in sheep but that recovery took 60 min. Even at these high doses, there were minimal cardiovascular changes except for a slight decrease in cardiac output and transient hypotension and an increase in vascular resistance. However, hypoventilation and a decrease in PaO_2 similar to that following IV xylazine/ketamine were reported. Hypoxemia in sheep is not usually observed if, following intubation, a high inspired concentration of O_2 is provided using oxygen supplementation (C. I. Dunlop, personal communication).

Maintaining anesthesia with diazepam/ketamine for prolonged periods (1.5 hr) causes a persistent depression of ventilation, oxygenation, and cardiac output and increases vascular resistance (Coulson *et al.,* 1991). Therefore, prolonged anesthesia with high doses of ketamine–diazepam would seem to offer few advantages over xylazine–ketamine.

3. Ketamine–Medetomidine

This mixture has been investigated for use in calves. Medetomidine is a far more specific α_2-adrenergic agonist than xylazine and is a more potent sedative and analgesic. Therefore the dose of ketamine needed to cause analgesia can be substantially reduced. In a report by Raekallio *et al.* (1991), calves became recumbent within 1 minute and remained sufficiently anesthetized to permit minor surgery for 20 to 40 min. Similar to the response of xylazine, medetomidine caused a reduction in heart rate, hypoventilation, and a more marked reduction in PaO_2. This anesthetic combination has the advantage that the sedation caused by medetomidine can be antagonized by atipamezole. In the report by Raekallio *et al.* (1991), atipamezole administered to calves following 15 to 118 min of surgery (7 to 30 min after the last dose of medetomidine or ketamine) caused them to become recumbent in an average of 2 min and to stand in an average of 3 min. There was no difference in recovery if the entire dose of atipamezole (0.06 mg/kg) was administered IV or if half was administered IM.

F. Tiletamine and Zolazepam

Tiletamine, a dissociative anesthetic, and zolazepam, a benzodiazepine, are available as a fixed combination (1:1 mixture; Telazol). This combination has been investigated over a long period of time (Conner *et al.,* 1974), but has only recently become available. This combination produces more profound anesthesia than ketamine–diazepam. Following 12 to 24 mg/kg IV administered to sheep (Lagutchik *et al.,* 1991), there was a rapid onset of general anesthesia sufficient for surgery that lasted 40 min, but recovery was prolonged with sheep unable to stand unassisted after 2 hr. At this high dose, cardiac output, systemic pressure, ventilation, and PaO_2 were depressed and systemic vascular resistance increased. For 20 min following a more typical dose in calves (4 mg/kg IV; Lin *et al.,* 1989), cardiopulmonary function changed in a similar fashion, but to a lesser degree. The duration of anesthesia was not reported, but calves remained recumbent for at least 60 min.

From a clinical perspective, this combination probably offers few advantages and potentially some disadvantages over ketamine–diazepam when used in healthy, calm ruminants. However, its one clinical advantage, which was not discussed in the earlier reports, is the rapid onset of anesthesia following IM administration. Occasionally there are situations where this is a desirable quality of an anesthetic drug. If adult cattle escape, Telazol can be administered IM via a capture dart to cause rapid heavy sedation or anesthesia.

G. Etomidate

Although it is not widely used for anesthesia of ruminants, Kumar *et al.* (1983) reported that 2 mg/kg IV etomidate in 40 sheep produced about 12 min of anesthesia and recovery to standing in 20 to 30 min, being considerably faster than that for thiopental, ketamine, or ketamine–xylazine. However, profuse salivation occurred in 20% of sheep and hemoglobinemia in 10% of sheep. Etomidate also causes pain on injection. One advantage of etomidate in other species is its minimal cardiovascular depression and low arrhythmogenicity. However, this is not documented for ruminants. Komar (1991) reported that combining etomidate and xylazine resulted in profound hypoxia, so this combination should be used with caution in ruminants not administered supplemental oxygen.

H. Guaifenesin

Guaifenesin is an effective, centrally acting muscle relaxant that causes minimal cardiopulmonary depression (Thurmon and Benson, 1993). Guaifenesin is most commonly prepared as a 5% solution in 5% dextrose and water, and at a dose of 75 to 100 mg/kg, large volumes are required (1 liter in cattle and around 150 ml in sheep). Guaifenesin is an irritant if administered perivascularly so it must be given IV, preferably via an indwelling catheter. It can cause recumbency when used alone, but does not alter central nervous system (CNS) consciousness and there is no evidence that it causes analgesia. Therefore, it should not be used as a sole restraint agent for animals undergoing surgical or painful manipulations. However, clinical experience suggests that guaifenesin potentiates the effects of other general anesthetic agents to a greater degree than other muscle relaxants (e.g., benzodiazepines), so it is a useful drug in anesthetic combinations. In adult horses, death from guaifenesin toxicity is seen at doses around 300 mg/kg (Funk, 1973). In small ruminants, this dose can easily be achieved with the amount of guaifenesin in a 1-liter bottle of 5% solution (total of 50 g per bottle, suitable for a 500-kg animal).

For IV induction of anesthesia, 60 to 100 mg/kg guaifenesin is commonly administered with 4 to 6 mg/kg thiopental, 2.2 mg/kg ketamine, or 3 mg/kg propofol. In sheep induced with one of these three combinations, that were then administered halothane for 1 hr, there was no difference in average recovery times to standing (propofol 28 min; ketamine 34 min; thiamylal 50 min; Crump *et al.,* 1994a) or recovery quality. A mixture of guaifenesin and thiopental (2 g added to 1 liter 5% guaifenesin) administered to effect (typically 150 to 200 ml/adult sheep) can be used to induce anesthesia in sheep for short procedures such as radiography or minor surgery. The advantage of this technique over thiopental alone is that a lesser dose of thiopental is required, resulting in faster and less excited recoveries from anesthesia.

I. Halothane or Isoflurane

In calves, lambs, sheep, and goats, induction of general anesthesia can easily be achieved with halothane or isoflurane. In larger calves (over 100 kg), sedation with 0.025 to 0.05 mg/kg IV xylazine will facilitate handling. Most commonly the animal is physically restrained and a mask is placed over the muzzle. Dog face masks are suitable for small calves, sheep, and goats. A large-size plastic funnel will work if a piece of rebreathing bag is taped over the wide opening and a hole is cut in the center. Plastic 1-gallon bottles with the bottom cut out and the cut edge padded with gauze and tape will also work. Alternatively, a lubricated 6- to 10-mm ID cuffed nasotracheal tube can be passed through the nostril, ventral nasal meatus, and nasopharynx and then through the larynx into the trachea. The cuff is then inflated. Mask inductions are slower than nasotracheal inductions and are associated with more occupational anesthetic gas exposure. Nasotracheal intubation is not a particularly easy technique and can be accompanied by some nasal bleeding which can cause significant problems in animals that receive anticoagulants during the surgical procedure. This technique also limits the diameter of the endotracheal tube that can be used.

Circle rebreathing systems found on most human and veterinary anesthetic machines are suitable for small ruminants. The carbon dioxide canister should be filled with fresh absorbent. For calves between 100 and 150 kg, the carbon dioxide absorber should have a volume of at least 2 liters. Halothane or isoflurane inductions using a nasotracheal tube take 3 to 5 min at fresh O_2 flows of 5 liters per minute with an out-of-circuit precision vaporizer set on 5%. Once anesthesia and orotracheal intubation are achieved, the O_2 flow can be reduced to 10 to 30 ml/kg/min and the vaporizer setting to between 1 and 2.5%. The carbon dioxide absorbent should be replaced if the color indicator shows that half has become exhausted. Inhalation anesthetic inductions limit the need for additional parenteral anesthetic agents, which limits cardiopulmonary dysfunction associated with a combination of drugs, limits the amount of drug metabolism and excretion, and results in a rapid recovery from anesthesia. These advantages make this a useful technique in inducing anesthesia in pregnant sheep, goats, and neonates.

IX. INTUBATION TECHNIQUES

Because ruminants profusely salivate during general anesthesia and have the potential to regurgitate at either light or deep levels of anesthesia, the authors advocate intubation with cuffed tubes. In small ruminants such as sheep, cuffed 10- to 16-mm ID vinyl plastic or silicone endotracheal tubes are used. For adult cattle, 22- to 26-mm ID, 80- to 100-cm-long tubes are necessary. Due to the fact that all ruminants have very sharp molar teeth, they should be adequately anesthetized such that they are neither chewing nor swallowing. It is also important that care be taken not to strike the edges of the molar teeth with the cuff of an endotracheal tube during insertion as it may rip the cuff. It is also important to adequately secure the endotracheal tube in place, and it is usally very helpful to use a rubber or wooden mouth gag that can be taped or tied in place to restrict the movement of the endotracheal tube throughout the anesthetic period. Commonly, these devices contain a large central hole in which the endotracheal tube passes fixing its position. The tube and mouth gag are then secured in place with tape and roller gauze. The patency of the tube cuff should be checked before it is placed into the animal. There are at least four useful techniques for the intubation of adequately anesthetized ruminant animals.

A. Direct Visualization

In sheep, goats, and calves, the oral cavity is small and short so a laryngoscope can be used to visualize the laryngeal opening. Conventional laryngoscope blades (such as a Miller No. 4) are too short, so usually a locally modified blade is required. A locally designed and manufactured 20-cm-long blade is used, which can simultaneously hold the tongue to one side and depress the epiglottis. The animal is best positioned in sternal recumbency, the head fully extended, and the mouth held open by an assistant using two strips of gauze (Fig. 2). After the epiglottis is visualized, 2% lidocaine can be applied to the laryngeal opening using a small spray bottle, then the endotracheal tube is passed and the cuff inflated.

The animal should not be swallowing or trying to regurgitate during intubation. It is easiest to pass the endotracheal tube during inspiration. Some endotracheal tubes are not sufficiently rigid to allow their tip to be easily directed, but this can be overcome by the use of a stylet (e.g., 3/8-inch malleable, aluminum rod). The end of the rod that protrudes from the endotracheal tube connector should be bent such that its other end does not protrude beyond the tip of the endotracheal tube. Once positioned, the endotracheal tube cuff should be positioned caudal to the larynx but not beyond the thoracic inlet. The endotracheal tube is then fastened to the lower jaw, using one of the gauze strips. A piece of PVC pipe may be placed over the endotracheal tube into the pharynx to prevent the tube from being lacerated by the molar teeth. This piece of pipe should be attached to the jaw to prevent it from being swallowed at the time of extubation. Occasionally, ruminants have to be intubated while they are regurgitating in order to minimize the chance of aspiration. For small ruminants, it is possible for a trained assistant to use the palm of their hand to apply firm pressure just caudal to the larynx, directing the force to the left side of the trachea, thus occluding the esophagus. This will then enable the larynx to be visualized and an endotracheal tube to be passed.

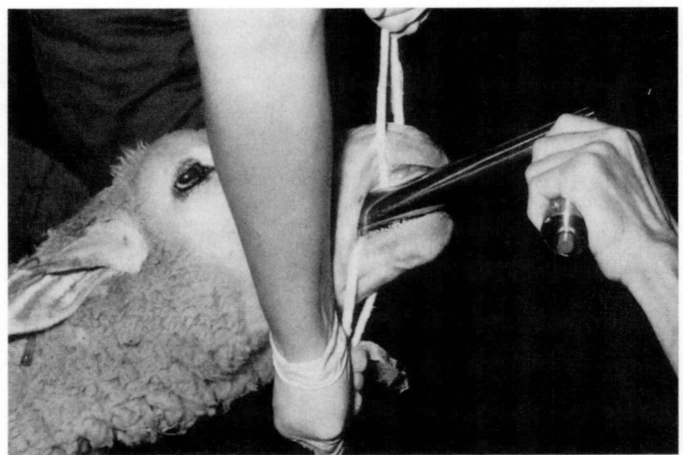

Fig. 2. Sheep and goats can be intubated positioned in sternal recumbency. The head is fully extended and the mouth is held open by an assistant using two strips of gauze. A laryngoscope can be used to visualize the larynx.

B. Blind Intubation

This method of intubation is suitable for sheep and goats. The animal should be positioned on its back with the head beyond the edge of a table. Therefore the head will be in full extension. A stiff, curved endotracheal tube can then be directed blindly through the mouth and into the larynx. The limitation of this technique is that usually a smaller tube than normal is passed. However, in a ''mass production'' system such as the multiple surgical embryo transfers, this can be a rapid method to simply attain a secured airway for a short procedure. This technique should be used with caution in pygmy goats because they can have atlantoaxial joint laxity, which can result in a dislocation and spinal cord injury. Cattle can also be blindly intubated, but they are positioned in sternal recumbency with the head and neck fully extended.

C. Digital Palpation

Adult cattle are easiest to intubate using digital palpation and manual placement of the tube into the larynx (Fig. 3). Either a dental wedge or a mouth gag must be used to keep the mouth open and prevent serious injury to the arm of the operator. After the head is fully extended, the hand is advanced between the dental arcades to the level of the epiglottis. The middle finger can be used to palpate the opening between the arytenoid cartilages. During this examination it is critical to observe swallowing or attempts to actively regurgitate. If either occurs, immediately withdraw the arm and administer a further dose of parenteral anesthetic. After the initial examination, withdraw the arm and surround the endotracheal tube cuff with the fingers and palm to protect it from the molar teeth. The middle finger should be most dorsal and again should locate the opening between the arytenoid cartilages. Using this finger to direct the tube tip ventrally, advance the tube by pushing the exteriorized end with the free hand. The least resistance to passing the tube is during inspiration, but the animal should be at a depth of anesthesia sufficient to prevent swallowing. When the tube has been successfully passed into the trachea, the cuff should be inflated immediately to effect a seal.

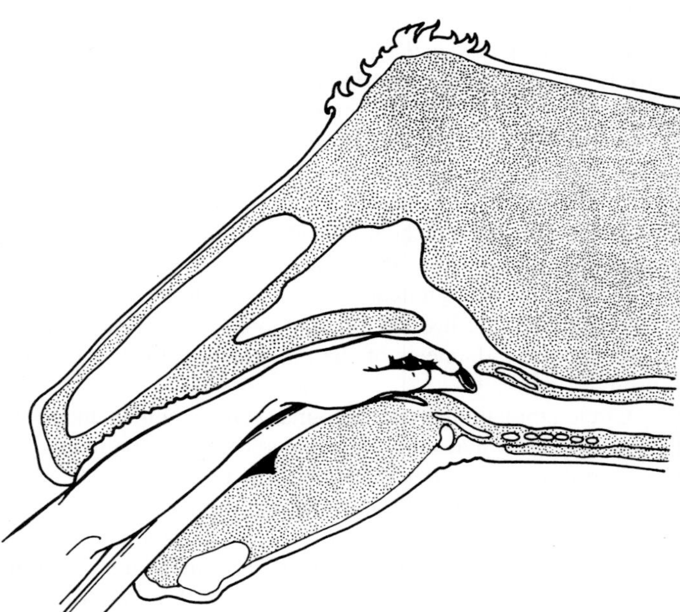

Fig. 3. Adult cattle are commonly intubated using digital palpation and manual placement of the tube into the larynx. The head is fully extended and the hand is advanced between the dental arcades to the level of the epiglottis. The fingers and palm are used to protect the endotracheal tube cuff from laceration by the molar teeth.

D. Use of Guides

This method is useful in small cattle and is similar to digital palpation except that instead of passing the endotracheal tube, a stiff tube such as an equine stomach tube is passed into the trachea. This is necessary because there is not sufficient room for both the endotracheal tube and the arm of the operator. The endotracheal tube is fed over this smaller tube which guides it into the trachea and the smaller tube is then removed. It is important that the outside diameter of the smaller tube be less than the internal diameter of the endotracheal tube and that the smaller tube be 2.5 to 3 times longer than the endotracheal tube. For this technique to work, the tubes must be constructed of materials that become sufficiently slippery from naturally occurring moisture to allow for them to slip over each other without binding.

X. MAINTENANCE OF ANESTHESIA

A. Inhalation Anesthetics

Because of their minimal metabolism and low solubility, which cause a rapid anesthetic recovery, halothane and isoflurane are the inhalation anesthetics most commonly used for the maintenance of anesthesia. The general pharmacology, physiological responses, and occupational health considerations associated with inhalation agents are discussed in Chapter 2. The relative potencies of the inhalation anesthetic drugs commonly used in ruminants are given in Table IV. Inhalation anesthetic agents depress arterial blood pressure, cardiac output, and ventilation in a dose-dependent fashion, which are useful in monitoring anesthetic depth (Runciman *et al.,* 1984b; Steffey and Howland, 1979). In other species, halothane is more arrhythmogenic and is metabolized to a greater degree than isoflurane (25% versus less than 2%). Cattle maintain higher heart rates with isoflurane (Greene *et al.,* 1988), which may result in higher cardiac outputs. This difference would be an advantage for neonatal ruminants, where cardiac output tends to be heart rate dependent. In sheep, distribution of blood flow to tissues is reduced by inhalation anesthesia (Runciman *et al.,* 1984b) and this effect is similar for both halothane and isoflurane (Abass *et al.,* 1988). This effect alters parenteral drug distribution and prolongs drug elimination. Interestingly, in sheep where anesthesia was induced with guaifenesin and thiamylal (a drug that is efficiently eliminated from ruminants and has a short duration of action), an end-expired halothane concentration of between 0.7 and 0.85% halothane was required to maintain anesthesia for bilateral stifle arthrotomy (Crump *et al.,* 1994b). This is a lower amount of halothane than would normally be expected for anesthesia of a large animal (vaporizer settings of 1% or less), suggesting that small ruminants may be overanesthetized if anesthetized as if they were a young, healthy dog on a circle anesthetic circuit.

TABLE IV

POTENCY OF COMMONLY USED INHALATION ANESTHETIC AGENTS IN SHEEP AND GOATS

	1 MAC[a] ± SE (vols %)		
Agent	Sheep (n = 6)	Goats (n = 6)[b]	Calves (n = 8)[c]
Halothane	0.69 ± 0.25 (5 pregnant ewes, MAC estimated from blood concentration)[d]	1.41 ± 0.11	0.76 ± 0.03
	0.73 ± 0.07 (pregnant ewes)[e]		
	0.97 ± 0.04[e]		
Isoflurane	1.01 ± 0.06 (pregnant ewes)[e]	1.63 ± 0.17	
	1.58 ± 0.07[e]		
Methoxyflurane	0.26 ± 0.02[e]		

[a]Minimum alveolar concentration of an inhalation anesthetic that prevents gross purposeful movement in response to a standard noxious stimulus; therefore MAC is an index of anesthetic potency.
[b]Alibhai *et al.* (1994).
[c]Steffey and Howland (1979).
[d]Gregory *et al.* (1983).
[e]Palahniuk *et al.* (1974).

The respiratory rate of adult ruminants anesthetized with isoflurane is lower than that seen with halothane, resulting in slightly greater hypoventilation ($PaCO_2$ range 70 to 85 mm Hg; Greene *et al.,* 1988). This is a greater level of hypoventilation than would be expected for similarly anesthetized horses, suggesting that adult cattle may benefit from intermittent positive pressure ventilation. However, excessive hypoventilation in cattle is not simply an effect of size, as calves at light levels of halothane anesthesia also had severe hypoventilation ($PaCO_2$ range 80 to 90 mm Hg at 1.0 to 1.5 MAC; Steffey and Howland, 1979). Halothane-anesthetized sheep do not hypoventilate to the same extent as cattle ($PaCO_2$ averaged 48 mm Hg; Crump *et al.,* 1994b). However, as compared to sheep anesthetized with pentobarbital, which causes minimal changes to pulmonary function, halothane anesthesia causes hypoventilation, decreases V/Q and functional residual capacity, and increases intrapulmonary shunting (Dueck *et al.,* 1984). Because of these responses, ventilatory function and oxygenation should be monitored during inhalation anesthesia.

B. Parenteral Anesthetics

Guaifenesin is particularly useful for maintenance of anesthesia via a constant infusion (Thurmon and Benson, 1993). A mixture consisting of 100 mg xylazine, 1 g ketamine, and 50 g guaifenesin is made by adding the xylazine and ketamine to a 1 liter bottle of guaifenesin. Anesthesia can be induced using a bolus of this mixture (typically 0.5 ml/kg) and/or maintained using 0.7 to 2.2 ml/kg/hr, administered at a constant rate via an IV infusion set. Recovery from anesthesia will be reasonably short if the total IV anesthesia time is kept to less than 1 hr. This anesthetic combination (''triple drip'') is associated with minimal cardiopulmonary depression (Thurmon *et al.,* 1986). However, it is prudent to intubate these animals and to administer supplemental oxygen, especially for longer procedures.

Other parenteral techniques such as administration of intermittent drug boluses are useful, but these are applications of the drugs and combinations that have already been discussed.

XI. MONITORING OF ANESTHETIZED RUMINANTS

The level of monitoring used for anesthetized ruminants being used for biomedical research depends on the nature, demands, and length of the procedure, as well as the age and physical status (e.g., pregnant vs. nonpregnant) of the animal. In addition, it depends on the skills of the technical staff, particularly their ability to observe physical signs, interpret these, and respond appropriately.

A. Physical Signs

The monitoring of anesthetic depth in adult cattle, sheep, and goats is most usefully achieved by observation of the following (in order of usefulness): (a) physical movement or jaw chewing in response to stimulation, (b) eye position and palpebral muscle tone, and (c) presence or absence of active or passive regurgitation.

Eye position and palpebral muscle tone certainly indicate sheep that are lightly anesthetized (i.e., dorso-lateral eye position and a relaxed palpebral fissure), but at deeper anesthetic levels, there can be a range from no eye signs to signs of very light anesthesia. Adequately to deeply anesthetized sheep have a centrally positioned eyeball with a dilated pupil. The cornea may appear to be dry, and there will be no palpebral reflex. The authors do not routinely test the corneal reflex because of the potential for injury. When such eye signs are present, it is advisable to attempt to lighten the anesthetic level, looking for signs such as physical movement to confirm that the animal is not overanesthetized.

The most reliable confirmation of anesthetic depth is the presence or absence of physical movement and either active or passive regurgitation (discussed under regurgitation). Active regurgitation is usually a sign of light anesthetic level. Passive regurgitation requires relaxation of the esophageal sphincter which usually occurs at moderate to deep levels of anesthesia. However, in ruminants not held off feed or water prior to anesthesia, regurgitation can occur continuously throughout anesthesia. When passive regurgitation occurs, the authors try decreasing the anesthetic level.

B. Cardiopulmonary Indicators of Anesthetic Depth

Systemic arterial blood pressure and ventilation are reduced in a dose-dependent fashion as anesthetic depth increases, particularly with inhalation agents.

1. Systemic Blood Pressure

Normal awake blood pressures for sheep are given in Table I. Typical mean arterial pressures during light surgical guaifenesin/thiobarbiturate/halothane anesthesia in sheep are 65 to 85 mm Hg (Crump *et al.,* 1994b). Determination of the direct systemic blood pressure is easily accomplished in sheep and goats. The auricular artery on the abaural surface of the ear can be catheterized prior to the induction of anesthesia. Blood pressure can then be determined directly in awake animals.

Alternatively, the blood pressure can be estimated indirectly using a blood pressure cuff with either auditory or Doppler detection techniques. In sheep and goats, the cuff should be between 30 and 60% of the circumference of the limb and should be placed on the forelimb at the midmetacarpus. In sheep, more accurate readings will be obtained if the wool is clipped. Systolic blood pressure is the most accurate indirect measure (i.e., not mean or diastolic pressure). As a rough guideline when mean systemic blood pressure falls below 60 mm Hg, the animal has likely become sufficiently hypotensive to have inadequate tissue perfusion which can result in the cessation of urine production and other physiological consequences.

2. Ventilation

Ventilatory function is best monitored using either intermittent arterial blood gas analysis or continuously monitoring $PaCO_2$ indirectly using capnography (end-expired CO_2). Capnography continuously samples gas using a small sample port placed between the endotracheal tube and the breathing circuit connection. The monitor determines the respiratory rate, inspired, and peak (end-expired) CO_2 levels. The end-expired CO_2 level is an estimate of the arterial CO_2 level. In anesthetized sheep positioned in dorsal recumbency, the difference between arterial and end-expired CO_2 averages 10 $\pm$ (SD) 4 mm Hg (Crump *et al.,* 1994b). Because of the potential for a large error, for longer procedures, or where the level of ventilation is important, it is advisable to analyze at least one arterial blood gas to know the accuracy of the end-expired CO_2 readings. In sheep at a light surgical level of guaifenesin/thiobarbiturate/halothane anesthesia, $PaCO_2$ averaged 48 mm Hg compared to awake levels around 32 mm Hg (Crump *et al.,* 1994b). In young animals with compliant chest walls or those undergoing a thoracotomy, the end-tidal CO_2 will be less accurate.

Arterial blood gas analysis still remains one of the most definitive methods of assessing the overall effectiveness of ventilation, especially during thoracic surgery where end-expired CO_2 values may be unreliable. Because many types of experimental surgical procedures for which small ruminants are used require arterial cannulation of peripheral arteries, this method of ventilation assessment is commonly used. Arterial blood gas analysis not only provides information on $PaCO_2$, but also measures oxygen tension (PaO_2), oxygen saturation of hemoglobin, and pH. Unlike capnography, the measurements are discontinuous.

C. Other Monitored Parameters

1. Heart Rate

It may be acceptable to monitor heart rate, pulse quality, and mucous membrane color for short procedures. For lengthy or more invasive procedures, monitoring cardiac function should include use of an electrocardiograph (ECG) to detect arrhythmias. Applying conventional human ECG leads that use stick-on skin electrodes can be difficult in small domestic ruminants (especially sheep). This is due primarily to the difficulty in removing hair completely from the area of application and adequately preparing the skin connection site using organic solvents such as alcohol to remove skin oils. These difficulties are compounded by the uneven and often small areas on the distal portion of the appendages of ruminants compared to the rather flat and fleshy portions of the forearm of humans. The use of clip-type skin electrodes that firmly grasp or even pierce the skin or band-type electrodes, as well as the application of adequate amounts of electrode gel to ensure conductivity, may provide more secure electrical connections.

2. Pulse Oximetry

The primary purpose of the cardiopulmonary system is to deliver oxygenated blood to tissues. Direct measurement of the arterial oxygen level by blood gas analysis is the most accurate method of determining arterial oxygenation.

Pulse oximetry is a valuable monitoring technique because it is simple, rapidly applied, noninvasive, and provides several useful pieces of information. A clothespin-type probe is usually clipped to the tongue and uses absorption of infrared light to detect blood flow and hemoglobin saturation with oxygen (SaO_2). As a general rule, one should strive to maintain a SaO_2 of 90% or greater during anesthesia. The pulse oximeter continuously calculates the pulse rate, emits a beep with each pulse, and estimates the tissue (tongue) saturation with oxygen. Most monitors have alarm systems that can be set for both pulse rate and inspired oxygen saturation levels.

3. Blood Glucose

Sheep that are anesthetized for long periods, are systemically ill, or are held off feed for more than 48 hr are likely to become dehydrated and hypoglycemic. In such cases, it is advisable to measure the blood glucose concentration prior to anesthetic induction, then repeat the measurement throughout anesthesia. The authors do not routinely administer intravenous glucose solutions such as 5% dextrose to anesthetized sheep unless the procedure is longer than 3 hr, if the blood glucose level drops markedly from the preanesthesia level, or if the blood glucose level falls below 90 mg/dl.

4. Blood Loss

Volume loss caused by hemorrhage or dehydration is not well tolerated. In addition, some ruminants have lower packed cell volumes (PCV) and hemoglobin concentrations. Healthy sheep and goats with a PCV over 30% and total serum protein (TSP) above 6.5 g/dl should be able to tolerate acute reductions in blood volume of 25% (blood volume normally is 7 to 8% of body weight or 70–80 ml/kg). This assumes appropriate volume replacement with polyionic solutions. When transfusions are needed (minimum PCV 20%, TSP 3.5 g/dl), whole blood is used that either has been collected prior to the procedure (a week or more) from the animal undergoing the surgery or from an appropriate donor animal of the same species. When blood from another animal is used, a major and minor cross match should be conducted. Although a number of blood types have been identified in each of the various species of ruminants commonly used in biomedical research, none have been shown to result in hemolysis in the absence of prior sensitization. It is possible, however, that more subtle reactions involving elevated core temperature and occasional decreases in hematocrit may occur without prior sensitization. Most commonly, blood for transfusion purposes is collected in a citrate–phosphate–dextrose–adenine solution (USP CPDA-1). Under refrigeration at 4°C, the shelf life of whole blood in humans has been documented to be as long as 35 days. Because similar information is not available in commonly used ruminant species, a shorter shelf life has been arbitrarily used that ranges from 7 to 21 days.

5. Body Temperature Maintenance

Thermal regulation is a complex integrated process involving many body systems and organs. The application of anesthetics disturbs thermal regulation, resulting in the body beginning to equilibrate with environmental temperatures. Generally this results in a decrease in core body temperature. As body temperature falls, metabolism is altered and with it the metabolism of anesthetics. In general, the amount of inhalant anesthetic necessary to sustain general anesthesia drops approximately 5% for every degree Celsius below normal body temperature in humans with similar changes likely occurring in animals.

The maintenance of normothermia is further compromised by the fact that more severe hypothermia can be expected to occur in anesthetized animals undergoing open body cavity procedures. In the case of small ruminants such as sheep, removing hair or wool from the surgical site and performing surgical site skin preparation using volatile aqueous and organic solutions also contribute to a drop in the body temperature of the animal through evaporation. Evaporative fluid loss from open body cavities coupled with the increase in surface area exposed to heat loss also magnifies the drop in core body temperature. For this reason, it is important to monitor core body temperatures during surgery and to use auxiliary heat sources such as circulating hot water blankets, warm water bottles, or other similar heat sources coupled with insulating materials to prevent loss into heat sinks such as stainless-steel surgical tables.

XII. SUPPORT DURING ANESTHESIA

A. Oxygenation

Because anesthesia depresses both ventilatory and circulatory function, anesthetized ruminants breathing room air are susceptible to hypoxemia. Therefore, the administration of at least 50% O_2 (balance N_2) to anesthetized ruminants is advisable. This is the single most important reason in considering the use of inhalation anesthesia, as these agents are administered from a breathing circuit in 50 to 95% O_2. If an anesthetic machine is not available or practical, such as with magnetic resonance imaging, O_2 can be insufflated via either a nasal tube or a tube placed deep into the endotracheal tube. For nasal insufflation to be effective, the tube should be inserted to the level of the nasopharynx, via a nasal passage. The fresh oxygen flow rate should be at least 3 liter/min if an inspired concentration of around 30% is desired. This is because the tidal volume of sheep and goats (around 500 ml) and the respiratory rate (20 to 40 breaths/min) result in less than 1 sec/breath of passive expiration, during which time 50 ml of oxygen would be added to the inspired gas. Thereapy in sheep with severe hypoxia may require insufflation flows of up to 6 liter/min or oxygen administration via a face mask attached to an anesthetic machine breathing circuit.

B. Ventilation

Anesthesia reduces the resting tone of the intercostal muscles and diaphragm. Lateral or dorsal positioning causes the abdominal viscera to be forced forward and ventral, reducing diaphragmatic excursions. Therefore, both large and small ruminants tend to hypoventilate during anesthesia. The approach of the authors is to allow small ruminants to ventilate spontaneously, unless they are undergoing a thoracotomy. However, the authors continuously monitor end-tidal CO_2 and intermittently monitor $PaCO_2$. Although not commonly encountered, if hypoventilation is detected ($PaCO_2 > 55$ mm Hg in anesthetized sheep) the authors first try to decrease the anesthetic depth, but if this is not successful in reducing the $PaCO_2$, they institute IPPV.

Most commonly for sheep, a ventilator with a 2-liter bellows is used, administering 12 to 18 breaths per minute, a tidal volume of 6 to 10 ml/kg, and a peak inspired airway pressure of 12 to 20 cm H_2O. As a general rule for ruminants, the initial tidal volume can be calculated as 10–15 ml/kg of body weight. If increased ventilation is required, in order to avoid barotrauma, the authors advocate increasing the respiratory rate rather than increasing the tidal volume or peak inspired pressure above 20 cm H_2O. Because anesthetized sheep are easily hypoventilated, either arterial or end-expired CO_2 should be monitored when IPPV is used. End-expired CO_2 concentrations ideally should range between 34 and 40 mm Hg. In spontaneously ventilating sheep, the authors have not observed unacceptable hypoventilation. Therefore, because positive pressure ventilation depresses cardiac function, the authors do not routinely ventilate sheep.

C. Fluid Therapy

In healthy ruminants, polyionic solutions should be routinely administered during anesthesia, such as lactated Ringer's solution or Normosol-R, at a rate between $4.8 \times$ body weight $kg^{0.75}$ (metabolic requirement, Lowe, 1981) and 10 ml/kg per hour, depending on the duration of anesthesia and the nature of the procedure. Fluids administered intravenously, especially those used to flush open body cavities, should be warmed to body temperature. Fluid support should be continued postanesthesia if marked hemorrhage occurred during surgery or if the animal does not resume eating or drinking.

D. Inotropic Therapy

Vasopressors and inotropes such as dopamine, dobutamine, and ephedrine are used in anesthetized ruminants to treat hypotension. Dobutamine is the most potent and effective of these drugs for increasing cardiac output and blood pressure. Ephedrine is less arrhythmogenic, is simplest to administer, and may better preserve uterine perfusion in pregnant sheep.

XIII. ANESTHETIC RECOVERY AND POSTOPERATIVE CARE

A. Personnel

Specific members of the research team should be identified to be responsible for observation and care of animals during anesthesia recovery and for several days following surgical procedures. In the latter case, these individuals should be responsible for either feeding the animals or observing the animals at this time, as behavior suggestive of pain perception is most likely to be evident at feeding times.

B. Facilities

Recovery from anesthesia should occur in a specific area or room that is dimly lit, quiet, well ventilated, and temperature controlled. For very young animals recovering from anesthesia, higher room temperatures are desirable. The floor should be dry and the surface be of a nonskid nature, such as rubber. The room should be of sufficient size to handle the maximum number of recumbent animals that would undergo a surgical procedure on each day's activity. In some situations, particularly with adult cattle, it may be necessary to recover animals in their stall, in which case fresh, dry straw should be used to cover the floor and pad the animal. Keeping the animal above a concrete surface will also help maintain body temperature.

In an experimental setting, recovery of small ruminants may occur in pens as well as in specially designed cages or slings that are appropriately padded and restrict the movement of the animal. During the recovery process, insulating materials such as blankets as well as supplemental heat sources, including circulating hot water blankets and heat lamps, can be used.

C. Airway Support

A patent airway should be maintained until ruminants are conscious and able to swallow. Intubated ruminants should not be left unobserved because they have very sharp molar teeth and sudden chewing can result in the orotracheal tube being damaged and possibly cut into two pieces, with one piece remaining in the airway. If the animal is recovering from anesthesia on straw bedding, the endotracheal tube should be positioned so that it is not lying in the straw to avoid straw and dust from being sucked into the airway during inspiration. If a large amount of ruminal regurgitation occurred during anesthesia and ruminal fluid drained from the nasal passages, upper airway obstruction could occur after extubation. Because sheep and goats have smaller nasal passages than adult cattle, this is a potential problem in these species. In such cases, lavage of the nasal passages with saline and suction of the pharynx can be done prior to ending anesthesia and deflating the endotracheal tube cuff to minimize this problem.

D. Positioning

Because ruminants have a quiet temperament and also sleep in a recumbent position, they tend to remain recumbent for prolonged periods after anesthesia and do not struggle. In order to improve reexpansion of the atelectic lung, improve ventilation, eructate ruminal gas, animals recovering from anesthesia should be maintained in sternal recumbency. The animals will remain in such a position for a long period unless stimulated, at which point they will suddenly stand.

E. Body Temperature

Despite their wool coat, anesthetized sheep are susceptible to hypothermia. During anesthesia, they have decreased muscular activity and do not conserve heat. Hypothermia causes peripheral vasoconstriction and bradycardia reduces cardiac output. These circulatory changes decrease tissue perfusion, which slows drug metabolism and excretion and prolongs anesthetic recovery. It is advisable to monitor body temperature during anesthesia. If the animal is hypothermic at the end of anesthesia, the body temperature should be increased by using circulating warm water blankets, infrared heating lamps, or other strategies. Heating lamps should be kept at least 4 feet from the animal to prevent burning the skin of an anesthetized animal. Use of electric heating pads can result in thermal injury and should be avoided.

XIV. PAIN PERCEPTION AND EVALUATION

Pain is a subjective modality resulting from the conscious perception of application of a noxious stimulus. In humans, somatic pain conducted by A-delta (fast) and C (slow) nerve fibers can be discretely localized. Visceral pain, conducted by C fibers, is not well localized. Pain perception depends on conduction via neural pathways from the peripheral nociceptors via the spinal cord to an intact cerebral cortex (via the brain stem and thalamus). Human beings verbally describe the type or quality of pain in terms such as acute or chronic; mild or severe; tolerable or intolerable; and localized or diffuse. Although animals cannot verbally describe or quantify the perceived response following noxious stimulation (i.e., pain), methodology for pain management must assume its perception for a given species of animal. Painful stimulation causes physical discomfort and stress which alters behavior, homeostasis, including feed and water intake, and recovery from tissue injury or trauma. Minimizing pain perception in animals improves their well-being and is desirable for humane reasons.

A. Behavioral Responses to Painful Stimulation

In order to ensure effective pain management, it is important that the researcher first be competent at evaluating both normal behavior and behavior indicative of pain in that particular animal species to avoid undertreating some animals, particularly if a generic analgesic drug "recipe" is used for all animals irrespective of the species of animal, the type, and the duration of noxious stimulation. Evaluation of an animal's response to noxious stimulation is mostly based on an interpretation of how humans would perceive that stimulus. Human beings are most likely to treat animals that vocalize following painful stimulation (Hanson and Hardie, 1993).

Assessment of behavior indicative of pain in ruminant animals is difficult. Domestic ruminants are herd animals that originated from species that were hunted. In such species, not displaying abnormal behavior, including painful behavior, is a survival advantage to avoid attracting predators (Grandin, 1989; Livingston, 1994). Signs of painful behavior in ruminants include an alteration in the feeding pattern, a reduction in feed intake, an animal separated from the herd or lagging behind, an abnormal gait or position, and a reluctance to stand and avoid humans entering the flight zone of the animal (Grandin, 1989; Molony and Wood, 1992; Morton and Griffiths, 1985; Randolph, 1994). Response to palpation of skin over a surgical site should also be considered when evaluating pain perception. Goats may vocalize, but sheep and cattle generally do not, except in immediate response to the application of a noxious stimulus. Sheep will tolerate painful stimulation, such as that caused by orthopedic surgery, with minimal change in their normal behavior, particularly when housed in a flock (Randolph, 1994; Wagner *et al.*, 1994). Altered feeding behavior may be unrelated to pain, such as withholding feed in ruminants prior to anesthesia and surgery which can alter ruminal motility and therefore appetite. Measurable clinical signs, including elevations in body temperature, respiratory rate, heart rate, and blood pressure, could be useful indicators of pain perception (Morton and Griffiths, 1985). Normal values for these parameters in sheep are given in Table I. These parameters and behaviors can be affected by sedative and anaesthetic drugs.

B. Biochemical Responses to Painful Stimulation

There are no reliable biochemical markers of pain perception described for ruminants. Cortisol, β-endorphin, and catecholamine levels are elevated in response to painful stimulation, stress, exercise, etc. (Grandin, 1989; Molony and Wood, 1992; Rushen, 1986; Shutt *et al.*, 1988). Following painful stimulation, there may be a delay in elevation of these neurohumoral agents, and the hormone levels in particular may remain elevated for a period following the cessation of painful stimulation. Painful stimulation causes stress, but stress such as that caused by physical restraint, withholding feed, or solitary housing of unconditioned ruminants should not cause pain (Grandin, 1989; Rushen, 1986; Shutt *et al.*, 1988). Because neurohumoral agents have limited usefulness in quantifying pain perception in ruminants, clinical assessment, particularly observations of behavior, is still the most reliable indicator of pain perception in ruminants.

C. Preemptive Pain Management

Analgesic agents may be administered to ruminants prior to the onset of painful stimulation, such as surgery, or in response to signs of perceived pain exhibited postoperatively. Numerous studies in humans suggest that the preemptive management of pain is more effective, actually reduces the amount of analgesic drug administered and the number of days of analgesic administration, and hastens postoperative recovery (McQuay *et al.*, 1988; Woolf and Chong, 1993).

There are several physiological explanations that justify the use of preemptive analgesia, which have been reviewed by Livingston (1994). (1) Altered sensitivity of peripheral nocicentors: tissue damage results in the local production and release of chemotactic agents that alter the threshold of high threshold nociceptors so they become low threshold nociceptors (Kitchell, 1987). (2) Altered excitability of neurons within the central nervous system: changes within neuronal networks, particularly within the spinal cord, result in sensitization so that stimuli previously perceived as touch or pressure become perceived as pain (Woolf, 1983). This type of hypersensitivity results from strong sensory input from the peripheral nociceptors at the time of the initial trauma, such as surgery, but the change persists for a long time after the insult (Waterman *et al.*, 1992).

D. Conclusions

From a clinical perspective, it is apparent that ruminants perceive pain but that evaluation of behavior indicative of pain is difficult, especially for the untrained or casual observer. Therefore, the routine use of analgesic therapy following surgical and painful procedures is more appropriate than treatment "based on patient need." Generally, prophylactic analgesic therapy is not associated with life-threatening untoward effects. Preemptive analgesic therapy is far more effective than management of an animal already perceiving pain. For surgical procedures, general anesthesia alone may not be sufficient to prevent neuronal hypersensitization and the use of postsurgical analgesic agents to treat painful behavior may be a suboptimal approach. A better approach may be to consider a combination of therapies that includes preoperative analgesic and anti-inflammatory drug administration, general anesthesia accompanied by local or regional nerve blocks, administration of parenteral analgesic drugs during surgery if the animal is moving in response to painful manipulations, and routine, regular administration of postoperative analgesic and anti-inflammatory drugs. Unfortunately, once hypersensitization has occurred, routine analgesic drug choices, dosages, and dosage intervals may be inadequate (Livingston, 1994; Waterman *et al.*, 1992).

XV. PAIN MANAGEMENT

A. Parenteral Analgesic Therapy

The analgesic properties of a wide variety of drugs have been extensively evaluated in sheep (Table V) compared to a lack of quantifiable information for goats and cattle. Therefore, comments pertaining to efficacy and clinical usage are specifically derived from their use in sheep. From a clinical perspective it should be of concern that many of the analgesic drugs reported in Table V (e.g., opioids) have a variable response and a relatively short duration of action following IV administration.

1. Opioid Drugs

These include the pure (mu,μ) agonists fentanyl and meperidine; mixed μ-agonist/antagonist drugs with action primarily on μ-receptors such as buprenorphine; and the κ-agonist/μ-antagonist drugs including butorphanol. Except for fentanyl, all of these drugs provide poor analgesia in sheep in response to mechanical stimulation applied to the thoracic limb (pressure from a blunt pin applied to the anterior aspect of the radius) as summarized in Table V. The most effective analgesia was provided by 0.01 mg/kg fentanyl, which had a rapid onset but a short duration of action. All these opioid drugs provide more reliable analgesia in response to cutaneous thermal stimulation (of the ear), although the duration of action varies between drugs. Buprenorphine provides a long duration of action (3.5 hr) compared to butorphanol (2 hr) and fentanyl (1 hr). Clinically, butorphanol administered to sheep at a dose of 0.1 to 0.2 mg/kg IV has been reported to cause agitation, bleating, and chewing (Waterman *et al.,* 1991). At 0.4 mg/kg IV, butorphanol caused ataxia. However, a study by O'Hair *et al.* (1988) found that 0.5 mg/kg SQ butorphanol caused sedation and analgesia (in response to skin, toe, or nose pinch) from 15 to 120 min following administration. Clinically, butorphanol seems to decrease the response to nasopharyngeal stimulation (e.g., passage of an endoscope) in adult cattle, which supports the findings of O'Hair *et al.* (1988) in sheep. The variability in analgesic and sedative responses to opiate drugs in sheep may be due to differences in the distribution and density of opiate receptor subtypes in the brain and spinal chord. Because of the unpredictable response, researchers should be cautious in assuming that all

TABLE V

CHARACTERISTICS OF ANALGESIC DRUGS IN SHEEP

Drug[a]	Sheep[b]	Dose (mg/kg IV)	Onset of analgesia (min)	Duration of analgesia (min): Mechanical stimulus	Duration of analgesia (min): Thermal stimulus	Comments	Ref.
Acepromazine	n	0.2	No analgesia observed			Caused heavy sedation, some sheep became recumbent	6
Xylazine	n	0.05	Rapid	30	45	Good analgesia, less pronounced for mechanical stimulus	6
	f	0.05	Rapid	25	None done	About three-fourths the analgesic potency of normal sheep	7
Clonidine	n	0.006	Rapid	90	120	Longer duration than xylazine	6
Medetomidine	n	0.005	Rapid	40–60	Not done		5
Fentanyl	n	0.01	Rapid	40	50 to 60	Intensity of analgesia less for mechanical stimulus	4
	f	0.01	Rapid	25	None done	Analgesic potency less than for normal sheep	7
Meperidine	n	5.0	Rapid	5	30	Similar to fentanyl, short duration	4
Buprenorphine	n	0.006	40 to 45	No analgesia observed	210	Slowest onset of action	4
Butorphanol	n	0.2	5 to 15	No analgesia observed	120	Slow onset of action, caused agitation, bleating, chewing	4
Fentanyl with droperidol	n	0.005 0.005	5 to 10	30	Not done	Caused agitation, locomotion activity, chewing	2
Midazolam	n	0.3	5 (rapid)	35	15	Caused marked sedation, some sheep became recumbent	3
Flunixin	n	2.2	10	150	Not done	About half the analgesic potency of xylazine or fentanyl with droperidol	1
	f	2.2			Not done	About half the analgesic potency of normal sheep	1
Dipyrone	n	25	10	150	Not done	Sheep struggled after IV injection, similar analgesic potency to flunixin	1
	f	25			Not done	About half the analgesic potency of normal sheep	1
Dexamethasone SP	n	0.1	No analgesia observed				1

[a]α_2-Adrenergic agonists (xylazine, clonidine, medetomidine): Effects abolished by pretreatment with 0.1 mg/kg IV idazoxan.

[b]n, normal; f, sheep with footrot following treatment.

[c]Key to references: (1) Chambers *et al.* (1995); (2) Kyles *et al.* (1993a); (3) Kyles *et al.* (1996); (4) Livingston *et al.* (1992b); (5) Muge *et al.* (1994); (6) Nolan *et al.* (1987a); (7) Waterman *et al.* (1992).

opiate drugs provide analgesia for all types of stimuli that result in pain perception in ruminants, particularly sheep.

2. α-Adrenergic Agonists

α_2-Adrenergic agonists, including xylazine, clonidine, and medetomidine, cause sedation and effective analgesia in sheep (Table V). There is a rapid onset of analgesia following IV administration, which is similar for both mechanical and thermal stimuli. The duration of action is longest following clonidine (1.5 to 2 hr) compared to that from medetomidine (1 hr) and xylazine (30 to 45 min). Nolan *et al.* (1987b) have shown that analgesia and sedation can be abolished by administration of a specific α_2-adrenergic antagonist, idazoxan, which could be clinically useful in situations where profound sedation might be a problem, such as animals recovering from anesthesia. The analgesic effects of α_2-adrenergic agonists are not simply due to their sedative properties, as acepromazine, a phenothiazine derivative, causes profound sedation in sheep without analgesia (Table V). Because binding sites for α_2-adrenergic agonists are present in the dorsal horn of the spinal cord (Bouchenafa and Livingston, 1987), it is likely that some of the analgesic effect is mediated at this level. From a clinical perspective, analgesia is more likely to be obtained in ruminants, particularly sheep, using α_2-adrenergic agonists as compared to opiate drugs.

3. Neuroleptic Drugs

Neuroleptic drugs combined with subanalgesic doses of fentanyl have been shown to cause analgesia in sheep (Kyles *et al.*, 1993a, Table V). This analgesia was of short duration and was associated with undesirable effects such as agitation. In sheep, neuroleptic drugs such as droperidol, which are dopamine antagonists, have minimal analgesic properties when administered alone. Although there may be better alternatives for pain management, neuroleptic drugs may warrant further investigation.

4. Nonsteroidal Anti-inflammatory Drugs

Nonsteroidal antiflammatory drugs (NSAIDs), including phenylbutazone, dipyrone, flunixin and aspirin, have been used clinically in ruminants, presumably to reduce pain caused by inflammation; NSAIDs inhibit cyclooxygenase, which facilitates the conversion of arachidonic acid to prostaglandins. These compounds sensitize peripheral nociceptors to bradykinin and histamine. Studies in sheep by Chambers *et al.* (1995) suggest that this might not be the only mechanism of NSAID-mediated analgesia. In healthy sheep without inflammation, both flunixin and dipyrone provided effective analgesia to mechanical stimulation. Although the analgesic potency of these drugs is probably less than that of the α_2-adrenergic agonists, it is longer lasting (up to 2.5 hr, Table V). Evidence for this effect being centrally mediated is (1) that similar analgesia was not obtained following dexamethasone administration (inhibits phospholipase A_2 and, therefore, arachidonic acid production) and (2) that this NSAID analgesia could be inhibited using naloxone or atipamezole. This suggests an interaction between these NSAIDs and central opioid and α_2-adrenergic pain modulation systems.

In humans, the preemptive administration of NSAIDs such as diclofenac sodium and ketorolac results in better postoperative pain relief and reduces the requirement for additional postoperative analgesics (Buckley and Brogden, 1990; Gillberg *et al.*, 1993). The preemptive use of NSAIDs is supported by similar studies in dogs using carprofen (Lascelles *et al.*, 1994). The use of NSAIDs for preemptive analgesia in ruminants is not routine practice, but has considerable potential value. Flunixin and dipyrone are effective analgesic agents in sheep, with a long duration of action. Clinically, phenylbutazone (2 to 6 mg/kg IV or PO) seems to be similarly effective in cattle and sheep and can be administered orally. Aspirin may be useful for oral medication (50 to 100 mg/kg PO). Welsh *et al.* (1992) reported the pharmacokinetics of carprofen in sheep. It is a novel NSAID that has a long plasma half-life in sheep (48 to 72 hr following 0.7 to 4.0 mg/kg IV), is less ulcerogenic than indomethacin or aspirin, and apparently causes analgesia that is at least as effective as flunixin or dipyrone (A. Livingston, personal communication from unpublished research, 1994). Because of their long duration of action, NSAIDs should have a place in pain management in ruminants.

5. Therapy for Chronic Pain

This has been investigated using sheep with footrot (Waterman *et al.*, 1992). These studies suggest that in sheep with chronic pain the duration of action of analgesic drugs is shorter and analgesia is less predictable and less effective compared to that in normal sheep (Table V). From a clinical perspective, the effective management of chronic pain in ruminants should entail more frequent administration rates and higher dosages of analgesic drugs. Using combinations of peripherally acting drugs such as NSAIDs or local anesthetic agents with parenteral centrally acting drugs should provide more effective analgesia.

B. Alternative Analgesic Therapies

1. Local and Regional Anesthesia

This has been discussed in terms of facilitating surgical procedures. However, these drugs and administration techniques can also be used to provide effective postoperative analgesia. Using bupivacaine, analgesia can be obtained for periods of up to 4 to 6 hr. In addition to local and regional nerve blocks, analgesia can be achieved from epidural administration (described under local anesthesia) and, following thoracotomies, via interpleural administration or intercostal nerve blocks. Interpleural administration of bupivacaine following intercostal thoracotomy in dogs was as effective as either systemic morphine administration or intecostal nerve blocks (Thompson and

Johnson, 1991). Bupivacaine (1.5 mg/kg) was administered via a pleural chest drain. Interpleural bupivacaine would be most useful following a sternal thoracotomy, where it would not be possible to block every intercostal space.

2. Spinal Administration of Analgesic Drugs

This is similar to the technique for administrating anesthetic drugs (spinal ''anesthesia'') previously described in the section on local and regional analgesia. However, instead of using local anesthetics, either opioids or α_2-adrenergic agonists are administered. The theoretical advantages of spinal analgesia are lower total drug doses, a long duration of action, and minimal loss of pelvic limb motor function. Epidural administration of either opioids or α_2-adrenergic agonists has been evaluated in ruminants.

Opioids administered spinally to ruminants include morphine, fentanyl, and etorphine. In normal sheep, spinal morphine caused only partial analgesia to mechanical stimulation of a distal limb (Livingston *et al.,* 1992a). In the same study, spinal fentanyl caused virtually no analgesia. In a study by Wagner *et al.* (1994), 0.1 mg/kg spinal morphine caused pelvic limb weakness and, in two sheep, agitation and chewing at the skin over the flank. In the same report, sheep receiving morphine were more lame after bilateral stifle surgery compared to nontreated sheep. In another study in sheep, chronic infusion of morphine via an epidural catheter caused hindlimb weakness and sterile granulomatous reactions in the epidural space (Coombs *et al.,* 1994). In a study of postcastration behavior in lambs, spinal etorphine did not alter behavior when compared to no analgesic therapy, although etorphine-treated lambs became sedate (Wood *et al.,* 1990). These studies suggest that spinal opioids do not cause reliable analgesia in sheep and may result in undesirable behavior or neurologic damage.

In goats recovering from pelvic limb orthopedic surgery, 0.1 mg/kg epidural morphine diluted to 0.13 ml/kg with saline resulted in quiet recoveries from anesthesia and less behavior suggestive of pain (less vocalization and teeth grinding) compared to nontreated goats (Pablo, 1993). The treated goats also appeared to be sedated. This study suggests that epidural morphine may be a useful analgesic technique in goats.

α_2-Adrenergic agonists, including xylazine and clonidine, have been administered spinally to sheep and cattle. In normal sheep, either 50 μg of xylazine or 12 μg of clonidine administered to either the L5 or C5 region via intrathecal catheters caused effective and prolonged analgesia to mechanical stimulation of a distal limb (Livingston *et al.,* 1992a; Kyles *et al.,* 1993b). Analgesia persisted for 2 to 2.5 hr. Autoradiographic studies suggest that the site of action of these drugs in the spinal cord is in the region of the substantia gelatinosa of the dorsal horn. The spread of the intrathecally administered drug is limited to spinal cord segments at the region of administration. Therefore, in clinical situations, administration of α_2-adrenergic agonists via lumbosacral injection might not result in analgesia at the level of the cervical spinal chord. Eisenach and Grice (1988) reported that epidural clonidine administered to sheep (17 to 25 μg/kg; which is 2.5 times the maximal antinociceptive epidural dose) caused minimal cardiovascular effects, except for a slight decrease in heart rate. No changes were seen in cardiac output, mean arterial pressure, or spinal cord blood flow. In these sheep there was a slight decrease in PaO_2.

In cattle, epidural xylazine (50 μg/kg diluted to a 5-ml volume) injected at the first coccygeal interspace caused a long duration (greater than 1.5 hr) of analgesia to the perineal region (Skarda and Muir, 1992). The onset of analgesia was about 15 to 20 min after administration. In addition, epidural xylazine caused sedation, mild ataxia, bradycardia, decreased ruminal motility, and mild bloat (in six of eight cattle). Xylazine has local anesthetic properties in addition to its central α_2-adrenergic agonist actions, which might explain the ataxia. The total dose and volume administered need to be carefully calculated to minimize the rostral spread of analgesia/anesthesia and therefore prevent femoral or sciatic nerve anesthesia. Increasing the xylazine dose above 60 μg/kg in cattle markedly increased pelvic limb weakness, ataxia, and recumbency. Epidural xylazine can cause effective and prolonged analgesia, but should be used carefully in cattle to avoid problems with pelvic limb ataxia and recumbency.

3. Electroanesthesia

Electroanesthesia has been investigated in sheep and cattle to provide ''analgesia'' (electroacupuncture) and to facilitate restraint for surgical procedures (electroimmobilization). Electroacupuncture uses the stimulation of traditional Chinese application points to cause cutaneous, regional analgesia. For procedures such as a flank laparotomy, analgesia will be present in the skin and body wall, but not in the peritoneal cavity. The use of acupuncture to provide postoperative analgesia has not been reported, but is certainly worth investigating.

Electroimmobilization for restraint has been investigated in sheep and cattle undergoing painful procedures (Kuchel *et al.,* 1990). In sheep, immobilization is achieved when a 35-mA current at 48 Hz is applied to subcutaneous submandibular and coccygeal needle electrodes. In cattle, a current of up to 120 mA is similarly applied (instructional video, Stockstill Cattle Immobilizer, Feenix Stockstill, Tarlee, South Australia). In electroimmobilized sheep, mean arterial pressure, heart rate, and cardiac output increase. Although there is an initial period of apnea, arterial carbon dioxide tension decreases as a result of hyperventilation. Claims have been made in some literature suggesting that electroimmobilization causes analgesia, therefore justifying its use to facilitate surgical procedures in sheep and cattle. However, scientific evidence contradicts these claims, apart from the cardiopulmonary stimulation that was observed in sheep. First, cattle that have experienced electroimmobilization have elevations in heart rate when they approach a chute where it was used and will choose to avoid that chute if given the choice of two chutes, suggesting that the procedure was unpleasant (Grandin, 1989). Second, studies in humans show that cutaneous pain is perceived when a current between 6 and 10 mA is applied to the skin, which is considerably less than the current used for immobilization. Humans that have been apprehended using ''Tasar'' units suggest that the experience is unpleasant. At this time there is a lack of evidence that indicates electroimmobilization causes analgesia in sheep or cattle. Furthermore, available evidence suggests that the procedure itself is associated with noxious or painful stimulation and does not provide analgesia. If this conclusion is correct, electroimmobilization should not be used in nonsedated animals and surgery should not be performed unless methods that are known to provide analgesia are also used.

ACKNOWLEDGMENTS

The authors acknowledge the generous assistance, advice, and provision of unpublished information in Section XV by Alex Livingston, Dean, Western College of Veterinary Medicine, University of Saskatchewan, Saskatoon, Canada.

REFERENCES

Abass, B. T., Weaver, B. M., and Staddon, G. E. (1988). The influence of halothane and isoflurane on the pharmacokinetics of pentobarbitone and thiopentone in sheep. *Proc. Int. Congr. Vet. Anesth., 3rd,* Brisbane, Australia, pp. 198–210.

Adetunji, A., Pascoe, P. J., McDonell, W. N., and Horney, F. D. (1984). Retrospective evaluation of xylazine/halothane anesthesia in calves. *Can. Vet. J.* **25,** 342–346.

Adetunji, A., McDonell, W. N., and Pascoe, P. J. (1985). Cardipulmonary effects of xylazine, acetylpromazine and chloral hydrate in supine cows (abstr). *Proc. Int. Congr. Vet. Anesth., 2nd,* Sacramento, California.

Alibhai, H. I., Clarke, K. W., and Hughes, G. (1994). Minimum alveolar concentration (MAC) in goats. *Proc. Int. Congr. Vet. Anesth., 5th,* Guelph, Ontario, Abstr., p. 87.

Bartels, H., and Harms, H. (1959). Oxygen dissociation curves of the bloods of mammals (man, rabbit, guinea pig, dog, cat, swine, ox and sheep). *Pflueger's Arch. Gesamte Physiol. Menschen Tiere* **268,** 334–365.

Baum, D., Halter, J. B., Taborsky, G. J., and Porte, D. (1985). Pentobarbital effects on plasma catecholamines: temperature, heart rate and blood pressure. *Am. J. Physiol.* **248,** E95–E100.

Benson, G. J., and Thurmon, J. C. (1993). Regional anesthesia. *In* ''Current Veterinary Therapy. III. Food Animal Practice'' (J. L. Howard, ed.), 3rd ed., pp. 77–88. Saunders, Philadelphia.

Bernard, K. W., Parham, G. L., Winkler, W. G., and Helmick, C. G. (1982). Q fever control measures: Recommendations for research facilities using sheep. *Infect. Control* **3,** 461–465.

Blaze, C. A., LeBlanc, P. H., and Robinson, N. E. (1988a). Effect of witholding feed on ventilation and the incidence of regurgitation during halothane anesthesia of adult cattle. *Am. J. Vet. Res.* **49,** 2126–2129.

Blaze, C. A., Holland, R. E., and Grant, A. L. (1988b). Gas exchange during xylazine-ketamine anesthesia in neonatal calves. *Vet. Surg.* **17,** 155–159.

Bonath, K. H., Nouh, S. R., Evans, J., Amelang, D. (1988). Antagonism of xylazine anesthesia with tolazoline in dromedary and sheep. *Proc. Int. Congr. Vet. Anesth.,* Sacramento, California, pp. 227–232.

Bouchenafa, O., and Livingston, A. (1987). Autoadiographic localization of alpha-2 adrenoceptor binding sites in the spinal cord of sheep. *Res. Vet. Sci.* **42,** 382–386.

Buckley, M. M., and Brogden, R. N. (1990). Ketorolac: A review of its pharmacodynamic and pharmacokinetic properties and therapeutic potential. *Drugs* **39,** 86–109.

Celly, C. S., McDonell, W. N., and Black, W. D. (1993). Comparative cardiopulmonary effects of four alpha-2 adrenoceptor agonists in sheep (abstr.). *Vet. Surgery* **22:** 545–546.

Centers for Disease Control/National Institutes of Health (CDC/NIH) (1993). ''Biosafety in Microbiological and Biomedical Laboratories,'' 3rd ed., HHS Publi. No. (CDC) 93-8395, pp. 102–104. CDC/NIH, Washington, D.C.

Chambers, J. P., Waterman, A. E., and Livingston, A. (1995). The effects of opioid and alpha-2 adrenergic blockade on NSAID analgesia in sheep. *J. Vet. Pharmacol. Ther.* **18,** 161–166.

Conner, G. H., Coppock, R. W., and Beck, C. C. (1974). Laboratory use of Cl-744. *Vet. Med.* **69,** 479–482.

Coombs, D. W., Colburn, R. W., DeLeo, J. A., Hoopes, P. J., and Twitchell, B. B. (1994). Comparative spinal neuropathology of hydromorphone and morphine after 9- and 30-day epidural administration in sheep. *Anesth. Analg.* **78,** 674–681.

Cottle, M. K., Van Petten, G. R., and Van Muyden, P. (1983). Maternal and fetal cardiovascular indices during fetal hypoxia due to cord compression in chronically cannulated sheep. *Am. J. Obstet. Gynecol.* **146,** 686–692.

Coulson, N. M., Januszkiewicz, A. J., Dodd, K. T., and Ripple, G. R. (1989). The cardiorespiratory effects of diazepam–ketamine and xylazine–ketamine anesthetic combinations in sheep. *Lab. Anim. Sci.* **39,** 591–597.

Coulson, N. M., Januszkiewicz, A. J., and Ripple, G. R. (1991). Physiological responses of sheep to two hours anesthesia with diazepam–ketamine. *Vet. Rec.* **129,** 329–332.

Crump, K. T., Dunlop, C. I., Hick, L. H., and Turner, A. S. (1994a). Recovery from anesthesia in sheep following induction with thiamylal, ketamine or propofol. *Proc. Int. Congr. Vet. Anesth., 5th,* Guelph, Canada, Abstr., p. 85.

Crump, K. T., Dunlop, C. I., Hick, L. H., and Turner, A. S. (1994b). A comparison of spontaneous and intermittent positive pressure ventilation in anesthetized sheep. *Proc. Int. Congr. Vet. Anesth., 5th,* Guelph, Canada, Abstr., p. 101.

Daunt, D. A., Link, R. E., Chruscinski, A. J., and Kobilka, B. K. (1994). Cloning of mouse alpha-2 adrenergic subtypes: molecular basis for inter-species variation in ligand binding. *Proc. Int. Congr. Vet. Anesth., 5th,* Guelph, Canada, Abstr., p. 79.

Dehghani, S., Behbodikhah, A., and Khorsand, N. (1991). Clinical, haematological and biochemical effects of xylazine, ketamine and their combination in cattle and sheep. *Proc. Int. Congr. Vet. Anesth., 4th,* Utrecht, The Netherlands, pp. 123–128.

Dueck, R., Rathbun, M., and Greenburg, A. G. (1984). Lung volume and V/Q distribution response to intravenous versus inhalation anesthesia in sheep. *Anesthesiology* **61,** 55–65.

Eisenach, J. C., and Grice, S. C. (1988). Epidural clonidine does not decrease blood pressure or spinal cord blood flow in awake sheep. *Anesthesiology* **68,** 335–340.

Eisenach, J. C., Rose, J. C. Dewan, D. M., and Grice, S. C. (1987). Intravenous clonidine produces fetal hypoxemia. *Anesthesiology* **67**(3A), A449 (abstr.).

Fayed, A. H., Abdalla, E. B., Anderson, R. R., Spencer, K., and Johnson, H. D. (1989). Effect of xylazine in heifers under thermoneutral or heat stress conditions. *Am. J. Vet. Res.* **50,** 151–153.

Flecknell, P. A. (1987). ''Laboratory Animal Anesthesia: An Introduction for Research Workers,'' pp. 42–74, 107–109, 131–132, 138–139. Academic Press, Orlando, Florida.

Floyd, J. G., and Randle, R. F. (1989). Endotracheal intubation of cattle under xylazine hydrochloride sedation in the field. *Compend. Contin. Educ.* **11,** 1302–1304.

Funk, K. A. (1973). Glyceryl guaiacolate: Some effects and indications in horses. *Equine Vet. J.* **5,** 15–19.

Garcia-Villar, R., Toutain, P. L., Alvinerie, M., & Ruckebusch, Y. (1981). The pharmacokinetics of xylazine hydrochloride: An interspecific study. *J. Vet. Pharmacol. Ther.* **4,** 87–92.

Gillberg, L. E., Harsten, A. S., and Stahl, L. B. (1993). Preoperative diclofenac sodium reduces post-laparoscopy pain. *Can. J. Anaesth.* **40,** 406–408.

Grandin, T. (1989). Behavioral principles of livestock handling. *Prof. Anim. Sci.* **5,** 1–11.

Gray, P. R., and McDonell, W. N. (1986). Anesthesia in goats and sheep. Part 11: General anesthesia. *Compend. Contin. Educ.* **8,** S127–S135.

Green, S. A., Tyner, C. L., Morris, D. L., and Hartsfield, S. M. (1988). Comparison of cardiopulmonary effects of isoflurane and halothane after atropine-guiafenesin-thiamylal anesthesia for rumenotomy in steers. *Am. J. Vet. Res.* **49,** 1891–1893.

Gregory, G. A., Wade, J. A., Beihl, D. R., Ong, B. Y., and Sitar, D. S. (1983). Fetal anesthetic requirement (MAC) for halothane. *Anesth. Analg.* **62,** 9–14.

Hales, J. R., Band, J. W., and Facade, A. A. (1976). Effects of cold exposure on the distribution of cardiac output in the sheep. *Pflueer's Arch.* **366,** 153–157.

Handel, I. G., Weaver, B. M., Staddon, G. E., and Cruz Mudorran, J. I. (1991). Observations on the pharmacokinetics of propofol in sheep. *Proc. Int. Congr. Vet. Anesth., 4th,* Utrecht, The Netherlands, pp. 143–147.

Hanson, B., and Hardie, E. (1993). Prescription and use of analgesics in dogs and cats in a veterinary teaching hospital: 258 cases (1983–1989). *J. Am. Vet. Med. Assoc.* **202,** 1485–1494.

Hikasa, Y., Takase, K., and Ogasawara, S. (1988). Antagonistic effects of alpha adrenoceptor blocking drugs on reticuloruminal hypomotility induced by xylazine in cattle. *Can. J. Vet. Res.* **52,** 411–415.

Hodgson, D. S., Dunlop, C. I., Chapman, P. L., and Smith, J. A. (1991). Effects of acepromazine and xylazine on uterine blood flow and oxygen delivery in pregnant cows. *Vet. Surg.* **20,** 157 (abstr.).

Hoyt, R. F., Jr. (1994). Intraoperative monitoring and equipment. *In* "Research Animal Anesthesia, Analgesia, and Surgery" (A. C. Smith and M. M. Swindle, eds.), pp. 137–146. Scientists Center for Animal Welfare, Greenbelt, Maryland.

Hsu, W. H., and Hummel, S. K. (1981). Xylazine-induced hyperglycemia in cattle: A possible involvement of alpha-adrenergic receptors regulating insulin release. *Endocrinology (Baltimore)* **109,** 825–829.

Hubbell, J. A., Hull, B. L., and Muir, W. W. (1986). Perianesthetic considerations in cattle. *Compend. Contin. Educ.* **8,** F92–F99.

Ilkiw, J. E., and Benthuysen, J. A. (1991). Comparative study of the pharmacokinetics of alfentanyl in rabbits, sheep and dogs. *Am. J. Vet. Res.* **52,** 581–584.

Ilkiw, J. E., Benthuysen, J. A., Ebling, W. F., and McNeal, D. (1991). A comparative study of the pharmacokinetics of thiopental in the rabbit, sheep and dog. *J. Vet. Pharmacol. Ther.* **14,** 134–140.

Jain, N. C. (1986). "Schalm's Veterinary Hematology," 4th ed., pp. 208–224. Lea & Febiger, Philadelphia.

Kariman, A., Sharifi, M., Teshfam, M., Nowrouzian, I., and Mojabi, A. (1994). The effects of detomidine hydrochloride on cardiovascular and hemodynamics in sheep. *Proc. Int. Congr. Vet. Anesth., 5th,* Guelph, Canada, Abstr., p. 119.

Kitchell, R. L. (1987). Problems in defining pain and peripheral mechanisms of pain. *J. Am. Vet. Med. Assoc.* **191,** 1195–1199.

Klein, L., Tomasic, M., and Nann, L. (1988). Flumazenil (RO 15- 1788) antagonism of midazolam sedation in sheep. *Proc. Int. Congr. Vet. Anesth., 3rd,* Brisbane, Australia, Abstr., p. 70.

Kolobow, T., Powers, T., Mandava, S., Aprigliano, M., Kawaguchi, A., Tsuno, K., and Mueller, E. (1994). Intratracheal pulmonary ventilation (ITPV): Control of positive end-expiratory pressure at the level of the carina through the use of a novel ITPV catheter design. *Anesth. Analg.* **78,** 455–461.

Komar, E. (1991). Intravenous anaesthesia in the sheep. *Proc. Int. Congr. Vet. Anesth., 4th,* Utrecht, The Netherlands, pp. 209–210.

Kuchel, T. R., Mather, L. E., Runciman, W. B., and Carapets, R. J. (1990). Physiological and biochemical consequences of electroimobilization in conscious sheep. *Aust. Vet. J.* **67,** 33–38.

Kumar, A., Thurmon, J. C., Nelson, D. R., Benson, G. J., and Tranquilli, W. J. (1983). Response of goats to ketamine hydrochloride with and without premedication of atropine, diazepam or xylazine. *Vet. Med.* **78,** 955–960.

Kyles, A. E., Waterman, A. E., and Livingston, A. (1993a). Antinociceptive effects of combining low doses of neuroleptic drugs and fentanyl in sheep. *Am. J. Vet. Res.* **54,** 1483–1488.

Kyles, A. E., Waterman, A. E., and Livingston, A. (1993b). The spinal antinociceptive activity of the alpha-2 adrenoceptor agonist, xylazine in sheep. *Br. J. Pharmacol.* **108,** 907–913.

Kyles, A. E., Waterman, A. E., and Livingston, A. (1996). Antinociceptive activity of midazolam in sheep. *J. Vet. Pharmacol. Ther.* (in press).

Lagutchik, M. S., Januszkiewicz, A. J., Dodd, K. T., and Martin, D. G. (1991). Cardiopulmonary effects of a tiletamine-zolazepam combination in sheep. *Am. J. Vet. Res.* **52,** 1441–1447.

Lascelles, B. D., Waterman, A. E., Livingston, A., Henderson, G., and Jones, A. (1994). Correlation of blood levels of carprofen given either pre- or post operatively with mechanical thresholds and humoral factors as indicators of post-operative pain in the dog. *Proc. Int. Congr. EAVPT, 6th,* France, Abstr., pp. 191–192.

Lin, H. C., Thurmon, J. C., Benson, G. J., Tranquilli, W. J., and Olson, W. A. (1989). The hemodynamic response of calves to tiletamine-zolazepam anesthesia. *Vet. Surg.* **18,** 328–334.

Livingston, A. (1994). Physiological basis for pain perception in animals. *Proc. Int. Congr. Vet. Anesth., 5th,* Guelph, Canada, pp. 1–6.

Livingston, A., Waterman, A. E., Bouchenafa, Ol, and Kyles, A. (1992a). Antinociceptive effects of intrathecal opioids and alpha-2 agonists in sheep. *In* "Animal Pain" (C. E. Short and A. Van Poznak, eds.), pp. 281–290. Churchill-Livingston, New York.

Livingston, A., Waterman, A. E., Nolan, A., and Amin, A. (1992b). Comparison of the thermal and mechanical antinociceptive actions of opioid and alpha-2 adrenoreceptor agonists in sheep. *In* "Animal Pain" (C. E. Short and A. Van Poznak, eds.). pp. 372–377. Churchill-Livingstone, New York.

Lowe, H. S., and Ernst, E. A., eds. (1981). "The Quantitative Practice of Anesthesia." William & Wilkins, Baltimore, Maryland.

McQuay, H. J., Carroll, D., and Moore, R. A. (1988). Postoperative orthopaedic pain—the effect of opiate premedication and local anesthetic blocks. *Pain* **33,** 291–295.

Meiklejohn, G., Reimer, L. G., Graves, P. S., and Helmick, C. (1981). Cryptic epidemic of Q fever in a medical school. *J. Infect. Dis.* **144,** 107–113.

Mirakhur, K. K., Khanna, A. K., and Kumar, V. R. (1985). Pulmonary ventilation during diazepam sedation in calves. *Proc. Int. Congr. Vet. Anesth., 2nd,* Sacramento, California, Abstr., p. 206.

Molony, V., and Wood, G. N. (1992). Acute pain from castration and tail docking lambs. *In* "Animal Pain" (C. E. Short and A. Van Poznak, eds.), pp. 285–395. Churchill-Livingstone, New York.

Morton, D. B., and Griffiths, P. H. (1985). Guidelines on the recognition of pain, distress and discomfort in experimental animals and an hypothesis for assessment. *Vet. Rec.* **116,** 431–436.

Muge, D. K., Chambers, J. P., Livingston, A., and Waterman, A. E. (1994). Analgesic effects of medetomidine in sheep. *Vet. Rec.* **135,** 43–44.

Muller, E. E., Kolobow, T., Mandava, S., Jones, M., Vitale, G., Aprigliano, M., and Yamada, K. (1993). How to ventilate lungs as small as 12.5% of normal: The new technique of intratracheal pulmonary ventilation. *Pediatr. Res.* **34,** 606–610.

Nolan, A., Livingston, A., and Waterman, A. E. (1987a). Antinociceptive actions of intravenous a_2 adrenoceptor agonists in sheep. *J. Vet. Pharmacol. Ther.* **10,** 141–144.

Nolan, A., Livingstone, A., and Waterman, A. (1987b). Antinociceptive actions of intravenous alpha-2 agonists in sheep. *J. Vet. Pharmacol. Ther.* **10,** 202–209.

Nolan, A. M., Reid, J., and Welsh, E. (1991). The pharmacokinetics of propofol in goats. *Acta Vet. Scand. Suppl.* **87,** 168–170.

O'Hair, K. C., Dodd, K. T., Phillips, Y. Y., and Beattie, R. J. (1988). Cardiopulmonary effects of nalbuphine hydrochloride and butorphanol tartrate in sheep. *Lab. Anim. Sci.* **38,** 58–61.

Pablo, L. S. (1993). Epidural morphine in goats after hindlimb orthopedic surgery. *Vet. Surg.* **22,** 307–210.

Palahniuk, R. J., Schider, S. M., and Eger, E. I., II (1974). Pregnancy decreases the requirement for inhaled anesthetic agents. *Anesthesiology* **41,** 82–83.

Pike, R. M. (1976). Laboratory-associated infections: Summary and analysis of 3921 cases. *Health Lab. Sci.* **13,** 105–114.

Pollack, K. L., Chestnut, D. H., Whiner, C. P., Thompson, C. S., and DeBruyn, C. S. (1990). Magnesium sulphate and promethazine do not interact to cause hypotension in gravid ewes. *Am. J. Obstet. Gynecol.* **163,** 787–794.

Rae, J. H. (1962). The fate of pentobarbitone and thiopentone in the sheep. *Res. Vet. Sci.* **3,** 399–407.

Raekallio, M., Kivalo, M., Jaianka, H., and Vainio, O. (1991). Medtomidine/ketamine sedation in calves and its reversal with atipamezole. *J. Vet. Anaesth.* **18,** 45–47.

Randolph, M. M. (1994). Post-operative care and analgesia of farm animals used in biomedical research. *AWIC Newsl.* **5,** 11–13.

Raptopoulos, D., and Weaver, B. M. (1984). Observations following intravenous xylazine administration in steers. *Vet. Rec.* **114,** 567–569.

Raptopoulos, D., Koutinas, A., Moustardas, N., and Papasteriadis, A. (1985). The effect of xylazine or xylazine plus atropine on blood gases in sheep. *Proc. Int. Congr. Vet. Anesth., 2nd.,* Sacramento, California, Abstr., p. 201.

Riebold, T. W. (1994). Ruminants: Anesthesia and analgesia. *In* "Research Animal Anesthesia, Analgesia, and Surgery" (A. C. Smith and M. M. Swindle, eds.), pp. 111–127. Scientists Center for Animal Welfare, Greenbelt, Maryland.

Rings, D. M., and Muir, W. W. (1982). Cardiopulmonary effects of intramuscular xylazine-ketamine in calves. *Can. J. Comput. Med.* **46,** 386–389.

Rosenberger, G., Hempel, E., and Baumeister, M. (1969). Contributions to the effect and applicability of Rompun in cattle. *Vet. Med. Rev.* **2,** 137–142.

Runciman, W. B., Ilsley, A. H., Mathew, L. E., Carapetis, R., and Rao, M. M. (1984a). A sheep preparation for studying interactions between blood flow and drug disposition. I. Physiological profile. *Br. J. Anaesth.* **56,** 1015–1027.

Runciman, W. B., Mather, L. E., Ilsley, A. H., Carapetis, R. J., and Upton, R. (1984b). A sheep preparation for studying interactions between blood flow and drug disposition: III. Effects of general and spinal anaesthesia on regional blood flow and oxygen tensions. *Br. J. Anaesth.* **56,** 1247–1257.

Runciman, W. B., Mather, L. E., Ilsley, A. H., Carapetis, R. J., and Upton, R. N. (1985). A sheep preparation for studying interactions between blood flow and drug disposition. IV. The effects of general and spinal anesthesia on blood flow and cefoxitin disposition. *Br. J. Anaesth.* **57,** 1239–1247.

Runciman, W. B., Mather, L. E., and Selby, D. G. (1990). Cardiovascular effects of propofol and of thiopentone anesthesia in the sheep. *Br. J. Anaesth.* **65,** 353–359.

Rushen, J. (1986). Some problems with the physiological concept of "stress." *Aust. Vet. J.* **63,** 359–361.

Rutten, A. J., Nancarrom, C., Mather, L. E., Isley, A. H., Runciman, W. B., and Upton, R. N. (1989). Hemodynamic and central nervous system effects of intravenous bolus doses of lidocaine, bupivacaine and ropivacaine in sheep. *Anesth. Analg.* **69,** 291–299.

Sandkuhler, J., Fu, Q-G., Helmchen, C., and Zimmermann, M. (1987). Pentobarbital in subanesthetic doses, depresses spinal transmission of nociceptive information but does not affect stimulation-produced descending inhibition in the cat. *Pain* **31,** 381–390.

Seifelnasr, E., Salch, M., and Soliman, F. A. (1974). In vivo investigations on the effect of rompun on the rumen motility in sheep. *Vet. Med. Rev.* **2,** 158–165.

Sharma, S. K., Nigam, J. M., Kumar, A., Varshney, A. C., and Singh, M. (1994). Preliminary studies on the use of medetomidine in calves and its reversal by atipamezole. *Proc. Int. Congr. Vet. Anesth., 5th,* Guelph, Canada, Abstr.), p. 121.

Short, C. R., Kappel, L. C., Ruhr, L. P., *et al.* (1985). Pharmacokinetics of pentobarbital and thiamylal as combined anesthetics in sheep. *J. Vet. Pharmacol. Ther.* **8,** 234–245.

Shutt, D. A., Fell, L. R., Connell, R., and Bell, A. K. (1988). Stress response in lambs docked and castrated surgically or by the application of rubber rings. *Aust. Vet. J.* **65,** 5–7.

Skarda, R. T. (1986). Techniques of local analgesia in ruminants and swine. *Vet. Clin. North Am. Food Anim. Pract.* **2,** 621–663.

Skarda, R. T., and Muir, W. W. (1992). Physiologic responses after caudal epidural administration of detomidine in horses and xylazine in cattle. *In* "Animal Pain" (C. E. Short and A. Van Poznak, eds.), pp. 292–302. Churchill-Livingstone, New York.

Spinelli, J., Ascher, M. S., and Brooks, D. L. (1981). Q fever crisis in San Francisco: Controlling a sheep zoonosis in a laboratory facility. *Lab. Anim.* **10,** 24–27.

Staub, N. C., Bland, R. D., Brigham, K. L., Demling, R., Erdmann, A. J., and Woolverton, W. C. (1975). Preparation of chronic lung lymph fistulas in sheep. *J. Surg. Res.* **19,** 315–320.

Steffey, E. P., and Howland, D. (1979). Halothane anesthesia in calves. *Am. J. Vet. Res.* **40,** 372–376.

Thompson, J. R., Hsu, W. H., and Kersting, K. W. (1989). Antagonistic effect of idazoxan on xylazine-induced central nervous system depression and bradycardia in calves. *Am. J. Vet. Res.* **50,** 734–735.

Thompson, S. E., and Johnson, J. M. (1991). Analgesia in dogs after intercostal thoracotomy: A comparison of morphine, selective intercostal nerve block and interpleural regional analgesia with bupivacaine. *Vet. Surg.* **20,** 73–77.

Thurmon, J. C., ed. (1986). Anesthesia. *Vet. Clin. North Am.: Food Anim. Pract.* **2,** 1–710.

Thurman, J. C., Benson, G. J. (1993). Anesthesia in ruminants and swine. *In* "Current Veterinary Therapy. III. Food Animal Practice" (J. L. Howard, ed.), 3rd ed., pp. 58–76. Saunders, Philadelphia.

Thurmon, J. C., Kumar, A., and Link, R. P. (1973). Evaluation of ketamine hydrochloride as an anesthetic in sheep. *J. Am. Vet. Med. Assoc.* **162,** 293–297.

Thurmon, J. C., Kumar, A., and Cawley, A. J. (1975). Changes in the acid base status of sheep anaesthetized with a combination of atropine sulphite, acepromazine and ketamine hydrochloride. *Aust. Vet. J.* **51,** 484–487.

Thurmon, J. C., Neison, D. R., Hartsfield, S. M., and Rushmore, C. A. (1978). Effects of xylazine hydrochloride in urine in cattle. *Aust. Vet. J.* **54,** 178–180.

Thurmon, J. C., Benson, G. J., Tranquilli, W. J., and Olson, W. A. (1986). Cardiovascular effects of intravenous infusion of guaifenesin, ketamine and xylazine in holstein calves (abstr). *Vet. Surg.* **15,** 463.

Toutain, P. L., Brandon, R. A., Alvinerie, M., and Baggot, J. D. (1983). Thiopentone pharmacokinetics and electrocortigram pattern in sheep. *J. Vet. Pharmacol. Ther.* **6,** 201–209.

Trim, C. M. (1981). Sedation and general anesthesia of ruminants. *Calif. Vet.* **4,** 29–35.

Tsuno, K., Prato, P., and Kolobow, T. (1990). Acute lung injury from mechanical ventilation at moderately high airway pressures. *J. Appl. Physiol.* **69,** 956–961.

Valverde, A., Diehard, T. J., Design, D., and Valiant, A. E. (1989). Evaluation of pentobarbital as a drug for standing sedation in cattle. *Vet. Surg.* **18,** 235–238.

Vesce, G., and Lucisano, A. (1991). Crossing of the placental barrier by propofol in small ruminants. *Proc. Int. Congr. Vet. Anesth.* 4th, Utrecht, The Netherlands, pp. 167–169.

Wagner, A. E., Dunlop, C. I., and Turner, A. S. (1994). Experiences with subarachnoid morphine in sheep (abstr.). *Proc. Int. Congr. Vet. Anesth, 5th,* Guelph, Canada, p. 103.

Waterman, A., and Livingston, A. (1978). Studies on the distribution and metabolism of ketamine in sheep. *J. Vet. Pharmacol. Ther.* **1,** 141–147.

Waterman, A. E., Nolan, A., and Livingston, A. (1987). Influences of idazoxan on the respiratory blood gas changes induced by alpha-2 adrenoceptor agonist drugs in conscious sheep. *Vet. Rec.* **121,** 105–107.

Waterman, A. E., Livingston, A., and Akin, A. (1991). Analgesic activity and respiratory effects of butorphanol in sheep. *Res. Vet. Sci.* **51,** 19–23.

Waterman, A. E., Livingston, A., Ley, S. J., and Brandt, S. (1992). Changes in nociceptive thresholds associated with chronic pain in sheep. *In* "Animal Pain" (C. E. Short and A. Van Poznak, eds.), pp. 378–384. Churchill-Livingstone, New York.

Welsh, E. M., Baxter, A. M., and Nolan, A. M. (1992). Pharmacokinetics of Carprofen administered intravenously to sheep. *Res. Vet. Sci.* **53,** 264–266.

Wood, G. N., Molony, V., Hodgson, J. C., and Fleetwood-Walker, S. M. (1990). Responses of young lambs to castration after intrathecal etorphine or xylazine. *Pain, Suppl.* **5,** S123 (abstr.).

Woolf, C. J. (1983). Guidance for a central component of post-injury pain hypersensitivity. *Nature (London)* **306,** 686–688.

Woolf, C. J., and Chong, M. S. (1993). Pre-emptive analgesia—treating post-operative pain by preventing the establishment of central sensitization. *Anesth. Analg.* **77,** 362–379.

Chapter 14

Anesthesia and Analgesia in Swine

Alison C. Smith, William J. Ehler, and M. Michael Swindle

In the past, swine have been considered one of the more difficult species to anesthetize in the laboratory. Much of the literature on anesthesia is more relevant to their agricultural use than to their use as an experimental animal, where the physiologic effects of anesthetics become of prime importance. Veterinary textbooks provide a complete listing of anesthetic and analgesic agents that have been used in swine (Riebold *et al.*, 1995; Short, 1987; Smith and Swindle, 1994). The Appendix at the end of this chapter presents data on drug administration and the protocols used in surgery.

ANESTHESIA AND ANALGESIA IN LABORATORY ANIMALS

ISBN 0-12-417570-8

I. ANATOMIC AND PHYSIOLOGIC CHARACTERISTICS

When using swine in research, the investigator should take into consideration that differences exist between domestic (commercial farm breeds) and miniature swine, as well as between breeds within both categories, even though all are classified as *Sus scrofa domestica.* Within the same breed and type of pig, differences related to age, weight, and sex may become apparent. All of these factors must be taken into consideration when considering the physiologic effects of an anesthetic agent during an experiment (Thurmon and Benson, 1987).

Miniature breeds are more mature than domestic swine at the same body weight. For example, at 12 weeks of age the comparative body weights of various types of pigs would be expected to be 15 kg, Yucatan miniature pig; 12 kg, Yucatan micropig; 19 kg, Hanford miniature pig; and 36 kg, a Yorkshire domestic pig. Both miniature and domestic pigs achieve sexual maturity between 4 and 6 months of age. In comparing the same pigs at 6 months of age, the comparative body weights would be expected to be even more diverse with a weight range of over 70 kg. Consequently, matching miniature and domestic stocks of pigs by body weight to compare pharmacologic actions could be misleading because of the potential age differences. When considering the hemodynamic effects of anesthetics on studies using different stocks, it is not appropriate to compare the pigs by body weight without indexing to body surface area (Smith *et al.,* 1990).

Within the same stock, differences in response to anesthetics can be age related (Crane *et al.,* 1975; Gootman *et al.,* 1986; Buckley *et al.,* 1979). In a study comparing the differences in various parameters among three stocks of miniature pigs, all of which were 4 months of age, significant differences were found between the heart weight: body weight ratio, peak systolic pressure, right ventricular peak pressure, mean arterial pressure, right atrial pressure, mean pulmonary artery pressure, and pulmonary vascular resistance when using equivalent anesthetic conditions (Smith *et al.,* 1990).

In a European study comparing Göttingen and Yucatan miniature pigs with farm pigs in an acute cardiovascular protocol, only minor differences between stocks were found when they were weight matched at 20–25 kg. The farm pigs tended to develop dysrhythmias and had a shorter stabilization period than either miniature breed. The Yucatan had higher absolute hemodynamic values. For their studies, these differences were considered insignificant (Benharkate *et al.,* 1993).

Consequently, comparisons of the response to various anesthetic agents in swine should be made with caution when the experimental animals are not the same stock, age, and weight (Buckley *et al.,* 1979; Smith *et al.,* 1990).

Swine are prone to vasospasm and many of the peripheral vessels are not readily visualized, which may make vascular access difficult for inexperienced personnel. Swine are also highly susceptible to ventricular arrhythmias, have fragile pulmonary tissue that may be damaged by overinflation with mechanical respirators, and have a left hemiazygous vein that drains the intercostal vessels into the coronary sinus. All of these factors should be considered when selecting anesthetic agents and their method of delivery to swine (Swindle, 1986; Swindle and Adams, 1988; Swindle, 1991; Smith *et al.,* 1989; Bobbie and Swindle, 1986).

II. ANESTHETIC DELIVERY

The most reliable sites for vascular access and injections of anesthetics in laboratory swine have been previously reviewed and illustrated (Swindle, 1983; Bobbie and Swindle, 1986). Intramuscular injections may be given in the major muscles of the thigh and in the neck. The semimembranous and semitendinosus muscles of the caudal thigh as well as the gluteal muscles of the cranial thigh are both acceptable sites for giving large volume IM injections. The muscle mass of the dorsolateral neck region is a good site for small volume injections.

Subcutaneous injections may be given in the neck or in the flank. The pig is a fixed-skin animal, like humans, and does not have a SC region as large or accessible as other animal species. Consequently, large volume SC injections are not as easily delivered.

The peripheral blood vessels are not easily visualized in most pigs. The ear veins are the most common site for IV injections. The veins are located on the lateral and medial dorsal ear margins and the artery is central.

The cephalic vein is located on the cranial aspect of the foreleg between the carpus and the elbow joint, is readily accessible, and can be palpated if a tourniquet is placed around the elbow joint. If blood flow through the vessel is occluded by the thumb and if the skin is rotated laterally, as is commonly done in the dog, the vessel will collapse and become inaccessible. The cephalic vein also crosses the medial aspect of the shoulder and can be distended in dorsal recumbency by putting digital pressure on the thoracic inlet.

In the rear leg the lateral and median saphenous veins are generally unreliable for anesthetic delivery because of their small size. The femoral vessels may be approached percutaneously but are more frequently used for vessel cannulation following surgical exposure.

The external jugular veins are deep in the pig and lie in the same plane as the internal jugular veins and carotid arteries. These vessels are best used for surgical cannulation or blood sampling and would probably only be used in an emergency situation for anesthetic delivery. The cranial vena cava may be cannulated by inserting a needle into the thoracic inlet from the right side between the shoulder and the manubrium sterni.

Restraint of the pig for injections or for delivery of inhalant anesthetics via a face mask may be very stressful. A humane restraint sling (Fig. 1) has been described (Panepinto *et al.*, 1983), and variations of the method have been developed by individual laboratories (Houpt, 1986). This restraint method is comfortable for the pig and most will readily accept suspended restraint after a few training sessions. Manual restraint by suspending the pig by the rear legs, use of snout snares, or tying to a V-trough is stressful and should be discouraged. Small pigs may be manually restrained and larger pigs may be herded into the corner of a pen using a panel.

Small needle sizes (20- to 22-gauge) can be used for most animals for IV, IM, or SC injections. Large gauge needles (14- to 16-gauge) used for venipuncture in domestic farm breeds are unnecessary and contraindicated in animals <60 kg. Teflon-coated catheters may lose their outer covering when being passed through the skin unless the skin is incised. As an alternative, polyethylene intravascular catheters or butterfly catheters may be used. Intramuscular or SC injections may be given by positioning a butterfly catheter into the injection site with a quick, open-handed slap technique while holding the wings of the butterfly between two fingers. This is performed by slapping the neck of the pig lightly with the hand several times prior to inserting the needle while simultaneously administering a slap. The pig will generally not notice the catheter and then can be approached, a syringe can be attached, and the injection administered without manual restraint.

Some of the chemical restraint methods discussed in this chapter may be necessary to reduce stress in swine prior to handling, venipuncture, or other invasive techniques such as catheterization. If combination injections are being used, administration of the tranquilizer/sedative portion may help reduce the stress. For example, if ketamine and acepromazine are used, administration of the acepromazine 10 min ahead of ketamine may reduce the stress associated with the more irritating ketamine injection.

A simple method of endotracheal intubation has been described for swine under 50 kg (Swindle, 1991). Swine are placed in dorsal recumbency and the epiglottis is visualized using a straight laryngoscope blade. A topical anesthetic, i.e., lidocaine, is sprayed on the larynx as prophylaxis against laryngospasm. The tip of the epiglottis is depressed with the tip of the laryngoscope against the base of the tongue and firm pressure with the blade of the laryngoscope presses the maxilla against the surface of the table. With the laryngoscope blade at a 45° angle, the larynx can now be visualized as it is pressed against the ventral surface of the neck and the vocal cords can be visualized by looking down the path of the blade. The tip of the endotracheal tube is passed into the larynx while still being visualized. Once the tube is at the larynx, further passage may be obstructed by the laryngeal diverticulum. Passage may be facilitated by gently rotating the tube 90° while simultaneously applying forward pressure. This forward corkscrew motion should facilitate passage without undue force (Fig. 2).

When using this method, it is contraindicated to have an assistant hold the jaw open because this causes the laryngeal–tracheal passage to bend at an angle that makes intubation difficult. Inexperienced anesthetists may find stylets to be helpful, but they are generally unnecessary. A standard laryngoscope blade with a length of less than 195 mm is usually indicated. Endotracheal tube sizes of 5–8 mm internal diameter, 22–30 cm long, can be used for swine under 50 kg. If more than minor force is required to pass the tube, then either the tube is too large or the angle of the laryngoscope blade is less than 45°.

The more conventional method of placing pigs in sternal recumbency and using specialized laryngoscopes and assistants to hold the mouth open is generally necessary in swine over 50 kg of body weight. Endotracheal tube sizes of 9–14 mm

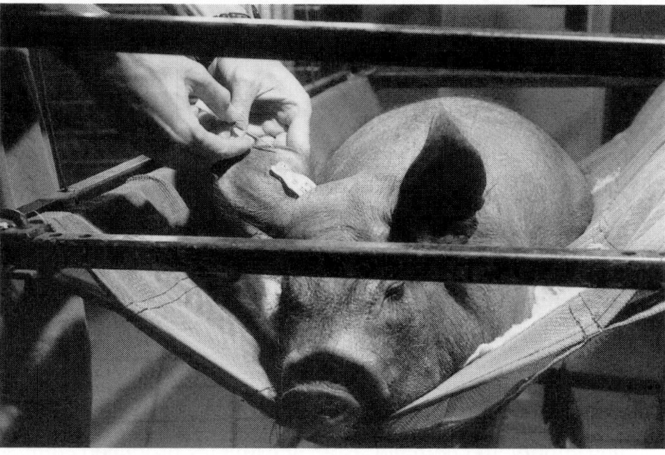

Fig. 1. Yucatan miniature pig in a Panepinto sling.

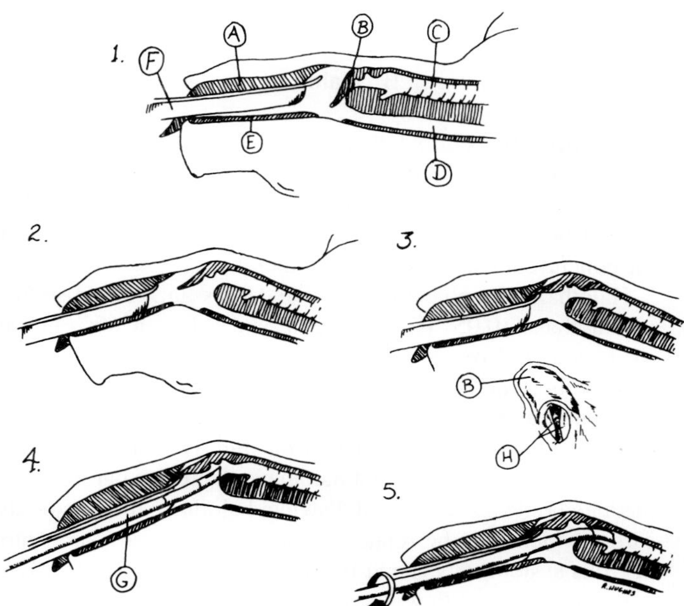

Fig. 2. Stepwise methods of endotracheal intubation in a pig in dorsal recumbency.

internal diameter are recommended for swine >50 kg. Methods of intranasal intubation (Riebold *et al.,* 1995), percutaneous transtracheal ventilation (Hudson and Gilroy, 1986), and tracheostomy have been described but are not commonly used and are seldom appropriate (Swindle, 1983).

III. INHALATIONAL ANESTHESIA

A. Nitrous Oxide

Use of nitrous oxide as a sole anesthetic agent in swine, as in other animal species, is contraindicated because its low potency results in the inability to maintain a surgical plane of anesthesia.

The minimum alveolar concentration (MAC) for nitrous oxide in swine is 195 volume % (Tranquilli *et al.,* 1985), similar to nitrous oxide MAC in other species (Steffey and Eger, 1985). Although Eisele *et al.* (1985) reported a lower value of 162 volume % for nitrous oxide MAC, both studies documented a quantitative difference between humans and swine in the ability of nitrous oxide to decrease the requirement for a concomitantly administered anesthetic agent. In swine, the combination of nitrous oxide and halothane or isoflurane causes less cardiovascular depression than equipotent levels of either agent used alone (Manohar and Parks, 1984a; Lundeen *et al.,* 1983).

The primary advantage of using nitrous oxide is derived from the ability to reduce the requirement of other volatile anesthetics, resulting in less cardiovascular depression than when greater concentrations of these agents are used alone with oxygen. However, because of its lesser potency in swine, the reduction in the requirement of other volatile agents is not as great as in humans.

B. Halothane

Determinations of halothane MAC in swine range from 0.91 (Tranquilli *et al.,* 1983) to 1.25% (Eisele *et al.,* 1985) with the majority of reports supporting the lower value (Tranquilli *et al.,* 1988; Cheng *et al.,* 1992). Halothane MAC can be reduced by the concurrent administration of other drugs.

Halothane is a potent myocardial depressant. Administration of halothane (0.05–1.7 volume % end tidal) in 60% nitrous oxide and 40% oxygen to swine produced dose-dependent decreases in aortic blood pressure, cardiac output, and indices of left ventricular function (Merin *et al.,* 1977). The heart rate is maintained (Wedel *et al.,* 1991) or slightly increased (Merin *et al.,* 1977), suggesting that halothane depresses baroreceptor-mediated reflexes. In neonatal swine, halothane induces bradycardia which may be related to the immaturity of the central nervous system (CNS) and baroreceptors (Crane *et al.,* 1975).

Coronary blood flow, oxygen consumption, and myocardial metabolism decrease with increasing halothane concentrations. The dose-dependent depression produced by halothane is not associated with tissue hypoxia and therefore is probably related to its negative inotropic effect (Merin *et al.,* 1977). In swine with normal hearts, 1.0 and 1.5 MAC halothane produced roughly equivalent reductions in myocardial blood flow in all regions of the cardiac ventricles. The ratio of perfusion of the subendocardium to the subepicardium was unchanged from that of unanesthetized animals, suggesting that oxygen delivery to the subendocardium kept pace with oxygen demand (Manohar and Parks, 1984a).

The effect of halothane on coronary vascular resistance is controversial. In humans and dogs, halothane has been reported to cause coronary vasodilation and vasoconstriction or to not affect coronary vascular resistance (Gilbert *et al.,* 1988). In the nondiseased porcine heart, halothane produced a dose-dependent decrease in the capacity of the coronary arteries to maximally vasodilate (coronary reserve) which could not be corrected by increasing coronary artery pressure. In contrast, isoflurane preserved coronary vascular reactivity and reserve at anesthetic concentrations that equaled or exceeded halothane concentrations that were associated with a markedly depleted coronary reserve (Gilbert *et al.,* 1988).

Because the response of the porcine heart to ischemia resembles that of humans (Thurmon and Tranquilli, 1986), the safety of anesthetic agents has been studied in swine that have been developed as animal models of humans with coronary artery disease. Following chronic coronary artery occlusion, a collateral-dependent region of myocardium develops that is potentially at increased risk for ischemia. Coronary steal results when coronary vasodilation distributes myocardial blood flow away from collateral-dependent zones to a normally perfused area resulting in ischemia to the flow-limited area. Anesthetic agents that produce significant coronary vasodilation at clinically useful concentrations therefore could potentiate myocardial ischemia by the mechanism of myocardial steal.

The effect of halothane and isoflurane on the distribution of myocardial blood flow was studied under controlled coronary perfusion pressures in swine with collateral development following chronic coronary occlusion. Isoflurane and halothane produced a dose-dependent decrease in myocardial blood flow with decreasing coronary perfusion pressure in both normal and collateral-dependent zones. Isoflurane was associated with higher endocardial blood flow than halothane in collateral-dependent myocardium. Neither agent produced significant coronary vasodilation or coronary steal (Cheng *et al.,* 1992).

Swine models of regional myocardial ischemia have been created to determine if anesthesia worsens myocardial performance following an ischemic insult. Such studies allow controlled comparisons to be made regarding the hemodynamic effects of various anesthetic agents. A 60% reduction in left anterior descending (LAD) coronary artery flow during halothane or fentanyl anesthesia compared the functional and metabolic effects of high (fentanyl) or low (halothane) myocardial oxygen supply and demand. Both anesthetics produced equivalent depression in ventricular pump function and degree of ischemia after stenosis. Following reperfusion, partial functional and metabolic recovery occurred, although cardiac output and indices of left ventricular contractile performance were higher with fentanyl. Neither agent was protective for myocardial ischemia (Merin *et al.,* 1982).

Halothane, compared with isoflurane, is more depressant to global ventricular function. Increasing concentrations of isoflurane in swine with a critical stenosis of the LAD coronary artery does not result in greater impairment of ventricular function than in animals without stenosis. In contrast, LAD stenosis exacerbated the reduction of cardiac output and stroke volume produced by increasing concentrations of halothane (Gilbert *et al.,* 1989).

Halothane has a lower margin of safety compared to isoflurane. Graded concentrations of halothane and isoflurane were compared in swine with and without major surgery, and with and without critical stenosis of the LAD coronary artery. The safety margin, defined as the ratio of the fatal anesthetic concentration to the 1 MAC level, for isoflurane was approximately twice that of halothane. The addition of surgical stress resulted in an additional 20% decrease in the safety margin for halothane, but did not affect that of isoflurane. Presence of a critical stenosis in addition to surgical stress did not affect the safety margin for either agent (Roberts *et al.,* 1987).

Halothane increases the sensitivity of the myocardium to the arrhythmogenic effects of epinephrine. The arrhythmogenic dose of epinephrine for swine at 1.66 MAC halothane was found to be 3.60 μg/kg, comparable to that reported for other species. Blockade with prazosin, an α_1-antagonist, increased the arrhythmia threshold approximately threefold in halothane–xylazine–anesthetized animals and was more protective than β_1-blockade with metoprolol (Tranquilli *et al.,* 1986). Although the mechanism by which halothane enhances arrhythmias is unknown, the protective effect of prazosin suggests that α_1-adrenergic receptors mediate halothane-induced arrhythmias.

C. Isoflurane

Isoflurane has a lower blood gas solubility than halothane, resulting in a more rapid induction and recovery from anesthesia as well as the ability to more quickly alter anesthetic depth.

MAC values for young swine of the age commonly used in research range from 1.45 (Lundeen *et al.,* 1983) to 2.04% (Eger *et al.,* 1988) with an average MAC value of 1.58%. Variations in reported values can be attributed to differences in experimental methodologies, age and breeds of swine, and body temperature. The reported MAC value for neonatal piglets ranging in age from 2 to 17 days is 1.2% (Schieber *et al.,* 1986).

Among all the currently used inhalation agents, isoflurane has the greatest margin of cardiovascular safety (Weiskopf *et al.,* 1989b). Isoflurane, like halothane, produces a dose-dependent depression of the cardiovascular system characterized by decreases in aortic pressure, cardiac output, and stroke volume. However, when compared to halothane, the depressant effects of isoflurane are less and occur at higher anesthetic concentrations (Gilbert *et al.,* 1989). Use of nitrous oxide as an adjunct to isoflurane anesthesia maintains heart rate, cardiac output and rate–pressure product, an indicator of myocardial oxygen uptake, near values for conscious animals (Lundeen *et al.,* 1983). Myocardial blood flow is decreased in a dose-related manner, especially in the endocardium. However, the ratio of endocardial to epicardial perfusion remains near 1 (Manohar and Parks, 1984b).

The hemodynamic changes produced by isoflurane are more profound in newborn piglets (2–17 days) than in more mature animals. Like human infants, neonatal piglets develop moderate to severe bradycardia and marked hypotension with preservation of cardiac output. The neonatal response has been attributed to the imbalance in the state of development of the autonomic nervous system characterized by a well-developed parasympathetic nervous system and an immature sympathetic nervous system. Although the margin of safety for isoflurane in neonatal swine is relatively low, it is greater than that for equipotent concentrations of halothane because it better preserves cardiac output (Schieber *et al.,* 1986).

The scientific literature is conflicting regarding the effects of isoflurane on the coronary circulation (Priebe, 1989). Much of the confusion may be due to differences in methodology as well as to species differences in coronary anatomy and pathophysiology (Cheng *et al.,* 1992). Although evidence suggests that isoflurane causes direct coronary vasodilation (Gilbert *et al.,* 1988), there is no convincing evidence that isoflurane, in swine with a normal coronary circulation, impairs myocardial oxygenation (Priebe, 1989). The contribution of isoflurane to myocardial ischemia by the mechanism of coronary steal in human patients with coronary artery disease will probably continue to be debated. Experimental and clinical evidence suggests that isoflurane is potentially dangerous to a subset of patients with coronary artery disease and ''steal-prone'' coronary anatomy (Priebe, 1989). Results of studies in a swine model of collateral-dependent myocardium indicate that neither isoflurane nor halothane at clinically useful concentrations produced myocardial ischemia by either intercoronary or transmural redistribution of regional myocardial blood flow (Cheng *et al.,* 1992).

Compared to halothane, significantly greater infusion rates of epinephrine are required to induce cardiac dysrhythmias in isoflurane-anesthetized swine (Weiskopf *et al.,* 1989a). Unlike enflurane, isoflurane does not produce seizure activity (Rampil *et al.,* 1988).

Isoflurane is preferred for liver surgery in swine, including transplantation. Isoflurane, compared to halothane, administered in 50% nitrous oxide and oxygen for swine undergoing liver transplantation was associated with higher intraoperative blood pressures, shorter anesthetic recovery times, and better long-range survival (Eisele *et al.,* 1986). In miniature swine, hepatic metabolism occurred at low alveolar concentrations of halothane but not with isoflurane (Halsey *et al.,* 1971). In swine undergoing laparotomy, the administration of either isoflurane or fentanyl in concentrations that decreased mean arterial pressure by ≤30% provided adequate hepatic oxygen supply. Halothane, at concentrations that produced a similar degree of hypotension, or isoflurane concentrations producing decreases in blood pressure exceeding 30% were associated with an inadequate hepatic oxygen supply (Gelman *et al.,* 1987).

Isoflurane does not cause hepatic injury. Determinations of plasma alanine aminotransferase (ALT), an indicator of hepatocellular necrosis, were not increased over baseline values after prolonged isoflurane anesthesia in swine (Holmes *et al.,* 1990). Although isoflurane administration results in a threefold increase in concentrations of plasma fluoride ions, the degree of biotransformation is minimal and not associated with renal toxicity (Koblin *et al.,* 1989).

Given the many advantageous properties of isoflurane, it should be considered to be the inhalation agent of choice for swine.

D. Enflurane

Enflurane is a halogenated ether derivative that is more soluble but less potent than isoflurane. Enflurane MAC for swine has been determined to be 1.66% and 0.82% with 50% nitrous oxide (Short, 1987). Enflurane, at high inspired concentrations, can produce muscle rigidity and seizure activity in swine. Enflurane is a more potent myocardial depressant than either isoflurane or halothane, producing decreases in mean arterial pressure, cardiac output, stroke volume, left ventricular work, systemic vascular resistance, and rate–pressure product (Thurmon and Benson, 1987). Because of its potential to induce seizures, use of enflurane in swine should be avoided.

E. Desflurane

Experimental studies with chronically instrumented swine have been utilized to obtain data regarding the anesthetic effects of desflurane. The effects of desflurane in swine appear to be similar to those in humans (Weiskopf *et al.,* 1992).

The minimum alveolar concentration for desflurane in swine, determined by using the tail clamp technique, was 8.28%. A more consistent and higher MAC value of 10.00% was obtained by clamping the dew claw. The difference in MAC values was attributed to the difference in the type of stimulus applied and that use of the tail clamp technique underestimated the anesthetic requirement of pigs because it was not a supramaximal stimulus (Eger *et al.,* 1988). The same authors later determined the MAC value for desflurane in humans to be 7%, which correlated with the observation that the MAC values for other anesthetic agents in swine are higher than those in humans (Weiskopf *et al.,* 1992). The cardiovascular effects of desflurane in swine are similar to those of isoflurane, producing vasodilation, hypotension, and dose-dependent myocardial depression. The cardiovascular effects of desflurane were studied at 0.8, 1.2, and 1.6 MAC and compared with data obtained while the animals were conscious and with equipotent concentrations of isoflurane. Desflurane produced decreases in mean arterial blood pressure and stroke volume in a dose-related manner despite increases in right and left heart-filling pressures and a dose-dependent decrease in systemic vascular resistance. Heart rate increased above the conscious state at 0.8 MAC, whereas the cardiac output was unchanged. At higher concentrations, heart rate decreased but was greater than that of conscious animals. Concentrations above 0.8 caused a dose-dependent decrease in cardiac output despite increased right and left heart preload. Dose-dependent decreases in oxygen consumption and left ventricular minute work, calculated as the product of systolic blood pressure, cardiac output, and a rate constant, were also observed (Weiskopf *et al.,* 1988).

The margin of safety for desflurane, defined as the ratio of fatal anesthetic concentration to MAC, is between that of isoflurane and other halogenated anesthetics. High concentrations of desflurane, similar to isoflurane, result in a progressive decrease in mean aortic pressure and cardiac output without affecting heart rate. However, neither parameter was a useful indicator of impending cardiovascular collapse. The most useful indices of imminent cardiovascular collapse were decreases in mixed venous PO_2, mixed venous oxyhemoglobin saturation, and the ratio of oxygen transport to oxygen consumption which occurred within 0.5 MAC of the fatal anesthetic concentration (Weiskopf *et al.,* 1989b).

Desflurane anesthesia does not produce adverse cardiovascular effects combined with the commonly used anesthetic adjuncts atropine, succinylcholine, atracurium, fentanyl, thiopental, naloxone, and nitrous oxide (Weiskopf *et al.,* 1990). Desflurane does not lower the myocardial threshold for epinephrine-induced dysrhythmias (Weiskopf *et al.,* 1989a).

Prolonged anesthesia with desflurane does not produce hepatic injury. Plasma ALT levels were determined before anesthesia, at the end of anesthesia, and 3–8 days after exposure to either desflurane or isoflurane. ALT activity was never increased above baseline levels at any time nor were there differences between the hepatic effects of isoflurane and desflurane (Holmes *et al.,* 1990).

Desflurane undergoes little to no metabolism. The plasma fluoride concentration is used as an index of desflurane metabolism. At the end of a total dose of 5.4 MAC hours of desflurane anesthesia, the plasma fluoride concentration did not change. A 17% increase in the plasma fluoride concentration occurred 4 hr after anesthesia, but this level was markedly below that associated with subclinical renal toxicity (Koblin *et al.,* 1989).

Desflurane has been identified as a trigger for malignant hyperthermia in those breeds that are susceptible (Wedel *et al.,* 1991).

Despite the many favorable characteristics of desflurane, additional experience with this agent is needed. Given its similarity in anesthetic effects and the substantially greater cost of desflurane compared to that of isoflurane, its choice over isoflurane cannot currently be justified.

IV. INJECTABLE ANESTHETICS AND TRANQUILIZERS

A. Tranquilizers, Sedatives, and Anticholinergics

The phenothiazine derivative tranquilizers are acceptable either for sedation as a sole agent or when combined with anesthetics. Acepromazine (0.11–1.1 mg/kg IM, IV, SC) (Riebold and Thurmon, 1986; Benson and Thurmon, 1979; Swindle, 1983) is the most commonly used agent in this class, although dosages have been reported for promazine (0.44–2.0 mg/kg IM, IV, SC) (Bolin and Runnels, 1992; Riebold and Thurmon, 1986; Benson and Thurmon, 1979; Swindle, 1983) and chlorpromazine (0.5–4 mg/kg IM, IV, SC) (Bolin and Runnels, 1992; Riebold and Thurmon, 1986; Benson and Thurmon, 1979; Swindle, 1983). This class of tranquilizers produces an α-adrenergic blockade in higher doses which may be a contraindication for their use in some cardiovascular studies.

Benzodiazepine tranquilizers used in swine include diazepam and midazolam. Diazepam (0.5–10 mg/kg IM or 0.44–2 mg/kg IV) is usually combined with other agents (Benson and Thurmon, 1979; Thurmon and Tranquilli, 1986; Bolin and Runnels, 1992; Schranz *et al.,* 1988). It has also been given as a continuous IV infusion (1 mg/kg/hr) (Riou *et al.,* 1988). The benzodiazepines provide good hypnosis and sedation in swine; however, they are associated with hypothermia with prolonged use (Danneberg *et al.,* 1986). In the experience of the authors, this hypothermia can be profound and life-threatening, especially if combined with other anesthetic agents. A specific benzodiazepine reversal agent, flumazenil, has been developed but its use has not been described in swine.

Midazolam (100–500 μg/kg IM or IV) is a water-soluble benzodiazepine that has been described as a sole agent providing deep sedation for 20 min with minimal depression of the cardiovascular and respiratory systems (Smith *et al.,* 1991). Higher dosages of midazolam (1 mg/kg IV) have been used in anesthetized swine (Ochs *et al.,* 1987). Continuous IV infusions have been administered at a rate of 0.6–1.5 mg/kg/hr in the authors' laboratory. Other benzodiazepines have been described for experimental use in swine and include brotizolam (1–10 mg/kg PO) (Danneberg *et al.,* 1986), lorazepam (0.1 mg/kg IV or intraosseous) (Pender *et al.,* 1991), climazolam (Becker, 1986), and flurazepam (2 mg/kg IV or PO) (Ochs *et al.,* 1987).

Butyrophenone tranquilizers used in swine include droperidol, fluanisone, and azaperone. Droperidol and fluanisone are not routinely used as sole agents in swine and will be discussed with combination drugs. Azaperone was introduced as an agent to prevent fighting in swine but has been used as a preanesthetic (2–4 mg/kg IM) or it may be used as an immobilizing agent in a higher dosage (5.3–8 mg/kg IM) with a minimal depression of cardiovascular function (Thurmon and Tranquilli, 1986; Portier and Slusser, 1985; Riebold and Thurmon, 1986).

The α_2-adrenergic agonist agents xylazine, (2 mg/kg IM), (Benson and Thurmon, 1979; Thurmon and Tranquilli, 1986; Riebold and Thurmon, 1986; Swindle, 1991) and medetomidine (0.2 mg/kg IV) have been described in swine usually in combination with other agents (Vainio *et al.,* 1992). None of these agents have been described as providing sedation in swine without combination with other agents. Xylazine has only transient analgesic activity in swine, produces hypotension and 1–3° heart block, and is not satisfactory as a sole agent (Benson and Thurmon, 1979; Thurmon and Tranquilli, 1986; Riebold and Thurmon, 1986; Swindle, 1991). Medetomidine can be reversed by atipamezole. α_2-agonists are not effective sedatives or analgesics when used alone in swine.

Reversal agents have not been routinely used in swine; however, a dose is available for yohimbine (1 mg/kg IV) (Armstead *et al.,* 1988).

Anticholinergics are useful in swine to dry secretions of the oral cavity and respiratory tract and to provide a vagolytic effect during endotracheal intubation or tracheal suctioning. Atropine (0.05 mg/kg IM or 0.02 mg/kg IV) or glycopyrrolate (0.004–0.01 mg/kg IM) may be used to produce these effects (Benson and Thurmon, 1979; Riebold *et al.,* 1995; Swindle, 1983). Anticholinergics are indicated to prevent bradycardia caused by vagal stimulation during intubation or secondary to induction or preanesthetic agents such as opioids. With experience, the routine use of these agents is not necessary and they are not essential unless the heart rate falls below 60 beats per minute. Sinus tachycardia can result from indiscriminate use.

B. Dissociative Agents

Dissociative anesthetic agents, specifically ketamine, are the most commonly used injectable anesthetic agents in swine. These agents provide rapid and safe immobilization with minimal depression of the cardiovascular system. However, because they provide poor muscle relaxation and little visceral analgesia, they are usually combined with other agents (Benson and Thurmon, 1979; Riebold *et al.,* 1995; Swindle, 1983; Loscher *et al.,* 1990).

Ketamine (11–33 mg/kg IM) as a sole agent provides approximately 30 min of immobilization characterized by catatonia. This muscle rigidity makes intubation difficult at best in swine that are anesthetized with ketamine alone. As a sole agent it is unsatisfactory for performing major surgery even if administered as an IV infusion of 3–10 mg/kg/hr because of inadequate analgesia and muscle relaxation (Benson and Thurmon, 1979; Bolin and Runnels, 1992; Short, 1987; Swindle, 1983; Tranquilli *et al.,* 1982; Worek *et al.,* 1988).

Ketamine has been combined with a wide variety of other agents either as a preanesthetic or to produce a surgical plane of anesthesia for minor procedures.

Ketamine (33 mg/kg IM) combined with acepromazine (1.1 mg/kg IM) provides chemical immobilization with muscle relaxation that lasts approximately 30 min and is useful as a preanesthetic but the combination is slightly cardiodepressant (Riebold *et al.,* 1995; Swindle, 1983; Benson and Thurmon, 1979). Ketamine (15 mg/kg IM) combined with diazepam (2 mg/kg IM) or azaperone (2 mg/kg IM) produces a similar state to the ketamine–acepromazine combination; however, because of the slow onset of action of the drugs, diazepam and azaperone are usually administered 15–20 min prior to the ketamine (Riebold and Thurmon, 1986). Ketamine (33 mg/kg IM) and midazolam (500 μg/kg IM) have been used in the authors' laboratory to provide up to 45 min of immobilization with a single injection; however, this combination leads to profound hypothermia and requires 1–4 hr to recover a righting reflex. Ketamine (10 mg/kg IV) and medetomidine (0.2 mg/kg IV) provide 30 min of deep sedation; however, it is accompanied by profound but reversible hemodynamic depression (Vainio *et al.,* 1992).

The combination of ketamine (20 mg/kg IM) and xylazine (2 mg/kg IM) has been used for immobilization and minor surgery. The analgesic effects of the xylazine are transient and last only approximately 5 min in swine. However, it is possible to intubate swine using this combination, and the cardiodepression associated with xylazine can be overcome with anticholinergics (Kyle *et al.,* 1979; Cantor *et al.,* 1981; Trim and Gilroy, 1985). Ketamine (2 mg/kg), xylazine (2 mg/kg), and oxymorphone (0.075 mg/kg) may be administered IV or the dose may be doubled for IM administration to provide short-term chemical restraint suitable for minor surgery (Breese and Dodman, 1984). Ketamine (20 mg/kg IM) and climazolam (0.5–1.0 mg/kg IM) provide similar immobilization to the ketamine–diazepam combination (Becker, 1986). A combination for continuous IV infusion that provides minimal cardiopulmonary depression in swine has been developed (Thurmon and Tranquilli, 1986). Ketamine (1 mg/ml), xylazine (1 mg/ml), and glycerol guaiacolate (5%) mixed in 5% dextrose solution are administered as a 1-ml/kg IV bolus followed by a continuous infusion of 1 ml/kg/hr. Meperidine (2.2 mg/kg IM) and azaperone (2.2 mg/kg IM) followed in 20 min by ketamine (22 mg/kg IM) and morphine (2.2 mg/kg IM) provide up to 1 hr of immobilization without splenic enlargement (Hoyt *et al.,* 1986).

A tiletamine–zolazepam combination (2–8.8 mg/kg IM) is commercially available and provides 20 min of immobilization suitable for minor surgery. However, the combination does produce hypothermia and cardiodepression and would not be recommended for use in cardiopulmonary-compromised animals (Bolin and Runnels, 1992; Swindle, 1991). This agent has been combined with xylazine (2.2 mg/kg IM) to produce a surgical plane of anesthesia; however, the combination can produce profound cardiopulmonary depression (Swindle 1991).

In a comparison of Telazol (4.4 mg/kg IM), Telazol (4.4 mg/kg IM)–ketamine (2.2 mg/kg IM), Telazol (4.4 mg/kg IM)–ketamine (2.2 mg/kg IM)–xylazine (2.2 mg/kg IM), and Telazol (4.4 mg/kg IM)–xylazine (2.2 mg/kg IM), all four induced chemical restraint. However, only the combinations with xylazine were suitable for intubation and short-term anesthesia. No significant difference was found in the effects of Telazol–xylazine when ketamine was added. However, no hemodynamic data were collected during this study, and the combinations were not recommended for cardiovascular studies unless hemodynamic studies have been conducted to determine its exact effects (Ko *et al.,* 1993).

Most of the IM combinations provide only 20–30 min of anesthesia suitable for minor procedures or for induction prior to maintenance with an inhalant anesthetic. Of all of the combinations discussed, only ketamine–xylazine, ketamine–midazolam, ketamine–xylazine–glycerol guaiacolate, and tiletamine– zolazepam–xylazine abolish the swallowing reflex adequately to make endotracheal intubation possible without the use of additional anesthetic agents.

C. Barbiturates

Barbiturates can be used for general anesthesia, and the pharmacologic effects in swine are similar to other species. The pig is prone to apnea secondary to their administration, and use of dilute solutions (1–2.5%) for IV bolus injections is recommended. The severe cardiopulmonary depression associated with the prolonged administration of these agents may be alleviated by using continuous IV infusions rather than repeated IV boluses. Dosages are affected by many factors, including age and the use of other chemical restraint agents. As in other species, the dosages are guidelines and barbiturates should be administered to effect with the careful monitoring of vital signs (Thurmon and Tranquilli, 1986; Riebold and Thurmon, 1986; Swindle, 1991).

The ultra-short-acting thiobarbiturates may be utilized either alone or following preanesthesia with ketamine, ketamine combinations, or other chemical restraint agents in order to induce a surgical plane of anesthesia or to abolish the swallowing reflex in order to facilitate endotracheal intubation. Thiopental (6.6–25 mg/kg IV) or thiamylal (6.6–25 mg/kg IV) is commonly used for these procedures (Bolin and Runnels, 1992; Riebold and Thurmon, 1986; Thurmon and Tranquilli, 1986; Swindle, 1991). Thiamylal or thiopental (3–6 mg/kg/hr) can be used as a continuous IV infusion to provide prolonged barbiturate anesthesia (Swindle, 1991; Swindle *et al.,* 1988; Worek *et al.,* 1988). Because the thiobarbiturates are rapidly eliminated by the kidney, they have a short recovery time.

Pentobarbital (20–40 mg/kg IV) is an intermediate-acting barbiturate that can be used for prolonged general anesthesia.

Pentobarbital is more cardiodepressant than the thiobarbiturates and may require prolonged recovery times postoperatively due to its metabolism by the liver (Bolin and Runnels, 1992; Riebold and Thurmon, 1986; Thurmon and Tranquilli, 1986; Swindle, 1991; Swindle *et al.,* 1988). Pentobarbital can be administered as a continuous IV infusion (5–15 mg/kg/hr), which may be useful, particularly for nonsurvival studies (Buckley *et al.,* 1979; Worek *et al.,* 1988).

D. Opioid Infusions and Combinations

Continuous IV infusions of opioids have been used with other anesthetics to enhance analgesia and produce balanced anesthesia and in higher doses to produce anesthesia for experimental cardiac surgery. The main advantages of using high-dose opioid anesthetic techniques are that they produce minimal depression of cardiac function and they provide protection against cardiac arrhythmias (Lunn *et al.,* 1979; Merin *et al.,* 1982; Swindle *et al.,* 1986; Ehler *et al.,* 1985; Schumann *et al.,* 1994).

Fentanyl and sufentanil are the most commonly used agents for opioid infusion techniques in swine. The agents may be used alone for induction or following light sedation to facilitate IV access. Fentanyl is given as an IV bolus (50 μg/kg) after initiating a continuous IV infusion of 30–100 μg/kg/hr (Ehler *et al.,* 1985; Swindle *et al.,* 1986; Gelman *et al.,* 1987). Sufentanil is more potent and is started as an IV infusion at 15–30 μg/kg/hr followed by an IV bolus of 7 μg/kg (Schumann *et al.,* 1994; Swindle *et al.,* 1993). Starting the IV infusion prior to the bolus prevents muscle rigidity, which is frequently associated with these drugs. Some animals may require supplemental anesthesia with other agents such as isoflurane and nitrous oxide. For this reason, paralytic agents should not be used until there is assurance that the animal is adequately anesthetized. Opioid infusion techniques produce a profound bradycardia during the IV bolus phase that can be controlled with anticholinergics. The heart rate stabilizes during the maintenance infusion. Monitoring heart rate and blood pressure are sensitive indicators of an adequate level of analgesia with these techniques.

Fentanyl (0.005 mg/kg IV) combined with thiopental (12.5 mg/kg IV) for induction followed by fentanyl (0.0025 mg/kg q20–30 min), thiopental (1 mg/kg q30 IM), and midazolam (0.25 mg/kg) has also been used. However, this combination would probably provide more stable anesthesia if given as a continuous infusion rather than as repeated boluses (Oldhafer *et al.,* 1993).

Fentanyl–droperidol is available commercially as a combination (0.4 mg fentanyl/ml and 20 mg droperidol/ml) for IM injection at a dose of 1 ml/13.5 kg or 0.25–0.5 ml/kg IV. This combination produces short-term immobilization suitable for preanesthesia or minor surgery. As in other species, it is associated with muscular movement, salivation, and hypersensitivity to sound (Benson and Thurmon, 1979; Thurmon and Tranquilli, 1986). It may be combined with ketamine (11 mg/kg IM) at a dose of 1 ml/13.5 kg IM of the fentanyl–droperidol mixture (Riebold and Thurmon, 1986; Cantor *et al.,* 1981; Benson and Thurmon, 1979). This combination provides 30 min of anesthesia suitable for minor surgery.

Naloxone (0.5–2.0 mg/kg IV) can be used to reverse opioids (Benson and Thurmon, 1979; Nishijima *et al.,* 1988; Trudeau *et al.,* 1988).

E. Miscellaneous Injectables

A mixture of the pregnanediones, alphaxalone and alphadolone, is commercially available in Europe. It may be administered at a dose of 6–8 mg/kg IM followed by 2–3 mg/kg IV. This combination is short acting, lasting only 4–10 min, and its use with barbiturates is contraindicated (Glen *et al.,* 1979; Bolin and Runnels, 1992; Flecknell, 1987).

Another combination available in Europe is etorphine–acepromazine (0.245 mg/10 kg IM). Etorphine can be reversed by diprenorphine (0.3 mg/kg IM) or naloxone 0.01–0.05 mg/kg IV. No particular advantage exists to using these combinations over anesthetic agents available in the United States, and cardiovascular depression and muscle rigidity may be encountered (Glen *et al.,* 1979; Bolin and Runnels, 1992).

α-Choralose has been used in the past for nonsurvival experimental procedures as a continuous infusion providing minimal cardiac depression and a sparing effect on the baroreceptors. Its use as a sole agent is discouraged because of its questionable analgesic value, especially if other methods such as high-dose narcotics or propofol infusions can be substituted (Silverman and Muir, 1993). α-Chloralose can be administered to swine at a dose of 55–86 mg/kg IV followed by a continuous IV infusion for maintenance (Thurmon and Tranquilli, 1986). For short-term acute cardiovascular studies, 2 mg/kg IM droperidol followed by 50 mcg/kg IV flunitrazepam for induction followed by 60 mg/kg IV α-chloralose with a maintenance infusion of 20 mg/kg/hr was deemed satisfactory. Hemodynamics were stable for 2–3 hr (Benharkate *et al.,* 1993).

Etomidate (4–8 mg/kg IV) has been used as a sedative in swine with minimal analgesia and muscle relaxation. When combined with azaperone (2 mg/kg IM), sedation and analgesia suitable for minor surgery or preanesthesia can be produced (Holzchuh and Cremonesi, 1991). It has also been used to induce anesthesia at 0.6 mg/kg IV followed by a ketamine infusion of 10 mg/kg/hr (Worek *et al.,* 1988).

Propofol is an IV hypnotic agent that has been described for use in swine after induction with azaperone and thiopental (Foster *et al.,* 1992). It provides effective sedation and muscle relaxation for anesthesia with a minimal depression of cardiovascular function. Analgesia is adequate for major surgery if infused at the higher dosages. Induction of anesthesia by bolus injections of 0.83–1.66 mg/kg IV was followed by incremental IV boluses approximating an infusion rate of 14–20 mg/kg/hr.

In a similar experiment, propofol was administered as an IV infusion of 12 mg/kg/hr to demonstrate its safety in malignant hyperthermia-susceptible swine (Raff and Harrison, 1989) along with its effectiveness for a teaching protocol in laparoscopic surgery at the same rate (Ramsey *et al.,* 1993).

Metomidate (4 mg/kg IV) has been given following preanesthesia with azaperone (2 mg/kg IM). The metomidate had to be repeated as bolus injections every 15–30 min but could probably be used as a continuous IV infusion (Svendsen and Carter, 1989). It has also been used as a premedication at 5 mg/kg IM with 6 mg/kg IM azaperone followed IV by 0.01 mg/kg atropine, 2.5 mg/kg metomidate, 0.01 mg/kg fentanyl, and 1 mg/kg droperidol. Analgesics were repeated as indicated (Kreimeier *et al.,* 1993). These combinations provided relatively stable cardiovascular function with minimal depression for short-term acute studies. However, others (Thurmon and Tranquilli, 1986) have reported that the metomidate dosage should be increased to 15 mg/kg to enhance the anesthetic state. Even at this dosage, metomidate has questionable anesthetic value when compared to other combinations.

V. REGIONAL ANESTHESIA

Regional anesthesia is the reversible loss of sensation to a limited body area without a loss of consciousness. Most surgical procedures performed in a research setting are performed under general anesthesia. However, regional anesthesia may be indicated for brief surgical procedures or when physiologic alterations due to general anesthesia are to be avoided. Knowledge of porcine anatomy and mastery of the technical skills are required for successfully providing regional anesthesia. To minimize stress and achieve immobilization, regional anesthesia should be performed using sedatives and with the animal physically restrained in a sling. The most commonly used local analgesic is lidocaine with a duration of action of 90–180 min. Longer-acting local analgesics, such as bupivacaine, provide from 180 to 300 min of analgesia (Short, 1987).

Injection of a local analgesic at the surgical site, or infiltration analgesia, places the agent in direct contact with nerve endings. In research, this technique is best suited for minor, minimally invasive procedures (e.g., auricular venipuncture). Any pathologic process involving the injection site, such as infection or decreased vascularization, would be contraindications for regional analgesia. The most commonly used form of regional analgesia in farm swine is epidural analgesia at the lumbosacral space. Descriptions of this technique for farm swine have been published previously (Swindle, 1986; Short, 1987). The depth of spinal needle penetration is dependent on the size of the animal. This technique provides analgesia by depositing the analgesic within the epidural space. It will produce analgesia caudal to the umbilicus suitable for laparotomy or cesarean section. Epidural analgesia is contraindicated in animals that are in shock or toxemic because of sympathetic blockade resulting in hypotension (Swindle, 1986). The economic and logistic factors that recommend epidural analgesia for farm swine generally do not apply to research settings. However, the technique may be useful in certain abdominal procedures that preclude the use of general anesthetics.

The primary indication for paravertebral analgesia in research involves the control of postoperative pain following thoracotomy. Although not formally described, one of the authors (MMS) has used this technique successfully in controlling thoracotomy-related pain prior to the availability of the longer-acting analgesics currently available.

VI. ANALGESIA

Most analgesics have a short half-life in swine, which limits their use as postoperative analgesics. Included in this category are fentanyl (0.05 mg/kg IM q2hr; 50–100 μg/kg/hr IV infusion), sufentanil (5–10 μg/kg IM q2hr; 15–30 μg/kg/hr IV infusion), meperidine (2–10 IM mg/kg q4hr), oxymorphone (0.15 mg/kg IM q4hr), and pentazocine (1.5–3 mg/kg IM q4hr) (Flecknell, 1984, 1987; Swindle, 1991; Blum, 1988). The short-acting opioids fentanyl and sufentanil can be administered as continuous IV infusions. Morphine has been reported to cause excitement in nonpainful swine similar to the adverse reaction reported for cats. Butorphanol (0.1–0.3 mg/kg IM q4-6hr) is longer acting and has few side effects in swine. In the authors' opinion, buprenorphine is the postoperative analgesic of choice because it is effective for 8–12 hr in higher doses. Low dosages of 0.005–0.01 mg/kg IM have been reported, but that dose has been proven to be inadequate for major surgical procedures such as organ transplantation; the drug should be administered in a dose range of 0.05–0.1 mg/kg IM q12h (Hermansen *et al.,* 1986). The authors have not found buprenorphine to be a significant respiratory depressant as determined by the postoperative monitoring of blood gases.

It is a good practice to administer the first dose of an analgesic preemptively prior to surgical trespass to diminish the afferent nerve impulses involved in the nociceptive process (Thurmon *et al.,* 1996). The initial dose may be the only one required, depending on the drug chosen and based on clinical assessment (Swindle *et al.,* 1988).

For musculoskeletal pain, phenylbutazone may be administered 10–20 mg/kg PO q12h. Aspirin (10 mg/kg q4-6hr) may also be used. Enteric-coated products are recommended due to the pigs predisposition for gastric ulcers. Recently, use of the nonsteroidal, antiinflammatory drugs, ketoralac (1 mg/kg IM or IV q12hr) and ketaprophen (1–3 mg-kg PO q12hr) have been reported to be effective as part of a balanced anesthetic regimen (Richard Fosse, oral communication). Oral medication is readily accepted in canned dog food or in chocolate syrup (Swindle, 1991; Swindle *et al.,* 1988).

VII. INTRAOPERATIVE SUPPORT AND MONITORING

The focus of intraoperative monitoring should be to support the cardiovascular and pulmonary systems and to maintain core body temperature. Careful attention to trends in the parameters that are monitored will help prevent anesthetic complications from developing. Interpretation and integration of vital signs and measured variables should be based on knowledge of the physiologic effects of the anesthetic agents that have been given and the health status of the animal. Selection of the parameters to be monitored should be tailored to the surgical procedure. For uncomplicated or brief surgeries, the monitoring of vital signs and ECG may be adequate. However, for more complex cases such as cardiovascular procedures, monitoring parameters may also include aortic blood pressure, central venous pressure, pulmonary artery pressure, and blood gas analysis.

ECG monitoring is recommended to detect dysrhythmias. Normal swine have a prolonged Q–T interval that may appear abnormal to those unfamiliar with interpreting porcine ECGs. Because miniature swine are practically hairless, no hair removal is necessary to establish good contact using ECG patches designed for human use. The anesthetist should remember that ECG monitoring only assesses the electrical activity of the heart, which can be completely normal in the face of inadequate circulatory function.

A. Cardiac Arrhythmias

Swine are prone to the development of fatal cardia arrhythmias secondary to manipulation of the heart during cardiac surgery or vagal stimulation following endotracheal intubation, suctioning of the trachea, or surgical manipulation of the pulmonary bronchus (Swindle *et al.,* 1986). Use of bretyllium (3.0–5.0 mg/kg IV) every 30 min prior to and during such manipulations will prevent the majority of the ventricular arrhythmias. If the injection is administered slowly IV and not repeated more often than every 30 min, cardiovascular hemodynamics will be minimally affected (Swindle, 1991; Swindle *et al.,* 1986; Schumann *et al.,* 1993).

Lidocaine infusions (0.3 mg/kg/hr) may also be used for the prevention of such arrhythmias. If ventricular fibrillation or cardiac asystole occurs, countershock [10 joules (J) internal paddles or 200–400 J external paddles] is more effective than chemical agents.

Pig skin has high electrical resistance, and minimal settings of external defibrillator paddles are ineffective (Schumann *et al.,* 1994).

A technique using calcium channel blockers and nitroglycerin in combination with bretyllium in order to enable the catheterization of coronary vessels without cardiac arrhythmias has been published (Rogers *et al.,* 1988).

Miniature swine seem to be less susceptible to cardiac arrhythmias than commercial breeds probably due to their greater maturity per kilogram of body weight (Swindle *et al.,* 1988).

Anesthetic depth is assessed by combining several clinical observations. The normal range for heart rate is between 70 and 150 beats per minute with less than 70 considered to be bradycardia. Generally, young animals tend to have faster heart rates than older animals. Heart rate will vary considerably with the different preanesthetic and anesthetic agents used. Fentanyl and xylazine will produce slower heart rates, whereas atropine and ketamine can cause tachycardia.

Ocular reflexes are not reliable indicators of anesthetic depth for swine. Instead, laxity of jaw tone, absence of a pedal reflex in response to painful stimulus, and a stable heart rate indicate that a surgical plane of anesthesia has been achieved.

The quality of the peripheral pulse is more difficult to assess in swine than other species. However, pulse pressure can be monitored using the saphenous, sublingual, and radial arteries. The color of the mucous membranes should be pink, which indicates adequate blood oxygen saturation. It does not assess blood carbon dioxide content.

The best means of assessing the adequacy of lung gas exchange is by performing blood gas analysis by obtaining an arterial blood sample or use of a pulse oximeter. The saphenous and dorsal auricular arteries can be catheterized percutaneously for this purpose.

The most accessible vessels for direct blood pressure monitoring are the external jugular vein, common carotid artery, and the femoral vessels. With experience, all but the common carotid artery can be catheterized percutaneously. However, catheterization can be accomplished most quickly by performing a vascular cut-down and is the recommended method.

Catheterization of an auricular vein is the usual route of fluid administration. A balanced electrolyte solution, such as lactated Ringer's solution, should be given at a maintenance rate of 5–10 ml/kg/hr. Administering warmed intravenous fluids will help maintain body temperature during prolonged surgeries.

General anesthetic agents, particularly inhalation agents, produce hypothermia which can be greatly exacerbated by neglecting to take measures in the operating room to reduce heat loss. Core body temperature should be monitored by placement of a probe either rectally or in the thoracic esophagus. Body temperature should be maintained using thermal support such as circulating water blankets.

B. Support

For prolonged anesthesia in all species, supporting homeostasis becomes of prime importance. The pig is very prone to hypothermia because it is relatively hairless and has a large body surface area. Hypothermia occurs in the pig within the first hour of anesthesia, especially if peripheral vasodilators are used. Thermal support is of vital importance for prolonged procedures. Swine are prone to pulmonary edema and pooling of blood in the abdominal viscera during anesthesia, and support of homeostasis by administrating IV fluids and by monitoring blood gases is important. Postoperative apnea can be a problem, especially if agents like pentobarbital are used. Swine should be weaned from respirators and their ability to breathe without assistance should be monitored.

VIII. POSTSURGICAL CARE

Frequent monitoring is the hallmark of good postoperative care. During recovery from anesthesia, vital signs and surgical incisions should be monitored and recorded frequently as needed. The greatest attention should be given to the cardiovascular and respiratory systems. Extubation should be performed only after a strong swallowing reflex is apparent. Transport cages designed for dogs are satisfactory for recovering swine up to 50 kg. Pigs weighing more than 50 kg can be recovered to the point of extubation by placing the animal in lateral recumbency atop a sling. Circulating water blankets can be used with either method. With large swine, water blankets can be placed both atop and below the animal.

Administrating analgesics preemptively prior to the perception of pain, will reduce the complications of hypoventilation, hypercapnia, and hypoxemia associated with major surgery. Use of an opioid agonist–antagonist will provide adequate analgesia but not cause further respiratory depression.

Facilities and equipment should be available to deal with complications encountered during recovery. Suction should be on hand to clear airway passages of excessive secretions. In the event of laryngospasm or airway obstruction, ready access to intubation supplies, including a laryngoscope, is necessary. A means of providing supplemental oxygen and continued fluid administration may also be necessary during recovery from anesthesia.

The pharmacokinetics of some of the newer broad-spectrum antibiotics have not been established in swine. As a rule of thumb, use of the human pediatric dose and dosing schedule should be followed when adapting human medications for swine. The authors have successfully used the following antibiotic regimen for swine for various cardiac and fetal surgical procedures: 50–100 mg IV ceftraxione perioperatively followed by 25–50 mg oral cephradine (mixed with canned dog food) every 12 hr.

IX. SPECIAL ANESTHETIC CONSIDERATIONS

A. Malignant Hyperthermia

Malignant hyperthermia (MH) has been encountered in the pig as an anesthetic complication during surgical procedures (Ehler *et al.,* 1985; Swindle *et al.,* 1988).

MH is a genetic condition occurring in certain breeds of commercial farm pigs (mainly Landrace, Pietrain, and Poland China breeds) but not in miniature pigs (Swindle *et al.,* 1988). MH is a "fulminant and hypermetabolic state of skeletal muscle" induced by volatile inhalational anesthetics, succinylcholine, and perhaps stress and exercise (Kaplan, 1991).

The mode of inheritance has long been felt to be polygenic, dominant with incomplete penetrance, or recessive (Harter and Wilsdorf, 1914; Britt *et al.,* 1978; Fuji *et al.,* 1991; Mickelson *et al.,* 1988). However, evidence suggests that it is transmitted as an autosomal dominant called the Hal genotype, which stands for halothane gene. The ryanodine receptor gene (*ryr-1* locus), which is also known as the calcium release channel, is the probable site. The defect is a C to T transition at nucleotide 1843 with T/T being the susceptibles and C/T being the carriers (Houde *et al.,* 1993; Geers *et al.,* 1992).

The pathophysiology of MH involves an inability of the affected animal to control intracellular ionized calcium levels, resulting in a rise in intracellular calcium (Endo, 1977; Putney, 1979). The earliest detectable changes in MH appear in the venous effluents from skeletal muscle and reductions in pH and PO_2, and increases in lactate, PCO_2, potassium, and temperature (Gronert, 1980).

Stress and exercise may be key considerations preoperatively. Potent preanesthetic regimes described elsewhere in this chapter are strongly advised. Ketamine, opioids, and benzodiazepines have all been successfully employed (Ehler *et al.,* 1985; Swindle, 1991). Past and present reviews of the literature suggest a common pathway of excessive myoplasmic Ca^{2+}. Lopez *et al.* (1988) reported that Ca^{2+} is primarily due to the excessive release of calcium from the sarcoplasmic reticulum.

Kaplan (1991) succinctly describes the metabolic chain of events. Excessive myoplasmic calcium leads to (1) activation of contractile elements; (2) hydrolysis of ATP; (3) heat production; (4) O_2 consumption; (5) CO_2 and lactate production; (6) uncoupling of oxidative phosphorylation; and (7) eventual cell breakdown and release of intracellular contents (creatine kinase (CK), K^+, Ca^{2+}, myoglobin, etc.).

1. Recognition of Malignant Hyperthermia

The first cardinal clinical sign of MH is an elevation in end-tidal CO_2. A rise of 5–10 mm Hg above an established baseline is highly suspect. Concurrent tachycardia and tachypnea and an increase in temperature (not always) should further alert the anesthetist/anesthesiologist. The increased metabolic rate results in an increased $PaCO_2 > 45–50$ mm, a decreased pH < 7.25, base deficit < -8, central venous desaturation, and hypercarbia which may occur within 5–10 min. If unchecked, fulminant MH will progress to rhabdomyolysis with severe hyperkalemia > 6, increased serum CK, myoglobinemia, myoglobinuria (a cola-colored urine), unstable blood pressure, and bradycardia with resulting asystole (from hyperkalemia).

It should be noted that an increase in end-tidal CO_2 and tachycardia may also result from excessive heating of the patient (i.e., heating blanket, ambient temperature, or covers), hypoventilation, light anesthesia, or other equipment malfunctions.

2. Triggering Agents

Kaplan (1991) has listed triggering drugs: succinylcholine, halothane, isoflurane, enflurane, and desflurane. All modern volatile inhalation anesthetics tested can trigger MH, although halothane seems to be the most potent. Barbiturates (i.e., pentothal) have the unique property of delaying MH episodes in animals and can be said to be the most safe. Other IV drugs (droperidol, diazepam, midazolam, etomidate, ketamine, propofol, narcotics), muscle relaxants (pancuronium, vecuronium, atracurium, doxacurium, mivacurium), ester local anesthetics, and N_2O are considered safe. Anticholinesterases, atropine, and glycopyrrolate are safe and can be used for the reversal of neuromuscular blockade if needed.

3. Prevention and Management of Malignant Hyperthermia

The successful prevention and avoidance of MH in susceptible swine have been previously described by appropriate breed selection and prophylactic treatment (Ehler *et al.,* 1985; Swindle *et al.,* 1986).

Dantrolene is the drug of choice for the treatment and prevention of MH. It is a lipid-soluble hydantoin derivative that acts distal to the end plate within the muscle fiber (Gronert *et al.,* 1976). It is most effective when administered while there is adequate muscle perfusion. It is for this reason that the immediate postbypass dose of dantrolene is administered at the maximum perfusion temperature of 34°C instead of earlier in the postbypass course (see later). Dantrolene attenuates calcium release without affecting uptake by action on the connections between the transverse tubules and the terminal cisternae of the sarcoplasmic reticulum (Flewellen and Nelson, 1982; Quintin *et al.,* 1981). The effectiveness of dantrolene has been documented in a variety of experimental porcine situations, including contracture studies of isolated muscle and in the intact pig (Harter and Wilsdorf, 1914; Rittenhouse *et al.,* 1974).

The porcine minimum effective dose of dantrolene of MH prophylaxis has been demonstrated to be in the range of 3.5–5 mg/kg (Flewellen and Nelson, 1982; Gronert, 1980). With this experimental information, the authors administer a prophylactic dose prebypass and attenuating doses postbypass.

At effective doses, dantrolene is not associated with serious toxicity. Even very high doses do not produce muscle paralysis, although weakness may result (Flewellen and Nelson, 1982; Gronert, 1980). This weakness appears to cause anorexia and decreased activity, which were ameliorated with lower doses of oral dantrolene (3 mg/kg vs 5 mg/kg). The only ill effect reported in the literature with the drug has been hepatic dysfunction in humans (Schneider and Mitchell, 1976), which has not been seen with oral administration of less than 3 weeks duration. This effect has not been observed in pigs.

Dantrolene, a mild muscle relaxant and antipyretic, is administered 5 mg/kg PO 10–12 hr prior to surgery. It is repeated 2 hr prior to surgery. An alternate administration is 3.5–5.0 mg IV 1 hr preoperatively with repeated hourly dosages (Ehler *et al.,* 1985). Inhalation and triggering agents are avoided.

4. Treatment and Management of Malignant Hyperthermia Crisis

In addition to the regular supplies and drugs, the drugs listed in Table I must be immediately available to manage a malignant hyperthermia episode. Although relatively expensive, dantrolene is the effective treatment for MH. It decreases the release of Ca^{2+} from the sarcoplasmic reticulum and rapidly reverses the MH episode. Enough dantrolene should be administered to completely normalize all signs of MH. Failure to do so may cause recrudescence (Flewellen and Nelson, 1982; Kaplan, 1991; Littleford *et al.,* 1991; Rosenberg, 1988).

Recommendations for treatment of a fulminant MH episode are listed in Table II as outlined by Kaplan (1991).

Cardiac arrests and dysrhythmias during MH are almost always the result of hyperkalemia compounded by acidosis. The heart muscle itself is not involved in the pathophysiology of MH. Aggressive treatment of the acidosis and hyperkalemia generally relieves the dysrhythmias. Bretyllium should also be considered. $CaC1_2$ should be considered only as a last resort. Finally, calcium channel blocking agents should not be used due to the synergistic effect of dantrolene leading to cardiovascular collapse (Kaplan, 1991).

B. Cardiopulmonary Bypass

This segment of the chapter is a general overview and guide to performing survival cardiopulmonary bypass (CPB) in swine. The following is a brief description of CPB, a stepwise description including equipment, anesthesia, flow rates, hemodynamics and hematologic parameters, adjunct pharmaceutical agents, and surgical connections. The authors discuss some of the anesthetic options they and others have successfully used.

CPB is by definition an extracorporal pumping system that temporarily replaces the physiological functions of the cardiopulmonary system. It the swine to receive CPB do not have a propensity for MH or cardiovascular compromise, the choice of anesthetic agent does not significantly affect the outcome. There appears to be no evidence that would justify one intravenous

TABLE I

DRUGS USED TO MANAGE MALIGNANT HYPERTHERMIA

Dantrolene
Sterile H_2O to reconstitute dantrolene
Sodium bicarbonate
Dextrose (50%) w/v
Insulin
Mannitol (25%) w/v
Furosemide

TABLE II

TREATMENT OF MALIGNANT HYPERTHERMIA EMERGENCY[a,b]

1. Stop all triggering agents immediately. Continue with safe agents if surgery cannot be stopped immediately.
2. Hyperventilate with 100% O_2. Use new circuit and soda lime. Change to "clean MH machine."
3. Administer 2.5 mg/kg dantrolene immediately. Continue until all signs normalize (total dose up to 10–20 mg/kg). Maintain IV dantrolene 1 mg/kg q 1hr.
4. Correct metabolic acidosis: HCO_3 1–2 mEq/kg stat. Follow ABGs.
5. Hyperkalemia: HCO_3 or glucose 0.5 g/kg and regular insulin 0.15 μ/kg.
6. Cooling: Iced saline, cooling blanket, body cavity lavage, extracorporeal circulation.
7. Arrhythmias: If persistent, 3 mg/kg procainamide to start; maximum of 15 mg/kg.
8. Monitor: ABG, VBG, (CVP OR PA), UOP, T° (central), end-tidal CO_2, K^+, Ca^{2+}, lactate, CK urine myoglobin, PT, PTT, platelets.
9. Maintain UOP >1 ml/kg/hr: Mannitol, furosemide, volume.
10. Transfer to ICU when stable.
11. Observe in ICU: Stable 24–28 hr. Monitor for recrudescence and late complications.
12. Convert to oral dantrolene when extubated and stable; 1 mg/kg PO q 6 hr × 24–28 hr.

[a]Malignant Hyperthermia Association of the United States 24-hr hotline: 209-634-4917. Modified from "Suggested Therapy for Malignant Hyperthermia Emergency" (MHAUS)

[b]ABG, arterial blood gas; VBG, venous blood gas; CK, creatine kinase; CVP, central venous pressure; PA, pulmonary artery; UOP, urine output; PT, prothrombin time; PTT, partial thromboplastin time.

or inhalational agent over the other in healthy normal animals. Ultimately the choice may be dictated by the individual protocol. Conversely, if MH is a deciding factor, there is a drastic difference in selecting "nontriggering" intravenous anesthetic agents and muscle relaxants. Many of the techniques used in the authors' surgical laboratory have had their initiation from techniques described in various texts and modified by the authors. If an inhalation agent is chosen, it can be delivered during CPB through the oxygen supply to the oxygenator.

CPB requires a team approach with effective communications. The perfusionist manages the composition of priming solutions, perfusion flows, supplemental fluids, and blood. The anesthetist/anesthesiologist manages drugs, i.e., anesthetics, muscle relaxants, and vasoactive agents. The surgeon controls cardioplegia and usually temperature regulation of the heart.

1. Fundamentals and Description of Cardiopulmonary Bypass

The system includes a pump, (either roller or pulsatile), an oxygenator to imitate lung function, (either bubble or membrane), PVC tubing, canulas, and filters (Fig. 3). For a more detailed description of bypass circuitry and the variations that are used, the reader is referred to several texts (Gravlee, 1993b; Reed and Stafford, 1989; Bain, 1988).

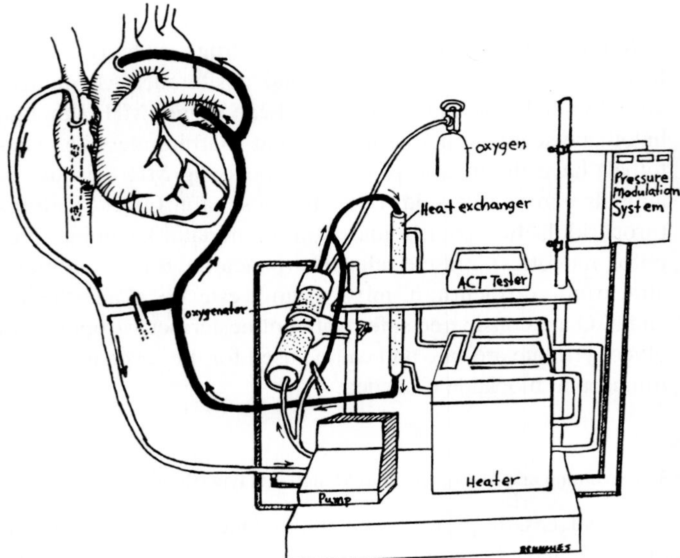

Fig. 3. (A) Diagram of a cardiopulmonary bypass circuit. Single cannulation of the right atrium is shown and is suitable for swine weighing 10–25 kg. Separate cannulation of the superior and inferior vena cavae should be performed for larger swine.

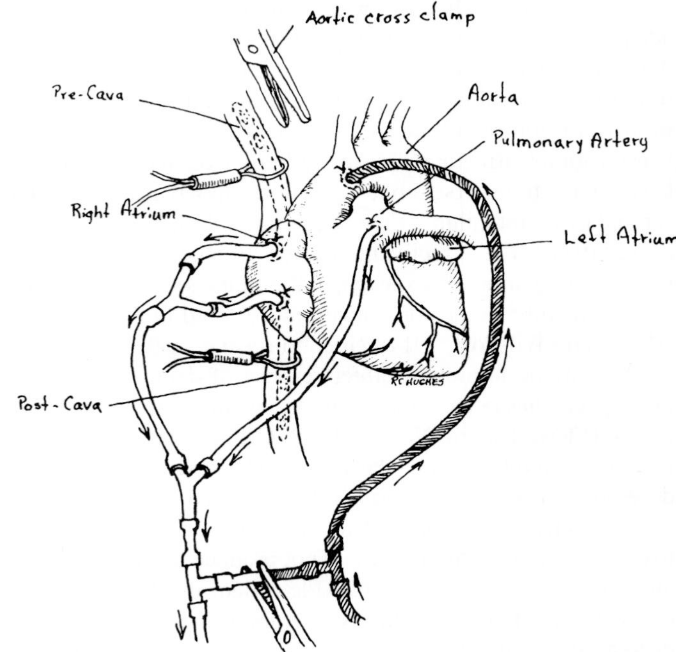

Fig. 3. (B) An alternate method to cannulation of swine for cardiopulmonary bypass with a pulmonary artery vent.

Blood is withdrawn before it reaches the heart, is oxygenated and CO_2 removed, and is then filtered and returned to the systemic arterial tree distal to the aortic valve. The step-by-step surgical procedure has been previously described (Ehler *et al.*, 1985). Venous blood is usually withdrawn through a single right atrial cannula for most of the CPB procedures (10- to 25-kg pigs). An alternate double cannula system placed retrograde into the superior and inferior vena cavae has been used in swine heavier than 40–50 kg. The double cannula system has the advantage of placing less tension on the friable right atrium.

Venous drainage is usually gravity driven into the venous reservoir. A Sarnes 7000 roller pump or Biomedicus pump Model 520 or 540 drives the blood through an arterial filter back into the aorta.

Two types of oxygenators are used: bubble and membrane. Generally, selection is based on time length of the run. Membrane oxygenators are now used for all survival procedures and bubble oxygenators are used for nonsurvival procedures. Discussion of the advantages and disadvantages of each type is beyond the scope and breadth of this chapter. The basic principle of artificially oxygenating blood is accomplished by both. Oxygenators are intrinsically not biocompatible. The blood–gas interface represents an unphysiological state. CPB procedures of less than 2 hr duration have shown no clinically significant difference between membrane or bubble oxygenators. Membrane oxygenators are preferred when longer perfusions (>2 hr) are to be performed. Both microemboli formation and blood trauma are reduced with membrane oxygenators.

2. Choice of Priming Solutions

For porcine CPB, fresh porcine blood from donors under general opioid or barbiturate anesthesia is used. The authors generally collect 2–3 units of blood to prime the bypass pump depending on the weight of the animal to be bypassed. Additional fresh whole blood may be needed post-CPB. A neuroendocrine stress response may be initiated in the recipient if the last unit of blood from the donor is used. This is due to the stress of terminal hemorrhage in the donor. Crystalloid solutions either alone or combined with blood have also been used, but care should be taken to keep the hematocrit >25. There is no consensus as to what constitutes an ideal prime.

3. Myocardial Protection

Myocardial protection during CPB is yet another topic of controversy and it too is beyond the scope of this chapter. The single most decisive factor is the effective application of myocardial preservation techniques during CPB by the surgeon and perfusionist. "The three components of myocardial protection by cardioplegia are the choice of an effective solution, the maintenance of electromechanical inactivity, and the use of profound topical hypothermia. Each need consideration if maximal protection is to be achieved" (Braimbridge, 1988).

Three basic methods are used in combination as indicated by the procedure:

1. Chemical cardioplegia, i.e., Buckberg's solution: 5% dextrose in 0.2% NaCl, 550 ml; KCl 20 mEq; THAM (Abbott Laboratory, North Chicago, IL 60064) 200 ml; CPD 50 ml. One part Buckberg:4 parts oxygenated blood. An alternative is cold Plegisol (Abbott Laboratory, North Chicago, IL 60064) slurry with 10 mEq biocarbonate/liter.
2. Electrical fibrillation, i.e., 1–6 V 60 cycle to cause continual ventricular fibrillation, applied through small electrodes applied directly to the heart.
3. Hypothermic cardioplegia: spontaneous cardiac arrest occurs when the myocardial temperature falls below 20°C. Saline slush is applied directly to the heart as it lies in the bowl-like tacked up pericardium. During this period of inactivity and myocardial temperature, metabolic rate and O_2 consumption are markedly reduced (Reed and Stafford, 1989).

4. Anticoagulation

Adequate anticoagulation cannot be overemphasized prior to the onset of CPB. Systemic heparinization is initiated with 300 units/kg IV. The adequacy of heparinization must be checked at this point by measuring the activated clotting time (ACT). The authors routinely maintain an ACT ⩾ 350 sec (Gravlee, 1991b).

Protamine sulfate is used as a reversal agent at the completion of CPB if needed. A 1.3:1 protamine to heparin reversal ratio is used to neutralize heparin administered during extracorporeal circulation. As a general rule, 1 mg protamine will neutralize approximately 100 units of heparin: specifically 90 units sodium heparin bovine lung, 100 units calcium heparin porcine intestinal mucosa, and 115 units sodium heparin porcine intestinal mucosa. Protamine should not be used if the ACT falls <180 sec and problems with bleeding have not been encountered.

5. Arterial Pump Flow Rates

Beginning arterial input from the pump should be 2.4 liters/m^2/min (Moffitt *et al.*, 1962). Higher flow rates do not improve tissue perfusion and oxygenation and are associated with more mechanical blood damage and a greater risk of embolism from microbubble formation (Bain, 1988). Moderate hypothermia allows for lower arterial pump flow rates (APFR), i.e., core temperature of 28°C results in 50% decrease in O_2 requirements. An APFR of 60–80 ml/kg/min is satisfactory in chilled larger swine (>25 kg) ⩽ 25°C. Because small swine (<10 kg) have an increased metabolic requirement, a higher perfusion rate of 80–100 ml/kg/min is required to maintain adequate tissue oxygenation. Flow rates are reduced as hypothermia progresses.

6. Measurement and Monitoring during Cardiopulmonary Bypass

Total management of CPB and the subsequent "weaning off the pump" and termination of CPB require the continuous monitoring and measurements of all physiological parameters. The primary purpose is to assure adequate O_2 and CO_2 exchange and tissue perfusion. Although the respirator is turned off during CPB, a slight positive pressure is advantageous to minimize pulmonary atelectasis postoperatively.

The following parameters should be monitored continuously:

ECG/HR	differs according to size and breed
Mean arterial pressure	50 mm Hg ± 5
Central venous pressure	0–5 mm Hg during CPB; 8–12 mm Hg pre/post-CPB
Arterial pump pressure	< 200 mm Hg (depending on circuit and cannula sizes)
Systemic venous O_2 saturation	70 ± 5%
Systemic vascular resistance (SVR)	800–1400 dyne/sec/cm^{-5}
Core temperature	esophageal, rectal probes, and/or with tympanic membrane probes (most sensitive)
Blood temperature	in both venous and arterial lines. Note: Arterial <38°C
Coolant temperature	(in heat exchanger) ≤ 40°C. Note: The temperature gradient in C° between the coolant in the heat exchanger and venous return should be ≤ 10–12°C.

The following parameters are monitored as needed:

PaO_2, $PaCo_2$	PaO_2 between 100 and 200 mm Hg; $PaCO_2$ 40 mm Hg ± 4
Arterial pH	7.38–7.42
Base excess	0 to −5
Hematocrit	>18% (prefer >25%)
K^+	4.5–5.5 mmol/liter
ACT	> 300 sec

Left atrial pressure and pulmonary wedge pressure are monitored as needed. A general rule is to match the pressure pre- and post-CPB.

The assessment of O_2 delivery during CPB, hypothermia, hemodilution, and deep anesthesia is difficult at best. During normothermic perfusion, flows exceeding 2.2 liter/m^2/min are usually sufficient, but might become inadequate in the presence of rewarming, "light" anesthesia, or especially low hematocrit levels (<18%). Hypothermia decreases O_2 consumption approximately 7%/°C, and pump flows can be reduced proportionately (Gravlee, 1991a). Thus reduced O_2 consumption allows one to safely reduce perfusion flow rates, hematocrit, or both without compromising O_2 delivery during hypothermia.

As a general rule of thumb, the percent hematocrit is approximately equal to temperature in degrees Celsius during hypothermia. Gravlee (1993a) suggests CPB hematocrit ranges at varying temperatures in humans (see Table III); however, these recommendations have not been subjected to rigorous scientific validation.

When CPB is initiated, the arterial BP falls and then slowly rises to normal or above normal levels during CPB. What the "ideal pressure" may be is controversial. Mean pressure <30 mm Hg is below the critical closing pressure of some vascular beds, and significant underperfusion of tissue may result (Bain, 1988).

TABLE III

SUGGESTED CPB HEMATOCRIT RANGE AT DIFFERENT TEMPERATURES IN HUMANS[a]

Temperature range (°C)	Hematocrit range (%)
30–37	25–30
23–30	20–25
15–22	15–20

[a]From Gravlee, G. P. (1993a). The anesthesiologist and management of cardiopulmonary bypass. *Am. Soc. Anesthesiol. 1993 Ann. Refresher Course Lec.* Washington, D.C., No. 126, pp. 1–7.

7. Central Venous Pressure

Normal central venous pressure (CVP) in normothermic anesthetized swine on intermittent positive pressure ventilation is approximately 4–10 mm Hg. During CPB, venous pressure drops to zero. When CPB is completed and normal circulation restored, the CVP is useful in determining volume replacement by adding 25- to 50-ml increments from the pump. CVP monitoring is beneficial through the early postoperative period to judge volume adjustments. Bain (1988) reported that the baseline CVP in a well-sedated patient on intermittent positive pressure ventilation is about 4–5 mm Hg higher postoperatively. Similar trends have been observed. CVP in these ranges are acceptable. It should be noted that a progressive sustained rise in CVP postoperatively is a hallmark for cardiac failure or peripheral vasoconstriction (↑ SVR) if colloids were not infused.

8. Left Atrial and Pulmonary Wedge Pressure Measurement

Left atrial (LA) pressure can be determined using a balloon-tipped Swan–Ganz catheter. The catheter is introduced into the femoral vein using the Seldinger technique (Gaymes *et al.*, 1995). An 18-gauge angiocath is introduced into the vein, the stylet is withdrawn and a guide wire is introduced through the angiocath. With the guide wire in place the angiocath is withdrawn over the guide wire. An appropriate size (8 French) catheter introducer is then inserted into the femoral vein over the guide wire and the guide wire is withdrawn. The introducer is sutured into place with 2-0 Prolene. The Swan–Ganz catheter is then advanced up the inferior vena cava through the right atrium, into the right ventricle, and out into the pulmonary artery, which measures pulmonary artery pressure (PAP); in the "wedged" position (balloon tip expanded) the pulmonary capillary wedge pressure (PWP) is measured and is an index of left atrial pressure. The Swan–Ganz catheter can also measure cardiac output by thermodilution measurement techniques. Normal left arterial pressure ranges from 5 to 15 mm Hg. Left ventricular function can be monitored. PWP is a useful index of myocardial function when "weaning off" or coming off CPB.

9. Termination of CPB or "Weaning off the Pump"

The core temperature should be stabilized at pre-CPB levels before warming is initiated. Generally, termination of CPB is uneventful. When CPB is terminated the heart will either function effectively or it will not function. Cardiac performance is the limiting factor in most instances. Hence it may be necessary to reinstitute CPB. Adequate cardiac output will not result until blood volume is restored in the pig. Usually, the authors slowly occlude the venous return line with a tubing clamp until the surgeon decides the right atrium and pulmonary artery are subjectively "back to normal." Arterial perfusion is stopped. Further volume expansion is administered in 25- to 50-ml increments to maintain a systolic pressure of 100 mm Hg ± 10. Conversely, overexpansion may put the heart into cardiac failure. Gradually titrating back toward pre-CPB pressures has worked best for our laboratory. Impatience can result in perilous premature termination of CPB. Subsequently, hearts that may have performed well with a little more recovery time may need to be treated with high doses of inotropes. This in turn can potentially delay mycardial recovery (Gravlee, 1991a). If cardiac output is inadequate, CPB should be resumed and the heart allowed a more reasonable recovery period. Continue at reduced CPB flows within normal tissue oxygenation limits as previously described (SvO_2 70 ± 5%). When the myocardium has rested for 5–10 min the "weaning process" is repeated. Once successful separation from CPB has been completed, the recovery phase begins.

During the recovery period, electrocardiographic, temperature, and arterial and venous pressure monitoring are continued. Arterial blood gas tensions, hematocrit, and serum electrolytes are obtained hourly or more frequently if clinical conditions dictate. Whole blood is administered to maintain the central venous pressure between 8 and 12 mm Hg. Once monitoring parameters indicate a return to homeostasis, the pigs are withdrawn from mechanical ventilation usually between 3 and 5 hr after completion of the surgical procedure, without the use of anticholinesterase agents to reverse neuromuscular blockade. Throughout this initial postoperative period a slow rewarming process via warming blanket and manipulation of environmental temperature is undertaken. Oxygen must be provided as required. Negative pressure is reestablished in the thoracic cavity through the use of a surgically implanted chest tube. Suction is applied either continuously or intermittently using a Heimlich valve, water trap, or similar device. The output of air or blood is recorded and replaced as necessary. Excessive bleeding through the chest tubes has not been a problem.

In the absence of complications, the pig is removed from the intensive care unit (ICU) cage the morning following surgery, at which time ambulation and oral alimentation are begun. The chest tubes are removed 24 hr after surgery.

This description of cardiopulmonary bypass has had routine success in the authors' laboratories; however, variations between laboratories may dictate changes in the described procedures.

For a more in-depth understanding of the complexities and physiological trespass of CPB, the reader is referred to texts by Gravlee (1993b), Reed and Stafford (1989a), and Taylor (1988).

10. Anesthetic Management

Anesthesia for CPB, for the most part, encompasses the traditional goals of anesthesia: adequate analgesia, hypnosis, and muscle relaxation. These can be met with minor changes in anesthetic techniques used for general surgery described elsewhere in this chapter. But the anesthetist must take into consideration the "physiologic trespass" one is inducing with CPB. Numerous investigators have addressed plasma levels of anesthetics pre-CPB, during, and post-CPB (Buylaert *et al.*, 1989; Holley *et al.*, 1982; Okutani *et al.*, 1988; Reves *et al.*, 1989). Plasma levels decrease proportionately to the volume of priming solution used. Gravlee (1991b) reports "while it might superficially appear that this would immediately lighten the anesthetic plane, the occurrence is potentially counterbalanced by an increase in the plasma ratio of free drug to protein-bound drug (enhancing drug available to cross the blood–brain barrier) and acute reduction in perfusion to some peripheral vascular beds (discouraging redistribution into peripheral compartments). In addition, drugs with relatively large apparent volumes of distribution (such as fentanyl) have redistributed before CPB, leaving a rather substantial drug "sink" in both central and peripheral compartments." Gravlee reports plasma levels of fentanyl rise after separating from CPB, probably because pulmonary reperfusion reestablishes plasma compartment access for a drug that has been sequestered in the lungs. This appears to hold true for other opioids and probably propofol. The synthetic opioids bind to the membrane circuit and tubing, hence reducing the anesthetic agent available.

It is important to recognize that the anesthetic regime will be quite different with hypothermic CPB versus normothermic CPB. Anesthetics, muscle relaxants, and vasoactive drug requirements are directly proportional to the core temperature of the patient (see Table IV).

On reviewing the veterinary and human anesthesia literature, one can be overwhelmed with the myriad of anesthetic protocols available. The bottom line is twofold: (1) the anesthetic regime must be consistent with the protocol's goal, and (2) the anesthetist/anesthesiologist must completely understand the agents selected and feel comfortable with their use.

Numerous studies have investigated the effect of anesthetics on the outcome of CPB. The common conclusion was that the choice of agents used—intravenous or inhalational—had no effect on the outcome of CPB (Slogoff and Keats, 1989; Tuman *et al.*, 1989). Taking into account that swine and humans appear to react pharmacokinetically similarly, one can interpolate if malignant hyperthermia is *not* a factor.

TABLE IV

CARDIAC INDEX AND HYPOTHERMIA TEMPERATURES FOR CARDIOPULMONARY BYPASS[a]

Temperature (°C)	Cardiac index
37	2.2–2.4
34–37	2–2.2
30–34	1.8–2
28–30	1.6–1.8
24–28	1.6–1.8
18–24	1.5–1.6

[a]Cardiac index (CI) × body surface area (BSA) = perfusion flow rate.

Techniques that have evolved at the authors' institution for swine with and without the malignant hyperthermia factor are described below.

11. Anesthesia for Normal (MH-Free) Pigs for CPB

Any of the anesthetic regimes described previously can be used. The inhalation agents are excellent selections alone or in combination with opioids, benzodiazepines, or propofol. In combination with a broad selection of muscle relaxants, swine anesthesia takes on all the benefits of "balanced anesthesia" described elsewhere in this chapter.

12. CPB Anesthesia for MH-Positive Pigs

Preoperatively the pigs are treated with a single oral dose of 5 mg/kg dantrolene the afternoon before surgery (Ehler *et al.*, 1985; Gronert *et al.*, 1976). This dose of dantrolene is repeated 2 hr before surgery.

Ketamine hydrochloride is administered at a dose of 30 mg/kg intramuscularly, after which a steel IV needle is inserted into an ear vein. Anesthetic induction is performed using 10 mg/kg sodium thiopental IV. The animals are intubated orotracheally with a cuffed endotracheal tube. Early maintenance of anesthesia consists of 100% oxygen, utilizing an oxygen-flushed anesthesia machine. (Note: Swine are subject to apnea with barbiturate induction. Ventilate immediately at 18–20 ml/kg at 20–25 cycles/min.)

With secure venous access in place, a "high-dose" opioid technique is initiated with fentanyl, 50–100 μg/kg/hr intravenously (Ehler *et al.*, 1985; Lunn *et al.*, 1979; Quintin *et al.*, 1981; Swindle *et al.*, 1986; Swindle, 1991). Pancuronium bromide (0.15 mg/kg) and dantrolene (4 mg/kg) are also administered intravenously at this time (Ehler *et al.*, 1985). Pancuronium bromide has been reported to have some protective effect over MH in MH-susceptible swine (Hall *et al.*, 1976; Short *et al.*, 1976). Sufentanil citrate (5.0–7.0 μg/kg IV) followed by 15–30 μg/kg/hr and vecuronium (0.1 mg/kg IV loading dose) followed by 30 mg/kg/hr is an alternative regime (Schumann *et al.*, 1994).

The authors select the high-dose narcotic anesthetic technique in an effort to avoid a stress response in the perioperative period. It is known that the use of fentanyl at a dose of 50 μg/kg can prevent the rise in cortisol, growth hormone, and glucose commonly associated with the stress response (Hall *et al.*, 1978).

During normothermic procedures, pigs have a propensity for fatal cardiac arrhythmias from cardiac manipulations. Bretylium (5 mg/kg IV) is an effective preventative agent when used in a two-dose regimen during thoracic surgery. It may be repeated by slow IV infusion every 30 min without affecting hemodynamics (Swindle, 1985; Swindle *et al.*, 1986; Swindle and Adams, 1988).

Depending on the procedure being performed, deep hypothermia and circulatory arrest may be used. Upon completion of the procedure, rewarming is begun. When the pig has been rewarmed to 34°C, ventilation with 100% oxygen is reinstituted, and the pig is slowly withdrawn from cardiopulmonary bypass. At this time additional pancuronium bromide (0.15 mg/kg) and dantrolene (1.0 mg/kg) are administered by slow intravenous infusion. It is not necessary to use temporary pacing in any of the experimental animals. Oral dantrolene (3 mg/kg q 8 hr) is continued for three doses. If deep hypothermia is not used, cold cardioplegia as previously described in this chapter is used for myocardial protection.

C. Pregnant Animals/Fetal Surgery

When using swine in fetal surgical experiments, special considerations are required. When placing the animal in dorsal recumbency, compression of the aorta by the gravid uterus may lead to vascular compromise of the uterus and distal extremities. Consequently, the position of the uterus should be noted. As in fetal surgery in other species, maintenance of a high environmental temperature is necessary to prevent fetal distress. The sow may be kept warm by using circulating water blankets both over and under the body or by using an airway heater. The room temperature should be kept above 28°C. The sow has a diffuse, epitheliochorial type of placentation and most anesthetics would be capable of crossing the placental barrier. The authors have successfully used isoflurane in nitrous oxide:oxygen (2:1) as the general anesthetic agent after induction with 33 mg/kg ketamine IM and 1.1 mg/kg acepromazine. Monitoring of arterial blood gas values and mean arterial blood pressure is recommended to prevent hypoxia and hypotension (Dyess *et al.*, 1994; Swindle *et al.*, 1996).

APPENDIX

Drug doses, routes of administration, and duration are presented in Table A.I. Tables A.II–A.IV include drug protocols for various types of surgery.

TABLE A.I

DRUG ADMINISTRATION FOR SWINE[a]

Drug name (generic)	Dose (mg/kg)	Route of administration	Approximate duration	Reference
Acepromazine	0.11–1.1	IM, IV, SC	6–8 hr	Riebold and Thurmon (1986), Benson and Thurmon (1979), Swindle (1983)
Promazine	0.44–2.0	IM, IV, SC	6–8 hr	Bolin and Runnels (1992), Riebold and Thurmon (1986), Benson and Thurmon (1979)
Chlorpromazine	0.5–4.0	IM, IV, SC	6–8 hr	Bolin and Runnels (1992), Riebold and Thurmon (1986), Benson and Thurmon (1979), Swindle (1983)
Diazepam	0.5–10	IM	2–4 hr	Benson and Thurmon (1979)
	0.44–2.0	IV	2 hr	Thurmon and Tranquilli (1986), Bolin and Runnels (1992), Schranz *et al.* (1988)
Midazolam	0.1–0.5	IM, IV	20 min	Oldhafer *et al.* (1993); Smith *et al.* (1991)
Brotizolam	1–10	PO		Danneberg *et al.* (1986)
Lorazepam	0.1	IV		Pender *et al.* (1991)
Climazolam	2.0	IV		Becker (1986)
Flurazepam	2.0	IV, PO		Ochs *et al.* (1987)
Azaperone	2.0–8.0	IM		Thurmon and Tranquilli (1986), Portier and Slusser (1985), Riebold and Thurmon (1986)
Xylazine	0.2	IM	5 min	Benson and Thurmon (1979), Thurmon and Tranquilli (1986), Riebold and Thurmon (1986), Swindle (1991)
Etomidate	4.0–8.0	IV		Holzchuh and Cremonesi (1991)
Metomidate	4.0	IV		Svendsen and Carter (1989)
Atropine	0.05	IM		Riebold *et al.* (1995)
	0.02	IV		Swindle (1983)
Glycopyrrolate	0.004–0.01	IM	30 min	Benson and Thurmon (1979), Riebold *et al.* (1995), Swindle (1983)
Ketamine	11–33.0	IM	30 min	Benson and Thurmon (1979), Bolin and Runnels (1992), Short (1987), Swindle (1983), Tranquilli *et al.* (1982), Worek *et al.* (1988)
Ketamine +	33.0	IM		
acepromazine	1.1	IM	30 min	Riebold *et al.* (1995), Swindle (1983), Benson and Thurmon (1979)
Ketamine +	15.0			
diazepam	2.0	IM	20 min	Riebold and Thurmon (1986)
Ketamine +	10.0	IV		
medetomidine	0.2	IV	30 min	Vainio *et al.* (1992)
Ketamine +	15.0			
azaperone	2.0	IM	20 min	Riebold and Thurmon (1986)
Ketamine +	33.0			
midazolam	0.5	IM	45 min	M. M. Swindle (oral communication)
Ketamine +	20.0			
xylazine	2.0	IM	20 min	Kyle *et al.* (1979), Cantor *et al.* (1981), Trim and Gilroy (1985)
Ketamine +	2.0			
xylazine +	2.0			
oxymorphone	0.075	IV, IM (2× dose)	20 min	Breese and Dodman (1984)
Ketamine +	20.0			
climazolam	0.5–1.0	IM	20 min	Becker (1986)
Ketamine +	22.0			
meperidine +	2.2			
azaperone +	2.2			
morphine	2.2	IM	1 hr	Hoyt *et al.* (1986)
Tiletamine +				
zolazepam (telazol)	2.0–8.8	IM	20 min	Bolin and Runnels (1992), Swindle (1991)
Tiletamine +				
zolazepam +	2.0–8.8			Ko *et al.* (1993)
xylazine	2.2	IM	20 min	Swindle (1991)

TABLE A.I

DRUG ADMINISTRATION FOR SWINE[a] *(continued)*

Drug name (generic)	Dose (mg/kg)	Route of administration	Approximate duration	Reference
Ketamine +	2.2	IM	20–30 min	
Telazol +	4.4	IM	20–30 min	Ko *et al.* (1993)
Ketamine +	2.2	IM		
telazol +	4.4	IM	20–30 min	
xylazine	2.2	IM	20–30 min	Ko *et al.* (1993)
Thiopental	6.6–25.0	IV	10–20 min	Bolin and Runnels (1992), Riebold and Thurmon (1986), Thurmon and Tranquilli (1986), Swindle (1991)
Thiamylal	6.6–25.0	IV	10–20 min	Bolin and Runnels (1992), Riebold and Thurmon (1986), Thurmon and Tranquilli (1986), Swindle *et al.* (1988), Swindle (1991)
Pentobarbital	20–40	IV	20–30 min	Bolin and Runnels (1992), Riebold and Thurmon (1986), Thurmon and Tranquilli (1986), Swindle (1991), Swindle *et al.* (1988)
Fentanyl +	1 ml/13.5 kg	IM	20 min	Benson and Thurmon (1979), Thurmon and Tranquilli (1986)
droperidol	0.25–0.5 ml/kg	IV		
Ketamine +	11.0			
fentanyl–droperidol	1 ml/13.5 kg		30 min	Benson and Thurmon (1979), Cantor *et al.* (1981), Riebold and Thurmon (1986)
Naloxone	0.5–2.0	IV		Benson and Thurmon (1979), Nishijima *et al.* (1988), Trudeau *et al.* (1988)
Alphaxolone +	6.0–8.0	IM		
alphadolone	2.0–3.0	IV	4–10 min	Glen *et al.* (1979), Bolin and Runnels (1992), Flecknell (1987)
Etorphine +				
acepromazine	0.025	IM	20 min	Glen *et al.* (1979), Bolin and Runnels (1992), Flecknell (1987)
Etomidate +	4.0–8.0	IV		
azaperone	2.0	IM		Holzchuh and Cremonesi (1991)
Propofol	0.83–1.66	IV		Foster *et al.* (1992), Raff and Harrison (1989), Ramsey *et al.* (1993)
Metomidate +	4.0–15.0	IV		
azaperone	2.0	IM	15–30 min	Thurmon and Tranquilli (1986), Svendsen and Carter (1989)
Yohimbine	1.0	IV		Armstead *et al.* (1988)
Fentanyl	0.05	IM	2 hr	Flecknell (1984), Flecknell (1987), Swindle (1991), Blum (1988)
Sufentanil	0.005–0.01	IM	2 hr	Flecknell (1984), Flecknell (1987), Swindle (1991)
Meperidine	2.0–10.0	IM	4 hr	Flecknell (1984), Flecknell (1987), Swindle (1991), Blum (1988)
Oxymorphone	0.15	IM	4 hr	Flecknell (1984), Flecknell (1987), Swindle (1991), Blum (1988)
Pentazocine	1.5–3.0	IM	4 hr	Flecknell (1984), Flecknell (1987), Swindle (1991), Blum (1988)
Butorphanol	0.1–0.3	IM	4–6 hr	Flecknell (1984), Swindle (1991), Flecknell (1987)
Buprenorphine	0.05–0.1	IM	8–12 hr	Hernansen *et al.* (1986), Swindle (1991)
Phenylbutazone	10–20	PO	12 hr	Swindle (1991), Swindle *et al.* (1988)
Aspirin	10	PO	4–6 hr	Swindle (1991), Swindle *et al.* (1988)
Bretyllium	3.0–5.0	IV	30 min	Swindle *et al.* (1986)

[a]Does not include indications and contraindications of the agents administered. It also excludes the use of continuous IV infusions of parentenal agents. Please refer to the text for details.

TABLE A.II

PROTOCOL FOR NONSURVIVAL TECHNIQUES USED FOR TEACHING LABORATORIES

Preanesthesia	Induction	Maintenance
Atropine (0.5 mg/kg IM) Ketamine (33.0 mg/kg IM) Acepromazine (1.1 mg/kg IM)	Pentobarbital (20–40 mg/kg IV) or thiopental (6.6–25.0 mg/kg IV)	Pentobarbital (5–15 mg/kg/hr IV) or thiopental (3.0–6.0 mg/kg/hr IV)

TABLE A.III

PROTOCOL FOR ROUTINE SURGERY WITHOUT PHYSIOLOGIC MEASUREMENTS

Preanesthesia	Induction	Maintenance
Atropine (0.05 mg/kg IM) Ketamine (33.0 mg/kg IM) Acepromazine (1.1 mg/kg IM)	Isoflurane via face mask (4–5%)	Isoflurane (0.5–2.0%); deliver in oxygen: nitrous oxide (1:2, v/v)

TABLE A.IV

OPIOID INFUSION PROTOCOL FOR CARDIOPULMONARY BYPASS WITH CARDIAC COMPROMISE

Preanesthesia	Induction	Maintenance
Fentanyl (30–50 μg/kg IV) or sufentanyl (7–15 μg/kg IV)	Fentanyl (50–100 μg/kg/hr IV) or sufentanyl (10–30 μg/kg/hr IV)	Supplement as required with 0.25–0.5% isoflurane

ACKNOWLEDGMENTS

The authors thank William D. McCoy, P.A., C.C.P., Rebecca L. Ochkie, P.A., C.C.P., and Robert M. Sharrow, P.A., C.C.P., for technical information and Ms. Rachel Montez, Ms. Velma Grantham, and Ms. Debra Allston for the preparation of this manuscript.

REFERENCES

Armstead, W. M., Leffler, C. W., Busija, D. W., Beasley, D. G., and Mirro, R. (1988). Adrenergic and prostanoid mechanisms in control of cerebral blood flow in hypotensive newborn pigs, *Am. J. Physiol.* **254** (Heart Circ. Physiol. 23), H671–H677.

Bain, W. H. (1988). Measurement and monitoring for cardiopulmonary bypass. *In* ''Cardiopulmonary Bypass'' (K. M. Taylor, ed.), pp. 13–28. Chapman & Hall, London.

Becker, M. (1986). Anesthesia in Gottingen miniature swine used for experimental surgery. *Lab. Anim. Sci.* **36**(4), 417–419.

Benharkate, M., Zanini, V., Blanc, R., Boucheix, O., Coyez, F., Genevois, J. P., and Pairet, M. (1993). Hemodynamic parameters of anesthetized pigs: A comparative study of farm piglets and Gottingen and Yucatan miniature swine. *Lab. Anim. Sci.* **43**(1), 68–72.

Benson, G. J., and Thurmon, J. C. (1979). Anesthesia of swine under field conditions. *J. Am. Vet. Med. Assoc.* **174,** 594–596.

Blum, J. R. (1988). Laboratory animal anesthesia. *In* ''Experimental Surgery and Physiology: Induced Animal Models of Human Disease'' (M. M. Swindle and R. J. Adams, eds.), pp. 329–345. Williams & Wilkins, Baltimore, Maryland.

Bobbie, D. L., and Swindle, M. M. (1986). Pulse monitoring, intravascular and intramuscular injection sites in pigs. *In* ''Swine in Biomedical Research'' (M. E. Tumbleson, ed.), Vol. 1, pp. 273–277. Plenum, New York.

Bolin, S. R., and Runnels, L. J. (1992). Chemical restraint and anesthesia. *In* ''Diseases of Swine'' (A. D. Leman, B. Straw, R. D. Glock, W. L. Mengeling, R. H. C. Penny, E. Scholl, eds.), 7th ed., pp. 933–942. Iowa State Univ. Press, Ames.

Braimbridge, M. V. (1988). Myocardial protection: Hypothermia and cardioplegia. *In* ''Cardiopulmonary Bypass'' (K. M. Taylor, ed.), pp. 375–392. Chapman & Hall, London.

Breese, C. E., and Dodman, N. H. (1984). Xylazine-ketamine-oxymorphone: An injectable anesthetic combination in swine. *J. Am. Vet. Med. Assoc.* **184,** 182–183.

Britt, B. A., Kalow, W., and Endrenyl, L. (1978). Malignant hyperthermia pattern of inheritance in swine. *In* ''Second International Symposium on Malignant Hyperthermia'' (J. A. Aldrete and B. A. Britt, eds.), pp. 195–211. Grune & Stratton, New York.

Buckley, N. M., Gootman, P. M., Brazeau, P., Matanic, B. P., Frasier, I. D., and Gentles, E. L. (1979). Cardiovascular function in anesthetized miniature swine. *Lab. Anim. Sci.* **29,** 200–208.

Buylaert, W. A., Herregods, L. L., Mortier, E. P., and Bogaert, M. G. (1989). Cardiopulmonary bypass and the pharmacokinetics of drugs: An update. *Clin. Pharmacokinet.* **17,** 10–26.

Cantor, G. H., Brunson, D. B., and Reibold, T. W. (1981). A comparison of four short-acting anesthetic combinations for swine. *VM/SMA, Vet. Med./Small Anim. Clin.,* pp. 715–720.

Cheng, D., Mayers, J., Knutson, R., Gomez, M., and Tinker, J. (1992). Dose-response relationship of isoflurane and halothane versus coronary perfusion pressures. *Anesthesiology* **76,** 113–122.

Crane, L., Gootman, N., and Gootman, P. M. (1975). Age-dependent cardiovascular effects of halothane anesthesia in neonatal pigs, *Arch. Int. Pharmacodyn. Ther.* **214,** 180–187.

Danneberg, P., Bauer, R., Boke-Kuhn, K., Hoefke, W., Kuhn, F., Lehr, E., and Walland, A. (1986). ''General Pharmacology of Brotizolam in Animals.'' Arzneim.–Forsch/Drug Res. 36(1), 540–551.

Dyess, D. L., Tacchi, E., Powell, R. Q., Ardell, J. L., Roberts, W. S., and Ferrara, J. J. (1994). Development of a protocol to provide prolonged general anesthesia to pregnant sows. *J. Invest. Surg.* **7**(3), 235–242.

Eger, E., Johnson, B., Weiskopf, R., Holmes, M., Yasuda, N., Targ, A., and Rampil, J. (1988). Minimum alveolar concentration of I-653 and isoflurane in pigs: Definition of a supramaximal stimulus. *Anesth. Analg.* **67,**1174–1176.

Ehler, W. J., Mack, J. W., Jr., Brown, D. L., and Davis, R. F. (1985). Avoidance of malignant hyperthermia in a porcine model for experimental open heart surgery. *Lab. Anim. Sci.* **35,** 172–175.

Eisele, P. H., Talken, L., and Eisele, J. H. (1985). Potency of isoflurane and nitrous oxide in conventional swine. *Lab. Anim. Sci.* **35,** 76–78.

Eisele, P. H., Woodle, E. S., Hunter, G. C., Talken, L., and Ward, R. E. (1986). Anesthetic, preoperative and postoperative considerations for liver transplantation in swine. *Lab. Anim. Sci.* **36,** 402–405.

Endo, M. (1977). Calcium release from the sarcoplasmic reticulum. *Physiol. Rev.* **57,** 71–108.

Flecknell, P. A. (1984). The relief of pain in laboratory animals. *Lab. Anim.* **18,** 147–160.

Flecknell, P. A. (1987). ''Laboratory Animal Anaesthesia.'' Academic Press, Orlando, Florida.

Flewellen, E. H., and Nelson, T. E. (1982). Masseter spasm induced succinylcholine in children: Contracture testing for malignant hyperthermia: Report of six cases. *Can. J. Anaesth.* **29,** 42–48.

Foster, P. S., Hopkinson, K. C., and Denborough, M. A. (1992). Propofol anaesthesia in malignant hyperpyrexia susceptible swine. *Clin. Exp. Pharmacol. Physiol.* **19,** 183–186.

Fuji, J., Ostiu, K., Zorzato, F., deLeon, S., Khanna, V. K., Weiler, J. E., O'Brien, P. J., and MacLennan, D. H. (1991). Identification of a mutation in porcine ryanodine receptor associated with malignant hyperthermia. *Science* **253,** 448–451.

Gaymes, C. H., Swindle, M. M., Gillette, P. C., Harold, M. E., and Schumann, R. E. (1995). Percutaneous serial catheterization in swine: A practical approach. *J. Invest. Surg.* **8**(2), 123–128.

Geers, R., Decanniere, C., Ville, H., Van Hecke, P., Goedseels, V., Bosschaerts, L., Deley, J., Janssens, S., and Nierynck, W. (1992). Identification of halothane gene carriers by use of in vivo 31P nuclear magnetic resonance spectroscopy in pigs. *Am. J. Vet. Res.* **53**(9), 1711–1714.

Gelman, S., Dillard, E., and Bradley, E. L. (1987). Hepatic circulation during surgical stress and anesthesia with halothane, isoflurane, or fentanyl. *Anesth. Analg.* **66,** 936–943.

Gilbert, M., Roberts, S., Mori, M., Blomberg, R., and Tinker, J. (1988). Comparative coronary vascular reactivity and hemodynamics during halothane and isoflurane anesthesia in swine. *Anesthesiology* **68,** 243–253.

Gilbert, M., Mori, M., and Myhre, E. (1989). Hemodynamic dose-responses to halothane and isoflurane are different in swine with and without critical coronary artery stenosis. *Anesth. Analg.* **68,** 752–758.

Glen, J. B., Davies, G. E., Thomson, D. S., Scarth, S. C., and Thompson, A. V. (1979). An animal model for the investigation of adverse responses to IV anesthetic agents and their solvents. *Br. J. Anaesth.* **51,** 819–827.

Gootman, N., Gootman, P. M., Buckley, B. J., and Buckley, N. M. (1986). Cardiovascular effects of catecholamine infusions in developing swine. *In* ''Swine in Biomedical Research'' (M. E. Tumbleson, ed.), Vol. 3, pp. 1615–1622. Plenum, New York.

Gravlee, G. P. (1991a). Management of cardiopulmonary bypass. *Am. Soc. Anesthesiol. 1991 Ann. Refresher Course Lec.* San Francisco, No. 176, pp. 1–7.

Gravlee, G. P. (1991b). Heparin monitoring for cardiopulmonary bypass. Does it really matter? *Anesthesiology* **5,** 2.

Gravlee, G. P. (1993a). The anesthesiologist and management of cardiopulmonary bypass. *Am. Soc. Anesthesiol. 1993 Ann. Refresher Course Lec.* Washington, D.C., No. 126, pp. 1–7.

Gravlee, G. P. (1993b). Cardiopulmonary bypass circuitry and cannulation techniques. *In* ''Cardiopulmonary Bypass Principles and Practice'' (G. P. Gravlee, D. D. Davis, and D. D. Utley, eds.), Williams & Wilkins, Baltimore, Maryland.

Gronert, G. A. (1980). Malignant hyperthermia. *Anesthesiology* 53(5), 395–423.

Gronert, G. A., Milde, J. A., and Theye, P. A. (1976). Dantrolene in porcine malignant hyperthermia. *Anesthesiology* **44,** 488–495.

Hall, G. M., Lucke, J. N., and Lister, D. (1976). Porcine malignant hyperthermia IV: Neuromuscular blockade. *Br. J. Anaesth.* **48,** 1135–1141.

Hall, G. M., Young, C., Holdcroft, A., and Alaghband-Zadeh, J. (1978). Substrate mobilization during surgery. A comparison between halothane and fentanyl anesthesia. *Anesthesiology* **33,** 924–931.

Halsey, M. J., Sawyer, D. C., Eger, E. I., Bahlman, S. H., and Impelman, D. M. K. (1971). Hepatic metabolism of halothane, methoxyflurane, cyclopropane, ethrane, and forane in miniature swine. *Anesthesiology* **35,** 43–47.

Harter, M., and Wilsdorf, G. (1914). Die bedeutung des schweines fur die fleischvrsorgung. *Berl., Arb. Dtsch. Landwirtschaftgesellsch.,* No. 270.

Hermansen, K., Pedersen, L. E., and Olesen, H. O. (1986). The analgesic effect of buprenorphine, etorphine and pethidine in the pig: A randomized double blind cross-over study. *Acta Pharmacol. Toxicol.* **59**(1), 27–35.

Holley, F. O., Ponganis, K. V., and Stanski, D. R. (1982). Effect of cardiopulmonary bypass in the pharmacokinetics of drugs. *Clin. Pharmacokinet.* **7,** 234–251.

Holmes, M., Weiskopf, R., Eger, E., Johnson, B., and Rampil, I. (1990). Hepatocellular integrity in swine after prolonged desflurane (I-653) and isoflurane anesthesia: Evaluation of plasma alanine aminotransferase activity. *Anesth. Analg.* **71,** 249–253.

Holzchuh, M. P., and Cremonesi, E. (1991). Anaesthesia in pigs. Analysis of azaperone and etomidate effects separately and in associations. *Proc. Int. Cong. Vet. Anaesth., 4th,* Utrecht, The Netherlands, *J. Vet. Anesth. Spec. Suppl.,* pp. 197–200.

Houde, A., Pommier, S. A., and Roy, R. (1993). Detection of the ryanodine receptor mutation associated with malignant hyperthermia in purebred swine populations. *J. Anim. Sci.* **71**(6), 1414–1418.

Houpt, T. R. (1986). The handling of swine in research. *In* ''Swine in Cardiovascular Research'' (H. C. Stanton and H. J. Mersmann, eds.), Vol. 1, pp. 25–38. CRC Press, Boca Raton, Florida.

Hoyt, R. F., Jr., Hayre, M. D., Dodd, K. T., and Phillips, Y. Y. (1986). Long acting intramuscular anesthetic regimen for swine. *Lab. Anim. Sci.* **36**(4), 413–416.

Hudson, L. C., and Gilroy, B. A. (1986). Precutaneous transtracheal ventilation in swine. *Lab. Anim. Sci.* **36**(4), 420–424.

Kaplan, R. F. (1991). Malignant hyperthermia. *Am. Soc. Anesthesiol., 1991 Annu. Refresher Course Lec.,* San Francisco, No. 231, pp. 1–7.

Ko, J. C. H., Williams, B. L., Smith, V. L., McGrath, C. J., and Jacobson, J. D. (1993). Comparison of Telazol, Telazol-Ketamine, Telazol-Xylazine and Telazol-Ketamine-Xylazine as chemical restraint and anesthetic induction combination in swine. *Lab. Anim. Sci.* **43**(5), 476–480.

Koblin, P., Weiskopf, R., Holmes, M., Konopka, K., Rampil, I., Eger, E., and Waskell, L. (1989). Metabolism of I-653 and isoflurane in swine. *Anesth. Analg.* **68,** 147–149.

Kreimeier, U., Brueckner, U. B., Gerspach, S., Veitinger, K., and Messmer, K. (1993). A porcine model of hyperdynamic endotoxinemia: Patterns of respiratory, macrocirculatory, and regional blood flow changes. *J. Invest. Surg.* **6**(2), 143–156.

Kyle, O. C., Novak, S., and Bolooki, H. (1979). General anesthesia in pigs. *Lab. Anim. Sci.* **29,** 123–124.

Littleford, J. A., Patel, L. R., Bose, D., Cameron, C. B., and McKillop, C. (1991). Masseter muscle spasm in children: implications of continuing the triggering anesthetic. *Anesth. Analg.* **72,** 151–160.

Lopez, J. R., Allen, P. D., Jones, D., and Streter, F. A. (1988). Myoplasmic free [Ca++] during a malignant hyperthermia episode in swine. *Muscle Nerve* **11,** 82–88.

Loscher, W., Ganter, M., and Fassbender, C. P. (1990). Correlation between drug and metabolite concentrations in plasma and anesthetic action of ketamine in swine. *Am. J. Vet. Res.* **51,** 391–398.

Lundeen, G., Manohar, M., and Parks, C. (1983). Systemic distribution of blood flow in swine while awake and during 1.0 and 1.5 MAC isoflurane anesthesia with or without 50% nitrous oxide. *Anesth. Analg.* **62,** 499–512.

Lunn, J. K., Stanley, T. M., Eisele, J., Webster, L., and Woodward, A. (1979). High dose fentanyl anesthesia for coronary surgery: Plasma fentanyl concentrations and influence of nitrous oxide on cardiovascular responses. *Anesth. Analg.* **58,** 390–405.

Manohar, M., and Parks, C. (1984a). Porcine regional brain and myocardial blood flows during halothane-O_2 and halothane-nitrous oxide anesthesia: Comparisons with equipotent isoflurane anesthesia. *Am. J. Vet. Res.* **45**(3), 465–473.

Manohar, M., and Parks, C. (1984b). Regional distribution of brain and myocardial perfusion in swine while awake and during 1.0 and 1.5 MAC isoflurane anaesthesia produced without or with 50% nitrous oxide. *Cardiovasc. Res.* **18,** 344–353.

Merin, R. G., Verdouw, P. D., and Jong, J. W. (1977). Dose-dependent depression of cardiac function and metabolism by halothane in swine (*Sus scrofa*). *Anesthesiology* **46,** 417–423.

Merin, R. G., Verdouw, P. D., and Jong, J. W. (1982). Myocardial functional and metabolic responses to ischemia in swine during halothane and fentanyl anesthesia. *Anesthesiology* **56,** 84–92.

Mickelson, J. R., Gallant, E. M., Litterer, L. A., Johnson, K. M., Rempel, W. E., and Louis, C. F. (1988). Abnormal sarcoplasmic reticulum ryanodine receptor in malignant hyperthermia. *J. Biol. Chem.* **263,** 9310–9315.

Moffitt, E. A., Kirklin, J. W., and Theye, R. A. (1962). Physiologic studies during whole body perfusion in Tetralogy of Fallot. *J. Thorac. Cardiovasc. Surg.* **44,** 180.

Nishijima, M. K., Breslow, M. J., Miller, C. F., and Traystman, R. J. (1988). Effect of naloxone and ibuprofen on organ blood flow during endotoxic shock in pig. *Am. J. Phys.* 255 (1 Pt 2): H177–H184.

Ochs, H. R., Greenblatt, D. J., Eichelkraut, W., Bakker, C., Gobel, R., and Hahn, N. (1987). Hepatic vs. gastrointestinal presystemic extraction of oral midazolam and flurazepam. *J. Pharmacol. Exp. Ther.,* **243**(3), 852–856.

Okutani, R., Philbin, D. M., Rosow, C. E., Koski, G., and Schneider, R. C. (1988). Effect of hypothermic hemodilutional cardiopulmonary bypass on plasma sufentanil and catecholamine concentrations in humans. *Anesth. Analg.* **67,** 667–670.

Oldhafer, K. J., Hauss, J., Gubernatis, G., Pichlmayr, R., and Spiegel, H. U. (1993). Liver transplantation in pigs: A model for studying reperfusion injury. *J. Invest. Surg.* **6**(5), 439–450.

Panepinto, L. M., Phillips, R. W., Norden, S. W., Pryor, P. C., and Cox, R. (1983). A comfortable minimum stress method of restraint for Yucatan miniature swine. *Lab. Anim. Sci.* **33,** 95.

Pender, E. S., Pollack, C. V., Woodall, B. N., and Parks, B. R. (1991). Intraosseous administration of lorazepam: Same-dose comparison with intravenous administration in the weanling pig. *J. MSMA,* pp. 365–368.

Portier, D. B., and Slusser, C. A. (1985). Azaperone: A review of a new neuroleptic agent for swine. *Vet. Med.* **80**(3), 88–92.

Priebe, H. J. (1989). Isoflurane and coronary hemodynamics. *Anesthesiology* **71,** 960–976.

Putney, J. W., Jr. (1979). Stimulus-permeability coupling: Role of calcium in the receptor regulation of membrane permeability. *Pharmacol. Rev.* **30,** 209–245.

Quintin, L., Whalley, D. G., Wynands, J. E., Morin, J. E., and Burke, J. (1981). High dose fentanyl anesthesia with oxygen for aorta-coronary bypass surgery. *Can. Anaesth. Soc. J.* **28,** 314–320.

Raff, M., and Harrison, G. (1989). The screening of propofol in MHS swine. *Anesth. Analg.* **68,** 750–751.

Rampil, I., Weiskopf, R., Brown, J., Eger, E., Johnson, B., Holmes, M., and Donegan, J. (1988). I653 and isoflurane produce similar dose-related changes in the electroencephalogram of pigs. *Anesthesiology* **69,** 298–302.

Ramsey, D. E., Aldred, N., and Power, J. M. (1993). A simplified approach to the anesthesia of porcine laparoscopic surgical subjects. *Lab. Anim. Sci.* **43**(4), 336–337.

Reed, C. C., and Stafford, T. B. (1989). Myocardial protection. *In* ''Cardiopulmonary Bypass,'' 2nd ed., pp. 415–426. Surgimedics/TMP, The Woodlands, Texas.

Reves, J. G., Croughwell, N., Jacobs, J. R., and Greeley, W. (1989). Anesthesia during cardiopulmonary bypass: Does it matter? *In* ''Cardiopulmonary Bypass: Current Concepts and Controversies'' (J. H. Tinker, ed.), pp. 69–98. Saunders, Philadelphia.

Riebold, T. W., and Thurmon, J. C. (1986). Anesthesia in swine. *In* ''Swine in Biomedical Research'' (M. E. Tumbleson, ed.), Vol. 1, pp. 243–254. Plenum, New York.

Riebold, T. W., Goble, D. O., and Geiser, D. R. (1995). ''Large Animal Anesthesia: Principles and Techniques,'' 2nd ed. Iowa State Univ. Press, Ames.

Riou, B., Rimailho, A., Galliot, M., Bourdon, R., and Huet, Y. (1988). Protective cardiovascular effects of diazepam in experimental acute chloroquine poisoning. *Intensive Care Med.* **4,** 610–616.

Rittenhouse, E. A., Mohri, H., Dillard, D. H., and Merendino, K. A. (1974). Deep hypothermia in cardiovascular surgery. *Ann. Thorac. Surg.* **17,** 63–98.

Roberts, S., Gilbert, M., and Tinker, J. (1987). Isoflurane has a greater margin of safety in swine with and without major surgery or critical coronary stenosis. *Anesth. Analg.* **66,** 485–491.

Rogers, G. P., Cromeens, D. M., Minor, S. T., and Swindle, M. M. (1988). Bretylium and diltiazem in porcine cardiac procedures. *J. Invest. Surg.* **1**(4), 321–326.

Rosenberg, H. (1988). Management of patients in whom trismus occurs following succinylcholine. *Anesthesiology* **68,** 653–654.

Schieber, R. A., Namnoum, A., Sugden, A., Shiu, G., Orr, R., and Cook, D. R. (1986). Hemodynamic effects of isoflurane in the newborn piglet: comparison with halothane. *Anesth. Analg.* **65,** 633–638.

Schneider, R., and Mitchell, D. (1976). Dantrolene hepatitis. *J. Am. Med. Assoc.* **235,** 1590–1601.

Schranz, D., Huth, R. G., Stopfkuchen, H., and Jungst, B. K. (1988). The effect of nifedipine alone or combined with low dose acetylsalicylic acid on endotoxin-induced pulmonary hypertension in the piglet. *Intensive Care Med.* **14,** 595–601.

Schumann, R. E., Harold, M., Gillette, P. C., Swindle, M. M., and Gaymes, C. H. (1993). Prophylactic treatment of swine with bretylium for experimental cardiac catheterization. *Lab. Anim. Sci.* **43**(3), 244–246.

Schumann, R. E., Swindle, M. M., Knick, B. J., Case, C. L., and Gillette, P. C. (1994). High dose narcotic anesthesia using sufentanil in swine for cardiac catheterization and electrophysiologic studies. *J. Invest. Surg.* **7**(3), 243–248.

Short, C. E., ed. (1987). ''Principles and Practice of Veterinary Anesthesia.'' Williams & Wilkins, Baltimore, Maryland.

Short, C. E., Paddleford, R. R., McGrath, C. J., and Williams, C. H. (1976). Preanesthetic evaluation and management of malignant hyperthermia in the pig experimental model. *Anesth. Analg.* **55,** 643–653.

Silverman, J., and Muir, W. W., III (1993). A review of lab animal anesthesia with chloral hydrate and chloralose. *Lab. Anim. Sci.* **43**(3), 210–216.

Slogoff, S., and Keats, A. S. (1989). Randomized trial of primary anesthetic agents on outcome of coronary artery bypass operations. *Anesthesiology* **70,** 179–188.

Smith, A. C., and Swindle, M. M., eds. (1994). ''Research Animal Anesthesia, Analgesia and Surgery.'' Scientists Center for Animal Welfare, Bethesda, Maryland.

Smith, A. C., Spinale, F. G., Carabello, B. A., and Swindle, M. M. (1989). Technical aspects of cardiac catheterization of swine. *J. Invest. Surg.* **2**(2), 187–194.

Smith, A. C., Spinale, F. G., and Swindle, M. M. (1990). Cardiac function and morphology of Hanford miniature swine and Yucatan miniature and micro swine. *Lab. Anim. Sci.* **40**(1), 47–50.

Smith, A. C., Zellner, J. L., Spinale, F. G., and Swindle, M. M. (1991). Sedative and cardiovascular effects of midazolam in swine. *Lab. Anim. Sci.* **41**(2), 157–161.

Steffey, E. P., and Eger, E. I. (1985). Nitrous oxide in veterinary practice and animal research. *In* ''Nitrous Oxide,'' (E. I. Eger, ed.) pp. 305–312. Elsevier, New York.

Svendsen, P., and Carter, A. M. (1989). Blood gas tensions, acid-base status and cardiovascular function in miniature swine anaesthetized with halothane and methoxyflurane or intravenous metomidate hydrochloride. *Pharmacol. Toxicol.* **64,** 88–93.

Swindle, M. M. (1985). ''Anesthesia in Swine,'' Charles River Technical Bulletin No. 3, p. 3. Charles River Laboratories, Wilmington, Massachusetts.

Swindle, M. M., ed. (1986). Surgery. *In* ''Swine in Biomedical Research'' (M. E. Tumbleson, ed.), pp. 233–433. Plenum, New York.

Swindle, M. M. (1991). ''Anesthetic and Perioperative Techniques in Swine,'' Charles River Technical Bulletin, Winter 1991. Charles River Laboratories, Wilmington, Massachusetts.

Swindle, M. M., and Adams, R. J. (1988). ''Experimental Surgery and Physiology: Induced Animal Models of Human Disease.'' Williams & Wilkins, Baltimore, Maryland.

Swindle, M. M. (1983). ''Basic Surgical Exercises Using Swine.'' Praeger, New York.

Swindle, M. M., Horneffer, P. J., Gardner, T. J., Gott, V. L., Hall, T. S., Stuart, R. S., Baumgartner, W. A., Borkon, A. M., Galloway, E., and Reitz, B. A. (1986). Anatomic and anesthetic considerations in experimental cardiopulmonary surgery in swine. *Lab. Anim. Sci.* **36**(4), 357–361.

Swindle, M. M., Smith, A. C., and Hepburn, B. J. S. (1988). Swine as models in experimental surgery. *J. Invest. Surg.* **1**(1), 65–79.

Swindle, M. M., Schumann, R. E., Johnson, W. S., Knick, B., Smith, A. C., Carabello, B. A. Zile, M., Case, C. L., and Gillette, P. C. (1991). High dose narcotic anesthesia using sufentanyl in swine and dogs. *Proc. Int. Cong. Vet. Anesth., 4th,* Utrecht, The Netherlands, *J. Vet. Anesth. Spec. Suppl.,* pp. 273–276.

Swindle, M. M., Wiest, D. B., Smith, A. C., Garner, S. S., Case, C. C., Thompson, R. P., Fyfe, D. A., and Gillette, P. C. (1996). Fetal surgical protocols in yucatan miniature swine. *Lab. Anim. Sci.* **46**(1), 90–95.

Taylor, K. M. (1988). ''Cardiopulmonary Bypass.'' Chapman and Hall, London.

Thurmon, J. C., and Benson, G. J. (1987). Special anesthesia considerations of swine. *In* ''Principles and Practice of Veterinary Anesthesia'' (C. E. Short, ed.), pp. 308–322. Williams & Wilkins, Baltimore, Maryland.

Thurmon, J. C., and Tranquilli, W. J. (1986). Anesthesia for cardiovascular research. *In* ''Swine in Cardiovascular Research'' (H. C. Stanton and H. J. Mersmann, eds.), Vol. 1, pp. 39–59. CRC Press, Boca Raton, Florida.

Thurmon, J. C., Tranquilli, W. J., and Benson, G. J. (1996). Perioperative pain and distress. *In* ''Lumb and Jones' Veterinary Anesthesia'' (J. C. Thurmon, W. J. Tranquilli, and G. J. Benson, eds.), 3rd ed. pp. 40–60. Williams and Wilkins, Baltimore, Maryland.

Tranquilli, W. J., Parks, C. M., Thurmon, J. C., Benson, G. J., Koritz, G. D., Manohar, M., and Theodorakis, M. C. (1982). Organ blood flow and distribution of cardiac output in nonanesthetized swine. *Am. J. Vet. Res.* **43,** 895–897.

Tranquilli, W. J., Thurmon, J. C., Benson, G. J., and Steffey, E. P. (1983). Halothane potency in pigs (*Sus scrofa*). *Am. J. Vet. Res.* **44,** 1106–1107.

Tranquilli, W. J., Thurmon, J. C., and Benson, G. J. (1985). Anesthetic potency of nitrous oxide in young swine (*Sus scrofa*). *Am. J. Vet. Res.* **46,** 58–60.

Tranquilli, W., Thurmon, J., and Benson, G. (1986). Halothane-catecholamine arrhythmias in swine (*Sus scrofa*). *Am. J. Vet. Res.* **47**(10), 2134–2137.

Tranquilli, W. J., Thurmon, J. C., Benson, G. J., and Steffey, E. P. (1988). Determination of halothane MAC in swine. *Anesth. Analg.* **67,** 597–598.

Trim, C. M., and Gilroy, G. A. (1985). Cardiopulmonary effects of a xylazine and ketamine combination in pigs. *Res. Vet. Sci.* **38,** 30–34.

Trudeau, V. L., Meijer, J. C., van de Wiel, D. F. M., and Bevers, M. M. (1988). Effects of morphine and naloxone on plasma levels of LH, FSH, prolactin and growth hormone in the miniature male pig. *J. Endocrinol.* **119,** 501–508.

Tuman, K. J., McCarthy, R. J., Spless, B. D., DaValle, M., Dabir, R., and Ivankovich, A. D. (1989). Does choice of anesthetic agent significantly affect outcome after coronary artery surgery? *Anesthesiology* **70,** 189–198.

Vainio, O. M., Bloor, B. C., and Kim, C. (1992). Cardiovascular effects of a ketamine-medetomidine combination that produces deep sedation in Yucatan mini swine. *Lab. Anim. Sci.* **42**(6), 582–588.

Wedel, D., Iaizzo, P., and Milde, J. (1991). Desflurane is a trigger of malignant hyperthermia in susceptible swine. *Anesthesiology* **74,** 508–512.

Weiskopf, R., Holmes, M., Eger, E., Johnson, B., Rampil, I., and Brown, J. (1988). Cardiovascular effects of I653 in swine. *Anesthesiology* **69,** 303–309.

Weiskopf, R., Eger, E., Holmes, M., Rampil, I., Johnson, B., Brown, J., Yasuda, N., and Targ, A. (1989a). Epinephrine-induced premature ventricular contractions and changes in arterial blood pressure and heart rate during I-653, isoflurane and halothane anesthesia in swine. *Anesthesiology* **70,** 293–298.

Weiskopf, R., Holmes, M., Rampil, I., Johnson, B., Yasuda, N., Targ, A., and Eger, E. (1989b). Cardiovascular safety and actions of high concentrations of I-653 and isoflurane in swine. *Anesthesiology* **70,** 793–798.

Weiskopf, R., Eger, E., Holmes, M., Yasuda, N., Johnson, B., Targ, A., and Rampil, I. (1990). Cardiovascular actions of common anesthetic adjuvants during desflurane (I-653) and isoflurane anesthesia in swine. *Anesth. Analg.* **71,** 144–148.

Weiskopf, R. B., Holmes, M. A., Eger, E., Yasuda, N., Rampil, I., Johnson, B., Targ, A., Reid, I., and Keil, L. (1992). Use of swine in the study of anesthetics. *In* ''Swine as Models in Biomedical Research'' (M. M. Swindle, ed.), pp. 96–117. Iowa State Univ. Press, Ames.

Worek, F. S., Blumel, G., Zeravik, J., Zimmerman, G. J., and Pfeiffer, U. J. (1988). Comparison of ketamine and pentobarbital anesthesia with the conscious state in a porcine model of *Pseudomonas aeriginosa* septicemia. *Acta Anaesth. Scand.* **32,** 509–515.

Chapter 15

Anesthesia and Analgesia in Nontraditional Laboratory Animal Species

Dorcas O. Schaeffer

ANESTHESIA AND ANALGESIA IN LABORATORY ANIMALS

ISBN 0-12-417570-8

I. INTRODUCTION

Nontraditional laboratory animal species are being used with increasing frequency in research environments. Many of these animals have only recently been understood from biological, husbandry, and medical perspectives. This chapter addresses anesthesia and analgesia of more commonly used unconventional species, including special physiologic considerations, structural peculiarities, and methods for anesthetic administration (see Appendix at the end of this chapter for a listing of preferred anesthetics and doses).

II. MAMMALS

Nontraditional mammalian species used in research include representatives of the orders Rodentia (chinchillas, ground squirrels, prairie dogs, woodchucks), Marsupialia (opossums), Edentata (armadillos), Insectivora (shrews), and Chiroptera (bats) (Table I).

A. Rodents

In many rodent species, the esophagus enters the stomach at an acute angle, thereby preventing regurgitation. Because of this feature and the high metabolic rate of rodents, many veterinarians do not fast rodents prior to induction of anesthesia. Others, however, are concerned with gastric distention and regurgitation that occur in some species, including hystrichomorphs, and therefore recommend fasting (Heard, 1993). If large rodents are fasted prior to induction of anesthesia, the fast should not exceed 6–12 hr to prevent hypoglycemia (Flecknell, 1987; Muir and Hubbell, 1989; Heard, 1993).

TABLE I

SELECTED MAMMALIAN SPECIES USED IN RESEARCH

Common name	Genus and species
Chinchilla	*Chinchilla laniger*
Richardson's ground squirrel	*Spermophilus richardsonii*
Thirteen-lined ground squirrel	*Spermophilus tridecemlineatus*
Golden-manteled ground squirrel	*Spermophilus lateralis*
Black-tailed prairie dog	*Cynomys ludovicianus*
Woodchuck	*Marmota monax*
North American opossum	*Didelphis virginiana*
Gray short-tailed opossum	*Monodelphis domestica*
Australian brush-tailed opossum	*Trichosurus vulpecula*
Nine-banded armadillo	*Dasypus novemcinctus*
Short-tailed shrew	*Blarina brevicauda*
Musk shrew	*Suncus murinus*
Giant Indian fruit bat	*Pteropus giganteus*
Big brown bat	*Eptesicus fuscus*
Little brown bat	*Myotis lucifugus*

1. Chinchillas

Chinchillas have been widely used in auditory research (Suarez *et al.*, 1989; Amoils *et al.*, 1992; Schachern *et al.*, 1992; Subramaniam *et al.*, 1993; Palombi *et al.*, 1994). Laryngeal and oropharyngeal anatomy in chinchillas is similar to that of other hystrichomorph rodents, thereby making endotracheal intubation difficult to accomplish (Hoefer, 1994). When attempting intubation, a small diameter (≤2.5 mm), noncuffed endotracheal tube should be used (Kujime and Natelson, 1981; Heard, 1993). The soft palate of the chinchilla is continuous with its tongue base, and the oropharynx and pharynx connect through the palatal ostium. Misplacement of the endotracheal tube down the side of the ostium and into the easily damaged soft palate may occur unless extreme care is taken when intubating chinchillas (Timm *et al.*, 1987).

Inhalant anesthetic agents can be safely administered via induction chamber and face mask. Isoflurane, delivered through a precision vaporizer and a nonrebreathing system, is the agent of choice for chinchillas. Induction and recovery rates are rapid and there is a wide margin of safety (Hoefer, 1994). Chinchillas can be induced at 3–4% and maintained at 1–2% isoflurane. Halothane has been administered, also through a precision vaporizer and nonrebreathing system, at 3.5–4% for induction and 1.5–2% for maintenance. This regimen has provided effective and safe surgical anesthesia for cesarean sections in chinchillas (Jones, 1990; Stephenson, 1990). Methoxyflurane has been used; however, induction times can be prolonged with this agent (Hoefer, 1994).

Injectable anesthetics have historically been used for research surgeries, especially those involving the auditory system or brain. These agents are usually given intraperitoneally or intramuscularly, using a 23- to 25-gauge needle and small injection volumes (<0.3 ml). Cephalic and saphenous veins are most commonly used for intravenous anesthetic administration (Hoefer, 1994); however, femoral veins have also been used (Tappa *et al.*, 1989). Twenty-five to 28-gauge needles are suggested for administration of IV anesthetic agents (Hoefer, 1994). Ketamine hydrochloride, sodium pentobarbital, α-chloralose, xylazine, tiletamine–zolazepam, diazepam, acetylpromazine, and urethane have been used alone or in various combinations to induce and maintain anesthesia for procedures ranging from vascular cut-downs to stereotaxic microelectrode implantation (Shaddock *et al.*, 1985; Kaltenbach and Saunders, 1987; Shirane and Harrison, 1987; Rybak and Whitworth, 1988; Sahley *et al.*, 1991; Paolini *et al.*, 1993; Henderson *et al.*, 1994).

Ketamine used alone does not provide adequate visceral analgesia; it should be combined with xylazine to enhance muscle relaxation and provide appropriate surgical anesthesia and analgesia (Heard, 1993). Ketamine–xylazine (40 mg/kg ketamine and 2 mg/kg xylazine), administered IM, will provide 2 hr of surgical anesthesia in the chinchilla (Hargett *et al.,* 1989). The xylazine can be reversed with yohimbine (2.1 mg/kg IP), thereby decreasing anesthesia time by half (Hargett *et al.,* 1989). Tiletamine–zolazepam has a more prolonged recovery time than ketamine. When tiletamine–zolazepam (30 mg/kg IM) followed by sodium pentobarbital (6.5 mg/ml in a constant IV infusion of 5 μl/minute) was used in chinchillas, an upward drift in auditory evoked potentials was observed (Sahley *et al.,* 1991).

The anesthetic effect of sodium pentobarbital is widely variable. Pentobarbital causes prolonged sedation and cardiopulmonary depression in some animals (Hall and Clarke, 1991). Additionally, analgesia is poor (Heard, 1993); therefore, pentobarbital is not generally recommended as an effective sole anesthetic agent. Although barbiturate solutions other than sodium pentobarbital are alkaline and therefore tissue irritants (Heard, 1993), thiopental at 40 mg/kg has been given IP to chinchillas for short-duration anesthesia (Collins, 1988).

The anesthetic and analgesic efficacy of α-chloralose has been reviewed. The current recommendation is to use this drug only after the animal has been preanesthetized with an opioid, barbiturate, α_2-adrenoceptor agonist, or phenothiazine tranquilizer. Depth of anesthesia may be difficult to assess, and animals may require mechanical ventilation. Because of its tendency to induce an inflammatory response when administered IP, α-chloralose should only be administered IV for survival surgeries (Silverman and Muir, 1993).

Depth of anesthesia in chinchillas and other rodents can be monitored by heart rate, respiratory rate, pedal withdrawal reflex, palpebral reflex, and jaw tone (Heard, 1993).

Analgesic doses for chinchillas have been adapted from those used in guinea pigs. Butorphanol (2mg/kg SC q 2–4 hr), buprenorphine (0.5 mg/kg SC q 8–12 hr), and morphine (2–5 mg/kg SC or IM q 4 hr) are among the agents more commonly used for guinea pigs (Heard, 1993).

2. Ground Squirrels, Prairie Dogs, and Woodchucks

Ground squirrels, prairie dogs, and woodchucks are also commonly used wild rodent species. Isoflurane anesthesia, as described for chinchillas, is the method of choice for these animals. Halothane has been used successfully to anesthetize the golden-manteled ground squirrel (*Spermophilus lateralis*). Five percent halothane was used to induce and 2% was used to maintain anesthesia (Osborne and Milsom, 1993).

Injectable anesthetics have been traditionally used in ground squirrels. Olson and McCabe (1986) demonstrated that ketamine–xylazine (86 mg/kg ketamine and 11 mg/kg xylazine IM) provided 20–30 min of surgical anesthesia in Richardson's ground squirrels (*Spermophilus richardsonii*). They also discovered that ketamine alone (86 mg/kg), fentanyl–droperidol (52 μg/kg fentanyl and 2.6 mg/kg droperidol), and sodium pentobarbital (50 mg/kg) did not provide surgical anesthesia in this species. Similarly, Carey and Cooke (1991) used 90 mg/kg ketamine and 10 mg/kg xylazine IM for abdominal surgery in the thirteen-lined ground squirrel (*Spermophilus tridecemlineatus*). Other investigators have used high sodium pentobarbital doses (60–75 mg/kg IP) for a variety of surgical manipulations (Forger *et al.,* 1986; Powley and Fox, 1986; Dark *et al.,* 1992; Hendriksen *et al.,* 1992; Beckman *et al.,* 1993; Bharma and Milsom, 1993). Caution should be exercised when attempting major surgical procedures, however, due to the cardiopulmonary depressive effects and lack of analgesic activity of sodium pentobarbital (Heard, 1993).

Woodchucks have been anesthetized with ketamine–xylazine combinations (25–40 mg/kg ketamine and 2.5–8 mg/kg xylazine (Leveille *et al.,* 1987; Sinha Hikim *et al.,* 1992). Prairie dogs have been anesthetized with ketamine/xylazine (100–150 mg/kg ketamine and 10–20 mg/kg xylazine IM) for abdominal surgery (Roslyn *et al.,* 1979; Pitt *et al.,* 1984; Miki *et al.,* 1993), whereas lower doses (50 mg/kg ketamine and 5 mg/kg xylazine IM) have been used for less invasive procedures (Krishnamurthy and Turner, 1988). Although pentobarbital (30 mg/kg IP) has been used as the sole anesthetic agent for abdominal surgery (Li *et al.,* 1991, 1994), low doses may not provide adequate surgical anesthesia and analgesia. Finally, α-chloralose has been used extensively as an anesthetic agent for prairie dogs (Lillemoe *et al.,* 1988; Webb *et al.,* 1988; Ahrendt *et al.,* 1992; Kaufman *et al.,* 1993; Tierney *et al.,* 1994). Based on the recent comprehensive review of chloral hydrate and α-chloralose, it is recommended that α-chloralose be administered only after induction with another agent and that α-chloralose should not be administered IP for survival surgeries (Silverman and Muir, 1993).

B. Marsupials

Both the North American opossum (*Didelphis virginiana*) and the South American gray short-tailed opossum (*Monodelphis domestica*) have been used extensively in biomedical research, especially in studies of sphincter motility, pancreatitis, and fetal development. Many of these investigations involve extensive surgical manipulations, and a variety of anesthetic agents have been used.

Halothane administered through a precision vaporizer into a chamber at an induction rate of 4% successfully anesthetized gray short-tailed opossums. Animals were subsequently maintained on face masks (Keller *et al.,* 1988). Halothane was found to affect motility of the sphincter of Oddi in Australian brush-tailed possums (*Trichosurus vulpecula*) (Liu *et al.,* 1993). Isoflurane has been safely and effectively used in marsupials, including the North American opossum. The marsupial can be induced with 50 mg/kg ketamine or 4% isoflurane (Liu *et al.,* 1992), then intubated and maintained on a nonrebreathing system at 1–2%.

If inhalants are not available, injectable anesthetics may be used. Ketamine (30 mg/kg IM) provided 10–15 min of restraint in North America opossums (Moore, 1984). Ketamine (40 mg/kg) and xylazine (5 mg/kg) provided sedation but not surgical anesthesia in gray short-tailed opossums (Keller *et al.,* 1988). Brush-tailed possums were anesthetized for jugular cannulation with 50 mg/kg ketamine and 10 mg/kg xylazine IM (Bathgate *et al.,* 1992). Ketamine (40 μg/g IM) has been used in conjunction with pentobarbital (50 μg/g SC) and atropine (2 μg IM), following induction with methoxyflurane, to provide 1–2 hr of surgical anesthesia in the gray short-tailed opossum (Robinson and VandeBerg, 1994). Administration of pentobarbital subcutaneously is not recommended. Alternatively, ketamine (50 mg/kg IM) has been used to induce anesthesia, followed by maintenance with halothane and N_2O (0.3–0.75% in N_2O:O_2, 1:2) in the Australian brush-tailed possum (Liu *et al.,* 1992). Ketamine (20 mg/kg IM) and xylazine (2 mg/kg IM) have also been used with halothane to induce and maintain surgical anesthesia in the North American opossum (Larsen *et al.,* 1991).

Sodium pentobarbital has been used as a sole anesthetic agent in North American opossums. It has been administered intraperitoneally at doses from 30 to 50 mg/kg (Runzi *et al.,* 1993; Anand and Paterson, 1994; Runkel *et al.,* 1995; Samuel *et al.,* 1995) and supplemented with intravenous doses of 5 mg/kg (Anand and Paterson, 1994) or 10–15 mg (Hanyu *et al.,* 1990) as needed. In gray short-tailed opossums, pentobarbital at $\leq$ 50 mg/kg did not provide adequate sedation or analgesia; a dose of 60 mg/kg proved fatal (Keller *et al.,* 1988). A thiobarbiturate derivative, 5-*sec*-butyl-5-ethyl-2-thiobarbituric acid (Inactin), has been used to anesthetize suckling and adult North American opossums, at doses of 35–80 mg/kg (doses increase with age) SC (sucklings) or IP (adults) (Farber, 1990, 1993a,b). Cranial and thoracic surgical procedures were performed on these animals. Animals were supplemented when necessary (25–33% of the original dose) to maintain abolition of withdrawal and palpebral reflexes. Because all barbiturates except sodium pentobarbital are alkaline and therefore irritating (Heard, 1993), the compounds should be given only intravenously.

Many investigators have induced anesthesia with sodium pentobarbital (40 mg/kg IP or IV) and maintained it with α-chloralose (20–25 mg/kg or 70–200 mg IV) (Culver and Rattan, 1986; Rattan and Culver, 1987; Nurko *et al.,* 1989; Paterson, 1989; Thatikunta *et al.,* 1993; Paterson and Kolyn, 1994; Shibamoto *et al.,* 1994). α-Chloralose is favored because it does not interfere with sympathetic nerve activity (Shibamoto *et al,* 1994). Sengupta *et al.* (1989) found that induction with 35–40 mg/kg IP pentobarbital followed by 25–50 mg/kg IV α-chloralose for maintenance provided good surgical anesthesia in the North American opossum. Recirculating water blankets should be used to maintain body temperature, and assisted ventilation may be required.

C. Edentates

Armadillos (edentates) are unique mammals that have been used extensively in leprosy and reproductive studies. They have a long, narrow snout which makes endotracheal intubation somewhat challenging. Isoflurane and halothane are the anesthetic agents of choice for armadillos. In the author's laboratory, armadillos are first immobilized with ketamine (10–20 mg/kg IM) and are then anesthetized with halothane or isoflurane delivered through a nonrebreathing system and a face mask. Induction rates are 3–5%; maintenance levels are 1–2%. Face masks for armadillos can be fashioned from syringe cases (this length adapts well to the edentate snout). A piece of latex glove with a central perforation is fitted over the end of the syringe case to decrease the escape of anesthetic gases (Fig. 1). Atropine (0.04 mg/kg IM) must be given to prevent copious salivation (Divers, 1986). Because armadillos can hold their breath for prolonged periods of time (Storrs, 1972), many feel intubation is necessary. The author, however, has found that after ketamine immobilization, the animals breathe spontaneously and steadily, using only a face mask.

A few injectable anesthetic agents have been used in armadillos. Ketamine (10–20 mg/kg IM) can be combined with diazepam (0.1 mg/kg IV or IM) or acetylpromazine (0.1 mg/kg IM) for deep sedation (Divers, 1986; Gillespie, 1993). Xylazine is not recommended for armadillos (Gillespie, 1993). Sodium pentobarbital (25 mg/kg IV) has been used to anesthetize armadillos (Divers, 1986). Fentanyl–droperidol (0.11 ml/kg, 1.0-ml total dose) has also been used for anesthesia (Storrs, 1972); however, other authors found this combination unsatisfactory for edentates (Gillespie, 1993).

D. Insectivores

Shrews have been used as animal models in reproductive, gastrointestinal, and metabolic studies. Because of their ex-

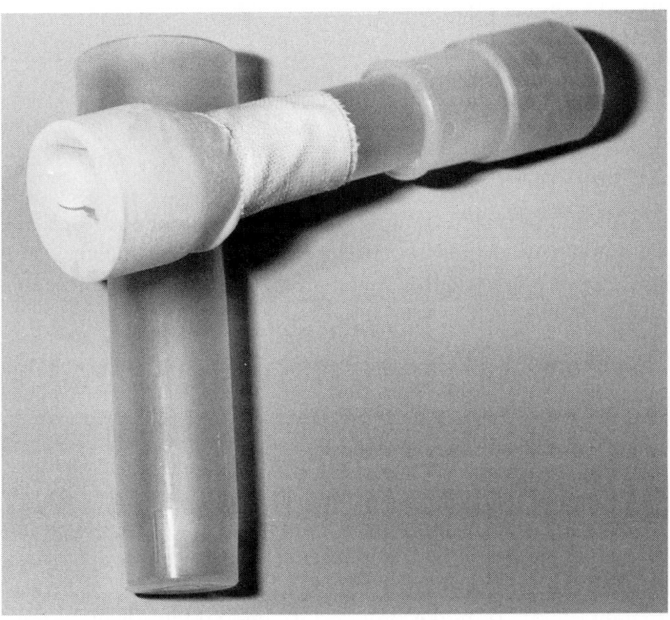

Fig. 1. A cut syringe case covered with perforated latex creates an ideal face mask for armadillos and other long-snouted species.

tremely high metabolic rates, they are prone to cardiogenic shock and death when stressed. Isoflurane is the anesthetic agent of choice for shrews. They can be induced in a chamber at 4% and maintained at 1.8–2.5%, delivered through a face mask (Meehan, 1993). Ketamine (1–2 mg/100 g) combined with xylazine (0.1 mg/100 g) will provide 20–30 min of anesthesia for minor procedures (Meehan, 1993). Sodium pentobarbital (0.4 mg/10 g) has been used for castrations in musk shrews (Rissman and Crews, 1989). Regardless of the agent used, shrews should be kept warm during anesthesia and recovery (Meehan, 1993).

E. Bats

Because of their unique echolocating abilities, bats have been used in auditory system and brain stem research. Furthermore, their unique forelimb anatomy (highly vascular, membranous wings) has made them popular for vascular studies. Ether has historically been used to anesthetize bats and has occasionally caused mortality (McCracken and Wilkinson, 1988). This, combined with the inherent dangers associated with handling ether, has led investigators to select other inhalant anesthetic agents. Isoflurane has provided good surgical anesthesia for castration of fruit bats (Barnard and Soloway, 1992). Halothane (1.5–2%) and methoxyflurane have been used successfully for experimental and surgical procedures including cranial surgery, muscle electrode implantation, and fracture repair (Racey, 1972; Constantine, 1986; Covey and Casseday, 1986; Wilson, 1988; Routh, 1991).

Ketamine (11 mg/kg IM) and acetylpromazine (1.1 mg/kg) have been used as an injectable anesthetic combination for restraint in giant fruit bats (Carpenter, 1986). Sodium pentobarbital has been administered IP in a range from 25 to 50 mg/kg, with most investigators using 30 mg/kg (Racey, 1972; Constantine, 1986; Sun and Jen, 1989; Sun *et al.,* 1990).

α-Chloralose has been used to produce light anesthesia in bats at a dose of 110–115 mg/kg (Bassett, 1987; Wilson, 1988). This drug should only be given IV and should follow administration of an opioid, barbiturate, or other induction agent (Silverman and Muir, 1993). Either pentobarbital or α-chloralose can cause hemodilution in bats (Bassett, 1987).

III. BIRDS

A list of avian species most commonly used in research is given in Table II. Several distinct anatomic features must be considered when dealing with anesthesia in avian species. Primary bronchi pass through the lungs and connect directly or indirectly to the caudal pulmonary (caudal thoracic and abdominal) air sacs. Additionally, the bronchi further subdivide into secondary then tertiary bronchi (parabronchi) which connect with the cranial pulmonary air sacs (cervical, clavicular, and cranial thoracic air sacs) and air capillaries. Air capillaries lie adjacent to blood capillaries and are the primary site for oxygen exchange (King and McLelland, 1984).

TABLE II

SELECTED AVIAN SPECIES USED IN RESEARCH

Common name	Genus and species
Chicken	*Gallus gallus*
Japanese quail	*Coturnix japonica*
Turkey	*Meleagris gallopavo*
Pigeon	*Columba livia*
Pekin duck/mallard	*Anas platyrhynchos*
Canvasback	*Aythya valisineria*
Greylag goose	*Anser anser*
Zebra finch	*Poephila castanotis*
House sparrow	*Passer domesticus*
Amazon parrot	*Amazona sp.*
Budgerigar	*Melopsittacus undulatus*
Cockatiel	*Nymphicus hollandicus*

Birds lack a functional diaphragm. Both inspiration and expiration require active muscular exertion. Thoracic volume is expanded and contracted by moving the sternum ventrally and dorsally. It is important when handling birds to allow the chest to move freely (S. E. Orosz, personal communication, 1996).

The trachea of birds is located caudal to the tongue, but in a much more cranial position than in mammals. In most species the glottis can be exposed by pulling the tongue rostrally. There is no epiglottis. Gentle forward pressure on the mandibular space or gular region will move the glottis into position for intubation (Flammer, 1989). Only noncuffed endotracheal tubes should be used; birds have complete tracheal rings, and overinflation of cuffs can cause pressure necrosis and subsequent tracheal damage (Sedgwick, 1988a; Flammer, 1989). If the bird is small, a tube may be fashioned from a red rubber feeding tube (Sinn, 1994) (Fig. 2). During anesthesia (of any type), the relatively rigid structure of the avian respiratory tract may prevent adequate ventilation, particularly if the bird is placed on its back. Careful intermittent positive pressure ventilation is therefore recommended for anesthetic procedures in birds (Sedgwick, 1988b; Heard, 1994; Sinn, 1994). Four to six breaths per minute is recommended (S. E. Orosz, personal communication, 1996).

The presence of air sacs allows administration of inhalant anesthetics and/or oxygen directly into them through surgical access as an alternate route for anesthesia maintenance. This is particularly useful for head surgery or to remove aspirated foreign bodies (Sinn, 1994). After induction, the bird is placed in right lateral recumbency with the left leg extended caudally. The skin is prepared and incised over the sternal notch, and sterile hemostats are used to bluntly enter the left caudal thoracic air sac. A short endotracheal tube is then introduced into this air sac and connected to the anesthetic machine (Sinn, 1994). Because manipulation of the bird can inadvertently dislodge the air sac tube, placement of a second tube in the trachea

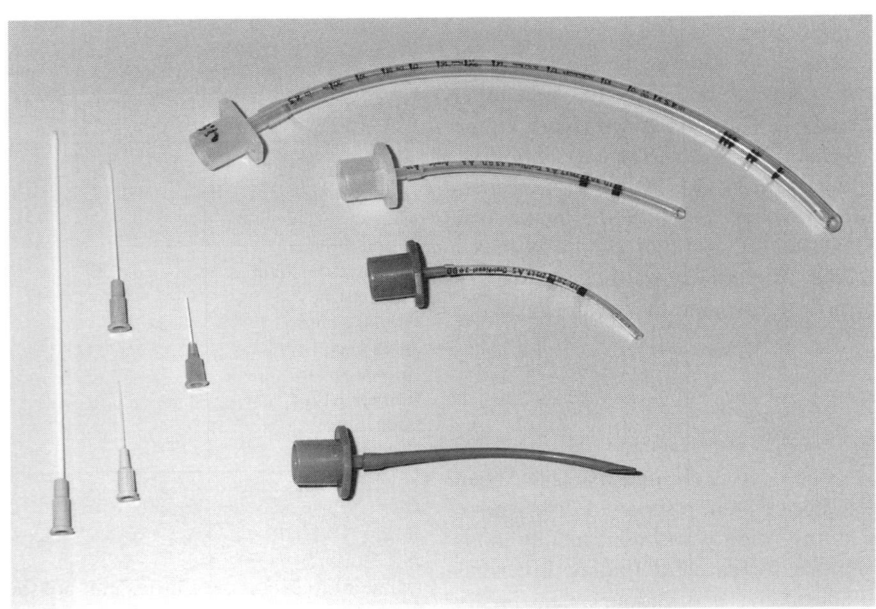

Fig. 2. Several types of small-diameter endotracheal tubes. A red rubber feeding tube can be cut, blunted, and attached to an adapter. Teflon intravenous catheters can also be used for very small patients.

is usually desirable as a backup measure (M. Jones, personal communication, 1996).

Because birds lose heat rapidly during anesthesia, supplemental heat should be provided with a recirculating water blanket, and temperature monitoring (normal body temperature is approximately 105–107°F) should be done, especially during prolonged procedures (Clutton, 1986). Electronic thermometers and tympanic scanners are recommended over traditional glass thermometers when monitoring avian cloacal temperatures (Sinn, 1994).

Other ways to conserve body heat in birds undergoing surgical procedures include plucking only the minimum amount of feathers necessary and taping the rest aside prior to scrubbing (Olsen *et al.,* 1992). Use of a povidone iodine **solution** to scrub will avoid evaporative heat loss that results from using alcohol preps (Sinn, 1994). This will also minimize removal of the birds' natural oils, which are necessary for waterproofing and insulation in many aquatic species (Olsen *et al.,* 1992).

Fasting prior to induction of anesthesia will vary with the species and size of bird. Some authors believe that fasting for prolonged periods will deplete glycogen stores and produce a fatal hypoglycemia (Flammer, 1989; Hall and Clarke, 1991). Others feel that overnight fasting is appropriate for larger species (Sinn, 1994). All agree that the bird should be fasted until the crop is empty. If this is not possible, crop contents should be aspirated or otherwise emptied during induction, taking care to prevent aspiration into the glottis or choana (Flammer, 1989; Sinn, 1994).

Depth of anesthesia in avian patients can be monitored by withdrawal reflex (wing or toe) and jaw tone. Additionally, when stimulated, the nictitating membrane moves rapidly in a light plane of anesthesia and slowly in a deep plane of anesthesia (M. Jones, personal communication, 1996).

Injectable anesthetics, analgesics, fluid therapy, and emergency drugs can be administered to avian species by a variety of routes. Intravenous agents are commonly injected into the basilic (wing) vein or metatarsal vein (Fig. 3). Intramuscular agents are almost always given in the pectoral (breast) muscle. If fluids are administered, isotonic solutions such as warmed lactated Ringer's solution are recommended. Fluid maintenance levels for birds are approximately 50 ml/kg/day (Redig, 1984). Fluids can be administered subcutaneously in the axillary or inguinal region. Avian skin is very thin and a bleb should be easily seen as subcutaneous fluids are administered. Care should be taken to avoid deep administration as this can result in fluid or drug placement directly into air sacs.

Because of the delicate nature of avian veins, maintenance of indwelling intravenous catheters can be difficult. As a result, many prefer to place intraosseous catheters into the ulna. To do this, the skin overlying the dorsal carpus is gently plucked and aseptically prepared. For a pigeon, an 18- to 20-gauge, 1.5-inch needle is inserted into the ulna dorsally on the medial side of the condyle with a gentle twisting motion. The needle is taped and sutured in place, and the wing is wrapped in a figure 8 bandage (S. E. Orosz, personal communication, 1996). Intraosseous catheters can be used for drug administration, fluid ther-

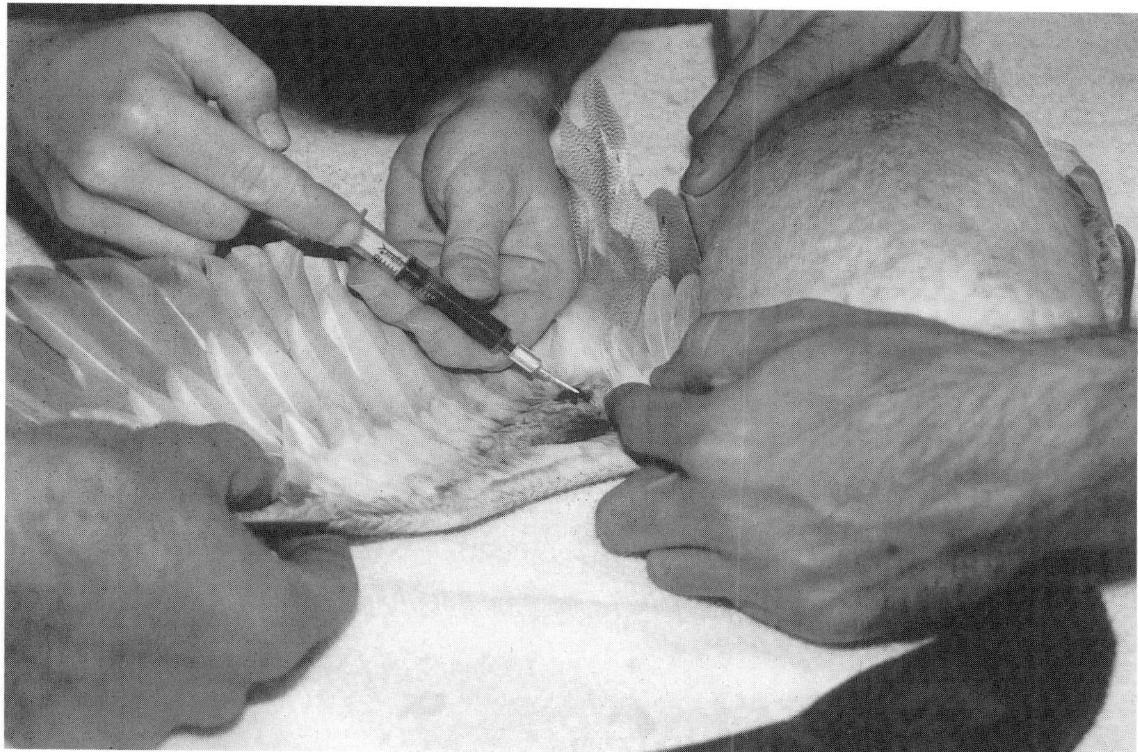

Fig. 3. The basilic vein of birds can be used for intravenous drug administration.

apy, blood transfusions, and total parenteral nutrition (Figs. 4a and 4b).

Birds recovering from anesthesia should be kept in a quiet, warm, oxygen-rich environment (Jenkins, 1993). The bird can also be wrapped in a towel to prevent wing flapping and self-injury during recovery (Sinn, 1994). Emergency drugs used for resuscitation in avian species include epinephrine (1:1000 solution), 0.1–0.3 ml/kg IV, IO, or intratracheally (Heard, 1994), doxapram, 5–10 mg/kg IM or IV, and prednisolone sodium succinate (Solu-Delta-Cortef, Upjohn, 10 mg/ml), 0.1–0.2 ml/100 g IV or IM (Clubb, 1986).

Some authors have advocated the use of allometric or metabolic scaling when calculating drug doses for avian and other species (Sedgwick, 1988b). This concept may have some application among mammalian species that metabolize, distribute, and excrete drugs by the same mechanism. In avian and poikilothermic species that are anatomically and physiologically distinct from mammals, allometric scaling has a much more limited application. Structural and functional differences in avian and reptilian kidneys can alter drug metabolism and distribution, thereby making allometric scaling inappropriate for these species (Fitzgerald and Cooper, 1990; Frazier *et al.,* 1995a,b).

The preanesthetic agents atropine and glycopyrrolate are not frequently used in avian species. Both drugs cause thickening of respiratory secretions, which could occlude endotracheal tubes. Additionally, atropine elevates heart rate, an undesirable effect in animals that normally have extremely high heart rates (Sinn, 1994). If atropine is given, the dose is 0.04–0.1 mg/kg IM (Clubb, 1986).

Inhalant anesthesia is the method of choice for birds. Inhalant anesthetics should be delivered to most avian species (birds <8 kg) through nonrebreathing systems.

Isoflurane is the preferred inhalant anesthetic agent (Sedgwick, 1988a; Muir and Hubbell, 1989; Doolen and Jackson, 1991; Sinn, 1994). Isoflurane can be delivered via a face mask or an anesthetic chamber to induce birds. Canine or feline anesthetic masks, disposable plastic cups, and syringe-case face masks have been used (Sinn, 1994); the choice of mask depends on the size and beak or bill structure of the species to be anesthetized (Fig. 5). Authors disagree over appropriate induction methodology. Some feel the bird should be masked with oxygen and the isoflurane levels slowly increased to 2.5–3%. Others believe that rapid induction at 5% with subsequent maintenance levels of 1–2% is preferable (Sinn, 1994). Both methods are used successfully. Induction may be smoother with the gradual introduction of isoflurane; however, in species that are already stressed and struggling, 5% provides rapid loss of consciousness. Halothane has been used successfully in many avian species. Induction usually occurs in 2–5 min at settings of 2–3%, maintenance levels are normally 1–1.5%, and recovery takes

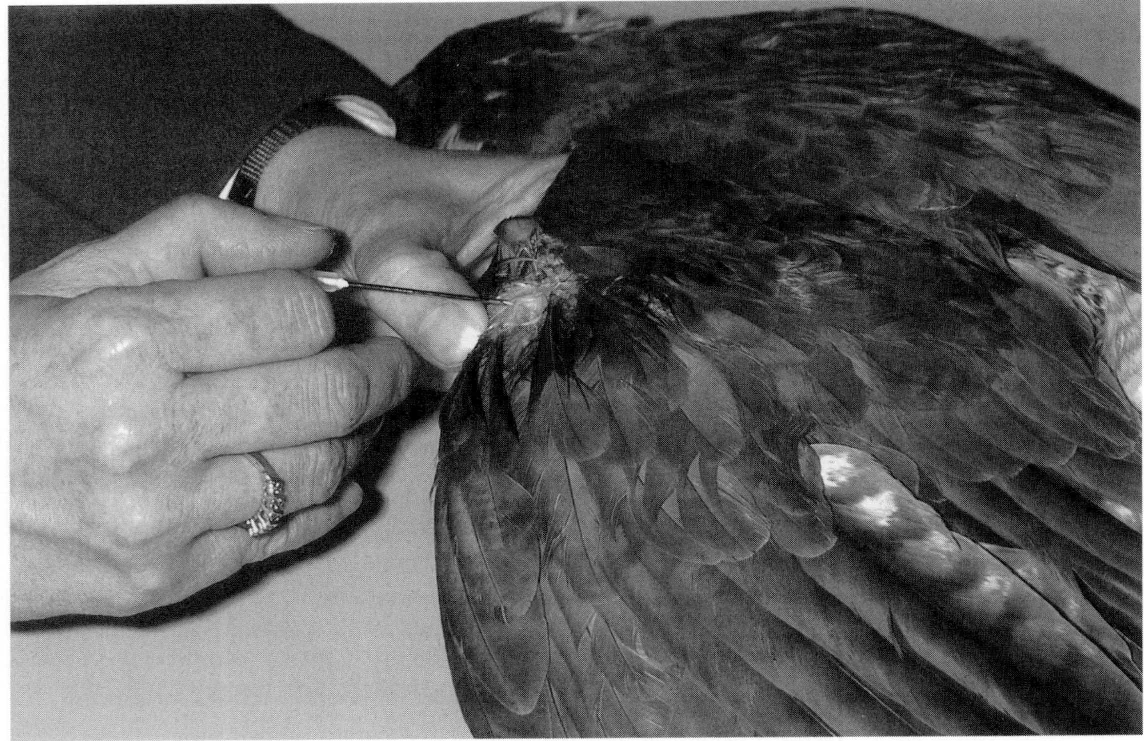

a

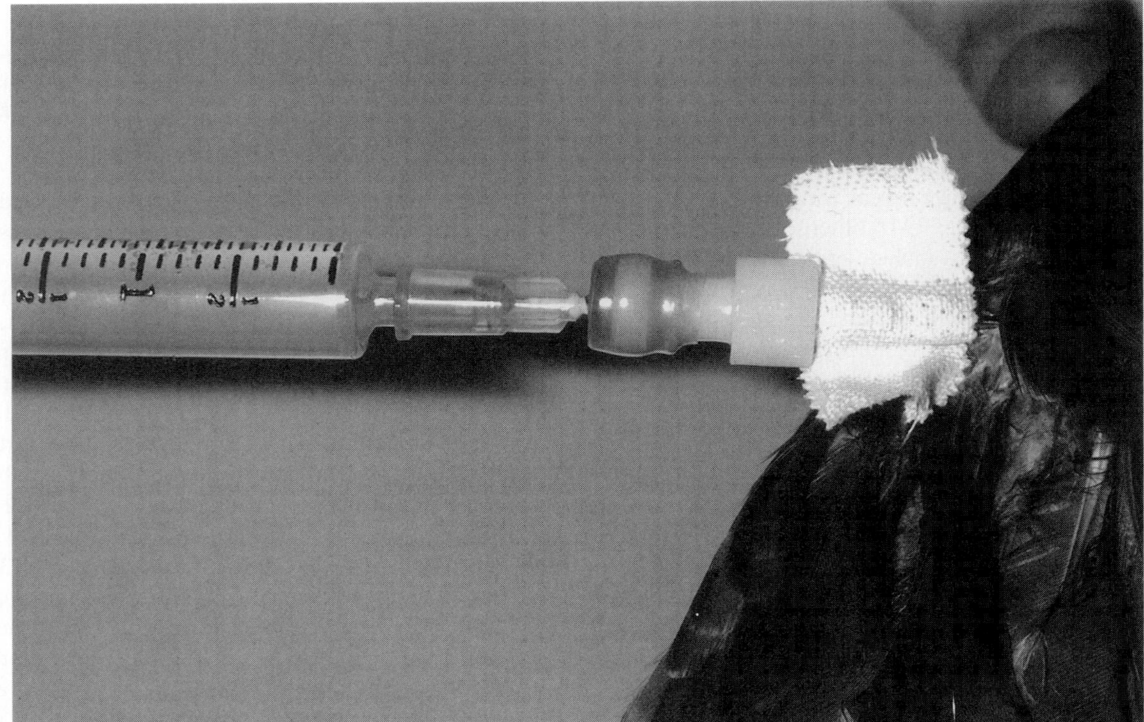

b

Fig. 4. (a and b) Cadaver demonstration of intraosseus catheter placement in the ulna of a bird. Photograph courtesy of Dr. S. E. Orosz.

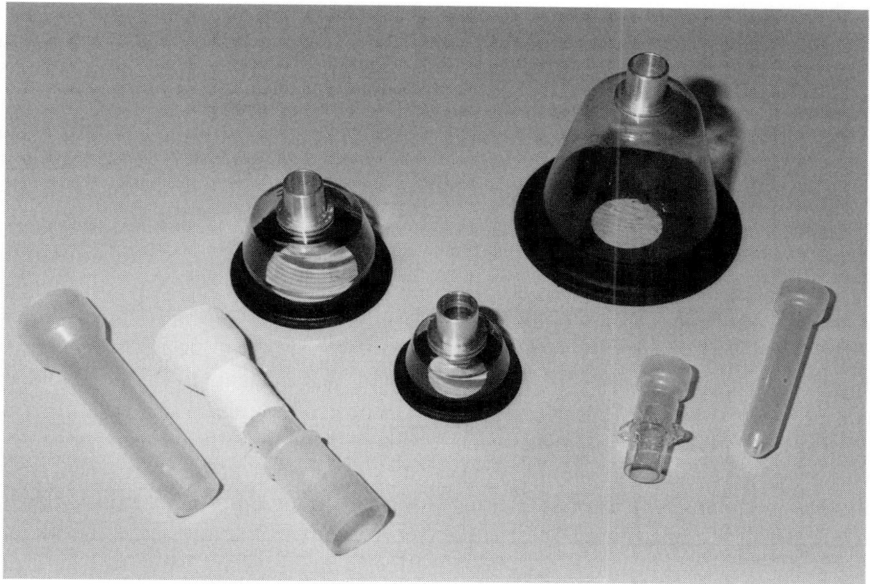

Fig. 5. Anesthetic masks should be selected based on a bill or beak structure of the avian patient.

approximately 15–20 min. Halothane can cause bradycardia, apnea with concurrent cardiac arrest, and sensitization of the heart to catecholamines, resulting in arrhythmias. Halothane, like isoflurane, must be delivered through a precision vaporizer (Sinn, 1994). Methoxyflurane has been used in birds. Rates of 3–4% induce birds within 5–15 min. Maintenance rates are 1–2% (Flammer, 1989). Although some authors appropriately warn against use of this agent in an open-drop system (Sinn, 1994), methoxyflurane has been used successfully for induction of light anesthesia in carefully controlled situations (Smith *et al.*, 1980; Rotella and Ratti, 1990).

Nitrous oxide has been used in birds; however, it can cause subcutaneous emphysema in diving birds with normal subcutaneous air pockets (Bednarski, 1985; Sinn, 1994).

A. Chickens, Quail, and Turkeys

Chickens, quail, and turkeys are galliform birds commonly used in research (Table II). As for other species, isoflurane is this author's anesthetic of choice for galliformes. Induction and recovery are smoother and more rapid, and fewer complications have been seen with isoflurane. Induction and maintenance rates of isoflurane are similar to other species. Atracurium, administered slowly IV at a dosage of 0.25 mg/kg to isoflurane-anesthetized chickens, induced clinically useful muscle relaxation for approximately 20 min (Nicholson and Ilkiw, 1992). Recovery from the atracurium was enhanced by edrophonium given slowly IV at 0.5 mg/kg. Heard (1994) has also used atracurium to achieve ocular relaxation and nonresponsiveness during ophthalmic surgery.

Isoflurane has been found to lower the threshold for electrical fibrillation in chickens (Greenlees *et al.,* 1990). The halothane minimum anesthetic dose in chickens (0.85%) is similar to minimum alveolar concentrations (MAC) in mammals; halothane also causes dose-dependent cardiopulmonary depression in chickens (Ludders *et al.,* 1988). Halothane at 1% was used to anesthetize chickens for thoracotomy (Ploucha and Fink, 1986). In another study, chicks were anesthetized with 3% halothane in 50% O_2/N_2O for serial daily intraocular injections (Rohrer and Stell, 1994). Both hatchling chicks and embryos have been anesthetized with halothane. Chicks received 3% halothane to produce deep anesthesia for cranial surgery and decerebration. Embryos were anesthetized by passing 1% halothane over their eggshells prior to drilling a hole and exposing the embryo. Withdrawal reflexes were used to gauge depth of anesthesia in both chicks and embryos (Valenzuela *et al.,* 1990).

Several injectable agents have been used in chickens. Ketamine may be administered IM, IV, or IO (intraosseously). In one study, ketamine was administered at 30 mg/kg IO; no significant gross or microscopic marrow abnormalities were observed postadministration (Valverde *et al.,* 1993). In another study, the median effective dose of ketamine was determined to be 14 mg/kg IV for restraint. The median lethal dose was 68 mg/kg IV, and the minimum dose for anesthesia was 60 mg/kg IV. Even at 60 mg/kg, analgesia was not adequate for surgical procedures. Furthermore, the extremely low margin of safety reinforces the conclusion that ketamine alone is inadequate for surgical procedures (McGrath *et al.,* 1984a). Long-term social stress has been shown to prolong ketamine sleep time in chickens (McGrath *et al.,* 1984b). Ketamine–xylazine (15 mg/kg ketamine and 5 mg/kg xylazine IM) has been used in chickens (Roder *et al.,*

1993). This study also found that chloramphenicol, ketaconazole, and cimetidine increased the duration of ketamine–xylazine anesthesia. Forty percent of the chickens pretreated with cimetidine died during the anesthetic procedure. Phenobarbital, 3-methylcholanthrene, or rifampin did not affect the duration of anesthesia.

Other injectable combinations have been used in chickens. Mohammad *et al.* (1993) found that a ketamine (10–20 mg/kg IM) and detomidine (0.3 mg/kg IM) combination induced anesthesia. Comb-pinch reflex was eliminated; however, no procedure was performed to assess the efficacy of this combination for surgical anesthesia. Additionally, respiratory depression was noted with these drugs. Christensen *et al.* (1987) examined several agents. Equithesin (sodium pentobarbital, chloral hydrate, magnesium sulfate, propylene glycol, ethanol, and water) alone (2.5 ml/kg IM) was insufficient for laparotomy. Equithesin (2.5 ml/kg IM) followed in 15 min by diazepam (2.5 mg/kg IV) provided suitable anesthesia for laparotomy. Increasing the diazepam dose to 5 mg/kg IV resulted in suitable anesthesia for thoracotomy. This combination has been used successfully by others (Morild, *et al.,* 1987). In contrast, metomidate (20 mg/kg IM) followed in 10 min by diazepam (2.5 mg/kg) provided varying anesthesia of short duration (< 15 min) that was marginally suitable for laparotomy. Ketamine (75 mg/kg IM) followed in 10 min by diazepam (2.5 mg/kg IV) was not suitable for laparotomy in the chicken (Christensen *et al.,* 1987).

Sodium pentobarbital (dose range 25–75 mg/kg IV or IM; 30 mg/kg average dose) has been used for a variety of procedures, ranging from assessment of sleep time (Bartlet *et al.,* 1990; Chanda *et al.,* 1990; Odom *et al.,* 1992) to extremely invasive intramuscular, intraabdominal, and intrathoracic procedures (Bottje and Holmes, 1989; Bottje *et al.,* 1989, 1990; DeSaedeleer *et al.,* 1989; Lang and Levy, 1989; Buhr, 1990; El-Fawal *et al.,* 1990; Estavillo, 1990; Estavillo *et al.,* 1990). Christensen *et al.* (1987) found that sodium pentobarbital alone or in combination with ketamine provided inconsistent anesthesia, poor muscle relaxation, poor analgesia, and death. This author, therefore, does not recommend pentobarbital as a surgical anesthetic agent for chickens.

Thiopental (20 mg/kg) has been administered intraosseously in the chicken (Valverde *et al.,* 1993). No observable damage occurred to bone marrow or cartilage as a result of injecting this highly alkaline drug. A brief initial period of apnea and persistent pale mucous membranes were noted postinjection. The animals then relaxed sufficiently to allow intubation. Based on this study, intraosseous administration of thiopental as an induction agent for general anesthesia in healthy chickens may be possible.

DIAL (diallylbarbituric acid, urethane, and monoethylurea) (1.5–2.5 ml/kg IM or IP) has been used to anesthetize chickens for procedures ranging from cannula placement to laparotomy (Wideman and Gregg, 1988; Glahn *et al.,* 1991; Nishimura *et al.,* 1994). Because most barbiturates are alkaline, they should be injected only IV.

In mammals, urethane produces anesthesia, with a minimal effect on blood pressure and respiration, an adequate analgesia (Flecknell, 1987). It has been used (0.75–2.0 g/kg IP) in a variety of studies involving chickens, turkeys, and quail (Domowicz *et al.,* 1988; Neya *et al.,* 1990b; Takei and Hasegawa, 1990; Takei and Watanabe, 1990; Campbell *et al.,* 1991; Saunders *et al.,* 1993). In one study (Takei and Watanabe, 1990), movement was noted even at the highest dose (2.0 g/kg). Urethane is cytotoxic, carcinogenic, and immunosuppressive, and is only recommended for nonsurvival procedures (Anonymous, 1993).

High doses of lidocaine can cause systemic toxicosis in birds; therefore, it is not generally recommended for avian species. Bekoff and Sabichi (1987) diluted lidocaine hydrochloride (2%) to a dose of 35 mg/kg in a volume of 0.2 ml. This was injected subcutaneously into the neck or thigh of chicks in a study of sensory control of hatching. The most effective dose was 35 mg/kg; 40 mg/kg or higher caused drowsiness.

A number of studies have investigated effects of opioid analgesics in chickens. Morphine [a mu(μ)-opioid agonist] and a kappa(κ)-opioid agonist (both infused at 0.1, 1.0, and 3.0 mg/kg IV) decreased the minimal anesthetic concentration of isoflurane in chickens in a dose-dependent manner. Neither drug significantly affected heart rate or mean arterial blood pressure (Concannon *et al.,* 1994, 1995). Morphine (30 mg/kg IM) produced analgesia in Rhode Island Red chickens, but hyperalgesia in White Leghorn and Cal-White strains, when these animals were subjected to tests of thermal nociception (Hughes, 1990a). This hyperalgesia is mediated by μ-receptors (Sufka and Hughes, 1990; Sufka *et al.,* 1991). Alternatively, codeine produced an analgesic effect when given to White Leghorns (Hughes, 1990b). An earlier study using vocalization as a response to a nociceptive stimulus suggested that high doses of morphine may simply impair movement and coordination, thereby inaccurately appearing to provide analgesia (Rager and Gallup, 1986). κ-opioid receptors may predominate in birds, therefore a κ-agonist such as butorphanol (1 mg/kg IM) may be useful in chickens (J. Ludders, personal communication, 1996).

Topical analgesic agents have been used in chicks post beak trimming (Glatz *et al.,* 1992). Both bupivicaine (0.5% mixed 50:50 with dimethyl sulfoxide gel) and phenylbutazone (Isobutalone gel, containing 50 mg/g phenylbutazone, 40 mg/g isopropylaminophenazone, and 80 mg/g dimethyl sulfoxide) provided analgesia when applied topically immediately after beak trimming.

B. Pigeons and Doves

Pigeons are widely used in research, particularly in psychology investigations. Pigeons have been successfully anesthetized with 2.5–3% isoflurane in a 1:1 O_2:N_2O mixture (Sap *et al.,* 1993). Fitzgerald (1993) found that the addition of N_2O decreased the minimum anesthetic dose of isoflurane from 1.51 to 1.02% in pigeons.

Propofol has been investigated for use in pigeons. An intravenous dose of 14 mg/kg provided smooth induction and muscle relaxation, but also caused severe respiratory depression and has a narrow safety margin. Duration of anesthesia was 2–7 min (Fitzgerald and Cooper, 1990).

Ketamine (20 mg/kg IM) and xylazine (10 mg/kg IM) were used to anesthetize pigeons for intraocular injections (Shimizu *et al.,* 1994). Pentobarbital and ketamine have been used (Dickman and Correia, 1989); however, neither of these agents provides good muscle relaxation or adequate visceral analgesia for invasive procedures. Urethane has also been used (Cabot and Bogan, 1987; Britto *et al.,* 1990; Necker, 1990; Neya *et al.,* 1990a); however, care must be taken to ensure adequate anesthesia during prolonged invasive procedures as well as appropriate personnel protection from this carcinogen. Chloral hydrate has been used with the local infiltration of lidocaine for occular surgery in pigeons (Fontanesi *et al.,* 1993). This agent should be preceded by administration of an opioid, barbiturate, or sedative, and should not be given IP for survival surgery (Silverman and Muir, 1993).

Gingicain (pantocain, 1.4 mg/bird) has been used to topically anesthetize the olfactory mucosa of pigeons (Kiepenheuer, 1985; Ganzhorn and Burkhardt, 1991).

C. Ducks and Geese

This group (Anseriformes) includes many species that are divers and have been used in baroreceptor studies.

Inhalant anesthetics have been used extensively in ducks. Olsen *et al.* (1992) used isoflurane for induction (4%) and maintenance (2–2.5%) of anesthesia during intraabdominal radiotransmitter implantation. The minimal anesthetic concentration of isoflurane in Pekin ducks was determined to be 1.3% (Ludders *et al.,* 1990). This study also demonstrated that isoflurane can cause significant hypoventilation in ducks as in other birds (Ludders *et al.,* 1990) and stressed the appropriateness of assisted ventilation in birds anesthetized with this agent. However, care must be taken not to overoxygenate, as hyperoxia (fraction of inspired oxygen > 40%) will contribute to hypoventilation in isoflurane-anesthetized ducks (Seaman *et al.,* 1994). Isoflurane (3.5% induction, 2–3% maintenance) was compared to halothane (3.5% induction, 2–2.5% maintenance) in terms of blood pressure, heart rate, and respiratory rate. Respiratory depression and cardiac arrhythmias were observed with halothane. Isoflurane had shorter induction and recovery times, was less of a cardiopulmonary depressant, and was not arrhythmogenic (Goelz *et al.,* 1990). Halothane has been used alone (Steeves *et al.,* 1987) or with nitrous oxide (3% induction and 1–2% maintenance with 20% N_2O) for cranial surgery in ducks and geese (Sholomenko *et al.,* 1991).

Injectables used in ducks include tiletamine–zolazepam (Telazol), 13 mg/kg IM, supplemented 3.0 mg/kg as necessary, for liver biopsy. Lidocaine (10 ml of a 1% solution) was also infiltrated intradermally and deeper to complete anesthesia. The authors suggested an even lower lidocaine concentration (0.5%) because of an anesthetic death associated with the administration of 2% lidocaine (Carp *et al.,* 1991). Other authors have also noted toxicity with lidocaine (Hall, 1987).

Sodium pentobarbital has been used in ducks at dose ranges of 25–40 mg/kg IV for a variety of invasive procedures (Adamson and Burger, 1986; Hempleman *et al.,* 1986; Hastings and Powell, 1987; Bickler *et al.,* 1989; Smith and Jones, 1990, 1992; Hempleman and Bebout, 1994). As stated for other avian species, pentobarbital does not provide good analgesia or muscle relaxation, except in dangerously high doses (60 mg/kg). Because of this, use of pentobarbital as a sole anesthetic should be discouraged.

D. Songbirds

Finches are among the species of songbirds most commonly used in research. Invasive and often complex investigations of neural control of song have been conducted on zebra finches, using a variety of anesthetic agents. Methoxyflurane has been used for induction followed by maintenance with 1–2 mg ketamine and 0.2 mg xylazine (Schariff and Nottebohm, 1991; Morrison and Nottebohm, 1993). Alternatively, 0.05 ml of a 50/50 ketamine–xylazine mixture (100 mg/ml ketamine and 20 mg/ml xylazine) has been used for induction, supplemented as necessary with methoxyflurane (Williams *et al.,* 1992). Prolonged recovery times and vacuolization of cortical neurons observed with ketamine–xylazine anesthesia (25 mg/kg ketamine and 50 mg/kg xylazine) were significantly decreased by administering yohimbine (1.25 mg/kg). Overdose of yohimbine (25 mg/kg) has proven fatal in zebra finches (Kilander and Williams, 1992). Other investigators have used ketamine (30 mg/kg) and xylazine (40 mg/kg) for stereotaxic surgery (Vicario, 1991; Vicario and Yohay, 1993). These ketamine–xylazine and yohimbine doses are much higher than the standard clinical dose used (5–30 mg/kg ketamine IM, 1.0–4.0 mg/kg xylazine IM, and 0.1 mg/kg yohimbine) (Degernes *et al.,* 1988; Sinn, 1994). Caution should be exercised when using high doses of these agents, especially xylazine, which causes hypotension and bradycardia.

Equithesin (0.04 ml total volume IM; 3.2 ml/kg IM) has been given to sparrows and zebra finches prior to cranial surgery (Tsutsui *et al.,* 1991; Nordeen and Nordeen, 1992; Richardson and Boswell, 1993). In two cases, reference is made to animals moving under anesthesia; one paper even states that birds under general anesthesia respond to toe pinch (Tsutsui *et al.,* 1991). In view of the findings by Christensen *et al.* (1987) that Equithesin alone is inadequate for laparatomy in chickens, it is unlikely that this drug provided adequate anesthesia for passerines; in fact, leg and body movement could be attributed to a light plane of anesthesia.

Urethane and DIAL have also been used in passerine species (Laverty and Alberici, 1987; Roberts and Dantzler, 1990; Margoliash and Fortune, 1992). The same precautions followed for other birds should be taken when using these agents in passerines.

E. Parrots and Budgies

Isoflurane is the anesthetic agent of choice for psittacines, including parrots, cockatiels, and budgerigars (parakeets) (Fig. 6). Ketamine–xylazine (40 mg/kg IM ketamine and 10 mg/kg IM xylazine) has been used to anesthetize budgerigars for neurosurgery (Heaton and Brauth, 1991). These investigators also demonstrated that yohimbine (0.275 mg/kg) was effective in hastening recovery from ketamine–xylazine anesthesia. Ketamine–xylazine (5–30 mg/kg IM ketamine and 1.0–4.0 mg/kg xylazine) has been used for surgical procedures in psittacines (Sinn, 1994). Ketamine–diazepam (5–30 mg/kg IM ketamine and 0.5–2.0 mg/kg IM diazepam) (Sinn, 1994) can be used for minor procedures. In addition to yohimbine, ketamine–xylazine anesthesia time in turkey vultures has been shortened by administration of the reversal agent, tolazoline (15 mg/kg IV) (Allen and Oosterhuis, 1986).

F. Analgesics for Birds

Several analgesic agents have been evaluated in chickens (see Section II, A). Butorphanol (3–4 mg/kg) has been studied in budgerigars (Bauck, 1990). Birds showed no significant differences in heart rate or respiratory rate. At this dose, mild motor deficits were observed for 2–4 hr. Bennett (1994) recommends starting with 0.2–0.4 mg/kg and working upward, as needed. Butorphanol has been used clinically at 0.5–2.0 mg/kg IM (Jenkins, 1993). Buprenorphine has also been used clinically at 0.01–0.05 mg/kg IM (Jenkins, 1993). Buprenorphine at 0.32 mg/kg IM has been shown to be a partial μ-agonist in pigeons; antinociceptive effects were not studied (Leander, 1988). Flunixin meglumine (1–10 mg/kg IM SID) can be used to treat painful inflammatory conditions, but is not usually effective for surgical pain (Bauck, 1990; Jenkins, 1993; Bennett, 1994). At 10 mg/kg, vomiting can occur (Bauck, 1990).

Aspirin (one 5-grain tablet dissolved in 250 ml drinking water) has also been used for inflammatory pain. Attention should be paid to potential gastrointestinal irritation and clotting deficits when using nonsteroidal agents (Bennett, 1994).

IV. REPTILES

Class Reptilia can be subdivided into four orders, three of which are commonly represented in research situations (Table III).

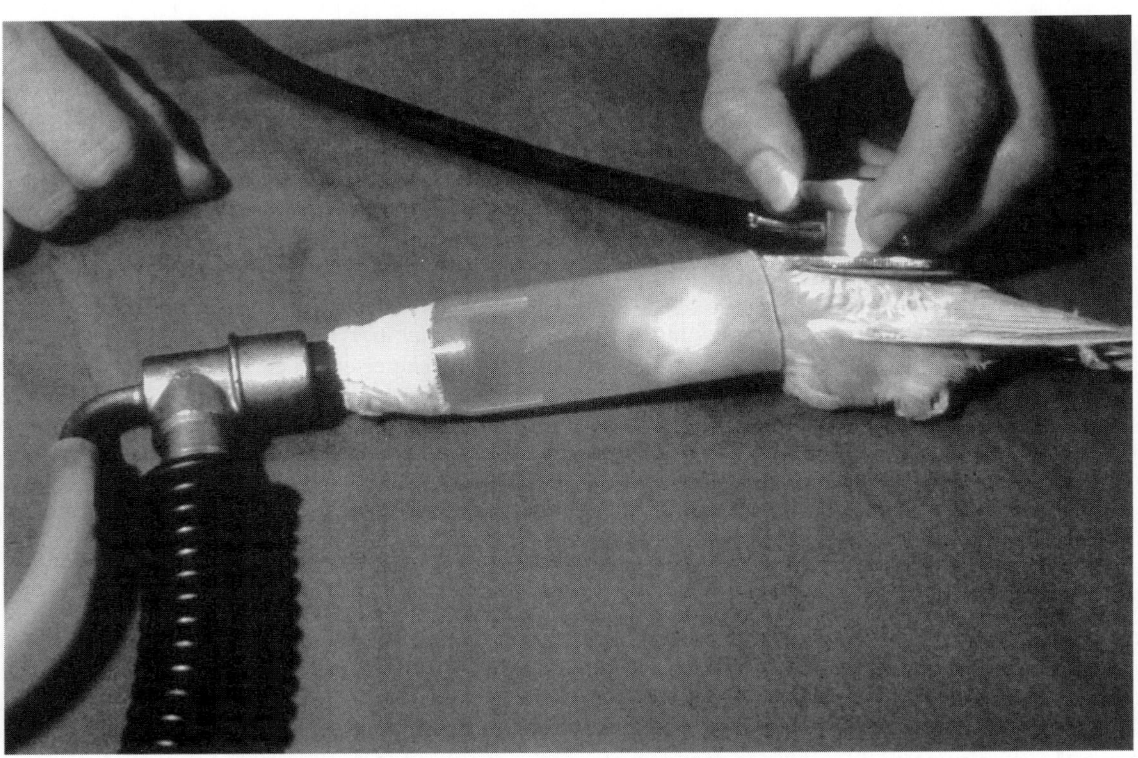

Fig. 6. Isoflurane anesthesia delivered via a face mask to a budgerigar. Photograph courtesy of Dr. S. E. Orosz.

TABLE III

SELECTED REPTILIAN SPECIES USED IN RESEARCH

Common name	Genus and species
Garter snake	*Thamnophis sirtalis*
Water snake	*Nerodia* sp.
Rat snake	*Elaphe obsoleta*
Corn snake	*Elaphe gutatta*
King snake	*Lampropeltis* sp.
Rattlesnake	*Crotalus* sp.
Copperhead	*Agkistrodon contortrix*
European adder	*Vipera berus*
Coral snake	*Micrurus fulvius*
Cobra	*Naja* sp.
Alligator	*Alligator mississippiensis*
Common caiman	*Caiman crocodilus*
Anole	*Anolis carolinensis*
Green iguana	*Iguana iguana*
Broad-head skink	*Eumeces laticeps*
Tokay gecko	*Gekko gecko*
Mud turtle	*Kinosternon subrubrum*
Slider	*Trachemys scripta*
Painted turtle	*Chrysemys picta*
Snapping turtle	*Chelydra serpentina*
Box turtle	*Terrepene carolina*
North American tortoise	*Gopherus* sp.
Herman's tortoise	*Testudo hermanni*

Squamata (snakes and lizards) and Chelonia (turtles and tortoises) are seen most frequently in research settings. Crocodilians (alligators, crocodiles, and caimans), although not used as extensively, are often the animal model chosen for neurobiological investigations.

Reptiles are distinct from mammals and birds in several respects. They are poikilothermic (''cold-blooded''), i.e., they require external heat sources to maintain their preferred body temperatures. This is accomplished most frequently by engaging in behavioral thermoregulation (moving to warmer or cooler areas as desired). A few species (e.g., female pythons incubating eggs) can actually raise their body temperature a few degrees by shivering. Maintaining the reptile at its optimal temperature is essential before, during, and after surgery. The immune system may be compromised if the animal is kept too cool. Doses required to achieve deep levels of anesthesia are decreased and recovery rates are prolonged at lower temperatures (Arena and Richardson, 1990). Additionally, if a reptile is anesthetized in a cool environment and subsequently recovered at a warmer temperature, many of the lipid-soluble agents (such as methoxyflurane) are released from fat, resulting in relapse into a deeper plane of anesthesia (Kaplan, 1969; Burke and Wall, 1970; Calderwood, 1971; Bennett, 1991). If the preferred temperature range of a specific reptile is not known, authors have recommended 22–24°C (Millichamp, 1988), 75–85°F (Heard, 1993), and 26–32°C (Page, 1993). The respiratory anatomy of reptiles is also different from that of homeotherms. The glottis is opened by the glottis dilator muscle only during respiration (Wood and Lenfant, 1976; Millichamp, 1988; Bennett, 1991). It is situated more rostrally in reptiles than in mammals, and its exact position varies with species. Generally, the glottis is readily visible immediately behind the tongue sheath in snakes. It is most accessible in this group (Millichamp, 1988; Bennett, 1991) (Figs. 7a and 7b). In turtles and crocodilians, the glottis lies at the base of a fleshy tongue and is more difficult to see. In crocodilians, the basihyal valve (epiglottis) is large and forms a seal with the soft palate (Calderwood, 1971; Davies, 1981; Bennett, 1991). This adaptation, which allows the animal to hold prey in its mouth while submerged, prevents easy visualization of the glottis (Figs. 8a and 8b). This author uses an oral speculum to keep the mouth open and a laryngoscope to depress the basihyal valve to allow visualization and intubation of the crocodilian glottis (Fig. 9).

Tracheal rings are complete in turtles and crocodilians and incomplete in lizards and snakes (Davies, 1981; Bennett, 1991). In species with complete tracheal rings, uncuffed endotracheal tubes are preferred because overinflation of cuffs can cause mucosal damage (Bennett, 1991). In turtles, the trachea is short; care must be taken not to pass the tracheal bifurcation and intubate only one lung (Bennett, 1991).

Most reptilian lungs are endothelium-lined sacs that are attached to the bronchi. Crocodilians are the exception, with a lung almost as complex as that of a mammal (Davies, 1981; Bennett, 1991). Turtles, lizards, and crocodilians have two, equal-sized, functional lungs. Primitive snakes (boas and pythons) have a well-developed right lung and a smaller left lung. In more advanced snakes, the left lung is vestigial (Millichamp, 1988; Bennett, 1991; Jacobson, 1993). The respiratory part of the lung in boids (boas and pythons) and colubrids (garter snakes, rat snakes, etc.) is situated between the heart and the liver. In viperids (rattlesnakes, adders, etc.) and elapids (coral snakes, cobras, mambas, etc.) the respiratory portion of the lung is cranial to the heart (Jacobson, 1993). Tracheal lungs (alveolar tissue in the dorsal part of the trachea) are present in some species. These are thought to have evolved for respiratory or behavioral functions (Davies, 1981; Bennett, 1991; Jacobson, 1993). The snake lung terminates in a large air sac, which is variable in length, thin-walled, and fragile. This air sac may serve as an air reservoir or have a buoyancy function. Care should be taken not to overinflate and rupture the lungs and air sacs during assisted ventilation (Millichamp, 1988; Bennett, 1991; Jacobson, 1993). Reptiles lack a true muscular diaphragm and therefore have a coelomic cavity. Turtles and crocodilians have membranous structures that function in a similar manner to a diaphragm (Bellairs, 1969; Davies, 1981; Millichamp, 1988; Bennett, 1991). Reptiles ventilate by movement of intercostal, trunk, and abdominal muscles. Additionally, the lung walls contain smooth muscle, which functions in ventilation (Davies, 1981; Bennett, 1991). These adaptations permit ventilation even if the coelomic cavity is opened (Bennett, 1991). Because turtle ribs are fused to the carapace, intercostal muscles

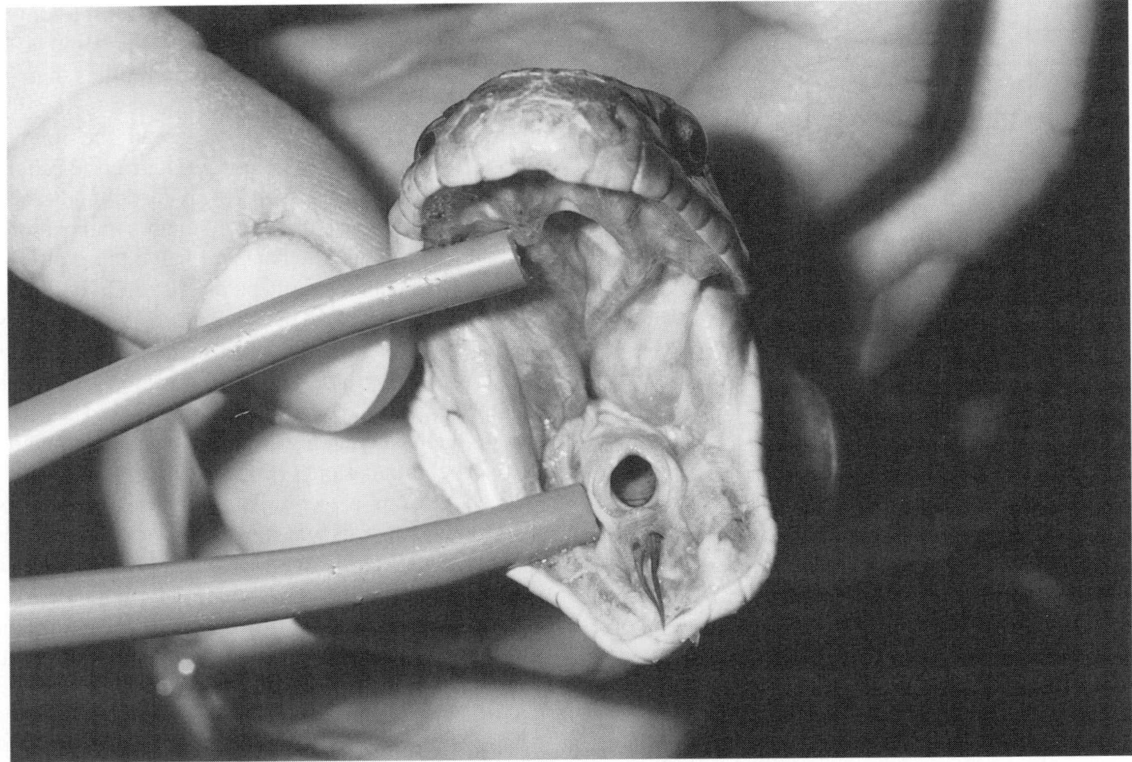

a

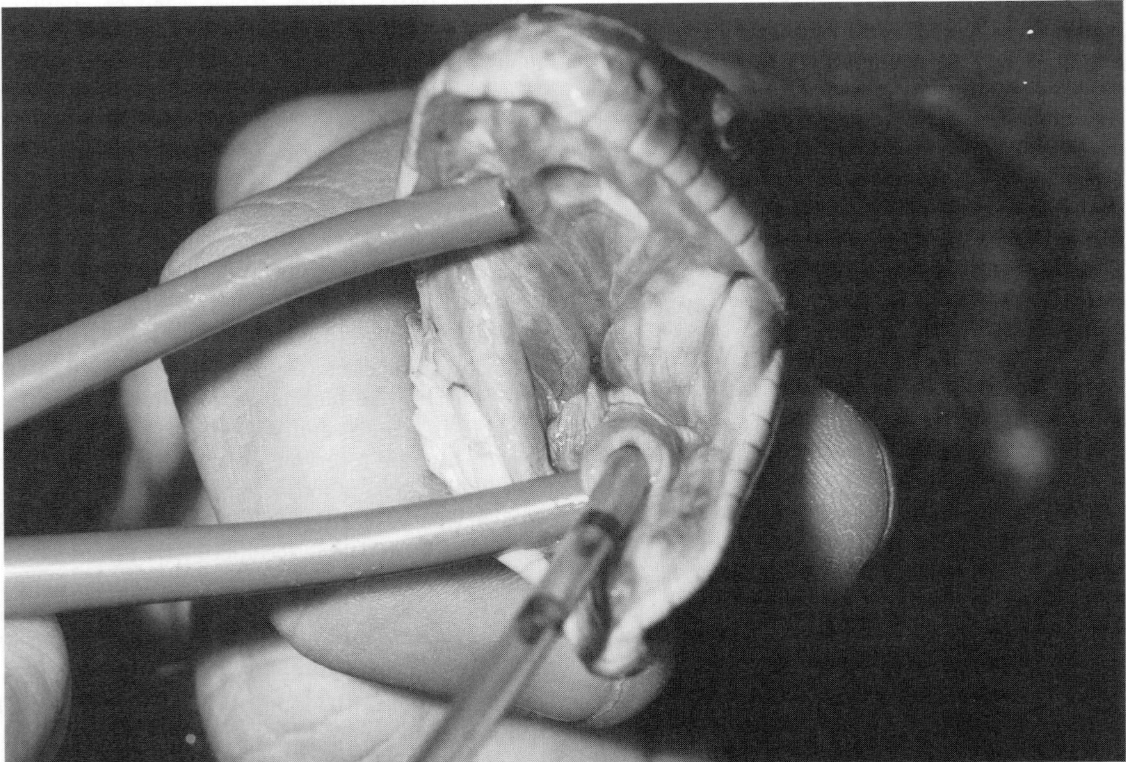

b

Fig. 7. (a) The glottis in snakes is immediately caudal to the tongue sheath. (b) Most snakes can be directly intubated, then manually ventilated until anesthetized.

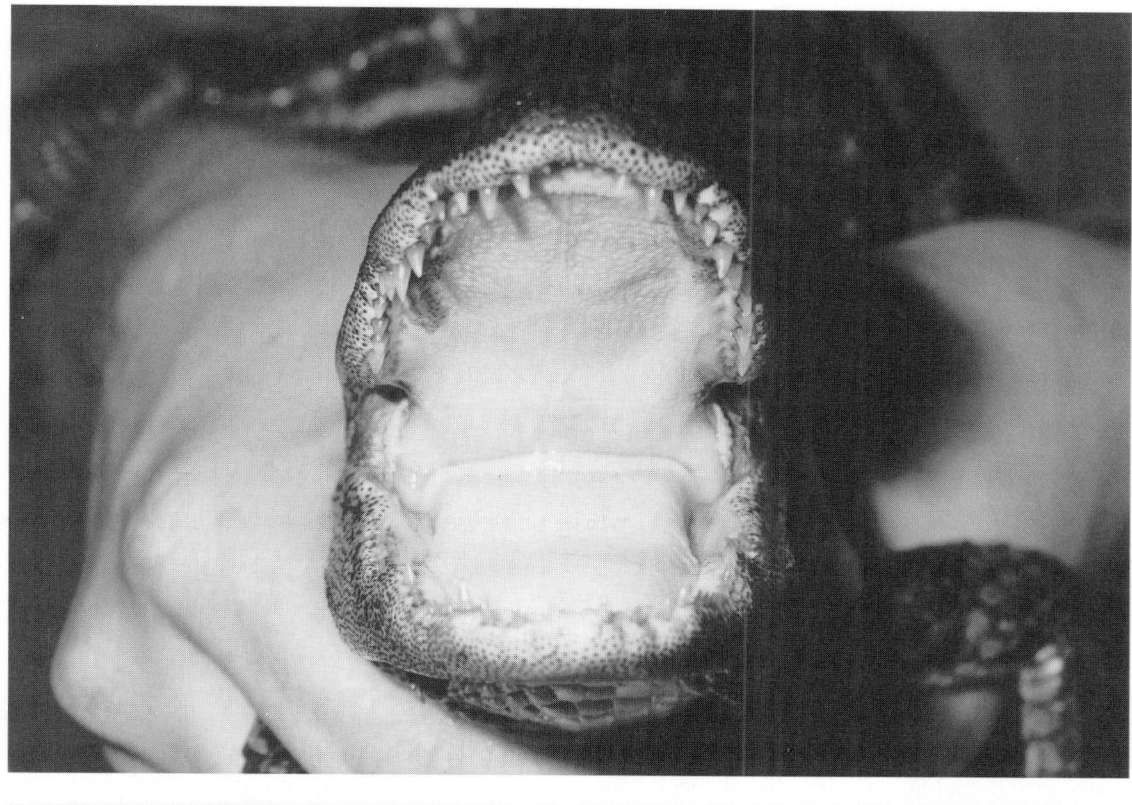

a

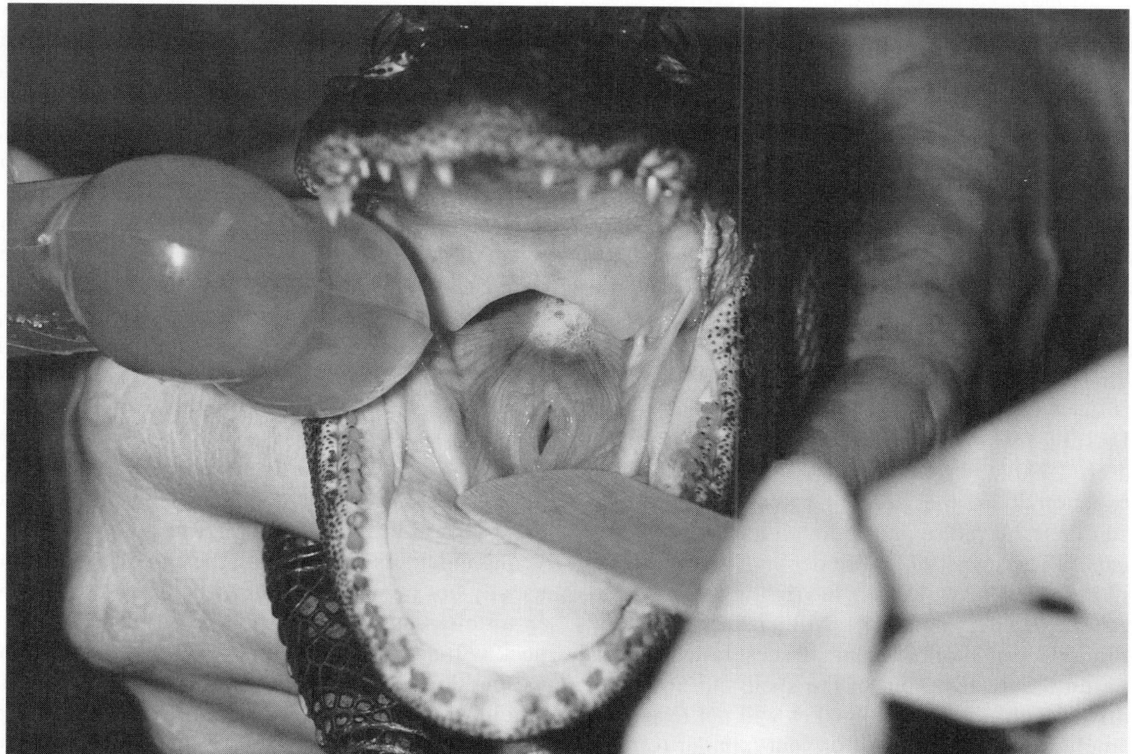

b

Fig. 8. (a) Alligators have a basihyal valve which, when elevated, forms a seal to prevent aspiration of water. (b) The basihyal valve must be depressed to visualize the glottis.

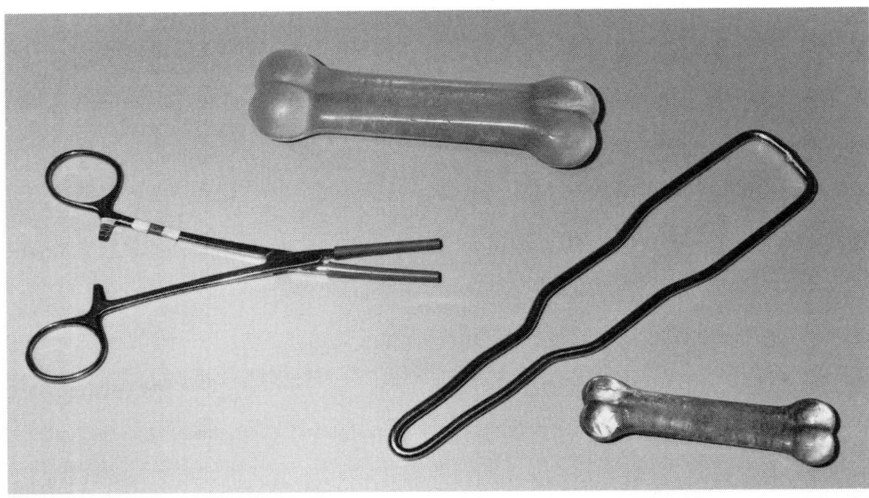

Fig. 9. A variety of atraumatic oral speculums can be used for birds and reptiles. Fitting part of a red rubber feeding tube over the tips of hemostats creates a speculum useful for most snake species.

cannot be moved; turtles ventilate by moving abdominal muscles and viscera, limbs, and the pelvic girdle (Calderwood, 1971; Davies, 1981; Bennett, 1991). Because turtle lungs are situated dorsally in the coelomic cavity, dorsal recumbency can cause viscera to compress the lungs and reduce tidal volume. Assisted ventilation (four to six breaths per minute) should be provided in these cases (Millichamp, 1988; Bennett, 1991).

Reptiles are resistant to hypoxia and convert easily to anaerobic metabolism (Millichamp, 1988; Bennett, 1991; Schaeffer, 1994). Some turtle species have survived in a 100% nitrogen atmosphere for 27 hr (Johlin and Moreland, 1933; Calderwood, 1971; Bennett, 1991). Iguanas can remain apneic for 4.5 hr (Moberly, 1968; Bennett, 1991).

Reptiles should be fasted before anesthesia to prevent regurgitation and putrefaction of food and to improve ventilation in snakes (Millichamp, 1988; Bennett, 1991; Page, 1993). A minimum fast of 24–48 hr is common (Page, 1993). Some authors recommend fasting for several days (Millichamp, 1988; Arena and Richardson, 1990).

During induction of anesthesia, reptiles relax anterior to posterior, and muscle tone returns posterior to anterior (Marcus, 1981; Bennett, 1991; Page, 1993). The depth of anesthesia can be monitored in several ways. Loss of the righting reflex occurs early in anesthesia (Calderwood, 1971; Millichamp, 1988; Bennett, 1991). In a deeper, surgical plane of anesthesia, the head and legs of a turtle will hang out of the shell and not retract when a painful stimulus is applied. Snakes will not respond to a painful stimulus applied to the tail or vent (similar to mammalian anal sphincter tone). Snakes fail to respond when a stimulus is run down the back or ventral scales (Calderwood, 1971; Bonath, 1979; Millichamp, 1988; Bennett, 1991). In crocodilians, lack of response to hind toe or tail pinch is a good indication of surgical anesthesia. Loss of the corneal reflex in most reptiles occurs in the deepest plane of anesthesia (Bonath, 1979; Millichamp, 1988; Bennett, 1991). This reflex should be present in a surgical plane of anesthesia. Tongue withdrawal reflex in snakes is reported by some authors to be present at a surgical plane of anesthesia (Betz, 1962; Bennett, 1991). Heard (1993) and Arena and Richardson (1990) state that the tongue withdrawal reflex is lost at a surgical plane of anesthesia. This reflex should be used in conjunction with others to ensure that the depth of anesthesia is appropriate.

Heart rate can be monitored by direct observation (snakes) or by an electrocardiograph monitor. In snakes, the left and right arm leads can be placed cranial to the heart and the left leg lead placed caudal to the heart (Bennett, 1991). A Doppler transducer placed over the heart is also useful for monitoring cardiac function (Heard, 1993; Page, 1993). The heart of snakes lies approximately one-fourth of the way down the body. This will vary with species; arboreal snakes have a more cranially located heart. The heart of some lizard species lies almost at the pectoral girdle (Heard, 1993). Crocodilian hearts are situated more caudally than in lizards. This author has often located the heart in young alligators by palpating the end of the sternum; the heart usually lies between the sternal tip and one scale row caudally. In tortoises, a Doppler pencil probe placed at the base of the neck will detect blood flow in the thoracic inlet (Heard, 1993). Respiration should also be monitored and assisted if necessary. A minimum of two to four breaths per minute is appropriate for most reptiles (Bennett, 1991). When ventilating snakes, adequate time must be allowed between respirations for the air sacs to empty. Turtles can be ventilated by moving their forelegs (Heard, 1993).

1. Inhalant Anesthetics

As for most species, inhalant anesthetics are the agents of choice for reptiles. Induction may be accomplished through the use of masks and chambers; however, in species that hold their breath (especially turtles and crocodilians), prior sedation may be necessary (Millichamp, 1988; Bennett, 1991). This author has directly intubated small- to medium-sized snakes, lizards, and crocodilians and induced them directly by manual ventilation. Topical anesthetic can be applied to the glottis before intubation (Heard, 1993). Venomous species of snakes can be induced to crawl into appropriately sized, clear plexiglass tubes (the diameter should be large enough for the snake to crawl easily into, but small enough to prevent its turning around once inside). The person providing restraint firmly holds the body of the snake and the tube so the animal cannot crawl forward or back out. The snake can then be induced by placing a mask over the front of the tube or, preferably, first induced by injecting an agent such as ketamine, followed by mask induction and eventual intubation (Fig. 10). Squeeze cages are also useful for injecting ketamine prior to induction in poisonous species.

Extreme caution should be used in handling *any* venomous reptile, even when anesthetized. Manual restraint should always be used in addition to chemical restraint, and a clear understanding should be established that complete immobilization is rarely achieved with injectable agents. Once a venomous snake is adequately immobilized, its fangs should be covered with cork or another protective substance. Depth of anesthesia should be constantly monitored in all poisonous reptiles by a designated, trained individual who has been assigned that responsibility only.

Isoflurane is the inhalant anesthetic agent of choice for reptiles. As with other species, it should be delivered through a precision vaporizer. Induction concentrations are generally 3–5%, and maintenance concentrations are 1–2% (Sedgwick, 1988a; Muir and Hubbell, 1989; Bennett, 1991; Frye, 1991; Jenkins, 1991; Heard, 1993). In species of turtles that hold their breath and cannot be directly intubated, 20–40 mg/kg IM ketamine may be given first to relax the head and permit opening of the mouth (Heard, 1993). Halothane has been used extensively in reptiles, with both open-drop and precision vaporizer systems. Because of its high vapor pressure and ability to reach lethal concentrations, most authors prefer to deliver halothane through a precision vaporizer (Millichamp, 1988; Bennett, 1991; Frye, 1991; Heard, 1993). Induction levels are usually 3–5% and maintenance levels are 1–2% (Jackson, 1970; Bonath, 1979; Green, 1979; Burke, 1986; Millichamp, 1988; Sedgwick, 1988a; Muir and Hubbell, 1989; Arena and Richardson, 1990; Bennett, 1991; Jenkins, 1991; Heard, 1993). Although halothane in an open-drop system has been used successfully, mortalities have been associated with this method (Bennett, 1991). This author has experienced both immediate and delayed deaths (within 24 hr) after apparent recovery from halothane, and strongly encourages use of a precision vaporizer when delivering this agent.

Methoxyflurane has been used in both open-drop systems and delivered through a precision vaporizer. Open-drop systems have used 1 ml/8400–19,000 cm^3 (Bennett, 1991). Jacobson

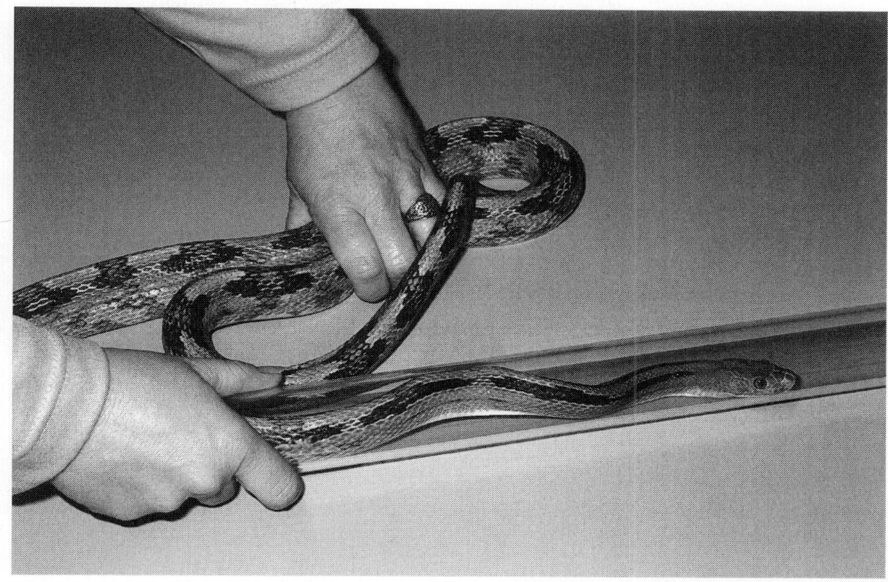

Fig. 10. A method used for restraining venomous or aggressive snakes. The animal can be encouraged to crawl into a clear plexiglass tube, then firmly held to prevent escape. Once restrained, injectable anesthesia can be readily administered.

(1984) used 10 ml in 42,800 cm^3. With a precision vaporizer, induction is achieved at approximately 3%; the maintenance concentration is 0.5% (Millichamp, 1988; Heard, 1993).

2. Injectable Anesthetics

If injectable agents are used, they may be delivered by several routes. Sites of injection vary, depending on species. Jugular veins can be catheterized in tortoises and turtles for intravenous injections. There is a large vein on the ventral midline of iguanas (Heard, 1993). Nile crocodiles have large ventrolateral vessels (Hocutt *et al.,* 1992). Other authors prefer to use the ventral venous sinus, which lies along the ventral midline of the tail in lizards and snakes. Turtles have a dorsal venous sinus, which can be located on the dorsal midline of the tail (Lawton, 1992; Haskell and Pokras, 1994). The dorsal vertebral sinus is sometimes used for bleeding and giving IV injections to crocodilians; however, incorrect needle placement will permit injection into the subarachnoid space (Heard *et al.,* 1988; Heard, 1993).

In large snakes, the palatine vein can be used. Intracoelomic injections in turtles can be given by entering the body cavity cranial to the hindlegs (Heard, 1993). Intracoelomic injections in lizards, snakes, and small crocodilians are similar to the IP method used for rodents.

Most injectable agents are delivered through the intramuscular route in reptiles. Reptiles have a renal portal system; veins in the caudal portion of the body drain directly to the kidneys, where drugs can be excreted before entering general circulation. Therefore, recommended sites for intramuscular injections include the forelegs of turtles, lizards, and crocodilians (Fig. 11) and the dorsal epaxial muscles in the cranial part of the body of snakes (Arena and Richardson, 1990; Boyer, 1992; Heard, 1993) (Fig. 12). Injections should be made between the scales rather than through them.

Some authors use atropine for premedication (0.01–0.04 mg/kg IM) to counteract bradycardia and decrease oral secretions (Frye, 1981). Others (Bennett, 1991; Heard, 1993) have not experienced these problems and do not routinely use atropine. Glycopyrrolate (10 μg/kg IM, IV, SQ) has also been used by some authors (Frye, 1991).

Acepromazine (0.1–0.5 mg/kg IM), given 1 hr prior to anesthesia, is reported to decrease the amount of barbiturate needed for anesthesia (Frye, 1981, 1991; Millichamp, 1988; Bennett, 1991). Some evidence suggests that certain groups of turtles (emydids) may lack benzodiazepine receptors; acepromazine would be ineffective in those species (Harvey-Clark, 1994).

3. Hypothermia

In many cases, hypothermia continues to be used as a sole anesthetic agent in reptiles (Calderwood, 1971; Peterson, 1987; Pritz and Stritzel, 1987, 1990, 1994; Janik and Menaker, 1990;

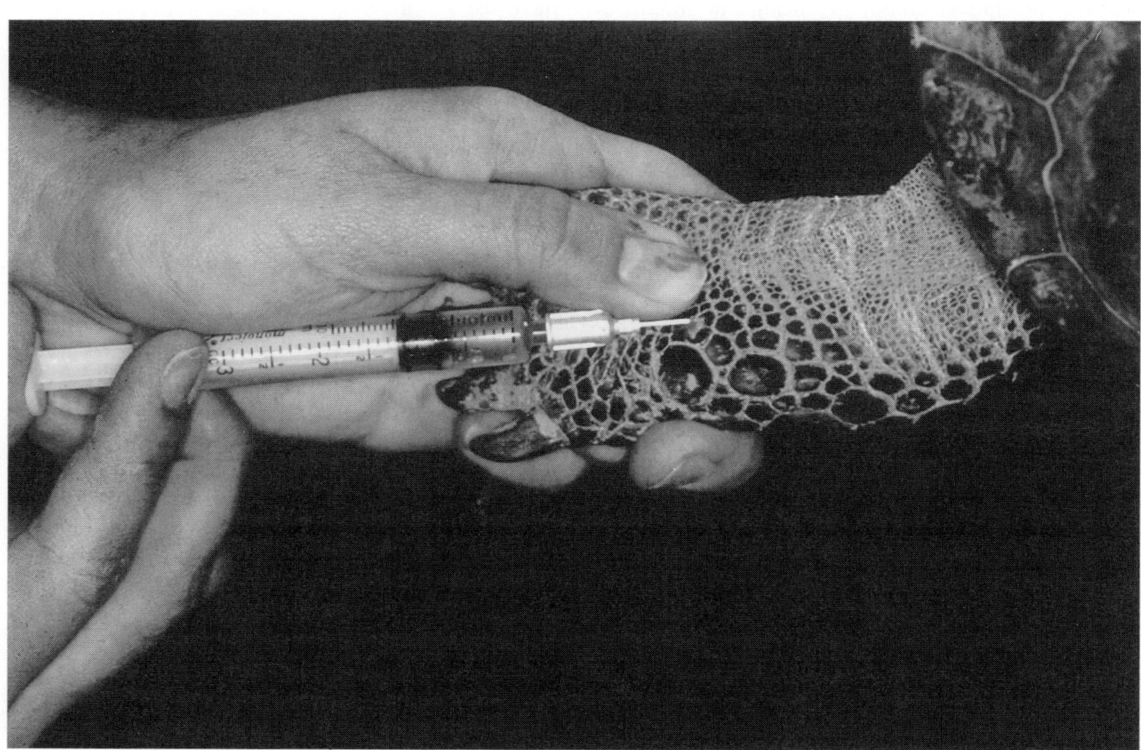

Fig. 11. Intramuscular injections should be given in the forelegs of turtles and tortoises.

Fig. 12. Intramuscular injections can be given in the dorsal epaxial muscles of snakes.

Wade and Crews, 1991; Roth and Jones, 1992; Shaw *et al.*, 1993). Rosenberg (1977) documented a nerve conduction velocity cold block below 3.5°C in warm acclimated tortoises and below 2.4°C in tortoises exposed to cold for 1 month (Rosenberg, 1978). He found that warm acclimated animals had faster conduction velocities during high acute temperatures when compared with cold acclimated tortoises. He also found that below 20°C, no change in conduction velocity was observed for any given temperature. He concluded that no evidence existed for a compensatory change in nerve transmission velocity during cold acclimation (Rosenberg, 1977) and that there was no change in conduction velocities of cold acclimated tortoises at the lower end of the thermal range (Rosenberg, 1978).

The conduction velocity papers by Rosenberg (1977, 1978) have been repeatedly cited as both condoning and condemning hypothermia as an effective anesthetic agent. Early investigations of frog sciatic nerve conduction velocity demonstrated slowing as the nerve was cooled to as low as 5°C (Lucas, 1908; Rosenberg and Sugimoto, 1925; Gasser, 1931). This, along with the findings of Rosenberg, indicates that poikilotherm nervous function persists at lower temperatures than mammals.

Studies on human patients indicate that peripheral nerve conduction decreases in a linear fashion with decreasing temperature until 20°C and that muscle action potentials presumably cease below 9°C (deJong *et al.*, 1966). All of these investigations addressed conduction velocity in large, primarily motor nerves. Bonath (1979) observed that lack of reaction to a painful stimulus could be mistaken for analgesia in a cold-immobilized animal. More accurate assessment of sensory input presumably would involve measuring the effects of sensory nerve stimulation.

Sensory nerve function during hypothermia has also been addressed. Investigators have studied electroencephalograms, visual evoked response, auditory evoked response, and somatosensory evoked response to hypothermia in a variety of species. In trout being slaughtered by removal from water (brain anoxia), Kestin *et al.* (1991) demonstrated that cooling the fish to 2°C on ice actually prolonged the presence of response to a visual stimulus. They further demonstrated that general anesthesia with MS222, unlike hypothermia, abolished visual evoked responses in trout. Thus, one can assume that fish continue to resist hypoxia and perceive stimulus at temperatures close to freezing. This indicates that hypothermia does not cause anesthesia as is reported for mammals.

Corwin *et al.* (1982) investigated brain stem auditory evoked responses in several vertebrate classes. They demonstrated an overall similar pattern of diminishing auditory evoked response with cooling across species. In leopard frogs, however, they observed an actual increase in amplitude of the initial biphasic peak with initial cooling. The slow waves remained essentially unchanged as the animals were cooled to 4°C (Corwin *et al.*,

1982). Again, this indicates that poikilotherm sensory input remains intact at almost freezing temperatures. Furthermore, sensitivity to stimulus may be actually increased during the early phase of the cooling process.

Corwin *et al.* (1982) also recorded almost total abolition of the auditory brain stem response in frogs when a local anesthetic (2% xylocaine) was injected directly into the brain stem, suggesting that auditory evoked responses are abolished by anesthesia. In contrast, Rossi and Britt (1984) recorded the effect of cooling on brain stem auditory evoked response in cats. They also noted an initial increase in amplitude during early rapid cooling (37–32°C). In cats, however, the brain stem auditory evoked response was nonrecognizable below 20°C. Metabolic acidosis was a serious side effect of hypothermia in cats.

Parsons and Huggins (1965) recorded cortical electroencephalograms in caimans with cloacal temperatures of 2–4°C. In alligators, Strain *et al.* (1987) observed that brain stem auditory evoked potentials were present from 0 to 36°C in contrast to mammals. They also noted that below 20°C, amplitude of the peaks decreased with decreasing temperature; latencies, however, decreased only slightly from controls. This indicates that alligators do not experience blocked sensory nerve transmission under hypothermic conditions.

Guérit *et al.* (1994) investigated the use of somatosensory-evoked potentials in humans undergoing hypothermia as an adjunct to cardiovascular surgery. They concluded that the brain stem is more resistant to hypothermia than the cerebral cortex; therefore, somatosensory evoked potentials (which measure brain stem activity) are more appropriate than electroencephalograms (which measure cortical activity) for monitoring. If this is the case, then somatosensory evoked potentials would be a better indicator in poikilotherms because cortical development is minimal to absent in these species. Guérit *et al.* (1994) determined that there is a wide variety of brain susceptibility to hypothermia among humans. Presumably this individual variation exists in other species as well. They also identified a number of problems associated with too deep a level of cooling, including decreased insulin release, increase in blood viscosity, and massive catecholamine release (Guérit *et al.,* 1994).

Stevens and Pezalla (1989) demonstrated that cold temperatures (4°C) cause a decrease in nociceptive thresholds and less sensitivity to the protective effects of endogenous opioids in leopard frogs. Other side effects of chilling a poikilotherm include suppression of its immune system and increased susceptibility to infection (Frye, 1991). Additionally, hypothermia can damage tissues (Northcutt and Butler, 1974; Wang *et al.,* 1977). Because of this data, most authors adamantly oppose the use of hypothermia as a sole anesthetic agent (Cooper, 1984; Burke, 1986; Millichamp, 1988; Bennett, 1991; Frye, 1991; Lawton, 1992; Heard, 1993; Cooper and Williams, 1995).

In all studies of hypothermia conducted on large, vocal mammals, the subject was induced with general anesthesia prior to cooling (deJong *et al.,* 1966; Rossi and Britt, 1984; Guérit *et al.,* 1994). Small species that are less able to audibly express their pain (neonatal rats, frogs, lizards, etc.) are usually immersed directly in crushed ice or ice water (Kaplan, 1969; Phifer and Terry, 1986). The onset of rapid chilling (immersion of a body part in ice water) is intensely painful for humans. It is reasonable to assume that poikilotherms also feel this same initial pain. Additionally, rewarming is painful, and humans and other mammals usually remain under anesthesia for this procedure. Therefore, this author is reluctant to recommend immersion in crushed ice or ice–water baths as an appropriate method of anesthesia induction in poikilotherms.

The studies discussed previously indicate that mammals may be more appropriate subjects for adjunct anesthesia by hypothermia. Individual and species variation, combined with our incomplete understanding of pain processing in the poikilothermic nervous system, make hypothermia an inappropriate anesthetic agent for poikilotherms.

A. Snakes

Inhalant anesthetics are widely used in snakes, and isoflurane is the author's preferred anesthetic agent. Snakes can be intubated directly (Bodri and Sadanaga, 1991) or induced first with ketamine (20–40 mg/kg IM) or tiletamine–zolazepam (10–20 mg/kg IM). Recovery will be prolonged if injectables are used (Heard, 1993; Frye, 1994).

Halothane has been used extensively in snakes (Millichamp *et al.,* 1986; Slip and Shine, 1988a,b; Furilla and Bartlett, 1989; Ragsdale *et al.,* 1993). It has been delivered through a precision vaporizer at 1.5–5% for induction and approximately 1% to maintain (Smith and Barker, 1988; Brown and Martin, 1990; Garrett and Harwell, 1991). Ketamine (20–60 mg/kg IM) has been used to induce snakes prior to maintenance with halothane (Millichamp *et al.,* 1983; Grain and Evans, 1984; Crawley *et al.,* 1988). Halothane has been administered by the open-drop method (Calderwood, 1971). Reinert and Cundall (1982) used halothane to anesthetize viperids in an open-drop system. Burke and Wall (1970) reported that viperids require higher induction concentrations of halothane than elapids. Cobras require lower levels than most venomous snakes (Bennett, 1991). Aird (1986) reported both high mortality and unpredictably rapid recoveries using halothane in an open-drop system. Custer and Bush (1980) observed respiratory depression and respiratory acidosis in gopher snakes administered halothane.

Methoxyflurane has been used in an open-drop system to safely anesthetize gravid garter snakes (Charland, 1991). Burke and Wall (1970) found methoxyflurane, induced by open-drop and maintained by precision vaporizer, to be lethal to cobras. It is possible that these animals, having undergone a prolonged procedure, stored methoxyflurane in fat; warming during recovery could have mobilized the anesthetic, thereby ''relapsing'' the snakes into a lethal plane of anesthesia (Calderwood, 1971; Aird, 1986; Bennett, 1991).

Ketamine is frequently used as an injectable anesthetic for snakes. Up to 30 min of sedation can be obtained using a dose of 50 mg/kg IM. Several hours will be needed for recovery. Anesthetic doses range from 50 to 130 mg/kg IM, and recovery may take days (Millichamp, 1988). Cooper (1974) and Lawton (1992) noticed no difference in effect between SC and IM injections. Glenn *et al.* (1972) determined the lethal dose in the Western diamondback rattlesnake (*Crotalus atrox*) to be 154 mg/kg. These investigators also found that larger (3.6–4.5 kg) snakes required higher anesthetic doses (88–110 mg/kg) than did smaller (<0.9 kg) snakes (22–44 mg/kg). Because of severe respiratory and cardiac depression seen at high doses, snakes given >110 mg/kg ketamine should be provided with ventilatory support (Glenn *et al.,* 1972; Bennett, 1991). The total cumulative dose of ketamine in venomous snakes should not exceed 99–132 mg/kg or respiratory failure and death can result (Glenn *et al.,* 1972).

Several species of crotalids (rattlesnakes, copperheads, moccasins, etc.), elapids (mambas, cobras, etc.), and viperids (gaboon viper, Russell's viper, etc.) were anesthetized with 60–100 mg/kg ketamine for venom gland isolation (Glenn *et al.,* 1973). Other authors have used ketamine (88 mg/kg IM) for celiotomy in a python (Bodri *et al.,* 1990). Ketamine does not provide effective visceral analgesia; consequently, it should not be used alone for coelomic (abdominal) procedures (Jacobson, 1984; Bennett, 1991). Movement and lack of muscle relaxation are also seen with even high doses of ketamine (Glenn *et al.,* 1972; Cooper, 1974; Bennett, 1991).

Ketamine has been combined with acepromazine (20 mg/kg IM ketamine; 0.1 mg/kg acepromazine) for removal of a subcutaneous mass in a milk snake (Ewing *et al.,* 1991). Ketamine anesthesia has also been reported to cause permanent aggression in some species of snakes (Lawrence and Jackson, 1983).

Other injectables used in snakes include metomidate at 10 mg/kg for surgery in a rat snake (Zwart *et al.,* 1986) and for induction (8 mg/kg) prior to halothane anesthesia in a boa (Zwart *et al.,* 1988). Lawton (1992) has used metomidate at 10–20 mg/kg IM for sedation in snakes. He emphasizes, however, that this agent has no analgesic properties (Burke, 1986; Lawton, 1992).

Barbiturates have been used to induce anesthesia for a variety of invasive procedures in snakes. The mechanism of barbiturate elimination in reptiles is unknown. Short-acting thiobarbiturates produce induction and recovery times equivalent to pentobarbital in these animals (Calderwood, 1971; Bennett, 1991). Induction times are long and unpredictable, and recovery times are quite prolonged (Bennett, 1991).

Methohexital is a short-acting barbiturate that has been used in snakes (Dickman *et al.,* 1987; Krohmer and Crews, 1987; Morris and Crews, 1990; Graves and Duvall, 1993). A dose of 5–15 mg/kg SC is recommended, but there is much species variation with this drug. Induction is rapid; recovery is slower. The concentration used must not exceed 0.125–0.5% due to the alkalinity of the compound. This drug is not recommended for young or debilitated animals or animals undergoing ecdysis (Wang *et al.,* 1977; Millichamp, 1988; Bennett, 1991). Methohexital is useful for short procedures or for induction of anesthesia followed by maintenance with inhalants (Bennett, 1991; Frye, 1991).

Sodium pentobarbital has been used by some authors (Haldar-Misra, 1986; Furilla and Bartlett, 1988). A suggested dose for snakes is 15–30 mg/kg IP (Millichamp, 1988). Lower doses may not provide adequate analgesia or muscle relaxation, and higher doses have a narrow margin of safety. Larger snakes require a higher dose than smaller snakes; venomous snakes require a higher dose than nonvenomous snakes (Calderwood, 1971; Bennett, 1991). Because of extreme variation in depth of anesthesia, prolonged recovery times, and cardiac and respiratory depression, this drug is not generally recommended for anesthesia in snakes (Lawton, 1992).

Alphaxalone–alphadolone has been used in snakes (Calderwood and Jacobson, 1979; Lawrence and Jackson, 1983; Harper, 1984; Millichamp, 1988). Lawton (1992) suggests doses of 6–9 mg/kg IV or 9–15 mg/kg IM. Lawrence and Jackson (1983) found unreliable results in snakes and relaxation without analgesia in several individuals. Harper (1984) found intramuscular administration of this drug to be ineffective. This drug may be useful as an induction agent prior to inhalant anesthesia (Lawrence and Jackson, 1983; Bennett, 1991).

Propofol (10 mg/kg IV) is recommended for anesthetic induction in snakes. It provides smooth induction, minimal cumulative effects, and rapid, smooth recovery. Additionally, it is not viscous and can be delivered through a small-gauge needle. If the snake has had previous metomidate sedation, 5 mg/kg IV propofol should be given. Propofol must be given IV (Lawton, 1992).

Local anesthetics have been used in snakes. Lidocaine (2% without epinephrine) or procaine (1%) can be infiltrated around the incision site for minor surgical procedures (Bennett, 1991; Page, 1993). Lidocaine has been used in conjunction with ketamine for venom gland removal (Kiel, 1984). It has been used with hypothermia (for immobilization) and an opioid (for analgesia) to repair a prolapsed cloaca in a python (George and Joseph, 1989). Local anesthetics have also been dropped onto mucous membranes of facial pits to anesthetize and block thermal stimulation of these organs (Chiszar *et al.,* 1986).

B. Lizards

Isoflurane is the preferred anesthetic agent for lizards. They can be induced via a face mask or in a chamber (Barten, 1993). Other authors (Faulkner and Archambault, 1993; Heard, 1993) prefer to directly intubate or premedicate as lizards can hold their breath. When intubating lizards, an oral speculum should be used because many species are capable of inflicting powerful bites (Barten, 1993; Heard, 1993). Isoflurane has been used successfully for a wide variety of surgical procedures in lizard species (Faggella and Raffe, 1987; Anderson, 1992; Blahak, 1994; Walker, 1994). Halothane has also been used (Flanders, 1988; Anderson, 1992; Gerwien and John-Alder, 1992; Faulkner and Archambault, 1993). The reptiles most susceptible to inhalation anesthetic overdose are lizards (Heard, 1993), especially with halothane (Page, 1993).

Low-dose ketamine (20–40 mg/kg IM) is the method recommended by many authors to anesthetize lizards (Cooper, 1974; Bennett, 1991; Heard, 1993). Ogunranti (1987) found 30 mg/kg to be the optimal dose for agamid lizards. Font and Schwartz (1989) used much higher doses in *Anolis equestris* (179–336 mg/kg) and *Iguana iguana* (70–80 mg/kg). Even at these elevated doses, some animals required supplementation with ketamine or an inhalant anesthetic. Dodd and Capranica (1992) were only able to achieve consistent surgical anesthesia in Tokay geckos by combining 440 mg/kg ketamine with 1.0 mg/kg oxymorphone. Kwantes (1992) used ketamine (50 mg/kg) in conjunction with hypothermia to perform coelomic surgery. Arena *et al.* (1988) found that ketamine produced anesthesia at 170–230 mg/kg at 30°C in two species of Australian skink.

Tiletamine–zolazepam (10–30 mg/kg IM) has been used for anesthesia in lizards (Millichamp, 1988; Frye, 1991). Lower doses (1–4 mg/kg IM) have been used for immobilization prior to intubation (Heard, 1993). Pentobarbital use (22–30 mg/kg IM) has been reported in lizards (Bonath, 1979; Douse and Mitchell, 1988). Dodd and Capranica (1992) found pentobarbital (30 mg/kg) combined with oxymorphone (1 mg/kg) to be appropriate for short, minor neurosurgical procedures.

Urethane has been used in lizards (Rose and Weiss, 1988; Eatock *et al.,* 1991). Because of previously referenced safety concerns and lack of pharmacokinetic data, this agent is generally not recommended for reptiles (Bennett, 1991).

C. Turtles

Isoflurane is the preferred anesthetic agent for this group of reptiles (Mautino and Page, 1993). Because turtles are very difficult to intubate directly, they can be given ketamine (20–40 mg/kg IM) first for sedation; the head and mouth can be manipulated usually within 30–45 min (Heard, 1993). Succinylcholine (0.3–1.5 mg/kg IM) is used in this same fashion in giant tortoises; equipment should be available to intubate and provide ventilation if necessary. Succinylcholine has no analgesic properties (Heard, 1993; Mautino and Page, 1993). Bickler (1992) used pentobarbital to induce light anesthesia, followed by isoflurane. Turtles have been directly induced with halothane (Norton *et al.,* 1990). Ketamine induction (40 mg/kg IM) followed by halothane anesthesia has been used in giant tortoises (Page, 1985). Gould *et al.* (1992) used tiletamine–zolazepam (2 mg/kg IM) for induction prior to intubation. Although turtles are prone to breath holding when placed in anesthetic chamber, Brannian *et al.* (1987) found that kinosternid turtles (mud and musk turtles) were easily anesthetized using halothane in an open-drop system. These same authors found that delivery through a precision vaporizer set at 5% halothane was unsuccessful in inducing anesthesia in this group of turtles. They were unable to anesthetize box turtles, sliders, and snapping turtles using an induction chamber. Once turtles are intubated, they should be manually ventilated at 6–10 breaths per minute (Mautino and Page, 1993).

Ketamine (20–40 mg/kg IM) and tiletamine–zolazepam (5–25 mg/kg IM) have been used to provide dissociative anesthesia for minor surgical procedures (Mautino and Page, 1993). As mentioned for other drugs in reptiles, and in contrast to the principles of allometric scaling, smaller species of turtles require lower doses (Mautino and Page, 1993). Dominguez *et al.* (1986) used 50 mg/kg IM ketamine to remove a mass from the tail of a snapping turtle; Serantes *et al.* (1987) used 80 mg/kg IM for otic surgery in a Greek tortoise; and Gould and Georgi (1991) used 50 mg/kg IM for maggot removal from the cervical area of two box turtles.

Pentobarbital has been used alone (Ishii and Dejours, 1986; Ramos, 1986; Feng *et al.,* 1988; Hicks and Malvin, 1992). Wood *et al.* (1982) compared three injectable anesthetics in green sea turtles and found ketamine IP adequate for short procedures. Pentobarbital was suitable for longer procedures; however, both it and thiopental gave an inconsistent depth of anesthesia. Some authors have used high doses of ketamine in combination with pentobarbital for celiotomies (Davies, 1989, 1991). Kennett *et al.* (1992) premedicated *Chelodina* with diazepam prior to IV ketamine anesthesia for celiotomy. Four of 17 animals failed to reach a deep plane of anesthesia with this method and did not receive surgery. Ketamine (20–40 mg/kg IM) and midazolam (2 mg/kg IM) produced adequate restraint in snapping turtles (Bienzle *et al.,* 1991). Care should be exercised in choosing the appropriate drug to combine with ketamine for invasive procedures. Because the benzodiazepenes do not provide analgesia, these combinations should be used only for restraint. Likewise, the analgesic efficacy of pentobarbital is variable; combining it with ketamine may not provide surgical anesthesia. Tiletamine–zolazepam has been used in turtles (Gray *et al.,* 1974; Bennett, 1991; Jacobson *et al.,* 1991). Although effective for sedation, tiletamine–zolazepam did not provide surgical anesthesia, and animals remained responsive to stimulation (Gray *et al.,* 1974; Bennett, 1991).

Alphaxalone–alphadolone at a 15-mg/kg total dose provided brief surgical anesthesia in turtles (Lawrence and Jackson, 1983; Bennett, 1991). Harper (1984) found that the drug was consistently effective only if given IV at 9–12 mg/kg.

Local anesthetic agents have been combined with cold narcosis for cranial surgery in turtles. Lidocaine (2%) infused subcutaneously and benzocaine (20%) applied topically to incised skin and periosteum provided local analgesia (Prechtl and

Bullock, 1992). The brain itself is insensitive and requires no topical analgesics.

D. Crocodilians

Isoflurane is the agent of choice for anesthetizing crocodilian species. This author has manually restrained, intubated, and induced alligators up to 3 feet in length with isoflurane. The animal is restrained behind the head and at the base of the tail. Gentle tapping on the snout will encourage it to open its mouth. A mouth speculum (PVC pipe, Nylabone, etc.) is inserted into one side of the mouth. A laryngoscope is used to depress the basihyal valve, and the glottis can be visualized. An endotracheal tube is passed into the glottis, and the animal is manually ventilated until anesthetized. The mouth is then secured with a tape or rubber band muzzle (speculum is left in place). Ventilation is assisted manually at two to four breaths per minute throughout the procedure. Induction takes approximately 2 to 3 min, and recovery occurs within 15 min. Halothane has been used in this same manner. Isoflurane has been used in young Nile crocodiles (Croce, 1989). Hatchling American alligators were anesthetized with halothane for coelomic surgery (Austin, 1990). Lumeij and Happé (1985) induced an adult caiman with ketamine, then maintained it on a halothane–oxygen–nitrous oxide mixture.

Loveredge and Blake (1987) were unable to immobilize Nile crocodiles with 18–45 mg/kg ketamine. Large crocodilians are normally first immobilized with succinylcholine at 3–9 mg/kg IM or IP (Bennett, 1991; Page, 1993). Alligators can require as little as 0.4–1.0 mg/kg IM (Page, 1993). Tiletamine–zolazepam and atracurium were evaluated in juvenile American alligators. Tiletamine–zolazepam (15 mg/kg IM) was effective for immobilization and minor procedures; atracurium caused prolonged apnea and was therefore considered unsuitable (Clyde *et al.,* 1994). Alphaxalone–alphadolone has been reported to cause deaths in young crocodilians (Calderwood and Jacobson, 1979; Bennett, 1991).

Several authors have used pentobarbital for invasive surgical procedures (Powell *et al.,* 1988; Douse *et al.,* 1989; Powell and Gray, 1989; Douse and Mitchell, 1991). Loveridge and Blake (1987) found pentobarbital to be unsuitable for crocodilian anesthesia. This author has also attempted to induce pentobarbital anesthesia in young American alligators. Results of IP and IM administration were extremely variable and unpredictable, ranging from no anesthesia to respiratory depression and death. Pentobarbital is not recommended as a routine anesthetic agent in crocodilians.

A few investigators have used local anesthetics in conjunction with neuromuscular blocking agents in field surgery in large crocodilians (Hocutt *et al.,* 1992). Others have used lidocaine infusion (Douse *et al.,* 1989) or spray (Powell *et al.,* 1988) for venous access. Unlike amphibians, the skin of reptiles is impermeable; topicals are effective only on mucous membranes.

E. Analgesics for Reptiles

Few analgesic studies have been conducted in reptiles. Kanui *et al.* (1990) determined that the Nile crocodile has a well-developed nociceptive system. They further demonstrated that morphine (0.5–1.0 mg/kg IP) increases the latency of response in the hot plate test and therefore provides analgesia to crocodiles (Kanui and Hole, 1992). Tonic immobility induced by manual restraint in the green anole (*Anolis carolinensis*) caused increased sensitivity to noxious stimuli. It should therefore not be used if painful procedures are to be performed (Mauk *et al.,* 1981). These authors, however, found that morphine (5 mg/kg IP) had an analgesic effect on *Anolis.*

Heard (1993) has clinically observed sedation in gopher tortoises given butorphanol (0.2 mg/kg IM). He has also listed flunixin (0.5–1.0 mg/kg IM q 24–72 hr), morphine (0.5–2.0 mg/kg SC or IM q 12–48 hr), meperidine (5.0–10.0 mg/kg IM q 12–24 hr), oxymorphone (0.05–0.2 mg/kg SC or IM q 12–48 hr), nalbuphine (1.0 mg/kg IM q 12 hr), pentazocine (2.0–5.0 mg/kg IM q 6–24 hr), and buprenorphine (0.005–0.02 mg/kg IM q 24–48 hr) as potential analgesic agents. These extrapolated doses can provide the veterinarian with a starting point until specific nociceptive and analgesic investigations are conducted and more appropriate analgesic doses are derived.

V. AMPHIBIANS

Representatives of class Amphibia commonly used in research include anurans (frogs and toads) and caudates (salamanders) (Table IV). Amphibians are poikilothermic and live in a variety of habitats. Frog species range from temperate (bullfrogs, leopard frogs) to tropical (*Xenopus*) climates. Temperate species should be kept at 20–25°C and tropical species at 25–30°C (Raphael, 1993). Salamanders in general prefer a cooler climate, with many of the temperate species having preferred

TABLE IV

SELECTED AMPHIBIAN SPECIES USED IN RESEARCH

Common name	Genus and species
Axolotl	*Ambystoma mexicanum*
Tiger salamander	*Ambystoma tigrinum*
Slimy salamander	*Plethodon glutinosus*
Red-spotted newt	*Notopthalmus viridescens*
Mudpuppy	*Necturus maculosus*
Leopard frog	*Rana pipiens*
Bullfrog	*Rana catesbeiana*
Tree frog	*Hyla* sp.
Marine toad	*Bufo marinus*
African clawed frog	*Xenopus laevis*

temperature ranges of 10–16°C (Jaeger, 1992). Tropical species require 15–20°C (Raphael, 1993). Amphibians represent a transition from an aquatic environment to a terrestrial one. Although many species are terrestrial, all need water to reproduce. They also require high environmental humidity (Crawshaw, 1993). Amphibian larvae are aquatic and breathe through gills, much like fish. They possess a two-chambered heart, also like fish. Most amphibians undergo metamorphosis, develop lungs, and begin a terrestrial existence. Some larval salamanders do not metamorphose; these retain gills and other larval characteristics, yet grow to adult size and are able to reproduce. This condition is referred to as neoteny. The axolotl (*Ambystoma mexicanum*), commonly used in research, is a neotenous salamander. Adult frogs breathe with lungs. Depending on the species, adult salamanders may breathe through lungs, gills, both, or neither (Fig. 13). The plethodontid salamanders are commonly known as ''lungless'' salamanders. Cutaneous respiration is the primary form of breathing in lungless salamanders and an important method in other species when exposed to low temperatures. Amphibians also breathe through the buccopharyngeal cavity (Crawshaw, 1992). Amphibian skin is extremely permeable, and water is absorbed across the skin rather than ingested (Crawshaw, 1992). This permeability allows amphibians to be anesthetized by immersion methods; it also makes these animals susceptible to dehydration and toxin absorption. Consequently, amphibian skin must always be kept moist, especially during surgery and recovery. Dechlorinated water with a pH of 6.5–8.5 should be used (Raphael, 1993). A protective mucous layer covers amphibian skin. When handling these species, care must be taken not to mechanically disrupt this barrier and permit entry of pathogens. Another unique modification of amphibians is the presence of lymph sacs in frogs. These structures lie subcutaneously on the caudodorsal surface of the frog and drain into the venous system. Injections may be made directly into the dorsal lymph sacs. Like reptiles, amphibians also have a renal portal system which drains venous blood from the caudal part of the body through the kidneys (Raphael, 1993).

When an amphibian is anesthetized, it loses its righting reflex. As depth of anesthesia increases, first abdominal respiration then gular movements are lost. At this point, the amphibian is in a surgical plane of anesthesia (Crawshaw, 1993). Response to noxious stimuli and heart rate are used to monitor anesthesia (Raphael, 1993). Tricaine methane sulfonate (MS222), a crystalline powder, is used as the anesthetic of choice for many amphibian species. Tadpoles and newts require 200–500 mg/liter of MS222 in deionized water. Five hundred milligrams to 2 g/liter is used for frogs and salamanders, and toads need 1–3 g/liter (Cooper, 1987b; Arena and Richardson, 1990; Crawshaw, 1993). Concentrations over 500 mg/liter should be buffered as MS222 is quite acidic (Heyer *et al.,* 1994). Sodium bicarbonate at 10–25 meq/liter (420–1050 mg/liter) can be used as a buffer (Crawshaw, 1992). Alternatively, 0.5 *M* dibasic sodium phosphate (Na_2HPO_4) can be used. Thirty-four milliliters of Na_2HPO_4 added to 2 liters of a 1-g/liter MS222 solution, will raise the pH from about 3.0 to approximately 7.0 (Downes, 1995). An unbuffered MS222 solution is predominantly ionized and nonabsorbable. Therefore, in unbuffered solutions, induction time is prolonged and the acidic solution is stressful to the amphibian (Downes, 1995). Induction may take up to 15 min using these concentrations (Crawshaw, 1992). Once the appropriate depth

Fig. 13. The tiger salamander is a terrestrial species that breathes through lungs, whereas some other species of salamanders are ''lungless.''

of anesthesia is reached, the amphibian can be removed for surgery (anesthesia should persist for 5–20 min), left in a shallow dish of the same solution, or removed to a dish containing a lower maintenance concentration (Arena and Richardson, 1990; Crawshaw, 1992; Downes, 1995).

Benzocaine has been used in a similar manner to MS222. Larvae can be anesthetized with 50 mg/liter, and adult frogs and salamanders require 200–300 mg/liter. Benzocaine must be dissolved in ethanol to render it more soluble in water (Crawshaw, 1993).

Ketamine has been administered at 50–150 mg/kg SC or IM for restraint and minor procedures. It does not provide adequate analgesia for major surgical procedures (Crawshaw, 1993).

Inhalant anesthetics have been used in amphibians. Terrestrial species can be induced with isoflurane or halothane (4–5%) in a chamber. Induction may take 20 min or longer. If desired (and feasible), the amphibian can be carefully intubated (the trachea is short) and maintained on lower concentrations with assisted ventilation (Crawshaw, 1993). Isoflurane and halothane can be bubbled through the water to anesthetize aquatic species (Raphael, 1993). Arena and Richardson (1990) used 0.5–1 ml methoxyflurane on a cotton ball in a 1-liter container to anesthetize adult frogs. The prepared chamber should sit for 20–30 min to produce a 3% vapor before introducing the frog. Induction takes approximately 2 min, and anesthesia lasts for approximately 40 min after the frog is removed from the chamber. Recovery can be prolonged.

Amphibians can be recovered postanesthesia by rinsing them with clean deionized water and placing them on a moist cloth. Doxapram administered to effect can be used to stimulate respiration (Raphael, 1993).

Hypothermia has been used by some investigators as an anesthetic for invasive intracoelomic procedures (Mihic *et al.*, 1994; Ueda *et al.*, 1994a; Bainton and Strichartz, 1994; Durieux, 1995). Until further research clarifies its analgesic properties, other methods should be used (see Section IV,3).

A. Frogs and Toads

Frogs and toads have been anesthetized with MS222 for a variety of surgical procedures of varying lengths. Initial immersion followed by dripping MS222 onto a cloth sling anesthetized bullfrogs throughout the 24-hr test period following impact injury (Fish *et al.*, 1991; Karnes *et al.*, 1992; Taylor *et al.*, 1992). This drug has been used for relatively short procedures such as physical examination, blood collection, and minor surgery (Stephens *et al.*, 1987; Ackermann and Miller, 1992; Wright, 1995).

MS222 provided anesthesia for procedures including gonadectomy, limb amputation, ophthalmic surgery, vascular surgery, intestinal surgery, and hypophysectomy (Muneoka *et al.*, 1986; Straznicky and McCart, 1987; Budtz, 1988; Murphy *et al.*, 1993; Preston *et al.*, 1993; Regnier and Herrera, 1993; Fournier *et al.*, 1994; Iela *et al.*, 1994; Zhao and Szaro, 1994). MS222 has also been used extensively for embryo and larval anesthesia (Grant and Ma, 1986; Ide *et al.*, 1987; O'Rourke and Fraser, 1989; Fritzsch and Sonntag, 1990; Tunstall and Roberts, 1991; Underwood and Ide, 1992; Underwood *et al.*, 1993). Benzocaine has also been used in embryos and tadpoles (Tiedeken and Rovainen, 1991).

MS222 has been administered as an injectable agent (Olesen, 1987; Letcher, 1992; Crawshaw, 1993). Letcher (1992) determined the effective injectable dose for leopard frogs (*Rana pipiens*) and bullfrogs (*Rana catesbeiana*). Leopard frogs required 100–250 mg/kg intracoelomically, and bullfrogs were anesthetized with 250–400 mg/kg intracoelomically. Significant interspecies differences in doses were noted with this agent. At necropsy, after multiple injections, no evidence of inflammation was detected, despite using an unbuffered, concentrated solution. Because MS222 desensitizes serosal tissues, Letcher (1992) also concluded the injection to be relatively painless. Other injectable agents used in frogs and toads include pentobarbital and methohexital (Wood, 1990; Sugai *et al.*, 1991) and urethane (Dietz, 1987; Tyml, 1987; Dietrich and Tyml, 1992).

Tonner *et al.* (1992) determined the EC_{50} of propofol in *R. pipiens* tadpoles to be 2.2 μmol, which is an order of magnitude more potent than other anesthetics investigated. Letcher and Durante (1995) found tiletamine–zolazepam to be extremely variable between species and unreliable as an anesthetic agent for frogs.

Lidocaine has been used alone and in conjunction with other anesthetic agents (Narins and Wagner, 1989; Brown, 1995). Due to the permeability of anuran skin and potential toxicity, this drug should be used with caution.

B. Salamanders

Halothane has been delivered through a precision vaporizer to induce and to maintain anesthesia in a newt undergoing nuclear magnetic resonance examination (Burt *et al.*, 1986). Open-drop halothane has also been used with salamanders (Malvin and Hlastala, 1989).

MS222 (0.02–1.0%) has been used as an immersion anesthetic in mudpuppies (*Necturus*), axolotls, and newts (Malik and Goodwin, 1988; Planelles *et al.*, 1991; Soybel *et al.*, 1992, 1993; Carlone *et al.*, 1993). Frangioni and Borgioli (1989) found barbiturates, ether, and chloroform to be inadequate immersion anesthetics for newts. They found 0.2% MS222, 0.2% chlorbutal, and 2.0% urethane to be adequate for surgical procedures. McPheeters and Roper (1985) submerged *Necturus* in urethane to achieve anesthesia. As stated earlier, the carcinogenic properties of this agent make it less favorable than MS222.

C. Analgesics for Amphibians

The pharmacology of antinociceptive drugs has been much more intensively studied in amphibians than in reptiles. Stevens

(1992) reviewed nonmammalian pain models and proposed an amphibian model to study the pharmacology of various agents. Amphibians possess endogenous opioids, including Met-enkephalin (Jackson *et al.,* 1980), β-endorphin (Jegou *et al.,* 1983), dynorphin (Cone and Goldstein, 1982), and demorphin (Castiglione *et al.,* 1981; Montecucchi *et al.,* 1981). Amphibian nociceptive pathways have been identified anatomically and histochemically (Stevens, 1988). Pezalla (1983) developed an acetic acid test (serial dilutions of acetic acid dropped onto a hind limb to elicit a ''wiping response'') to determine nociceptive threshold in frogs. This method has been used to evaluate various opioids.

Stevens *et al.* (1994) determined relative analgesic potency of several μ- and κ-opioids in *R. pipiens,* the leopard frog. They found that fentanyl was the most potent (ED_{50} = 1.4 n*M*/g SC), followed by levorphanol (7.5 n*M*/g SC), methadone (19.9 n*M*/g SC), morphine (86.3 n*M*/g SC), buprenorphine (99.1 n*M*/g SC), meperidine (128.1 n*M*/g SC), codeine (140.3 n*M*/g SC), and nalorphine (320.9 n*M*/g SC). Amphibians required a much higher dose than rodents when the drugs were administered subcutaneously. The opioids reached their peak effect over a 60- to 90-min period, and analgesia lasted at least 4 hr. Stevens and Kirkendall (1993) also demonstrated that morphine tolerance to daily injections in *R. pipiens* developed after 3 days of treatment. Benyhe *et al.* (1989) found oxymorphazone (200 mg/kg) to be a relatively weak analgesic with a long duration of action (48 hr). Stevens *et al.* (1987) found that dynorphin (5, 50, and 500 ng) administered spinally to *R. pipiens* produced profound antinociception without the motor dysfunction seen in dynorphin-treated mammals.

The analgesic potency of α-adrenergic compounds in *R. pipiens* was studied. The most potent was dexmedetomidine (0.1–3.0 n*M*/g), followed by epinephrine (1–30 n*M*/g), norepinephrine (10–300 n*M*/g), and clonidine (100–1000 n*M*/g). Analgesia induced by epinephrine and dexmedetomidine took effect within 15 min and lasted more than 8 hr (Brenner *et al.,* 1994).

VI. FISH

Fish are gaining popularity as laboratory animals, particularly in toxicology and carcinogenicity research (Battalora *et al.,* 1990) (Table V). The majority of fish used are in the class Osteichthys, the ''bony fishes.'' Representatives from this group include warm-water (catfish, medaka, guppy) and cold-water (trout), freshwater species; marine species (fathead minnow), and species that live in both fresh and salt water (salmon). Cartilaginous fish, class Chondrichthys (sharks, rays, and skates), are also used in the laboratory, as are lampreys and hagfish from class Agnatha (jawless fishes).

Because fish breathe through gills rather than lungs, anesthesia must be delivered through an aquatic medium. Fish gills are composed of a series of gill arches, with primary and secondary lamellae arising off of these arches. Capillaries are located in the secondary lamellae. Blood flows countercurrent to the water as it passes over the lamellae, and oxygen exchange takes place here. Ammonium ions are also excreted via the gills. The gills in bony fishes are covered by an operculum, or bony flap. Opercular movements help direct water, which enters through the mouth, over the gills, and out (Sedgwick, 1986). Most fish are induced by adding the anesthetic agent to the tank water. A recirculating system which passes water into the mouth and out over the gills has been described (Anders and Ostrow, 1986; Brown, 1987; Lewbart *et al.,* 1995). Briefly, a recirculating system is set up to allow anesthetic in water to pass into the mouth and over the gills. The solution is then pumped back into the delivery tank and recirculated in this fashion. The fish is usually positioned in a special V-shaped trough during maintenance of anesthesia. It is very important, when anesthetizing fish, to have two separate containers of tank water. One serves as an anesthetic tank and the other is reserved for use as a recovery tank. Water containing anesthetic agents should be well-aerated to provide extra oxygen required during induction (induction stress causes increased respiration). Fish should not be disturbed before anesthesia. Food should be withheld for several hours prior to induction of anesthesia (Brown, 1993).

Fish epidermis is protected by a mucous layer and is very delicate. If the protective mucus is wiped away, the animal's skin abrades quite easily, and bacterial or fungal infections often result. When fish are handled, the handler should wear wet latex gloves to prevent abrading the skin of the fish. Several stages, planes, and categories of anesthesia have been described

TABLE V

SELECTED FISH SPECIES USED IN RESEARCH

Common name	Genus and species
Channel catfish	*Ictalurus punctatus*
Rainbow trout	*Oncorhynchus mykiss*
Pacific salmon	*Oncorhynchus gorbuscha*
Guppy	*Poecilia reticulata*
Medaka	*Oryzias latipes*
Molly	*Poecilia latipinna*
Three-spined stickleback	*Gasterosteus aculeatus*
Goldfish	*Carassius auratus*
Mozambique tilapia	*Tilapia mossambica*
Zebrafish	*Brachydanio rerio*
Swordfish	*Xiphophorus helleri*
Fathead minnow	*Pimephales promelas*
Gulf killifish	*Fundulus grandis*
Spiny dogfish shark	*Squalus acanthias*
Atlantic stingray	*Dasyatis sabina*
Common skate	*Raja batis*
River lamprey	*Lampetra fluviatilus*
Atlantic hagfish	*Myxine glutinosa*

in fish (Stoskopf, 1985; Brown, 1993). Briefly, Stage I, Plane 2 (deep sedation) is characterized by the cessation of voluntary swimming, a slight decrease in respiratory rate, normal equilibrium, and loss of response to stimuli. Stage II, Plane 2 (deep narcosis) is characterized by decreased muscle tone, equilibrium loss, and is an appropriate level for fin and gill biopsies, skin scrapings, and other external sampling. Prior to Stage II, Plane 2, the fish may undergo an excitement phase identified by increased respiration. Stage III, Plane 2 (surgical anesthesia) is characterized by slow respiration and heart rate, and a total loss of reactivity to stimuli. If gill movement ceases completely, cardiac arrest will follow shortly. The fish should immediately be removed from the anesthetic solution and placed in a well-aerated recovery tank. The fish should be manually moved gently forward through the water, forcing water through the mouth and out over the gills. This should be carefully repeated until the fish begins moving its opercula and breathing again on its own (Brown, 1993).

Immersion anesthesia is the method of choice for fish. For short-term procedures, the fish can be induced in a tank, lifted out for the procedure, and returned to a well-aerated, clean tank for recovery (the fish should be kept moist throughout the procedure) (Fig. 14). For longer or more invasive procedures, the fish can be induced in a tank and maintained on a recirculating anesthetic system.

If immersion anesthesia is inappropriate for the situation or species, injectable anesthetics can be used. Agents can be administered via intraperitoneal or intramuscular routes. Smaller fish generally require a 25- to 28-gauge, 0.25- to 0.50-inch needle; 18- to 23-gauge, 2- to 4-cm needles can be used for larger fish. Intraperitoneal injections are made on the ventral midline, aiming the needle craniodorsally to avoid the spleen. Intramuscular injections are usually given in the epaxial musculature below the dorsal fin. This route is not preferred because swimming movements can force the drug back out the needle track. Additionally, large injection volumes can result in necrosis at the injection site (Brown, 1993). For large bony and cartilagenous fishes, remote intramuscular injection using an underwater dart gun has been described (Harvey *et al.,* 1988). Intravenous injections can be given via the cannulated aorta in larger fish.

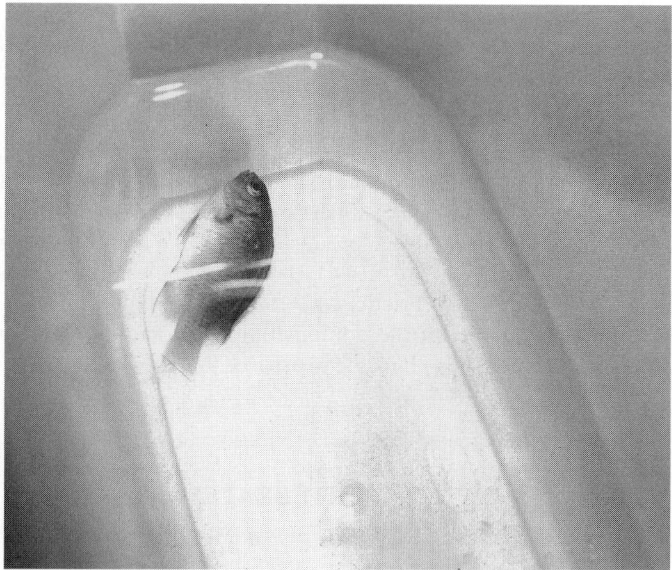

Fig. 14. Loss of righting reflex in a fish being anesthetized with MS222.

Electronarcosis has been described in fish (Gunstrom and Bethers, 1985; Barham *et al.,* 1987; Orsi and Short, 1987; Mitton and McDonald, 1994a,b). This method has been used primarily for collecting wild species in field situations and is rarely used in the laboratory. Hypothermia has been used in fish as a method of restraint. Kestin *et al.* (1991) observed that visual evoked responses were present in fish at 2°C. This suggests that hypothermia does not suppress nerve transmission, and it should not be used as a sole anesthetic agent.

A. Bony Fish

The largest body of anesthesia data has been collected from bony fishes, primarily those species of economic importance. Fish producers and processors have extensively investigated drugs for immobilization and transport.

Tricaine methane sulfonate, a derivative of benzocaine, is the most widely used fish anesthetic agent. Because this compound is acidic, it should be buffered to neutral pH before use. Doses vary according to species, but in general, 1:20,000 (50 mg/liter) dilutions can be used for tranquilization and 1:10,000 (100 mg/liter) for surgical anesthesia (Brown, 1993). MS222 has been used for invasive procedures in catfish (Burleson and Smatresk, 1989; Goodwin and Grizzle, 1991; Plakas *et al.,* 1992; Valentincic *et al.,* 1994), glass catfish (Bever and Borgens, 1991), mollies (Cohen and Kriebel, 1989; Cornbrooks and Parsons, 1991), goldfish (Anders and Ostrow, 1986; Sloley *et al.,* 1992), and *Tilapia* (Jurss *et al.,* 1984). Trout and salmon have also been successfully anesthetized with MS222 (Gingerich and Drottar, 1989; Hinton and Lauren, 1989; Bolinger and Kennish, 1992; Roberts *et al.,* 1992; McKim *et al.,* 1993; Sanz *et al.,* 1993; Wooster *et al.,* 1993).

MS222 has been associated with hypoxia and bradycardia (Brown, 1993). Lewis *et al.* (1985) noted the destruction of olfactory epithelial cilia in catfish exposed to 0.5 mg/liter MS222. Lesions resolved in 28 days. Taylor (1988) concluded that MS222 affected memory in juvenile salmon. Quinn *et al.* (1988) found that the olfactory homing behavior in salmon was not affected by MS222 at doses of 100 mg/liter. Other reported effects of MS222 include a decrease in nonesterified fatty acids (Harrington *et al.,* 1991) and a decrease in liver glycogen (Palace *et al.,* 1990). Kleinow *et al.* (1986) found buffered MS222 (50 mg/liter) to have no observable effect on P450 levels, and Barton and Peter (1982) concluded that buffered and unbuffered MS222 did not decrease stress hormone levels.

MS222 has been used to anesthetize saltwater fish (Bourne, 1984; Robertson *et al.,* 1988; Whitaker, 1991). Gabryelak *et al.* (1989) found that MS222 affected peroxide metabolism enzymes in both freshwater and marine fish species.

Benzocaine, the parent compound of MS222, is also used as a fish anesthetic agent. Because benzocaine is relatively insoluble in water, a stock solution is usually made in acetone or ethanol. This stock solution should be stored in a brown glass bottle in the dark (Brown, 1993). Because of its high lipid solubility, benzocaine administration may result in prolonged recovery in older or gravid fish (Brown, 1993). A benzocaine stock solution (100 g/liter) can be prepared, stored in a brown glass bottle, and added dropwise to effect (Ross and Ross, 1984). Arinc and Sen (1994) found benzocaine to have a prolonged (20 hr) inhibitory effect on mixed function oxidase activity in sea bream and warned against its use in xenobiotic transformation and steroid hormone studies. Benzocaine (25–45 mg/liter) has been used for restraint and anesthesia in salmon and trout. Large fish and low water temperatures required doses at the higher end of the range for both species (Gilderhus, 1989). Lucas (1989) used benzocaine for intracoelomic transmitter implantation. Fredricks *et al.* (1993) found minimal cardiovascular alterations with benzocaine and MS222.

Two-phenoxyethanol has been reported by some to be effective for restraint (Gilderhus and Marking, 1987; Josa *et al.,* 1992; Fredricks *et al.,* 1993) found phenoxyethanol to decrease heart rate and blood pressure and to alter EKG. Mattson and Riple (1989) found phenoxyethanol to be ineffective as an anesthetic for cod. Additionally, 2-phenoxyethanol has caused permanent neurologic damage in workers exposed chronically to this drug (Morton, 1990).

Quinaldine (2-methylquinoline) has been used for fish restraint (Davis *et al.,* 1986; Gilderhus and Marking, 1987; Hocutt, 1989). Quinaldine provides only minimal analgesia (Tytler and Hawkins, 1981). It is ineffective at a pH <5 (Brown, 1993) and is lethal at water temperatures higher than 29°C (Schramm and Black, 1984).

Metomidate is useful for tranquilizing fish at 2.5–5 mg/liter (Brown, 1993); this dose varies widely with species. Alphaxalone–alphadolone has been used occasionally in fish (Brown, 1993). Ueda *et al.* (1994b) cautioned that with this highly lipid-soluble steroid anesthetic, all the drug in the water would be absorbed by the fish. Consequently, the same concentration of drug in a larger tank would result in a larger amount being absorbed by the fish.

Ketamine is the most commonly used injectable anesthetic agent in fish. Williams *et al.* (1988) suggested 14–18 mg/kg IM for a variety of species. Whitaker (1991) used 15 mg/kg IM for marine species. Graham and Iwama (1990) administered 30 mg/kg ketamine IV into trout and salmon. The drug produced a long duration of anesthesia. Ventilation decreased immediately but returned within 10–30 sec. Acidosis was noted in both species.

Inhalant anesthetics have been used in fish. Halothane or isoflurane administered through a precision vaporizer can be bubbled into tank water to effect. Care must be taken to assure that the anesthetic is evenly distributed throughout the tank. Halothane (0.5–2 ml/liter) can also be dissolved in the tank for use as an immersion anesthetic agent (Brown, 1993).

B. Cartilaginous Fish

MS222 (60 mg/liter) has been used to anesthetize stingrays for cranial surgery (Livingston and Leonard, 1990). Skates were deeply anesthetized with 1 g/gal (265 mg/liter) MS222 prior to euthanasia (Malchow *et al.,* 1993). Ketamine–xylazine (15–25 mg/kg ketamine and 6–12 mg/kg xylazine) was effective in immobilizing sharks; injectable tiletamine–zolazepam, carfentanil, and metomidate were ineffective (Stoskopf, 1986).

Tonic immobility has been used in captive sharks, rays, and skates to facilitate examination and minor procedures (Henningsen, 1994). The animals are gently inverted in the water. Induction usually occurs within 15–50 sec, and duration ranges from approximately 30 sec to 4 min. This method is useful for restraint; it provides no analgesia for invasive procedures.

C. Jawless Fish

Lampreys have been anesthetized with MS222 (100 mg/liter) prior to isolation of brain and spinal cord (Yamamura *et al.,* 1990).

D. Analgesics for Fish

Fish do not appear to have a spinothalamic tract; however, another area of the brain could have assumed the function of the thalamus (Brown, 1985). Fish do respond to noxious stimuli (Brown, 1993). Bony fishes produce opiate-like hormones (Ng and Chan, 1990; Stoskopf, 1994). Benzodiazepine-binding proteins are found in fish (Friedl *et al.,* 1988). Although little information is available on the administration of opioids or other analgesics to fish, it is plausible to assume that these agents may be effective.

VII. INVERTEBRATES

Invertebrates have been studied in the laboratory for many years; however, since several of the species used are classified as "pests," little effort has been made to provide them with the same standards of care as for vertebrate species. Laboratory invertebrates represent a wide array of taxonomic phyla, including Mollusca (octopus, squid, *Aplysia,* snails), Arthropoda (*Limulus,* shrimp, crabs, crawfish, spiders, insects), Echinodermata (sea urchins, starfish), and Annelida (worms) (Table VI).

Respiratory systems of invertebrates are as widely diversified as the species. Aquatic invertebrates breathe through gills. In marine and freshwater snails, water is drawn into the mantle cavity (the mantle is a fleshy structure that envelopes the visceral mass) and passed over the gills. As in fish, both oxygen exchange and nitrogenous waste elimination take place here. Additionally, gametes are expelled via this route (Fig. 15).

In species with a shell, gills are lost on the side to which the shell coils. In species without a shell (*Aplysia*), the gills are bilateral finger-like projections on the back of the animal. Octopus and squid also have a mantle which encloses the gills and other viscera. A funnel protrudes from beneath the head and assists respiration by circulating water through the mantle cavity. The octopus has a very well-developed nervous system.

Horseshoe crabs (*Limulus*) respire through book gills, which are thin, vascular plates located on the appendages of the second segment (opisthosoma). Movement of these appendages creates a water current over the book gills; this is where oxygen exchange occurs. Crabs, shrimp, and crawfish have a chitinous carapace which protects their gills. A pulmonary vascular plexus lies inside the respiratory cavity in air-breathing snails. Terrestrial arthropods (spiders) breathe through book lungs located on the ventral surface of the abdomen.

Invertebrates possess hemolymph rather than blood. Hemolymph contains hemocyanin and carries oxygen throughout the body. Cooper and Knowler (1991) present an excellent review of snail anatomy, husbandry, and diseases.

TABLE VI

SELECTED INVERTEBRATE SPECIES USED IN RESEARCH

Common name	Genus and species
Horseshoe crab	*Limulus polyphemus*
Opossum shrimp	*Mysidopsis* sp.
Sea hare	*Aplysia californica*
Crawfish	*Procambarus clarkii*
Octopus	*Octopus* sp.
Common squid	*Loligo pealei*
Common garden snail	*Helix aspersa*
Land snail	*Cepaea nemoralis*
Medicinal leech	*Hirudo medicinalis*
Slug	*Arion ater*
Atlantic sea urchin	*Arbacia punctulata*
Earthworm	*Lumbricus terrestris*
Roundworm	*Caenorhabditis elegans*
German cockroach	*Blattella germanica*
Cockroach	*Periplaneta americana*
Fruit fly	*Drosophila melanogaster*

A. Anesthetics

Relatively few anesthetic agents have been used in invertebrates. Terrestrial species can be anesthetized in a chamber with 5–10% halothane (Cooper and Knowler, 1991). Cooper (1987a) anesthetized spiders with 4–8% halothane or 4% methoxyflurane in a chamber. Halothane induction took 5–10 min; methoxyflurane induction took 10–30 min. Animals recovered in 5–20 min after removal from the chamber.

More commonly, invertebrates such as *Drosophila* and *Caenorhabditis* are used to study molecular activity, mutagenicity, and fertility effects of inhalant anesthetics. Thus, indirectly, several inhalant agents have been shown to be safe and effective when administered to invertebrates. These include halothane, isoflurane, and enflurane (Kundomal and Baden, 1985a,b; Morgan *et al.,* 1988; Campbell and Nash, 1994; Sedensky *et al.,* 1994). Carbon dioxide has been used to anesthetize several species of invertebrates (Keeley and McKercher, 1985; Kono *et al.,* 1987; Comer *et al.,* 1988; Baptista, 1989; Cooper, 1990; Gupta *et al.,* 1990; Joachim and Curtsinger, 1990; Perino *et al.,* 1990; Costa and Francescoli, 1991; Royer and McNeil, 1992; Shimizu and Masaki, 1993). Valles and Koehler (1994) found that repeated carbon dioxide anesthesia caused mortality in the German cockroach. Cooper and Knowler (1991) suggested 10–20% carbon dioxide for light anesthesia of invertebrates, but warned against using CO_2 for painful procedures, as its analgesic effects are not known. Aquatic invertebrates can be anesthetized by immersion with 100 mg/liter MS222 or benzocaine (benzocaine must first be dissolved in ethanol or acetone). Once anesthetized, the invertebrate can be removed from water for up to 15 min. It is placed in fresh water to recover (Cooper and Knowler, 1991). Other authors have found these standard doses of MS222 to be inadequate in *Aplysia* (Clark *et al.,* 1996). A combination of MS222, urethane, and magnesium chloride ($MgCl_2$) has been used in *Aplysia* (Fredman, 1988).

CO_2 has been bubbled into aquatic systems to anesthetize invertebrates for minor procedures (Cooper and Knowler, 1991). Halothane and isoflurane could probably be administered in a similar fashion to induce a surgical plane of anesthesia.

Triethylamine (''Flynap'') has been used in an open-drop system to anesthetize *Drosophilia* and *Culicoides* (Work *et al.,* 1990; Seiger and Kink, 1993). Kramer *et al.* (1990), however, found this compound to be fatal to mosquitoes.

Many researchers have used $MgCl_2$ as an invertebrate anesthetic (Rowley, 1989; Geusz and Page, 1991; Small *et al.,* 1992; Nagahama *et al.,* 1993, 1994; Teyke *et al.,* 1993). Others have combined $MgCl_2$ with succinylcholine, a neuromuscular blocking agent (Adamo and Chase, 1990a,b; Balaban and Maksimova, 1993; Morand and Faliex, 1994). Succinylcholine provides no analgesia; $MgCl_2$ is also a neuromuscular blocking agent. Although the *Aplysia* nervous system is characterized as a group of ganglia (Kandel, 1979), until further studies are done to confirm that sensory as well as motor transmission is blocked, these agents should not be used alone for painful procedures in *Aplysia*.

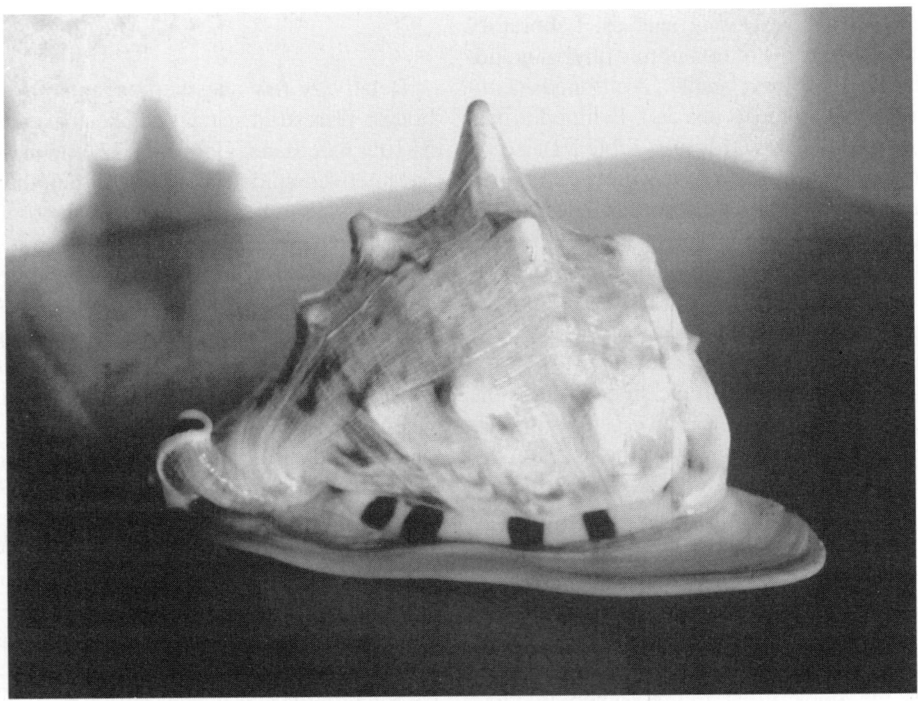

Fig. 15. Aquatic mollusks are being used with increasing frequency as research animals. Mantle cavity can not be seen in this picture.

Ethanol (5%) was used to anesthetize earthworms for 30–45 min prior to surgery (Cooper and Roch, 1992). Oliver *et al.* (1991) found that steroid anesthetics are not effective in invertebrates. Hypothermia has been used extensively to immobilize and presumably anesthetize invertebrates for a variety of invasive procedures (MacIver and Roth, 1987; Colwill *et al.,* 1988; Mellon and Quigley, 1988; Mellon *et al.,* 1989; Mimura, 1990; Weaver and Edwards, 1990; Chaudhury, 1991; Murchison *et al.,* 1993). Cooper and Knowler (1991) recommend hypothermia as a means of restraint but warn, as for other species, that it should not be used for invasive procedures.

Several investigators have ''anesthetized'' invertebrates by placing them in a water or saline solution (Sevala and Davey, 1991; Mikolajczyk and Cymborowski, 1993; Muszynska-Pytel *et al.,* 1993). It is doubtful that this method provides appropriate anesthesia or analgesia for any invasive procedure.

Unfortunately, many researchers conduct invasive and potentially painful studies on invertebrates using no anesthetic at all (Carlson and O'Gara, 1983; Schmid *et al.,* 1986; Grossfeld and Hansen, 1987; Colwell and Page, 1989; Stocker and Gendre, 1989; Barrera-Mera and Block, 1990; Ramirez and French, 1990; Salgado, 1990; Casagrand and Ritzmann, 1991; Sevala and Davey, 1991; Sink and Whitington, 1991; Foster *et al.,* 1992). Invertebrates possess functional nervous systems and pain perception analogous to that of vertebrates; therefore, invasive procedures performed without appropriate anesthesia should not be tolerated in any species.

B. Analgesics

Opioid-like compounds have been described in many invertebrates, including butterflies, blowflies, cockroaches, earthworms, snails, and octopus. This suggests that these neuropeptides have been conserved throughout evolution (Leung and Stefano, 1987; Kavaliers, 1988). Sakharov *et al.* (1993) described Met- and Leu-enkephalin-like substances in the brain of the snail, *Cepaea*. Kavaliers (1987) demonstrated that warm stress-induced analgesia in *Cepaea* is opioid mediated whereas cold stress-induced analgesia is not, similar to the situation in mammals. Aversive behavior and endogenous opioid analgesia has been described in leeches (Cooper, 1990), snails (Kavaliers and Tepperman, 1988), slugs (Dalton and Widdowson, 1989), crabs (Valeggia *et al.,* 1989), and cockroaches (Brown *et al.,* 1994). Kavaliers and Ossenkopp (1991) described phylogenetic continuity in magnetic field effects on endogenous opioid systems. Additionally, morphine-induced analgesia and tolerance have been described for crickets (Zabala and Gomez, 1991) and snails (Kavaliers and Hirst, 1986). In snails, morphine (10 mg/kg) was injected through the side of the foot into the hemocoel of the mantle cavity. Based on these investigations, it is reasonable to assume that most, if not all, invertebrates possess opioid receptors and pain processing similar to vertebrates; these creatures, therefore, should be provided with appropriate anesthetics and analgesics when subjected to painful procedures.

APPENDIX

Table A.I lists preferred anesthetics and doses for nontraditional laboratory animals.

TABLE A.I

PREFERRED ANESTHETICS AND DOSES

Animal	Anesthetic	Dose
Chinchillas	Isoflurane	3–4% induction
		1–2% maintenance
	Ketamine + xylazine	40 mg/kg 2 mg/kg
Ground squirrels	Isoflurane	3–4% induction
		1–2% maintenance
	Ketamine + xylazine	90 mg/kg 10 mg/kg
Woodchucks	Isoflurane	3–4% induction
		1–2% maintenance
	Ketamine + xylazine	35 mg/kg 5 mg/kg
Prairie dogs	Isoflurane	3–4% induction
		1–2% maintenance
	Ketamine + xylazine	100 mg/kg 10 mg/kg
North American opposums	Isoflurane	4% induction
		1–2% maintenance
Armadillos	Isoflurane	3–4% induction
		1–2% maintenance
	Ketamine	10–20 mg/kg IM for restraint or induction
	Atropine	0.04 mg/kg IM
Shrews	Isoflurane	4% induction
		2% maintenance
	Ketamine + xylazine	1–2 mg/100 g 0.1 mg/100 g
Bats	Isoflurane	3–4% induction
		1–2% maintenance
	Ketamine + acepromazine	11 mg/kg IM 1.1 mg/kg IM for restraint
Chickens	Isoflurane	4–5% induction
		1–2% maintenance
Pigeons	Isoflurane	4–5% induction
		1–2% maintenance
	Ketamine + xylazine	20 mg/kg 10 mg/kg
Ducks	Isoflurane	4% induction
		2–2.5% maintenance (mask)
		1–2% maintenance (intubation)
Finches, parrots	Isoflurane	4–5% induction
		1–2% maintenance
	Ketamine + xylazine	5–30 mg/kg 1–4 mg/kg
Snakes	Isoflurane	3–5% induction (direct intubation)
		1–2% maintenance
	Ketamine	50 mg/kg IM for restraint
Lizards	Isoflurane	3–5% induction
		1–2% maintenance
	Telazol	10–30 mg/kg IM
	Ketamine	20–40 mg/kg IM for restraint
Turtles	Ketamine	20–40 mg/kg IM to sedate prior to intubation
	Isoflurane	1–2% maintenance
	Ketamine	20–40 mg/kg IM for restraint
	Telazol	5–25 mg/kg IM for restraint
Crocodilians	Isoflurane	3–5% induction (direct intubation)
		1–2% maintenance
	Telazol	15 mg/kg IM for minor procedures
Tadpoles and newts	MS222	200–500 mg/liter immersion
Frogs and salamanders	MS222	0.5–2 g/liter immersion
		250 mg/kg intracoelomically (frogs)
Toads	MS222	1–3 g/liter immersion
	Isoflurane	3–5% induction
		1–2% maintenance (mask)
Bony fish	MS222	50 mg/liter tranquilization
		100 mg/liter anesthesia
Cartilaginous fish	MS222	60 mg/liter
Invertebrates	Halothane	5–10% terrestrial species
	MS222	100 mg/liter aquatic species

REFERENCES

Ackermann, J., and Miller, E. (1992). Chromomycosis in an African bullfrog, *Pyxicephalus adspersus*. *Bull. Assoc. Reptilian Amphibian Vet.* **2,** 8–9.

Adamo, S. A., and Chase, R. (1990a). The ''Love Dart'' of the snail *Helix aspersa* injects a pheromone that decreases courtship duration. *J. Exp. Zool.* **225,** 80–87.

Adamo, S. A., and Chase, R. (1990b). Dissociation of sexual arousal and sexual proclivity in the garden snail, *Helix aspersa*. *Behav. Neural Biol.* **54,** 115–130.

Adamson, T. P., and Burger, R. E. (1986). Sodium bicarbonate infusion increases discharge frequency of intrapulmonary chemoreceptors only at high CO_2. *Respir. Physiol.* **66,** 83–93.

Ahrendt, S. A., Ahrendt, G. M., Lillemoe, K. D., and Pitt, H. A. (1992). Effect of octreotide on sphincter of Oddi and gallbladder motility in prairie dogs. *Am. J. Physiol.* **262,** G909–G914.

Aird, S. D. (1986). Methoxyflurane anesthesia in *Crotalus:* Comparisons with other gas anesthetics. *Herpetoli Rev.* **17,** 82–84.

Allen, J. L., and Oosterhuis, J. E. (1986). Effect of tolazoline on xylazine-ketamine-induced anesthesia in turkey vultures. *J. Am. Vet. Med. Assoc.* **189,** 1011–1012.

Amoils, C. P., Jackler, R. K., Kelly, K. E., and Cao, K. (1992). An animal model of chronic tympanic membrane perforation. *Otolaryngol. Head Neck Surg.* **106,** 47–55.

Anand, N., and Paterson, W. G. (1994). Role of nitric oxide in esophageal peristalsis. *Am. J. Physiol.* **266,** G123–G131.

Anders, J. J., and Ostrow, M. E. (1986). Goldfish in research: Use and maintenance. *Lab Anim.,* July/August, pp. 33–41.

Anderson, N. L. (1992). Diseases of *Iguana iguana*. *Compendium* **14,** 1335–1343.

Anonymous (1993). Anesthesia. *In* ''Guide to the Care and Use of Experimental Animals'' (E. D. Olfert, B. M. Cross, and A. A. McWilliam, eds.), 2nd ed., pp. 129–139. Canadian Council on Animal Care, Ottawa, Ontario.

Arena, P. C., and Richardson, K. C. (1990). The relief of pain in cold-blooded vertebrates. *ACCART News* **3,** 1–4.

Arena, P. C., Richardson, K. C., and Cullen, L. K. (1988). Anaesthesia in two species of large Australian skink. *Vet. Rec.* **123,** 155–158.

Arinc, E., and Sen, A. (1994). *In vivo* effects of the anesthetic, benzocaine, on liver microsomal cytochrome P450 and mixed-function oxidase activities of gilthead seabream (*Sparus aurata*). *Comp. Biochem. Physiol. C* **107C,** 399–404.

Austin, H. B. (1990). The effects of estradiol and testosterone on Mullerian-duct regression in the American alligator (*Alligator mississippiensis*). *Gen. Comp. Endocrinol.* **76,** 461–472.

Bainton, C. R., and Strichartz, G. R. (1994). Concentration dependence of lidocaine-induced irreversible conduction loss in frog nerve. *Anesthesiology* **81,** 657–667.

Balaban, P. M., and Maksimova, O. A. (1993). Positive and negative brain zones in the snail. *Eur. J. Neurosci.* **5,** 768–774.

Baptista, F. C. (1989). A portable carbon dioxide anesthetizer for spiders. *Southwest. Nat.* **34,** 288–289.

Barham, W. T., Schoonbee, H. J., and Visser, J. G. J. (1987). The use of electronarcosis as anaesthetic in the cichlid, *Oreochromis mossambicus* (peters). I. General experimental procedures and the role of fish length on the narcotizing effects of electric currents. *Onderstepoort J. Vet. Res.* **54,** 617–622.

Barnard, S. M., and Soloway, S. M. (1992). A note on the castration of two giant indian fruit bats (*Pteropus giganteus*). *Anim. Keepers' Forum* **19,** 398–399.

Barrera-Mera, B., and Block, G. D. (1990). Photocerebral circadian pacemakers in crayfish: Evidence for mutually coupled pacemakers. *Brain Res.* **522,** 241–245.

Barten, S. L. (1993). The medical care of iguanas and other common pet lizards. *Vet. Clin. North Am.: Small Anim. Pract.* **23,** 1213–1249.

Bartlet, A. L., Harvey, S., and Klandorf, H. (1990). Contrasting effects of nitrofurans on plasma corticosterone in chickens following administration as a bolus or diet additive. *J. Vet. Pharmacol. Ther.* **13,** 261–269.

Barton, B. A., and Peter, R. E. (1982). Plasma cortisol stress response in fingerling rainbow trout, *Salmo gairdneri* Richardson, to various transport conditions, anaesthesia, and cold shock. *J. Fish Biol.* **20,** 39–51.

Bassett, J. E. (1987). Hemodilution with anesthesia in the bat, *Antrozous pallidus. J. Mammal.* **68,** 378–381.

Bathgate, R. A. D., Sernia, C., and Gemmell, R. T. (1992). Brain content and plasma concentrations of arginine vasopressin in an Australian marsupial, the brushtail possum *Trichosurus vulpecula. Gen. Comp. Endocrinol.* **88,** 217–223.

Battalora, M. S. J., Hawkins, W. E., Walker, W. W., and Overstreet, R. M. (1990). Occurrence of thymic lymphoma in carcinogenesis bioassay specimens of the Japanese medaka (*Oryzias latipes*). *Cancer Res.* **50,** 5675s–5678s.

Bauck, L. (1990). Analgesics in avian medicine. *Proc. Assoc. Avian Vet.,* pp. 239–244.

Beckman, A. L., Beaver, T. A., and Lewis, F. A. (1993). Morphine physical dependence in the hibernator: Central nervous system mechanisms underlying the development of dependence remain functional during depression induced by pentobarbital anesthesia. *Life Sci.* **52,** 1079–1086.

Bednarski, R. M. (1985). Isoflurane-nitrous oxide-oxygen anesthesia in an Andean condor. *J. Am. Vet. Med. Assoc.* **187,** 1209–1210.

Bekoff, A., and Sabichi, A. L. (1987). Sensory control of the initiation of hatching in chicks: Effects of a local anesthetic injected into the neck. *Dev. Psychol.* **20,** 489–495.

Bellairs, A. (1969). ''The Life of Reptiles.'' Weidenfeld & Nicholson, London.

Bennett, R. A. (1991). A review of anesthesia and chemical restraint in reptiles. *J. Zoo Wildl. Med.* **22,** 282–303.

Bennett, R. A. (1994). Surgical considerations. *In* ''Avian Medicine: Principles and Applications'' (B. W. Ritchie, G. J. Harrison, and L. R. Harrison, eds.), pp. 1093–1095. Wingers Publishing, Lake Worth, Florida.

Benyhe, S., Hoffman, G., Varga, E., Hosztafi, S., Toth, G., Borsodi, A., and Wollemann, M. (1989). Effects of oxymorphazone in frogs: Long lasting antinociception in vivo, and apparently irreversible binding in vitro. *Life Sci.* **44,** 1847–1857.

Betz, T. W. (1962). Surgical anesthesia in reptiles, with special reference to the water snake, *Natrix rhombifera. Copeia,* pp. 284–287.

Bever, M. M., and Borgens, R. B. (1991). The regeneration of electroreceptors in *Kryptopterus. J. Comp. Neurol.* **309,** 200–217.

Bharma, S., and Milsom, W. K. (1993). Acidosis and metabolic rate in golden mantled ground squirrels (*Spermophilus lateralis*). *Respir. Physiol.* **94,** 337–351.

Bickler, P. E. (1992). Effects of temperature and anoxia on regional cerebral blood flow in turtles. *Am. J. Physiol.* **262,** R538–R541.

Bickler, P. E., Hoh, S. O., and Severinghaus, J. W. (1989). Effects of hypoxia and hypocapnia on brain redox balance in ducks. *Am. J. Physiol.* **257,** R132–R135.

Bienzle, D., Boyd, C. J., Valverde, A., and Smith, D. A. (1991). Sedative effects of ketamine and midazolam in snapping turtles (*Chelydra serpentina*). *Vet. Surg.* **20,** 154 (abstr).

Blahak, S. (1994). Urolithiasis beim Grunen Leguan (*Iguana iguana*). *Tieraerztl. Parx.* **22,** 187–190.

Bodri, M. S., and Sadanaga, K. K. (1991). Circumcostal cloacapexy in a python. *J. Am. Vet. Med. Assoc.* **198,** 297.

Bodri, M. S., Hendrick, M. J., O'Brien, R. T., and Sadanage, K. K. (1990). Retained caseous yolk sac in a Burmese python (*Python molurus bivittatus*). *J. Wildl. Dis.* **26,** 564–566.

Bolinger, R. A., and Kennish, J. M. (1992). The effect of naphthoflavone and phenobarbital on UDP-glucuronosyltransferase in king salmon (*Oncorhyncus tshawaytscha*). *Mar. Environ. Res.* **34,** 227–231.

Bonath, K. (1979). Halothane inhalation anesthesia in reptiles and its clinical control. *Int. Zoo Yearb.* **19,** 112–125.

Bottje, W. G., and Holmes, K. R. (1989). Effect of hemorrhage on hepatic tissue blood flow in the domestic fowl (*Gallus domesticus*). *Comp. Biochem. Physiol.* **93,** 725–727.

Bottje, W. G., Holmes, K. R., Neldon, H. L., and Koike, T. K. (1989). Relationships between renal hemodynamics and plasma levels of arginine vasotocin and mesotocin during hemorrhage in the domestic fowl (*Gallus domesticus*). *Comp. Biochem. Physiol. A* **92A,** 423–427.

Bottje, W. G., Wang, S., Kinzler, S., Neldon, H. L., and Koike, T. I. (1990). Plasma concentrations of arginine vasotocin and mesotocin following pentobarbital anaesthesia and carotid cannulation in domestic fowl. *Br. Poult. Sci.* **31,** 189–195.

Bourne, P. K. (1984). The use of MS 222 (tricaine methanesulphonate) as an anaesthetic for routine blood sampling in three species of marine teleosts. *Aquaculture* **36,** 313–321.

Boyer, T. H. (1992). Clinical anesthesia of reptiles. *Bull. Assoc. Reptilian Amphibian Vet.* **2,** 10–13.

Brannian, R. E., Kirk, C., and Williams, D. (1987). Anesthetic induction of kinosternid turtles with halothane. *J. Zoo Anim. Med.* **18,** 115–117.

Brenner, G. M., Klopp, A. J., Deason, L. L., and Stevens, C. W. (1994). Analgesic potency of *alpha* adrenergic agents after systemic administration in amphibians. *J. Pharmacol. Exp. Ther.* **270,** 540–545.

Britto, L. R. G., Gasparotto, O. C., and Hamassaki, D. E. (1990). Visual telencephalon modulates directional selectivity of accessory optic neurons in pigeons. *Visual Neurosci.* **4,** 3–10.

Brown, C. S. (1995). Rear leg amputation and subsequent adaptive behavior during reintroduction of a green tree frog, *Hyla cinerea. Bull. Assoc. Reptile Amphibian Vet.* **5,** 516–7.

Brown, C. W., and Martin, R. A. (1990). Dystocia in snakes. *Compendium* **12,** 361–367.

Brown, G. E., Anderson, C. I., and Scruggs, J. L. (1994). Shock induced analgesia in the cockroach (*Periplaneta americana*). *Psychol. Rep.* **74,** 1051–1057.

Brown, L. A. (1985). Pain in fish. *In* ''Pain in Animals.''(Br. Vet. Assoc. Anim. Welfare Found. Symp., London.

Brown, L. A. (1987). Recirculation anaesthesia for laboratory fish. *Lab. Anim.* **21,** 210–215.

Brown, L. A. (1993). Anesthesia and restraint. *In* ''Fish Medicine'' (M. K. Stoskopf, ed.), pp. 79–90. Saunders, Philadelphia.

Budtz, P. E. (1988). Epidermal tissue homeostasis. III. Effect of hydrocortisone on cell pool size, cell birth rate and cell loss in normal toads and in toads deprived of the pars distalis of the pituitary gland. *Cell Tissue Kinet.* **21,** 259–270.

Buhr, R. J. (1990). Selective denervation of the *Musculus pectoralis* muscle in the chicken. *Poult. Sci.* **69,** 124–132.

Burke, T. J. (1986). Reptile anesthesia. *In* ''Zoo and Wild Animal Medicine'' (M. E. Fowler, ed.), 2nd ed., pp. 153–155. Saunders, Philadelphia.

Burke, T. J., and Wall, B. E. (1970). Anesthetic deaths in cobras (*Naja naja* and *Ophiophagus hannah*) with methoxyflurane. *J. Am. Vet. Med. Assoc.* **157,** 620–621.

Burleson, M. L., and Smatresk, N. J. (1989). The effect of decerebration and anesthesia on the reflex responses to hypoxia in catfish. *Can. J. Zool.* **67,** 630–635.

Burt, C. T., Roberts, J., Lojeski, E., and Hitzig, B. M. (1986). Phosphorus and fluorine NMR examination of the anesthetized newt (*Notopthalamus viridescens*). *Comp. Biochem. Physiol. C.* **85C,** 409–412.

Cabot, J. B., and Bogan, N. (1987). Light microscopic observations on the morphology of sympathetic preganglionic neurons in the pigeon, *Columba livia. Neuroscience* **20,** 467–486.

Calderwood, H. W. (1971). Anesthesia for reptiles. *J. Am. Vet. Med. Assoc.* **159,** 1618–1625.

Calderwood, H. W., and Jacobson, E. (1979). Preliminary report on the use of saffan on reptiles. *Proc. Annu. Meet. Am. Assoc. Zoo Vet.,* pp. 23–26.

Campbell, B., Garner, A., Dimaline, R., and Dockray, G. J. (1991). Hormonal control of avian pancreas by gastrin-releasing peptide from the proventriculus. *Am. J. Physiol.* **261,** G16–G21.

Campbell, D. B., and Nash, H. A. (1994). Use of *Drosophila* mutants to distinguish among volatile general anesthetics. *Proc. Natl. Acad. Sci. U.S.A.* **91,** 2135–2139.

Carey, H. V., and Cooke, H. J. (1991). Effect of hibernation and jejunal bypass on mucosal structure and function. *Am. J. Physiol.* **261,** G37–G44.

Carlone, R. L., Boulianne, R. P., Vijh, K. M., Karn, H., and Fraser, G. A. D. (1993). Retinoic acid stimulates the synthesis of a novel heat shock protein in the regenerating forelimb of the newt. *Biochem. Cell Biol.* **71,** 43–50.

Carlson, J. R., and O'Gara, B. A. (1983). The ecdysis of the cricket, *Teleogryllus oceanicus:* Generation of the pharyngeal air-swallowing motor program by the isolated frontal ganglion. *Comp. Biochem. Physiol. A.* **75A,** 579–587.

Carp, N. Z., Saputelli, J., Halbherr, T. C., Mason, W. S., and Jilbert, A. R. (1991). A technique for liver biopsy performed in pekin ducks using anesthesia with telazol. *Lab. Anim. Sci.* **41,** 474–475.

Carpenter, R. E. (1986). Old world fruit bats. *In* ''Zoo and Wild Animal Medicine'' (M. E. Fowler, ed.), 2nd ed., pp. 634–644. Saunders, Philadelphia.

Casagrand, J. L., and Ritzmann, R. E. (1991). Localization of ventral giant interneuron connections to the ventral median branch of thoracic interneurons in the cockroach. *J. Neurobiol.* **22,** 643–658.

Castiglione, R., Farao, F., Perseo, G., and Piani, S. (1981). Synthesis of dermophin, a new class of opioid-like peptides. *Int. J. Pept. Protein Res.* **17,** 263–272.

Chanda, A., Ghosh, R., Chakravarty, A. K., and Debnath, P. K. (1990). Possible role of a 5-hydroxytryptamine in pentobarbitone induced hypnosis in chicks. *Indian Vet. J.* **67,** 411–414.

Charland, M. B. (1991). Anesthesia and transmitter implantation effects on gravid garter snakes (*Thamnophis sirtalis* and *T. elegans*). *Herpetol. Rev.* **22,** 46–47.

Chaudhury, M. F. B. (1991). Evidence of hormonal control of sexual maturation, receptivity and mating in the male, *Glossina pallidipes. Proc. Int. Sci. Counc. Trypanosomiasis Res. Control,* pp. 205–209.

Chiszar, D., Dickman, D., and Colton, J. (1986). Sensitivity to thermal stimulation in prairie rattlesnakes (*Crotalus viridis*) after bilateral anesthetization of the facial pits. *Behav. Neural Biol.* **45,** 143–149.

Christensen, J., Fosse, R. T., Halvorsen, O. J., and Morild, I. (1987). Comparison of various anesthetic regimens in the domestic fowl. *Am. J. Vet. Res.* **48,** 1649–1657.

Clark, T. R., Nossov, P. C., Apland, J. P., and Filbert, M. G. (1996). Chemical restraint methods in the invertebrate sea snail, *Aplysia californica.* Submitted for publication.

Clubb, S. (1986). Therapeutics: Individual and flock treatment regimens. *In* ''Clinical Avian Medicine and Surgery Including Aviculture'' (G. J. Harrison and L. R. Harrison, eds.), pp. 327–355. Saunders, Philadelphia.

Clutton, R. E. (1986). Prolonged isoflurane anesthesia in the golden eagle. *J. Zoo Anim. Med.* **17,** 103–105.

Clyde, V. L., Cardeilhac, P. T., and Jacobson, E. R. (1994). Chemical restraint of American alligators (*Alligator mississippiensis*) with atracurium or tiletamine-zolazepam. *J. Zoo Wildl. Med.* **25,** 525–530.

Cohen, S. L., and Kriebel, R. M. (1989). Terminal processes of serotonin neurons in the caudal spinal cord of the molly, *Poecilia latipinna,* project to the leptomeninges and urophysis. *Cell Tissue Res.* **255,** 619–625.

Collins, B. R. (1988). Common diseases and medical management of rodents and lagomorphs. *In* ''Exotic Animals'' (E. R. Jacobson and G. V. Kollias, eds.), pp. 261–316. Churchill-Livingstone, New York.

Colwell, C. S., and Page, T. L. (1989). The electroretinogram of the cockroach *Leucophaea maderae. Comp. Biochem. Physiol. A* **92A,** 117–123.

Colwill, R. M., Absher, R. A., and Roberts, M. L. (1988). Conditional discrimination learning in *Aplysia californica. J. Neurosci.* **8,** 4440–4444.

Comer, C. M., Dowd, J. P., and Stubblefield, G. T. (1988). Escape responses following elimination of the giant interneuron pathway in the cockroach, *Periplaneta americana. Brain Res.* **445,** 470–475.

Concannon, K. T., Hellyer, P. W., and Dodam, J. R. (1994). Influence of a mu- and kappa- opioid agonist on isoflurane minimum alveolar concentration in chickens. *Vet. Surg.* **23,** 80 (abstr.).

Concannon, K. T., Dodam, J. R., and Hellyer, P. W. (1995). Influence of a mu- and kappa-opioid agonist on isoflurane minimal anesthetic concentration in chickens. *Am J. Vet. Res.* **56,** 806–811.

Cone, R. I., and Goldstein, A. A. (1982). A dynorphin-like opioid in the central nervous system of an amphibian. *Proc. Natl. Acad. Sci. U.S.A.* **79,** 3345–3349.

Constantine, D. G. (1986). Insectivorus bats. *In* ''Zoo and Wild Animal Medicine'' (M. E. Fowler, ed.), 2nd ed., pp. 650–653. Saunders, Philadelphia.

Cooper, E. L., and Roch, P. (1992). The capacities of earthworms to heal wounds and to destroy allografts are modified by polychlorinated biphenyls (PCB). *J. Invertebr. Pathol.* **60,** 59–63.

Cooper, J. E. (1974). Ketamine hydrochloride as an anaesthetic for east African reptiles. *Vet. Rec.* **95,** 37.

Cooper, J. E. (1984). Anaesthesia of exotic animals. *Anim. Technol.* **35,** 13–20.

Cooper, J. E. (1987a). A veterinary approach to spiders. *J. Small Anim. Pract.* **28,** 229–239.

Cooper, J. E. (1987b). Veterinary work with non-domesticated pets IV. Lower verebrates. *Br. Vet. J.* **143,** 193–202.

Cooper, J. E. (1990). A veterinary approach to leeches. *Vet. Rec.* **127,** 226–228.

Cooper, J. E., and Knowler, C. (1991). Snails and snail farming: An introduction for the veterinary profession. *Vet. Rec.* **129,** 541–549.

Cooper, J. E., and Williams, D. L. (1995). Veterinary perspectives and techniques in husbandry and research. *In* ''Health and Welfare of Captive Reptiles'' (C. Warwick, F. L. Frye, and J. B. Murphy, eds.), pp. 102–111. Chapman & Hall, London.

Cornbrooks, E. B., and Parsons, R. L. (1991). Sexually dimorphic distribution of a galanin-like peptide in the central nervous system of the teleost fish *Poecilia latipinna. J. Comp. Med.* **304,** 639–657.

Corwin, J. T., Bullock, T. H., and Schweitzer, J. (1982). The auditory brain stem response in five vertebrate classes. *Electroencephalogr. Clin. Neurophysiol.* **54,** 629–641.

Costa, F. G., and Francescoli, G. (1991). Analyse expérimentale de l'isolement reproductif entre deux espèces jumelles et sympatriques d'araignées: Le *Lycosa thorelli* (Keyserling) et le *Lycosa carbonelli* Costa et Capocasale. *Can. J. Zool.* **69,** 1768–1776.

Covey, F., and Casseday, J. H. (1986). Connectional basis for frequency representation in the nuclei of the lateral lemniscus of the bat *Eptesicus fuscus. J. Neurosci.* **6,** 2926–2940.

Crawley, G. R., Bishop, C. R., and Woody, B. J. (1988). Gastrotomy in a burmese python. *Calif. Vet.* 42(3), 7–8.

Crawshaw, G. J. (1992). Amphibian medicine. *In* "Current Veterinary Therapy. XI. Small Animal Practice" (R. W. Kirk and J. D. Bonagura, eds.), pp. 1219–1230. Saunders, Philadelphia.

Crawshaw, G. J. (1993). Amphibian medicine. *In* "Zoo and Wild Animal Medicine" (M. E. Fowler, ed.), 3rd ed., pp. 131–139. Saunders, Philadelphia.

Croce, A. (1989). Un coccodrillo goloso. *Sci. Vet.* **8,** 9–11.

Culver, P. J., and Rattan, S. (1986). Genesis of anal canal pressures in the opossum. *Am. J. Physiol.* **251,** G765–G771.

Custer, R. S., and Bush, M. (1980). Physiologic and acid-base measures of gopher snakes during ketamine or halothane-nitrous oxide anesthesia. *J. Am. Vet. Med. Assoc.* **177,** 870–874.

Dalton, L. M., and Widdowson, P. S. (1989). The involvement of opioid peptides in stress-induced analgesia in the slug *Arion ater. Peptides (N. Y.)* **10,** 9–13.

Dark, J., Ruby, N. F., Wade, G. N., Licht, P., and Zucker, I. (1992). Accelerated reproductive development in juvenile male ground squirrels fed a high-fat diet. *Am. J. Physiol.* **262,** R644–R650.

Davies, D. G. (1989). Distribution of systemic blood flow during anoxia in the turtle, *Chrysemys scripta. Respir. Physiol.* **78,** 383–390.

Davies, D. G. (1991). Chemical regulation of cerebral blood flow in turtles. *Am J. Physiol.* **260,** R382–R384.

Davies, P. M. C. (1981). Anatomy and physiology. *In* "Diseases of the Reptilia" (J. E. Cooper and O. F. Jackson, eds.), pp. 9–73. Academic Press, New York.

Davis, K. B., Goudie, C. A., Simco, B. A., MacGregor, R., and Parker, N. C. (1986). Channel catfish (*Ictalurus punctatus*). *Physiol. Zool.* **59,** 717–724.

Degernes, L. A., Kreeger, T. J., Mandsager, R., and Redig, P. T. (1988). Ketamine-xylazine anesthesia in red-tailed hawks with antagonism by yohimbine. *J. Wildl. Dis.* **24,** 322–326.

deJong, R. H., Hershey, W. N., and Wagman, I. H. (1966). Conduction velocity during hypothermia in man. *Anesthesiology* **27,** 805–810.

DeSaedeleer, V., Wechsung, E., and Houvenaghel, A. (1989). The influence of substance P on oviductal, duodenal and blood pressure in the anaesthetized domestic hen. *Vet. Res. Commun.* **13,** 183–191.

Dickman, J. D., Colton, J. S., Chiszar, D., and Colton, C. A. (1987). Trigeminal responses to thermal stimulation of the oral cavity in rattlesnakes (*Crotalus viridis*) before and after bilateral anesthetization of the facial pit organs. *Brain Res.* **400,** 365–370.

Dickman, J. D., and Correia, M. J. (1989). Responses of pigeon horizontal semicircular canal afferent fibers. II. high-frequency mechanical stimulation. *J. Neurophysiol.* **62,** 1102–1111.

Dietrich, H. H., and Tyml, K. (1992). Microvascular flow response to localized application of norepinephrine on capillaries in rat and frog skeletal muscles. *Microvasc. Res.* **43,** 73–86.

Dietz, F. R. (1987). Effect of peripheral nerve on limb development. *J. Orthop. Res.* **5,** 576–585.

Divers, B. J. (1986). Edentata. *In* "Zoo and Wild Animal Medicine" (M. E. Fowler, ed.), 2nd ed., pp. 622–630. Saunders, Philadelphia.

Dodd, F., and Capranica, R. R. (1992). A comparison of anesthetic agents and their effects on the response properties of the peripheral auditory system. *Hear. Res.* **62,** 173–180.

Dominguez, J., Hernandez-Jauregui, P., Grycuk, J. R., Maria, J. S., and Hori, C. S. (1986). Surgical excision of a keratinizing epidermal papilloma from the tail of a snapping turtle. *J. Zoo Anim. Med.* **17,** 105–109.

Domowicz, M. S., Kivatinitz, S. C., Caputto, B. L., and Caputto, R. (1988). Synthesis and translocation of gangliosides and glycoproteins during urethane anesthesia. *J. Neurochem.* **50,** 1369–1374.

Doolen, M. D., and Jackson, L. (1991). Anesthesia in caged birds. *Iowa State Univ. Vet.* **53,** 76–80.

Douse, M. A., and Mitchell, G. S. (1988). Temperature effects on CO-sensitive intrapulmonary chemoreceptors in the lizard, *Tupinambis nigropunctatus. Respir. Physiol.* **72,** 327–342.

Douse, M. A., and Mitchell, G. S. (1991). Time course of temperature effects on arterial acid-base status in *Alligator mississippiensis. Respir. Physiol.* **83,** 87–102.

Douse, M. A., Powell, F. L., Milsom, W. K., and Mitchell, G. S. (1989). Temperature effects on pulmonary receptor responses to airway pressure and CO in *Alligator mississippiensis. Respir. Physiol.* **78,** 331–334.

Downes, H. (1995). Tricaine anesthesia in amphibia: A review. *Bull. Assoc. Reptilian Amphibian Vet.* **5,** 11–16.

Durieux, M. E. (1995). Halothane inhibits signaling through m1 muscarinic receptors expressed in *Xenopus oocytes. Anesthesiology* **82,** 174–182.

Eatlock, R. A., Weiss, T. F., and Otto, K. L. (1991). Dependence of discharge rate on sound pressure level in cochlear nerve fibers of the alligator lizard: Implications for cochlear mechanisms. *J. Neurophysiol.* **65,** 1580–1597.

El-Fawal, H. A. N., Jortner, B. S., and Ehrich, M. (1990). Modification of phenyl saligenin phosphate-induced delayed effects by calcium channel blockers: *In vivo* and *in vitro* electrophysiological assessment. *Neurotoxicology* **11,** 573–590.

Estavillo, J. A. (1990). Cardiac receptors evoke ventilatory response with increased venous PCO at constant arterial PCO. *J. Appl. Physiol.* **68,** 369–373.

Estavillo, J. A., Adamson, T. P., and Burger, R. E. (1990). Middle cardiac nerve section alters ventilatory response to PaCO in the cockerel. *Respir. Physiol.* **81,** 349–358.

Ewing, P. J., Setser, M. D., Stair, E. L., Waurzyniak, B., and Cowell, R. L. (1991). Myxosarcoma in a Sinaloan milksnake. *J. Am. Vet. Med. Assoc.* **199,** 1775–1776.

Fagella, A. M., and Raffe, M. R. (1987). The use of isoflurane anesthesia in a water monitor lizard and a rhino iguana. *Companion Anim. Pract.,* pp. 52–53.

Farber, J. P. (1990). Effects on breathing of rostral pons glutamate injection during opossum development. *J. Appl. Physiol.* **69,** 189–195.

Farber, J. P. (1993a). Maximum discharge rates on respiratory neurons during opossum development. *J. Appl. Physiol.* **75,** 2040–2044.

Barber, J. P. (1993b). GABAergic effects on respiratory neuronal discharge during opossum development. *Am. J. Physiol.* **264,** R331–R336.

Faulkner, J. E., and Archambault, A. (1993). Anesthesia and surgery in the green iguana. *Semin. Avian Exotic Pet Med.* **2,** 103–108.

Feng, Z.-C., Rosenthal, M., and Sick, T. J. (1988). Suppression of evoked potentials with continued ion transport during anoxia in turtle brain. *Am. J. Physiol.* **255,** R478–R484.

Fish, D. R., Mendel, F. C., Schultz, A. M., and Gottstein-Yerke, L. M. (1991). Effect of anodal high voltage pulsed current on edema formation in frog hind limbs. *Phys. Ther.* **71,** 724–730.

Fitzgerald, G. (1993). Effet du protoxyde d'azote sur ladose anesthésique minimale de l'isoflurane chez le pigeon et la buse à queue rousse. *Méd. Vet. Quebec* **23,** 132.

Fitzgerald, G., and Cooper, J. E. (1990). Preliminary studies on the use of propofol in the domestic pigeon (*Columba livia*). *Res. Vet. Sci.* **49,** 334–338.

Flammer, K. (1989). Update on avian anesthesia. *In* "Current Veterinary Therapy. X" (R. W. Kirk and J. D. Bonagura, eds.), pp. 776–780. Saunders, Philadelphia.

Flanders, M. (1988). Head movement co-ordination in the African chameleon. *Neuroscience* **24,** 511–517.

Flecknell, P. A. (1987). "Laboratory Animal Anesthesia." Academic Press, Toronto.

Font, E., and Schwartz, J. M. (1989). Ketamine as an anesthetic for some squamate reptiles. *Copeia* **2,** 484–486.

Fontanesi, G., Traina, G., and Bagnoli, P. (1993). Somatostatin-like immunoreactivity in the pigeon visual system: Developmental expression and effects of retina removal. *Visual Neurosci.* **10,** 271–285.

Forger, N. G., Dark, J., Barnes, B. M., and Zucker, I. (1986). Fat ablation and food restriction influence reproductive development and hibernation in ground squirrels. *Biol. Reprod.* **34,** 831–840.

Foster, M. C., Castiglia, C. M., and Saubermann, A. J. (1992). Effects of serotonin and carbachol on glial and neuronal rubidium uptake in leech CNS. *Brain Res.* **597,** 181–188.

Fournier, P. A., Nadeau, A., and Guderley, H. (1994). The pattern of catecholamine response to burst activity in leopard frogs, *Rana pipiens. Gen. Comp. Endocrinol.* **95,** 125–132.

Frangioni, G., and Borioli, G. (1989). Relationships between spleen and respiration in the newt. *J. Exp. Zool.* **252,** 118–125.

Frazier, D. L., Jones, M. P., and Orosz, S. E. (1995a). Pharmacokinetic considerations of the renal system in birds. Part I. Anatomic and physiologic principles of allometric scaling. *J. Avian Med. Surg.* **92,** 92–103.

Frazier, D. L., Jones, M. P., and Orosz, S. E. (1995b). Pharmacokinetic considerations of the renal system in birds: Part II. Review of drugs excreted by renal pathways. *J. Avian Med. Surg.* **92,** 104–121.

Fredman, S. M. (1988). Recovery of escape locomotion following a CNS lesion in *Aplysia. Behav. Neural Biol.* **49,** 261–279.

Fredricks, K. T., Gingerich, W. H., and Fater, D. C. (1993). Comparative cardiovascular effects of four fishery anesthetics in spinally transected rainbow trout, *Oncorhynchus mykiss. Comp. Biochem. Physiol. C* **104C,** 477–483.

Friedl, W., Hebebrand, J., Rabe, S., and Propping, P. (1988). Phylogenetic conservation of the benzodiazepine binding sites: Pharmacological evidence. *Neuropharmacology* **27,** 163–170.

Fritzsch, B., and Sonntag, R. (1990). Oculomotor (N III) motoneurons can innervate the superior oblique muscle of *Xenopus* after larval trochlear (N IV) nerve surgery. *Neurosci. Lett.* **114,** 129–134.

Frye, F. L. (1981). "Biomedical and Surgical Aspects of Captive Reptile Husbandry." Veterinary Medicine Publ., Edwardsville, Kansas.

Frye, F. L. (1991). Anesthesia, *In* "Reptile Care: An Atlas of Diseases and Treatments" (F. L. Frye, ed.), pp. 420–439. T. F. H. Publications, Neptune City, California.

Frye, F. L. (1994). Colo-rectal atresia and its surgical repair in a juvenile amelanistic Burmese python (*Python molurus bivittatus*). *J. Small Exotic Anim. Med.* **2,** 149–150.

Furilla, R. A., and Bartlett, D. (1988). Intrapulmonary receptors in the garter snake (*Thamnophis sirtalis*). *Respir. Physiol.* **74,** 311–322.

Furilla, R. A., and Bartlett, D. (1989). Intrapulmonary CO_2 inhibits inspiration in garter snakes. *Respir. Physiol.* **78,** 207–218.

Gabryelak, T., Zalesna, G., Roche, H. and Peres, G. (1989). The effect of MS 222 an anaesthetic on the peroxide metabolism enzymes in erythrocytes of freshwater and marine fish species. *Comp. Biochem. Physiol. C* **92C,** 5–8.

Ganzhorn, J. U., and Burkhardt, J. F. (1991). Pigeon Homing: New airbag experiments to assess the role of olfactory information for pigeon navigation. *Behav. Ecol. Sociobiol.* **29,** 69–75.

Garrett, C. M., and Harwell, G. (1991). Loreal pit impaction in a black speckled palm-pitviper (*Bothriechis nigroviridis*). *J. Zoo Wildl. Med.* **22,** 249–251,

Gasser, H. S. (1931). Nerve activity as modified by temperature changes. *Am. J. Physiol.* **92,** 254.

George, P. O., and Joseph, J. (1989). Prolapse of cloaca in a python (*Python molurus*). *Indian Vet. J.* **66,** 648–649.

Gerwien, R. W., and John-Alder, H. B. (1992). Growth and behavior of thyroid-deficient lizards (*Sceleoporus undulatus*). *Gen. Comp. Endocrinol.* **87,** 312–324.

Geusz, M. E., and Page, T. L. (1991). An opsin-based photopigment mediates phase shifts of the *Bulla* circadian pacemaker. *J. Comp. Physiol. A* **A168,** 565–570.

Gilderhus, P. A. (1989). Efficacy of benzocaine as an anesthetic for salmonid fishes. *North Am. J. Fish. Manage.* **9,** 150–153.

Gilderhus, P. A., and Marking, L. L. (1987). Comparative efficacy of 16 anesthetic chemicals on rainbow trout. *North Am. J. Fish. Manage.* **7,** 288–292.

Gillespie, D. S. (1993). Edentata: diseases. *In* "Zoo and Wild Animal Medicine, Current Therapy" (M. E. Fowler, ed.), 3rd. ed., pp. 304–309. Saunders, Philadelphia.

Gingerich, W. H., and Drottar, K. R. (1989). Plasma catecholamine concentrations in rainbow trout (*Salmo gairdneri*) at rest and after anesthesia and surgery. *Gen. Comp. Endocrinol.* **73,** 390–397.

Glahn, R. P., Beers, K. W., Bottje, W. G., Wideman, R. F., Huff, W. E., and Thomas, W. (1991). Aflatoxin alters avian renal function, calcium, and vitamin D metabolism. *J. Toxicol. Environ. Health* **34,** 309–321.

Glatz, P. C., Murphy, L. B., and Preston, A. P. (1992). Analgesic therapy of beak-trimmed chickens. *Aust. Vet. J.* **69,** 18.

Glenn, J. L., Straight, R., and Snyder, C. C. (1972). Clinical use of ketamine hydrochloride as an anesthetic agent for snakes. *Am. J. Vet. Res.* **33,** 1901–1903.

Glenn, J. L., Straight, R., and Snyder, C. C. (1973). Surgical technique for isolation of the main venom gland of viperid, crotalid and elapid snakes. *Toxicon* **11,** 231–233.

Goelz, M. F., Hahn, A. W., and Kelley, S. T. (1990). Effects of halothane and isoflurane on mean arterial blood pressure, heart rate, and respiratory rate in adult pekin ducks. *Am. J. Vet. Res.* **51,** 458–460.

Goodwin, A. E., and Grizzle, J. M. (1991). Granulomatous inflammation and monstrous giant cells in response to intraperitoneal hormone implants in channel catfish (*Ictalurus punctatus*). *J. Comp. Pathol.* **104,** 147–160.

Gould, W. J., and Georgi, M. E. (1991). Myiasis in two box turtles. *J. Am. Vet. Med. Assoc.* **199,** 1067–1068.

Gould, W. J., Yaegar, A. E., and Glennon, J. C. (1992). Surgical correction of an intestinal obstruction in a turtle. *J. Am. Vet. Med. Assoc.* **200,** 705–706.

Graham, M. S., and Iwama, G. K. (1990). The physiologic effects of the anesthetic ketamine hydrochloride on two salmonid species. *Aquaculture* **90,** 323–331.

Grain, E., and Evans, J. E. (1984). Egg retention in four snakes. *J. Am. Vet. Med. Assoc.* **185,** 679–681.

Grant, P., and Ma, P. M. (1986). Optic fibers follow aberrant pathways from rotated eyes in *Xenopus laevis. J. Comp. Neurol.* **250,** 364–376.

Graves, B. M., and Duvall, D. (1993). Reproduction, rookery use, and thermoregulation in free-ranging, pregnant *Crotalus v. viridis. J. Herpetol.* **27,** 33–41.

Gray, C. W., Bush, M., and Beck, C. C. (1974). Clinical experience using CI-744 in chemical restraint and anesthesia of exotic specimens. *J. Zoo Anim. Med.* **5,** 12–21.

Green, C. J. (1979). "Animal Anesthesia." Laboratory Animals. London.

Greenlees, K. J., Clutton, R. E., Larsen, C. T., and Eyre, P. (1990). Effect of halothane, isoflurane, and pentobarbital anesthesia on myocardial irritability in chickens. *Am. J. Vet. Res.* **51,** 757–758.

Grossfeld, R. M., and Hansen, D. B. (1987). Long-term persistence of GAD activity in injured crayfish CNS tissue. *Neurochem. Res.* **12,** 977–983.

Guérit, J. -M., Verhelst, R., Rubay, J., El Khoury, G., Noirhomme, P., Baele, P., and Dion, R. (1994). The use of somatosensory evoked potentials to determine the optimal degree of hypothermia during circulatory arrest. *J. Cardiac Surg.* **9,** 596–603.

Gunstrom, G. K., and Bethers, M. (1985). Electrical anesthesia for handling large salmonids. *Prog. Fish-Cult.* **47,** 67–69.

Gupta, R. K., Rutledge, L. C., and Korte, D. W. (1990). Portable tray for anesthetizing and sorting insects. *J. Med. Entomol.* **27,** 406–408.

Halder-Misra, C. (1986). Methods of pinealectomy in some vertebrates. *Indian J. Exp. Biol.* **24,** 319–322.

Hall, L. W., and Clarke, K. W. (1991). "Veterinary Anesthesia." Baillière Tindall, London.

Hall, M. R. (1987). External stimuli affecting incubation behavior and prolactin secretion in the duck (*Anas platyrhynchos*). *Horm. Behav.* **21,** 269–287.

Hanyu, N., Dodds, W. J., Layman, R. D., Hogan, W. J., Chey, W. Y., and Takahashi, I. (1990). Mechanism of cholecystokinin-induced contraction of the opossum gallbladder. *Gastroenterology* **98,** 1299–1306.

Hargett, C. E., Record, J. W., Carrier, M., Bordwell, K. C., and Patterson, J. H. (1989). Reversal of ketamine-xylazine anesthesia in the chinchilla by yohimbine. *Lab. Anim.* **18,** 41–43.

Harper, R. C. (1984). Anaesthetising reptiles. *Vet. Rec.,* 115(18), 475–476.

Harrington, A. J., Russell, K. A., Singer, T. D., and Ballantyne, J. S. (1991). The effects of tricaine methanesulfonate (MS-222) on plasma nonesterified fatty acids in rainbow trout, *Oncorhynchus mykiss. Lipids* **26,** 774–775.

Harvey, B., Denny, C., Kaiser, S., and Young, J. (1988). Remote intramuscular injection of immobilising drugs into fish using a laser-aimed underwater dart gun. *Vet. Rec.* **122,** 174–177.

Harvey-Clark, C. (1994). Research aspects of anesthesia of reptiles. *CALAS/ACTAL* **29,** 25–29.

Haskell, A., and Pokras, M. A. (1994). Nonlethal blood and muscle tissue collection from redbelly turtles for genetic studies. *Herpetol. Rev.* **25,** 11–12.

Hastings, R. H., and Powell, F. I. (1987). High-frequency ventilation of ducks and geese. *J. Appl Physiol.* **63,** 413–417.

Heard, D. J. (1993). Principles and techniques of anesthesia and analgesia for exotic practice. *Vet. Clin. North Am.: Small Anim. Pract.* **23,** 1301–1327.

Heard, D. J. (1994). Avian anesthesia. *Proc. Avian Surg. Symp.,* Abstr., pp. 13–17.

Heard, D. J., Jacobson, E. R., Clemmons, R. E., and Campbell, G. A. (1988). Bacteremia and septic arthritis in a West African dwarf crocodile. *J. Am Vet. Med. Assoc.* **192,** 1453–1454.

Heaton, J. T., and Brauth, S. E. (1992). Effects of yohimbine as a reversing agent for ketamine-xylazine anesthesia in budgerigars. *Lab. Anim. Sci.* **42,** 54–56.

Hempleman, S. C., and Bebout, D. E. (1994). Increased venos PCO enhanced dynamic responses of avian intrapulmonary chemoreceptors. *Am. J. Physiol.* **266,** R15–R19.

Hempleman, S. C., Adamson, T. P., and Burger, R. E. (1986). Sensitivity of avian intrapulmonary chemoreceptors to venous CO_2 load. *Respir. Physiol.* **66,** 53–60.

Henderson, D., Subramaniam, M., Papazian, M., and Spongr, V. P. (1994). The role of middle ear muscles in the development of resistance to noise induced hearing loss. *Hear. Res.* **74,** 22–28.

Hendriksen, J. H., Bachelor, P. A., Newman, R. J., and Stanton, T. L. (1992). Thyrotropin-releasing hormone action in the preoptic/anterior hypothalamus decreases thermoregulatory set point in ground squirrels. *Regul. Pept.* **38,** 55–69.

Henningsen, A. D. 1994). Tonic immobility in 12 elasmobranchs: Use as an aid to captive husbandry. *Zoo Biol.* **13,** 325–332.

Heyer, W. R., Donnelly, M. A., McDiarmid, R. W., Hayek, L. -A. C., and Foster, M. S. (1994). ''Measuring and Monitoring Biological Diversity Standard Methods for Amphibians.'' Smithsonian Institution Press, Washington, D. C.

Hicks, J. W., and Malvin, G. M. (1992). Mechanism of intracardiac shunting in the turtle *Pseudemys scripta. Am. J. Physiol.* **262,** R986–R992.

Hinton, D. E., and Lauren, D. J. (1989). *In vivo* microscopy of liver microvasculature in rainbow trout (*Oncorhynchus mykiss*). *Mar. Environ. Res.* **28,** 407–410.

Hocutt, C. H. (1989). Seasonal and diel behavior of radio-tagged *Clarias gariepinus* in Lake Ngazi, Zimbabwe (Pisces: Clariidae). *J. Zool.* **219,** 181–199.

Hocutt, C. H., Loveridge, J. P., and Hutton, J. M. (1992). Biotelemetry monitoring of translocated *Cocodylus niloticus* in Lake Ngezi, Zimbabwe. *J. Zool.* **226,** 231–242.

Hoefer, H. L. (1994). Chinchillas. *Vet. Clin. North Am.: Small Anim. Pract.* **24,** 103–107.

Hughes, R. A. (1990a). Strain-dependent morphine-induced analgesic and hyperalgesic effects on thermal nociception in domestic fowl. (*Gallus gallus*). *Behav. Neurosci.* **104,** 619–624.

Hughes, R. A. (1990b). Codeine analgesic and morphine hyperalgesic effects on thermal nociception in domestic fowl. *Pharmacol. Biochem. Behav.* **35,** 567–570.

Ide, C. F., Wunsh, L. M., Lecat, P. J. C., Kahn, D., and Noelke, E. L. (1987). Healing modes correlate with visuotectal pattern formation in regenerating embryonic *Xenopus retina. Dev. Biol.* **124,** 316–330.

Iela, L., D'Aniello, B., DiMeglio, M., and Rastogi, R. K. (1994). Influence of gonadectomy and steroid hormone replacement therapy on the gonadotropin-releasing hormone neuronal system in the anterior preoptic area of the frog (*Rana esculenta*) brain. *Gen. Comp. Endocrinol.* **95,** 422–431.

Ishii, K., and Dejours, P. (1986). Activity of vagal afferent fibers innervating CO_2-sensitive receptors in the tortoise, *Testudo hermanni. Jpn. J. Physiol.* **36,** 1015–1026.

Jackson, I. M. D., Bolaffi, J. L., and Guillemen, R. (1980). Presence of immunoreactive-endorphin and eukephalin-like material in the retina and other tissues of the frog, *Rana pipiens. Gen. Comp. Endocrinol.* **42,** 505–508.

Jackson, O. F. (1970). Snake anesthesia. *Br. J. Herpetol.* **4,** 172–175.

Jacobson, E. R. (1984). Immobilization, blood sampling, necropsy techniques, and diseases of crocodilians: A review. *J. Zoo Anim. Med.* **15,** 38–45.

Jacobson, E. R. (1993). Snakes. *Vet. Clin. North Am.: Small Anim. Pract.* **23,** 1179–1212.

Jacobson, E. R., Buergelt, C., Williams, B., and Harris, R. K. (1991). Herpesvirus in cutaneous fibropapillomas of the green turtle *Chelonia mydas. Dis. Aquat. Organisms* **12,** 1–6.

Jaeger, R. G. (1992). Housing, handling and nutrition of salamanders. *In* ''The Care and Use of Amphibians, Reptiles and Fish in Research'' (D. O. Schaeffer, K. M. Kleinow, and L. Krulisch, eds.), pp. 25–29. S.C.A.W.

Janik, D. S., and Menaker, M. (1990). Circadian locomotor rhythms in the desert iguana. *J. Comp. Physiol., A* **166,** 803–810.

Jegou, S., Tonon, M.C., Leroux, P., Leboulenger, F., Pelletier, G., Cote, J., Ling, N., and Vaudry, H. (1983). Immunological characterizations of endorphins, adrenocorticotropin and melanotropins in the frog hypothalamus. *Gen. Comp. Endocrinol.* **51,** 246–254.

Jenkins, J. R. (1991). Medical management of reptile patients. *Compend. Contin. Educ.* **13,** 980–988.

Jenkins, J. R. (1993). Postoperative care of the avian patient. *Semin. Avian Exotic Pet Med.* **2,** 97–102.

Joachim, D., and Curtsinger, J. W. (1990). Genotype and anesthetic determine mate choice in *Drosophila melanogaster. Behav. Genet.* **20,** 73–79.

Johlin, J. M., and Moreland, F. B. (1933). Studies of the blood picture of the turtle after complete anoxia. *J. Biol. Chem.* **103,** 107–114.

Jones, A. K. (1990). Caesarean section in a chinchilla. *Vet. Rec.,* 126(17), 441.

Josa, A., Espinosa, E., Cruz, J. I., Gil, L., Falceto, M. V., and Lozano, R. (1992). Use of 2-phenoxyethanol as an anaesthetic agent in goldfish (*Cyprinus carpio*). *Vet. Rec.* **131,** 468.

Jurss, K., Bittorf, T., Vokler, T., and Wacke, R. (1984). Biochemical investigations into the influence of environmental salinity on starvation of the tilapia, *Oreochromis mossambicus. Aquaculture* **40,** 171–182.

Kaltenback, J. A., and Saunders, J. C. (1987). Spectral and temporal response patterns of single units in the chinchilla dorsal cochlear nucleus. *Exp. Neurol.* **96,** 406–419.

Kandel, E. R. (1979). *Aplysia* among the molluscs II: The nervous system. *In* ''Behavioral Biology of *Aplysia,* a Contribution to the Study of Opisthobranch Molluscs'' (E. R. Kandel, ed.), pp. 107–148. Freeman, San Francisco.

Kanui, T. I., and Hole, K. (1992). Morphine and pethidine antinociception in the crocodile. *J. Vet. Phamacol. Ther.* **15,** 101–103.

Kanui, T. I., Hole, K., and Miaron, J. O. (1990). Nociception in crocodiles: Capsaicin instillation, formalin and hot plate tests. *Science* **7,** 537–540.

Kaplan, H. M. (1969). Anesthesia in amphibians and reptiles. *Fed. Proc., Fed. Am. Soc. Exp. Biol.* **28,** 1541–1546.

Karnes, J. L., Mendel, F. C., and Fish, D. R. (1992). Effects of low voltage pulsed current on edema formation in frog hind limbs following impact injury. *Phys. Ther.* **72,** 273–278.

Kaufman, H. S., Ahrendt, S. A., Pitt, H. A., and Lillemoe, K. D. (1993). The effect of erythromycin on motility of the duodenum, sphincter of oddi, and gallbladder in the prairie dog. *Surgery* (*St. Louis*) **114,** 543–548.

Kavaliers, M. (1987). Evidence for opioid and non-opioid forms of stress-induced analgesia in the snail, *Cepaea nemoralis. Brain Res.* **410,** 111–115.

Kavaliers, M., and Hirst, M. (1986). Environmental specificity of tolerance to morphine-induced analgesia in a terrestrial snail: Generalization of the behavioral model of tolerance. *Pharmacol., Biochem. Behav.* **25,** 1201–1206.

Kavaliers, M., and Ossenkopp, K. -P. (1991). Opioid systems and magnetic field effects in the land snail, *Cepaea nemoralis. Biol. Bull.* (*Woods Hole, Mass.*) **180,** 301–309.

Kavaliers, M., and Tepperman, F. S. (1988). Exposure to novel odors induces opioid-mediated analgesia in the land snail, *Cepaea nemoralis. Behav. Neural Biol.* **50,** 285–299.

Kavaliers, M. (1988). Evolutionary and comparative aspects of nociception. *Brain Res. Bull.* **21,** 923–931.

Keeley, L. L., and McKercher, S. R. (1985). Endocrine regulations of ovarian maturation in the cockroach *Blaberus discoidalis. Comp. Biochem. Physiol. A* **80A,** 115–121.

Keller, L. S. F., Drozdowicz, C. K., Rice, L., Bowman, T. A., and Lang, C. M. (1988). An evaluation of three anaesthetic regimes in the gray short-taled opossum (*Monodelphis domestica*). *Lab. Anim.* **22,** 269–275.

Kennett, R., Christian, K., and Prichard, D. (1992). Underwater nesting by the tropical freshwater turtle, *Chelodina rugosa* (Testudinate: Chelidae). *Aust. J. Zool.* **41,** 47–52.

Kestin, S. C., Wotton, S. B., and Gregory, N. G. (1991). Effect of slaughter by removal from water on visual evoked activity in the brain and reflex movement of rainbow trout (*Oncorhynchus mykiss*). *Vet. Rec.* **128,** 443–446.

Kiel, J. L. (1984). Long-term effects of venom gland adenectomy of snakes. *Avian/Exotic Pract.* **1,** 35–41.

Kiepenheuer, J. (1985). Can pigeons be fooled about the actual release site position by presenting them information from another site? *Behav. Ecol. Sociobiol.* **18,** 75–82.

Kilander, K., and Williams, H. (1992). Yohimbine reduces neuropathology induced by ketamine/xylazine anesthesia. *Physiol. Behav.* **51,** 657–659.

King, A. S., and McLelland, J. (1984). "Birds: Their Structure and Function," 2nd ed. Baillière Tindall, London.

Kleinow, K. M., Haasch, M. L., and Lech, J. J. (1986). The effect of tricaine anesthesia upon induction of select P-450 dependent monooxygenase activities in rainbow trout (*Salmo gairdneri*). *Aquat. Toxicol.* **8,** 231–241.

Kono, Y., Narai, N., Numata, M., and Yoshioka, M. (1987). Effects of venom from the joro spider on insects. *Appl. Entomol. Zool.* **22,** 392–394.

Kramer, L. D., Presser, S. B., Houk, E. J., and Hardy, J. L. (1990). Effect of anesthetizing agent triethylamine on western equine encephalomyelitis and St. Louis encephalitis viral titers in mosquitoes (Diptera: Culcidae). *J. Med. Entomol.* **27,** 1008–1010.

Krishnamurthy, G. T., and Turner, F. E. (1988). Gallbladder filling and post-ceruletide emptying in prairie dogs and rabbits. *Nucl. Med. Commun.* **9,** 382–388.

Krohmer, R. W., and Crews, D. (1987). Temperature activation of courtship behavior in the male red-sided garter snake (*Thamnophis sirtalis parietalis*): Role of the anterior hypothalamus-preoptic area. *Behav. Neurosci.* **101,** 228–236.

Kujime, K., and Natelson, B. H. (1981). A method for endotracheal intubation of guinea pigs (*Cavia porellus*). *Lab. Anim. Sci.* **31,** 715–716.

Kundomal, Y. R., and Baden, J. M. (1985a). Inhaled anaesthetics have no effect on fertility in *Drosophilia melanogaster. Br. J. Anaesth.* **57,** 900–903.

Kundomal, Y. R., and Baden, J. M. (1985b). Mutagenicity of inhaled anesthetics in *Drosophila melanogaster. Anesthesiology* **62,** 305–309.

Kwantes, L. J. (1992). Surgical correction of cystic urolithiasis in an iguana. *Can. Vet. J.* **33,** 752–753.

Lang, S. A., and Levy, M. N. (1989). Effects of vagus nerve on heart rate and ventricular contractility in chicken. *Am. J. Physiol.* **256,** H1295–H1302.

Larsen, F., Schlarman, D., Andrus, C. C., and Kaminski, D. L. (1991). The effect of the CCK receptor antagonist CR 1409 on bile reflux pancreatitis in the opossum. *Pancreas* **6,** 291–297.

Laverty, G., and Alberici, M. (1987). Micropuncture study of proximal tubule pH in avian kidney. *Am. J. Physiol.* **253,** R587–R591.

Lawrence, K., and Jackson, O. F. (1983). Alphaxalone/alphadolone anaesthesia in reptiles. *Vet. Rec.* **112,** 26–28.

Lawton, M. P. C. (1992). Anaesthesia. *In* "British Small Animal Veterinary Association Manual of Reptiles" (P. H. Beynon, M. P. C. Lawton, and J. E. Cooper, eds.), pp. 170–183. Br. Small Anim. Vet. Assoc., Gloucestershire.

Leander, J. D. (1988). Buprenorphine is a potent k-opioid receptor antagonist in pigeons and mice. *Eur. J. Pharmacol.* **151,** 457–461.

Letcher, J. (1992). Intracoelomic use of tricaine methanesulfonate for anesthesia of bullfrogs (*Rana catesbeiana*) and leopard frogs (*Rana pipiens*). *Zoo Biol.* **11,** 243–251.

Letcher, J., and Durante, R. (1995). Evaluation of use of tiletamine/zolazepam for anesthesia of bullfrogs and leopard frogs. *J. Am. Vet. Med. Assoc.* **207,** 80–82.

Leung, M. K., and Stefano, G. B. (1987). Comparative neurobiology of opioids in invertebrates with special attention to senescent alterations. *Prog. Neurobiol.* **28,** 131–159.

Leveille, C. R., Center, S. A., Baldwin, B. H., Hornbuckle, W. R., and Tennant, B. C. (1987). Serum bile acid determination for assessment of hepatic injury in the woodchuck. *Lab. Anim. Sci.* **37,** 224–228.

Lewbart, G. A., Stone, E. A., and Love, N. E. (1995). Pneumocystectomy in a midas cichlid. *J. Am. Vet. Med. Assoc.* **207,** 319–321.

Lewis, D. H., Tarpley, R. J., Marks, J. E., and Sis, R. F. (1985). Drug induced structural changes in olfactory organ of channel catfish *Ictalurus punctatus,* Rafinesque. *J. Fish Biol.* **26,** 355–358.

Li, Y. F., Wesibrodt, N. W., and Moody, F. G. (1991). Effect of bile diversion and sphincterotomy on gallbladder muscle contractility and gallstone formation. *Am. J. Surg.* **162,** 31–35.

Li, Y. F., Russell, D. H., Myers, S. I., Weisbrodt, N. W., and Moody, F. G. (1994). Gallbladder contractility in aspirin- and cholesterol-fed prairie dogs. *Gastroenterology* **106,** 1662–1667.

Lillemoe, K. D., Webb, T. H., and Pitt, H. A., (1988). Neuropeptide Y: A candidate neurotransmitter for biliary motility. *J. Surg. Res.* **45,** 254–260.

Liu, Y.-F., Saccone, G. T. P., Thune, A., Baker, R. A., Harvey, J. R., and Touli, J. (1992). Sphincter of Oddi regulates flow by acting as a variable resistor to flow. *Am. J. Physiol.* **263,** G683–G689.

Livingston, C. A., and Leonard, R. B. (1990). Locomotion evoked by stimulation of the brain stem in the atlantic stingray, *Dasyatis sabina. J. Neurosci.* **10,** 194–202.

Loveridge, J. P., and Blake, D. K. (1987). Crocodile immobilization and anesthesia. *In* "Wildlife Management: Crocodiles and Alligators" (G. J. W. Webb, S. C. Manolis, and P. J. Whitehead, eds.), pp. 259–267. Beatty and Sons, Ltd., Surrey.

Lucas, K. (1908). The temperature coefficient of the rate of conduction in nerve. *J. Physiol.* (*London*) **37,** 112.

Lucas, M. C. (1989). Effects of implanted dummy transmitters on morality, growth and tissue reaction in rainbow trout, *Salmo gairdneri* Richardson. *J. Fish Biol.* **35,** 577–587.

Ludders, J. W., Mitchell, G. S., and Schaefer, S. L. (1988). Minimum anesthesic dose and cardiopulmonary dose respone for halothane in chickens. *Am. J. Vet. Res.* **49,** 929–932.

Ludders, J. W., Mitchell, G. S., and Rode, J. (1990). Minimal anesthetic concentration and cardiopulmonary dose response of isoflurane in ducks. *Vet. Surg.* **19,** 304–307.

Lumeij, J. T., and Happé, R. P. (1985). Endoscopic diagnosis and removal of gastric foreign bodies in a caiman (*Caiman crocodilus crocodilus*). *Vet. Q.* **7,** 234–236.

MacIver, M. B., and Roth, S. H. (1987). Anesthetics produce differential actions on the discharge activity of a single neuron. *Eur. J. Pharmacol.* **139,** 43–52.

Malchow, R. P., Qian, H., and Ripps, H. (1993). Evidence for hemi-gap junctional channels in isolated horizontal cells of the skate retina. *J. Neurosci. Res.* **35,** 237–245.

Malik, R., and Goodwin, M. (1988). Removal of gastric foreign bodies from and axolotl. *Aust. Vet. Practitioner* **18,** 95–96.

Malvin, G. M., and Hlastala, M. P. (1989). Effects of environmental O_2 on blood flow and diffusing capacity in amphibian skin. *Respir. Physiol.* **76,** 229–241.

Marcus, L. C. (1981). ''Veterinary Biology and Medicine of Captive Amphibians and Reptiles.'' Lea & Febiger, Philadelphia.

Margoliash, D., and Fortune, E. S. (1992). Temporal and harmonic combination-sensitive neurons in the zebra finch's HVc. *J. Neurosci.* **12,** 4309–4326.

Mattson, N. S., and Riple, T. H. (1989). Metomidate, a better anesthetic for cod (*Gadus morhua*) in comparison with benzocaine, MS 222, chlorobutanol, and phenoxyethanol. *Aquaculture* **83,** 89–94.

Mauk, M. D., Olson, R. D., LaHoste, G. J., and Olson, G. A. (1981). Tonic immobility produces hyperalgesia and antagonizes morphine analgesia. *Science* **213,** 353–354.

Mautino, M., and Page, C. D. (1993). Biology and medicine of turtles and tortoises. *Vet. Clin. North Am.: Small Anim. Pract.* **23,** 1251–1269.

McCracken, G. F., and Wilkinson, G. S. (1988). Allozyme techniques and kinship assessment in bats. *In* ''ecological and Behavioral Methods for the Study of Bats'' (T. H. Kunz, ed.), pp. 141–156. Smithsonian Institution Press, Washington, D. C.

McGrath, C. J., Lee, J. C., and Campbell, V. L. (1984a). Dose-response anesthetic effects of ketamine in the chicken. *Am. J. Vet. Res.* **45,** 531–534.

McGrath, C. J., Gross, W. B., and Lee, J. C. (1984b). Effects of social stress on response to ketamine anesthesia in chickens. *Am. J. Vet. Res.* **45,** 1421–1424.

McKim, J. M., Naumann, S., Hammermeister, D. E., Hoffman, A. D., and Klaassen, C. D. (1993). *In vivo* microdialysis sampling of pheno and phenyl glucuronide in the blood of unanesthetized rainbow trout: Implications for toxicokinetic studies. *Fundam. Appl. Toxicol.* **20,** 190–198.

McPheeters, M., and Roper, S. D. (1985). Amiloride does not block taste transduction in the mudpuppy, *Necturus maculosus. Chem. Senses* **10,** 341–352.

Meehan, T. P. (1993). Medical problems of shrews. *In* ''Zoo and Wild Animal Medicine'' (M. E. Fowler, ed.), 3rd ed., pp. 302–303. Saunders, Philadelphia.

Mellon, D., and Quigley, M. M. (1988). Disruption of muscle reorganization by lesions of the peripheral nerve in transforming claws of snapping shrimps. *J. Neurobiol.* **19,** 532–551.

Mellon, D., Tuten, H. R., and Redick, J. (1989). Disruption of radioactive leucine following uptake by olfactory sensory neurons in normal and heteromorphic crayfish antennules. *J. Comm. Neurol.* **280,** 645–662.

Mihic, S. J., McQuilkin, S. J., Eger, E. I., Ionescu, P., and Harris, R. A. (1994). Potentiation of gamma-aminobutyric acid type A receptor-mediated chloride currents by novel halogenated compounds correlates with their ability to induce general anesthesia. *Mol. Pharmacol.* **46,** 851–857.

Miki, S., Mosbach, E. H., Cohen, B. I., Mikami, T., Infante, R., Ayyad, N., and McSherry, C. K. (1993). Metabolism of muricholic acid in the hamster and prarie dog. *J. Lipid Res.* **34,** 1709–1716.

Mikolajczyk, P., and Cymborowski, B. W. (1993). Lower temperature influences developmental rhythms of the wax moth *Galleria mellonella:* Putative role of ecdysteroids. *Comp. Biochem. Physiol. A* **105A,** 57–66.

Millichamp, N. J. (1988). Surgical techniques in reptiles. *In* ''Exotic Animals'' (E. R. Jacobson and G. V. Kollias, eds.), pp. 49–73. Churchill-Livingstone, New York.

Millichamp, N. J., Lawrence, K., Jacobson, E. R., Jackson, O. F., and Bell, D. A. (1983). Egg retention in snakes. *J. Am. Vet. Med. Assoc.* **11,** 1213–1218.

Millichamp, N. J., Jacobson, E. R., and Dziezyc, J. (1986). Conjunctivalostomy for treatment of an occluded lacrimal duct in a blood python. *J. Am. Vet. Med. Assoc.* **189,** 1136–1138.

Mimura, K. (1990). Developmental process of visual pattern discrimination in the fly. *Brain Res.* **512,** 75–80.

Mitton, C. J. A., and McDonald, D. G. (1994a). Consequences of pulsed DC electrofishing and air exposure to rainbow trout (*Oncorhynchus mykiss*). *Can. J. Fish. Aquat. Sci.* **51,** 1791–1798.

Mitton, C. J. A., and McDonald, D. G. (1994b). Effects of electroshock, air exposure, and forced exercise on swim performance in rainbow trout *(Oncorhynchus mykiss). Can. J. Fish. Aquat. Sci.* **51,** 1799–1803.

Moberly, W. R. (1968). The metabolic responses of the common iguana, *Iguana iguana,* to walking and diving. *Comp. Biochem. Physiol.* **27,** 21–32.

Mohammad, F. K., Al-Badrany, M. S., and Al-Hasan, A. M. (1993). Detomidine-ketamine anaesthesia in chickens. *Vet. Rec.,* 133(8), 192.

Montecucchi, P. C., DeCastiglione, R., Piana, S., Gozzini, L., and Erspamer, V. (1981). Amino acid composition and sequence of dermorphin, a novel opiate-like peptide from the skin of *Phylomedusa sauagei. Int. J. Pept. Protein Res.* **17,** 275–277.

Moore, D. M. (1984). A simple technique for blood collection in the opossum (*Didelphis virginiana*). *Lab. Anim.* **18,** 52–54.

Morand, S., and Faliex, E. (1994). Study on the life cycle of a sexually transmitted nematode parasite of a terrestrial snail. *J. Parasitol.* **80,** 1049–1052.

Morgan, P. G., Sedensky, M. M., Meneely, P. M., and Cascorbi, H. F. (1988). The effect of two genes on anesthetic response in the nematode *Caenorhabditis elegans. Anesthesiology* **69,** 246–251.

Morild, I., Christensen, J. A., Halvorsen, O. J., and Farstad, M. (1987). Effect of volume depletion on the afferent arterioles in the avian kidney. *Virchows Arch. A: Pathol. Anat. Histol.* **411,** 149–155.

Morris, Y. A., and Crews, D. (1990). The effects of exogenous neuropeptide Y on feeding and sexual behavior in the red-sided garter snake (*Thamnophis sirtalis parietalis*). *Brain Res.* **530,** 339–341.

Morrison, R. G., and Nottebohm, F. (1993). Role of a telencephalic nucleus in the delayed song learning of socially isolated zebra finches. *J. Neurobiol.* **24,** 1045–1064.

Morton, W. E. (1990). Occupational phenoxyethanol neurotoxicity: A report of three cases. *J. Occup. Med.* **32,** 42–45.

Muir, W. W., and Hubbell, J. A. E. (1989). ''Handbook of Veterinary Anesthesia.'' Mosby, St. Louis, Missouri.

Muneoka, K., Holler-Dinsmore, G., and Bryant, S. V. (1986). Intrinsic control of regenerative loss in *Xenopus laevis* limbs. *J. Exp. Zool.* **240,** 47–54.

Murchison, D., Chrachri, A., and Mulloney, B. (1993). A separate local pattern-generating circuit controls the movements of each swimmeret in crayfish. *J. Neurophysiol.* **70,** 2620–2631.

Murphy, S., Li, Z.-S., Furness, J. B., and Campbell, G. (1993). Projections of nitric oxide synthase- and peptide-containing neurons in the small and large intestines of the toad *(Bufo marinus). J. Auton. Nerv. Syst.* **46,** 75–92.

Muszynska-Pytel, M., Trzcinska, R., Aubry, M., Pszczolkowski, M. A., and Cymborowski, B. (1993). Regulation of prothoracic gland activity in diapausing larvae of the wax moth, *Galleria mellonella* L. (Lepidoptera). *Insect Biochem. Mol. Biol.* **23,** 33–41.

Nagahama, T., Weiss, K. R., and Kupfermann, I. (1993). Effects of cerebral neuron C-PR on body postural muscles associated with a food-induced arousal state in *Apylsia. J. Neurophysiol.* **70,** 1231–1243.

Nagahama, T., Weiss, K. R., and Kupfermann, I. (1994). Body postural muscles active during food arousal in *Aplysia* are modulated by diverse neurons that receive monosynaptic excitation from the neuron C-PR. *J. Neurophysiol.* **72,** 314–325.

Narins, P. M., and Wagner, I. (1989). Noise susceptibility and immunity of phase locking in amphibian auditory-nerve fibers. *J. Acoust. Soc. Am.* **85,** 1255–1265.

Necker, R. (1990). Sensory representation of the wing in the spinal dorsal horn of the pigeon. *Exp. Brain Res.* **81,** 403–412.

Neya, T., Yamasato, T., Mizutani, M., and Nakayama, S. (1990a). Regulation of bile duct motility by vagus and sympathetic nerves in the pigeon. *Acta Med. Okayama* **44,** 113–115.

Neya, T., Matsuo, S., and Nakayama, S. (1990b). Intrinsic reflexes mediated via peptidergic neurons in the smooth muscle esophagus of the hen. *Acta Med. Okayama* **44,** 129–133.

Ng, T. B., and Chan, S. T. H. (1990). Adrenocorticotropin-like and opiate-like materials in the brain of the red grouper, *Epinephelus akara* (Telostei: Serranidae). *Comp. Biochem. Physiol. C* **95C,** 159–162.

Nicholson, A., and Ilkiw, J. E. (1992). Neuromuscular and cardiovascular effects of atracurium in isoflurane-anesthetized chickens. *Am. J. Vet. Res.* **53,** 2337–2341.

Nishimura, H., Walker, O. E., Patton, C. M., Madison, A. B., Chiu, A. T., and Keiser, J. (1994). Novel angiotensin receptor subtypes in fowl. *Am. J. Physiol.* **36,** R1174–R1181.

Nordeen, K. W., and Nordeen, E. J. (1992). Auditory feedback is necessary for the maintenance of stereotyped song in adult zebra finches. *Behav. Neural Biol.* **57,** 58–66.

Northcutt, R. G., and Butler, A. B. (1974). Retinal projections in the northern water snake, *Natrix sipedon sipedon. J. Morphol.* **142,** 117–135.

Norton, T. M., Jacobson, E. R., and Sundberg, J. P. (1990). Cutaneous fibropapilomas and renal myxofibroma in a green turtle, *Chelonia mydas. J. Wildl. Dis.* **26,** 265–270.

Nurko, S., Dunn, B. M., and Rattan, S. (1989). Peptide histidine isoleucine and vasoactive intestinal polypeptide cause relaxation of opossum internal anal sphincter via two distinct receptors. *Gastroenterology* **96,** 403–413.

Odom, A., Gross, W. B., and Ehrich, M. (1992). Role of socialization, stress and sex of chickens on response to anesthesia and on response to an organophosphate neurotoxicant. *Vet. Hum. Toxicol.* **34,** 134–137.

Ogunranti, I. O. (1987). Some physiological observations on ketamine hydrochloride anaesthesia in the agamid lizard. *Lab. Anim.* **21,** 183–187.

Olesen, S.-P. (1987). Free oxygen radicals decrease electrical resistance of microvascular endothelium in brain. *Acta Physiol. Scand.* **129,** 181–187.

Oliver, A. E., Deamer, D. W., and Akeson, M. (1991). Evidence that sensitivity to seroid anesthetics appears late in evolution. *Brain Res.* **557,** 298–302.

Olsen, G. H., Dein, F. J., Haramis, G. M., and Jorde, D. G. (1992). Implanting radio transmitters in wintering canvasbacks. *J. Wildl. Manage.* **56,** 325–328.

Olson, M. E., and McCabe, K. (1986). Anesthesia in the Richardson's ground squirrel: Comparison of ketamine, ketamine and xylazine, droperidol and fentanyl, and sodium pentobarbital. *J. Am. Vet. Med. Assoc.* **189,** 1035–1037.

O'Rourke, N. A., and Fraser, S. E. (1989). Gradual appearance of a regulated retinotectal projection pattern in *Xenopus laevis. Dev. Biol.* **132,** 251–265.

Orsi, J. A., and Short, J. W. (1987). Modifications in electrical anesthesia for salmonids. *Prog. Fish-Cult.* **49,** 144–146.

Osborne, S., and Milsom, W. K. (1993). Ventilation is coupled to metabolic demands during progressive hypothermia in rodents. *Respir. Physiol.* **92,** 305–318.

Page, C. D. (1985). Soft tissue celiotomy in an aldabra tortoise. *J. Zoo Anim. Med.* **16,** 113–114.

Page, C. D. (1993). Current reptilian anesthesia procedures. *In* "Zoo and Wild Animal Medicine" (M. Fowler, ed.), 3rd ed., pp. 140–143. Saunders, Philadelphia.

Palace, V. P., Majewski, H. S., and Klaverkamp, J. F. (1990). Effects of sampling and storage conditions on the stability of biochemical parameters measured in rainbow trout *(Oncorhynchus mykiss)* liver. *Comp. Biochem. Physiol. B* **95B,** 51–55.

Palombi, P. S., Backoff, P. M., and Caspary, D. M. (1994). Paired tone facilitation in dorsal cochlear nucleus neurons: A short-term potentiation model testable in vivo. *Hear. Res.* **75,** 175–183.

Paolini, R. V., Rossman, J. E., Patel, V., and Stanievich, J. F. (1993). A reliable method for large volume blood collection in the chinchilla. *Lab. Anim. Sci.* **43,** 524–525.

Parsons, L. C., and Huggins, S. E. (1965). Effects of temperature on electroencephalogram of the caiman. *Proc. Soc. Exp. Biol. Med.* **120,** 422–426.

Paterson, W. G. (1989). Electrical correlates of peristaltic and nonperistaltic contractions in the opossum smooth muscle esophagus. *Gastroenterology* **97,** 665–675.

Paterson, W. G., and Kolyn, D. M. (1994). Esophageal shortening induced by short-term intraluminal acid perfusion in opossum: A cause for hiatus hernia? *Gastroenterology* **107,** 1736–1740.

Perino, L. J., Wright, R. E., Hoppe, K. L., and Fulton, R. W. (1990). Bovine leukosis virus transmission with mouthparts from *Tabanus abactor* after interrupted feeding. *Am. J. Vet. Res.* **51,** 1167–1169.

Pezalla, P. (1983). Morphine-induced analgesia and explosive motor behavior in an amphibian. –Brain Res. *273,* **297**305.

Phifer, C. B., and Terry, L. M. (1986). Use of hypothermia for general anesthesia in preweanling rodents. –Physiol. Behav. *38,* **887**890.

Pitt, H. A., Lewinski, M. A., Muller, E. L., Porter-Fink, V., and DenBesten, L. (1984). Ileal resection-induced gallstones: Altered bilirubin or cholesterol metabolism? *Surgery (St. Louis)* **96,** 154–162.

Plakas, S. M., Stehly, G. R., and Khoo, L. (1992). Pharmacokinetics and excretion of phenol red in the channel catfish. *Xenobiotica* **22,** 551–557.

Planelles, G., Anagnostropoulos, T., Cheval, L., and Doucet, A. (1991). Biochemical and functional characterization of H^+-K^+-ATPase in distal amphibian nephron. *Am. J. Physiol.* **260,** F806–F812.

Ploucha, J. M., and Fink, G. D. (1986). Hemodynamics of hemorrhage in the conscious rat and chicken. *Am. J. Physiol.* **251,** R846–R850.

Powell, F. L., and Gray, A. T. (1989). Ventilation-perfusion relationships in alligators. *Respir. Physiol.* **83**94.

Powell, F. L., Milsom, W. K., and Mitchell, G. S. (1988). Effects of intrapulmonary CO and airway pressure on pulmonary vagal afferent activity in the alligator. *Respir. Physiol.* **74,** 285–298.

Powley, T. L., and Fox, E. A. (1986). Vagotomy does not alter circannual body weight cycles in the hibernator *Citellus lateralis. Brain Res.* **364,** 159–161.

Prechtl, J. C., and Bullock, T. H. (1992). Barbiturate sensitive components of visual ERPs in a reptile. *NeuroReport* **3,** 801–804.

Preston, E., Carlo-Stella, R., and Courtice, G. P. (1993). Desensitisation of the cardiac response to somatostatin in isolated toad atria but not in whole toads. *J. Auton. Nerv. Syst.* **42,** 151–158.

Pritz, M. B., and Stritzel, M. E. (1987). Percentage of intrinsic and relay cells in a thalamic nucleus projecting to general cortex in reptiles, *Caiman crocodilus. Brain Res.* **409,** 146–150.

Pritz, M. B., and Stritzel, M. E. (1990). Thalamic projections from a midbrain somatosensory area in a reptile, *Caiman crocodilus. Brain Behav. Evol.* **36,** 1–13.

Pritz, M. B., and Stritzel, M. E. (1994). Anatomical identification of a telencephalic somatosensory area in a reptile, *Caiman crocodilus. Brain Behav. Evol.* **43,** 107–127.

Quinn, T. P., Olson, A. F., and Konecki, J. T. (1988). Effects of anaesthesia on the chemosensory behaviour of Pacific salmon. *J. Fish Biol.* **33,** 637–641.

Racey, P. A. (1972). Bats. *In* "Handbook on the Care and Management of Laboratory Animals" (Universities' Federation for Animal Welfare, ed.), 4th ed., pp. 298–310. Williams & Wilkins, Baltimore, Maryland.

Rager, D. R., and Gallup, G. G. (1986). Apparent analgesic effects of morphine in chickens may be confounded by motor deficits. *Physiol. Behav.* **37,** 269–272.

Ragsdale, F. R., Imel, K. M., Nilsson, E. E., and Ingermann, R. L. (1993). Pregnancy-associated factors affecting organic phosphate levels and oxygen affinity of garter snake red cells. *Gen. Comp. Endocrinol.* **91,** 181–188.

Ramirez, J.-M., and French, A. S. (1990). Phentolamine selectivity affects the fast sodium component of sensory adaptation in an insect mechanoreceptor. *J. Neurobiol.* **21,** 893–899.

Ramos, J. G. (1986). Aminoacids and oscillatory electrocortical activity. *Acta Physiol. Pharmacol. Latinoam.* **36,** 289–301.

Raphael, B. L. (1993). Amphibians. *Vet. Clin. North Am.: Small Anim. Pract.* **23,** 1271–1286.

Rattan, S., and Culver, P. J. (1987). Influence of loperamide on the internal anal sphincter in the opossum. *Gastroenterology* **93,** 121–128.

Redig, P. T. (1984). Fluid therapy and acid base balance in the critically ill avian patient. *Proc. Int. Conf. Avian Med.,* pp. 59–73.

Regnier, M., and Herrera, A. A. (1993). Differential sensitivity to androgens within a sexuality dimorphic muscle of male frogs *(Xenopus laevis). J. Neurobiol.* **24,** 1215–1228.

Reinert, H. K., and Cundall, D. (1982). An impoved surgical implantation method for radio-tracking. *Copeia* **3,** 702–705.

Richardson, R. D., and Boswell, T. (1993). A method for third ventricular cannulation of small passerine birds. *Physiol. Behav.* **53,** 209–213.

Rissman, E. F., and Crews, D. (1989). Effect of castration on epididymal sperm storage in male musk shrews (*Suncus murinus*) and mice (*Mus musculus*). *J. Reprod. Fertil.* **86,** 219–222.

Roberts, B. L., Van Rossem, A., and DeJager, S. (1992). The influence of cerebellar lesions on the swimming performance of the trout. *J. Exp. Biol.* **167,** 171–178.

Roberts, J. R., and Dantzler, W. H. (1990). Micropuncture study of the avian kidney: Infusion of mannitol or sodium chloride. *Am. J. Physiol.* **258,** R869–R875.

Robertson, L., Thomas, P., and Arnold, C. R. (1988). Plasma cortisol and secondary stress responses of cultured red drum (*Sciaenops ocellatus*) to several transportation procedures. *Aquaculture* **68,** 115–130.

Robinson, E. S., and VandeBerg, J. L. (1994). Blood collection and surgical procedures for the laboratory opossum *(Monodelphis domestica). Lab. Anim. Sci.* **44,** 63–68.

Roder, J. D., Akkaya, R., Amouzadeh, H. R., Sangiah, S., Burrows, G., and Qualls, C. W. (1993). Effects of hepatic P-450 enzyme inhibitors and inducers on the duration of xylazine + ketamine anesthesia in broiler chickens and mice. *Vet. Hum. Toxicol.* **35,** 116–118.

Rohrer, B., and Stell, W. K. (1994). Basic fibroblast growth factor (bFGF) and transforming growth factor beta (TGF-) act as stop and go signals to modulate postnatal ocular growth in the chick. *Exp. Eye Res.* **58,** 553–562.

Rose, C., and Weiss, T. F. (1988). Frequency dependence of synchronization of cochlear nerve fibers in the alligator lizard: Evidence for a cochlear origin of timing and non-timing neural pathways. *Hear. Res.* **33,** 151–166.

Rosenberg, H., and Sugimoto, T. (1925). Uber die physikochemischenbedingungeh der energungsleitung. *Biochem. Z.* **156,** 262.

Rosenberg, M. E. (1977). Temperature and nervous conduction in the tortoise. *J. Physiol. (London)* **270,** 50.

Rosenberg, M. E. (1978). Thermal relations of nervous conduction in the tortoise. *Comp. Biochem. Physiol. A* **60A,** 57–63.

Roslyn, J., Thompson, J. E., and DenBesten, L. (1979). Anesthesia for prairie dogs. *Lab. Anim. Sci.* **29,** 542–544.

Ross, L. G., and Ross, B. (1984). ''Anaesthetic and Sedative Techniques for Fish.'' Institute of Aquaculture, Scotland.

Rossi, G. T., and Britt, R. H. (1984). Effects of hypothermia on the cat brainstem auditory evoked response. *Electroencephalogr. Clin. Neurophysiol.* **57,** 143–155.

Rotella, J. J., and Ratti, J. T. (1990). Use of methoxyflurane to reduce nest abandonment of mallards. *J. Wild. Manage.* **54,** 627–628.

Roth, J. R., and Jones, R. E. (1992). A single ovary of *Anolis carolinensis* responds more to exogenous gonadotropin if the contralateral overy is absent. *Gen. Comp. Endocrinol.* **85,** 486–492.

Routh, A. (1991). Bats in the surgery: A practitioner's guide. *Vet. Rec.,* 128(14), 316–318.

Rowley, R. J. (1989). Settlement and recruitment of sea urchins (*Strongylocentrotus* spp.) in a sea-urchin barren ground and a kelp bed: Are populations regulated by settlement or post-settlement processes? *Mar. Biol.* **100,** 485–494.

Royer, L., and McNeil, J. N. (1992). Evidence of a male sex pheromone in the european corn borer, *Ostrinia nubilalis* (Hubner) (Lepidoptera: Pyralidae). *Can. Entomol.* **124,** 113–115.

Runkel, N. S. F., Rodriguez, L. F., and Moody, F. G. (1995). Mechanisms of sepsis in acute pancreatitis in opossums. *Am. J. Surg.* **169,** 227–232.

Runzi, M., Saluja, A., Lerch, M. M., Dawra, R., Nishino, H., and Steer, M. L. (1993). Early ductal decompression prevents the progression of biliary pancreatitis: An experimental study in the opossum. *Gastroenterology* **105,** 157–164.

Rybak, L. P., and Whitworth, C. (1988). Quinine reduces noxious cochlear effects of furosemide and ethacrynic acid. *Am. J. Otolaryngol.* **9,** 238–243.

Sahley, T. L., Kalish, R. B., Musiek, F. E., and Hoffman, D. W. (1991). Effects of opioid be drugs on auditory evoked potentials suggest a role of lateral olivocochlear dynorphins in auditory function. *Hear. Res.* **55,** 133–152.

Sakharov, D., Nezlin, L., Moroz, L., and Elofsson, R. (1993). Patterns of enkephalin immunolabeling in the pulmonate snail *Cepaea nemoralis* and related molluscs. *Brain Res.* **620,** 114–121.

Salgado, V. L. (1990). Mode of action of insecticidal dehydropyrazoles: Selective block of impulse generation in sensory nerves. *Pestic. Sci.* **28,** 389–411.

Samuel, I., Wilcockson, D. P., Regan, J. P., and Joehl, R. J. (1995). Ligatino-induced acute pancreatitis in opossums: Acinar cell necrosis in the absence of colocalization. *J. Surg. Res.* **58,** 69–74.

Sanz, A., Cardenete, G., Hidalgo, F., and Garcia-Gallego, M. (1993). Study of biliary secretion in trout (*Oncorhynchus mykiss*) by cannulation of bile duct. *Comp. Biochem. Physiol. A* **104A,** 525–529.

Sap, R., van Wandelen, R. M., and Hellebrekers, L. J. (1993). Spontane ademhaling versus IPPV BIJ De Duif. *Tijdschr. Diergeneeskd.* **118,** 402–404.

Saunders, J. C., Torsiglieri, A. J., and DeDio, R. M. (1993). The growth of hearing loss in neonatal chicks exposed to intense pure tones. *Hear. Res.* **69,** 25–34.

Schachern, P. A., Paparella, M. M., Hybertson, R., Sano, S., and Duvall, A. J. (1992). Bacterial tympanogenic labyrinthitis, meningitis, and sensorineural damage. *Arch. Otolaryngol. Head Neck Surg.* **118,** 53–57.

Schaeffer, D. O. (1994). Miscellaneous species: Anesthesia and analgesia. *In* ''Research Animal Anesthesia, Analgesia and Surgery'' (A. C. Smith and M. M. Swindle, eds.), pp. 129–136. Scientists Center for Animal Welfare, Greenbelt, Maryland.

Schariff, C., and Nottebohm, F. (1991). A comparative study of the behavioral deficits following lesions of various parts of the zebra finch song system: Implications for vocal learning. *J. Neurosci.* **11,** 2896–2913.

Schmid, H., Gendre, N., and Stocker, R. F. (1986). Surgical generation of supernumerary appendages for studying neuronal specificity in *Drosophila melanogaster. Dev. Biol.* **113,** 160–173.

Schramm, H. L., and Black, D. J. (1984). Anesthesia and surgical procedures for implanting radio transmitters into Grass carp. *Prog. Fish-Cult.* **46,** 185–190.

Seaman, G. C., Ludders, J. W., Erb, H. N., and Gleed, R. D. (1994). Effects of low and high fractions of inspired oxygen on ventilation in ducks anesthetized with isoflurane. *Am. J. Vet. Res.* **55,** 395–398.

Sedensky, M. M., Cascorbi, H. F., Meinwald, J., Radford, P., and Morgan, P. G. (1994). Genetic differences affecting the potency of stereoisonomers of halothane. *Proc. Natl. Acad. Sci. U.S.A.* **91,** 10054–10058.

Sedgwick, C. J. (1986). Anesthesia for fish. *Vet. Clin. North Am.: Food Anim. Pract.* **2,** 737–742.

Sedgwick, C. J. (1988a). Anesthesia for reptiles, birds, primates, and small exotic animals. *In* ''Manual of Small Animal Anesthesia'' (R. R. Paddleford, ed.), pp. 309–321. Churchill-Livingstone, New York.

Sedgwick, C. J. (1988b). Anesthesia for small nondomestic mammals. *In* ''Exotic Animals'' (E. R. Jacobson and G. V. Kollias, eds.), pp. 209–229. Churchill-Livingstone, New York.

Seiger, M. B., and Kink, J. F. (1993). The effect of anesthesia on the photoresponses of four sympatric species of *Drosophila. Behav. Genet.* **23,** 99–104.

Sengupta, N., Kauvar, D., and Goyal, R. K. (1989). Characteristics of vagal esophageal tension-sensitive afferent fibers in the opossum. *J. Neurophysiol.* **5,** 1001–1010.

Serantes, P. A., Barreiro, A. L., Diaz, C. V., Espinel, J. L. P., Diez, A. E., and Gonzalo, J. M. (1987). Surgical treatment of otic granuloma in a greek tortoise. *An. Fac. Vet. Leon, Univ. Oviedo* **33,** 93–95.

Sevala, V. L., and Davey, K. G. (1991). The transparent accessory reproductive gland secretes a polypeptide into the hemolymph of male *Rhodnius prolixus. Insect Biochem.* **21,** 215–221.

Shaddock. L. C., Wright, C. G., and Hamernik, R. P. (1985). A morphometric study of microvascular pathology following experimental rupture of Reissner's membrane. *Hear. Res.* **20,** 119–129.

Shaw, A. P., Collazo, C. R., Easterling, K., Young, C. D., and Karwoski, C. J. (1993). Circadian rhythm in the visual system of the lizard *Anolis carolinensis. J. Biol. Rhythms* **8,** 107–124.

Shibamoto, T., Chakder, S., and Rattan, S. (1994). Role of hypogastric nerve activity in opossum internal anal sphincter function: Influence of surgical and chemical denervation. *J. Pharmacol. Exp. Ther.* **271,** 277–284.

Shimizu, T., and Masaki, S. (1993). Injury causes microptery in the ground cricket, *Dianemobius fascipes*. *J. Insect Physiol.* **39,** 1021–1027.

Shimizu, T., Cox, K., Karten, H. J., and Britto, L. R. G. (1994). Cholera toxin mapping of retinal projections in pigeons (*Columbia livia*), with emphasis on retinohypothalamic connections. *Visual Neurosci.* **11,** 441–446.

Shirane, M., and Harrison, R. V. (1987). The effects of deferoxamine mesylate and hypoxia on the cochlea. *Acta Oto-laryngol.* **104,** 99–107.

Sholomenko, G. N., Funk, G. D., and Steeves, J. D. (1991). Avian locomotion activated by brainstem infusion of neurotransmitter agonists and antagonists. *Brain Res.* **85,** 659–673.

Silverman, J., and Muir, W. W. (1993). A review of laboratory animal anesthesia with chloral hydrate and chloralose. *Lab. Anim. Sci.* **43,** 210–216.

Sinha Hikim, A. P., Woolf, A., Bartke, A., and Amador, A. G. (1992). Further observations on estrus and ovulation in woodchucks (*Marmota monax*) in captivity. *Biol. Reprod.* **46,** 10–16.

Sink, H., and Whittington, P. M. (1991). Early ablation of target muscles modulates the arborisation pattern of an identified embryonic *Drosophila* motor axon. *Development (Cambridge, UK)* **113,** 701–707.

Sinn, L. C. (1994). Anesthesiology. *In* ''Avian Medicine: Principles and Applications'' (B. W. Ritchie, G. J. Harrison, and L. R. Harrison, eds.), pp. 1066–1080. Wingers Publishing, Lake Worth, Florida.

Slip, D. J., and Shine, R. (1988a). Reptilian endothermy: A field study of thermoregulation by brooding diamond pythons. *J. Zool.* **216,** 367–378.

Slip, D. J., and Shine, R. (1988b). Thermophilic response to breeding of the diamond python, *Morelia s. spilota* (Serpentes: Boidae). *Comp. Biochem. Physiol. A* **89A,** 645–650.

Sloley, B. D., Kah, O., Trudeau, V. L., Dulka, J. G., and Peter, R. E. (1992). Amino acid neurotransmitters and dopamine in brain and pituitary of the goldfish: Involvement in the regulation of gonadotropin secretion. *J. Neurochem.* **58,** 2254–2261.

Small, S. A., Cohen, T. E., Kandel, E. R., and Hawkins, R. D. (1992). Identified FMRFamide-immunoreactive neuron LPL16 in the left pleural ganglion of *Aplysia* produces presynaptic inhibition of siphon sensory neurons. *J. Neurosci.* **12,** 1616–1627.

Smith, D. A., and Barker, I. K. (1988). Healing of cutaneous wounds in the common garter snake *(Thamnophis sirtalis). Can. J. Vet. Res.* **52,** 111–119.

Smith, F. M., and Jones, D. R. (1990). Effects of acute and chronic baroreceptor denervation on diving responses in ducks. *Am. J. Physiol.* **258,** R895–R902.

Smith, F. M., and Jones, D. R. (1992). Baroreflex control of arterial blood pressure during involuntary diving in ducks (*Anas platyrhynchos* var.). *Am. J. Physiol.* **263,** R693–R702.

Smith, L. M., Hupp, J. W., and Ratti, J. T. (1980). Reducing abandonment of nest-trapped gray partridge with methoxyflurane. *J. Wildl. Manage.* **44,** 690–691.

Soybel, D. T., Davis, M. B. E., and West, A. B. (1992). Effects of aspirin on pathways of ion permeation in *Necturus antrum:* Role of nutrient HCO. *Gastroenterology* **103,** 1475–1485.

Soybel, D. I., Davis, M. B. E., and Cheung, L. Y. (1993). Characteristics of basolateral Cl^- transport by gastric surface epithelium in *Necturus* antral mucosa. *Am. J. Physiol.* **264,** G910–G920.

Steeves, J. D., Sholomenko, G. N., and Webster, D. M. S. (1987). Stimulation of the pontomedullary reticular formation initiates locomotion in decerebrate birds. *Brain Res.* **401,** 205–212.

Stephens, L. C., Cromeens, D. M., Robbins, V. W., Stromberg, P. C., and Jardine, J. H. (1987). Epidermal capillariasis in South African clawed frogs *(Xenopus laevis). Lab. Anim. Sci.* **37,** 341–344.

Stephenson, R. S. (1990). Caesarean section in a chinchilla. *Vet. Rec.,* 126(15), 370.

Stevens, C. W. (1988). Opioid antinociception in amphibians. *Brain Res. Bull.* **21,** 959–962.

Stevens, C. W. (1992). Alternatives to the use of mammals for pain research. *Life Sci.* **50,** 901–912.

Stevens, C. W., and Kirkendall, K. (1993). Time course and magnitude of tolerance to the analgesic effects of systemic morphine in amphibians. *Life Sci.* **52,** 111–116.

Stevens, C. W., and Pezalla, P. D. (1989). Endogenous opioid system down-regulation during hibernation in amphibians. *Brain Res.* **494,** 227–231.

Stevens, C. W., Pezalla, P. D., and Yaksh, T. L. (1987). Spinal antinociceptive action of three representative opioid peptides in frogs. *Brain Res.* **402,** 201–203.

Stevens, C. W., Klopp, A. J., and Facello, J. A. (1994). Analgesic potency of *Mu* and *Kappa* opioids after systemic administration in amphibians. *J. Pharmacol. Exp. Ther.* **269,** 1086–1093.

Stocker, R. F., and Gendre, N. (1989). Courtship behavior of *Drosophila* genetically or surgically deprived of basiconic sensilla. *Behav. Genet.* **19,** 371–385.

Storrs, E. E. (1972). Armadillos. *In* ''Handbook on the Care and Management of Laboratory Animals'' (Universities' Federation for Animal Welfare, ed.), pp. 229–239. Williams & Wilkins, Baltimore, Maryland.

Stoskopf, M. K. (1985). ''Manual for the Aquatic Animal Workshop.'' American Association for Laboratory Animal Science National Capitol Area Branch, Washington, D. C.

Stoskopf, M. K. (1986). Immobilization of captive sharks. *Annu. Meet. Am. Assoc. Zoo Vet.,* Abstr., pp. 106–108.

Stoskopf, M. K. (1994). Pain and analgesia in birds, reptiles, amphibians, and fish. *Invest. Ophthalmol. Visual Sci.* **35,** 774–780.

Strain, G. M., Tucker, T. A., Graham, M. C., and O'Malley, N. A. (1987). Brainstem auditory evoked potentials in the alligator. Effects of temperature and hypoxia. *Electroencephalogr. Clin. Neurophysiol.* **67,** 68–76.

Straznicky, C., and McCart, R. (1987). The effects of tectal lesion on the survival of isthmic neurones in *Xenopus. Development (Cambridge, UK)* **101,** 869–876.

Suarez, C., Honrubia, V., Gomez, J., Lee, W. S., and Newman, A. (1989). Primary vestibular projections in the chinchilla. *Arch. Oto-Rhino-Laryngol.* **246,** 242–248.

Subramaniam, M., Henderson, D., and Spongra, V. (1993). Effect of low-frequency ''conditioning'' on hearing loss from high-frequency exposure. *J. Acoust. Soc. Am.* **93,** 952–956.

Sufka, K. J., and Hughes, R. A. (1990). Dose and temporal parameters of morphine-induced hyperalgesia in domestic fowl. *Physiol. Behav.* **47,** 385–387.

Sufka, K. J., Hughes, R. A., and Giordao, J. (1991). Effects of selective opiate antagonists on morphine-induced hyperalgesia in domestic fowl. *Pharmacol. Biochem. Behav.* **38,** 49–54.

Sugai, T., Sugitani, M., and Ooyama, H. (1991). Effects of activation of the divergent efferent fibers on the spontaneous activity of vestibular afferent fibers in the toad. *Jpn. J. Physiol.* **41,** 217–232.

Sun, D., Sun, X., and Jen, P. H.-S. (1990). The influence of the auditory cortex on acoustically evoked cerebellar responses in the CF-FM bat, *Rhinolopus pearsoni chinesis. J. Comp. Physiol.* **166,** 477–487.

Sun, X., and Jen, P. H.-S. (1989). Electrical stimulation of bat superior colliculus influences responses on inferior collicular neurons to acoustic stimuli. *Brain Res.* **497,** 214–222.

Takei, Y., and Hasegawa, Y. (1990). Vasopressor and depressor effects of native antiotensins and inhibition of these effects in the Japanese quail. *Gen. Comp. Endocrinol.* **79,** 12–22.

Takei, Y., and Watanabe, T. X. (1990). Vasodepressor effect of atrial natriuretic peptides in the quail, *Coturnix coturnix japonica. Zool. Sci.* **7,** 435–441.

Tappa, B., Amao, H., and Takahashi, K. W. (1989). A simple method for intravenous injection and blood collection in the chinchilla *(Chinchilla laniger). Lab. Anim.* **23,** 73–75.

Taylor, K., Fish, D. R., Mendel, F. C., and Burton, H. W. (1992). Effect of a single 30-minute treatment of high voltage pulsed current on edema formation in frog hind limbs. *Phys. Ther.* **72,** 63–68.

Taylor, P. B. (1988). Effects of anaesthetic MS 222 on the orientation of juvenile salmon, *Oncorhynchus tschawytscha* Walbaum. *J. Fish Biol.* **32,** 161–168.

Teyke, T., Rosen, S. C., Weiss, K. R., and Kupfermann, I. (1993). Dopaminergic neuron B20 generates rhythmic neuronal activity in the feeding motor circuitry of *Aplysia. Brain Res.* **630,** 226–237.

Thatikunta, P., Chakder, S., and Rattan, S. (1993). Nitric oxide synthase inhibitor inhibits catecholamines release caused by hypogastric sympathetic nerve stimulation. *J. Pharmacol. Exp. Thera.* **267,** 1363–1368.

Tiedeken, J. J., and Rovainen, C. M. (1991). Fluorescent imaging in vivo of developing blood vessels on the optic tectum of *Xenopus laevis. Microvasc. Res.* **41,** 376–389.

Tierney, S., Qian, Z., Burrow, C., Lipsett, P. A., Pitt, H. A., and Lillemoe, K. D. (1994). Estrogen inhibits sphincter of Oddi motility. *J. Surg. Res.* **57,** 69–73.

Timm, K. I., Jahn, S. E., and Sedgwick, C. J. (1987). The palatal ostium of the guinea pig. *Lab. Anim. Sci.* **37,** 801–802.

Tonner, P. H., Poppers, D. M., and Miller, K. W. (1992). The general anesthetic potency of propofol and its dependence on hydrostatic pressure. *Anesthesiology* **77,** 926–931.

Tsutsui, K., Wingfield, J. C., and Bottjer, S. W. (1991). Adenohypophysectomy in the zebra finch. *Gen. Comp. Endocrinol.* **81,** 163–173.

Tunstall, M. J., and Roberts, A. (1991). Longitudinal coordination of motor output during swimming in *Xenopus* embryos. *Proc. R. Soc. London* **244,** 27–32.

Tyml, K. (1987). Red cell perfusion in skeletal muscle at rest and after mild and severe contractions. *Am. J. Physiol.* **252,** H485–H493.

Tytler, P. and Hawkins, A. D. (1981). Vivisection, anaesthetics and minor surgery. *In* ''Aquarium Systems'' (A. D. Hawkins, ed.), pp. 247–278). Academic Press, London.

Ueda, H., Miyamae, T., Fukushima, N., and Misu, Y. (1994a). Protein kinase inhibitor potentiates opioid α-receptor currents in *Xenopus oocoytes. NeuroReport* **5,** 1985–1988.

Ueda, I., Tatara, T., Chiou, J.-S., and Krishna, P. R. (1994b). Structure-selective anesthetic action of steroids: Anesthetic potency and effects on lipid and protein. *Anesth. Analg.* **78,** 718–725.

Underwood, L. W., and Ide, C. F. (1992). An autoradiographic time study during regeneration in fully differentiated *Xenopus* eyes. *J. Exp. Zool.* **262,** 193–201.

Underwood, L. W., Carruth, M. R., Vandecar-Ide, A., and Ide, C. F. (1993). Relationship between local cell division and cell displacement during regeneration of embryonic *Xenopus* eye fragments. *J. Exp. Zool.* **265,** 165–177.

Valeggia, C., Fernandez-Duque, E., and Maldonado, H. (1989). Danger stimulus-induced analgesia in the crab *Chasmagnathus granulatus. Brain Res.* **481,** 304–308.

Valentincic, T., Wegert, S., and Caprio, J. (1994). Learned olfactory discrimination versus innate taste responses to amino acids in channel catfish *(Ictalurus punctatus). Physiol. Behav.* **55,** 865–873.

Valenzuela, J. I., Hasan, S. J., and Steeves, J. D. (1990). Stimulation of the brainstem reticular formation evokes locomotor activity in embryonic chicken (in ovo). *Dev. Brain Res.* **56,** 13–18.

Valles, S. M., and Koehler, P. G. (1994). Influence of carbon dioxide anesthesia on chlorpyrifos toxicity in the german cockroach (Dictyoptera: Blattellidae). *J. Econ. Entomol.* **87,** 709–713.

Valverde, A., Bienzle, D., Smith, D. A., Dyson, D. H., and Valliant, A. E. (1993). Intraosseous cannulation and drug administration for induction of anesthesia in chickens. *Vet. Surg.* **22,** 240–244.

Vicario, D. S. (1991). Contributions of syringeal muscles to respiration and vocalization in the zebra finch. *J. Neurobiol.* **22,** 63–73.

Vicario, D. S., and Yohay, K. H. (1993). Song-selective auditory input to a forebrain vocal control nucleus in the zebra finch. *J. Neurobiol.* **24,** 488–505.

Wade, J., and Crews, D. (1991). The relationship between reproductive state and ''sexually'' dimorphic brain areas in sexually reproducing and parthenogenetic whiptail lizards. *J. Comp. Neurol.* **309,** 507–541.

Walker, N. D. (1994). Caesarean section in chameleons. *Vet. Rec.* **134,** 508.

Wang, R. T., Kubie, J. L., and Halpern, M. (1977). Brevital sodium: An effective anesthetic agent for performing surgery on small reptiles. *Copeia,* pp. 738–743.

Weaver, R. J., and Edwards, J. P. (1990). The role of the corpora allata and associated nerves in the regulation of ovarian cycles in the oviparous cockroach *Periplaneta americana. J. Insect Physiol.* **36,** 51–59.

Webb, J. H., Lillemoe, K. D., and Pitt, H. A. (1988). Gastrosphincter of Oddi reflex. *Am. J. Surg.* **155,** 193–198.

Whitaker, B. R. (1991). Common disorders of marine fish. *Compendium* **13,** 960–966.

Wideman, R. F., and Gregg, C. M. (1988). Model for evaluating avian renal hemodynamics and glomerular filtration rate autoregulation. *Am. J. Physiol.* **254,** R925–R932.

Williams, H., Crane, L. A., Hale, T. K., Esposito, M. A., and Nottebohm, F. (1992). Right-side dominance for song control in the zebra finch. *J. Neurobiol.* **23,** 1006–1020.

Williams, T. D., Christiansen, J., and Nygren, S. (1988). A Comparison of intramuscular anesthetics in teleosts and elasmobranchs. *Int. Assoc. Aquatic Anim. Med. Proc.* **19,** 148.

Wilson, D. E. (1988). Maintaining bats for captive studies. *In* ''Ecological and Behavioral Methods for the Study of Bats'' (T. H. Kunz, ed.), pp. 247–263. Smithsonian Institution Press, Washington, D.C.

Wood, F. E., Critchley, K. H., and Wood, J. R. (1982). Anesthesia in the green sea turtle, *Chelonia mydas. Am. J. Vet. Res.* **43,** 1882–1883.

Wood, S. C. (1990). Effect of hematocrit on behavioral thermoregulation of the toad *Bufo marinus. Am. J. Physiol.* **258,** R848–R851.

Wood, S. C., and Lenfant, C. J. M. (1976). Respiration: Mechanics, control and gas exchange. *In* ''Biology of the Reptilia'' (C. Gans and W. R. Dawson, eds.), pp. 223–274. Academic Press, New York.

Wooster, G. A., Hsu, H.-M., and Bowser, P. R. (1993). Nonlethal surgical procedures for obtaining tissue samples for fish health inspections. *J. Aquat. Anim. Health* **5,** 157–164.

Work, T. M., Sawyer, M. M., Jessup, D. A., Washino, R. K., and Osburn, B. I. (1990). Effects of anesthetization and storage temperature on bluetongue virus recovery from *Culicoides varipennis* (Deptera: Ceratopogonidae) and sheep blood. *J. Med. Entomol.* **27,** 331–333.

Wright, K. (1995). Blood collection and hematological techniques in amphibians. *Assoc. Reptilian Amphibian Vet.* **5,** 8–10.

Yamamura, T., Harada, K., Okamura, A., and Kemmotsu, O. (1990). Is the site of action of ketamine anesthesia the n-methyl-d-aspartate receptor? *Anesthesiology* **72,** 704–710.

Zabala, N. A., and Gomez, M. A. (1991). Morphine analgesia, tolerance and addiction in the cricket *Pteronemobius* sp. (Orthoptera, Insecta). *Pharmacol., Biochem. Behav.* **40,** 887–891.

Zhao, Y., and Szaro, B. G. (1994). The return of phosphorylated and nonphosphorylated epitopes of neurofilament proteins to the regenerating optic nerve of *Xenopus laevis. J. Comp. Neurol.* **343,** 158–172.

Zwart, P., Volker, V., Wijnands, M., and Gerritsen, R. (1986). Corpus alienum in de maag bij een slang. *Tijdschr. Diergeneeskd.* **111,** 925–927.

Zwart, P., Kik, K. J. L., and Das, A. H. A. (1988). Dysfunctie van de eileider en salpingectomie bij een boa *(C. constrictor). Tijdschr. Diergeneeskd.* **113,** 494–497.

Chapter 16

Postoperative Support and Intensive Care

Steve C. Haskins and Pamela H. Eisele

I. INTRODUCTION

Effective monitoring and the provision of needed supportive care for postoperative animals contribute to the success of research projects. The overall health and well-being of the subject are critical to the welfare of the animal and to the validity of the results of the experiment. (Exceptions might be in cases where specific adverse effects are being studied as part of an approved protocol.) Although some animals may not require special care after recovery from anesthesia, others may require intensive monitoring and support for successful recovery and a satisfactory outcome to the procedure.

Postoperative care is one facet of adequate veterinary care addressed by regulations and guidelines concerning the care and use of research and teaching animals. The most familiar of these are the Animal Welfare Act regulations (U.S. Department

ANESTHESIA AND ANALGESIA IN LABORATORY ANIMALS

ISBN 0-12-417570-8

of Agriculture, 1992) and the ''Guide for the Care and Use of Laboratory Animals'' [Institute of Laboratory Animal Resources (ILAR), 1985]. Final authority for the provision of adequate veterinary care, including presurgical and postsurgical care, rests with the attending veterinarian of the institution. According to the Animal Welfare Act regulations (U.S. Department of Agriculture, 1992), the provision of postprocedural care for research animals should be ''in accordance with current established veterinary medical and nursing procedures.'' Guidance in the proper use of anesthetic, analgesic, and tranquilizing drugs is a related part of the required veterinary care program. A report by Brown *et al.* (1993) provides more detailed guidelines for animal surgery in the research and teaching environment and generally addresses principles of postoperative care.

Guidelines for postoperative care and monitoring, personnel training, and record keeping should be developed by the institution. Responsibilities for postoperative care can be designated at the time that a protocol is approved by the Institutional Animal Care and Use Committee (IACUC). Personnel providing postoperative and intensive care for research animals should work closely with the attending veterinarian (or an appointee) and with the investigator to assure that the animal receives adequate veterinary care, that the care given corresponds to the level of care required by the approved protocol, and that treatments and other interventions do not render the animal unsuitable for the intended project. Because of this last concern and to minimize pain and distress, the investigator and veterinary staff may elect to have an animal euthanized if it is not recovering from surgery as expected instead of undertaking major supportive measures. However, provisions for extensive supportive care may be written into protocols involving complex surgical procedures, such as major organ transplants, because the need for this care is considered a normal consequence of the procedure.

The purpose of this chapter is to discuss aspects of postoperative monitoring and support of importance to research scientists and to veterinary and animal care personnel responsible for research animals. The monitoring and support of major organ functions, assessment of postoperative pain, facility and equipment requirements, personnel training, general nursing care protocols, record keeping, management of wounds and indwelling catheters, and fluid and nutritional support are all discussed. The focus of the discussion will be the more common research animals: dogs, cats, small ruminants, pigs, and small laboratory mammals such as rabbits and rodents.

II. FACILITIES AND EQUIPMENT FOR POSTOPERATIVE AND INTENSIVE CARE

An animal should be recovered from anesthesia in an area that is warm, safe, quiet, comfortable, and appropriate for the needs of the individual animal and species. In general, animals should be kept separate according to species and health status. A separate anesthesia recovery area, close to the surgery suite and consisting of cages or runs, may be used in centralized facilities. In this case, animals can be transported to other housing after they are fully ambulatory. Another approach is to recover the animals in rooms, runs, or cages where they will remain for part or all of the postoperative period. Finally, animals can be recovered in their home cages, if precautions are made to provide for adequate surveillance, warmth, and animal safety during recovery. Some components of intensive care can be provided in any of these scenarios. Alternatively, where there is frequent need for intensive care, an intensive care unit (ICU) can be incorporated into the veterinary clinic of the institution.

Postoperative housing should allow for adequate surveillance, easy access to the animal for physical examination and treatment, and confinement if this is needed. The proper environment should minimize animal anxiety, fear, and social stresses and optimize comfort and well-being. Factors such as feed palatability and availability, watering systems, light intensity and light cycle, noise levels, social isolation, caging, flooring, bedding, and environmental temperature should be considered and modified to satisfy both the requirements of the individual animal and those of the protocol.

For most species, except smaller rodents, equipment useful for routine postoperative monitoring includes a stethoscope and thermometer. The ability to monitor such parameters as the electrocardiogram, arterial blood pressure, blood gases and electrolytes, hematology and biochemical parameters, food and water intake, and animal weight may be required in some cases. Radiographic equipment and the services of a diagnostic laboratory should be available for cases requiring more extensive follow-up. Equipment required for supportive care, such as warming blankets, heat lamps, special bedding, oxygen cages, and intravenous fluid pumps, should be available as needed. Table I lists equipment and supplies that may be needed in the postoperative or ICU setting.

III. GENERAL NURSING CARE PROTOCOLS

The development of written protocols for procedures commonly performed is desirable for standardizing the delivery of nursing care and preventing procedural errors. In the research environment where proper conduct of these procedures may be important to the data generated, standardized written protocols are even more important and may be required by reviewing agencies.

Standard operating procedures define how to perform a particular activity (Peterson, 1989). The degree of detail provided will be somewhat determined by the training of personnel performing the task and the complexity of the procedure, but should be adequate to prevent inadvertent mistakes. One approach is to designate who is responsible for performing the task, list the materials needed, and then chronologically list the steps to be taken to carry out the activity successfully (Peterson,

TABLE I

EQUIPMENT, SUPPLIES, AND DRUGS USED IN POSTOPERATIVE CARE OR ICU SETTINGS

Equipment	Supplies	Drugs
Means of providing oxygen	Full assortment of syringes, needles, and intravenous catheters	Epinephrine
Means of providing positive-pressure ventilation	Sterile, isotonic, polyionic, intravenous fluids (e.g., lactated Ringer's)	Atropine
Laryngoscope and assortment of blades	Wide selection of concentrates (sodium bicarbonate; potassium chloride; 50% dextrose in water; mannitol)	Dopamine; dobutamine
Assortment of endotracheal tubes	Heparin	Phenylephrine; norepinephrine
Stethoscope	Fluid administration sets: regular, pediatric, and blood filter sets	Calcium gluconate
Sterile surgical instruments and suture material	Extension sets, three-way stopcocks, and injection caps	Lidocaine
Sterile gloves, drapes, and gauze sponges	Urinary catheters	Glucocorticoids
Hair clippers	Tape, cotton, gauze, and other bandaging materials	Diuretics (furosemide, mannitol, dopamine)
Surgical lights	Fleece pads for cage padding	Antipyretic (aspirin, dipyrone)
Assortment of stomach tubes	Sandbags for positioning	Anesthetics of choice
Electrocardiograph		Analeptics [opioid reversing agent (naloxone, nalbuphine); doxapram]
Anesthetic machine		Tranquilizers/sedatives of choice
Hematocrit tubes and centrifuge		Analgesics (opioid agonist or agonist–antagonist; antiprostaglandin)
Refractometer		Antiemetics (metoclopramide; thorazine)
Microscope		Emetics (apomorphine; hydrogen peroxide)
Suction apparatus and catheters		Peristaltics (metoclopramide; cisipride)
Chemistry strips for blood glucose and urea nitrogen		Broad-spectrum antibiotics
Ophthalmoscope		
Coagulation testing method		
Blood pressure measuring equipment		
Radiographic equipment		
Thermometers		
Circulating warm water blanket		
Infrared heat lamp		
Blood gas analyzer		
Fluid infusion pump		
Nebulizer (ultrasonic)		

1989). These protocols should be written so that they are easy to follow and should be easily accessible to personnel performing the described care. At the University of California, Davis, small animal ICU, written guidelines have been developed for a number of commonly performed procedures. Some of these include placement of jugular and peripheral venous catheters, indwelling catheter maintenance, monitoring of arterial blood pressure and central venous pressure, placement and care of indwelling urinary catheters, care of chest drainage systems, and care of tracheostomy tubes. The provision of written guidelines should be tailored to the needs of the particular facility.

IV. PERSONNEL TRAINING

Institutional responsibilities for personnel training have been addressed (Brown *et al.,* 1993; ILAR, 1985; U.S. Department of Agriculture, 1992). Ideally, trained veterinary personnel should be responsible for postoperative monitoring and for supportive and intensive care. However, it is not uncommon for research personnel to perform some of these duties. Personnel should be trained and competent in the aspects of postoperative monitoring and care utilized in the project. General skills required would include knowledge of the proper handling and restraint of the species involved; the ability to perform the required monitoring tasks such as taking of the body temperature, pulse rate, and respiratory rate; and the ability to perform required treatments properly. Additionally, personnel should be trained and experienced in the detection of deviations from normal in the species of concern and, if possible, familiar with normal for the individual being evaluated. They should work closely with the facility veterinary staff, reporting any abnormalities found or problems encountered so that they can be evaluated by the veterinarian. Personnel involved in the use of sophisticated monitoring and ICU equipment should be trained in its proper use. Personnel training can be achieved via a number of approaches, including formal classes and individual or small group practical training sessions under close supervision. Personnel should be able to demonstrate proficiency in the procedures required for the particular postoperative care situation in which they are involved.

V. RECORD KEEPING

Records of postoperative and intensive care unit care should be maintained as part of the animal's medical record which may be reviewed on occasion by regulatory agencies. In addition to documenting adequate veterinary care, accurate records are important for ongoing assessment of the patient's progress and for completeness of the medical record. Information contained in these records may be needed in the future medical management of the patient and is often helpful in the management of other animals with similar conditions or subject to similar experimental protocols. These records may also be relevant to the individual research project and, in some cases, are a necessary part of the data.

The type of postoperative records in use at a particular institution will depend somewhat on institutional policies. Policies regarding research animal medical records should address the documentation needs of the veterinary staff, the researchers, and the relevant regulatory officials, and fulfill any special institutional requirements. Several different types of records may be required to satisfy the needs of various projects involving different species of animals. For example, an institution may require that individual postoperative records be kept for animals the size of guinea pigs and larger. The same institution may allow the postoperative records for multiple small rodents of the same species, undergoing the same procedure on the same day, to be kept on one ''batch'' postoperative form. A general guideline would be that the type of record used should be appropriate for the species and procedure of concern and fulfill current regulatory requirements.

Individual postoperative and ICU records typically include the animal's identification number and species; the investigator's name; the protocol number; the date of surgery; and a description of the surgical procedure or medical problem for which the animal is being treated. Other useful information includes the phone numbers of personnel to contact in case of an emergency, the animal's weight and age, and any special instructions to follow when working with the animal. Timed and dated entries should be made for all treatments administered, observations made, and physiological parameters monitored. Results of diagnostic testing performed, as well as the rest of the animal's medical record, should be accessible.

The use of standardized forms is helpful for maintaining both postoperative and ICU records. Figure 1 is an example of an individual postoperative record that provides space for recording the results of frequent observations in the immediate postoperative period and for less intensive follow-up in subsequent days. This type of record may be adequate for many routine surgeries that do not require extensive monitoring, supportive care, and documentation. In certain situations where animals may require intensive care and close monitoring for recovery and for the success of a project, a more detailed flowchart record may be helpful.

Overall, the style of the records used for a particular research project may not be of paramount importance. However, it is imperative that all relevant information is recorded in a logical fashion that can be easily read and understood by the research scientist, by the veterinary staff, and by outside reviewers. The record should tell the whole story, from beginning to end: what happened, when it happened, and what was done about it. Nothing should be left to the imagination and nothing should be left unrecorded. One never knows what bit of information might become important sometime in the future.

LABORATORY ANIMAL HEALTH CLINIC
UNIVERSITY OF CALIFORNIA, DAVIS
SURGERY POST-OPERATIVE RECORD DATE ______ LOG# ______

POST-ANESTHETIC RECOVERY RECORD

INVESTIGATOR	PROTOCOL #	SPECIES	ANIMAL ID #	PROCEDURE
PERSON TO CALL	PHONE #	LOCATION OF SURGERY SUITE		DATE OF SURGERY
Anesthetic Drugs Administered				

IMMEDIATE POST-OP PERIOD

TIME	TEMP	HR	RR	CRT	TURNED	OBSERVATIONS AND TREATMENTS	SIG.

SUBSEQUENT DAYS POST-OP

ALL ENTRIES MUST BE DATED, TIMED, AND SIGNED (LAST NAME, FIRST INITIAL)

DATE TIME	TEMP	APPETITE	BM	URINE	INCISION CONDITION	OBSERVATIONS AND TREATMENTS	SIG.

POST-ANESTHETIC RECOVERY RECORD

D3822-1 (4/90) **SURGERY UNIT COPY**

Fig. 1. Postoperative record used by the Laboratory Animal Health Clinic, University of California, Davis.

VI. MONITORING

A. General Guidelines

Guidelines for postoperative care should be developed by the institution so that veterinary personnel, animal care staff, and researchers know what care is required as an animal passes through the various stages of postoperative recovery. These guidelines, usually developed by the attending veterinarian and adopted by the IACUC, are based on current standards of veterinary practice, adapted for the various laboratory animal species and for the research environment.

One approach for developing general guidelines considers the postoperative period in three overlapping phases: recovery from anesthesia, the immediate postoperative period, and the long-term postoperative period (Brown *et al.,* 1993). During the first two phases, the major concern is to provide for needed monitoring and supportive care, for proper patient positioning and safety, and for analgesic therapy if indicated.

In general, an animal with an endotracheal tube in place should not be extubated until it is able to control its airway. Once extubated, a recovering animal should be positioned so that the nose and mouth are not obstructed, to facilitate breathing. Partially recovered animals should not be housed together in the same primary enclosure because smothering and injuries can occur. Most rodents can be placed sternally or on their sides for recovery from anesthesia. If unable to maintain sternal recumbency, animals such as rabbits, dogs, cats, and small pigs should be turned from side to side periodically to prevent atelectasis in the dependent lung. Ruminants should be maintained in sternal recumbency with the head and neck propped up after the endotracheal tube has been removed to minimize the risk of bloat and passive regurgitation.

Supplemental heat is often required for the support of body temperature during recovery from anesthesia. This can be provided with warm water heating pads, heat lamps, and heated recovery cages. Care should be taken to prevent accidental burns and overheating when supplemental heat sources are used.

The rationale for postoperative monitoring is to assure that the animal recovers safely and comfortably from anesthesia and surgery and that any complications are identified so that they can be addressed. Consequently, parameters that should be monitored and monitoring frequency will be somewhat dependent on the severity of the procedure and individual animal variability. Usually, the most intensive and frequent monitoring will be required during recovery from anesthesia. In very small animals such as rats and mice, extensive monitoring may not be possible. Valuable observations that can be made in these animals include the color of the mucous membranes and nonpigmented areas of the body, the respiratory rate and pattern, the return of reflexes after anesthesia, and the assessment of body temperature. Evidence of cyanosis, respiratory distress or depression, prolongation of the return of reflexes, or hypothermia necessitates supportive measures such as the provision of supplemental oxygen and heat. Monitoring of rodents is further addressed in the chapter on anesthesia and analgesia in rodents found elsewhere in this volume. Under normal conditions, useful parameters to monitor in larger animals include the body temperature, heart rate and rhythm, respiratory rate and rhythm, mucous membrane color and capillary refill time, jaw tone, and the return of reflexes. In cases where surgery has been extensive or the animal is not recovering as expected, more intensive monitoring and diagnostic testing may be indicated and should be instituted based on the perceived need. The specific indications and approaches to intensive monitoring and the associated supportive care are addressed in the relevant sections of this chapter. The general rule would be to act on any significant problems identified so that body temperature and major organ system functions are supported and required analgesics, antibiotics, and other medications are provided.

Once the animal has recovered from anesthesia and satisfactory progress in the immediate postoperative period has been confirmed, monitoring guidelines for the extended postoperative period may be applicable. Daily postoperative monitoring should continue until the animal is behaviorally and physiologically stable and skin sutures, staples, or clips have been removed, usually at about 10 to 14 days postoperatively. Ongoing monitoring and nursing care are required for animals with chronic catheters and other external appliances for as long as these devices are in place. Parameters assessed and monitoring frequency should be sufficient to assure that recovery is progressing well and to identify any complications that may arise. Examinations should be as detailed as needed to assess patient health and well-being, but should not disturb the animal unnecessarily. Baseline presurgical physical findings and diagnostic test results may be helpful when attempting to determine the significance of postoperative findings.

Useful observations include assessment of overall attitude and activity, hydration, food and water consumption, urination and defecation, and the condition of the operative site. Temperature, pulse, and respiration, along with other parameters discussed in the section on early postoperative monitoring, should be evaluated where indicated either by the condition of the animal or by the specific procedure. Care of the incision and required supportive care should be provided as discussed elsewhere in this chapter.

B. Postoperative Pain

Pain assessment and analgesia in the different species are addressed elsewhere in this volume. However, a discussion on postoperative monitoring and care would be incomplete without some reference to these topics.

The monitoring and assessment of animal pain and the need for analgesic treatment can be approached in a number of ways, the most important of which is observation of the animal. A number of guidelines have been proposed to assist those working with animals in the assessment of pain, distress, and discomfort [Association of Veterinary Teachers and Reseach Workers (AVTRW), 1986; Beynen *et al.,* 1987; ILAR, 1992; Laboratory Animal Science Association (LASA), 1990; Morton and Griffiths, 1985; Soma, 1987; Wright *et al.,* 1985]. Additionally, knowledge of the expected severity of experimental procedures (LASA, 1990) and familiarity with the type of postoperative pain typical of various veterinary surgical procedures (ILAR, 1992; Soma, 1987; Wright *et al.,* 1985) may aid in patient evaluation and contribute to improved management of postoperative pain and distress. However, knowledge of the procedure alone should not preclude careful assessment of the individual due to significant biological variability in the perception and signs of pain (Crane, 1987; Wright *et al.,* 1985).

Signs of pain and distress may differ in different species and between individuals within a species; however, some generalizations may be made regarding typical findings of pain and distress in animals. Nonspecific signs of pain, distress, and discomfort which may be observed include postprocedural temperament or behavioral changes such as increased aggression or escape behavior; an apprehensive, dejected, or uncomfortable appearance; restlessness; inability to rest or sleep normally; decreased activity; reluctance to move; abnormal posture; depression; decreased food and water intake; dehydration; weight loss; an ill-groomed appearance and rough unkempt hair coat; excessive salivation; dilated pupils; abnormal respiratory patterns or sounds such as expiratory grunting; tooth grinding; and vocalization that is excessive in response to a given stimulus (AVTRW, 1986; ILAR, 1992; Morton and Griffiths, 1985; Soma, 1987; Wright *et al.,* 1985). More specific signs of pain include lameness or nonuse of an extremity, guarding or bothering the injured part, stretching or tucking up of the abdomen with abdominal discomfort, and behavioral aggression, withdrawal, and vocalization when the affected site is approached or manipulated (AVTRW, 1986; LASA, 1990; Soma, 1987; Wright *et al.,* 1985). Frantic vocalization, rolling, and thrashing about not associated with excitation during recovery from anesthesia generally indicate severe pain such as that observed during acute thromboembolic ischemic episodes. Physiological indicators of pain or distress may include signs of decreased peripheral circulation, increased or decreased body temperature, increased respiratory rate and hyperventilation, increased heart rate and tachyarrhythmias, hypertension or hypotension, and hyperglycemia (AVTRW, 1986; Crane, 1987; Haskins, 1987; Morton and Griffiths, 1985; Wright *et al.,* 1985). Deviations from baseline values of parameters such as white blood cell counts and circulating cortisol and catecholamines may also occur with pain and distress (Benson *et al.,* 1991; Morton and Griffiths, 1985). More detailed descriptions of species-specific signs of pain and distress are available in the literature (AVTRW, 1986; Gibson and Paterson, 1985; ILAR, 1992; LASA, 1990; Morton and Griffiths, 1985) and elsewhere in this text. Likewise, pain management for the different species is addressed in the relevant chapters of the current work.

C. Pulmonary Function

An animal experiencing respiratory difficulty may exhibit an increased breathing rate, exert an extraordinary amount of effort during each breath, appear to have a diminished tidal volume, exhibit an irregular breathing pattern, exhibit cyanotic mucous membrane color, or make snoring or wheezing sounds when it breathes (McKiernan, 1984). Lung sounds may be difficult to auscultate if there is a pleural space filling disorder, if the lungs are consolidated, or if the animal is too small. Lungs may sound crepitant if there is an accumulation of airway fluid.

Arterial PCO_2 ($PaCO_2$) is a measure of the ventilatory status of the patient (Shapiro, 1989). The normal $PaCO_2$ is in the range of 35–45 mm Hg; a $PaCO_2 < 35$ mm Hg indicates hyperventilation whereas a $PaCO_2 > 45$ mm Hg indicates hypoventilation. Hypercapnia may be caused by hypoventilation due to medullary, cervical, or neuromuscular disease; upper or lower airway obstruction; pleural filling disorders; pulmonary parenchymal disease; and abdominal or thoracic restrictive disorders. A $PaCO_2$ in excess of 60 mm Hg is associated with severe respiratory acidosis and hypoxemia if the animal is breathing room air. Hypocapnia may be due to excitement, pain, anemia, hypoxemia, or metabolic acidosis. A $PaCO_2$ below 20 mm Hg may be associated with severe respiratory alkalosis and an inappropriate decrease in cerebral blood flow. Venous PCO_2 ($PvCO_2$) is a reflection of tissue PCO_2, which represents a combination of arterial PCO_2 and tissue metabolism. $PvCO_2$ is approximately 3 to 6 mm Hg higher than $PaCO_2$ in stable states and is variably higher in transition states, during anemia and carbonic anhydrase inhibitor therapy. $PaCO_2$ may be estimated by measuring the carbon dioxide in a sample of gas taken at the end of an exhalation.

The arterial PO_2 (PaO_2) is a measure of the oxygenating efficiency of the lungs (Shapiro, 1989). An animal may have a normal PaO_2 and yet have a reduced blood oxygen content (anemia, methemoglobinemia). The normal PaO_2 is considered to range between 90 and 100 mm Hg. A PaO_2 of 50–60 mm Hg is a commonly selected minimum value. *In vitro* oxygen–hemoglobin saturation analyzers are commercially available and economical, and are easy to operate. Pulse oximeters attach to an animal externally, on the tongue, toe, tail, or lip, and are good for monitoring trends. An oxygen saturation of 90% is equivalent to a PaO_2 of 60 mm Hg. Venous PO_2 reflects tissue PO_2 but bears no correlation to arterial PO_2. Jugular or vena caval blood PvO_2 is usually between 40 and 50 mm Hg. Values below 30 mm Hg may be caused by anything that decreases the delivery of oxygen to the tissues (hypoxemia, low cardiac output, vasoconstriction); values above 60 mm Hg (while breathing room air) suggest reduced tissue uptake of oxygen (shunting, septic shock, metabolic poisons).

Venous admixture is the collective term for all of the ways in which blood passes from the right to the left side of the circulation without being properly oxygenated. There are four causes of venous admixture: (1) low ventilation/perfusion units, (2) no ventilation/perfusion units (small airway and alveolar collapse), (3) abnormal anatomic shunts, and (4) diffusion impairment. If the measured PaO_2 and $PaCO_2$ values add to a value less than 120 mm Hg, lung oxygenating ability is decreased (venous admixture). The lower the added value, the greater the magnitude of the venous admixture. This rule of thumb only applies if the barometric pressure is close to 760 mm Hg, the inspired oxygen is 21%, and the respiratory quotient is around 0.9

D. Cardiovascular Function

There are three distinctly different aspects of cardiovascular function which must be evaluated in order to obtain an overall assessment of cardiovascular effectiveness. (1) The electrical activity of the heart: Is it beating too fast, too slow, or too arrhythmically? Tachycardia is the common finding in most stressed animals (hypovolemia, pain, excitement). Bradycardia may be caused by excessive vagal tone, atrioventricular conduction disturbances, severe hypothermia, terminal hypoxia, or drugs (opioids, xylazine, cholinergics). Arrhythmias should be identified electrocardiographically (Tilley, 1985). (2) The mechanical activity of the heart: Does the heart auscultate strong with robust sounds or weak and distant? Does the pulse quality feel sharp, strong, and full or is it weak and thready? Pulse quality reflects stroke volume. (3) Peripheral perfusion: Vasoconstriction decreases peripheral perfusion and is manifested by pale mucous membrane color, prolonged capillary refill time, anuria, and a decrease in peripheral (skin, muscle, or toe web) temperature. Vasoconstriction may be caused by any stress-induced increase in sympathetic tone: hypovolemia, low cardiac output, excitement, fear, pain, or the excitement phase of recovery; by α-receptor agonist therapy or smooth muscle agonists (vasopressin, calcium, oxytocin).

Central venous pressure (CVP) is the luminal pressure of the intrathoracic anterior vena cava. The measurement technique has been previously described (Haskins, 1981). Peripheral venous pressure is variably higher than CVP and is not a reliable indicator of CVP. Measurements should be made between ventilatory excursions (during either spontaneous or positive pressure ventilation) because changes in pleural pressure affect the luminal pressure within the anterior vena cava. The normal CVP in dogs and cats is 0 to 10 cm H_2O, and there is considerable overlap in hypovolemic patients depending on other cardiovascular changes. The CVP must be interpreted in the context of other cardiovascular parameters. The CVP is a measure of the relative ability of the heart to pump the venous return and therefore is a useful monitor in animals with a heart failure component. It is also an estimate of the relationship between the blood volume and the blood volume capacity and therefore is useful when it is necessary to administer large volumes of fluids rapidly.

Arterial blood pressure is determined by blood volume, cardiac output, and vascular resistance. Adequate cerebral and coronary perfusion is considered to require a mean systemic blood pressure of 50–60 mm Hg. Arterial blood pressure can be measured in animals by indirect and direct techniques (Haskins, 1981). Indirect sphygmomanometry involves the application of an occlusion cuff over an artery in a cylindrical appendage. The occlusion cuff should be placed snugly around the leg. Inflation of the cuff applies pressure to the underlying tissues and will occlude blood flow when the pressure exceeds the systolic blood pressure. As the cuff pressure is gradually decreased, blood will begin to flow intermittently when the cuff pressure falls below the luminal systolic pressure. The return of blood flow can be assessed by the appearance of needle oscillations on the manometer (systolic blood pressure), digital palpation of a pulse distal to the cuff (systolic blood pressure), and Doppler detection of blood flow distal to the cuff (systolic blood pressure).

Oscillometry is the measurement of changes in intracuff pressure caused by the changes in appendage size associated with each pulse wave as the cuff is slowly deflated (systolic, mean, and diastolic blood pressure, and heart rate are digitally displayed). These instruments are popular in human medicine, but many do not work well in small animals. All external techniques are least accurate when vessels are small, when the blood pressure is low, and when the vessels are constricted. Direct measurements of arterial blood pressure are accurate and continuous compared to indirect methods, which require the introduction of a catheter into an artery. In most species, normal systolic, diastolic, and mean blood pressures are approximately 100 to 160, 60 to 100, and 80 to 120 mm Hg, respectively. Hypotension may be due to hypovolemia, peripheral vasodilation, or reduced cardiac output. Hypovolemia is the most common cause (Haskins, 1986). Extracellular fluid deficits may result from inadequate intake, diarrhea, vomiting, diuresis, or third-space fluid accumulations. Vascular volume deficiencies may, in addition, be due to whole blood or plasma losses, or hypoproteinemia.

E. Renal Function

Oliguria or anuria may be prerenal (hypovolemia, vasoconstriction, thrombosis), renal (nephrosis, nephron destruction), or postrenal (urinary tract rupture or obstruction). Urine output is often used as an indirect indicator of the adequacy of renal blood flow (in an animal with normal renal function). Adequate renal blood flow is often used as an indirect indicator of visceral blood flow. Urine output, then, is a valuable monitoring tool in assessing the adequacy of hydration, circulation volume, and visceral perfusion.

If urine can be obtained by expressing the bladder, cathterization, or cystocentesis, a urinalysis (including sediment evaluation) should be performed. High specific gravity, and osmolality, and low urine sodium concentrations suggest that the animal is dehydrated. Isosthenuria, casts, or tubular epithelial cells are indicative of renal disease. If urine output is not evident (by bladder palpation or catheterization) after fluid therapy is sufficient to eliminate any prerenal component, renal disease should be suspected, and diuresis therapy should be instituted. Anuric animals that are not responsive to fluid loading are predisposed to fluid overloading and interstitial edema. Although fluid therapy must be aggressive early in the management of renal failure, it may be subsequently necessary to be conservative if the anuria or oliguria persists. Furosemide (1 to 5 mg/kg IV) is a renal vasodilator or natriuretic agent. Mannitol (0.25 to 0.5 g/kg IV) increases blood volume and renal perfusion, and is an osmotic diuretic. Dopamine (1 to 3 μg/kg/min IV), via dopaminergic receptors, causes renal afferent arterial vasodilation which increases renal blood flow. Dopamine may also have a natriuretic effect.

1. Urinary Catheter Placement and Care

Intermittent or indwelling urinary catheters may be indicated for quantified urine collection for renal function tests or pharmacokinetic studies. Most large animals are easy to catheterize. Exceptions are male ruminants (goats, sheep, cattle) and swine because of anatomical features which make urethral catheterization difficult to impossible (Garrett, 1987).

Indwelling urinary catheters predispose to mechanical and septic cystitis and urethritis. Urinary catheters must be soft and must be introduced aseptically. Hair around the prepuce or vulva should be clipped and the area washed with a povidone–iodine soap solution. The prepuce or vestibule should be flushed with a dilute povidone–iodine solution. In male animals, the prepuce should be retracted and the penis gently washed with the povidone–iodine solution. In female animals the urethra is visualized with the aid of a laryngoscope, otoscope, or vaginal speculum that has been disinfected. In larger animals the urethral papilla may be palpated with the finger of a sterile gloved hand.

A sterile catheter of sufficient length is generously lubricated and inserted into the urethra and advanced until urine can be obtained. The introduction of excessive lengths of catheter into the bladder should be avoided because it enhances the mechanical trauma to the bladder and may predispose to knotting of the catheter within the bladder. The catheter should be sutured in place (to the prepuce or vulva) to minimize its sliding in and out which predisposes to infection and mechanical trauma.

The catheter is attached to a closed drainage system. The collection system should be positioned so that it drains downhill. Draining the collection reservoir or needle puncture of the collection tube to obtain a urine sample must be accomplished aseptically. A 3% hydrogen peroxide solution or a 0.25% acetic acid solution may be placed in the collection reservoir to help prevent bacterial growth. The collection tubing should be taped to the hind leg or abdomen to prevent accidental traction on the catheter and suture sites. Enough slack should be provided in the tubing to allow the patient a full range of motion. Irrigation of the catheter with antiseptic or antibiotic solutions has not been shown to prevent urinary infections, only to delay them and to select for resistant infections. The prepuce or vestibule should be flushed with a dilute iodine solution three times daily to prevent the migration of infectious agents into the bladder along the outside of the catheter.

F. Central Nervous System

The mental status, activity, and behavior of the patient should be monitored as a general index of animal welfare. Light-reactive, mydriatic pupils are sympathetically mediated and represent a nonspecific response to stress. Light-reactive miotic or unequal pupils, which exhibit physiologic nystagmus, are suggestive of cerebral disease. Bilaterally unresponsive miotic (pontine injury) or mydriatic (midbrain injury) pupils and the absence of physiological nystagmus are indicative of brain stem involvement, as are irregular breathing patterns. Agonal gasps and apnea indicate severe terminal respiratory center dysfunction. Multiple cranial nerve dysfunction is indicative of brain stem injury. Decerebrate rigidity indicates midbrain to medullary injury. Unless intracranial injury is the subject of the study, these signs of brain stem injury signal sufficient reason to terminate the experiment.

If cerebral edema is suspected, intracranial pressure should be minimized by avoiding hypoxia, hypercapnia, hypertension, excitement, hyperthermia, anemia, head-down positioning, venous outflow occlusion, and drugs such as ketamine, xylazine, sympathomimetics, phenothiazines, and inhalational anesthetics. Unconscious animals should be intubated and moderately hyperventilated. Mannitol (0.25–0.5 g/kg IV over 20 min; may be repeated, if efficacious, in 2 to 4 hr), furosemide (5 mg/kg IV), and dexamethasone (0.5 mg/kg IV) may be beneficial in minimizing the cerebral edema.

G. Abdominal Function

The abdomen should be palpated for pain, fluid accumulation, and the presence, size, and location of abdominal viscera. Pain may be attributed to trauma to the abdominal wall, lumbar musculature, vertebral column, pelvis, or visceral organs. Abdominal fluid accumulation may be attributed to urine or blood, transudate, or exudate. Paracentesis and cytology of any fluid obtained should be performed. An area just lateral to the umbilicus on the right side is clipped and asceptically prepared. A 20- to 22-gauge needle is inserted perpendicularly into the abdomen and the plunger is gently retracted.

VII. POSTOPERATIVE CARE

A. Incision Care

The postoperative management of uncomplicated surgical incisions includes monitoring of the wound for signs of infection or dehiscence, ongoing assessment of the healing process, and protection of the incision from injury or self-trauma. This latter may necessitate the use of bandaging and some form of restraint, such as an Elizabethan collar, body stockinette, or protective jacket, to prevent the patient from damaging the incision. Some animals, particularly nonhuman primates, rabbits, and rodents, will remove conventional skin sutures, causing incisional trauma, dehiscence, and associated complications including infection. In these animals, the use of a buried continuous intradermal skin closure can reduce self-trauma to the incision and minimize the need for postoperative incision care (Smeak, 1992). In some circumstances, skin clips or staples may be used in research animals. Tissue adhesive may be useful for facilitating the closure of small incisions and elsewhere, where indicated.

Specific incision care varies with the surgical procedure and wound characteristics. Generally, incisions should be kept clean and dry postoperatively. This may require periodic gentle surface cleansing with sterile isotonic saline or with antiseptics such as povidone–iodine or chlorhexidine diacetate diluted as for wound lavage (Swaim and Henderson, 1990) and protection of the wound with a nonadherent bandage (Swaim, 1990). Bandages that become too tight, or are wet or otherwise damaged, should be changed. Other topical treatment is not generally required in the postoperative management of clean, uncomplicated surgical incisions. For special cases, literature on topical medications, specialized bandaging techniques, and the use of drains should be consulted (Bauer, 1989; Degernes, 1992; Swaim, 1990; Swaim and Henderson, 1990; Swaim and Lee, 1987; Swaim and Wilhalf, 1985; Vasseur, 1985).

Wound clips, skin staples, and skin sutures that are not buried are usually removed at 10 to 14 days postoperatively. In cases of delayed healing, sutures are removed when the incision is judged to be healed and of sufficient strength to prevent dehiscence.

B. Care of Chronic Implants

The laboratory animal science literature is replete with descriptions of research techniques employing implanted devices in animals. Examples of such implants include catheters, recording leads, telemetry devices, flow probes, cephalic recording devices, and orthopedic implants. Often, regimens for the care of implants have been developed and reported as part of the model. In some cases, the veterinary and investigative staff can work together to improve care of these animals and longevity of the implants. An in-depth discussion of the care of implanted animals is beyond the scope of this chapter, but some general considerations will be presented.

Ongoing monitoring and nursing care is required for animals as long as they have chronic catheters and other external appliances in place. Implanted materials that are completely buried may also require ongoing attention, depending on what they are, because implanted foreign materials may incite tissue reactions or become infected over time (Johnston, 1990).

Procedures should be developed and carefully followed to maintain the patency, sterility, and function of indwelling catheters and to protect skin exit sites and implanted devices from trauma and bacterial contamination. The care of implanted catheters, including infection control, anticoagulation, and clot removal, has been addressed extensively in the literature (Dennis *et al.,* 1984, 1986, 1989, 1990; Desjardins, 1986; Kamal *et al.,* 1991; Palm *et al.,* 1991; Penner *et al.,* 1994; Scalese *et al.,* 1990). A regimen for the care of external skeletal fixators has also been described and can be adapted for care of similar appliances that exit through the skin (Aron and Dewey, 1992).

In addition to caring for the implanted devices, periodic monitoring of such things as body temperature, the hemogram, food intake, and body weight may be necessary to confirm that the animal is healthy and tolerating the implant well. Other more specific assessments, including evaluation of the need for analgesics, may be required, depending on the model.

C. Identification and Treatment of Infections

The identification and treatment of postoperative infections are important parts of patient monitoring and care. Although there is a need for more information regarding the frequency and characteristics of surgical wound infections in different species, an overall postoperative wound infection rate of 5.1% was reported in one survey of canine and feline patients at a teaching hospital (Vasseur *et al.,* 1988).

Wound contamination may occur from bacteria that are exogenous or endogenous to the animal. The risk of infection increases with a number of factors, many of which may apply to surgery in research animals. Examples of risk factors include lengthy surgical procedures; the presence of foreign bodies, such as implanted devices; underlying disease states such as diabetes mellitus; and immunodeficiencies (Daly, 1985; Orsini, 1985; Romatowski, 1989). As a preventative measure, animals at high risk to develop postoperative infections should be identified ahead of time, considered for prophylactic antibiotic treatment, and carefully observed for signs of infection. Guidelines for case selection and treatment, and descriptions of prophylactic perioperative antibiotic regimens used in veterinary patients have been presented elsewhere (Haven *et al.,* 1992; Klein and Firth, 1988a,b; La Regina *et al.,* 1992; Orsini, 1985; Penwick, 1985; Romatowski, 1989; Rosin, 1989; Wilcke, 1990).

Clinical signs of postoperative infection generally develop 3 to 6 days after surgery and may be associated with a concurrent elevation in body temperature (Orsini, 1985). The signs of infection include heat, pain, swelling, and redness around the infected site and loss of function (Daly, 1985). Although the presence of pus, either as an exudate or accumulated as in an abscess, is a principal sign of wound infection, some infections may present as a paraincisional cellulitis with no obvious accumulation of pus (Daly, 1985; Orsini, 1985; Rosin, 1989).

Infections that are localized may not be associated with significant systemic signs. Infections that are more generalized, such as peritonitis, cellulitis, and septicemia, may be associated with a variety of clinical signs including fever, depression, anorexia, and signs specific to the organ systems affected. Fever, anorexia, and weight loss were observed in chronically catheterized rabbits that developed *Pseudomonas* bacteremia 2 to 6 weeks after surgery (Penner *et al.,* 1994). A hemogram from an animal with systemic involvement will likely have a leukocytosis with a neutrophilia and left shift, often with signs of toxic change (Dennis *et al.,* 1986; Orsini, 1985).

Appropriate cultures should be done to identify the offending pathogens in a particular case. Specimens for bacterial culture and antibiotic sensitivity testing should be taken from the wound and/or other foci of infection prior to initiation of antibiotic therapy. Specialized culture techniques, such as blood cultures in chronically catheterized animals suspected to be septicemic, should be applied where indicated. A gram stain from the sample submitted for culture may be useful in the selection of the initial antibiotic regimen.

Localized wound infections that manifest as paraincisional cellulitis generally respond to hot packing three to four times daily and appropriate antibiotic therapy (Rosin, 1989). Wounds that become abscessed should be treated locally with a combination of gentle debridement; establishment of adequate drainage, which may include some removal of sutures and/or placement of a drain; flushing with sterile physiologic saline or appropriately diluted wound antiseptic; and protective bandaging if indicated. Detailed management of contaminated and infected wounds is described elsewhere (Swaim and Henderson, 1990). A short (5–14 day) course of systemic antibiotics may be prescribed in some cases. In others, local treatment may be adequate for the resolution of uncomplicated local wound infections.

In more serious cases, where signs of systemic infection are present, systemic antibiotic treatment is indicated. Antibiotic treatment should be initiated prior to receiving results of the bacterial culture and sensitivity. Selection of the initial antibiotic regimen is based on many considerations, including the results of the gram stain, animal species, case history, wound characteristics, results of the physical exam and diagnostic testing, antibiotic spectra, drug side effects and toxicities, dosing regimen (including route and frequency of administration required), facility experience, and protocol constraints. The initial antibiotic regimen should be changed if needed based on the results of the culture and sensitivity or if the desired response to therapy is not observed within 24 to 72 hr.

If possible, infected implants should be removed, as these can be persistent niduses of infection, even during treatment (DaRif and Rush, 1983; Johnston, 1990). If an infected implant must be retained, the infection should be treated aggressively. Catheter disinfection techniques may help in clearing catheter-related infections (Palm *et al.,* 1991). It is recommended that antibiotics be continued for at least 1 week after cultures become negative (Dennis *et al.,* 1986). In some cases, antibiotic treatment may be required for as long as the implants remain in place.

Consideration should be given to species intolerances of particular antibiotic regimens. In rabbits and some rodents, the use of antibiotics with predominantly gram positive and anaerobic spectra may upset normal gastrointestinal flora, leading to diarrhea, enterotoxemia, and death. Guinea pigs and hamsters are among the most sensitive to antibiotic-induced gastrointestinal disturbances. Drugs that have reportedly caused gastrointestinal problems in these two rodent species include penicillin, lincomycin, and erythromycin (Harkness and Wagner, 1989). In the rabbit, oral administration of ampicillin (10 mg/kg for 6 days), low doses of lincomycin (5–30 mg), and the administration of erythromycin may be associated with enteric disturbances and death (Harkness and Wagner, 1989). Other drugs with similar activities may also adversely affect gastrointestinal function in these species. To minimize adverse side effects, the use of broad-spectrum antibiotics, parenteral routes of administration, and short treatment periods, and the avoidance of overdosing, is recommended. Enrofloxacin, a fluoroquinolone antibiotic, is currently used at 5 to 10 mg/kg IM, SC, or PO q 12 hr for the treatment of pasteurellosis in the rabbit (Broome *et al.,* 1991; Jenkins and Brown, 1993). Further investigation may confirm its usefulness for treating postoperative infections in rabbits and rodents.

VIII. NUTRITIONAL SUPPORT

Adequate nutrition of the intensive care or postoperative veterinary patient is an important component of supportive care as malnutrition may be associated with prolonged recovery, poor wound healing, and increased susceptibility to infection (Remillard and Martin, 1990). Formulation of the nutritional plan requires an understanding of the dietary requirements of an animal based on the species, age, physiologic state, activity level, and medical history. Other considerations include the status of the gastrointestinal tract and other organ systems, including dysfunctions associated with the present illness or surgical procedure; the behavioral needs of the animal as they relate to feeding activities; the advantages and disadvantages of various diets and routes of administration; and facility, equipment, personnel, and protocol constraints.

Basic nutritional guidelines for various species (including dogs, cats, rabbits, rats, mice, gerbils, guinea pigs, hamsters, nonhuman primates, mink and foxes, sheep, goats, swine, poultry, fishes, cattle, and horses) are addressed in the NRC series on the nutrient requirements of domestic animals (Committee on Animal Nutrition) and in a number of other publications (Cheeke, 1987; Fowler, 1986; Fox *et al.,* 1984; Robbins, 1983) which can be consulted when formulating patient diets. Different species have different requirements for protein, carbohydrate, fat, fiber, vitamins, minerals, and essential amino and fatty acids. Nutritional support is formulated to fulfill patient requirements and spare tissue protein, primarily by providing adequate exogenous calories and protein, with the long-term goal of maintaining optimal body weight and condition (Donoghue, 1992).

A number of equations are used to determine the energy needs of animals under differing conditions of activity, development, and disease (Armstrong and Lippert, 1988; Lewis *et al.,* 1987; Moser, 1990; Remillard and Martin, 1990; Remillard and Thatcher, 1989; Sweeney and Divers, 1990). In general, the resting energy requirement (RER) for eutherian mammals can be estimated using the following allometric equation adapted from the work of Kleiber (1961) relating metabolic rate to body mass (Remillard and Thatcher, 1989):

$$\text{RER (kcal of metabolizable energy/day)} = 70[\text{body weight in kg(BW)}]^{0.75}.$$

For dogs and cats weighing between 2 and 45 kg, this has been approximated with the linear equation (Remillard and Thatcher, 1989):

$$\text{RER} = 30(\text{BW}) + 70.$$

Significant deviations from the former equation are reported for a number of mammals. Some adult desert rodents, desert carnivores, and fossorial rodents have lower metabolic rates, whereas marine mammals, many mustelids, wild lagomorphs, shrews, voles, and young animals have higher metabolic rates than those predicted for placental mammals as a group (Robbins, 1983). Allometric equations relating metabolic rate to body size have been derived for marsupials and birds (Quesenberry *et al.*, 1989; Schmidt-Nielsen, 1984). Marsupial metabolic rates are approximately 30% lower than those of the placental mammals. Separate equations have been developed for passerine and non-passerine birds (Schmidt-Nielsen, 1984).

Maintenance energy requirements (MER), determined for normal dogs and cats, are twice RER for the dog and 1.5 times RER for the cat. However, it has been suggested that energy requirements of sick or injured animals usually fall between RER and MER (Remillard and Thatcher, 1989). Disease factors (DF) for various conditions are used to calculate total energy requirements (TER), sometimes referred to as illness/injury requirements (IER), for patients using the following formula (Armstrong *et al.*, 1990; Moser, 1990; Remillard and Thatcher, 1989):

$$\text{TER} = \text{IER} = \text{RER} \times \text{DF}.$$

As an example, for dogs and cats, respectively, the DF used to calculate TER are 1.2 and 1.1 for medical patients, and 1.5 and 1.3 for surgical patients. A DF of 2 (dogs) or 1.5 (cats) is generally considered maximal and is used for burn or head trauma patients (Remillard and Thatcher, 1989). Maintenance requirements and adjustments to maintenance for various physical conditions in birds have been presented (Quesenberry *et al.*, 1989). The amount to feed is based on the calculated TER and caloric density of the chosen diet, which should be balanced for the particular animal and species of concern. Frequent weighing, especially of birds and small mammals, may be required to be sure that caloric intake is adequate to maintain acceptable body weight and growth.

Various methods of nutritional support are available, including enteral and parenteral. In most cases where the gastrointestinal tract is functional, enteral nutrition is preferred because it is more physiologic and less expensive than parenteral nutrition. Methods for enteral nutrition include voluntary intake of commercial or homemade diets, force feeding, and tube feeding. Although voluntary intake is preferred as the least stressful and easiest method, some patients are at least partially anorectic. These animals require more aggressive nutritional support if they are unable to consume enough of the diet to maintain body weight and condition, have reduced oral food intake for 3 to 5 days, or meet other criteria indicative of inadequate nutrition (Abood and Buffington, 1992; Donoghue, 1992; Remillard and Thatcher, 1989). Voluntary intake can be encouraged by minimizing environmental stressors that may be contributing to the inappetence of the animal and improving palatability of the diet. The latter can be done in a number of ways; however, species and individual requirements and preferences, as well as neophobic behaviors in guinea pigs and other colony animals, should be considered. Appetite may be stimulated in many species by offering small amounts of treats and feeding highly palatable commercial or homemade diets. Warming or moistening the food for dogs and cats and adding flavor enhancers, such as small amounts of baby foods, broth, or canned tuna for carnivores and peanut butter for nonhuman primates, may encourage eating. Force feeding, hand or syringe feeding, and the use of pharmacologic appetite stimulants such as diazepam may be of use in selected cases (Crowe, 1989). However, tube feeding is generally indicated if enteral support is required for more than 2 days (Remillard and Martin, 1990).

Indications, routes, and diets for tube feeding and methods of tube placement have been discussed in detail (Armstrong and Lippert, 1988; Armstrong *et al.*, 1990; Crowe, 1985, 1989; Donoghue, 1992; Lewis *et al.*, 1987; Remillard and Martin, 1990). Orogastric, nasogastric, or nasoesophageal, pharyngostomy, gastrostomy, or enterostomy tubes may be used. To minimize handling stress, placement of an indwelling tube is preferred to repeated orogastric gavage in patients requiring tube feeding for more than a day or two. A general principle would encourage use of the least invasive method which allows use of as much of the functional gastrointestinal tract as possible (Armstrong *et al.*, 1990). The potential need for postoperative enteral nutritional support should be considered early and written into the animal use protocol where indicated, with tube placement ideally performed at the time of surgery. Improved techniques for tube placement in small animals including nasogastric intubation in the dog (Abood and Buffington, 1991) and esophagostomy and gastrostomy tube placement in dogs and cats (Armstrong *et al.*, 1990; Crowe, 1989; Fulton and Dennis, 1992; Mathews and Binnington, 1986) have been described. A needle catheter duodenostomy technique has been developed for use in birds (Goring *et al.*, 1986), and placement of an esophagostomy tube has been recommended as an alternative to periodic orogastric intubation in tortoises requiring prolonged nutritional support (Page and Mautino, 1990). Long-term enteral nutrition via gastrostomy has been described in rats (Muller *et al.*, 1992; Tsukamoto *et al,* 1984). Diets useful for tube feeding include slurries of high quality complete commercial animal diets or balanced homemade diets and commercial enteral liquid diets of varying composition manufactured for human or veterinary use. Diet descriptions, indications for use in various disease states, and guidelines for the formulation of feeding plans are discussed elsewhere (Abood and Buffington, 1992; Armstrong and Lippert, 1988; Armstrong *et al.,* 1990; Crowe, 1985, 1989; Donoghue, 1992; Lewis *et al.,* 1987; Remillard and Martin, 1990). In general, nutritional support should be introduced gradually and overfeeding avoided to prevent gastrointestinal complications. Initially, tube feeding volumes of 3 to 5 ml/kg or 1 to 3 ml/kg/hr are used for dogs and cats at the University of California, Davis, small animal ICU. Feeding volumes (up to 30 ml/kg) and rate are gradually increased as tolerated to fulfill the daily nutritional requirements of the patient (Haskins, 1989a).

Parenteral nutrition is considered in cases where the gastrointestinal tract is nonfunctional, as in gastrointestinal obstruction and severe small bowel disease, or when it requires resting, as in some cases of severe pancreatitis, and in cases where required nutritional support cannot be provided enterally (Lippert, 1989; Remillard and Thatcher, 1989). If possible, at least some nutrition should be administered enterally to supply nutritional substrates to the enterocytes, prevent hypoplasia of the intestinal mucosa, and maintain enteric immunity (Armstrong *et al.,* 1990; Remillard and Martin, 1990; Sweeney and Divers, 1990). The term total parenteral nutrition (TPN) refers to the provision of all essential nutrients parenterally. Solutions for parenteral nutrition are compounded from dextrose, amino acids and electrolytes, and lipid emulsions. Because the resultant formulations are hypertonic, they are administered via a dedicated catheter placed in a large central vein. Partial parenteral nutrition, used to spare endogenous proteins, may be provided with isotonic amino acid solutions and electrolytes administered via a peripheral venous catheter (Crowe, 1989). Lipid emulsions do not add to the osmolality of the solution and may be added to the peripheral infusion to provide additional calories and essential fatty acids (Crowe, 1989). Amino acids, dextrose, and vitamins have been given via the intraosseous route in patients with central vessels that are difficult to access (Donoghue, 1992; Otto *et al.,* 1989; Ritchie *et al.,* 1990). Vitamin and trace mineral supplementation is provided as dictated by species and individual requirements and by the duration of the parenteral regimen.

Parenteral nutrition is associated with increased costs as well as other practical disadvantages including requirements for personnel training, strict asepsis during compounding and administration of parenteral solutions, and increased patient monitoring. Potential adverse effects to the patient include infection, metabolic disturbances, and intestinal atrophy (Armstrong and Lippert, 1988). However, parenteral nutrition has been beneficial in many species, including neonatal large animals and wildlife (Donoghue, 1992; Sweeney and Divers, 1990). The administration of TPN using an implantable venous access device has been investigated for use in birds (Degernes *et al.,* 1994). Guidelines for patient selection, diet formulation and administration, vitamin and mineral supplementation, patient monitoring, and other aspects of parenteral nutrition have been extensively discussed elsewhere (Armstrong and Lippert, 1988: Haskins, 1989b; Remillard and Thatcher, 1989; Sweeney and Divers, 1990).

IX. VASCULAR CATHETER PLACEMENT AND CARE

The placement of an indwelling vascular catheter may be required for the short-term administration of intravenous fluids and medications in the postoperative period. If the need for intravenous access can be anticipated, a catheter is most easily placed, secured, and bandaged while the animal is still under anesthesia. In larger species with accessible peripheral veins (such as dogs, cats, rabbits, monkeys, and small ruminants), catheters are often placed percutaneously and are protected by bandaging. However, if a catheter is accessible to the animal, it may need to be removed once the animal regains consciousness or close surveillance and/or special protective measures (Elizabethan or other types of restraint collars) may be required to prevent the animal from pulling the catheter out. Common sites for venous catheter placement in the larger species are listed in Table II.

In swine, there are no good peripheral sites for percutaneous venous catheterization. Ear vein catheters are useful for administrating fluids and medications while the animal is anesthetized. These catheters are usually removed during the recovery period as most pigs will not tolerate them once they have awakened from anesthesia. If long-term vascular access is required in the pig, chronic catheters are usually placed into the jugular vein via cutdown under anesthesia and tunneled subcutaneously to an exit point in the interscapular area, inaccessible to the animal. A customized jacket can be used to protect the catheter exit site. Because most nonhuman primates will also pull out peripheral catheters, techniques have been developed for these species using similarly tunneled chronic catheters protected by jackets. An alternative technique uses catheters and injection

TABLE II

Suggested Sites for Placement of Peripheral Venous Catheters in Different Animal Species

Species	Location
Dog	Cephalic, jugular, lateral saphenous
Cat	Cephalic, jugular, medial saphenous
Goat, sheep, calf	Jugular
Rabbit	Ear, cephalic
Ferret	Jugular, cephalic, lateral saphenous
Monkey	Saphenous, cephalic

ports that are buried subcutaneously, and thus protected from the animal. Humane restraint devices, designed for use in nonhuman primates, may be used to protect peripheral catheters required for several hours of intravenous therapy, while the animal is under supervision. The use of a light cast may also be considered to protect peripheral catheters for relatively short periods of time.

In very small species, such as mice and rats, the intravenous route for routine fluid administration is rarely practical and fluids are generally administered either subcutaneously or intraperitoneally. In the guinea pig, which lacks substantial peripheral veins, these same administration routes are also commonly used. The intraosseous route can be applied for administering fluids in some of the smaller species. The placement of intraosseous catheters in the femurs of rabbits and ferrets has been described (Jenkins and Brown, 1993).

Percutaneous vascular catheter insertions are generally associated with lower infection rates compared to cutdown insertions. A cutdown should be implemented when the vessel is small and/or difficult to catheterize percutaneously. Indwelling vascular catheters expose the patient to the risk of local and systemic infections. Appropriate aseptic safeguards during the insertion and maintenance of an indwelling catheter decrease the risk of untoward mechanical or infectious complications.

A wide area of hair over the vessel to be catheterized should be clipped. The skin is prepared wtih antiseptic solutions in a manner similar to that used for any surgical procedure. One to 2% tincture of iodine is the antiseptic solution of choice, but chlorhexidine, iodophors, and 70% (v/v) isopropyl alcohol could be used. Sterile drapes should be used if it is necessary or helpful to extend the aseptic field beyond the margins of the prepared area. Hands should be well washed and sterile gloves should be worn, if necessary, to prevent contamination of either the catheter or the skin puncture site during insertion. Vessels are most effectively immobilized by stretching them between two points of traction. Insufficient traction allows the vessel to roll laterally or wrinkle longitudinally. Excessive traction may collapse the vessel. Digital pressure against the side of the vessel will usually collapse it rather than effectively immobilize it. It may be helpful to make a relief opening in the skin to minimize skin catheter resistance and to help prevent tearing of the end of the catheter. This is easily accomplished with the cutting edge of the bevel of a hypodermic needle.

Once the catheter is in place, it should be capped with an infusion plug and flushed with heparinized saline (1000 units of heparin per 250–500 ml of saline). Catheters that are not continuously in use for fluid administration should be periodically (approximately every 4 hr) flushed with heparinized saline. Fifty percent glucose and dextran 70 will also prevent a catheter from clotting. Pure heparin (1000 units/ml) could be used to fill the catheter if prolonged (24 hr) anticoagulation is desired. Care must be taken not to heparinize the animal when the heparin is first instilled into the catheter (fill only the dead space of the catheter) and when the next fluid or drug administration occurs (aspirate the heparin into a scavenging syringe prior to the fluid/drug administration). Arterial catheters should be continuously flushed with heparinized saline if possible (continuous, low-volume flush devices are available from Abbott and Baxter Medical supply companies).

The position of the catheter must be fixed close to the skin puncture site to prevent it from sliding in and out. On an appendage, fixation may be accomplished by taping around the circumference of the catheter and then taping around the circumference of the leg. On flat body surfaces the catheter must be sutured. An antibiotic–antifungal ointment should be placed on the skin puncture site and the site should be wrapped occlusively. Catheters on appendages or in a jugular vein should be sufficiently bound by the wrap so that accidental extraction of the catheter is prevented. Indwelling catheters should be redressed and inspected every 24–48 hr and all soiled bandage material should be discarded.

The skin puncture site and the vessel should be inspected at each redressing. Normally there is a very small ring of inflammation at the puncture site. If the puncture site appears healthy, it should be cleaned with antiseptic solutions and then fresh antibiotic–antifungal ointment and the occlusive wrap reapplied. Signs of untoward effects of the indwelling catheter may dictate removal. Cellulitis is characterized by warm, erythematous and tender skin around the insertion site. Phlebitis may be caused by mechanical, chemical, or infectious irritation of the vein and is recognized as warm, erythematous skin overlying a tender, indurated vessel. Purulent thrombophlebitis is heralded by all of the signs of simple phlebitis plus free exudate which may drain exteriorly with or without palpation or may drain internally resulting in a severe septicemia. Thrombotic occlusion of the vessel is recognized by hardening or ropiness of the vessel. It may be associated with an impaired ability to infuse fluids or withdraw blood samples. An occult IV site infection may also occur in the absence of any local inflammatory, thrombotic, or exudative signs. Unexplained fever and leukocytosis may be the only warning signs. The diagnosis may be confirmed by culturing the catheter tip.

Infusion fluids and administration tubing must be sterile. Connections should not be disconnected unless absolutely necessary and then this must be done aseptically. All injection caps should be cleaned well with an antiseptic solution prior to needle insertion. The primary catheter or infusion line should not be used for the collection of blood samples except in emergencies. The fluid bottles and all administration tubing should be changed every 24–48 hr, and tubing should be changed after blood or colloid infusion. The person, time, and date that a new bottle is started should be indicated on the bottle. The fluid bottle should be clearly marked if any drug or concentrate has been added. In-line filters are not necessary for infection control for routine fluid administration.

As alluded to earlier in the discussion on catheterization of swine and nonhuman primates, chronic indwelling catheters are often used in the research environment. These devices may be placed as part of the research surgery and used throughout the project for the administering of test substances, for gathering sequential blood samples, and for the measurement of study parameters. If such catheters are in place, they often may be used for administrating fluids and medications during the postoperative period. Various techniques for placement and maintenance of chronic catheters have been described in the literature (Dennis *et al.,* 1984, 1986; Desjardins, 1986).

X. FLUID THERAPY

Naturally occuring fluid loss can be divided into four broad categories: (a) whole blood loss, (b) predominantly extracellular fluid losses that have electrolyte compositions similar to that of the extracellular fluid compartment (ECF) (vomiting, diarrhea, diuresis, and some "third-space" fluid accumulations). (c) ECF fluid losses that are high in protein (exudates, modified transudates, some third-space accumulations and some diarrhea), and (d) pure water losses (evaporation from the respiratory and skin surfaces). The clinical signs of a vascular volume deficit are those associated with the normal compensatory responses to hypovolemia: tachycardia and vasoconstriction (pale color, prolonged capillary refill time, cool appendages by touch or actual temperature measurement, oliguria, high urine specific gravity, low urine sodium concentration). Pulse quality and amplitude, and central venous and arterial blood pressure measurements are usually low. The clinical sign of an interstitial fluid deficit is an increased tendency for the skin to retain its position after it has been lifted into a fold. An interstitial fluid deficit is expected in all categories of fluid loss except pure whole blood loss. A patient that is dehydrated is, by definition, hypovolemic because the vascular and interstitial compartments are in complete crystalloid and fluid equilibrium. The clinical sign of an intracellular water deficit is depressed mentation. The diagnosis of an intracellular water deficit is based on a measured hypernatremia or hyperosmolality. An intracellular water deficit occurs when free water loss (insensible and urinary losses, and most vomition, diarrhea, and diuresis contain free water) exceeds free water intake (water deprivation).

A. Blood Volume Restoration

If a patient is judged to be vascular volume depleted (see earlier discussion), blood volume restoration therapy should be instituted. An ECF replacement solution (a solution with about normal extracellular concentrations of sodium, potassium, chloride, and bicarbonate, lactate, gluconate, or acetate) is a good fluid to begin blood volume restoration therapy. These solutions are economical, readily available, and can be safely administered in large volumes to normal animals. Quantities of 20 to 80 ml/kg, in most species, may be required to achieve adequate blood volume restoration. Hypertonic saline has been reported to be efficacious when time is a limiting factor in fluid therapy. Hypertonic saline provides a greater blood volume expansion effect with considerably less infused volume compared to isotonic saline.

The end point of fluid administration is determined by the alleviation of the peripheral vasoconstriction, the restoration of an acceptable pulse quality, and a return of urine output. Objective measurements of the adequacy of blood volume restoration include an increase of the central venous pressure to 5 to 10 cm H_2O, or an arterial blood pressure of $>$ 60 mm Hg mean, or $>$80 mm Hg systolic. Only about 20% of crystalloid solutions remain in the vascular fluid compartment 30 min following their administration; the remainder is redistributed into the interstitial fluid compartment. The volume restoration achieved by crystalloid fluids may be fleeting, and if hypotension or vasoconstriction recurs, colloids or whole blood may be indicated. Excessive hemodilution is a common limitation to crystalloid fluid administration. Excessive hemodilution is often defined as a packed cell volume below 20% or a hemoglobin below 7 g/dl, and/or a total protein below 3.5 g/dl or an albumin below 1.5 g/dl.

Although subcutaneous, oral and intraperitoneal routes of fluid administration may be appropriate in simple dehydration, these routes are not direct enough for a life-threatening shock situation where fluids must be administered intravenously or intraosseously.

If the total protein is below about 3.5 g/dl or is likely to be reduced below this level with crystalloid therapy, plasma or a plasma substitute (dextran, hetastarch) should be administered. Colloids are more effective blood volume expanders than are crystalloids (Dawidson *et al.,* 1980; Falk *et al.,* 1989; Griffel and Kaufman, 1992; Rachow *et al.,* 1983) and should be considered when the patient does not appear to be responding appropriately to the crystalloid fluid infusion or when edema develops prior to adequate blood volume restoration. Plasma can serve as an amino acid source and can provide platelets (if fresh) and coagluation precursors (if fresh or fresh frozen). Dextran 70 is a large molecular polysaccharide (number-averaged molecular weight 49,000) with a duration of effect of about 8 to 24 hr. Molecular sizes below 50,000 are rapidly excreted by the kidneys; larger sizes are slowly metabolized by the liver. Large-dose dextran 70 infusions may change coagulation parameters to a greater extent than saline-induced hemodilution (Concannon *et al.,* 1992). Large infusions can interfere with red cell antigen testing. The authors administer dextran 70 at dosages of up to 20 to 40 ml/kg/day or at a rate of 1 to 2 ml/kg/hr. Hydroxyethyl starch is a branched chain polysaccharide (number-averaged molecular weight 69,000) with a duration of effect of about 12 to 36 hr. Hetastarch may be less likely to cause coagulopathies and interfere with red cell testing. Hetastarch can be administered in volumes similar to those suggested for dextran 70.

If the packed cell volume is below 20% or is likely to be reduced below 20% with crystalloid fluid therapy, packed red cells or whole blood should be administered as part of the fluid therapy program. The blood can be anticoagulated with heparin if the blood is to be immediately retransfused (1 unit of heparin/ml of whole blood) or with acid–citrate–dextrose (ACD) or citrate–phosphate–dextrose (CPD) (0.14 ml of solution/ml of whole blood) if the blood is to be stored.

Blood types and the problems of blood transfusions vary from species to species and the reader is referred to specific references which detail the current knowledge for each of the common species: dog (Pichler and Turnwald, 1985; Turnwald, 1985), cat (Giger and Bucheler, 1991; Giger *et al.,* 1989), and other species of interest.

The amount of blood or plasma to administer may be calculated as 10 to 40 ml/kg; for a small to large transfusion, respectively. The rate of whole blood or plasma administration should be conservative so as to minimize the clinical manifestations of transfusion or foreign protein histamine-mediated reactions. Infused blood should be filtered. Immunologic transfusion reactions may be manifested by restlessness, acute collapse, hypotension, dyspnea, wheezing, urticaria, hemoglobinemia or hemoglobinuria, or fever.

B. Daily Planning

Once life-threatening hypovolemia and electrolyte problems have been corrected, a fluid plan for the remaining fluid and electrolyte abnormalities can be developed. There are three categories of consideration. (1) The deficit: How much fluid volume and of what type will it take to restore the patient to normal. (2) The maintenance: How much (and what type) fluid will it take to accommodate the normal ongoing losses. (3) The ongoing losses: How much (and what type) fluid will take to accommodate the abnormal ongoing losses.

1. Deficits

An animal may be qualitatively determined to be dehydrated by the signs of dry skin and mucous membranes, oliguria, hemoconcentration, and signs referable to compensatory peripheral vasoconstriction. An acute change in body weight could be used as a quantitative guide to the volume of the deficit; however, large volumes of fluid can accumulate in the intestinal lumen or the peritoneal or pleural cavities, resulting in an effective diminution of the extracellular fluid volume without a change in body weight. Skin pliability is often used to determine the magnitude of the deficit. On the average, an animal is 5% dehydrated if the skin returns a little slowly to its resting position, after it has been lifted into a fold. An animal is about 12% dehydrated if the skin stands in a fold. Intermediate variations are extrapolated between 5 and 12%. An ECF replacement solution should be administered in an amout equivalent to the estimated percent dehydration times the body weight. It may be necessary to supplement the solution with additional potassium (20 to 50 mEq/liter) because hypokalemia is a common occurrence. Potassium should not be administered faster than 0.5 mEq/kg/hr.

2. Maintenance (Normal Ongoing Losses)

The volume of fluids required to replace the normal ongoing losses can be determined from predictive charts (Kirk, *et al.,* 1986). For dogs, cats, and rabbits the maintenance volume can be assumed to range between 50 and 100 ml/kg per day. The sodium concentration of a maintenance solution should be about 40 mEq/liter and the potassium concentration should be about 20 mEq/liter.

3. Abnormal Ongoing Loss

If the ongoing losses occur via transudation into one of the major body cavities, into the tissues, or via burn wounds, they are electrolytically in equilibrium with the extracellular fluid compartment and should be replaced with an ECF solution. If the ongoing losses occur via vomition, diarrhea, or diuresis, they should be replaced with an ECF solution and supplemented with postassium to a concentration of 10 mEq/liter.

4. Administration of Fluid Plan

Fluids may be combined into one large mixing container or may be administered separately, in series, in the order in which they have been discussed. Fluids may be administered intravenously or intraosseously, or they may be administered orally, subcutaneously, and even intraperitoneally, in several divided daily dosages.

XI. CONCLUSION

This chapter has discussed the major aspects of postoperative monitoring of common research animals and has made some recommendations with regard to specific support procedures. It is the charge of the research scientist and the animal care team to assure the well-being of the animal subject. Animals should be treated with the respect that they deserve, they should never be allowed to suffer, and all discomforts, in the name of science, must be minimized.

A great deal of intensive monitoring and support can be accomplished. The authors are, however, not suggesting that all things should be done in all animals, but in general, the more complex the research protocol or the more debilitated the animal, the greater the need for intensive interventions. There may become a point in the deterioration of an animal when the "benefit ratio" of such extensive monitoring and support is not warranted. The welfare of the animal and the result of the experiment are at stake. The veterinary staff and the research scientist should be alert to this development, and in such situations, the merits of euthanasia should be considered.

The complexities of some research protocols are clearly beyond the scope of this chapter. Consultation with institutional veterinary staff and animal care committees, specific reference books and previous, similar protocols in the research literature, and individual experts in the field is essential.

REFERENCES

Abood, S. K., and Buffington, C. A. (1991). Improved nasogastic intubation technique for administration of nutritional support in dogs. *J. Am. Vet. Med. Assoc.* **199,** 577–579.

Abood, S. K., and Buffington, C. A. (1992). Enteral feeding of dogs and cats: 51 cases (1989–1991). *J. Am. Vet. Med. Assoc.* **201,** 619–622.

Armstrong, P. J., and Lippert, A. C. (1988). Selected aspects of enteral and parenteral nutritional support. *Semin. Vet. Med. Surg. (Small Anim.)* **3,** 216–226.

Armstrong, P. J., Hand, M. S., and Frederick, G. S. (1990). Enteral nutrition by tube. *Vet. Clin. North Am.: Small Anim. Pract.* **20,** 237–275.

Aron, D. N., and Dewey, C. W. (1992). Application and postoperative management of external skeletal fixators. *Vet. Clin. North Am.: Small Anim. Pract.* **22,** 69–97.

Association of Veterinary Teachers and Research Workers (AVTRW) (1986). Guidelines for the recognition and assessment of pain in animals. *Vet. Rec.* **118,** 334–338.

Bauer, M. S. (1989). Management of deadspace and principles of drains. *Proc. 17th Annu. Vet. Surg. Forum,* Chicago, pp. 26–28.

Benson, G. J., Wheaton, L. G., Thurmon, J. C., Tranquilli, W. J., Olson, W. A., and Davis, C. A. (1991). Postoperative catecholamine response to onychectomy in isoflurane-anesthetized cats: Effects of analgesics. *Vet. Surg.* **20,** 222–225.

Beynen, A. C., Baumans, V., Bertens, A. P. M. G., Havenaar, R., Hesp, A. P. M., and Van Zutphen, L. F. M. (1987). Assessment of discomfort in gallstone-bearing mice: A practical example of the problems encountered in an attempt to recognize discomfort in laboratory animals. *Lab. Anim.* **21,** 35–42.

Broome, R. L., Brooks, D. L., Babish, J. G., Copeland, D. D., and Conzelman, G. M. (1991). Pharmacokinetic properties of enrofloxacin in rabbits. *Am. J. Vet. Res.* **52,** 1835–1841.

Brown, M. J., Pearson, P. T., and Tomson, F. N. (1993). Special report: Guidelines for animal surgery in research and teaching. *Am. J. Vet. Res.* **54,** 1544–1559.

Cheeke, P. R. (1987). ''Rabbit Feeding and Nutrition.'' Academic Press, New York.

Committee on Animal Nutrition. Veterinary Medicine. ''Nutrient Requirements of Domestic Animals Series.'' Board on Agriculture and Renewable Resources, National Research Council, National Academy of Sciences, Washington, D.C.

Concannon, K. T., Haskins, S. C., and Feldman, B. F. (1992). Hemostatic defects associated with two infusion rates of dextran 70 in dogs. *Am. J. Vet. Res.* **53,** 1369–1375.

Crane, S. W. (1987). Perioperative analgesia: A surgeon's perspective. *J. Am. Vet. Med. Assoc.* **191,** 1254–1257.

Crowe, D. T., Jr. (1985). Tube feeding diets for nutritional support of the critically ill or injured patients. *J. Vet. Emerg. Crit. Care* **1,** 8–18.

Crowe, D. T., Jr. (1989). Nutrition in critical patients: Administering the support therapies. *Vet. Med.,* February, pp. 152–180.

Daly, W. R. (1985). Wound infections. *In* ''Textbook of Small Animal Surgery'' (D. H. Slatter, ed.), pp. 37–51. Saunders, Philadelphia.

DaRif, C. A., and Rush, H. G. (1983). Management of septicemia in Rhesus monkeys with chronic indwelling venous catheters. *Lab. Anim. Sci.* **33,** 90–94.

Dawidson, I., Gelin, L. E., and Haglund, E. (1980). Plasma volume, intravascular protein content, hemodynamic and oxygen transport changes during intestinal shock in dogs; Comparison of relative effectiveness of various plasma expanders. *Crit. Care Med.* **8,** 73–80.

Degernes, L. A. (1992). Soft tissue wound management in avian patients. *Proc. Annu. Conf. Assoc. Avian Vet.,* New Orleans, Louisiana, pp. 476–483.

Degernes, L., Davidson, G., Flammer, K., Kolmstetter, C., and Munger, L. (1994). Administration of total parenteral nutrition in pigeons. *Am. J. Vet. Res.* **55,** 660–665.

Dennis, M. B., Jr., Cole, J. J., Jensen, W. M., and Scribner, B. H. (1984). Long-term blood access by catheters implanted into arteriovenous fistulas of sheep. *Lab. Anim. Sci.* **34,** 388–392.

Dennis, M. B., Jr., Cole, J. J., and Scribner, B. H. (1986). Vascular access in large laboratory animals. *In* ''Methods of Animal Experimentation'' (W. I. Gay and J. E. Heavner, eds.), Vol. VII, Pt. A, pp. 195–241. Academic Press, Orlando, Florida.

Dennis, M. B., Jr., Jones, D. R., and Tenover, F. C. (1989). Chlorine dioxide sterilization of implanted right atrial catheters in rabbits. *Lab. Anim. Sci.* **39,** 51–55.

Dennis, M. B. Jr., Jones, D. R., and Tenover, F. C. (1990). Sepsis of vascular catheters. II: *In vitro* disinfection of colonized tubing. *J. Biomed. Mater. Res.* **24,** 95–105.

Desjardins, C. (1986). Indwelling vascular cannulas for remote blood sampling, infusion, and long-term instrumentation of small laboratory animals. *In* ''Methods of Animal Experimentation'' (W. I. Gay and J. E. Heavner, eds.), Vol. VII, Pt. A, pp. 143–194. Academic Press, Orlando, Florida.

Donoghue, S. (1992). Nutritional support of hospitalized animals. *J. Am. Vet. Med. Assoc.* **200,** 612–615.

Falk, J. L., Rackow, E. C., and Weil, M. H. (1989). Colloid and crystalloid fluid resuscitation. *In* ''Textbook of Critical Care'' (W. C. Shoemaker, S. Aayres, A. Grenvik, *et al.,* eds.), 2nd ed., pp. 1060–1069. Saunders, Philadelphia.

Fowler, M. E., ed. (1986). ''Zoo and Wild Animal Medicine,'' 2nd ed. Saunders, Philadelphia.

Fox, J. G., Cohen, B. J., and Loew, F. M., eds. (1984). ''Laboratory Animal Medicine.'' Academic Press, New York.

Fulton, R. B., and Dennis, J. S. (1992). Blind percutaneous placement of a gastrostomy tube for nutritional support in dogs and cats. *J. Am. Vet. Med. Assoc.* **201,** 697–700.

Garrett, P. D. (1987). Urethral recess in male goats, sheep, cattle, and swine. *J. Am. Vet. Med. Assoc.* **191,** 689–691.

Gibson, T. E., and Paterson, D. A., eds. (1985). ''The Detection and Relief of Pain in Animals,'' *Proc. Anim. Welfare Found. 2nd Symp.* Br. Vet. Assoc. Anim. Welfare Found., London.

Giger, U., and Bucheler, J. (1991). Transfusion of type-A and type-B blood to cats. *J. Am. Vet. Med. Assoc.* **198,** 411–418.

Giger, U., Kilran, C. G., Filippich, L. J., and Bell, K. (1989). Frequencies of feline blood groups in the United States. *J. Am. Vet. Med. Assoc.* **195,** 1230–1232.

Goring, R. L., Goldman, A., Kaufman, K. J., Roberts, C., Quesenberry, K. E., and Kollias, G. V. (1986). Needle catheter duodenostomy: A technique for duodenal alimentation of birds. *J. Am. Vet. Med. Assoc.* **189,** 1017–1019.

Griffel, M. I., and Kaufman, B. S. (1992). Pharmacology of colloids and crystalloids. *In* ''Fluid Resuscitation of the Critically Ill'' (B. S. Kaufman), pp. 235–253. Critical Care Clinics, Saunders, Philadelphia.

Harkness, J. E., and Wagner, J. E. (1989). ''The Biology and Medicine of Rabbits and Rodents,'' 3rd ed. Lea & Febiger, Philadelpha.

Haskins, S. C. (1981). Standards and techniques of equipment utilization. *In* ''Veterinary Critical Care'' (F. P. Sattler, R. P. Knowles, and W. G. Whittick, eds.), pp. 60–110. Lea & Febiger, Philadelphia.

Haskins, S. C. (1986). Postoperative care. *In* ''Methods of Animal Experimentation'' (W. I. Gay and J. E. Heavner, eds.), Vol. VII, Part A, pp. 113–142. Academic Press, Orlando, Florida.

Haskins, S. C. (1987). Use of analgesics postoperatively and in a small animal intensive care setting. *J. Am. Vet. Med. Assoc.* **191,** 1266–1268.

Haskins, S. C. (1989a). Enteral nutrition—products and techniques. *Proc. 17th Annu. Vet. Surg. Forum,* Chicago, pp. 434–436.

Haskins, S. C. (1989b). Parenteral nutrition—products and techniques. *Proc. 17th Annu. Vet. Surg. Forum,* Chicago, pp. 437–439.

Haven, M. L., Wichtel, J. J., Bristol, D. G., Fetrow, J. F., and Spears, J. W. (1992). Effects of antibiotic prophylaxis on postoperative complications after rumenotomy in cattle. *J. Am. Vet. Med. Assoc.* **200,** 1332–1335.

Institute of Laboratory Animal Resources (ILAR) Committee on Care and Use of Laboratory Animals (1985). ''Guide for the Care and Use of Laboratory Animals,'' NIH Publ. No. 86–23. U. S. Department of Health and Human Services, Public Health Service, National Institutes of Health, Bethesda, Maryland.

Institute of Laboratory Animal Resources (ILAR) (1992). ''Recognition and Alleviation of Pain and Distress in Laboratory Animals.'' Committee on Pain and Distress in Laboratory Animals, Institute of Laboratory Animal Resources, Commission on Life Sciences, National Research Council, National Academy of Sciences, National Academy Press, Washington, D.C.

Jenkins, J. R., and Brown, S. A. (1993). ''A Practitioner's Guide to Rabbits and Ferrets.'' The American Animal Hospital Association, Lakewood, Colorado.

Johnston, D. E. (1990). Wound healing in skin. *Vet. Clin. North Am.: Small Anim. Pract.* **20,** 1–25.

Kamal, G. D., Pfaller, M. A., Rempe, L. E., and Jebson, P. J. R. (1991). Reduced intravascular catheter infection by antibiotic bonding: A prospective, randomized, controlled trial. *JAMA, J. Am. Med. Assoc.* **265,** 2364–2368.

Kirk, R. W., Bistner, S. I., and Ford, R. B. (1986). ''Handbook of Veterinary Procedures and Emergency Treatment.'' Saunders, Philadelphia.

Kleiber, M. (1961). ''The Fire of Life, An Introduction to Animal Energetics,'' Part V. Wiley, New York.

Klein, W. R., and Firth, E. C. (1988a). Infection rates in clean surgical procedures with and without prophylactic antibiotics. *Vet. Rec.* **123,** 542–543.

Klein, W. R., and Firth, E. C. (1988b). Infection rates in contaminated surgical procedures: A comparison of prophylactic treatment for one day or four days. *Vet. Rec.* **123,** 564–566.

Laboratory Animal Science Association (LASA) (1990). The assessment and control of the severity of scientific procedures on laboratory animals. *Lab. Anim.* **24,** 97–130.

La Regina, M. C., Frank, P. A., Vogler, G. A., and Baudendistel, L. J. (1992). The use of combined prophylactic antibiotics in chronically instrumented goats—a case report. *Contemp. Top. Lab. Anim. Sci.* **31,** 18–21.

Lewis, L. D., Morris, M. L., and Hand, M. S. (1987). ''Small Animal Clinical Nutrition,'' 3rd ed., Chapters 1, 5. Mark Morris Associates, Topeka, Kansas.

Lippert, A. C. (1989). Enteral and parenteral nutritional support in dogs and cats with gastrointestinal disease. *Semin. Vet. Med. Surg. (Small Anim.)* **4,** 232–240.

Mathews, K. A., and Binnington, A. G. (1986). Percutaneous incisionless placement of a gastrostomy tube utilizing a gastroscope: Preliminary observations. *J. Am. Anim. Hosp. Assoc.* **22,** 601–610.

McKiernan, B. C. (1984). Initial management of wounds and injuries. *In* ''Veterinary Trauma and Critical Care'' (I. M. Zaslow), pp. 245–286. Lea & Febiger, Philadelphia.

Morton, D. B., and Griffiths, P. H. M. (1985). Guidelines on the recognition of pain, distress, and discomfort in experimental animals and an hypothesis for assessment. *Vet. Rec.* **116,** 431–436.

Moser, E. (1990). Dietary principles in disease management. *Semin. Vet. Med. Surg. (Small Anim.)***5,** 145–153.

Muller, G., Schaarschmidt, K., and Stratmann, U. (1992). A new device for long-term continuous enteral nutrition of rats by elementary diet via gastrostomy, following extensive oesophageal or lower gastrointestinal surgery. *Lab. Anim.* **26,** 9–14.

Orsini, J. A. (1985). Wound infections and antimicrobial therapy. ***In* ''General Small Animal Surgery'' (I. M. Gourley and P. B. Vasseur, eds.), pp. 121–144. Lippincott, Philadelphia.**

Otto, C. M., Kaufman, G. M., and Crowe, D. T., Jr. (1989). Intraosseous infusion of fluids and therapeutics. *Compend. Contin. Educ. Pract. Vet.* **11,** 421–430.

Page, C. D., and Mautino, M. (1990). Clinical management of tortoises. *Compend. Contin. Educ. Pract. Vet.* **12,** 221–228.

Palm, U., Boemke, W., Bayerl, D., Schnoy, N., Juhr, N., and Reinhardt, H. W. (1991). Prevention of catheter-related infections by a new catheter-restricted antibiotic filling technique. *Lab. Anim.* **25,** 142–152.

Penner, J., St. Claire, M., Beckwith, C., Besselsen, D., Wright, J., Fish, R., and Friskey, S. (1994). *Pseudomonas* sp. bacteremia associated with chronic vascular catheterization in rabbits. *Lab. Anim. Sci.* **44,** 642–644.

Penwick, R. C. (1985). Use of antimicrobial drugs in surgery. *In* ''Textbook of Small Animal Surgery'' (D. H. Slatter, ed.), pp. 52–70. Saunders, Philadelphia.

Peterson, W. A. (1989). FDA/GLP regulations. *In* ''Good Laboratory Practice Regulations'' (A. F. Hirsch, ed.), pp. 15–89. Dekker, New York.

Pichler, M. E., and Turnwald, G. H. (1985). Blood transfusion in the dog and cat. Part I. Physiology, collection, storage, and indications for whole blood therapy. *Compend. Contin. Educ. Pract. Vet.* **7,** 64–71.

Quesenberry, K. E., Mauldin, G., and Hillyer, E. (1989). Nutritional support of the avian patient. *Proc. Annu. Conf. Assoc. Avian Vet.* Seattle, Washington. pp. 11–19.

Rackow, E. C., Falk, J. L,, Fein, I. A., Siegel, J. S., Packman, M. I., Haupt, M. T., Kaufman, B. S., and Putnam, D. (1983). Fluid resuscitation in circulatory shock: A comparison of the cardiorespiratory effects of albumin, hetastarch, and saline solutions in patients with hypovolemic and septic shock. *Crit. Care Med.* **11,** 839–850.

Remillard, R. L., and Martin, R. A. (1990). Nutritional support in the surgical patient. *Semin. Vet. Med. Surg. (Small Anim.)* **5,** 197–207.

Remillard, R. L., and Thatcher, C. D. (1989). Parenteral nutritional support in the small animal patient. *Vet. Clin. North Am.: Small Anim. Pract.* **19,** 1287–1306.

Ritchie, B. W., Otto, C. M., Latimer, K. S., and Crowe, D. T., Jr. (1990). A technique of intraosseous cannulation for intravenous therapy in birds. *Compend. Contin. Educ. Pract. Vet.* **12,** 55–58.

Robbins, C. T. (1983). ''Wildlife Feeding and Nutrition.'' Academic Press, New York.

Romatowski, J. (1989). Prevention and control of surgical wound infection. *J. Am. Vet. Med. Assoc.* **194,** 107–114.

Rosin, E. (1989). The selection and use of antibiotics in small animal general surgery. *Proc. 17th Annu. Vet. Surg. Forum,* Chicago, pp. 11–13.

Scalese, R. J., DeForrest, J. M., Hammerstone, S., Parente, E., and Burkett, D. E. (1990). Long term vascular catheterization of the Cynomolgus monkey. *Lab. Anim. Sci.* **40,** 530–532.

Schmidt-Nielsen, K. (1984). ''Scaling: Why is Animal Size so Important?,'' Chapter 6. Cambridge Univ. Press, New York.

Shapiro, B. A. (1989). ''Clinical Application of Blood Gases,'' 4th ed. Year Book Medical Publishers, Chicago.

Smeak, D. D. (1992). Buried continuous intradermal suture closure. *Compend. Contin. Educ. Pract. Vet.* **14,** 907–919.

Soma, L. R. (1987). Assessment of animal pain in experimental animals. *Lab. Anim. Sci.* **37,** 71–74.

Swaim, S. F. (1990). Bandages and topical agents. *Vet. Clin. North Am.: Small Anim. Pract.* **20,** 47–65.

Swaim, S. F., and Henderson, R. A. (1990). ''Small Animal Wound Management.'' Lea & Febiger, Philadelphia.

Swaim, S. F., and Lee, A. H. (1987). Topical wound medications: A review. *J. Am. Vet. Med. Assoc.* **190,** 1588–1593.

Swaim, S. F., and Wilhalf, D. (1985). The physics, physiology, and chemistry of bandaging open wounds. *Compend. Contin. Educ. Pract. Vet.* **7,** 146–156.

Sweeney, R. W., and Divers, T. J. (1990). The use of parenteral nutrition in calves. *Vet. Clin. North Am.: Food Anim. Pract.* **6,** 125–131.

Tilley, L. P. (1985). ''Essentials of Canine and Feline Electrocardiography: Interpretation and Treatment,'' 2nd ed. Lea & Febiger, Philadelphia.

Tsukamoto, H., Reidelberger, R. D., French, S. W., and Largman, C. (1984). Long-term cannulation model for blood sampling and intragastric infusion in the rat. *Am. J. Physiol.* **247,** R595–R599.

Turnwald, G. H. (1985). Blood transfusions in dogs and cats. Part II: Administration, adverse effects and component therapy. *Compend. Contin. Educ. Pract. Vet.* **7,** 115–122.

U. S. Department of Agriculture (1992). Title 9, Code of Federal Regulations, Subchapter A, part 2. USDA, Washington, D. C.

Vasseur, P. B. (1985). Principles of operative surgery. *In* ''General Small Animal Surgery'' (I. M. Gourley and P. B. Vasseur, eds.), pp. 75–104. Lippincott, Philadelphia.

Vasseur, P. B., Levy, J., Dowd, E., and Eliot, J. (1988). Surgical wound infection rates in dogs and cats, data from a teaching hospital. *Vet. Surg.* **17,** 60–64.

Wilcke, J. R. (1990). Use of antimicrobial drugs to prevent infections in veterinary patients. *Probl. Vet. Med.* **2,** 298–311.

Wright, E. M., Jr., Marcella, K. L., and Woodson, J. F. (1985). Animal pain: Evaluation and control. *Lab Anim.* **14,** 20–36.

Chapter 17

Current Research in Anesthesia and Trends in Clinical Applications

Mervyn Maze, David A. Daunt, and Markku Salonen

I. INTRODUCTION

Recent investigations have provided valuable insight into the molecular mechanisms of anesthetic and analgesic action. A variety of classes of injectable anesthetic agents have been shown to enhance chloride conductance through the $GABA_A$ (γ-aminobutyric acid, GABA) receptor-gated chloride channel, thereby enhancing inhibitory neurotransmission and producing the anesthetic state. Glutamate receptors, another family of ligand-gated ion channel, are modulated by the dissociative anesthetic ketamine. Selective and potent α_2-adrenergic agonists have been shown to produce an anesthetic state by activating the α_2-adrenoceptor which couples to second messenger or effector system via the pertussis toxin-sensitive guanine nucleotide-binding (G) proteins. Most of the putative anesthesia- and analgesia-mediating receptors, effectors (including ion channels), and G proteins have been isolated and cloned. Investigations of the molecular mechanism of anesthetic action using these molecular genetic reagents in *in vitro* systems are now being pursued. Such studies may result in anesthetic strategies which are designed to precisely modulate each of the different molecular components of the signal transduction pathway to produce the anesthetic state. Finally, knowledge of the site(s) of anesthetic and analgesic action has resulted in the implementation of drug delivery strategies designed to promote the optimal anesthetic/analgesic drug concentration in the biophase, thereby using less drug with fewer side effects.

ANESTHESIA AND ANALGESIA IN LABORATORY ANIMALS

ISBN 0-12-417570-8

II. NOVEL SITES FOR ANESTHETIC DRUG ACTION

A. $GABA_A$ Receptor Macromolecular Complex

1. Membrane Components

When the $GABA_A$ receptor proteins were initially purified from bovine cerebral cortex (Sigel *et al.,* 1983) they were found to contain at least two different subunits. The α subunit is thought to harbor the binding site for benzodiazepines whereas the β subunit is primarily the binding site for GABA.

Subsequent molecular biologic techniques have contributed greatly to our understanding of the structure of the benzodiazepine receptor. Based on amino acid homology, the subunits discovered by cloning techniques are classified as α, β, γ, δ, ϵ, and ζ (Pritchett *et al.,* 1988). There is an abundance of each subunit, and it is likely that the expression of different combinations of subunits in transfected cells will lead to further discovery of new benzodiazepine receptor subtypes. So far it is not understood if all or only certain subunits are necessary to form a receptor. The α_2 subunit is known to be important for high-affinity benzodiazepine binding, whereas the benzodiazepine receptor subtype is determined by the β subunit. Each receptor consists of five glycoprotein subunits, and each subunit consists of four membrane-spanning loops, an extracellular drug-accepting region, and an intracellular loop subject to phosphorylation by cyclic AMP-dependent protein kinase. Phosphorylation may be a mechanism of modulating the function of the receptor by other neurotransmitters and second messengers, although its importance is still unconfirmed (Tehrani *et al.,* 1989). A schematic diagram of the benzodiazepine receptor is presented in Fig. 1.

2. Transmembrane Signaling

The benzodiazepine–$GABA_A$ receptor complex gates chloride conductance across the neuronal cell membrane. GABA activates chloride currents in a concentration-dependent manner. The Hill coefficient of this effect is 2, indicating that two molecules of GABA are required to activate a single receptor. Benzodiazepines increase the potency of GABA. Chloride ions may flow in either direction, depending on the resting potential of the cell, but in most cases the current will be inward, leading to cell hyperpolarization. Opening chloride channels makes cell depolarization more difficult, thus GABAergic activity results in inhibition of neuronal excitability (Bormann, 1988).

An interesting question relates to the mechanism whereby the chloride channel is selective for only chloride ions. It has been argued that the channel pore size (6 Å) is not sufficient to discriminate among sodium, potassium, and calcium ions. Bormann *et al.* (1987) demonstrated ion selectivity by showing that anions must interact with positively charged residues in the channel pore of the receptor complex to be able to pass through the pore.

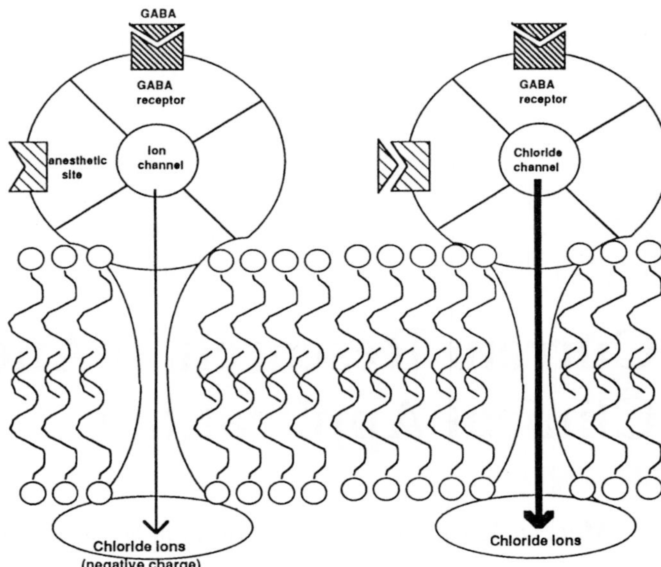

Fig. 1. Mechanism of action of drugs acting at the $GABA_A$ receptor macromolecular complex. After GABA, the inhibitory neurotransmitter, binds to its receptor on the $GABA_A$ receptor–chloride ionophore heteropentamer, chloride ions flow along the concentration gradient into the cell. The net result of this ionic conductance is to hyperpolarize the neuronal membrane and to make the cell less ''excitable.'' Modulating ligands such as benzodiazepines, barbiturates, and propofol[1], bind to separate and individual sites on the complex and enhance the GABA-mediated chloride conductance through the membrane (right-hand side).

3. Modulation by States

The $GABA_A$/benzodiazepine receptor complex is allosterically modulated by many drugs, including barbiturates, steroid anesthetics, etomidate, propofol, and ''cage convulsants.'' Barbiturates preferentially increase the likelihood of the GABA receptor being in the high-affinity state. Benzodiazepines increase the frequency of chloride channel openings (Study and Barker, 1981) whereas barbiturates increase the mean channel open time. In addition to benzodiazepine receptor agonists which increase chloride flux, inverse agonists bind to the same receptor but reduce the GABA-mediated chloride flux (Haefely *et al.,* 1985). Steroids enhance GABA-stimulated chloride conductance in rat brain by prolonging the open time of the chloride channel. In addition, these compounds enhance the binding affinity of $GABA_A$ and benzodiazepine ligands, but inhibit binding of cage convulsants to their binding site (Gee, 1988). Picrotoxinin and some bicyclic cage compounds antagonize GABA-induced chloride conductance responses by binding to a site that sterically hinders the entry of chloride ions into the ion channel pore (Gee, 1988). Dose-dependent propofol[1] potentiates GABA-activated currents and, at higher doses, directly increases the chloride conductance of cellular membranes in a

[1]Diprivan (Stuart Pharmaceuticals, Wilmington, DE).

bicuculline-sensitive fashion. In addition, propofol enhances the binding of GABA agonists but has no effect on the binding of [^{3}H] flunitrazepam or on pentobarbital-enhanced [^{3}H] flunitrazepam binding, suggesting that propofol binds at a site that is distinct from either the benzodiazepine or the barbiturate-binding sites on the $GABA_A$ receptor macromolecular complex (Concas *et al.,* 1991). Adenosine agonists and antagonists have some effects on the benzodiazepine responses as well. Caffeine and theophylline, methylxanthines that antagonize the hydrolysis of cAMP by phosphodiesterase, can antagonize binding to the benzodiazepine receptor as well as the sedation caused by benzodiazepines (Scolnik *et al.,* 1978). It is possible that the increase in cAMP-dependent protein kinase activity may phosphorylate the receptor and produce both the biochemical and the behavioral effects that were noted following the administration of phosphodiesterase inhibitors. GABA modulates its own effects and probably those of benzodiazepines by desensitizing its own receptor. The receptor is subject to glycosylation and phosphorylation, the roles of which are not understood at present.

4. Anesthetic Action and GABA–Chloride Ionophore

Enhancement of inhibitory synaptic transmission via the $GABA_A$ receptor–chloride channel has been proposed as a mechanism for the general anesthetic action of a variety of compounds. Among the intravenous anesthetic agents, benzodiazepines, barbiturates, steroids (Longoni and Olsen, 1992), and propofol (Concas *et al.,* 1990) enhance $GABA_A$ receptor function as reflected by an increase in $^{36}Cl^-$ flux. An excellent correlation exists between the clinical potency of benzodiazepines and their affinity for the "central" (flunitrazepam) binding site (Haefely *et al.,* 1985). Volatile anesthetic agents also enhance GABA-mediated chloride conductance (Yeh *et al.,* 1991). However, unlike the intravenous anesthetic agents which may exert their effect by allosteric interactions, volatile agents appear to act indirectly and are critically dependent on changes in calcium concentration (Mody *et al.,* 1991). The $GABA_A$ receptor–chloride ionophore macromolecular complex has been suggested as the final common pathway for general anesthetic action (Tanelian *et al.,* 1993).

B. Glutamate Receptors

The excitatory amino acid neurotransmitter, glutamate, mediates its effects in the central nervous system (CNS) by activation of at least three main types of ligand-gated ion channels. These channels are distinguished on the basis of their response to the agonists kainate, quisqualate (α-amino-3-hydroxy-5-methyl-4-isoxazole propionate or AMPA) and *N*-methyl-D-aspartate (NMDA). An expanded classification involving two additional pharmacologic receptor subtypes, namely 2-amino-4-phosphonobutyrate (L-AP4) and *trans*-1-aminocyclopentane 1,3-dicarboxylate (ACPD or metabotropic glutamate receptor), has been added (Gasic and Heinemann, 1991). Others have suggested a more simple functional classification, with AMPA receptors mediating fast synaptic currents and NMDA receptors mediating currents that are slower and longer lasting.

1. Transmembrane-Signaling Mechanisms

Kainate, NMDA, and AMPA receptors belong to the ligand-gated ion channel family (iGluR) which are activated in a millisecond time span. The glutamate receptor is the only neurotransmitter receptor that regulates a voltage-dependent channel directly linked to the second messenger Ca^{2+}. The L-AP4 and ACPD receptors utilize a G protein in their signal transduction mechanism and act slower (hundreds of microseconds). The L-AP4 receptor has been shown to increase the hydrolysis of cGMP in some cells, which leads to closure of an inward current (Nawy and Jahr, 1990). The ACPD is a G-protein-coupled receptor linked to inositol phosphate/diacylglycerol formation (Nicoletti *et al.,* 1986).

2. Glutamate Receptors and Anesthetic Action

General anesthetics reduce the amplitude of excitatory postsynaptic potentials and responses to excitatory amino acids (Krnjevic, 1991). This is especially true for the dissociative anesthetic ketamine[2] (Oshima and Richards, 1988; Davies *et al.,* 1988). However, the potency of other anesthetics is enhanced by noncompetitive blockers of the glutamate receptor–ionophore complex (Daniell, 1990, 1991, 1992). Ethanol inhibits NMDA receptor-mediated responses over the concentration range that produces intoxication (Weight *et al.,* 1992).

C. α_2-Adrenergic Receptors

α_2-Adrenergic agonists have been introduced into the clinical practice of human anesthesiology for their anxiolytic, sedative, analgesic, anesthetic-sparing, and hemodynamic-stabilizing properties (Maze and Tranquilli, 1991). In veterinary medicine, α_2-agonists such as xylazine[3] have been used for sedation, analgesia, and surgery since the mid-1970s (Maze and Tranquilli, 1991). More potent and α_2-specific agonists such as detomidine[4] and medetomidine[5] and antagonists such as yohimbine[6] and atipamezole[7] have been introduced to both human and veterinary medicine. The agonists produce somnolence, sedation, hypnosis, analgesia, and anesthetic-sparing effects by activating

[2]Vetalar (Parke-Davis, Morris Plains, NJ).
[3]Rompun (Haver Bayvet Division, Miles Laboratories, Shawnee, KS).
[4]Dormosedan (Smith-Kline-Beecham, Westchester, PA).
[5]Domitor (Farmos, Turku, Finland).
[6]Yobine (Lloyd Laboratories, Shenandoah, IA).
[7]Antisedan (Farmos, Turku, Finland).

α_2-adrenergic receptors. Through molecular cloning, ligand-binding studies, and autoradiographic techniques, we now appreciate the presence of at least three α_2-adrenergic receptor subtypes (Kobilka, 1992). Investigations are underway to determine the physiological significance of receptor subtype diversity and specificity.

1. Signal Transduction and Molecular Mechanisms of α_2-Adrenergic Agonists

α_2-Adrenergic receptors are members of the G-protein-coupled family of plasma membrane receptors (Gilman, 1987). From the amino acids in the polypeptide chain, a theoretical ''hydropathy'' plot can be deduced according to the hydrophobicity or hydrophilicity of each residue. From these plots, the G-protein-coupled receptors are proposed to have seven transmembrane regions (Fig. 2). The transmembrane domains form the ligand-binding pocket, whereas the cytoplasmic loops confer specificity for G protein coupling (Kobilka *et al.,* 1988). When agonist binds receptor, this activates an α_2-specific G protein which in turn regulates the effector mechanism(s) (Fig. 3).

Although relatively little is understood regarding the molecular mechanisms of anesthesia for many drugs, the effects of α_2-adrenergic agonists are known to be mediated through a pertussis toxin-sensitive G protein (Fig. 3). Animal studies have shown that by inhibiting the activation of G proteins by ADP-ribosylation of the G protein α subunit with pertussis toxin, the anesthetic effects of α_2-adrenergic agonist are prevented (Correa-Sales *et al.,* 1992a). In addition to the identified molecular components that induce these effects, α_2-adrenergic receptors in the locus coeruleus appear to be a major site for the hypnotic action in the CNS (Correa-Sales *et al.,* 1992b).

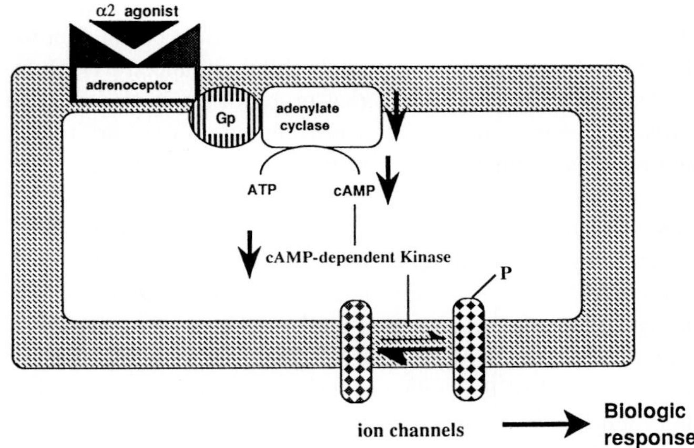

Fig. 3. Signal transduction mechanism for α_2-adrenergic agonists in the locus coeruleus. α_2-Adrenoceptors in the locus coeruleus mediate the hypnotic–anesthetic action of α_2-agonists. The initial postreceptor components involved in the hypnotic action of α_2-agonists are a pertussis toxin-sensitive G protein (Gp), coupled in an inhibitory fashion to adenylate cyclase. The authors propose that the resulting decrease in cAMP and cAMP-dependent protein kinase alters the phosphorylation state, and thereby the conductance, through ion channels. The end result is a hyperpolarized membrane that is less likely to be rendered ''excitable.''

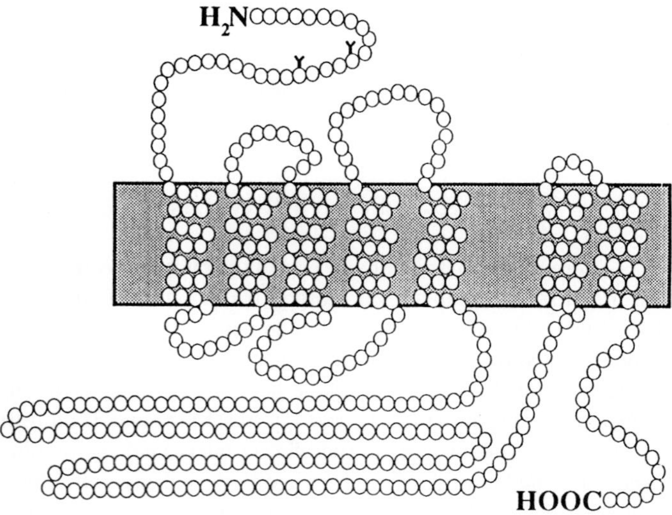

Fig. 2. Diagram of the proposed membrane topology of α_2-adrenergic receptors. These structural features are representative of all G-protein-coupled receptors. Circles represent amino acid residues. Seven hydrophobic regions are proposed to traverse the plasma membrane (shown as the shaded box) with the amino terminus orientated extracellularly and the carboxyl terminus intracellularly. These transmembrane regions form the ligand-binding pocket and confer ligand specificity. The third intracellular loop and parts of intracellular loop two and the amino end of the cytoplasmic tail confer G-protein-coupling specificity. Several asparagine-linked glycosylation sites (shown as Y) are located on the extracellular surface near the amino terminus.

In veterinary medicine, α_2-adrenergic agonists have been used for sedation and analgesia or in combination with other drugs as anesthetics in exotic, domestic, and farm animals (Jalanka, 1991; Maze and Tranquilli, 1991; Short, 1986). Specific α_2-antagonists have also been investigated and used clinically to reverse anesthetic states. More selective and potent α_2-adrenergic agonists (e.g., detomidine, medetomidine, and dexmedetomidine) and antagonists (e.g., atipamezole) have been investigated and used in clinical veterinary medicine. Several of these newer agents are being studied for their usefulness in human anesthesia. These agents may be particularly useful in humans where α_2-adrenergic agonists not only have anesthetic-sparing effects, but also reduce increases in heart rate and blood pressure associated with strong stimuli such as intubation, surgery, and emergence from anesthesia (Aho *et al.,* 1991; Engelman *et al.,* 1989; Ghignone *et al.,* 1987, 1988). In both human and veterinary anesthesia, α_2-agonists are also administered by the epidural route for their profound analgesic properties (Eisenach *et al.,* 1987. α_2-Agonists have been used as continuous infusions in conjunction with other injectable or inhalation anesthetics (Aho *et al.,* 1992; Dunlop *et al.,* 1991).

Not surprisingly, the activation of adrenergic receptors modulates the autonomic nervous system and affects multiple organ systems. α_2-Adrenergic receptors may be located presynaptically or postsynaptically and are found in tissues such as brain, spinal cord, kidney, vasculature, platelets, pituitary, spleen, lung, liver, and skeletal muscle (Lorenz *et al.,* 1990). Because of this wide tissue distribution, many physiologic changes are observed following the administration of α_2-specific drugs. Changes such as hypoventilation, hypoxemia, bradycardia, second-degree atrioventricular block, transient hypertension, hypotension, reduction in cardiac output, hypothermia, and diuresis have been reported in a number of species (Maze and Tranquilli, 1991). Prolonged gastrointestinal transit time is also produced by α_2-agonists; however, this probably is not clinically significant following acute drug administration for anesthesia. Other gastrointestinal effects following systemic α_2 administration include emesis which is common in dogs and cats (Tranquilli and Benson, 1992). α_2-Adrenergic agonists also produce mydriasis (Hsu *et al.,* 1981) and reduce intraocular pressure in both normal and glaucomatous eyes (Jin *et al.,* 1991; Vartiainen *et al.,* 1992). In addition, certain species such as ruminants appear to be much more sensitive to α_2-adrenergic agonists, exhibiting more profound side effects with a prolonged duration of action (Short, 1986). We do not appreciate the mechanism(s) responsible for these species differences in sensitivity for this class of drugs.

The newer, more potent, and selective α_2-agonists and antagonists are being investigated in a variety of laboratory animal species. Because of their ease of use, relatively wide margin of safety, and availability of specific antagonists, these potent and selective α_2-agonists will prove very useful in laboratory animal medicine.

D. Voltage-Gated Calcium Channels

Voltage-gated calcium channels regulate the entry of Ca^{2+} ions across the surface of membranes. The influx of Ca^{2+} ions along its electrochemical gradient generates both electrical and chemical messages for Ca^{2+}-sensitive mechanisms, including those that control ion channel gating, transmitter release, and cellular kinase activity (Tsien *et al.,* 1988). The channels are characterized according to their gating, ionic conductance, and pharmacologic properties as T (transient, small conductance, activated briefly following depolarizations), N (neuronal coupled to transmitter release), and L (long-lasting, large conductance from strong depolarizations) types (Tsien *et al.,* 1988).

The calcium channel is a complex of five subunits with a central α_1 component (Catterall, 1988). Considerable homology exists between this subunit and the α subunit of the sodium channel. Each subunit has four replicating domains that contain six hydrophobic α helices or membrane-spanning sequences. The transmembrane regions presumably contribute to the ion pore.

Volatile anesthetic agents inhibit a voltage-dependent calcium current at concentrations in the clinical range (Krnjevic and Puil, 1988; Study, 1994). A high threshold form of the calcium channel has been shown to be especially sensitive to volatile anesthetics (Herrington *et al.,* 1991). When dihydropyridine-sensitive calcium channels are blocked, the sensitivity of dogs to anesthesia with halothane is enhanced, suggesting a mediating role for voltage-operated calcium channels in its action (Maze *et al.,* 1983). Also, several species of voltage-operated calcium channels are mechanistically involved in the hypnotic action of α agonists (Horvath *et al.,* 1992; Nacif-Coelho *et al.,* 1994).

E. Voltage-Gated Potassium Channels

In the brain, voltage-gated K^+ channels are essential for terminating the action potential, repolarizing the neuron, setting the resting potential, and regulating neurotransmitter release. These functions are mediated by a diverse number of potassium channels. The K^+ channels can be classified into delayed rectifiers, inward rectifiers, and fast activation/inactivation channels, according to electrophysiologic criteria based on channel opening and closing kinetics (Rehm and Tempel, 1991). The availability of selective toxins is proving invaluable in the task of characterizing the many subtypes of potassium channels (Castle *et al.,* 1989). Voltage-gated K^+ channels have structural features in common with other voltage-gated channels, but the central core is composed of four subunits rather than a single polypeptide chain with repetitive domains.

In a mollusc neuron preparation, Franks and Lieb (1988) have reported a direct correlation between the presence of a volatile anesthetic-induced current and anesthetic inhibition of spontaneous firing in a given neuron. However, there is not uniformity in the manner in which volatile anesthetic agents affect the various species of potassium channels (Urban, 1993). Using selective toxins, Nacif-Coelho and colleagues (1994) were able to document the involvement of at least two species of potassium channels in the hypnotic action of α_2-adrenergic agonists.

III. NOVEL DRUGS

A. Telazol

Telazol[8] is an anesthetic approved for intramuscular administration in dogs and cats. Telazol is a combination anesthetic containing equal parts by weight of tiletamine hydrochloride and zolazepam hydrochloride in a water-soluble formulation. Tiletamine is an arylcycloalkylamine similar to ketamine and produces a dissociative anesthetic state characterized by rapid induction, retained pharyngeal and laryngeal reflexes, and

[8]Telazol (A. H. Robins, Richmond, VA).

analgesia. However, tiletamine does not provide good skeletal muscle relaxation. Zolazepam is a water-soluble benzodiazepine similar to midazolam[9] and has sedative-hypnotic, muscle relaxant, anxiolytic, and anticonvulsant properties (Rall, 1990). When used by itself, Telazol provides satisfactory analgesia and anesthesia for minor surgical procedures in primates and sheep (Ward *et al.,* 1974), dogs and cats (Ilkiw, 1992), rats (Silverman *et al.,* 1983), gerbils (Hrapkiewicz *et al.,* 1989), and ferrets (Payton and Pick, 1989). It also provides effective sedation and restraint for painless procedures in many other species. When combined with an α_2-adrenergic agonist such as xylazine, Telazol also produces surgical anesthesia in rabbits (Popilskis *et al.,* 1991), horses (Hubbell et al., 1989), and swine (Thurmon *et al.,* 1988).

Dose requirements for tranquilization and surgical anesthesia vary considerably among species. The recommended dose in cats is 9.7 to 15.8 mg/kg IM or SC and in dogs is 6.6 to 13.2 mg/kg IM or SC. However, these doses result in a prolonged and sometimes unpleasant recovery period (Ilkiw, 1992). Therefore, smaller doses (e.g., 2.5 mg/kg IM) are usually initially administered for restraint, and induction is completed either by an inhalation anesthetic delivered by face mask or by intravenous administration of Telazol at 1 mg/kg. Prolonged recovery has been reported in other species when surgical anesthesia doses are administered intramuscularly (Hrapkiewicz *et al.,* 1989). Induction of anesthesia is usually smooth with an onset of approximately 2 to 5 min following IM administration (Ilkiw, 1992).

The cardiorespiratory effects of Telazol are minimal in healthy animals. Because tiletamine and zolezepam are both excreted by the kidneys, their use is not recommended in patients with renal insufficiency (Ilkiw, 1992). Nephrotoxicity has been reported following relatively large doses (32 and 64 mg/kg, IM) of Telazol in rabbits (Brammer *et al.,* 1991) and is due to the tiletamine component (Doerning *et al.,* 1992). Lower doses of tiletamine in rabbits (7.5 mg/kg, IM) were not associated with serum chemical (creatinine or blood urea nitrogen) or urine abnormalities; however, mild nephrosis was evident with histopathology (Doerning *et al.,* 1992). Telazol causes hypersalivation, but this can be reduced or eliminated by treatment with an anticholinergic agent such as atropine or glycopyrrolate (Ward *et al.,* 1974).

B. Anesthetic Reversal Agents

Following the development of anesthetic drugs that have a specific target (e.g., a receptor) through which the anesthetic effects are mediated, the stage is set for the development of anesthetic reversal agents that displace the anesthetic drug from its target, thereby restoring the organism to its preanesthetic behavioral state. This situation should be analogous to that at the neuromuscular junction, where the anesthesiologist has exquisite control at inducing and reversing muscle relaxation.

1. Flumazenil

Flumazenil[10] is a centrally active (Persson *et al.,* 1985) benzodiazepine antagonist that effectively blocks the CNS effects of benzodiazepine agonists, including sedation, impaired cognition, and anesthesia (Brogden and Goa, 1991). Flumazenil also inhibits the benzodiazepine-induced decrease in plasma catecholamine levels (Duka *et al.,* 1986). Flumazenil does exhibit weak agonist properties, providing an anticonvulsant action at high doses (Scollo-Lavizzari, 1988); however, it also displays inverse agonist properties that provoke panic attacks in subjects with panic disorders (Nutt *et al.,* 1990). Following intravenous flumazenil administration, the onset of action is rapid, achieving maximal brain concentrations within 8 min; its elimination half-life is relatively short (0.8–1.16 hr) (Breimer *et al.,* 1991) with a duration of antagonism lasting less than 3 hr (McKay *et al.,* 1990). Because of these pharmacokinetic features, residual sedation is consistently observed and thus patients should be monitored following reversal with this drug.

Flumazenil administered in doses up to 1 mg was significantly better than placebo at reversing benzodiazepine-induced sedation, with earlier awakening and orientation in time and space and improved psychomotor performance. These reversal properties are not associated with any adverse cardiovascular effect (Marty *et al.,* 1991). It is especially useful in the differential diagnosis of coma with its ability to rapidly reverse a benzodiazepine overdose-induced coma (Brogden and Goa, 1988). It is ineffective at reversing coma from barbiturates, polycyclic antidepressants, phenothiazines, and alcohol.

2. Yohimbine/Atipamezole

At this time, there are no selective α_2-antagonists available for clinical human use. Phentolamine, a nonselective α-adrenoceptor blocker, is currently available; however, this compound does not easily penetrate the blood–brain barrier and thus will not antagonize the centrally active α_2-agonist responses. The following selective α_2-antagonists are currently being developed or are in clinical use in veterinary practice. Yohimbine, an indole alkaloid, is approximately 60 times more selective for the α_2 than the α_1 receptor (Clarke *et al.,* 1986). Idazoxan is a relatively specific antagonist at α_2-adrenoceptors (Doxey *et al.,* 1983) but does exert some partial α_1-agonist properties (Paciorek and Shepperson, 1983). Atipamezole is at least 100 times more selective for α_2-adrenoceptors than idazoxan (Virtanen *et al.,* 1989) and has no behavioral or neurochemical effects other than to increase the central turnover of norepinephrine (Scheinin *et al.,* 1988). The continuing development of these and other novel

[9]Versed (Hoffmann-La Roche, Nutley, NJ).

[10]Mazicon (Roche Laboratories, Nutley, NJ).

compounds for reversal of α_2-agonist-mediated responses will be a further useful adjunct in anesthetic practice.

C. New Inhalational Anesthetics

Sevoflurane and desflurane are two inhalational anesthetics recently introduced in clinical anesthesia. Both agents are significantly less soluble than halothane or isoflurane (Table I), allowing rapid changes in anesthetic depth and faster recovery from anesthesia. These characteristics are due to the exclusion of chlorine or bromine in their molecular structure, but instead are halogenated with fluorine (Eger, 1994). Other relevant physical characteristics of these newer inhalational anesthetics include lower potency than halothane or isoflurane, with minimal alveolar concentrations (MAC) values of 2 and 6% for sevoflurane and desflurane, respectively (Table I). The vapor pressure of desflurane is much greater than all other commonly used inhalation anesthetics (Table I), approaching 1 atm. Because the physical properties of desflurane are so unique, its use requires unconventional vaporizer technology (Weiskopf *et al.,* 1994). Desflurane vaporizers are relatively large and require electrical power as they heat desflurane to a gas which is blended with a diluent fresh gas to produce a regulated concentration. The lower vapor pressure of sevoflurane allows conventional vaporizer technology (Eger, 1994). Desflurane is also more pungent than other currently used inhalation anesthetics, resulting in airway irritation, salivation, breathholding, coughing, and laryngospasm during induction when inspired desflurane exceeds 6% (Eger, 1994).

This airway irritation has been associated with hypoxemia in children during inhalation induction (Zwass *et al.,* 1992) and therefore is not recommended for pediatric inhalation induction (Eger, 1994). The pungency of sevoflurane is similar to that of halothane, making it an acceptable agent for induction (Eger, 1994).

Molecular stability is enhanced with fluorination, therefore, both desflurane and sevoflurane are resistant to biodegradation (Eger, 1994). However, soda lime or Baralyme degrade sevoflurane, producing an olefin, a toxic breakdown product that can be lethal in rodents (Morio *et al.,* 1992). Although lethal concentrations of this toxin do not accumulate in the anesthetic circuit with normal use, low-flow or closed-circuit techniques accumulate the olefin, albeit at sublethal concentrations (Eger, 1994).

TABLE I

PHYSICAL CHARACTERISTICS OF NEW INHALATION ANESTHETICS[a]

Drug	Blood/gas partition coefficients	Vapor pressure at 20°C (mm Hg)	MAC[b] (%)
Halothane	2.4	244	0.75
Isoflurane	1.4	240	1.15
Desflurane	0.45	669	6.0
Sevoflurane	0.65	170	2.0

[a]From Eger, E. I. (1994). New inhaled anesthetics. Anesthesiology **80,** 906–922.
[b]MAC values for 30- to 60-year-old human patients (Eger, 1994).

At comparable MAC multiples, the effects of desflurane and sevoflurane on most vital organs are comparable to halothane and isoflurane (Eger, 1994). Desflurane and sevoflurane depress ventilation in a dose-dependent manner. Most circulatory changes also parallel those of halothane and isoflurane, which are characterized by dose-dependent decreases in blood pressure, total peripheral resistance, and myocardial contractility (Cahalan *et al.,* 1991). Although sevoflurane provides a relatively stable heart rate, at concentrations greater than 1 MAC, desflurane increases the heart rate (Cahalan *et al.,* 1991). Induction of anesthesia with desflurane can be associated with transient increases in heart rate and blood pressure (Zwass *et al.,* 1992). These effects may be undesirable in patients with coronary artery disease or diseases where cardiac reserve is limited, as increases in myocardial work increase myocardial oxygen consumption.

Neither desflurane nor sevoflurane sensitize the heart to the arrhythmic effects of epinephrine (Eger, 1994). Both agents can decrease cerebral vascular resistance and cerebral metabolic rate and dose dependently increase intracranial pressure (Lutz *et al.,* 1990). Electroencephalogram (EEG) changes are similar to those produced by isoflurane (Eger, 1994). Although neither agent causes convulsive activity, induction of anesthesia with sevoflurane can be associated with transient skeletal muscle rigidity (Morisaki *et al.,* 1988). Evidence also shows that neither agent causes hepatic or renal damage.

IV. ANESTHETIC DELIVERY

A. Computer-Controlled Drug Delivery

The precise titration of intravenous drugs is needed to rapidly achieve and maintain a steady plasma concentration for studies designed to assess drug concentration–effect relationships. Commonly, intravenous drugs are delivered by bolus administration, resulting in an initial overdose, followed by a rapidly declining blood level below therapeutic concentrations. Alternatively, administration of intravenous drugs by continuous infusion will only produce a steady plasma concentration after four to five elimination half-lives have elapsed (e.g., after more than 15 hr) (Shafer *et al.,* 1988). Schwilden (1981) calculated an infusion regimen for obtaining and maintaining a desired plasma level of a drug using algorithms that depended on the standard polyexponential disposition characteristics of the drug. This regimen lends itself to automation, thus computer-controlled infusion pumps have become a cornerstone in the delivery of total intravenous anesthesia (Shafer *et al.,* 1988). Pumps programmed according to pharmacokinetic data obtained from multicompartmental models infuse the drug at an exponentially decreasing rate. Although individual pharmacokinetic variability precludes knowing the precise plasma drug concentration that is achieved in "real-time," the ability to rapidly achieve and maintain a steady state allows the anesthesiologist to make proportional adjustments in plasma drug concentration in order to achieve a stable anesthetic depth. These principles have now been applied to animals (Dunlop *et al.,* 1991; Gustafsson *et al.,* 1992). It is anticipated that in the future these systems will be coupled to a direct measurement of pharmacodynamic effect (e.g., computerized EEG analysis), which would complete the feedback control loop (Shafer *et al.,* 1988).

B. Epidural/Subarachnoid Anesthesia

Epidural anesthesia is defined as the injection of local anesthetic into the epidural space, which is the space between the spinal cord dura and the vertebral canal. Spinal anesthesia is the term used when local anesthetic is administered in the subarachnoid space (i.e., intrathecal) where the cerebrospinal fluid is located. Because intrathecal administration of an anesthetic bathes the spinal cord directly, smaller doses are used compared to epidural administration. The vertebral levels at which the spinal cord and dura end are variable both within and between species. For instance, in cats the spinal cord usually ends at L7–S1, whereas the dural sac terminates at S1–S2. In dogs, these levels are usually one vertebral space cephalad.

Local anesthetic agents administered by the epidural or subdural route have been used to produce surgical anesthesia in both humans and animals for many years. Agents such as lidocaine and bupivicaine are two commonly administered local anesthetics used for this purpose. Other classes of anesthetic agents whose actions are mediated by distinctive receptors located in the neuraxis have been administered by this route. To date, opiates and α_2-adrenergic agonists have been the most extensively studied for these effects. Initially, these agents were administered by this route to provide postoperative analgesia; however, they have also been shown to reduce anesthetic requirement and are currently under investigation as components of "balanced" anesthesia techniques. Compared to systemic administration, relatively low doses of opiates and α_2-adrenergic agonists are used when delivered by the epidural route, thus reducing possible systemic side effects. In addition, compared to the epidural administration of local anesthetics, opiates do not alter sympathetic tone, whereas both opiates and α_2-agonists preserve motor nerve function. Therefore, the incidence of profound hypotension is reduced and patients can ambulate postoperatively while analgesia is maintained.

1. Local Anesthetics

Local anesthetics can block sympathetic, sensory, or motor nerves, depending on the dose administered (Ritchie and Greene, 1990). Because they affect motor function, local anesthetics are ideal when muscle relaxation is important for surgical exposure or repair.

Local anesthetic agents such as 2% lidocaine[11] or 0.5% bupivicaine[12] (with or without a vasoconstrictor such as 1:200,000 epinephrine to prolong their duration) provide anesthesia by interrupting the conduction of sensory nerve impulses by inhibiting voltage-sensitive sodium channels (Ritchie and Greene, 1990). The duration of surgical anesthesia using lidocaine without epinephrine is 60 to 90 min, whereas bupivicaine provides 4 to 6 hr of surgical anesthesia (Heath, 1992; Brown and Wedel, 1990). Because local anesthetics administered by this method do not affect the mental status of animals undergoing surgical procedures, light-to-moderate doses of neuroleptic–sedative agents (such as oxymorphone[13]–acepromazine[14]) are recommended (Heath, 1992).

Local anesthetics are potentially toxic, especially if large amounts are administered to small animals. Depending on the mass of drug injected and patient positioning, the epidural block can extend cranially and paralyze the intercostal motor nerves which impedes chest wall excursion. It is also possible for the block to ascend further and involve the phrenic nerve (C5-7), resulting in respiratory arrest. Because local anesthetics functionally sympathectomize the anesthetized region, vascular tone is reduced, resulting in vasodilation. Because the entire caudal body region is affected, this can result in significant hypotension. For this reason, epidural or intrathecal anesthesia with local anesthetics is inappropriate in hypovolemic or debilitated patients. If profound hypotension is encountered, intravenous fluids (e.g., lactated Ringer's solution bolused at 10 to 90 mg/kg IV) with sympathomimetic agents (e.g., ephedrine or phenylephrine) would be appropriate therapy. Finally, accidental intravenous injection of local anesthetic can cause neurotoxicity and cardiovascular collapse (Ritchie and Greene, 1990).

2. Opiates and α_2-Agonists

Opiates and α_2-adrenergic agonists have been administered intrathecally or epidurally to improve analgesia during surgical procedures on caudal body regions. These classes of drugs were initially given by these routes for postoperative analgesia or for chronic cancer pain; however, investigators who administered these drugs before surgery observed reduced anesthetic requirements (Valverde *et al.,* 1989a). Subsequently it was demonstrated that halothane MAC in dogs was reduced by about 40% following epidural morphine (Valverde *et al.,* 1989b). Specific receptors for these drugs were known to exist in the spinal cord and function in pain pathways. Therefore, by applying these agents epi- or subdurally, analgesia was provided at lower doses. Delivery of these agents by this route resulted in more prolonged analgesia with diminished side effects compared to equipotent intravenous administration (Eisenach *et al.,* 1987;

[11]Xylocaine (Astra Pharmaceutical Products, Inc., Worchester, MA).
[12]Marcaine (Winthrop Laboratories, New York, NY).
[13]Numorphan (Pitman-Moore Co., Washington Crossing, NJ).
[14]PromAce (Fort Dodge, Fort Dodge, IA).

Fisher *et al.,* 1991). It should be noted that drug absorption rates are comparable following intramuscular or epidural morphine administration; however, for equipotent analgesia, lower doses are used epidurally and are therefore associated with lower plasma drug concentrations and reduced side effects (Chauvin *et al.,* 1982).

α_2-Adrenergic agonists have been administered intrathecally in animals (Yaksh and Reddy, 1981) and humans (Filos *et al.,* 1992) to provide analgesia mediated by an opiate-independent mechanism (Yaksh, 1985). α_2-Adrenergic receptors are present in the dorsal horn of the spinal cord (Fleetwood-Walker *et al.,* 1985). Studies in sheep have shown that intrathecal clonidine[15] binds to specific sites in the spinal cord (Bouchenafa and Livingston, 1987) and produces analgesia (Eisenach *et al.,* 1987; Waterman *et al.,* 1988) without producing hypoxemia, bradycardia, or hypotension which occur following intravenous administration (Eisenach *et al.,* 1987). Human studies have shown that intrathecal administration of the α_2-adrenergic agonist clonidine enhances spinal anesthesia with local anesthetics (Racle *et al.,* 1987).

Opiates produce their analgesic effects by interacting with opioid receptors that normally bind naturally occurring, endogenous endorphins, enkephalins, and dynorphins (Jaffe and Martin, 1990). Both endogenous and exogenously administered opiates act by inhibiting pain transmission in the dorsal horn, inhibiting somatosensory afferents at supraspinal levels, and activating descending inhibitory pathways (Jaffe and Martin, 1990). Substance P (SP) is an important neurotransmitter in the nociceptive pathway. Substance P (Arg-Pro-Lys-Pro-Cln-Gln-Phe-Phe-Gly-Leu-Met-NH_2) was the first discovered member of structurally related peptides known as tachykinins (Helke *et al.,* 1991). The G-protein-coupled receptor which mediates the SP responses is referred to as NK-1. Upon activation of δ opioid receptors, SP release is inhibited (Collin *et al.,* 1991). Others have suggested that the μ receptor subtype may also be involved in modulating SP release (Aimone and Yaksh, 1989). Glutamate is also an important neurotransmitter of the nociceptive pathways at these sites in the dorsal columns of the spinal cord, and its release and function may be altered by activated opioid receptors (Foldes, 1991).

Intrathecal injection of opioids results in a profound increase in the pain threshold without significantly affecting motor or sensory functions (Jaffe and Martin, 1990). Epidural or subdural opiates have been used in conjunction with local anesthetics for spinal anesthesia for cesarean section in women (Gaffud *et al.,* 1986; Preston *et al.,* 1988) or as adjuncts to systemic anesthetics in animals (Pascoe, 1992a). An additional use has been to reduce tourniquet pain by intrathecal morphine during spinal anesthesia with bupivicaine (Tuominen *et al.,* 1988). Epidural morphine is reported to provide up to 24 hr of analgesia postoperatively (Pascoe, 1992b).

Although relatively rare, side effects have been reported following epidural use of these agents. The most commonly reported adverse effects include nausea and vomiting, sedation, respiratory depression, bradycardia, hypotension, urine retention, pruritis, and transient paresis (Morgan, 1989; Pascoe, 1992b).

Selective spinal analgesia has been advocated for the relief of postoperative, obstetrical, and chronic pain. It was thought that neuraxially administered opioids could relieve pain without side effects. However, after several hours both intrathecal- and epidural-delivered morphine spreads rostrally and can be recovered in the fourth ventricle (Bromage *et al.,* 1982). Opioid receptors in the brain stem can be activated resulting in the same side effects (e.g., respiratory depression, nausea, vomiting, and pruritis) as can be seen following parenterally administered opiates (Pfeiffer *et al.,* 1988). In an attempt to mitigate these side effects, highly lipophilic opiate narcotics have been administered neuraxially to facilitate penetration into the spinal cord with lesser amounts available for rostral spread. However, with the more lipophilic compounds the opioid dose, plasma concentration, degree of pain relief, and side effects are indistinguishable from systemic administration (Glass *et al.,* 1992). Therefore, morphine remains the drug of choice for epidural administration for acute postoperative pain, but attention to the potential of bothersome and even dangerous side effects must be given in the clinical setting.

[15]Catapres (Boehringer Ingelheim Pharmaceuticals, Ridgefield, CT).

REFERENCES

Aho, M., Lehtinen, A.-M., Erkola, O., Kallio, A., and Korttila, K. (1991). The effect of intravenously administered dexmedetomidine on perioperative hemodynamics and isoflurane requirements in patients undergoing abdominal hysterectomy. *Anesthesiology* **74,** 997–1002.

Aho, M., Erkola, O., Kallio, A., Scheinin, H., and Korttila, K. (1992). Dexmedetomidine infusion for maintenance of anesthesia in patients undergoing abdominal hysterectomy. *Anesth. Analg.* **75,** 940–946.

Aimone, L. D., and Yaksh, T. L. (1989). Opioid modulation of capsaicin-evoked release of substance P from rat spinal cord in vivo. *Peptides* **10,** 1127–1131.

Bormann, J. (1988). Electrophysiology of $GABA_A$ and $GABA_B$ receptor subtypes. *Trends Neurosci.* **1,** 112–116.

Bormann, J., Hamill, O. P., and Sakmann, B. (1987). Mechanisms of anion permeation through channels gated by glycine and γ-aminobutyric acid in mucose cultured spinal neurons. *J. Physiol. (London)* **385,** 243–286.

Bouchenafa, O., and Livingston, A. (1987). Autoradiographic localisation of a_2 adrenoceptor binding sites in the spinal cord of the sheep. *Res. Vet. Sci.* **42,** 382–386.

Brammer, D. W., Doerning, B. J., Chrisp, C. E., and Rush, H. G. (1991). Anesthetic and nephrotoxic effects of Telazol® in New Zealand white rabbits. *Lab. Anim. Sci.* **41,** 432–435.

Breimer, L. T. M., Burm, A. G. L., Danhof, M., Hennis, P. J., and Vletter, A. A. (1991). Pharmacokinetic-pharmacodynamic modelling of the interaction between flumazenil and midazolam in volunteers by aperiodic EEG analysis. *Clin. Pharmacokinet.* **20,** 497–508.

Brogden, R. N., and Goa, K. L. (1988). Fluamzenil: A preliminary review of its benzodiazepine antagonist properties, intrinsic activity and therapeutic use. *Drugs* **35,** 448–467.

Brogden, R. N., and Goa, K. L. (1991). Flumazenil: A reappraisal of its pharmacological properties and therapeutic efficacy as a benzodiazepine antagonist. *Drugs* **42,** 1061–1089.

Bromage, P. R., Camporesi, E. M., Durant, P. A. C., and Nielsen, C. H. (1982). Rostral spread of epidural morphine. *Anesthesiology* **56,** 431–436.

Brown, D. L., and Wedel, D. J. (1990). Spinal, epidural, and caudal anesthesia. *In* "Anesthesia" (R. D. Miller, ed.), 3rd ed., pp. 1377–1405. Churchill-Livingstone, New York.

Cahalan, M., Weiskopf, R., Eger, E. I., II, Yasuda, N., Ionescu, P., Rampil, I., Lockhart, S., Friere, B., and Peterson, N. (1991). Hemodynamic effects of desflurane/nitrous oxide anesthesia in volunteers. *Anesth. Analg.* **73,** 157–164.

Castle, N. A., Haylett, D. G., and Jenkinson, D. H. (1989). Toxins in the characterization of potassium channels. *Trends Neurosci.* **12,** 59–65.

Catterall, W. A. (1988). Structure and function of voltage-sensitive ion channels. *Science* **242,** 50–61.

Chauvin, M., Samii, K., Schermann, J. M., Sandouk, P., Bourdon, R., and Viars, P. (1982). Plasma pharmacokinetics of morphine after IM, extradural and intrathecal administration. *Br. J. Aneasth.* **54,** 843–846.

Clarke, R. D., Michel, A. D., and Whiting, R. L. (1986). Pharmacology and structure-activity relationships of a_2-adrenoceptor antagonists. *Prog. Med. Chem.* **23,** 1–39.

Collin, E., Mauborgne, A., Bourgoin, S., Chantrel, D. Hamon, M., and Cesselin, F. (1991). In vivo tonic inhibition of spinal substance P (-like material) release by endogenous opioid(s) acting at delta receptors. *Neurosciences* **44,** 725–731.

Concas, A., Santoro, G., Mascia, M. P., Serra, M., Sanna, E., and Biggio, G. (1990). The general anesthetic propofol enhances the function of g-aminobutyric acid-coupled chloride channel in the rat cerebral cortex. *J. Neurochem.* **55,** 2135–2138.

Concas, A., Santoro, G., Serra, M., Sanna, E., and Biggio, G. (1991). Neurochemical action of the general anaesthetic propofol on the chloride ion channel coupled with $GABA_A$ receptors. *Brain Res.* **542,** 225–232.

Correa-Sales, C., Reid, K., and Maze, M. (1992a) Pertussis toxin mediated ribosylation of G proteins blocks the hypnotic response to an a_2 agonist in the locus coeruleus of the rat. *Pharmacol. Biochem. Behav.* **43,** 723–727.

Correa-Sales, C., Rabin, B. C., and Maze, M. (1992b). A hypnotic response to dexmedetomidine, and a_2 agonist, is mediated in the locus coeruleus in rats. *Anesthesiology* **76,** 948–952.

Daniell, L. C. (1990). The non-competitive N-methyl-D-aspartate antagonists, MK 801, phencyclidine and ketamine increase the potency of general anesthetics. *Pharmacol. Biochem. Behav.* **36,** 111–115.

Daniell, L. C. (1991). Effect of Cgs 19755 a competitive N-methyl-D-aspartate antagonist on general anesthetic potency. *Pharmacol. Biochem. Behav.* **40,** 767–769.

Daniell, L. C. (1992). Alteration of general anesthetic potency by agonists and antagonists of the polyamine binding site of the N-methyl-D-aspartate receptor. *J. Pharmacol. Exp. Ther.* **261,** 304–310.

Daunt, D. A., Dunlop, C. T., Chapman, P. L., Shafer, S. L., Ruskoaho, H., Vakkuri, O., Hodgson, D. S., Tyler, L. M., Maze, M. Cardiopulmonary and behavioral responses to computer-driven infusion of detomidine in standing horses. *Am. J. Vet. Res.* **54,** 2075–82, 1993.

Davies, S. N., Alford, S. T., Coan, E. J., Lester, R. A. J., and Collingridge, G. L. (1988). Ketamine blocks an NMDA receptor-mediated component of synaptic transmission in rat hippocampus in a voltage-dependent manner. *Neurosci. Lett.* **92,** 213–217.

Doerning, B. J., Brammer, D. W., Chrisp, C. E., and Rush, H. G. (1992). Nephrotoxicity of tiletamine in New Zealand white rabbits. *Lab. Anim. Sci.* **42,** 267–269.

Doxey, J. C., Roach, A. G., and Smith, C. F. (1983). Studies on RX 781094 a selective, potent, and specific antagonist of a_2-adrenoceptors. *Br. J. Pharmacol.* **78,** 489–505.

Duka, T., Ackenheil, M., Nederer, J., Doenicke, A., and Dorow, R. (1986). Changes in noradrenaline plasma levels and behavioral responses induced by benzodiazepine agonists and the benzodiazepine antagonist Ro 15-1788. *Psychopharmacology* **90,** 351–357.

Eger, E. I. (1994). New inhaled anesthetics. *Anesthesiology* **80,** 906–922.

Eisenach, J. C., Dewan, D. M., Rose, J. C., and Angelo, J. M. (1987). Epidural clonidine produces antinociception, but not hypotension, in sheep. *Anesthesiology* **66,** 496–501.

Engelman, E., Lipszyc, M., Gilbart, E., Van der Linden, P., Bellens, B., Van Romphey, A., and de Rood, M. (1989). Effects of clonidine on anesthetic drug requirements and hemodynamic response during aortic surgery. *Anesthesiology* **71,** 178–187.

Filos, K. S., Goudas, L. C., Patroni, O., and Polyzou, V. (1992). Intrathecal clonidine as a sole analgesic for pain relief after cesarean section. *Anesthesiology* **77,** 174–267.

Fisher, B., Zornow, M. H., Yaksh, T. L., and Peterson, B. M. (1991). Antinociceptive properties of intrathecal dexmedetomidine in rates. *Eur. J. Pharmacol.* **192,** 221–225.

Fleetwood-Walker, S. M., Mitchell, R., Hope, P. J., Molony, V., and Iggo, A. (1985). An a_2 receptor mediates the selective inhibition by noradrenalin of nociceptive responses of identified dorsal horn neurons. *Brain Res.* **334,** 243–254.

Foldes, F. F. (1991). Pain control with intrathecally and epidurally administered opioids and other drugs. *Anaesthesiol. Reanim.* **16,** 287–298.

Franks, N. P., and Lieb, W. R. (1988). Volatile general anesthetics activate a novel neuronal K^+ current. *Nature (London)* **333,** 662–664.

Gaffud, M. P., Bansal, P., Lawton, C., Velasquez, N., and Watson, W. A. (1986). Surgical anesthesia for cesarean delivery with epidural bupivacaine and fentanyl. *Anesthesiology* **65,** 331–334.

Gasic, G. P., and Heinemann, S. (1991). Receptors coupled to ionic channels: The glutamate receptor family. *Curr. Opin. Neurobiol.* **1,** 20–26.

Gee, K. W. (1988). Steroid modulation of the GABA/benzodiazepine receptor-linked chloride ionophore. *Mol. Neurobiol.* **2,** 291–317.

Ghignone, M., Cavillo, O., and Quintin, L. (1987). Anesthesia and hypertension: The effect of clonidine on perioperative hemodynamics and isoflurane requirements. *Anesthesiology* **67,** 3–10.

Ghignone, M., Noe, C., Cavillo, O., and Quintin, L. (1988). Anesthesia for ophthalmic surgery in the elderly: The effects of clonidine on intraocular pressure, perioperative hemodynamics, and anesthesia requirements. *Anesthesiology* **68,** 707–716.

Gilman, A. G. (1987). G proteins: Transducers of receptor-generated signals. *Annu. Rev. Biochem.* **56,** 615–649.

Glass, P. S. A., Estok, P., Ginsberg, B., Goldberg, J. S., and Sladen, R. N. (1992). Use of patient-controlled analgesia to compare the efficacy of epidural to intravenous fentanyl administration. *Anesth. Analg.* **74,** 345–351.

Gustafsson, L. L., Ebling, W. F., Osaki, E., Harapat, S., Stanski, D. R., and Shafer, S. L. (1992). Plasma concentration clamping in the rat using a computer-controlled infusion pump. *Pharm. Res.* **9,** 800–807.

Haefely, W., Kyburz, E., Gerecke, M., and Mohler, H. (1985). Recent advances in the molecular pharmacology of benzodiazepine receptors and in the structure-activity relationships of their agonists and antagonists. *Adv. Drug Res.* **14,** 165–322.

Heath, R. B. (1992). Lumbosacral epidural management. *Vet. Clin. North Am.: Opin. Small Anim. Anesth.* **22,** 417–419.

Helke, C. J., Krause, J. E., Mantyh, P. W., Couture, R., and Bannon, M. J. (1991). Diversity in mammalian tachykinin peptidergic neurons: Multiple peptides, receptors, and regulatory mechanisms. *FASEB J.* **4,** 1606–1615.

Herrington, J., Stern, R. C., Evers, A. S., and Lingle, C. J. (1991). Halothane inhibits two components of calcium current in clonal (GH_3) pituitary cells. *J. Neurosci.* **11,** 2226–2240.

Horvath, G., Szikszay, M., and Benedek, G. (1992). Calcium channels are involved in the hypnotic-anesthetic action of dexmedetomidine in rats. *Anesth. Analg.* **74,** 884–888.

Hrapkiewicz, K. L., Stein, S., and Smiler, K. L. (1989). A new anesthetic agent for use in the gerbil. *Lab. Anim. Sci.* **39,** 338–341.

Hsu, W. H., Lee, P., and Betts, D. M. (1981). Xylazine-induced mydriasis in rats and its antagonism by a-adrenergic blocking agents. *J. Vet. Pharmacol. Ther.* **4,** 97–101.

Hubbell, J. A. E., Bednarski, R. M., and Muir, W. W. (1989). Xylazine and tiletamine-zolazepam anesthesia in horses. *Am. J. Vet. Res.* **50,** 737–741.

Ilkiw, J. E. (1992). Other potentially useful new injectable anesthetic agents. *Vet. Clin. North Am.: Opin. Small Anim. Anesth.* **22,** 281–289.

Jaffe, J. H., and Martin, W. R. (1990). Opioid analgesics and antagonists. *In* "The Pharmacological Basis of Therapeutics" (A. G. Gilman, T. W. Rall, A. S. Nies, and P. Taylor, eds.), 8th ed., pp. 485–521. Pergamon, New York.

Jalanka, H. H. (1991). Medetomidine, medetomidine-ketamine combinations and atipamezole in nondomestic mammals: A clinical, physiological and comparative study. Academic dissertation, pp. 28–282, College of Veterinary Medicine, Helsinki, Finland.

Jin, Y., Wilson, S., Elko, E., and Yorio, T. (1991). Ocular hypotensive effects of medetomidine and its analogs. *J. Ocul. Pharmacol.* **7,** 285–296.

Kobilka, B. K. (1992). Adrenergic receptors as models for G protein-coupled receptors. *Annu. Rev. Neurosci.* **15,** 87–114.

Kobilka, B. K., Kobilka, T. S., Daniel, K., Regan, J. W., Caron, M. G., and Lefkowitz, R. J. (1988). Chimeric a_2- b_2 adrenergic receptors: Delineation of domains involved in effector coupling and ligand binding specificity. *Science* **240,** 1310–1316.

Krnjevic, K. (1991). Cellular mechanisms of anesthesia. *Ann. N.Y. Acad. Sci.* **625,** 1–16.

Krnjevic, K., and Puil, E. (1988). Halothane suppresses slow inward currents in hippocampal slices. *Can. J. Physiol. Pharmacol.* **66,** 1570–1575.

Longoni, B., and Olsen, R. W. (1992). Studies on the mechanism of interaction of anesthetics with $GABA_A$ receptors. *In* "GABAergic Synaptic Transmission" (G. Biggio, A. Concas, and E. Costa, eds.), pp. 365–378. Raven Press, New York.

Lorenz, W., Lomasney, J. W., Collins, S., Regan, J. W., Caron, M. G., and Lefkowitz, R. J. (1990). Expression of three a_2 adrenergic receptor subtypes in rat tissues: Implications for a_2 receptor classification. *Mol. Pharmacol.* **38,** 599–603.

Lutz, L., Milde, J., and Milde, L. (1990). The cerebral functional, metabolic, and hemodynamic effects of desflurane in dogs. *Anesthesiology* **73,** 125–131.

Marty, J., Nitenberg, A., Philip, I., Foult, J. M., and Joyon, D. (1991). Coronary and left ventricular hemodynamic responses following reversal of flunitrazepam-induced sedation with flumazenil in patients with coronary artery disease. *Anesthesiology* **74,** 71–76.

Maze, M., and Tranquilli, W. (1991). a_2 Adrenoceptors agonists: Defining its role in clinical anesthesia. *Anesthesiology* **74,** 581–605.

Maze, M., Mason, D. M., and Kates, R. E. (1983). Verapamil decreases MAC for halothane in dogs. *Anesthesiology* **59,** 327–329.

McKay, A. C., McKinney, M. S., and Clarke, R. S. J. (1990). Effect of flumazenil on midazolam-induced amnesia. *Br. J. Anaesth.* **65,** 190–196.

Mody, I., Tanelian, D. L., and MacIver, M. B. (1991). Halothane enhances tonic neuronal inhibition by elevating intracellular calcium. *Brain Res.* **538,** 319–323.

Morgan, M. (1989). The rational use of intrathecal and extradural opioids. *Br. J. Anaesth.* **63,** 165–188.

Morio, M., Jujii, K., Satoh, N., Imai, M., Kawakami, U., Mizuno, T., Kawai, Y., Ogasawara, Y., Tamura, T., Negishi, A., Kumagai, Y., and Kawai, T. (1992). Reaction of sevoflurane and its degradation products with soda lime: Toxicity of the byproducts. *Anesthesiology* **77,** 1155–1164.

Morisaki, H., Suzuki, G., Miyazawa, N., Kiichi, Y., Misaki, T., and Suzuki, A. (1988). A clinical trial of sevoflurane in children for herniorraphy. *J. Anesth.* **2,** 94–97.

Nacif-Coelho, C., Correa-Sales, C., Chang, L. L., and Maze, M. (1994). Perturbation of ion channel conductance alters the hypnotic response to the a_2 adrenergic agonist dexmedetomidine in the locus coeruleus of the rat. *Anesthesiology.* 81: 1527–1534, 1994.

Nawy, S., and Jahr, C. E. (1990). Suppression by glutamate of cGMP-activated conductance in retinal bipolar cells. *Nature (London)* **346,** 269–271.

Nicoletti, F., Iadarola, M. J., Wroblewski, J. T., and Costa, E. (1986). Excitatory amino acid recognition sites coupled with inositol phospholipid metabolism: Developmental changes and interaction with a_1 adrenoceptors. *Proc. Natl. Acad. Sci. U.S.A.* **83,** 1931–1935.

Nutt, D. J., Glue, P., Lawson, C., and Wilson, S. (1990). Flumazenil provocation of panic attacks. *Arch. Gen. Psychiatry* **47,** 917–925.

Olsen, R. W., Sapp, D. M., Bureau, M. H., Turner, D. M., and Kokka, N. (1991). Allosteric actions of central nervous system depressants including anesthetics on subtypes of the inhibitory δ-aminobutyric $acid_A$ receptor-chloride channel complex. *Ann. N.Y. Acad. Sci.* **625,** 145–154.

Oshima, E., and Richards, C. D. (1988). An in vitro investigation of the action of ketamine on the excitatory synaptic transmission in the hippocampus of the guinea pig. *Eur. J. Pharmacol.* **144,** 25–33.

Paciorek, P. M., and Shepperson, N. B. (1983). a_1 Adrenoceptor agonist activity of a_2 adrenoceptor antagonists in the pithed rat preparation. *Br. J. Pharmacol.* **79,** 12–14.

Pascoe, P. J. (1992a). Epidural morphine. *In* "Animal Pain" (C. E. Short and A. Van Poznak, eds.), pp. 269–274. Churchill-Livingstone, New York.

Pascoe, P. J. (1992b). Advantages and guidelines for using epidural drugs for analgesia. *Vet. Clin. North Am.: Opin. Small Anim. Anesth.* **22,** 421–423.

Payton, A. J., and Pick, J. R. (1989). Evaluation of a combination of tiletamine and zolezepam as an anesthetic for ferrets. *Lab. Anim. Sci.* **39,** 243–246.

Persson, A., Ehrin, E., Ericksson, L., Farde, L., and Hedstrom, C.-G. (1985). Imaging of [^{14}C]Ro 15–1788 binding to benzodiazepine receptors in the human brain by positron emission tomography. *J. Psychiatr. Res.* **19,** 609–622.

Pfeiffer, B. L., Sernaker, H. L., and Horst, U. M. (1988). Pain scores and ventilatory and circulatory sequelae of epidural morphine in cancer patients with and without prior narcotic therapy. *Anesth. Analg.* **67,** 838–842.

Popilskis, S. J., Oz, M. C., Gorman, P., Florestal, A., and Kohn, D. F. (1991). Comparison of xylazine with tiletamine-zolazepam (Telazol®) and xylazine-ketamine anesthesia in rabbits. *Lab. Anim. Sci.* **41,** 51–53.

Preston, P. G., Rosen, M. A., Hughes, S. C., Glosten, B., Ross, B. K., Daniels, D., Shnider, S. M., and Dailey, P. A. (1988). Epidural anesthesia with fentanyl and lidocaine for cesarean section: Maternal effects and neonatal outcome. *Anesthesiology* **68,** 938–943.

Pritchett, D. B., Sontheimer, H., Gorman, C. M., Kettenmann, H., Seeburg, P. H., and Schofield, P. R. (1988). Transient expression shows ligand gating and allosteric potentiation of $GABA_A$ receptor subunits. *Science* **242,** 1306–1308.

Racle, J. P., Benkhadra, A., Poy, J. Y., and Gleizal, B. (1987). Prolongation of isobaric bupivacaine spinal anesthesia with epinephrine and clonidine for hip surgery in the elderly. *Anesth. Analg.* **66,** 442–446.

Rall, T. W. (1990). Hypnotics and sedatives: Ethanol. *In* "The Pharmacological Basis of Therapeutics" (A. G. Gilman, T. W. Rall, A. S. Nies, and P. Taylor, eds.), 8th ed., pp. 345–382. Pergamon, New York.

Rehm, H., and Tempel, B. L. (1991). Voltage-gated K^+ channels of the mammalian brain. *FASEB J.* **5,** 164–170.

Ritchie, J. M., and Greene, N. M. (1990). Local anesthetics. *In* "The Pharmacological Basis of Therapeutics" (A. G. Gilman, T. W. Rall, A. S. Nies, and P. Taylor, eds.), 8th ed., pp. 311–331. Pergamon, New York.

Scheinin, H., MacDonald, E., and Scheinin, M. (1988). Behavioral and neurochemical effects of atipamezole, a novel a_2-adrenoceptor antagonist. *Eur. J. Pharmacol.* **151,** 35–42.

Schwilden, H. (1981). A general method for calculating the dosage scheme in linear pharmacokinetics. *Eur. J. Clin. Pharmacol.* **20,** 379–383.

Scollo-Lavizzari, G. (1988). The anticonvulsant effect of the benzodiazepine antagonist, Ro 15–1788: An EEG study in 4 cases. *Eur. J. Anaesthesiol. Suppl.* **2,** 129–138.

Scolnik, P., Marangos, P. J., Goodwin, F. K., Edward, S. M., and Paul, S. (1978). Identification of inosine and hypoxanthine as endogenous inhibitors of [^{3}H] Diazepam binding in the central nervous system. *Life Sci.* **23,** 1473–1480.

Shafer, S. L., Siegel, L. C., Cooke, J. E., and Scott, J. C. (1988). Testing computer-controlled infusion pumps by simulation *Anesthesiology,* **68,** 261–266.

Short, C. E. (1986). Preanesthetic medications in ruminants and swine. *Vet. Clin. North Am.: Food Anim. Pract.: Anesth.* **2,** 553–566.

Sigel, E., Stephenson, F. A., Mamalaki, C., and Barnard, E. A. (1983). A gamma-aminobutyric acid/benzodiazepine receptor complex of bovine cerebral cortex: Purification and partial characterization. *J. Biol. Chem.* **258,** 6965–6971.

Silverman, J., Huhndorf, M., Balk, M., and Slater, G. (1983). Evaluation of a combination of tiletamine and zolazepam as an anesthetic for laboratory rodents. *Lab. Anim. Sci.* **33,** 457–460.

Study, R. E. (1994). Isoflurane inhibits multiple voltage-gated calcium currents in hippocampal pyramidal neurons. *Anesthesiology* **81,** 104–116.

Study, R. E., and Barker, J. L. (1981). Diazepam and (-)-pentobarbital: Fluctuation analysis reveals different mechanisms for potentiation of gamma-aminobutyric acid responses in cultured central neurons. *Proc. Natl. Acad. Sci. U.S.A.* **78,** 7180–7184.

Tanelian, D. L., Kosek, P., Mody, I., and MacIver, M. B. (1993). The role of the $GABA_A$ receptor/chloride channel complex in anesthesia. *Anesthesiology* **78,** 757–776.

Tehrani, M. H. J., Hablitz, J. J., and Barnes, E. M. (1989). cAMP increases the rate of $GABA_A$ receptor desensitization in chick cortical neurons. *Synapse* **4,** 126–131.

Thurmon, J. C., Benson, G. J., Tranquilli, W. J., Olson, W. A., and Tracy, C. H. (1988). The anesthetic and analgesic effects of Telazol® and xylazine in pigs: Evaluating clinical trials. *Vet. Med.* **83,** 841–845.

Tranquilli, W. J., and Benson, G. J. (1992). Advantages and guidelines for using a_2 agonists as anesthetic adjuvants. *Vet. Clin. North Am.: Opin. Small Anim. Anesth.* **22,** 289–293.

Tsien, R. W., Lipscombe, D., Madison, D. V., Bley, K. R., and Fox, A. P. (1988). Multiple types of neuronal calcium channels and their selective modulation. *Trends Neurosci.* **11,** 431–438.

Tuominen, M., Valli, H., Kalso, E., and Rosenberg, P. H. (1988). Efficacy of 0.3 mg morphine intrathecally in preventing tourniquet pain during spinal anesthesia with hyperbaric bupivacaine. *Acta Anaesthesiol. Scand.* **32,** 113–116.

Urban, B. W. (1993). Differential effects of gaseous and voltale anaesthetics on sodium and potassium channels. *Br. J. Anaesth.* **71,** 25–38.

Valverde, A., Dyson, D. H., McDonell, W. N., and Pascoe, P. J. (1989a). Use of epidural morphine in the dog for pain relief. *Vet. Comp. Orthop. Trauma* **2,** 55–58.

Valverde, A., Dyson, D. H., and McDonell, W. N. (1989b). Epidural morphine reduces halothane MAC in the dog. *Can. J. Anaesth.* **36,** 629–632.

Vartiainen, J., MacDonald, E., Urtti, A., Rouhiainen, H., and Virtanen, R. (1992). Dexmedetomidine-induced ocular hypotension in rabbits with normal or elevated intraocular pressures. *Invest. Ophthalmol. Visual Sci.* **33,** 2019–2023.

Virtanen, R., Savola, J. M., and Saano, V. (1989). Highly selective and specific antagonism of central and peripheral a_2-adrenoceptors by atipamezole. *Arch. Int. Pharmacodyn. Ther.* **297,** 190–204.

Ward, G. S., Johnson, D. O., and Roberts, C. R. (1974). The use of CI744 as an anesthetic for laboratory animals. *Lab. Anim. Sci.* **24,** 737–742.

Waterman, A., Livingston, A., and Bouchenafa, O. (1988). Analgesic effects of intrathecally-applied a_2-adrenoceptor agonists in conscious, unrestrained sheep. *Neuropharmacology* **27,** 213–216.

Watkins, J. C., and Evans, R. H. (1981). Excitatory amino acid transmitters. *Annu. Rev. Pharmacol. Toxicol.* **21,** 165–204.

Weight, F. F., Aguayo, L. G., White, G., Lovinger, D. M., and Peoples, R. W. (1992). GABA- and glutamate-gated ion channels as molecular sites of alcohol and anesthetic action. *In* "GABAergic Synaptic Transmission" (G. Biggio, A. Concas, and E. Costa eds.), pp. 335–347. Raven Press, New York.

Weiskopf, R. B., Sampson, D., and Moore, M. A. (1994). The desflurane (Tec 6) vaporizer: Design, design considerations and performance evaluation. *Br. J. Anaesth.* **72,** 474–497.

Yaksh, T. L. (1985). Pharmacology of spinal adrenergic systems which modulate spinal nociceptive processing. *Pharmacol., Biochem. Behav.* **22,** 845–858.

Yaksh, T. L., and Reddy, S. V. R. (1981). Studies in the primate on the analgesic effects associated with intrathecal actions of opiates, a adrenergic agonists, and baclofen. *Anesthesiology* **54,** 451–467.

Yeh, J. Z., Quandt, F. N., Tanguy, J., Nakahiro, M., Narahashi, T., and Brunner, E. A. (1991). General anesthetic action on g-aminobutyric acid activated channels. *Ann. N.Y. Acad. Sci.* **625,** 155–173.

Zwass, M., Fisher, D., Welborn, L., Cote, C., Davis, P., Dinner, M., Hannallah, R., Liu, L., Sarner, J., McGill, W., Alifimoff, J., Embree, P., and Cook, D. (1992). Induction and maintenance characteristics of anesthesia with desflurane and nitrous oxide in infants and children. *Anesthesiology* **76,** 373–378.

Appendix

Generic and Trade Names and Sources of Selected Drugs

Generic name	Trade name	Manufacturer/source[a]
Acepromazine maleate	PromAce	Fort Dodge Laboratories
Alphaxalone-alphadolone	Saffan[b]	Glaxo-Wellcome, Inc.
Amitriptyline	Elavil	Zeneca Pharmaceuticals
Atipamezole	Antisedan[b]	Pfizer and Company, Inc.
Atracurium	Tracrium	Glaxo-Wellcome, Inc.
Azaperone	Stresnil	Mallinckrodt Veterinary
Bupivacaine	Marcaine	Sanofi-Winthrop Pharmaceuticals
Buprenorphine	Buprenex, Temgesic	Reckitt & Colman, Norwich-Eaton,
Butorphanol	Stadol, Torbugesic,	Bristol-Myers Squibb, Fort Dodge
Carfentanil	Wildnil	Wildlife Pharmaceuticals
α-Chloralose	—	Sigma Chemical Company
Chloral hydrate	—	Sigma Chemical Company
Cimetidine	Tagamet	SmithKline Beecham
Clonidine	Catapres	Boehringer Ingelheim
Dantrolene	Dantrium	Procter & Gamble Company
Desflurane	Suprane	Ohmeda Pharmaceuticals, Inc.
Detomidine	Dormosedan	Pfizer and Company, Inc.
Diazepam	Valium	Roche Laboratories
Dopamine	Intropin	Dupont Pharmaceuticals
Doxacurium	Nuromax	Glaxo-Wellcome, Inc.
Enflurane	Ethrane	Ohmeda Pharmaceuticals, Inc.
Ephedrine	Ephedrine	Abbott Laboratories
EMTU (Ethylmalonylthiourea)	Inactin	BYK Gulden (Germany)
Etomidate	Amidate	Abbott Laboratories
Etorphine	M-99	Lemmon Company
Fentanyl citrate	Sublimaze	Janssen Pharmaceutica
Fentanyl-droperidol	Innovar-Vet	Mallinckrodt Veterinary
Fentanyl-fluanisone	Hypnorm[b]	Janssen Pharmaceutica
Fentanyl transdermal	Duragesic	Janssen Pharmaceutica

Generic and Trade Names and Sources of Selected Drugs *(continued)*

Generic name	Trade name	Manufacturer/source[a]
Fentanyl-azeperone	Fentaz	Janssen Pharmaceutica
Flunixin meglumine	Banamine	Schering-Plough, Inc.
Gallamine	Flaxedil	Davis and Geck, Inc.
Glycopyrrolate	Robinul-V	Fort Dodge Laboratories
Guaifenesin	Gecolate	Summit Hill Laboratories
Halothane	Fluothane	Wyeth-Ayerst Laboratories
Isoflurane	AErrane, Isoflo	Ohmeda, Inc., Solvay Animal Health
Isoproterenol	Isuprel	Sanofi-Winthrop Pharmaceuticals
Ketamine	Ketaset, Vetalar	Fort Dodge, Parke-Davis Company
Lidocaine	Xylocaine	Astra Pharmaceutical Products
Medetomidine	Domitor	Farmos, Inc. (Finland)
Meperidine	Demerol	Sanofi-Winthrop Pharmaceutical
Mepivacaine	Carbocaine	Sanofi-Winthrop Pharmaceutical
Methadone	Dolophine	Eli Lilly and Company
Methohexital	Brevital, Brevane	Eli Lilly, Corn State Labs.
Methotrimeprazine	Levoprome	Lederle Laboratories
Methoxamine	Vasoxyl	Glaxo-Wellcome, Inc.
Methoxyflurane	Metofane	Mallinckrodt Veterinary
Metomidate	Hypnodil	Janssen Pharmaceutical
Midazolam	Versed	Roche Products
Morphine	Duramorph	Elkins-Sinn, Inc.
Nalbuphine	Nubain	Dupont Pharma
Nalorphine	Naline	Merck and Company
Naloxone	Narcan, P/M Naloxone	DuPont Pharma, Mallinckrodt
Neostigmine	Prostigmin	Hoffmann-LaRoche, Inc.
Oxymorphone	Numorphan	DuPont Pharma
Pancuronium	Pavulon	Organon, Inc.
Pentazocine	Talwin-V	Upjohn Company
Pentobarbital	Nembutal	Abbott Laboratories
Phenylephrine	Neo-Synephrine	Sanofi-Winthrop Pharmaceutical
Phenytoin	Dilantin	Parke-Davis Company
Pipecuronium	Arduan	Organon, Inc.
Procaine	Novocain	Sanofi-Winthrop Pharmaceutical
Propofol	Diprivan	Zeneca Pharmaceuticals
Propranolol	Inderal	Wyeth-Aherst Laboratories
Protamine	Lyphomed	Eli Lilly and Company
Pyridostigmine	Regonol	Organon, Inc.
Sevoflurane	Ultane	Abbott Laboratories
Succinylcholine	Anectine	Glaxo-Wellcome, Inc.
Sufentanil	Sufenta	Janssen Pharmaceutica
Tetracaine	Pontocaine	Sanofi-Winthrop Pharmaceutical
Thiopental	Pentothal	Abbott Laboratories
Tiletamine-zolazepam	Telazol	Fort Dodge Laboratories
Tribromoethanol		Aldrich Chemical Company
Tricaine methane sulfonate (MS222)	Finquel	Argent Chemical Laboratories
Tubocurarine	Tubocurarine	Bristol-Myers Squibb
Urethane (ethyl carbamate)		Sigma Chemical, Aldrich Chemical
Vecuronium	Norcuron	Organon, Inc.
Xylazine	Rompun	Miles Animal Health
Yohimbine	Yobine	Lloyds Laboratories

[a]List of manufacturers/sources is not inclusive, and inclusion of trade names does not imply endorsement.

[b]Not currently available in United States.

Index

Entry followed by t or f denotes table or figure.

B

D

G

H

I

J

K

N

O

P

Q

W

X

Y

Z